AF774721

Boiler Operator's Handbook

3rd Edition

RIVER PUBLISHERS SERIES IN ENERGY ENGINEERING AND SYSTEMS

Series Editors

BOBBY RAUF
Semtrain
USA

The "River Publishers Series in Energy Engineering and Systems" is a series of comprehensive academic and professional books focussing on the theory and applications behind various energy-related technologies and control systems. The series features handbooks for related technology, as well as manuals on the fundamentals and theoretical aspects of energy engineering projects.

The main aim of this series is to be a reference for academics, researchers, managers, engineers, and other professionals in related matters with energy engineering and control systems.

Topics covered in the series include, but are not limited to:

- Energy engineering;
- Machinery;
- Testing;
- HVAC;
- Electricity and electronics;
- Water systems;
- Industrial control systems;
- Web based systems;
- Automation;
- Lighting controls;
- Difference between AC and DC;
- Batteries;
- Electrical Power;
- VFD's and Motors.

For a list of other books in this series, visit www.riverpublishers.com

Boiler Operator's Handbook

3rd Edition

By
Carl Bozzuto

River Publishers

Routledge
Taylor & Francis Group
LONDON AND NEW YORK

Published 2021 by River Publishers
River Publishers
Alsbjergvej 10, 9260 Gistrup, Denmark
www.riverpublishers.com

Distributed exclusively by Routledge
4 Park Square, Milton Park, Abingdon, Oxon OX14 4RN
605 Third Avenue, New York, NY 10017, USA

Library of Congress Cataloging-in-Publication Data

Boiler operator's handbook / Carl Bozzuto. -- Third edition.

©2021 River Publishers. All rights reserved. No part of this publication may be reproduced, stored in a retrieval systems, or transmitted in any form or by any means, mechanical, photocopying, recording or otherwise, without prior written permission of the publishers.

Routledge is an imprint of the Taylor & Francis Group, an informa business

ISBN 978-87-7022-591-5 (print)
ISBN 978-87-7022-590-8 (online)
ISBN 978-1-003-20736-8 (ebook master)

While every effort is made to provide dependable information, the publisher, authors, and editors cannot be held responsible for any errors or omissions.

Table of Contents

Chapter 1—**Operating Wisely** . . . 1
- Why Wisely? . . . 1
- Prioritizing . . . 1
- Safety . . . 5
- Measurements . . . 6
- Trends And Charting . . . 12
- Flow . . . 13
- Pressure Drop And Flow . . . 14
- What Comes Naturally . . . 15
- Water, Steam, and Energy . . . 15
- The Steam and Water Cycle . . . 19
- Combustion . . . 22
- The Central Boiler Plant . . . 28
- Electricity . . . 29
- Documentation . . . 34
- Standard Operating Procedures . . . 36
- Disaster Plans . . . 38
- Logs . . . 40

Chapter 2—**Boiler Plant Operations** . . . 47
- Operating Modes . . . 47
- Valve Manipulation . . . 47
- New Startup . . . 50
- Dead Plant Startup . . . 62
- Normal Boiler Startup . . . 64
- Emergency Boiler Startup . . . 66
- Normal Operation . . . 68
- Safety Testing . . . 70
- Idle Systems . . . 71
- Superheating . . . 74
- Switching Fuels . . . 75
- Standby Operation . . . 78
- Rotating Boilers . . . 78
- Bottom Blow Off . . . 79
- Continuous (Surface) Blowdown . . . 80
- Annual Inspection . . . 82
- Operating During Maintenance and Repair . . . 83
- Code Repairs . . . 84
- Pressure Testing . . . 84
- Lay Up . . . 86
- Tune Ups . . . 88
- Auxiliary Turbine Operation . . . 89
- Power Turbine Operation . . . 93

Chapter 3—**What The Wise Operator Knows** . . . 97
- Know The Load . . . 97
- Know The Plant . . . 101
- Matching Equipment to the Load . . . 101
- Efficiency . . . 104
- Performance Monitoring . . . 109
- Modernizing and Upgrading . . . 111

Chapter 4—**Special Systems** 113
Special Systems 113
Vacuum Systems 113
Hydronic Heating 114
Boiler Water Circulating Pumps 118
HTHW Boiler Plants 119
Organic Fluid Heaters and Vaporizers 121
Service Water Heating 122
Waste Heat Service 127
Once-Through Boilers 128
Engines and Emergency Generators 128
Gas Turbines 129
HRSGS and Combined Cycle Plants 131

Chapter 5 —**Refrigeration And Air Conditioning** 135
Refrigerants 135
The Refrigeration Cycle 137
Refrigerant Superheat and Sub-Cooling 138
Evaporators 139
Freezing and Ice Storage 141
Compressors 142
Condensers 148
Throttling Devices 152
Miscellaneous Components of a Refrigeration System 156
Cooling Towers 166
Air Conditioning 168

Chapter 6— **Maintenance** 185
Maintenance 185
Cleaning 186
Instructions and Specifications 187
Lock Out, Tag Out 188
Lubrication 189
Insulation 191
Refractory 193
Packing 195
Controls and Instrumentation 197
Lighting and Electrical Equipment 198
Voltage and Current Imbalance 201
Eddy Current Testing 202
Miscellaneous 202
Replacements 202
Maintaining Efficiency 207
Records 207

Chapter 7— **Consumables** 213
Fuels 213
Fuel Gases 215
Fuel Oil 217
Coal 221
Other Solid Fuels 223
Water 225
Treatment Chemicals 227
Miscellaneous 228

Chapter 8 —**Water Treatment** 231
Water Treatment 231
Water Testing 232
Pretreatment 236
Boiler Feed Tanks and Deaerators 239
Blowdown 243
Chemical Treatment 244
Preventing Corrosion 246
Preventing Scale Formation 247

Chapter 9 —**Strength Of Materials** 251
Strength of Materials 251
Stress 251
Cylinders Under Internal Pressure 253
Cylinders Under External Pressure 255
Piping Flexibility 255

Chapter 10— **Plants And Equipment** 257
Types of Boiler Plants 257
Boilers 258
Heat Transfer In Boilers 259
Circulation 261
Boiler Construction 264
Fire Tube Boilers 266
Water Tube Boilers 270
Trim 283
Heat Traps 294
Burners 298
Pumps 314
NPSH 319
Fans and Blowers 332
Cogeneration 342
Gas Turbines, Engines, and HRSGs 347

Chapter 11 — **Controls** 351
Controls 351
Self-Contained Controls 365
Control Linearity 368
Steam Pressure Maintenance 368
Fluid Temperature Maintenance 372
Fluid Level Maintenance 373
Burner Management 378
Firing Rate Control—General 380
Firing Rate Control—Low Fire Start 381
Firing Rate Control—High/Low 381
Firing Rate Control—Burner Cut Out 381
Firing Rate Control—Jackshaft 382
Establishing Linearity 384
Startup Control 389
Firing Rate Control— Parallel Positioning 390
Firing Rate Control—Add Air Metering 391
Firing Rate Control— Inferential Metering 392
Firing Rate Control— Steam Flow/Air Flow 393
Firing Rate Control—Full Metering 393
Firing Rate Control—Dual Fuel Firing 395
Firing Rate Control—Choice Fuel Firing 395

Firing Rate Control—Oxygen Trim....396
Draft Control....401
Feed Water Pressure Controls....403
Instrumentation....405

Chapter 12— **Why They Fail**....413
Why They Fail....413
Low Water....413
Thermal Shock....415
Corrosion and Wear....416
Operator Error and Poor Maintenance....416

Appendix A— **Properties of Water and Steam**....419
Properties of Water and Steam....419

Appendix B —**Water Pressure Per Foot Head**....423

Appendix C —**Nominal Capacities of Pipe**....424

Appendix D —**Properties of Pipe**....426
Properties of Pipe....426

Appendix E —**Secondary Ratings**....434
Secondary Ratings of Joints, Flanges, Valves, and Fittings....434

Appendix F —**Pressure Ratings for Various Pipe Materials**....437

Appendix G —**Square Root Flow Curve**....438

Appendix H —**Square Root Graph Paper**....439

Appendix I —**Viscosity Conversions**....440

Appendix J —**Thermal Expansion of Materials**....442

Appendix K —**Value Conversions**....443

Appendix L —**Excess Air/O_2 Curve**....444

Appendix M —**Properties of Dowtherm A**....445

Appendix N —**Properties of Dowtherm J**....446

Appendix O —**Chemical Tank Mixing Table**....447

Appendix P —**Suggested Mnemonic Abbreviations**....449

Appendix Q —**Specific Heats of Some Common Materials**....451

Appendix R —**Design Temperatures and Degree Days**....452

Appendix S —**Code Symbol Stamps**....455
Bibliography....456

Abbreviations....457

Index....461

Biography....471

Chapter 1

Operating Wisely

WHY WISELY?

There are many books with the title "Boiler Operator Handbook." Most of these books talk about plant and equipment, but do not really talk about operating a boiler, and often fail to explain why some things should be done while others should not. It is said that any automatic operating control system "will revert to the level of competence of the operator." Engineers can design all kinds of neat gadgets and algorithms, but they will not work any better than an operator allows. The key factor is to explain to the operator what that gadget or algorithm is supposed to do and how to make sure that it does, in fact, do it. Lacking that information, the operator will revert to a strategy that basically keeps the plant running. Hopefully, this book will provide an operator with a way to figure out what the engineer wanted to accomplish so that the operator can make the gadget or the algorithm really work, presuming that it actually does a better job. New gadgets and algorithms can be put to good use. There will be times when they do not work as advertised, but they should not be dismissed out of hand.

Operators can easily do things that are unwise and, perhaps, even dangerous. Most of these actions result from instructions for situations that no longer exist or from an operator's misunderstanding of what was going on. The key to wise operation is to know why certain things are done and what happens when the wrong things are done. This book will try to cover both keys. With an understanding of why things are done, the chances of doing the wrong things are minimized. When there is an opportunity to make a mistake, it is always helpful to know how someone else made a similar mistake. As Sam Levenson once said, "You must learn from the mistakes of others. You cannot possibly live long enough to make them all yourself." Many types of mistakes will be described in this book so that an operator can learn from the experience of others and, hopefully, not repeat them.

There are two other important factors to be covered: the environment and economics. If every boiler operator applied a few of the wise actions described in this book, there would be a significant reduction in energy consumption at their plants which would subsequently lead to a cleaner environment and more profitable operations. Proper operation will keep the fuel, electricity, and water costs as low as possible, while still providing the necessary thermal energy to buildings and processes.

PRIORITIZING

The first step in operating wisely is to recognize the proper priorities. If a poll was taken of most boiler operators asking them to name the most important thing to be done, the result would be keeping the steam pressure up. This result is most likely due to the fact that complaints from the plant or management are most often heard when the pressure is lost and production suffers or the building is too cold. Nevertheless, keeping up the steam pressure under all circumstances can be extremely dangerous.

History is replete with stories of boiler operators doing stupid things because their first priority was continued operation. There were operators that literally held down old, lever acting safety valves to get the steam pressure higher so that their boat would beat another in a race. Many did not live to tell the tale. In one case, the chief engineer aboard a steamship gave instructions to hit a safety valve with a hammer when a signal was given so that the safety valve would pop at the right pressure. The objective was to convince the Coast Guard inspector that the safety valve opened when it was supposed to. A close look at that safety valve indicated that hitting it with a hammer was not a good thing to do. Thankfully, the valve opened at the right pressure of its own accord. That was an example of self-endangerment to achieve a purpose that, quite simply, was not worth risking someone's life.

It is regrettable that keeping the pressure up is the priority of many operators. Several of them now sit alongside Saint Peter because they were influenced by the plant manager or others who put the wrong things at the top of their list of priorities. Another operator followed his chief's instructions to hit a safety valve so that it would pop several years ago. The valve cracked and ruptured, relieving the operator of his head. Without doubt, the superintendents and plant managers who demanded their now-dead operators to blindly meet selected objectives are still asking themselves why they contributed to their operator having the wrong impression. Despite how it may seem, the boss does not want risking one's life to keep the pressure up. It is just that the boss does not necessarily realize the safety situation and just loses sight of the priorities. The wise operator does not list pressure maintenance, or other events, as having priority over safety concerns.

So what is at the top of the list? The operator, of course. An operator's top priority should always be personal safety. Despite the desire to be a hero, personal safety should take priority over the health and wellbeing of other people. It simply makes sense. A boiler plant is attended by a boiler operator to keep it in a safe and reliable operating condition. If the operator is injured, or worse, he or she cannot control the plant to prevent it from becoming a hazard to other people.

For several years, a major industrial facility near Baltimore had an annual occurrence. An employee entered a storage tank without following proper entry procedures and subsequently succumbed to fumes or lack of oxygen. Now that is bad enough, but, invariably, his buddy went into the tank in a failed effort to remove him, and they both died. Rushing to rescue an ignorant person is neither heroic nor the right thing to do. Calling 911 and then maintaining control of the situation is the proper procedure. That way, no one else gets hurt. Abandoning responsibility to maintain control of a situation and risking one's life is getting the priorities out of order. While preventing or minimizing injury to someone else is important, it is not as important as protecting oneself.

There are occasions when the life or wellbeing of other people is dependent on a boiler operator's actions. There are many stories of cold winters in the north where operators kept their plants going through unusual means to keep a population from freezing. A favorite one is the school serving as a shelter when gas service was cut off to a community. When the operator ran out of oil, he started burning the furniture to keep the heat up. That form of ingenuity comes from the skill, knowledge, and experience that belong to a boiler operator and allows that operator to help other people.

Next in the proper list of priorities are the **equipment and facilities**. Keeping the pressure up is not as important as preventing damage to the equipment or the building. A short-term outage to correct a problem is less disrupting and easier to manage. It is better than a long-term outage because a boiler or other piece of equipment was run to destruction. The wise operator does not permit continued operation of a piece of equipment that is failing. Plant operations might be halted for a day or week while parts are manufactured or the equipment is overhauled. It is preferable to running the equipment until it fails and then having to wait for nine months to obtain a replacement. One can counter complaints from fellow employees that a week's layoff is better than nine months. There are several elements of operating wisely that consider the priority of the equipment.

Many operators choose to bypass an operating limit to keep the boiler on line and avoid complaints about pressure loss. Even worse, they bypass the limit because it was a nuisance. "That thing is always tripping the boiler off line; so I fixed it." The result of that fix is frequently a major boiler failure. Operator error and improper maintenance account for more than 34% of boiler failures. In the case of a large, coal fired, utility boiler, the plant operators short circuited the flame scanners. Flame scanners are safety devices that provide proof of flame. During a strike by the plant operators, the supervisory personnel attempted to start up the boiler. They did not realize that the flame scanners had been disconnected. They got the boiler into an unsafe condition and blew up the boiler.

The **environment** has taken a new position on the operator's list of priorities within the last half century. Reasons are not only philanthropic but also economic. Regularly during the summer, the notices advise that the air quality is marginal. Sources of quality water are dwindling dramatically. The wrong perception in the minds of the company's customers can reduce revenue (in addition to the costs of a cleanup) and the combination is capable of eliminating a source of income for the company and its employees. Several of the old rules have changed as a result. It is no longer appropriate to maintain an efficiency haze because it contributes to the degradation of the environment. The light brown haze that was once thought to be a mark of efficient operation when firing heavy fuel oil has become an indication that there are particulate emissions. Once upon a time, an oil spill was considered nothing more than a nuisance. Insurance for environmental damage has gotten so expensive that many firms cannot afford the insurance to cover the risk. Today, a single oil spill can destroy a company.

One must understand that operation of the plant always has a detrimental effect on the environment. There is no such thing as zero emissions. However, it is possible to reduce the impact of the plant's operation on the environment. The wise operator has a concern for the environment and keeps it appropriately placed on the list of priorities.

Those four priorities should precede **continued operation of the plant** on the list of priorities. Despite what the boss may say when the plant goes down, he or she does not mean or intend to displace them. Most operators manage to develop the perception that continued operation of the plant is on the top of the boss's list of priorities. That impression is formed when the boss is upset and feels threatened and not when he or she is conscious of all ramifications. Continued operation is important and dependent on the skill and knowledge of the operator only after the more important things are covered.

Since continued operation is so important, the operator has an obligation that many never think of and some avoid. The wise operator is always **training a replacement**. If the plant is going to continue to operate, there must be someone waiting to take over the operator's job when the operator retires or moves up to management. Producing a skilled replacement is simply one of the more important ways that the wise operator ensures for continued operation of the plant.

There is a common fear of being replaceable. As such, many operators refuse to tell fellow employees how they solved a problem or managed a situation, believing that they are protecting their job. That first priority is not the job. It is the operator's safety, health, and welfare. Note that protecting the position is not even on the list. When an employer becomes aware of an employee's act to protect the job (and they will notice it), they will question whether they need that position or that person. If the person is not replaceable, any opportunity for promotion is lost. Many operators have been bypassed for promotion simply because there was no one to replace them. Training a replacement is part of the job.

Preserving historical data is a responsibility of the operator. The major way an operator preserves data is by keeping an operator's log. If information is retained only in the operator s mind, the operator's replacement will not have it and neither will the other personnel and contractors. Lack of information can have a significant impact on the cost of a plant operation and on recovery in the event of a failure. Equipment instructions, parts lists, logs, maintenance records, and even photographs can be and are needed to operate wisely.

Operating the plant economically is the last priority and the one that involves the most time. The priorities discussed so far are covered quickly by the wise operator. An operator is paid a wage that respects the knowledge, skill, and experience necessary to maintain the plant in a safe and reliable operating condition. That pay is earned by operating the plant economically. One can make a difference equal to a multiple of wages in most cases. Note that the word efficiency does not fall on the list of priorities. It can be said that operating efficiently is operating economically, but that is not necessarily true. For example, fuel oil is utilized more efficiently than natural gas. However, gas historically costs less than oil. The wise operator knows what it costs to operate the plant and operates it accordingly. Efficiency is just a measure used by the wise operator to determine how to operate the plant economically. Frequently, the operator finds this task daunting because the boss will not provide the information necessary to make the economic decisions. The employer considers the cost data to be a confidential material that should only be provided to senior management personnel. However, if an operator does not know the true cost of the fuel burned, the water and chemicals consumed, the electrical power that runs the pumps and fans, etc., the operator will make judgments in operation based on perceived costs. Frequently, those perceptions are flawed. This point has been demonstrated many times in the past. Regrettably for the employer, it was after a lot of dollars went up the stack.

It is not necessary to know what the boss's or fellow employee's wages are. They are not subject to operator activities. However, one should know what it costs to keep oneself on the job. Taxes and fringe benefits can represent more than 50% of a person's wages. Many of the extra costs, but not all, for a union employee appear on the check because the funds are transferred to the union. Non-union employers should also inform the operators what is contributed on their behalf. Even if the employer does not allow the operator to have that information, the wise operator should know that the paycheck is only a part of what it costs to put a person on the job. In addition to retirement funds, health insurance, vacation pay, and sick pay, there is the employer's share of Social Security and Medicare. The employer has to contribute a match to what the employee has withheld from salary. There are numerous taxes and insurance elements as well. An employer pays health insurance, State Unemployment Taxes, Federal Unemployment Taxes, and Workmen's Compensation Taxes. As a guess, it really costs the employer, for all those extras, around 50%–100% of base salary. Economic operation requires utilizing a balance of resources, including manpower, in an optimum manner so that the total cost of operation is as low as possible.

To summarize, the wise operator keeps priorities in order and they are as follows:

1. the operator's personal safety, health, and welfare;
2. the safety and health of other people;
3. the safety and condition of the equipment operated and maintained;
4. minimizing damage to the environment;
5. continued operation of the plant;
6. training a replacement;
7. preserving historical data;
8. economic operation of the plant.

Prioritizing in the Real World

Prioritizing activities and functions is simply a matter of keeping the above list in mind. Every activity of an operator should contribute to the maintenance of those priorities. Only by documenting them can it be proven that they are done, and done according to priority. Following the list of priorities makes it possible to decide what to do and when.

Changes in the scope of a boiler plant operator's activities make the maintenance of that order important. Modern controls and computers that are used to make things such as building automation systems have relieved boiler plant operators of some of the more mundane activities. Huge strides have been taken from shoveling coal into the furnace to what is almost a white collar job today. As a result, operators find themselves being assigned other duties. Examples include a variety of duties which, when listed on a resume, would appear to outweigh the actual activity of operating a boiler. Today, a boiler plant operator may serve as a watchman, receptionist, mechanic, and receiving clerk in addition to operating the boiler plant. As mentioned earlier, maintenance functions can be performed by an operator or the operator can supervise contractors in their performance. The trend to assuming or being assigned other duties will continue and a wise operator will be able to handle that trend.

Many operators simply complain when assigned other tasks. They also frequently endeavor to appear inept at them, hoping the boss will pass them off on someone else. Note that if one intentionally appears inept at that other duty, it may give rise to a question of one's ability to be an operator. An operator has an opportunity to handle the concept of additional assignments in a professional manner. One can view the new duty as something that can be fit into the schedule, in which case it increases the operator's value to the employer. A wise operator will have developed systems that grant him (or her) plenty of time to handle other tasks. If, however, the duty does not fit, one can demonstrate that the new duty will take away from the work that must be done to maintain the priorities and, pleasantly, inform the boss of the increased risk of damage or injury that could occur if one takes on the new requirements. Should the boss insist upon the assumption of the duties that will alter the priorities, it should be opposed. Every place of employment should have a means for employees to appeal a boss's decision to a higher authority. Seek out that option and use it when necessary, but always be pleasant about it. It is during such contentious conditions that the value of documentation is demonstrated. A wise operator with a documented schedule, standard operating procedures (SOPs), and a to-do list will have no problem demonstrating that an additional task will have a negative effect on the safety and reliability of the boiler plant. On the other hand, documentation that is evidently self-serving will disprove a claim. The wise operator will always have supporting and qualifying documentation to support his or her position.

Another situation that produces contentious conditions in a boiler plant involves the work of outside contractors. Frequently, the contractor was employed to work in the plant with little or no input from the operators. When a contractor is working in the plant, it changes the normal routine and regularly interferes with the schedule an operator has grown accustomed to. The wise boss will have the contractor reporting to the operator. Even if someone is just visiting a plant, it is still important to make certain that they report to the operator on duty and check out as well. Regardless of the reporting requirements, the operator and contractor will have to work together to ensure the priorities are maintained.

The wise operator will be able to work reasonably with the contractor to facilitate the contractor getting its work done. Many operators have expressed an attitude that a contractor is only interested in profits and treat all contractors accordingly. However, the wise operator wants the contractor to make a profit. If the contractor is able to perform the work without hindrance or delay, he or she will be able to finish the work on time and make a profit. If the contractor perceives no threat to the profit that was contemplated when starting the job, he or she will do everything that was intended, including doing a good job. If the operator stalls and blocks the contractor's activity so that the contractor's costs start to run over, the contractor will attempt to protect its profit. If the contractor perceives the operator is intentionally making life difficult, the contractor may complain to the operator's boss as well as start cutting corners to protect his/her profit. A contractor can understand the list of priorities and work with the operator who understands the contractor's needs.

Dealing with fellow employees also requires demonstrative use of the list of priorities. The problem is not usually associated with swing shift operation because the duties are balanced over time. When operators remain on one shift, it is common for one shift to complain that another shift has less work to do. Another common problem is the one operator who, in the minds of the rest, does not do anything or does not do it right. If the priority order is right, it is already known that priority number 6 applies: train that operator.

There is nothing on the list about pride, convenience, or free time. Self-interest is not a priority when it comes to any job. One can be proud of how one does the job. It may be convenient to do something in a different way (but make sure the boss knows of and approves the way). One should always have a certain amount of free time during a shift to attend to the unexpected situations that arise, but it should be no more than an hour per shift. Keep in mind that one is not employed to further self-interests or simply occupy space. One can, and should, provide value to the employer in exchange for that salary.

Most employers understand an employee's need to handle a few personal matters during the day. They will tolerate some time spent on the phone, reading personal documents, and simply fretting over a problem at home. They will not, however, accept situations where the employee places personal interests ahead of the job. Some employers allow their employees to use the plant tools to work on personal vehicles, repair home appliances, make birdhouses, and so on during the shift. Other employers would not allow their people to make personal calls, locking up the phone. Limiting personal activity as much as possible and never allowing it to take priority over that priority list should prevent those situations where, because the boss's good nature was abused, the employer suddenly comes down hard restricting personal activity on the job.

The operator's health and wellbeing is at the top of the list primarily because the operator is the one responsible for the plant. Keep the priorities straight. Maintain the priorities in the specified order. That should always make it possible to resolve any situation.

SAFETY

One of the worst accidents in the United States was the result of a boiler explosion. In 1863, the boilers aboard the steamship *Sultana* exploded and killed almost eighteen hundred people. The most expensive accident was a boiler explosion at the River Rouge steel plant in February 1999. Six men died and the losses were measured at more than $1 billion. Boiler accidents are rare compared to figures near the first part of the 20th century when thousands were killed and millions injured by boiler explosions. Today, less than 20 people die each year as a result of a boiler explosion. Do not be one of them. Safety rules and regulations were created after an accident with the intent of preventing another. A simple rule like "always hold the handrail when ascending and descending the stair" was created to help prevent injury. Falls on stairs in office buildings are one of the most common accidents. Follow those safety rules and go home to the family healthy at the end of the shift. There are many simple rules that the macho boiler operator chooses to ignore and, in doing so, risks life and limb. Make every effort to comply with all of them. That is operating wisely.

Hold on to the handrail. Wear the face shield, boots, gloves, and leather apron when handling chemicals. Do not smoke near fuel piping and fuel oil storage tanks. Read the material safety data sheets, concentrating on the part about treatment for exposure. Connect that grounding strap. Do a complete lock-out and tag-out before entering a confined space and follow all the other safety rules that have been handed down at the place of work. Remember who is on the top of the priority list.

PPE stands for personal protective equipment and was an acronym that was often ignored. Someday, it will be nice to be able to see and hear those grandchildren when they come along. It would be even better if they do not have to observe that grandpa (or grandma as applicable) has deformed hands, burns, or other injuries that can only be described as displeasing. Wear that PPE.

Prevention of explosions in boilers has come a long way since the Sultana went down. The modern safety valve and the strict construction and maintenance rules applied to it have reduced pressure vessel explosions to less than 1% of the incidents recorded in the U.S. each year, always less than two. On the other hand, furnace explosions seem to be on the increase and that, generally, is due to the lack of training and knowledge on the part of the installer which results in inadequate training of the operator.

It is an absolute necessity to know what the rules are and make sure that everyone else abides by them. A new service technician, who is sent to the plant by a trusted contractor, could be poorly trained and unwittingly expose the plant to danger. Even experienced hands can make a mistake and create a hazard. Part of the lesson is to seriously question anything new and different, especially when it violates a rule.

There are lots of rules and some will not apply to every boiler plant. Some rules are covered by qualified inspectors and applied on inspection. There will be rules for the facility that were generated as a result of an accident or an analysis by a qualified inspector. The last time there was a boiler rattling BOOM in the furnace, a rule was created that basically said, don't do that again! State and local jurisdiction (city or county) may also have rules regarding boiler operation. They need to be checked as well. Here is a list of the published rules that every operator should be aware of and, when applicable, should know.

ASME Boiler and Pressure Vessel Codes (BPVC):

- Section I—Rules for Construction of Power Boilers[a]
- Section IV—Rules for Construction of Heating Boilers[a]
- Section VI—Recommended Rules for Care and Operation of Heating Boilers[b]
- Section VII—Recommended Rules for Care and Operation of Power Boilers[b]
- Section VIII—Pressure Vessels, Divisions 1 and 2[c] (rules for construction of pressure vessels including deaerators, blowoff separators, softeners, etc.)
- Section IX—Welding and Brazing Qualifications (the section of the Code that defines the requirements for certified welders and welding)
- B-31.1—Power Piping Code
- CSD-1—Controls and Safety Devices for Automatically Fired Boilers (applies to boilers with fuel input in the range of 400 thousand and less than 12.5 million Btu/hr (British thermal unit per hour) heat input)

National Fire Protection Association (NFPA) Codes

- NFPA—30—Flammable and Combustible Liquids Code
- NFPA—54—National Fuel Gas Code
- NFPA—58—Liquefied Petroleum Gas Code
- NFPA—70—National Electrical Code
- NFPA—85—Boiler and Combustion Systems Hazards Code (applies to boilers over 12.5 million Btu/hr heat input)

[a]Requires inspection by an authorized inspector so one does not have to know all these rules.
[b]This section is being revised. Some recommendations may be out of date.
[c]Requires inspection by an authorized inspector so one does not have to know all these rules.

There are volumes of codes and rules. It is impossible to know them all. They are typically revised every three years and would be out of date before one finished reading them all. It is not important to know everything—only that they are there for reference. Flipping through them at a library that has them or checking them out on the Internet will allow one to learn what applies. CSD-1 or NFPA-85, whichever applies to the plant's boilers, is a must read. Sections VI and VII of the ASME Code are good reads. The rest of the ASME Codes apply to construction and not operation. Even so, one should be aware that they exist. Many rules were produced as a result of accidents. A problem today is that many rules have been lost to history because they were not passed along with the reason for them fully explained. Keep a record of the rules. If there is no record, develop it. Many lives can be saved, including the boiler operator's life.

MEASUREMENTS

Imagine going to a gas station to get gas with no reading on the pump. The attendant might ask for $20, but one would have no way of knowing if that was correct. An argument might ensue. Yet, plants are operated without knowing the amount of fuel the plant burns every day. One wonders how much fuel has been wasted with no regard for the cost. Any boiler large enough to warrant a boiler operator burns hundreds, if not thousands, of dollars each day in fuel. Measurements and appropriate readouts and records are a must. Often boiler operators either do not understand measurements or they have a wrong impression of them.

First, there are two types of measures: measures of quantity and measures of a rate. There is about 100 miles between Baltimore and Philadelphia. That is a quantity. If one were to drive from one city to the other in 2 hrs, one would average 50 miles per hour. That is a rate. Rates and time determine quantities and vice versa. If the boiler burns 7–1/2 gpm (gallons per minute) of oil, it will drain that full 8000-gallon oil tank in less than 19 hrs. Quantities are fixed amounts and rates are quantity per unit of time.

The most important element in describing a quantity or rate is the unit. The term "unit" comes from the Latin word "uno" meaning one. Units are defined by a standard. One talks about the height in feet and inches using those units without thinking of their origin. A foot two centuries ago was defined as the length of the king's foot. Since there were several kings in several different countries, there was always a little variation in actual measurement. Today, a foot is accepted as determined by a ruler, yardstick, or tape measure all of which are based on a piece of metal maintained by the National

Bureau of Standards. That piece of metal is defined as the standard for that measure having a length of precisely one foot. They also have a chunk of metal that is the standard for one pound. There are also units that are based on the property of natural things. The meter, for example, is defined as one ten millionth of the distance along the surface of the earth from the equator to one of the poles. Regrettably, that is no longer a true value because a few years ago, it was discovered that the earth is slightly pear shaped. The distance from the equator to the pole depends on which pole is used. Many units have a standard that is a property of water.

Unless a unit reference is used for a measurement, nobody will know what one is talking about. How would it be handled if someone was asked how far it was to the next town and they said "about a hundred?" Did they mean miles, yards, furlongs, or football fields? Unless the units are tacked on, one cannot relate to the number. With few exceptions, there are multiple standards (units) of measure that can be used. The choice of which one to use is dependent on a trade or occupation. Frequently, relating one to the other should be possible because different trades are being dealt with. Conversion factors will be needed. One can think of a load of gravel as weighing a few hundred pounds, but the truck driver will think of it in tons. An 8-ton truck load may have to be converted to pounds because one may have no concept of tons. One might understand what 16,000 pounds are like. Another example is a cement truck delivery of 5 yards of concrete. That is not 15 feet of concrete. It is 135 cubic feet (there are 27 cubic feet in a cubic yard, 3 × 3 × 3). One needs to understand the type of measurement that is being dealt with to be certain that its value is understood. Also, as with the cement truck driver, one has to understand trade shorthand.

When measuring objects or quantities, there are three basic types of measurement: distance, area, and volume. There are only three dimensions. That is the extent of the types. Distances are taken in a straight line or the equivalent of a straight line. One will drive 100 miles between Baltimore and Philadelphia but will not travel between those two cities in a straight line. If one were to lay a string down along the route and then lay it out straight when done, it would be 100 miles long. The actual distance along a straight line between the two cities would be less, but that way cannot be taken.

Levels are distance measurements. Level measurements that are the distance between two levels are always used because one never talks about a level of absolute zero. If there was such a thing, it would probably refer to the absolute center of the earth. Almost every level is measured from an arbitrarily selected reference. The water in a boiler can be one to hundreds of feet deep, but the bottom is not used as a reference. When one talks about the level of the water in a boiler, inches and negative numbers are always used at times. That is because the reference everyone is used to is the center of the gauge glass which is almost always the normal water line in the boiler. The level in a 12-inch gauge glass is described as being in the range of –6 inches to +6 inches. For level in a tank, the bottom of the tank is normally used for a reference so that the level is equal to the depth of the fluid and the range is the height of the tank.

With so many choices for level, it could be difficult to relate one to the other. That could be important if one wants to know if condensate will drain from another building in a facility to the boiler room. There is one standard reference for level, but it is not called level. It is called "elevation," which is normally understood to be the height above mean sea level (MSL) and labeled "feet MSL" to indicate that is the case. In facilities at lower elevations, it is common to use that reference. A plant in Baltimore, MD, USA will have elevations normally in the range of 10–200 feet, unless it is a very tall building. When the facility is a thousand feet or more above mean sea level, it gets clumsy with too many numbers. The normal procedure is to indicate an elevation above a standard reference point in the facility. A plant in Denver, CO, USA would have elevations of 5200–5400 feet if sea level is used as a reference so that plant references would be used there. It is common for elevations to be negative. They simply refer to levels that are lower than the reference. It happens below sea level or when the designers decide to use a point on the main floor of the plant as the reference elevation of zero. Anything in the basement would be negative. The choice of zero at the main floor is a common one. Note that a point on the main floor is used. All floors should be sloped to drains. One cannot arbitrarily pull a tape measure from the floor to an item to determine its precise elevation.

An area is the measurement of a surface as if it were flat. A good example is the floor in the boiler plant, which would be described in units of square feet. One square foot is an area that is one foot long on each side. It is said "square" foot because the area is the product of two linear dimensions—one foot times one foot. The unit square foot is frequently written as ft^2, meaning feet two times or feet times feet. That is relatively easy to calculate when the area is a square or rectangle. If it is a triangle, the area is one half the overall width times the overall length. If it is a circle, the area is 78.54% of a square with length and width identical to the circle's diameter. A diameter is the longest dimension that can be measured across a circle, the distance from one side to a spot on the opposite side. In some cases, one uses the radius of a circle and says the area is equal to the radius squared times Pi (3.1416). When dealing with odd-shaped areas, laying graph paper over it and counting squares plus estimating the parts of squares at the borders is another way to determine an area. A complex-shaped area can also be broken up into squares, rectangles, triangles, and circles, adding and subtracting them to determine the total area.

Volume is a measure of space. A building's volume is described as cubic feet, abbreviated as ft^3, meaning multiply the width times the length times the height. One cubic foot is the space that is one foot wide by one foot long by one foot high. For volumetric measurements, there is also the gallon. It takes 7.48 of them to make a cubic foot.

Note that the volumetric measure of gallons does not relate to any linear or area measure. It is only used to measure volumes. That is some help because many trades use unit labels that are understood by them to mean area or volume. A painter may say there is another thousand feet to do. That is not a straight line. That means one thousand square feet. The cement hauler that uses the word "yards," which means cubic yards, is already mentioned. Always make sure to understand what the other person is talking about.

When talking, or even describing measurements, descriptions of direction will be used to aid in explaining them. While most people understand north, south, east, and west plus up and down, other terms require some clarification. Perpendicular is the same as perfectly square. When one looks for a measurement perpendicular to something, it is as if a square is set on it so that the distance being measured is along the edge of the square. An axial measurement is one that is parallel to the central axis or the center of rotation of something. On a pump or fan, it is measured in the same direction as the shaft. Radial is measured from the center out. On a pump or fan, it is from the centerline of the shaft to whatever is being measured. When saying tangentially or tangent to, one is describing a measurement to the edge of something round at the point where a radial line is perpendicular to the line being measured along.

Another measure that confuses operators is mass. Mass is the amount of matter in an object. It is what one weighs at sea level under the influence of the earth's gravity. If one is sent to Cape Kennedy, loaded into the space shuttle, sent up in space, and is then asked to stand on a scale and tell what it reads, it would be zero. With nearly zero gravity, there is no weight. However, one still has the same amount of mass that was weighed at sea level. There is a difference in weight with a change in altitude. One would weigh less in Denver, CO, USA because it is a mile higher. For all practical purposes, the small difference is not important to boiler operators. Once the fact that mass and weight have the same number on the surface of the earth is accepted (with some adjustment required for precision at higher elevations), one can accept that a pound mass weighs a pound and let it go at that.

Volume and mass are not consistently related. A pound mass is a pound mass despite its temperature or the pressure applied to it. One cubic foot of something can contain more or less mass depending on the temperature of the material and the pressure it is exposed to. Materials expand when heated and contract when cooled (except for ice which does just the opposite). A fluid like water can be put on a scale to determine its mass, but the weight will depend on how much is put on the scale. If a one-gallon container of 32° water is put on a scale, it will weigh 8.33 pounds. If a cubic foot of that water is put on a scale, it will weigh 62.4 pounds.

Density is the mass per unit volume of a substance. That is pounds per cubic foot. So, water must have a density of 62.4 pounds per cubic foot. Things are not always that simple. Pure clean water weighs that much. Sea water weighs in at about 64 pounds per cubic foot. Heat water up and it becomes less dense. When it is necessary to be precise, one can use the steam tables (page 419) to determine the density of water at a given temperature. Keep in mind that its density will also vary with the amount of material dissolved in it.

In many cases, water is the reference. The term specific gravity or specific weight means the comparison of the weight of the liquid to water (unless it is a gas when the reference is air). Knowing the specific gravity of a substance allows one to calculate its density by simply multiplying the specific gravity by the typical weight of water (or air if it is a gas). If the specific gravity is less than 1, it is lighter than water (or air), and if it is greater than 1, it will sink.

Gases, such as air, can be compressed. More and more pounds of air can be packed into a compressed air storage tank. As the air is packed in, the pressure increases. When the compressor is off and air is consumed, the tank pressure drops as the air in the tank expands to replace what leaves. The compressor tends to heat the air as it compresses it and that hot air will cool off while it sits in the tank and the pressure will drop. The pressure and temperature of a gas must be known to determine the density. The steam tables list the specific volume (cubic feet per pound) of steam at saturation and some superheat temperatures. Specific volume is equal to 1 divided by the density. To determine density, divide 1 by the specific volume.

Liquids are normally considered non-compressible. Only the temperature must be known to determine the density. The specific volume of water is also shown on the steam tables for each saturation temperature. Water at that temperature occupies the volume indicated regardless of the pressure.

Force is also measured in pounds. Just like a weight of, say, ten pounds can bear down on a table when that weight is set down, the table can be tipped up with its feet against a wall and one can push on it to produce a force of ten pounds with the same effect. Weights can only act down toward the center of the earth, but a force can be applied in any direction. Just like a weight with a scale can be measured, one can put the scale (if it is a spring loaded type) in any position and measure force. They are both measured in pounds.

Rates are invariably one of the measures of distance, area, volume, weight, or mass traversed, painted, filled, or moved per unit of time. Common measurements for a rate are feet per minute, feet per second, inches per hour, feet per day, gallons per minute, cubic feet per hour, miles per hour, and its equivalent of knots (nautical miles per hour). Take any quantity and any time frame to determine a rate. Which one to use is normally determined according to the trade discussing it or the size of the number. One normally drives at 60 miles per hour although it is also correct to say the vehicle is covering 88 feet per second. One would not say the speed is 316,800 feet per hour. Be conscious of the units used in trade magazines and by various workmen to learn which units are appropriate to use. Values can always be converted to units that are more meaningful. The appendix contains a list of common conversion factors.

There are common units of measure used in operating boiler plants. Depending on what is being measured, units of pounds, cubic feet, or gallons will be used when discussing volumes of water. Steam generated is measured in pounds (mass) per hour but feed the water to the boiler in gallons per minute. Oil is burned in gallons per hour, gas in thousands of cubic feet per hour, and coal in tons per hour. One uses a measure that is shared with the plumbing trade which is called pressure, normally measured in pounds per square inch. Occasionally, calling it "head" is confusing to everyone.

Normally, the rate of steam generation is in pounds per hour and abbreviated as "pph." The typical boiler plant can generate thousands of pounds of steam per hour. Thus, the numbers get large. A problem arises when using the abbreviations for large quantities because the industry is not consistent and uses a multitude of symbols. Common use is "kpph" to mean thousands of pounds of steam per hour. From the oil and gas industry, "MBtu/hr" is used to describe a thousand Btu's per hour (M being Latin for 1000). A measure of a million Btu's per hour can be labeled as "MMBtu/hr." The ASME is trying to be consistent in using only lower case letters for the units. It will be some time before that is accepted.

Pressure exists in fluids, gases, and liquids and has an equivalent called "stress" in solid materials. Most of the time, the unit is in pounds per square inch for both, but there are occasions when pounds per square foot are used. Pounds per square inch are abbreviated psi. The units imply force per unit area. It is not hard to imagine a square inch. It is an area measuring one inch wide by one inch long. Then, if one pound of water were put on top of that area, the pressure on that surface would be one pound per square inch. If 100 pounds of water were placed over each square inch, the pressure on the surface would be 100 psi. It is not necessary for the fluid to be on top of the area because the pressure is exerted in every direction. A square inch on the side of a tank or pipe centered so that there is 100 pounds of water on top of every square inch above it sees a pressure of 100 psi. The air in a compressed air storage tank is pushing down, up, and out on the sides of the tank with a force, measured in pounds, against each square inch of the inside of the tank. That is the pressure. Under very low pressure conditions, like the pressure of the wind on the side of a building, pounds per square foot could be used, but it is more common to use inches of water. A manometer with one side connected to the outside of the building and another to the inside would show two different levels of water, and the pressure difference between the inside and outside of the building is identified in inches of water, the difference in the water level. That is the common measure for air pressures in the air and flue gas passages of the boiler and the differential of flow measuring instruments.

There is another measure of pressure called "head." Head is the height of a column of liquid that can be supported by a pressure. A column of water one foot high will bear down on one square foot at a pressure of 62.4 pounds per square foot. Divide that by 144 square inches per square foot to get 0.433 pounds in a column of water one inch square and one foot high. Thus, one foot of water produces a pressure of 0.433 psi. Dividing that number by 1 gives a column of water 2.31 feet tall to produce a pressure of 1 psi. The reason to use head is because pumps produce a differential pressure, which is a function of the density of the liquid being pumped. Head in feet and inches of water (abbreviated as "in. W.C." for inches of water column) are both head measurements even though a value for head is normally understood to mean feet.

Often, pressure units are reported as psig and psia. They stand for pounds per square inch gauge and pounds per square inch absolute. The difference is related to atmospheric pressure. The air in the atmosphere has weight (mass). There is a column of air on top of every square inch that is more than 30 miles high. That may sound like a lot, but if one wanted to simulate the atmosphere on a globe (one of those balls with a map of the earth wrapped around it), the best way is to pour some water on it. After the excess has run off, the wet layer that remains is about right for the thickness of the atmosphere, about three one-hundredths of an inch on an 8-inch globe. The column of air over any square inch of the earth's surface, located at sea level, is about 15 pounds. Therefore, the atmosphere exerts a pressure of 15 pounds per square inch on the earth at sea level under normal conditions. (The actual standard value is 14.696 psi, but the value 15 is close enough for most purposes.) If all the air were taken away, there would be no pressure. The pressure would be zero. A pressure gauge actually compares the pressure in the connected pipe or vessel to atmospheric pressure. When the gauge is connected to nothing, it reads zero. There is atmospheric pressure on the inside and outside of the gauge's sensing element. When the gauge is connected to a pipe or vessel containing a fluid at pressure, the gauge is indicating the difference between atmospheric pressure and the pressure in the pipe or vessel. The reading is referred to as gauge pressure (psig). Absolute pressure is a combination of the pressure in the pipe or vessel and atmospheric pressure and is referred to as psia. Add 15 to gauge pressure to get absolute pressure, the pressure in the vessel above absolutely no pressure. For more precision, use 14.696 instead of 15. Atmospheric pressure varies a lot in any case, which obviates the need for more precision in most cases. Thermodynamic calculations may require more accuracy, but pressure trends in an operating plant can just rely on gauge pressure readings.

Viscosity is a measurement of the resistance of a fluid to flowing. All fluids have a viscosity that varies with their temperature. Normally, a fluid's viscosity decreases with increasing temperature. The phrase "slow as molasses in January" implies little or no flow. Cold molasses has a high viscosity because it takes a long time for it to flow through a standard tube (a viscometer). The normal measure of viscosity is the time taken for a certain volume of fluid to flow through the viscometer, and, thus, the viscosity is described in terms of seconds. A chart for conversion of viscosities is included in the appendix along with the viscosity of some typical fluids found in a boiler plant.

It is important to understand that, in any formula or equation, the dimensions on both sides of the equal sign are the same (something called dimensional analysis). Formulas that engineers use are checked for units matching on both sides of the equation to ensure the formula is correct in its dimensions (measurements). It ensures that inches are on both sides of an equation and not feet on one side and inches on the other. Units have to be consistent when making calculations. This is also true when using or ordering materials. A state side jet liner fueled up in Montreal. It ran out of jet fuel half way across Canada because it purchased liters of fuel instead of gallons of fuel (1 gal = 3.79 liters). Always check the units to ensure they are right.

Another value used in boiler plants is "turndown." Turndown is another way of describing the operating range of a piece of equipment or system. Instead of stating the boiler will operate between 25% and 100% of capacity, the manufacturer will claim it has a four to one turndown ratio. The full capacity of the equipment or system is described as multiples of the minimum rate it will operate at. The minimum operating rate is determined by dividing the larger number into 1. Divide the large number (4) into 1 and multiply by 100 to get the minimum firing rate in percent (25%).

The term "load" is used when describing equipment operation. Load usually refers to the demand the facility served places on the boiler plant, but, within the correct context, it also implies the capacity of a piece of equipment to serve that load. A boiler that is operating at a full load means that it is at its maximum design capacity; half load is 50%, etc.

More difficult measures to address are "implied" measures. Some are subtle and others are very apparent. A common implied measure in a boiler plant is half the range of the pressure gauge. Engineers normally select a pressure gauge or thermometer so that the needle is pointing straight up when the system is at its designed operating pressure or temperature. The water level in a boiler should be at the center of the gauge glass. That is another implied measurement. In other cases, the extreme of the device is used to imply the capacity of a piece of equipment. Steam flow recorders are typically selected to match the boiler capacity, even though they should not be. The problem with implied measurements is that situations can arise where the reading is assumed to be correct when it is not. Keep in mind that someone could have replaced that pressure gauge with something that was in stock but a different range. That gets back to good record keeping and documentation.

Probably one of the most common mistakes that are made is not getting something square. All too often, the layout is simply "eyeballed" or an improper instrument is used. The typical carpenter's square, a piece of steel consisting of a 2-foot length and 16-inch length of steel connected at one end and accepted as being connected at a right angle, works well for small measurements. Using it to lay out something larger than 4 feet can create problems. Also note that "accepted as being square" may not be so. Drop a carpenter's square on concrete any way but flat and it is surprising how it can be bent. On any job that is critical, always check the square by scribing a line with it and flipping it over to see if it shows the same line. Of course, the one side that is being dealt with has to be straight. Eyeballing (looking along the length of an edge with one's eye close to it) is the best way to check to confirm an edge is straight.

For measures larger than something that can be checked with that square, use a 3 by 4 by 5 triangle, which is the same thing the Egyptians used to build the pyramids. Lay it out by making three arcs as indicated in Figure 1-1. One frequently also needs a straight edge as the reference that one is going to be square to, in which case 3 units are marked off along that edge to form the one side that is drawing the arc to find the point B by measuring from point A. An arc is made with 4 units on the side at point C by measuring from point A and then another arc of 5 units is made by measuring from point B and laying down an arc at D. Where the A to C and B to

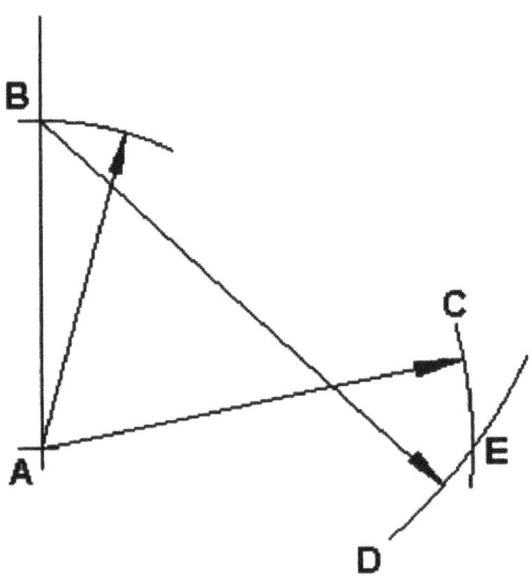

Figure 1-1. Creating a right angle.

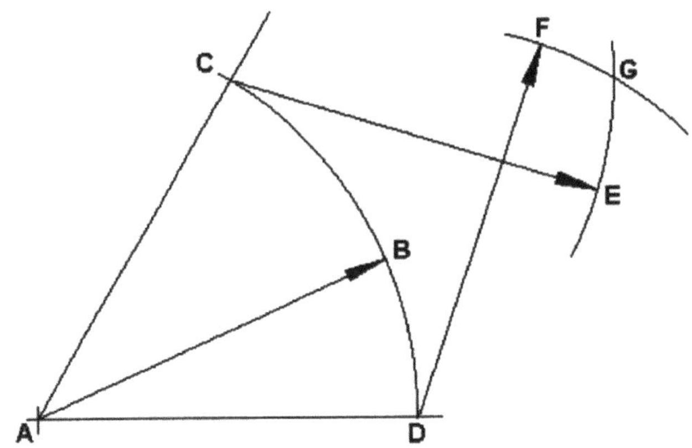

Figure 1-2. Dividing an angle.

D arcs cross (point E) is the other corner of the 3 by 4 by 5 triangle, and side A to B is square to A to E. The angle in between them is precisely 90°.

The beauty of the 3 by 4 by 5 triangle is that the units can be anything as long as the ratio is 3 to 4 to 5. Use inches, or even millimeters, on small layouts and feet on larger ones. For laying out a new storage shed, the triangle might be 30 feet, 40 feet, and 50 feet. It is difficult to get more precise, even with a transit.

Another challenge is finding a 45° angle. The best solution for that is to lay out a square side to get that 90° angle and then divide the angle in half. Figure 1-2 shows the arrangement for finding half an angle. Simply measure from the corner of the angle out to two points (C and D) the same distance (A to B) and then draw two more arcs, measuring from points C and D a distance E, and F identical to E to locate a point where the arcs cross at G. A line from A to G will be centered between the two sides, splitting the angle. In order to start with a 90° angle and split it into three 30's, measure off F at twice the length of E. Then shift around to get two points that are at 30° and 60°. The same scheme will allow the creation of any angle.

It is Always Done That Way

This language is often used as an excuse for why a certain action was taken. Often times, the underlying set of circumstances have changed since the original decision was made. It means that the correct action today is not always "the way we always did it."

In one case, a consultant was called upon to assist in the design of an installation of new burners plus burner management and controls on three existing boilers in a central heating plant. Observation of the plant's operation and a review of the fuel records revealed that the maximum load was less than two-thirds of the capacity of one boiler.

It was suggested that the design include only two boilers at reduced capacity to reduce turndown losses. The operator's instant response was to reject the suggestion because "We always run two boilers in the winter." After several exchanges, the plant decided to prove their point and did so by collecting data. They recorded outdoor air temperature and oil consumption for several hours. They produced that data at the next meeting, indicating that it proved their point. A plot of that data is shown in Figure 1-3. Note that an increase in fuel usage with lower temperatures is obvious. Then note the difference between the markers. The filled symbols are for readings with two boilers in operation, but the open ones are for one boiler in operation. Not only were they able to operate one boiler at colder temperatures, but they also burned less fuel doing so. Note that on the coldest hour, 16°F, they only burned 300 gallons of fuel while operating one boiler. However, on a warmer hour, 22°F, they burned 420 gallons. A difference in fuel use of 120 gallons per hour adds up to a lot of money in the winter season. This is only one example of how operators can make a difference—when they operate wisely.

TRENDS AND CHARTING

Figure 1-3 and the accompanying story is only one example of several situations where graphing data managed to prove a point which owners and operators simply did not understand until they looked at a graph. The graph or chart of the data gives a visual picture that, often times, is easier to comprehend. Modern control systems and data acquisition systems typically have graphic capability. They are typically referred to as trends. Take advantage of that capability. Trending data which produces a graph on the monitor or display in the plant on a regular basis simply provides one more visual aid that can be used to catch problems early or detect changes that would not have been noticed otherwise. Lack of fancy instrumentation does not restrict the creation of graphics that serve as a visual aid. Leaf through the logbook and enter the data into one of the spreadsheet programs that are readily available. Then use that program to produce a graph. That is how the graphic of Figure 1-3 was produced. Hopefully, the discussion about the situation that produced the graphic in Figure 1-3 is a clue. If the plant data is not graphed by the plant's operator, someone else can come up with data that contradicts the operator's perception of the facility's operation. Always remember the saying "if you always do what you've always done, you'll always get what you always got." Make the next step in operating wisely. Spend time looking at trends and graphs of one's data.

According to the dictionary, a trend is "the general course or prevailing tendency" and it is the graphic that is most readily produced because it is nothing more than a record of data values, frequently called data points, all taken at specific times. Today's instrumentation systems will produce a graphic showing data for different time intervals, and the time intervals to be displayed on a monitor or printed on paper can be selected. Only print the graphic when it is meaningful, showing something that has changed. Look at that displayed graphic on a regular basis to detect significant changes or undesirable trends. Typical indications include a sudden drop in condensate returns (sometimes indicated by sudden increase in makeup water) which frequently allows connecting that change to another event that occurred at the same time. Undesirable trends can include a constant decrease in evaporation rate, an increasing stack temperature, and a gradual decrease in condensate temperature, all of which are indicative of something going wrong and a need to discover what is causing it.

There are, however, load-related reasons for undesirable trends and situations where changes in load

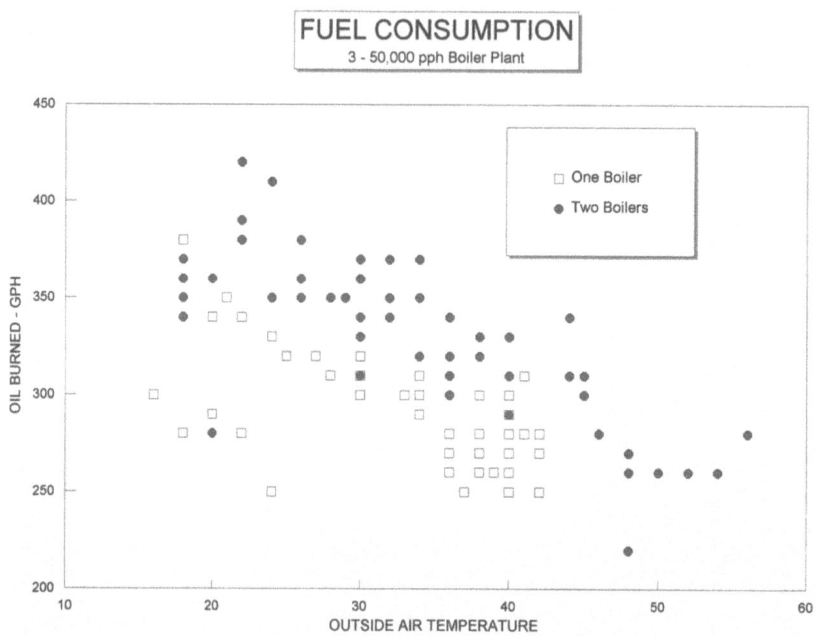

Figure 1-3. A comparison of boiler operation.

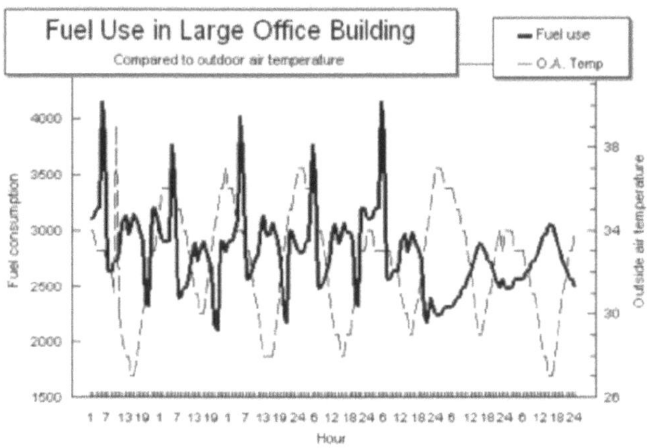

Figure 1-4. Fuel use for a building.

cloud the issue, producing oscillations in a trend graphic which prevent detecting trends. Many times, those deviations can be resolved by filtering the data so that only the trend graphic for a specific load on a boiler, plant, or other specific equipment or system is being displayed. At other times, it may be desirable to actually plot the data relative to the load. Figure 1-4 is one example of a trend graph for outside air temperature and fuel use for a large office building during a cool week. The fuel use obviously increases as the temperature drops. The effect of night setback when the building is closed from 7 pm to 6 am and during weekends reveals the reasons for the fuel use not being consistent for changes in outdoor air temperature alone. The peaks of fuel use around 6 and 7 in the morning of the first five days are due to recovering from the night setback. They are missing the last two days because of the weekend. The setback also reduces the fuel use over the weekend. Notice the increases in load associated with people entering and leaving the building when air infiltration is the greatest. There are many activities that can have an effect on load that will have to be identified to understand the operation. In addition to the ones mentioned here, there is the effect of wind, solar gain (sunny or cloudy), unusual activities (such as a conference that increases use of doors), hot water heating, and such things as kitchens, laundries, and other heat users.

FLOW

Many key variables are not directly controlled. What is controlled is the flow. Pressure and temperature are maintained by controlling the flow of fuel and air. Drum level is maintained by controlling the flow of feed water. Pressure, temperature, level, and other measures will increase or decrease only with a change in flow. An increase in flow will increase or decrease the value being measured depending on the direction of the flow. By understanding the concept of flow and how it affects the particular measure of interest, the operator can determine what is causing the problem. It is a fundamental that, once grasped, will always serve an operator in determining the cause of, and solution to, a problem with control.

Now with that concept in hand, the next issue is to see how control of flow can maintain all those desirable conditions in the boiler plant. There are two means for controlling flow. One can turn it on and off. Or the flow rate can be varied. Changing the flow rate is called "modulating." The method is called "modulation." To restore the level in a chemical feed tank, open a valve and shut it when the level is near the top. Then add chemicals to restore the concentration. That is "on/off" control. A float valve on a makeup water tank opens as the level drops to increase water flow and closes to decrease flow as the level rises. That is flow modulation. There is, of course, more to know and understand about these two methods of control, but they will be addressed in the chapter on controls.

Accepting the premise that what can be controlled is the flow makes it a lot simpler to understand the operation of a boiler plant. Every pound of steam that leaves the boiler plant must be matched by a pound of water entering it or the levels in the plant will have to change. Water wasted in blowdown and other uses like softener regeneration must also be replaced by water entering the plant.

The energy in the steam leaving the boiler plant requires energy to enter the plant in the form of fuel flow. If the steam leaving contains more energy than that is supplied by the fuel entering, then the steam pressure will fall. Some of the energy in the fuel ends up in the flue gases going up the stack. The energy in the fuel has to match the sum of the energy lost up the stack and leaving in the steam. The sum of everything flowing into the boiler plant has to match what is flowing out or plant conditions will change. An operator is something of a juggler. There is always a balancing act in controlling flows into the plant to match what is going out.

A boiler operator basically controls the flow of fluids. The energy added to heat water or make steam comes from the fuel. The amount of energy released in the boiler is adjusted by controlling the flow of the fuel. Gas and oil are both fluids because they flow naturally. Operators in coal fired plants could argue that they are controlling the flow of a solid, but when they look at it, they will realize that they are treating that coal the same way they would treat a fluid. The only other flow an operator controls is the flow of electrons in electrical circuits, electricity. Controlling those flows requires an understanding of what makes them flow and how the flow affects the pressures and temperatures that are needed.

All fluids have mass. Fuel oil normally weighs less than water. Natural gas weighs less than air but still has mass. They can all be treated the same way in general terms because what happens when they flow is about the same. Gas and air are a little more complicated because they are compressible (their volume changes with pressure). In practice, the relationship between flow and pressure drop is consistent regardless of the fluid. Flow metering using differential pressure is based on the Bernoulli principle. Bernoulli discovered the relationship between pressure drop and flow back in the 17th century. In order for air to flow from one spot to another, the pressure at spot one has to be higher than the pressure at spot two, like water flowing downhill. The higher the pressure differential, the faster a fluid will flow. Consider the wind. Small changes in atmospheric pressure cause the wind. It does not take a lot of difference in pressure to really get that air moving. Bernoulli discovered that the total pressure in the air does not change except for friction and that total pressure can be described as the sum of static pressure and velocity pressure.

The measurement of static pressure, velocity pressure, and total pressure is described using Figure 1-5. The static pressure is the pressure in the fluid measured in a way that is not affected by the flow. Note that the connection to the gauge is perpendicular to the flow. The gauge measuring total pressure is pointed into the flow stream so that the static pressure and the velocity pressure are measured on the gauge. What really happens at that nozzle pointed into the stream is that the moving liquid slams into the connection, converting the velocity to additional static pressure sensed by the gauge. There is no flow of fluid up the connecting tubing to the gauge. The measurement of velocity pressure requires a special gauge that measures the difference between static pressure and total pressure. With that measurement, one can determine the velocity of the fluid independent of the static pressure. A velocity reading in a pipe upstream of a pump, where the pressure is lower, would be the same as that in a pipe downstream of the pump (provided the pipe size is the same).

Observe the water flow in a creek or stream. Notice the level of water leaving a still pool and flowing over and between some rocks. Put a large rock in one of the gaps, and the flow through the gap is reduced. Yet, that water has to go somewhere. The level in the pool will go up. It may not be because the water flow that was blocked is shared by all the other gaps and the only way more water can flow is to have more cross section to flow through. Some real insights into fluid flow can be gained by spending some time observing a creek (that is a creek, not a large deep river). All the education is acquired by seeing how the water flows over and through the rocks and relating what is seen to the concepts of static, velocity, and total pressure.

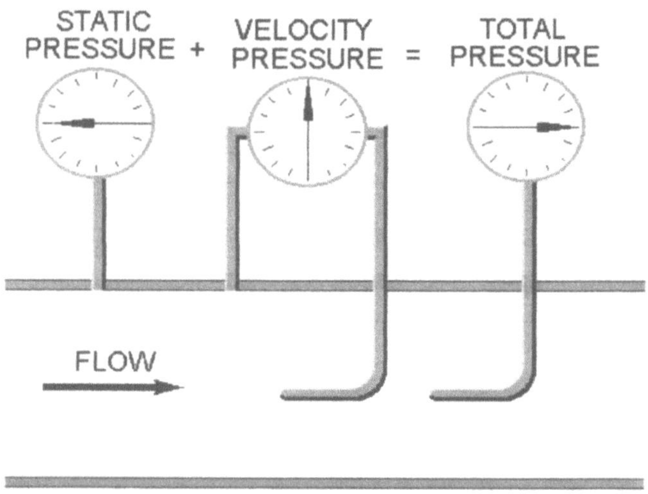

Figure 1-5. Static, velocity, and total pressure measurements.

PRESSURE DROP AND FLOW

There is another thing about flow that is important to understand. A change in pressure drop is proportional to the square of a change in flow. The diagram (Figure 1-6) of a sprinkler on the lawn hooked up to a

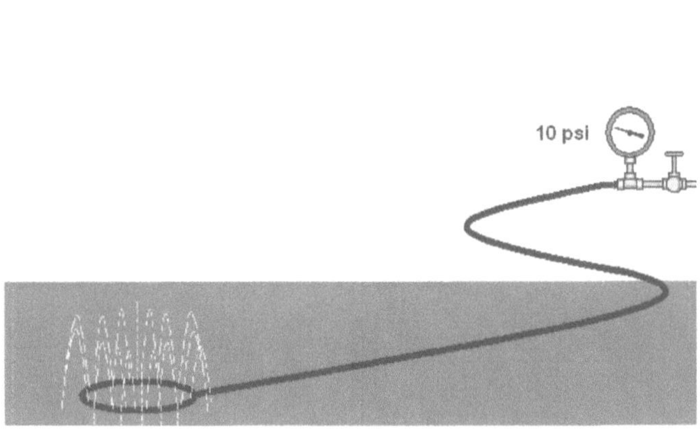

Figure 1-6. A lawn sprinkler example.

hose after a tee with a pressure gauge that reads 10 psig provides an example. When the flow is twice as much as shown, the pressure drop has to increase by four times (two squared is four) and the gauge has to read 40 psig.

Fractions can be used to compare different situations. Imagine a gas fired boiler where the gas head pressure is 16 inches of water column at full load. What is the pressure going to be at three-fourths load? Squaring three-fourths gives nine-sixteenths which provides the answer of 9 inches of water column. Note that nine-sixteenths is much closer to one-half than it is to three-fourths. If the firing rate is decreased to one-half, then the pressure drop is one-fourth and the gas head pressure would be only 4 inches of water column. At one-tenth load, the pressure drop is one-hundredth of full load. That should help explain why many flow instruments fail to record accurately at flows of less than 10%. The pressure drop is so small that its value is lower than the pressure fluctuations associated with noise in the piping. The square root chart in Appendix G provides a curve to obtain more precise values. Divide the full load pressure drop noted on gauges between two points in a system (actually using the same gauge connected at two points would be better to eliminate gauge calibration error) to get a 100% (full load) pressure drop value. For any other pressure drop, divide it by that full load value to get a percent pressure drop. The result of the division will always be less than 1 and more than 0. Then draw a horizontal line from that value to the intersection of the curve and a vertical line down to the bottom scale to read the percent of full load flow. To compare one state to another, use the one with the higher pressure drop as if it was full load and perform the math again to determine the flow at the lower pressure drop in percent of the value of flow at the higher pressure drop.

WHAT COMES NATURALLY

Observing what happens in nature helps to understand what is going on in the boiler plant. Engineering practice is based on learning about what happens naturally and then using it to accomplish the desired purposes like making steam. The formation of clouds, fog, and dew all conform to rules set up by nature. Observing them, cause and effect can be learned and made to work for everyone. See how it works and then relate it to what is happening in the boiler plant.

Many natural functions occur in the boiler plant. Steam is generated and condensed by nature. Notice that rain is falling and observe that puddles disappear when it is dry. Fire occurs naturally. See what happens when the fuel and air are mixed efficiently (as in a raging forest fire) and not so efficiently (a smoldering campfire). Observe the hawks spinning in close circles in a rising column of air heated by a hot spot on the ground or air deflected by wind hitting a mountain. Even though the air is not visible, the buoyancy can be demonstrated or how an air stream is diverted. Buoyancy is also evident in a block of wood floating on water. The wood is not as dense as the water. Thus, it is lifted up. The hot air the hawks ride is not as dense as cold air, and so it floats up in the sea of colder air around it. The movement of air and gases of different densities is important in a boiler plant. "Natural draft" refers to the movement of air that naturally occurs because air or gas of higher temperatures is lighter than colder surroundings and rises. The leaves and twigs in a stream spin off to the side, indicating the water is deflected by a rock in the stream. The level of the water increase beside the rock revealing the increase in static pressure as the velocity pressure is converted when it hits the rock. That conversion of velocity pressure to static pressure is how centrifugal fans and pumps work. When something happens that does not make sense, try to relate it to what is observed to happen in nature.

WATER, STEAM, AND ENERGY

At almost every hearing for the installation or expansion of a new boiler plant, there is always someone claiming the plant is not needed because it is much easier and cleaner to use electricity. People do not realize that almost all electricity is generated using boilers—even nuclear power. If it were not for the development of boilers, homes would still be heated with a fireplace in each room. Imagine the environmental consequences of that! Most people know so little about the use of water and steam for energy that it is important to establish an understanding of the very simple basics. The basics are important enough to go over because there are some simple shortcuts described here that can be helpful.

Water is the basis for heat energy measurement. The measure of heat energy, the British thermal unit (Btu for short), is defined as the amount of heat required to raise the temperature of water by 1°F. Engineers know that it is not precisely true at every condition of water temperature, but it is good enough for the boiler operator. As for the energy in steam, it depends on the pressure and temperature of the steam. For most practical applications, it takes 1000 Btu to make a pound of steam.

Likewise, upon condensation, the steam will give up that 1000 Btu. That is called the latent heat of vaporization. For more precision (particularly at high temperatures and pressures), use the steam tables. A few words on using those steam tables are appropriate. Engineers use the thermodynamic term "enthalpy" to describe the amount of heat in a pound of water or steam. A reference point is needed where the energy is zero. That reference point is the temperature of ice water, 32°F. By definition, that water has no enthalpy even though it has energy and energy could be removed from it by converting it to ice. Thus, the enthalpy of water or steam is the amount of energy required to get a pound of water at freezing temperature up to the temperature of the water or steam at the outlet of the boiler. Since freezing water is the reference point, the difference in enthalpy is always equal to the amount of heat required to get one pound of water from one condition to the other.

Chemically, steam and water are the same molecule. In vapor form, it is steam. In liquid form, it is water. It is the same H_2O molecules that have absorbed so much energy, heated up, that they are bouncing around so frantically that they now look like a gas. The form of the water changes as heat is added. It gets hotter until it reaches saturation temperature. Then it converts to steam with no change in temperature and finally superheats. There is, for each pressure, a temperature where both water and steam can exist and that is what is called the saturation point or saturation condition.

It is common knowledge that water boils at 212°F. That is only true at sea level. In Denver, CO, USA, it boils at about 203°F. Under a nearly pure vacuum, 29.75 inches of mercury, it boils at 40°F. The steam tables list the relationships between temperature and pressure for saturated conditions. Since a boiler operator does not need to be concerned with the small differences in atmospheric pressure, the table shows temperatures for inches of mercury vacuum and gauge pressure. If the plant altitude is a mile high (like Denver), then subtract about 3 psi from the table data. Any steam table used by an engineer will relate the temperatures to absolute pressure (psia).

As long as the temperature of water is always less than the saturation temperature at the pressure the water is exposed to, the water will remain a liquid. The enthalpy of the water can be calculated by subtracting 32 from the temperature in degrees Fahrenheit. For example, boiler feed water at 182°F would have an enthalpy of 150 Btu. It takes 970 Btu to convert one pound of water at 212°F to steam at the same temperature. Assuming steam at one atmosphere has an enthalpy of 1150 Btu (212–32+970) provides reasonable accuracy. If the 182°F feed water were sent to a boiler to convert it to steam, it would need to add 1000 Btu to each pound. Just remembering 32°F water has 0 Btu and it takes 970 Btu to convert water to steam from and at 212°F is about all it takes to handle the math of saturated steam problems.

There are other measures of energy that are unique to the industry. One is the boiler horsepower (BHP). With 1000 Btu to make a pound of steam and the ability to generate several hundred pounds of it, the numbers get large and cumbersome. The term boiler horsepower was standardized to equal 34.5 pounds of steam per hour from and at 212°F. Knowing that one pound requires 970 Btu at those conditions, a BHP is also about 33,465 Btu per hour (34.5 × 970). More precisely, it is 33,472 Btu. It is important here to note the distinction that a BHP is a rate value (quantity per hour) and Btu's are quantities. Btu's per hour is abbreviated as "Btu/hr" to identify the number as representing a rate of flow of energy.

Another unique measure of energy, but not used much anymore, is Sq. Ft. E.D.R., meaning square feet of equivalent direct radiation. It is also a rate value. It was used to determine boiler load by calculating the heating surface of all the radiators and baseboards in a building. There are two relative values of Sq. Ft. E.D.R. depending on whether the radiators are operating on steam or hot water. It is 240 Btu/hr for steam and 150 Btu/hr for water. While used rarely, it can relate what happens with heating surface. If a steam installation were converted to hot water, it would need an additional 60% (240/150 = 1.6) of heating surface to heat the same as the steam. Flooded radiators cannot produce the same amount of heat as one with steam in it even though the water is at the same temperature.

The rate of heat transfer from a hot metal to steam and vice versa is always greater than heat transfer from a hot metal to water. That is because of the change in volume more than anything else. Take a simple steam heating system operating at 10 psi (240°F). Check the steam tables to find that a pound of water occupies 0.01692 cubic feet and a pound of steam occupies 16.6 cubic feet. As the steam is created, it takes up almost 1000 times as much space as the liquid did. That rapid change in volume creates turbulence so that the heating surface always has water and steam rushing along it. It is about the same effect as experienced when skiing or riding in a convertible. The skin feels cooler because the air is sweeping over it. When the steam is condensing, it collapses into a space one-thousandth of its original volume and more steam rushes in to fill the void. That is the mechanism that improves heat transfer with steam. Note the fact that steam has more heat on a per-pound basis.

Steam may have more heat per pound, but those pounds take up a lot more space. One cubic foot of water at 240°F contains 12,234 Btu. One cubic foot of steam only contains 69.88 Btu. Why not use only hot water systems because water can hold more heat? The best answer is because a lot of water would have to be moved around to deliver the heat. To deliver the heat provided by one pound of steam would require about 200 pounds of water. Steam, as a gas, naturally flows from locations of higher pressure to those of lower pressure. It does not have to be pumped. The rate of water flow is restricted to about 10 feet per second to keep down noise and erosion. Steam can flow at ten times that speed. Nominal design for a steam system is a flowing velocity of about 6000 feet per minute.

Hot water is a little easier to control when there are many low temperature users. Hot water systems have a minimal change in water volume at all operating temperatures. For that reason, the cost of pumping water around a hot water system will be paid in exchange for avoiding the dramatic volume changes in steam systems. Never forget that there is a change in volume in a hot water system. To forget that is to invite a disaster. Water changes volume with changes in temperature at a greater rate than anything else, almost ten times as much as the steel used in most boiler systems. Unlike steam, it does not compress as the pressure rises. The system must allow the expanded water to go somewhere. The normal means for the expansion of the water in a hot water system is an expansion tank—a closed vessel containing air or nitrogen gas in part of it. Modern versions of expansion tanks have a rubber bladder in them to separate the air and water. The bladder prevents absorption of the air into the water. The air or nitrogen compresses as the water expands, making room for the water with a little increase in overall system pressure. Tanks without bladders normally have a gauge glass that shows the level of the water in the tank so that the condition can be known.

A hot water system will also have a means to add water, usually directly from a city water supply. Most systems have a water pressure regulator that adds water as needed to keep the pressure above the setting of the regulator. A relief valve (not the boiler's safety valve) is also provided to drain off excess water. Older systems can be modified and added to the extent that the expansion tank is no longer large enough to handle the full range of expansion of a system. In some newer installations, there are tanks that were not designed to handle the full expansion of the system. Those systems require automatic pressure regulators to keep the pressure in the system as the water shrinks when it cools and the relief valve to dump water as it expands while the system heats up. The tank should be large enough, however, to prevent the constant addition and draining of water during normal operation. A good tight system with a properly sized expansion tank should retain its initial charge of water and water treatment chemicals to simplify system maintenance.

All hot water systems larger than a residential unit should have a meter in the makeup water line in order to determine if water was added to the system and how much. Lacking that meter, a hot water system can operate with a small leak for a long period of time. Scale and sludge formation will occur until someone finally notices the stack temperature getting higher or some other indication of permanent damage to the boiler or system.

Steam compresses. There is seldom a problem of expansion with steam boilers unless the system is flooded. However, since steam temperature and pressure are related, when using steam at low temperatures, a vacuum can result and air from the atmosphere will leak in. The atmospheric air is at a higher pressure and will flow into the vacuum. In those cases where there is a tight system, the vacuum formed as steam condenses will approach absolute zero. The weight of the air outside the system will produce a differential pressure of 15 psi which can be enough to crush pressure vessels in the system. To prevent that from happening, low temperature steam systems usually have vacuum breakers to allow air into the system. Check valves make good vacuum breakers because they can let the air in but do not let the steam out. Thermostatic steam traps and air vents are required to let the air out when steam is admitted to the system. If installed and operated properly, low pressure steam systems can work well because the metal in the system will be hot and dry when the air contacts it. Corrosion is minimal.

To know how much heat is delivered per hour, determine the difference in enthalpy of the water or steam going to the facility and what is returning. Then multiply that difference by the rate of water or steam flowing to the process. The basic formula is: enthalpy in minus enthalpy out times pounds per hour of steam or water. Be sure to check the units. Water flow is often in gallons per minute. That can be converted by a simple rule of thumb: 1 gpm times 500 equals pounds per hour. One gallon of water weighs about 8.33 pounds and 1 gpm would be 60 gallons per hour. Thus, 8.33×60 equals 499.8, or 500, close enough. Since the difference in enthalpy is about the same as the difference in temperature for water, heat transferred in a hot water system can be calculated as temperature in minus temperature out multiplied by gpm times 500.

For steam systems, it is simply 1000 times the steam flow in pounds per hour if the condensate is returned. There are times when the condensate is not returned. Perhaps, a condensate line or pump broke or the condensate is contaminated. That happens in a lot of industrial plants because it is too easy for the condensate to get contaminated and is wasted intentionally. In those circumstances, the heat lost in the condensate that would have been returned has to be added. What is really being delivered to the plant under those conditions is the heat to convert the water to steam plus the energy required to heat it from makeup temperature to steam temperature. There are also applications where the steam is mixed with the process, becoming part of the production output. An example is heating water by injecting steam into it. The amount of heat that must be added to make the steam is the same as that in the previous example, but the heat delivered to the process is all the energy in the steam.

One problem many boiler operators have is the concept of saturation. Steam cannot be generated until the water is heated to the temperature corresponding to the saturation pressure. Once the water is at that temperature, the temperature cannot go any higher as long as water is present. At the saturated condition, any addition of heat will convert water to steam and any removal of heat will convert steam to condensate. The temperature cannot change as long as steam and water are both present and the pressure remains constant. When the heat is only added to the steam, the steam temperature will rise because there is no water to be converted to steam. Whenever the steam temperature is above the saturation temperature, it is called superheated steam.

Superheated steam does not just require addition of heat. In an insulated vessel containing nothing but saturated steam and when the pressure is reduced, the saturation temperature drops. The energy in the steam does not change. The temperature cannot drop and the steam is superheated. When high pressure steam is delivered through a control valve to a much lower pressure in a process heater, the superheat has to be removed before the steam can start to condense. The heat transfer is from gas to the metal. The turbulence associated with steam condensing to a liquid is eliminated. It is not as efficient as the heat transfer for condensing steam. Process heaters can be choked by superheated steam when poor gas to metal heat transfer leaves much of the surface of the heat exchanger unavailable for the higher rates of condensing heat transfer.

So why superheat the steam? Steam is superheated so that it will stay dry as it flows through to a steam turbine or engine. Without superheating, some water would form as soon as energy is extracted. The water droplets would impinge on the moving parts of the turbine (a familiar concept would be spraying water into the spinning wheel of a windmill) damaging the turbine blades. In an engine, it would collect in the bottom of the cylinder. In electric power generating plants, it is common to pipe the steam out of the turbine, raise its temperature again (reheating it), and then return it to the turbine just to maintain the superheat in the last stages of the turbine. From a thermodynamic point of view, the higher the temperature of the steam to a heat engine such as a steam turbine, the more efficient the heat engine will be.

When generating superheated steam, some of it can be used for purposes other than the turbine, in which case it should not be superheated. In that case, the steam can be desuperheated. Heat is removed or water is added to the superheated steam for desuperheating. When water is added, it absorbs the heat required to cool the steam by boiling into steam. In most applications, superheat cannot be eliminated entirely because some small amount of superheat is needed to detect the difference between that condition and saturation. As long as there is a little superheat, its condition is all steam. When the steam is at saturation conditions, the amount of water in the steam is not known.

Understanding saturation is the key to understanding steam explosions. When water is heated to saturation conditions higher than 212°F, as in a boiler, it cannot exist as water at that temperature if the pressure vessel containing it fails. Under those circumstances, the saturated condition becomes one atmosphere and 212°F as the water leaks out. A portion of the water is converted to steam to absorb the heat required to reduce the temperature of the remaining water to 212°F. How much steam is generated is determined by the original boiler water temperature. Every pound of water converted to steam expands to 26.8 cubic feet. The rapid expansion of the steam is the steam explosion. For a heating boiler operating at 10 psig, the 240°F water has to cool to 212°F, releasing 28 Btu per pound. It can only do so by generating steam at 212°F which contains 1150 Btu per pound. One pound of steam can cool 41 pounds of water (1150 ÷ 28). The volume of 42 pounds of 240°F water at 0.01692 cubic feet per pound (0.71 cubic feet) becomes 41 pounds of water at 212°F (0.01672 × 41 = 0.685 cubic feet) and one pound of steam (26.8 cubic feet). The original volume of water expanded 38.71 times (0.685 + 26.8 = 27.48 ÷ 0.71) and it happens almost instantly.

THE STEAM AND WATER CYCLE

For commercial, industrial, or institutional boiler plants, the concept of the steam and water cycle may be a foreign concept. Understanding a steam cycle becomes important when power is generated with the steam. Generating power means generating electricity or powering a mechanical device normally with a steam turbine. For almost any plant, there is a steam and water cycle. The exception is hot water heating boiler where steam is not generated. In hot water heating plants, the cycle is simply heat in and heat out. It is added to the water in the plant boilers, the water is transferred by circulating pumps (or natural circulation) to radiators, convectors, kitchen equipment, etc., where the water is cooled by the users of the heat, and then the water is returned to the boiler to be heated again.

Cycle diagrams show the flows of steam and water to the various components in the system. Valves, both isolating and control, are not shown, nor are steam traps always shown. They are understood to exist in components that use the steam and where required to isolate systems for maintenance.

The simplest steam and water cycle exists in a low pressure heating system. Water is pumped into the boiler, heated to saturation temperature, and then converted to steam. Almost all of the energy added to the water and steam in the boiler is latent heat (the latent heat of evaporation, the energy used to convert the water to steam). The steam then leaves the boiler and flows through piping and control valves to radiators, convectors, kitchen equipment, etc., where the steam is condensed. Those users of the heat primarily use the latent heat of condensation. There may be some heat transferred to the user by cooling the condensate. The condensate is then returned to the boiler completing the cycle. Figure 1-7 is an example of a steam cycle diagram for a conventional heating plant, perhaps a school, a restaurant, or an apartment building. When steam is used to heat things at temperatures close to, at, or above 212°F, the temperature of the condensate from that heating equipment is much higher and any drop in pressure will result in some of the condensate flashing into steam.

That is very true for condensate that is formed in the main steam piping. The cycle diagram in Figure 1-7 is provided to explain how some little changes in the cycle can make a difference. Steam is represented by two lines, sort of like along a highway, where the steam flows down the middle of the lines. Water, being much heavier, is shown in black between the two lines. The boxes containing a capital "T" represent the steam traps for

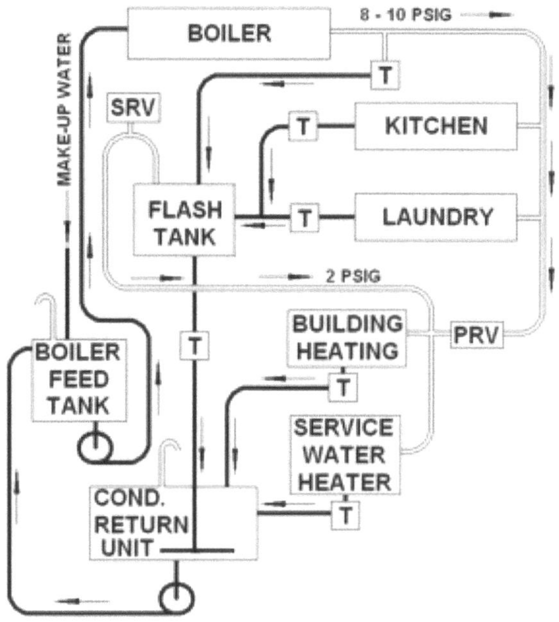

Figure 1-7. Low pressure steam cycle.

the systems. "SRV" represents a safety relief valve (not a safety valve). It is one that relieves a fluid proportional to the pressure above its set point. "PRV" represents a pressure reducing valve or a complete pressure reducing station.

What is going on here is a system that does not waste heat. It contains a flash tank that does what it is supposed to do—unlike so many that are installed and do nothing or simply separate the flash steam from the condensate and throw it away. Some of the condensate that is drained from the steam piping will flash into steam. Condensate from the main header drains, kitchen, and laundry is bound to be hotter than 212°F, and once it leaves the steam traps, some of it will flash into steam. By piping that condensate to the flash tank, that flash steam is recovered at a lower pressure and can be used for heating the building in the winter and service water (domestic hot water at 120°F) in the winter and summer. The condensate from different pressure sources is always piped independently to the flash tank so that flashing condensate from the higher pressure sources does not block the flow of condensate from lower pressure sources. If the flash steam is more than the demand for heating and service water, the SRV (in this system set at 2.5 psig) will open and waste the excess steam because there is no other place to use it. Under normal operating conditions, there is no waste to atmosphere because all the flash steam from flashing higher pressure condensate is used in the 2-psig system. Condensate from the flash tank would produce more flash steam when entering the condensate return unit, and that is why it is piped to a subsurface diffuser (pipe with holes drilled in the bottom) so that it can heat colder condensate from the heating and hot water systems. The steam condenses as it heats the low pressure condensate. The system practically eliminates any steam venting from the condensate unit and boiler feed tank (unless a trap has failed).

Why try to save a little steam? Every pound of vented steam represents about 1000 Btu wasted. Some systems waste condensate that has to be replaced by cold makeup water and heated to saturated steam in the boiler. In this low pressure plant example, that is 1180 Btu per pound. Part of the wise operator's job is to constantly monitor vents for wasted steam and try to do something about it.

In power plants where the steam is primarily used to generate power, the steam and water cycle becomes more complex. Some of these are referred to as supercritical boiler plants because the steam is generated at pressures exceeding 3190 psig where the density of water and steam are the same. Density is simply a value representing the weight of a substance in pounds per cubic foot. Because there is no difference in density, those boilers do not have a level indicator or gauge glass as there is no water level. After the water is converted to steam in the boiler, it passes through the superheater, more tubes exposed to the boiler flue gases or radiant heat from the fire where it is heated further, thus increasing its temperature. Steam heated to temperatures higher than the saturation temperature is superheated steam. The superheated steam is piped to a steam turbine or steam engine where the energy in the steam is converted to power. The conversion of energy to create power is associated with a drop in steam pressure and temperature. Each horsepower of mechanical energy is produced by removing 2545.1 Btu per hour from the steam plus an allowance for losses due to inefficiency.

Many industrial plants produce superheated steam in higher pressure boilers to generate power with steam turbines that exhaust at a nominal high pressure for use in the facility. These turbines are referred to as "back pressure turbines." In many plants, that exhaust steam is at 15 psig after a turbine where the steam is used mostly for evaporators in the process plant. Other facilities have turbine exhaust pressures as high as 150 psig for distribution to all users such as in a chemical plant.

Now compare that low pressure plant with an electric utility plant like the one depicted in Figure 1-8. The plant consists of two boilers and two turbine sets. Each set of turbines (H.P. for high pressure, I.P. for intermediate pressure, and L.P. for low pressure) in this plant are all driving one shaft connected to their respective electricity generator. Steam generated in the boiler is piped to headers, either external to or within the boiler that connects to superheater tubes which raise the temperature of the steam from saturation to the temperature required by the steam turbine. The maximum temperature

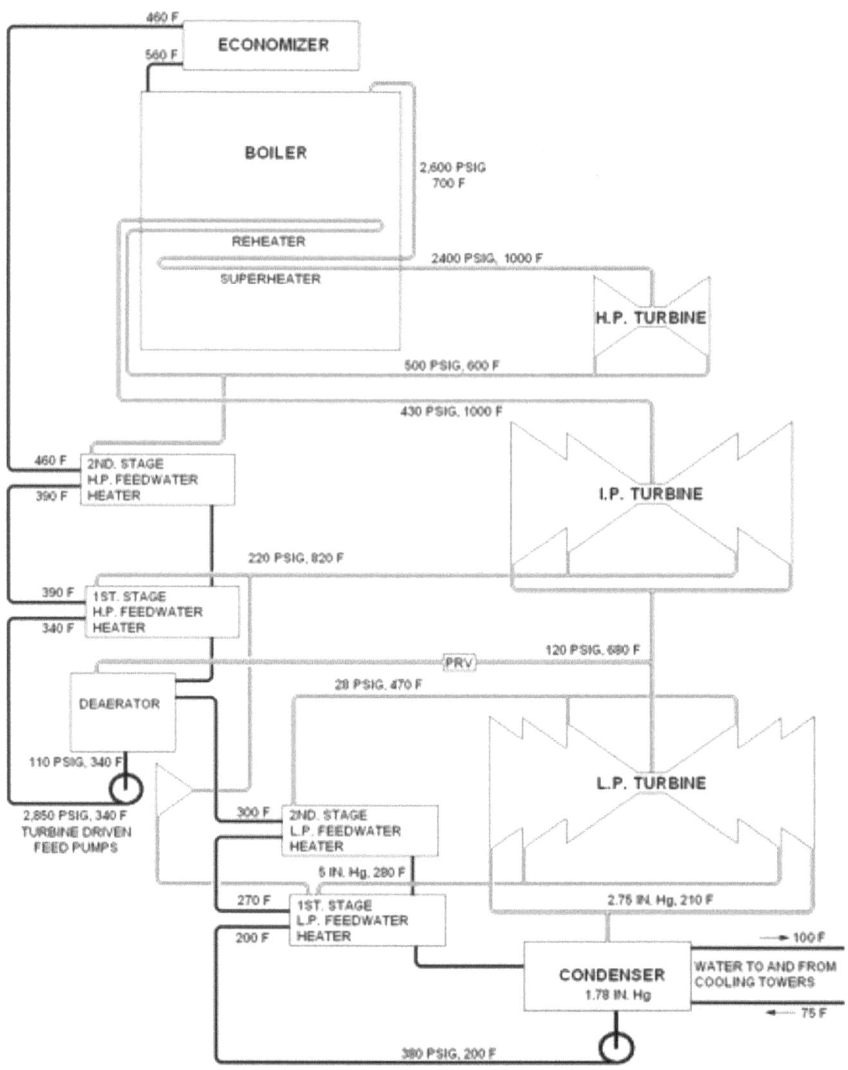

Figure 1-8. Typical utility cycle.

is normally determined by the strength of the metal that the superheater tubes are made of. Utility plant superheaters are typically exposed to radiant energy from the fireball located inside the boiler. Several methods are used to control the superheat temperature. An "interpass" desuperheater could consist of redirecting a portion of the superheated steam between two sections of the superheater through coils in the steam drum where the boiler water is heated by the superheated steam. The steam is cooled and then mixed with the remainder before passing through the next section of superheater tubes. Another interpass method is injecting boiler feed water into the stream of superheated steam between sections where the steam is cooled by evaporating the added boiler water. The spray station that does this is called the desuperheater or spray station. Controlling the flow of the steam through the drum desuperheater or the water to the spray station controls the steam temperature of the steam at the superheater outlet.

In a typical electric utility plant, the energy extracted in the high pressure turbine reduces the temperature of the steam so much that generating any additional power with that steam would result in the steam reaching saturated conditions where droplets of water would form, strike the rapidly rotating turbine blades, and damage the turbine. Thus, utility plants typically have reheaters—additional tubes in the boiler that are exposed to the flue gases before they reach the boiler economizer section. There, the steam from the high pressure turbine is heated again to increase its superheat before continuing through the intermediate pressure turbine. The reheater is "convective" because the superheater shields it from the radiant heat of the furnace and it is heated only by the flowing flue gas. Each "stage" of a turbine has different inlet and outlet pressures. The reheater operates at a lower pressure than the boiler and superheater, but can have an outlet temperature as high as the superheater.

As the pressure decreases, the volume of steam increases. To accommodate the increasing volume of the steam, the turbine casing would have to grow in size substantially so much so that it would become extremely large and extremely expensive to build. Economics in place when the plant was designed determine an optimal size for the casing and turbine blades for each stage of the turbine. At selected stages, steam is piped from the turbine for auxiliary uses, removing enough steam to keep the turbine from becoming excessively large. That steam is normally referred to as extraction steam or bleed steam. In some industrial plants, steam supplied to the facility is extracted with the remaining steam continuing through the turbine to the condenser. An additional feature of some turbines in that application incorporate what is referred to as a goggle plate that throttles the flow of steam to the rest of the turbine to maintain the pressure of the steam supplied to the facility. In electric utility plants, extracted steam is used to power auxiliary turbines, comfort heating of the facility, and heat the feed water at different stages between the condenser and the boiler inlet.

Steam leaving the last stage of the turbine enters the condenser where the latent heat is removed and the steam is converted to water. The condenser can be air cooled which normally consists of multiple heat exchangers with fan forced atmospheric air used to cool them. Air cooled condensers are only used where water comes at a premium or is in very limited supply. Most condensers are water cooled. Water cooled condensers can use water from cooling towers or from a river, lake, or the ocean which normally provides the highest plant efficiency because that water is the coldest. The colder cooling water reduces the pressure on the steam side of the condenser because the colder surfaces lower the saturation temperature of the steam. The pressure in the condenser is always less than atmospheric so that pressure is referred to as a vacuum. The vacuum, typically expressed in inches of mercury, can be estimated from the steam tables by adding 10°–20° to the cooling water temperature. In the example of Figure 1-8, note that the temperature of the condensate is higher at the discharge of the condensate pumps. That is because the condensers are designed to sweep the steam leaving the turbine over the droplets of condensate before turning up to contact the tubes that condense the steam.

The steam condensate is removed from the condenser by condensate pumps that force the water to flow from the condenser to the deaerator. On its way, the condensate will first pass through low pressure heat exchangers that use extraction steam and operate at a partial vacuum, draining their condensate into the condenser. Condensate from auxiliary systems that heat equipment at low temperatures, such as space heating, is typically drained into one of the low pressure heat exchangers. When the condensate reaches the deaerator, it is mixed with some extracted steam to heat it to saturation conditions. That steam source is commonly augmented with steam from a pressure reducing station supplied by a higher pressure source to maintain the plant's heat balance (explained later). When water is heated to saturation, any gases that are absorbed in the water cannot remain absorbed because the water is about to boil. The gases form bubbles in the water, which is agitated to force the bubbles out of the water and out a vent at the top of the deaerator. The heated, deaereated water is collected in a storage tank which is either part of the deaerator or as a separate vessel normally designed to hold enough feed water to keep the plant boilers operating for 20 minutes. The deaerator is a critical piece of equipment, as oxygen dissolved in the water is corrosive to many types of steel. The deaerator serves the purpose of heating the return water from the condenser and removing oxygen that might be dissolved in the water.

Boiler feed pumps draw water from the deaerator and supply the system pressure to the feed water. The water is pumped through the high pressure feed water heaters, the feed water control valves, and the economizer and then to the boiler circulating system. For subcritical boilers, there will be a steam drum to separate the steam from the water. For super critical boilers, the fluid is pumped directly to the lower headers of the boiler. Newer plants may operate without feed water control valves by using variable speed control of the boiler feed pumps to change the feed water flow. Some plants, especially supercritical, may use booster pumps to help increase feed water pressure. Controlling flows of extraction steam is one of the ways a utility plant boiler operator earns a salary. Maximizing electricity generation by ensuring maximum flow of steam through each turbine stage produces more Kwhr output and that is the product the plant is selling to its customers. The system of feed water heaters, pumps, and deaerators is called the feed water train. The cycle that uses the feed water train is called a regenerative cycle. A regenerative cycle has a better efficiency because the latent heat of vaporization of the extraction steam is being utilized, whereas sending that steam to the condenser basically rejects that heat to the environment (cooling tower or once through cooling).

At least most of the water, sometimes a liquid and sometimes steam, is circulated within the system leaving the boiler and then returning to it. That constitutes a cycle. In particular, this is a closed cycle. Assuming no contamination, the water is 100% recycled back to the boiler. The wise operator should be able to describe the cycle for the plant being operated. If there is no diagram for the plant lying around or buried in a file somewhere, try making one to be displayed. Displaying the cycle diagram in the plant not only informs visitors of the scope of the system but also serves as a reminder as to what an operator should monitor to make the cycle as efficient as possible.

COMBUSTION

Most of the fuel being used is called "fossil fuel" because its origin is animal and vegetable matter that was trapped in layers of the earth where it became fossilized, breaking down, for the most part, into hydrocarbons. Hydrocarbons are materials made up principally of hydrogen and carbon atoms. It is the hydrocarbon portion of fossil fuels that generates more than 90% of the energy in use today, from the propane that fires up a barbecue to the coal burned in a large utility boiler to make electricity. The normal everyday boiler plant also burns hydrocarbons. The four main forms are natural gas, light oil, heavy oil, and coal. The principal difference in these fuels is the hydrogen/carbon ratio and the amount of other elements that are in the fuel. Despite the fact that typical hydrocarbons vary from a gas lighter than air to a solid, they all burn by combining with oxygen from the air to release energy in the form of heat. It is not necessary to know exactly how it does it, only to understand that certain relationships exist. What happens depends on changes the operator makes or changes that are imposed on the operator by the system. Most fuels consist of many compounds. In order to better understand the properties of the fuel and the products of combustion, a chemical analysis of the fuel is needed .This analysis provides a breakdown of each element contained in the fuel. An analysis broken down by the elements is called an "ultimate analysis." An inspection of the "ultimate analysis" reveals that the fuel gets heavier with the amount of carbon in the fuel and lighter as hydrogen increases.

The Boy Scouts teach the fire triangle. To create a fire, three things are needed: a fuel, air, and enough heat to get the fire going. Wood, about four inches in diameter and over a foot long, can be stacked up to prepare a campfire, but even though there is a lot of fuel there, with air all around it, the fire will not start with just a match. Obviously, there is fuel and air. The problem is not enough heat. To get that fire going, kindling is needed—smaller and lighter pieces of fuel that will continue to burn once heated with a match. They produce more heat to light those big sticks that were stacked up. Once the fire gets going, the heat generated by those big sticks burning is enough to keep them going and light more big sticks as they are added to the fire. If the big sticks are pulled apart from each other, the fire will go out. This provides a very good lesson on the relationship of fuel and heat in a fire. As the fuel burns, it generates heat. Some of that heat is used to keep the fuel burning and some is used to start the added fuel burning. When the fire is compact, where a good portion of the heat generated is only exposed to the fire and more fuel, the fire will be self-supporting. If the fire is spread out where all its heat radiates out to cold objects, the fire will go out.

The fuel in the furnace of a boiler burns at temperatures in the range of 1200–3200° F, which is usually more than enough to keep it burning and heat up any new fuel that is added to the fire. Modern furnaces, however, are almost entirely composed of water cooled walls, which absorb most of the radiant heat of the fire to boil water to make steam. Despite that high temperature, a fire in a modern boiler is barely holding on and it does not take much to put it out. That is one of the reasons for deploying flame detectors.

All common fuels are principally hydrocarbons—material containing atoms of hydrogen and carbon in various combinations with varying amounts of other elements. The reason hydrocarbons are important is that they release energy in the form of heat when they burn. This burning of the fuel is called the "process of combustion."

Different adjectives can be used to describe different combustion results, including partial, perfect, complete, and incomplete. Partial combustion means part, but not all, of the fuel has been burned. Incomplete combustion is basically the same. The difference is that partial combustion is the intended result and incomplete combustion is an undesirable result. Perfect combustion is an ideal condition that is almost never achieved. It is when all of the fuel is burned with the precise amount of air necessary to do so. This condition is also referred to as "stoichiometric" combustion. Complete combustion burns all of the fuel, but there is usually some air left over.

Every fuel has its stoichiometric air/fuel ratio. That is the number of pounds of air required to perfectly burn one pound of fuel. The air/fuel ratio of a fuel is dependent on the ratio of carbon to hydrogen in the fuel, the amount of hydrocarbon in the fuel, and, to a lesser degree, the air required to combine with other elements in the fuel. Note that this is a mass ratio, which is not related to volumes. It can be converted to a volumetric ratio (cubic feet of air per cubic foot of fuel), provided the conditions of pressure and temperature are specified to define the density of the fuel and air. The air/fuel ratio for a fuel can be determined from an ultimate analysis of the fuel (Appendix L).

The air required for the fuel is not fully consumed. Only part of the air is used, the oxygen. Atmospheric air contains about 21% oxygen by volume. A more precise value is 20.9%, but for all practical purposes, 21% is close enough. What is in the other 79%? It is all nitrogen. For the most part, nitrogen is "inert." It does not do much of anything except hang around in the atmosphere. When various air emissions are considered when operating a boiler, it is not entirely inert. That other tenth of a percent contains a lot of gases, mostly noble gases and carbon dioxide, that do not really do anything in the process of combustion either. They are also considered to be inert. The rest of the air is 79% nitrogen. It absorbs a lot of the heat generated in the fire and limits how fast that oxygen can get to the fuel. It is considered a moderator in the process of combustion because it keeps the fuel and oxygen from going wild. Without it, everything would burn to a crisp very quickly.

In the early space program, an accident where three astronauts were burned to death in a capsule during a test while sitting on the ground was due to a small electrical fire in a pure oxygen environment. Without the nitrogen to moderate the rate of combustion, the inside of the capsule was consumed by fire in seconds. There are flames that burn fuel with pure oxygen. The space shuttle's engines do it and the typical metalworker's cutting torch uses it. But those applications have a limit on their burning imposed by consumption of all the fuel and the moderating effect of the nitrogen in air surrounding those operations. Keeping those cutting torch oxygen tanks properly strapped down in the boiler plant is important because they are a source of pure oxygen that could produce a rapid, essentially explosive, fire in the plant which it is not prepared for.

The appropriate title for this part should be combustion chemistry. However, mention the word "chemistry" and a boiler operator's eyes glaze over. Nevertheless, it is important to understand what is happening in that fire to know how to operate a boiler properly. Understand the chemistry and use that understanding to become a wiser operator.

Any fossil fuel has only three elements in it that will combine with the oxygen in the air and release heat. Actually, there are only four reactions of interest. (Combining of materials to produce different materials is a reaction.) The first one is hydrogen. Hydrogen in the fuel combines with oxygen in the air to produce dihydrogen oxide (H_2O). That is really what is called H-two-O, and it is water. Of course, the heat generated by the process produces water so hot that it is steam. There is no liquid water dripping from a fire. Hydrogen is very reactive. It will react with one of the other products of combustion if necessary to get its oxygen. To date, nobody has been able to find any hydrogen left over from a combustion process because it always gets its oxygen to make water. All of the hydrogen in the fuel will burn to water if combustion is complete. If it is not complete, the hydrogen will still be combined with some carbon atoms to produce an "unburned hydrocarbon." Sometimes it is not any of the hydrocarbons that the fuel started out with. It can be an entirely different one.

Carbon, in complete combustion, combines with the oxygen in the air to make carbon dioxide—CO_2 for short. That is one atom of carbon and two atoms of oxygen. The fizz in soda pop is carbonated water (CO_2 dissolved in water). Breathing produces carbon dioxide. Actually, human bodies convert hydrocarbons to water and carbon dioxide. The process is slower and at much lower temperatures than those in a boiler furnace. Since carbon is the major element in fuel, combustion produces lots of carbon dioxide in a boiler. Next in quantity is water. For the most part, firing a boiler is natural and it produces mostly CO_2 and H_2O which are not harmful to health.

If there is incomplete combustion, the carbon will not burn completely. Instead of forming CO_2, it forms CO—carbon monoxide. That is the colorless and odorless gas that kills. A person deciding to commit suicide by sitting in a running car in a closed garage dies because the car engine generates CO. That CO is trying to find another oxygen atom to become CO_2, and it will strip it from human bodies if it can. That is what happens. It robs the blood of oxygen and the person dies of asphyxiation.

The last flammable (stuff that burns) constituent in fuel is sulfur. Sulfur combines with the oxygen in the air to form SO_2—sulfur dioxide. There is not a lot of sulfur in fuel, but what is present burns. Those are the three elements: carbon, hydrogen, and sulfur combine with oxygen to produce CO_2, water, and SO_2. Heat is generated in the process. When numbers are used in subscript (small and slightly below normal), that indicates the numbers of atoms (represented by the letter in front of the number) in a molecule. Numbers in normal case indicate the number of molecules. Atoms, represented by the letters, combine to form molecules. Many gases, oxygen is one of them, are diatomic. That means it takes two atoms to make a molecule of that gas. All fuels are made up of atoms of hydrogen and carbon. It is the mix of atoms to form the molecules of the fuel that produces the different fuels in use. In other words, it is the combination of hydrogen and carbon atoms in the molecules that determines if the fuel is a gas, an oil, or a solid material like coal. Here is the list of basic combustion chemistry equations.

$$C + O_2 \Rightarrow CO_2$$
+ 14,096 Btu for each pound of carbon burned.

$$2H_2 + O_2 \Rightarrow 2\,H_2O$$
+ 61,031 Btu for each pound of hydrogen burned

$$S + O_2 \Rightarrow SO_2$$
+ 3,894 Btu for each pound of sulfur burned

$$2C + O_2 \Rightarrow 2CO$$
+ 3,960 Btu for each pound of carbon burned

C is carbon, one atom.
CO is a molecule of carbon monoxide, containing two atoms.
CO_2 is carbon dioxide, one molecule containing three atoms.
H_2 is a molecule of hydrogen, consisting of two atoms.
H_2O is a molecule of water, consisting of three atoms.
O_2 is a molecule of oxygen, consisting of two atoms.
S is an atom of sulfur.
SO_2 is a molecule of sulfur dioxide, three atoms.

The rules of the equations are rather simple. The same number of atoms must be on both sides of the equation. Matter is neither created nor destroyed in a chemical reaction. It is merely converted. Nothing is destroyed as a result of combustion. It is one of the natural laws of thermodynamics, which is called the law of conservation of mass. It can also be expressed as input equals output at steady state. It may appear that the wood in the campfire disappeared, but the truth is that it combined with the oxygen in the air to form gases that disappeared into the atmosphere along with the smoke. Every pound of carbon is still there. It is just combined with oxygen in the CO and CO_2.

There is another natural rule which states that all gas molecules at any particular pressure and temperature take up the same amount of space. When CO is produced, the number of flue gas molecules is doubled and the gas volume increases. The increased gas volume produces more pressure drop through the boiler which restricts flue gas from flowing out. Since the gas cannot get out as fast, less air can get in and there is less oxygen. That makes more CO. The result is a generous generation of CO until the heat input has dropped to where there is a balance between the pressure drop from more CO and the reduced generation of CO, as the air input is decreased. If the air flow is decreased at a constant fuel flow rate, the steam flow will start to drop when the CO starts forming.

At this point, it will be helpful to discuss how fuels are produced. Coal is not necessarily made but is simply dug up from the ground. However, some of it is put through a water washer to reduce the amount of ash. Some of it may be treated by exposure to superheated steam. Some of it may be ground up fine and mixed with fuel oil to create another fuel (synthetic fuel). Natural gas and fuel oil also go through preparation processes.

Natural gas is normally put through a scrubber system after it is extracted from the ground to remove undesirable carbon dioxide and sulfur compounds. Oil is refined into different grades.

For all practical purposes, the gas flowing up the large pipelines from Louisiana and Texas to consumers on the east coast does not have any sulfur in it to speak of. If it did, the sulfur might react with the oils in the big compressors that the pipeline companies use to pump the gas north and make those oils acidic. Once the gas arrives at a gas supplier in the northeast, a tiny amount of sulfur is added back into the gas in the form of mercaptans—chemical compounds that give gas its odor so that leaks can be detected.

Fuel oils, whether they are number 1 (kerosene), 2 (diesel), or any of the heavier grades (4, 5, and 6), all come from crude oil—the oil that is pumped from the earth or gushes when it is under pressure. The crude oil is "refined" in a refinery to separate the different fuels, and a lot more, from the material that comes out of the ground. One big fraction of crude oil is gasoline. In fact, there is such a big demand for gasoline that some of the other products are converted by different processes to make more gasoline to satisfy the driving demand. The basic process of separating the different components from crude oil is distillation, where the oil is heated until the lighter portions including naphtha, gasoline, and others evaporate.

A good portion of kerosene and light fuel oil (number 2) is produced by distillation. Some of that and heavier parts of the crude oil are heated further and exposed to catalysts (materials that help a reaction occur) to "crack" them, breaking more complex hydrocarbons down into lighter, less complex ones. That is what happens when the fuel is exposed to the heat of a fire. It is distilled and cracked until it becomes very simple hydrocarbons that readily react with air to burn. It is argued, with some degree of accuracy, that only gases burn and the heat has to convert the fuel to a gas before it will burn. All that distillation and cracking takes some time and that is why a fuel does not burn instantly once it is exposed to air.

The primary component of natural gas is methane, which is represented by the formula CH_4. The same rules for formulas apply. To burn the methane requires two oxygen molecules, O_2, from the air. One molecule of the O_2 combines with the carbon to form CO_2 and the other combines with the four hydrogen atoms to make two molecules of H_2O. The equation is as follows:

$$\underset{16}{CH_4} + \underset{64}{2\,O_2} \Rightarrow \underset{44}{CO_2} + \underset{36}{2H_2O}$$

The numbers under the groups of molecules in the equation represent the molecular weights of the different molecules. Atomic weights have no units that refer to the weights assigned to the atoms in the periodic table (C = 12, H = 1, etc.). They are all relative to oxygen assigned an atomic weight of 16 as the reference (equal to the sum of the protons and neutrons in the nucleus of the oxygen atom). Hydrogen has an atomic weight just slightly more than 1 because there is more than one type of nucleus for a hydrogen atom. For the accuracy of this example, the atomic weight of 1 is sufficient. Carbon has an atomic weight of 12. The molecular weight of a compound is equal to the sum of the atomic weights of all of the atoms in that compound. For methane, there is one atom of carbon and four atoms of hydrogen, making the molecular weight 16. One carbon plus four hydrogen atoms give methane a molecular weight of 16 (12 + 4), the number beneath the formula. One molecule of oxygen consists of two atoms. Its molecular weight is 32 (2 × 16). Two molecules of oxygen are needed to react with one molecule of methane. The total reacting weight is 64, the number beneath the formula. The CO_2 is 12 + 2 × 16, giving 44. The two water molecules are twice (2 × 1 + 16) giving 18 each, or 36 in total. The law of conservation of mass means that the mass should be the same on each side of the equation. The totals on each side amount to 80 and total mass is conserved. Matter is neither created nor destroyed in an ordinary chemical reaction. The reactants are converted into products, but the total amount of matter remains constant. Note that it is the molecules that are the reacting species. The specific amounts are dictated by the numbers of molecules that are involved in the reactions. Getting back to units of mass, when pounds are the common unit, the molecular weight expressed in pounds is called the pound mole. If grams are being used, it is the gram molecular weight and that is a gram mole. In the combustion of methane, one lb-mole of methane reacts with 2 lb-moles of oxygen to form one lb-mole of carbon dioxide and 2 lb-moles of water. The molecular weight of a substance that is a gas occupies the same volume. That happens to be convenient because volumes are easier to measure in flowing systems than weights. In the case of methane combustion, one volume of methane reacts with two volumes of oxygen to form one volume of carbon dioxide and two volumes of water vapor. Volumetric flow rates can be used to provide the proper amounts of fuel and oxygen for reaction (combustion).

Weights can be used to determine the amount of energy in the fuel. Recall that carbon will produce 14,096 Btu for every pound of carbon burned. In the case of methane 12/16ths of it is carbon, and that will provide

10,572 Btu per pound of CH_4 (12 ÷ 16 × 14,096). Similarly, the 4/16ths of hydrogen will produce 15,257 Btu (4 ÷ 16 × 61,031). Add the two values to get a higher heating value of methane of 25,829 Btu per pound. That figure is a little high for natural gas, as some energy is needed to break the chemical bonds of the starting molecules. The measured value is 23,879 Btu/lb. More commonly, natural gas is valued at about 1000 Btu per cubic foot. That is because it is always measured by volume, in cubic feet. However, the measurement is also always corrected for the actual weight of the gas because it is the mass that determines the heating value and not the volume.

In order to calculate exactly how much flue gas will be produced by a fuel, precisely what the air to fuel ratio is for that fuel, and how much energy will be released from the fuel, an "ultimate analysis" of the fuel is required. That analysis gives precisely how much carbon, hydrogen, sulfur, etc., is in the fuel. An ultimate analysis also includes a measure of the actual heating value. The worksheet in the appendix on page 419 is used to determine the amount of air required to burn a pound of fuel and some other useful information.

Getting back to the campfire, the big sticks did not start burning right away. In addition to the fact that a big heavy stick sucks up all the heat from the match without its temperature going high enough for it to burn, there is also the flammability limit. If enough heat is added to any mixture of air and fuel, some of it will burn. What is needed is to come up with a mixture of air and fuel that will not only burn but also will produce enough heat in that process that it will continue to burn. There are two rules. First, the fuel and air mixture has to be in the flammable range. The second requirement is a fuel-rich condition to start. The hard part for the design of boiler plants is to make certain that those conditions exist on startup. The flammable range and the explosive range are basically the same. They are a range of mixtures of fuel and air within which a fire will be self-supporting, not requiring added heat to keep the process of combustion going. Every burner produces an explosive mixture of fuel and air. It does not explode because it burns as fast as it is being created. If it does not burn and production of that mixture continues, the story is a lot different. Eventually, something will produce a spark or add enough heat to start it burning. Then the mixture burns almost instantly and it is that rapid burning and heating to produce rapidly expanding gases that creates an explosion.

A graphic of a typical fuel's flammability range is shown in Figure 1-9. At the far left of the graph is where the mixture is all air and no fuel. On the far right, there is all fuel and no air. The quantities of fuel and air in the mixture vary proportionally along the graph, as indicated by the two triangles. The thin line in the middle of that band is the stoichiometric point, the mixture that would produce perfect combustion. Mixtures to the left of the stoichiometric point are called lean mixtures because they have less fuel than required for perfect combustion. They can also be called air rich. Mixtures to the right are called fuel rich because there is more fuel in the mixture than that required for perfect combustion. The flammability range is the shaded area, and it is only within that narrow range of mixtures that a flame will be self-sustaining.

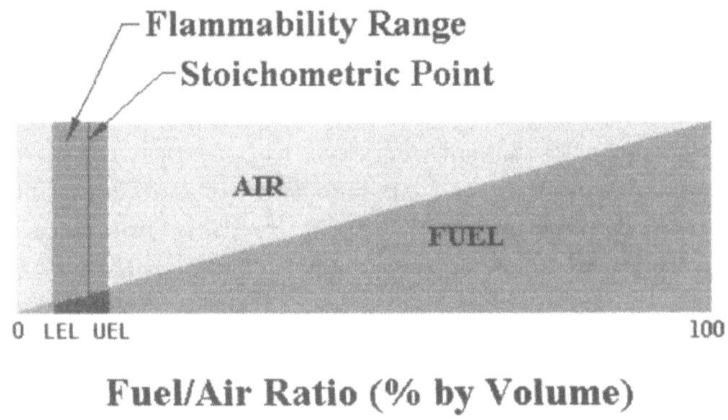

Figure 1-9. Flammability range.

At either end of the flammable range are the two limits of flammability. The one where flammability will be lost if more air is added is called the lower explosive limit—LEL for short. The one where too much fuel prevents sustained combustion is called the upper explosive limit (UEL). It is essential that the flammability range exists. Otherwise, the sticks would burn as they were carried to put on the campfire. Actually, everything would burn up. It is not that easy to get a fire going in a furnace considering that the fuel and air mixture has to be within that narrow range. The designer has taken this into consideration to assure that the system works before the operator ever gets near it.

Getting the mixture in the flammable range is not the only criteria when it comes to combustion in a boiler furnace. The only way that flame will burn steadily and stably is if it begins at the UEL. In other words, the point where ignition begins is where the fuel and air mixture passes from a really fuel-rich condition into the explosive range. Many a boiler failed to light because there was not that fuel-rich edge right where the ignitor added the heat to light it off. Usually, it is due to the mixture being too fuel rich and the ignitor not reaching the point where the UEL is to get things started. In other situations, the fire is lit and the heat from the fire manages to force ignition into the fuel and air entering the furnace until the fire reaches a point that is way too fuel rich and the fire goes out. Then, because the furnace has some heat, the fuel and air mix again to reach the flammable range and the mixture lights again and burns back toward the burner again. This situation is unstable and unsafe.

Sometimes, it is helpful to watch a service engineer tune up a burner. However, it is always important to be careful. It is fairly standard practice for service engineers to manually control the fuel going into the furnace when lighting a burner they just adjusted. They do it because they are not certain about the mixture and have their hand on the valve to control it, usually shutting the burner down faster than the flame safety system would. Once they get it right, they usually let it light off the automatic valves. Any roughness on light off is just another form of explosion and should not be tolerated. Sometimes, a contractor who placed the equipment in operation will not be able to establish a smooth light off. The disturbance may be referred to as "just a puff" that occurred as the burner started. Every puff is an explosion that did no or limited damage. Every puff should be considered a warning and is not to be tolerated. Sooner or later, whatever is causing the problem will get worse and then a serious explosion will occur that does cause some serious damage.

What causes explosions, including puffs? It is the direct result of an accumulation of a flammable mixture. When a fuel is being burnt, an explosive mixture is always created because there is no difference between a mixture of fuel and air that will burn and one that will explode. The reason that a boiler can be safely fired is that the explosive mixture burned at the same rate that it was created. It is only when the mixture does not burn and accumulates that an explosion results. The combustion is controlled by controlling the rate of burning. When an accumulation ignites, it burns at a rate dictated by nature. That is typically a lot faster than a normal fire. It is so fast that the products of combustion expanding can create a pressure wave which will create a force of 18–70 psig. The explosions that are called a puff were simply small accumulations of an explosive mixture which did not produce a pressure high enough to rupture the furnace.

It is not always possible to avoid a puff or rough light off. They occur when burner systems fail to repeat the conditions established when they were set up. The material can plug orifices. Linkages can slip. Regulator springs can soften. Many times, a combination of minimal factors can combine to prevent a smooth light off or burner operation. A puff should be considered a warning sign that something is going wrong and that something should be done about it. Keep in mind that more than 34% of boiler explosions are attributed to operator error or poor maintenance. Only make adjustments when there is a full understanding of what is causing the delayed ignition. If there is uncertainty, it is much wiser to call for a service technician who has experience with burner adjustments.

For a typical round burner with a refractory lined quarl, it is important that the flame begin within the throat of the burner where heat radiating from the refractory throat provides the ignition energy. A boiler just south of Baltimore had a furnace explosion in 1993 that was due to the improper adjustment of the burner such that the UEL was established so far out in front of the burner that it would not light in the first two or three tries. An accumulation of unburned fuel brought the mixture into the explosive range on the next attempt and the boiler room walls flew out into the parking lot. That incident and several others justified the following rule of thumb. Push the reset button on the flame detector chassis two times and only two times. Never take a chance on strike three.

There is another type of firing system under the trade name of "tangential firing." In this system, the fuel and air are injected into the furnace from the corners of the furnace and are aimed at an imaginary circle at the center of the furnace. The line of the flame front will be tangent to this imaginary circle. This system creates a single flame envelope. Intensive mixing occurs where the flames meet. A rotating motion is imparted to the flame body inside the furnace, which eventually spreads out to fill the furnace area. This firing system has moderate turbulence and mixing intensities compared to a typical round burner system firing from one wall. Instead of radiation from the refractory lined quarl of the round burner, radiation from the entire flame envelope provides the heat for ignition and stable flame conditions.

Finally, combustion optimization has gotten the US Environmental Protection Agency's (EPA) attention and, therefore, every State's department of air quality. The Industrial Boiler Maximum Achievable Control Technology (MACT) rules typically require an annual tune-up for most boilers over 10 MMBtu/hr. Combustion optimization is simply the process of adjusting the air to fuel ratio on a boiler to get the most heat out of the fuel. The environmental regulators also want it to include generating the smallest amount of emissions. For many a small plant, a service technician comes in once or twice a year (the typical state regulation requires a combustion analysis at least once a year) and "tunes up" the boiler. Plants with more sophisticated controls and oxygen trim have automatic combustion optimization. The controls are constantly adjusting the fuel to air ratio to maintain an oxygen set point in the flue gas. The objective is to assure that there is always enough oxygen in the system to minimize the formation of CO and burn all of the fuel. At the same time, the control system looks to minimize the stack losses that arise from having too much excess air. Any air that is fed to the boiler that is more than what is needed to completely burn the fuel and minimize the formation of CO will be heated up to stack temperature and pumped up the stack, resulting in a greater heat loss and a reduction in boiler efficiency.

THE CENTRAL BOILER PLANT

Steam and hot water are used for building and process heating because the conversion of fossil fuels (coal, oil, and natural gas) and biomass (like wood and bagasse) to heat is not a simple process. Water and steam are clean and inexpensive and are excellent for transferring energy from one location to another. It is also relatively easy and inexpensive to extract the heat from the steam or hot water once it has been delivered to where the heat is required. Boilers made it possible to centralize the process for converting fuel to heat so that the heat could be distributed throughout a facility for use. One boiler plant in a large commercial or industrial facility can serve hundreds or even thousands of heat users. The central plant concept is typically the most efficient way to deliver heat to a facility.

Many will question that statement. Many facilities are installing local boilers and doing away with the central plant. This could be a case of false economy. Many central plants are at the age where all the equipment and piping are well past their original design life and should be replaced. Replacing the central plant with several small local boilers can be seen as a way to reduce capital expenditure in any given year. The installation of one gas pipe distributing fuel to all those local boilers can be done at a much lower cost than installing insulated steam and condensate or hot water supply and return piping. However, the cost of several small boilers with a combined capacity exceeding that of the central plant will be higher than the cost for the one main boiler due to economies of scale. Admittedly, a small boiler operating at a low steam pressure or with hot water below 250°F will not need a boiler operator present. Installing many small boilers will reduce the cost of qualified operators.

In a recent study by Servidyne Systems Inc. and the California Energy Commission, there are claims that "a well trained staff and good PM program has the potential of 6% to 19% savings in energy." If the staff is eliminated, then an increase in cost of 6.3%–23.4% is possible because the staff are not there to maintain that savings. A little plant with a 500-horsepower boiler load could see energy cost increases in terms of 2013 dollars of $46 to $171 per day in fuel alone. That is considerably less than the numbers quoted in the first edition of this book because fracking has increased the production of natural gas dramatically, thereby forcing the price of gas down. However, fuel prices tend to vary considerably. Fuel prices in January 2001 were triple the cost in 1999. During the polar vortex, gas prices in the northeast went to more than $90/MMBtu for a brief period on the spot market. Thus, decentralizing almost any existing plant will save on labor but potentially burn those savings up in fuel. That does not consider the additional cost of maintaining several boilers instead of two or three. By the time all those local boilers start needing regular maintenance, the people who decided to eliminate the central plant will have claimed success and left. The facility maintenance bill starts to climb to join the high fuel bills associated with all those local boilers.

There is a misconception that the local boilers are more efficient because they are operating at low pressure. That is not true. Nothing prevents a high pressure steam plant with economizers from generating steam more efficiently than a low pressure boiler when the feed water temperature is less than the saturation pressure of the heating boiler. A typical central plant in an institution will have 227°F feed water to cool the flue gases, but local heating boilers will be about 238°F. Since the flue gases can be cooled more by the high pressure plant, the central plant boiler efficiency will likely be higher.

Along with the higher efficiency of a central plant, the ability to burn oil as well as gas provides additional flexibility when fuel prices vary. Today's time of use pricing has almost eliminated the deals that were available for interruptible gas. In the 1990s, when firm gas was about $5/MMBtu, interruptible gas was about $3.50/MMBtu. A plant could save 30% on the price of gas by allowing the supplier to call for the plant to stop burning that fuel at any time. The ability to burn fuel oil allowed the plant to take advantage of an interruptible gas contract. Today, it is not interruptible, but the price for gas can be much higher than oil when gas is in short supply (pipeline constraints, weather, etc.).

Running fuel oil supply and return piping to a lot of local boilers is usually abandoned as a first cost savings. With time of use pricing, someone needs to compare them for oil and gas to decide when to fire oil. In the winter of 2001, one customer capable of firing oil actually fired gas at prices of $10–$11/MMBtu, when oil cost only about $7.50/MMBtu. In less than two months, the difference in fuel cost would have paid a boiler operator's salary for a year. The only way a central plant may cost more to operate than a lot of local boilers is if the heat loss from the distribution piping is excessively high. However, it takes a lot of low quality installed distribution piping to produce enough heat loss to justify a lot of local boilers.

Another reason to maintain a central facility is cogeneration. The cost differential for a boiler capable of operating at 600 psig instead of 150 psig is not that great compared to the value of the potential for adding a superheater and converting the boiler for generating electricity later. A plant that generates power with the same steam that is used in the facility can produce that electricity at a lower cost than an electric generating station. The main saving is the cost of transmission and distribution, which now costs as much or more than the generation cost. However, care needs to be taken when throwing numbers around about the efficiency of cogeneration. Many references claim that cogeneration is 80% efficient, while the power grid is only 30% efficient. These two numbers are not comparable. The heat energy portion of the cogeneration plant is the same as the plant without cogeneration. The same temperature and pressure steam are produced and utilized in the plant. The 80% figure is comparable to the boiler efficiency. It is the percentage of the fuel energy that is used by the plant. For electric generation, a heat engine is required. This engine could be a steam turbine, a gas turbine, a diesel engine, or a gasoline engine. These engines all convert heat into work (i.e., mechanical rotation). That work can be used to turn an electric generator to produce an electric current which can be utilized as electricity. The difference in efficiency is all due to the difference in the electric generation portion of the cogeneration plant. In a power plant, the engines reject heat to the environment. It might be exhaust gas from a gas turbine or diesel engine. Or it might be steam going to a condenser. In the cogeneration plant, the exhaust from the heat engine is used to generate steam. Since that heat is utilized, the overall efficiency is better for the cogeneration plant. The improvement is related to the marginal improvement in heat rate for the cogeneration plant and applies to the electrical portion of the plant and not the heating portion. A modern, gas fired, combined cycle plant can generate electricity with a full load heat rate of 6500 Btu/Kwhr (the amount of heat energy needed to generate one kilowatt-hour of electricity). The marginal heat rate for a typical cogeneration plant is on the order of 5000 Btu/Kwhr. That 23% reduction in heat rate applies to the electrical power that is produced in the cogeneration plant. If, perhaps, 40% of the output is electricity and 60% of the output is heat energy, the net difference in fuel use between the two systems is about 12% in favor of the cogeneration plant. That is a much smaller difference than 80% compared to 30%. Nevertheless, that difference is real and can result in overall fuel savings. All of the facilities that dumped their central plants for a multitude of little boilers also dumped their ability to make power economically.

Distributed generation is a new buzzword that basically means electricity is generated in many locations (instead of large centrally located power plants that are often long distances from the users of the power). By having several small plants distributed throughout an area, transmission lines lose less power and do not have to be so big. The same problems apply. A small engine that is just used to generate electricity is less efficient and more costly than a larger central station power plant. However, there may be advantages in transmission and distribution costs, reliability issues, and local control which may be more valued than the generation cost.

ELECTRICITY

If there is anything that boiler operators pretend to know nothing about, it is electricity. Electrification is being proposed as one means of reducing greenhouse gas emissions. Heating and cooling can be accomplished with electric heat pumps. Cogeneration provides distributed generation, where every decent sized boiler plant can be generating electricity. It will be essential that the boiler operators of tomorrow know enough about electricity to use it, generate it, and occasionally troubleshoot a circuit. The current trend is toward engine and gas turbine cogeneration. That is where the fuel that is normally burned in the boiler is fired in the engine or gas turbine instead. The engine or turbine generates electric power and the steam or hot water is generated by the heat from the exhaust of the engine or turbine. Whether it is an engine, a gas turbine, a fuel cell, an electric boiler, or a very conventional steam turbine driving an electric generator, an operator will eventually encounter one of these because all plants will have them.

Electricity is different but it is not a dark and mysterious thing that is beyond the understanding of a competent boiler operator. Electric current is nothing more than the flow of electrons through a circuit. There are only two basic things that must be known about electricity, and the rest will fall into place. For electricity to work, there has to be a closed circuit. A circuit is a path that the electricity flows through. Break the circuit anywhere so that it is not a closed path and electric current cannot flow through it. The second thing is that there has to be something in that circuit that produces electrical current. If electric current is not flowing through the circuit, the circuit is not doing anything. That is it—create a circuit to make electricity work and break the circuit to stop it. When the path is complete so that current can flow, it is called a closed circuit. Whenever there is a break in the circuit, it is called an open circuit. A "break" is typically undesirable, whereas the "open" circuit is a normal interruption in the circuit.

Unplugging a toaster opens the circuit and stops the toaster from working. Turning the light switch off opens the circuit and the light goes out. In most cases, opening a circuit consists of moving a piece of metal so that there is a gap between it and the rest of the metal that forms the circuit. In almost every case where electricity is used, there is a metal wire and metal parts to form the circuit. Metals are used because they provide low resistance to the flow of electricity. Sometimes, as with the toaster plug, the opening can be seen in the form of a plug removed from the socket. In other situations, as with the light switch, the opening cannot be seen because it is enclosed in plastic to protect the user and the circuit.

Electrons can be stored to some extent. Rubbing a glass rod with a wool cloth will transfer some electrons to the glass. If the rod is brought near a piece of metal (like a door knob), a spark will jump from the glass to the metal. The glass rod is said to have built up static electricity (i.e., electrons that are not flowing). When the glass gets close enough to the metal, the electrons jump across the gap showing a visible spark. The glass rod is said to have discharged this static electricity. These electrons came from the wool cloth. The electrons will eventually move through the door knob and into the ground. When the wool cloth comes in contact with the ground (or something touching the ground), it will recover those electrons and complete the circuit. Lightning is just a giant example of the same process. A battery is like having stored electrons. The difference is that a battery contains chemicals that react to replace the electrons that are used when it starts discharging. A battery can be discharged by running the electrons through a light bulb, as in a flashlight. Some batteries are rechargeable. The chemical process is reversed by using electricity to restore the charge. A battery will keep restoring the charge until the chemicals are all changed. Then the battery is "dead." There is not much difference between a dead battery and a dead electrical circuit except that the dead battery just cannot produce enough electrons to raise the voltage to move the electrons and the dead circuit can have full voltage someplace but not be able to move the electrons because the circuit is open. It is important to realize that an electrical circuit that is not doing anything can still have a charge of electrons stored someplace. The problem with electric circuits is that they have the capacity to store a lot of electrons. When the human body becomes the piece that completes the circuit, it is the current that kills.

The flow of electricity is analogous to the flow of fluids. Controlling the flow of the electrons controls the flow of electricity. Voltage is, in effect, an electrical pressure that is used to force the flow of electrons. The generation of electricity must produce enough electron flow to keep the voltage up in the same way that the production of enough steam flow can keep the pressure up. Most electric flow control is on–off. Close the switch and open it to control the flow. A dimmer changes the resistance to current flow on one or more lights in order to dim the lights. At other times, special equipment is designed to automatically control the flow.

In the mid-1800s, it was learned that moving a conducting wire through a magnetic field will induce a current in the wire. An electric generator uses a coil of wire that is rotated in a magnetic field. The rotation comes from the heat engine, such as a gas turbine or a steam turbine. The ends of the wire are set up to slip inside metal rings so that the rotation does not keep twisting the wire. In this way, when the wire is horizontal, the current flows in one direction. As the wire rotates, it flips 180°. Now the direction of current flow is reversed in the wire. When the wire rotates the full 360°, the current flow in the wire is back in the original direction. This mechanism produces alternating current (AC). By contrast, a battery produces direct current (DC). AC became the standard for the electrical grid system in the late 1800s and the early 1900s. The AC allowed the filament inside a light bulb to last a little longer. It also allowed the electric current to be transmitted longer distances at the time. Nevertheless, some rules that apply to DC circuits can be helpful for AC. For example, Ohm's law can be applied to AC circuits to get an idea of what is going on. It is not a perfectly correct analysis, but it does provide some insight into what is going on in an electrical circuit. Ohm's law relates the voltage to the current and the resistance to current flow. The voltage between any two points in a circuit is equal to the value of the current flowing through the circuit times the resistance of the circuit between the two points.

$$V = I \times R$$

where V stands for voltage, I stands for current in amperes, and R represents resistance in ohms.

If any two of the values are known, the third can be calculated. Ohm's law can be a lot of help when troubleshooting electronic control circuitry. Most control circuits today use a standard range of 4–20 milliamps to represent the measured values. For example, a steam pressure transmitter set at a range of 0–150 psig will produce a current of 12 milliamps when the measured pressure is 75 psig. If the device is not getting a 75-psig indication on the control panel, a voltmeter can be used to measure the voltage at several points in the circuit in an effort to find the problem. Start with the power supply. It should be about 24 volts if it is a typical one. That provides a starting point. One side of the power supply can be used, whenever possible, to check for voltage at other points in the circuit. The voltage drop across the transmitter should be more than half that of the power supply because all the transmitter does is increase or decrease its resistance to control the current so that it relates to the measured steam pressure. If there is not much voltage drop across the transmitter, then there is a problem elsewhere in the circuit. Check for a voltage drop between each wire before it is connected to the transmitter terminal and a spot past the screw that holds the wire because poor connections are frequently a problem. The 24 volts DC cannot push current through a loose or corroded connection. Corrosion products provide additional resistance to the flow of electrons. Corrosion is always a problem in the humid atmosphere of a boiler plant. Many faulty circuits can be fixed by just tightening screws without even checking the voltage.

A voltmeter, or even a light bulb in a socket with two wires extended, can be used to check the typical 120-volt control circuit. Make sure not to touch those test leads on the light to anything that could be of higher or lower voltage. If the resistance between two points is zero, or nearly zero, then there is no voltage and the meter or test light will show nothing. If the circuit is open between the two points, there will be a reading or the light will shine. The circuit will not operate because the meter or light does not pass enough electrical current.

In the days of electro-mechanical burner management systems, as an aid to the operator, a light was added to the control panel, down in the bottom corner, and labeled "test." The light was connected to the grounded conductor, and a piece of wire long enough to reach anywhere in the panel was connected to the light and left coiled up in the bottom. All an operator had to do was to pick up the coiled wire and touch its end to any terminal or other wire in the panel to find out if the wire or terminal was "hot." The idea was to allow the operator to pick up that lead and troubleshoot the system when there was a problem. The need for troubleshooting burner management systems has decreased considerably with the introduction of microprocessor-based control systems. Many of them include a display that will tell the operator what is not working. They are also more reliable than all those relays and that extensive wiring involved in the electro-mechanical system. Just the same, the test light is a good idea. Read the drawings and sequence of operation until it is well understood how the system works. Then review them on a regular basis so that they can be recalled when the need to solve a problem comes up. Figure 1-10 is an example of a circuit test. For some reason, the fan motor starter (FMS) coil is not energizing to start the fan. The test light is connected to the grounded conductor. That upside-down Christmas tree symbol is normally used to represent a ground wire connection. The clip on the test light allows it to be connected to any grounded metal part of the boiler plant for testing. There can be times when something accessible and metal is not grounded even if it should be. Check it first. The other side of the light is connected to a test lead with an insulated handle and extended metal probe. Always check the test light by tapping a hot lead to ensure that it is working before testing. In the figure, the test lead is indicated to be touching terminal 4 in the panel. To quickly isolate a problem, start testing at the middle of the circuit. Terminal 4 is in the middle in this example.

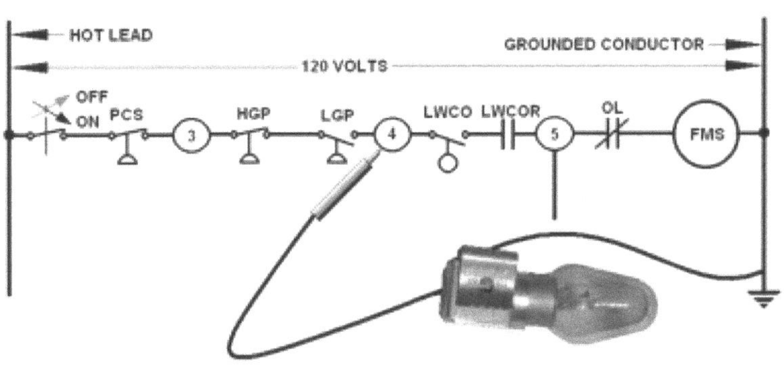

Figure 1-10. Testing a control circuit.

If the light works, then the break or open is closer to the grounded conductor. This eliminates checking everything before terminal 4. If the light does not work, then the break or open has to be between the hot lead and terminal 4. From this information, test at a point near the middle of the part of the circuit that was just isolated. Pick terminal 3 or 5 depending on which end of the circuit must have the break because they are conveniently in the panel. If the light did not work at 4 or 3, then check the wire connected at the on/off switch because it is also conveniently in the panel. The problem is isolated when the location where the light does not work when checking after a device and does work before it. Of course, it does not mean that there is not a problem with something else in the circuit that is between the isolated problem and the left connection of the FMS. It is slightly more complicated because other parts of a circuit must be activated first. However, once it has been determined where the circuit is open, it provides a lead to finding out what is preventing the operation.

If a fuel safety shutoff valve should open but does not, check its terminal (when the burner management system indicates it should be energized) to see if it is getting power (light on). If it is not, then check back through the panel circuitry to find what is open. Timing is important. There are only about 10 or 15 seconds to do that most of the time before other problems occur. It may take several burner cycles until the problem is found. If the output terminal is energized, then check the power at the valve to be certain that there is not a loose or broken wire between panel and valve motor. Always check to be certain that a burner management system is properly grounded. Lack of a ground can produce some very unusual and weird conditions. Anytime there are lights that are about half bright or equipment running that is noisy and just not normal, look for lack of a ground or an additional one. Exactly what is a ground? It is anything that is connected to a closed circuit to the earth. In most plants, there is a ground grid, an arrangement of wires laid out in a grid underground and all interconnected to each other, and the steel of the building to produce a grounded circuit. A ground wire is any wiring connected to the ground.

Do not confuse a ground wire with a grounded conductor. Ground wires are there to bleed stray voltage to ground and not to carry current. A grounded conductor is a wire that carries electrical current but is connected to a ground wire. All the white wires in a typical home should be grounded conductors. Examination of the circuit breaker panel should show that they are all connected together in there and also connected to a wire that is attached to the water line (the ground wire). All the steel in a building, the boilers, pumps, piping, etc., should all be connected to a ground. In cases like the building steel or pumps and piping, the electricians will call them "bonded." Bonding (or grounding) is the process of attaching everything that could carry electrical current (but should not) to the ground below the building.

With everything connected to a ground, the difference in voltage between any wire and ground should indicate the voltage of the system the wire is in. System voltages do vary. The common 120-volt system can vary from a low of 98 to a high of 132 although they typically fall in the 115–120 range. Higher voltage systems, such as 480-volt systems, usually range from 440 to 460 volts between leads at the motor. It is also a good idea to always know another location where the power can be disconnected to a circuit. For a simple device like a toaster, the plug can be pulled from the wall socket to disconnect the device. The reason to pull the plug out of the wall was that the toaster control did not work. There is usually a button or lever that can be pushed or flipped to release the toast and turn the toaster off, but, sometimes, it gets jammed. Unplugging the toaster stops all current flow to the toaster and allows it to be safely inspected and unjammed. Just like the toaster, another means of shutting down every piece of electrical equipment in the plant should be identified. Usually, there is a "panic" button labeled "stop" for that purpose. The stop button moves a metal bar away from two contacts to open the control circuit that stops current from flowing through a coil that holds the motor starter contacts closed. The coil releases the motor starter contacts and the motor stops. The question is, what happens when a) the push button contacts do not open? b) the insulation on the two wires leading to the push button in a conduit placed too low over a boiler melts and the wires touch each other (what is called a short)? c) a screwdriver somebody left in the motor control center (MCC) dropped onto the terminal board for the starter shorting out that same push button circuit? d) humidity in the electrical room promoted corrosion on the metal core of the coil until the portion holding the motor contacts rusted to it so that the motor contacts stay closed even when there is no power to the coil? e) two or more of the motor starter contacts fused together and will not release even though the coil is not holding them shut? f) something else happens? Make sure to know where to flip a circuit breaker or throw a disconnect switch in case something like that happens.

Keep in mind that disconnects are not normally used to break circuits. They are normally used to isolate a circuit when conducting maintenance. They are the devices that have copper bars that are hinged at one end and slip between two other pieces of copper that press against the bar to produce a closed circuit. If one of those is opened to shut a motor down, expect some sparks. Those copper bars are not designed for arcing and they will melt a little wherever the arc forms. A circuit breaker is the proper device for opening an operating circuit. If necessary, during an emergency, open the disconnect device as fast as possible. Every motor starter and circuit breaker is fitted with an "arc chute." It is constructed with insulating material and designed to help break the arc that forms when opening a circuit. They are not common on 120 volt or lower circuitry because there is not enough voltage and seldom enough current to produce a sizable arc. Normally, the arc chute has to be removed to see, let alone get at, the main circuit contacts to inspect and maintain them. Whenever the arc chute is removed, make certain it is put back! When somebody leaves the arc chutes off, the arc that forms when the contacts open lasts longer and does serious damage to the contacts because all the current in the arc tends to leave through one point. That point gets so hot that the metal melts and tries to follow the current, producing a high spot on the contacts. The next time the contacts close, that high spot is the only place contact is made and the metal is overheated because all the current for the motor has to go through that one little point. It melts and the coil pressure pushes the contacts together squeezing that melted part out until enough metal is touching on the contact to reduce the heat. Then the contact is fused closed and it will not necessarily open when the coil is de-energized. That is when it is necessary to find another way to shut the motor down!

If only two of the contacts fuse together or something happens to one of the three circuit wires for a three-phase motor, it then runs on only one phase. That is called single phasing because current can only flow one way at a time between two wires. Three-phase motors can operate on one phase if the load is low enough, but it will destroy the motor in a short period of time.

Due to the considerable number of injuries from arc flashes, there are currently new regulations for maintaining electrical equipment to prevent injuries from arc flashes. If the opportunity to take a course in arc flash protection arises, it is advisable to do so. There are requirements for PPE to be worn when working with energized electrical components. While it is improbable to have a problem with high voltage equipment in normal operation, it is very advisable to leave the handling of high voltage (600 volts and over) to electricians trained in arc flash prevention.

Figure 1-11. Motor control center.

Three-phase motors use three electrical currents that flow between the wires. If they are not balanced, the motors can run hot and fail early. Motor starter terminals should be checked regularly (every two or three years) and after any maintenance to be certain that the voltage is balanced. Use a meter to measure the voltage on each pair of leads: L1 to L2, L2 to L3, and L3 to L1. That big L, by the way, stands for "line" meaning line voltage, the supply voltage. The difference between the average difference and the lowest or highest measurement should not exceed 5%. If there is a big difference in voltage, an electrician should be called in to check everything in the plant.

Most boiler plants use MCCs (Figure 1-11) to control and distribute electrical power around the plant. Some plants have more than one MCC for improved reliability. These are more economical to install than local motor starters at each motor, but they can also be a source of unique problems in operation. Individual motor starters use their own power source for control power. Control power is electricity used to feed the motor starter coil that, when energized, pulls in the contacts that close the circuit to feed power to the motor. A typical MCC will have one control power transformer that supplies control power for all the motor starters in the MCC. It is necessary to rotate or pull the handle on the door of a particular section of the MCC because the handle operates and disconnects for the three-phase power before the door can be opened.

In some MCCs, that will also disconnect the control power, but in others, it will not. To access the starter, it will normally pull out because they are on a drawer assembly and that will disconnect the control power. Because the control power can still be on and the three-phase power disconnected, it is possible that everything indicates the motor is running when it is not. Always be aware that indicating lights on any electric motor starter or MCC can provide false indications. Also, a boiler can try to start without the fan running. Disconnect and tag out control power for any boiler before disengaging the starter at the MCC. The typical MCC indicating light arrangement provides a red light to indicate a motor is running and a green light to indicate it is shut down. This can be confusing as normally green means it is going. Accept the fact that the electrical engineers use that color arrangement to indicate that it is necessary to stop when the red light is on before doing anything that would shut down that motor unless, of course, it is meant to be shut down.

Motors can be destroyed if not treated properly. The common method is starting and stopping one. Motors are rated for "continuous duty," "intermittent duty," and "severe duty." Continuous duty motors are designed to operate continuously but only be started once or twice an hour. Intermittent duty motors are designed to start and stop a little more frequently. Severe duty motors are designed to be started and stopped all the time. Thus, a small boiler with a level-controlled feed pump that starts and stops all the time should have an intermittent or severe duty motor. When a motor is started, the electricity has to bring it from a dead stop up to speed, and that takes a lot of energy. A motor has what is called a high inrush current. In other words, a lot of electricity flows through it when it starts. All that energy heats up the motor because it is not as efficient as it is when it is up to speed. If stopped and then started up again right away, the heat is still there and added to. Do not start and stop continuous duty motors a lot. Sometimes there are some problems getting a boiler started and attempts to repeatedly start and stop the burner blower. If there is a selector switch on the panel that allows the fan to run constantly, that is a better thing to do than letting it continually start and stop. One operating technique for starting a centrifugal pump is to keep the discharge valve shut. It will not damage the pump, at least not right away, and preventing any fluid flow reduces the load on the pump while the motor is coming up to speed. Once the motor is up to speed, open the discharge valve so that fluid can flow. That only works on centrifugal pumps. It is also possible to overload a motor. Some plants installed oversized motors so that no matter what the operating conditions were, it could not be overloaded. However, oversized motors are very inefficient. As energy efficiency has taken on more importance (to reduce greenhouse gas emissions), there will be more opportunities to burn up a motor.

DOCUMENTATION

The importance of a boiler plant log, SOPs, and disaster plans has already been stressed. They are also becoming more of a legal requirement, both for safety and environmental compliance reasons. The US EPA imposes more fines for lack of required paperwork than for actual emission violations. Documentation is all the paperwork that is required. For most operator requirements, it is filling out the logs. Since the logs are the proof of what has been done, they are always part of an operator's job. SOPs, disaster plans, and the rest are primarily one-time activities with maintenance as required. They are prepared once and then revised when necessary. However, should an Occupational Safety and Health Agency (OSHA) or EPA inspector arrive at the plant and ask to see these plans, they must readily be produced and shown to the inspector. It means that the operator should know where they are kept. It also means that these plans must be followed. Emissions are to be minimized during startup by following the proper procedures. The EPA inspector will want some proof that the procedures are being followed. That gets back to the plant logs. Maintaining documentation can make a big difference in plant operation. When a piece of equipment fails, it would be helpful to know what company made that piece of equipment, what size it is, and where and when another one can be obtained. Of course, that will be difficult if the nameplate on the piece of equipment is (1) covered with eight layers of paint, (2) scratched and hammered until it is beyond recognition, or (3) simply missing and the plant has not one piece of paper that describes it. It is wise practice to look around the plant at every piece of equipment and imagine what is going to happen if it falls apart when it is needed the most. There is an old saying in the construction industry that applies to everyone. It is short and sure—"Document or Disaster."

Documentation is not much help if it is disorganized and cannot be located when needed. Every plant should have an equipment list and a bill of materials. Make sure that they are correct. When that job is done, those documents become the index for the operating and maintenance (O&M) instruction manuals. A good practice is to assign every piece of equipment in the plant a three-digit equipment number beginning with 101. Create a file with the important information. Drawing number 02 for every job is the equipment list where every piece of equipment is described along with a common name, manufacturer's information (including shop order, invoice, and serial number), and performance requirements. Drawing number 01 is a list of the drawings. When equipment or systems are added to the plant, the 02 drawing for that job becomes an extension of the first, and so on. When they are properly prepared on 8½ × 11 paper, equipment lists are invaluable, single, and readily accessible for use or inspection. It is also a good idea to produce an alphabetical index for equipment that references the number so that the information can be found in the equipment list.

Material is identified by a bill of material number that consists of a drawing number and the bill of material item number from that drawing. That way, the item can be found on a drawing (the drawing number) and where it is described (in the bill of material on the drawing). If there is not a drawing describing some material (for example, there is no creating a drawing of water chemicals), make up a drawing that is nothing but a list of those materials. Equipment typically needs instruction manuals. Materials can be described in one or two lines on a drawing. The equipment number should be marked on the equipment, and some materials, to facilitate reference. Stamp every page of the O&M instructions with the number before they are put into binders. Everything is then arranged and stored by the numbers. Numbers are generally preferable, particularly as plants get larger. Every time something is added to the plant, it gets the next consecutive number and goes in the last space in the last drawer of the file cabinet. Numbering devices and using an index to locate the number is just easier to manage.

Each equipment file also needs to have references to repairs and maintenance history, spare parts, and other pertinent information. Since repairs and maintenance are ongoing, the easiest way in a paper system is to have a sheet in each equipment file which has a line for each activity. The sheet might look something like this:

101—Boiler 1—Maintenance and Repair History

Original installation and startup complete—October 11, 1993
Annual Inspection—July 18, 1994
Annual Inspection—July 22, 1995
Replaced fan motor—August 12, 1995
Annual Inspection—June 30, 1996
Annual Inspection—July 11, 1997
Annual Inspection—July 17, 1998
Annual Inspection—June 23, 1999
Annual Inspection—July 21, 2000
Replaced burner—October 11, 2000
Plugged three tubes—January 22, 2001
Annual Inspection—June 30, 2001
Replaced probe on low water cutoff—August 21, 2001
Replaced steam pressure switches—August 30, 2001

This brief history of repairs and maintenance can easily fit on one sheet of paper to cover several years. To know more about why the three tubes were plugged on January 22, 2001, simply look at the maintenance and repair logs for January 22, 2001. It is also obvious that this requires some discipline. The item has to be added to the equipment record. It is so much easier with a computerized system and equipment numbers. Of course, all of this information can be stored in computer files. The problem comes in when the inspector wants to see them and have a paper copy for verification. Today it is easiest to use a computer to maintain the records. Just be sure to back up the files. The location of the instruction manual can be identified by file number and drawer number or other reference. The digital processing allows the insertion of information for a piece of equipment in a record without having to move everything about. Actually, it is moved. It is just that the computer takes care of it. Maintenance and repair information and other data related to a piece of equipment can be found by simply searching those files for an equipment number.

Even though the matter of filing is facilitated by the computer, equipment numbers should still be used. A number is unique to the computer. The machine cannot always pick out differences in alpha references when they are used. For example, data files could have references to Boiler No. 1, boiler #1, Blr. 1, boiler 1, and Number one boiler all entered by different people and sometimes even by the same person. The computer does not realize that all those references mean boiler 1, and some information could be lost in the depths of the data files.

In smaller plants, it is convenient to have everything stored together (the original specification, the manufacturer's paperwork, maintenance and repair records, parts lists, record of parts on hand, and where they are stored). When all the documentation for a piece of equipment is stored in one spot, information can be found quickly. In addition, when the equipment is removed from the plant, the paperwork can be removed from only one spot to discard it or move it. If the equipment was replaced, the documentation can be replaced readily as well. This may be particularly important for property tax purposes as many jurisdictions continue to collect taxes on property as long as the paperwork says that it still exists. When a piece of equipment is replaced and the old piece is discarded, it is just as important to get it off the tax inventory as well as the plant equipment list. Make it a point to have a second copy of all of the documentation stored in a separate location. Information that is kept only on a computer should have backup files stored somewhere else in a separate remote building, on an isolated server, or in the cloud. Some companies have data centers in two widely different physical locations so that critical data will not be lost in case of a natural disaster.

Here is a list of equipment items that is regarded as reasonably complete. All of the information will not always need everything, but none are unnecessary. The best thing to do is to keep everything. The value of a piece of information is greater when it cannot be found.

- An equipment list, arranged in numerical order with a description of each piece of equipment, a name for the equipment and manufacturer, manufacturer's model number, a copy, rubbing or photo of the nameplate, model number, serial number, National Board and State numbers for boilers and pressure vessels, capacity, maximum allowable pressure, maximum operating temperature, minimum operating temperature, maximum and minimum ambient temperatures for operation and storage, voltage requirements, power or amp draw, weight dry, weight operating, and overall dimensions.

- Original specification and/or purchase order for the equipment.

- Manufacturer's data report forms and all repair forms (boilers and pressure vessels).

- The manufacturer's operating and maintenance manual.

- Process and instrumentation diagrams (P&IDs): These drawings show the intended flow of all the process fluids (water, steam, gas, oil, etc.) in the plant and the instruments that are used to measure, indicate, and record the values of those fluid flows. Frequently, they will have the range of flow for each fluid. A steam line may show values like "0 to 25,000 pph" so that the range of flows is known. It can also show pipe sizes.

- Lubrication records, what lubricants are required, and when the equipment was lubricated or lubricant was changed. Include tribology reports.

- Maintenance and repair records. Either a reference to the date of repair (see above) so that details can be found in the maintenance and repair log or a description of the work and when it was done.

- Spare parts list furnished by manufacturer (including updated lists when they change part numbers and prices).

- List of spare parts on hand and the location where they are stored.

Of course, a material list will be needed. In small facilities, that can just be the bills of materials on the drawings. When there are more than 10–20 drawings for the plant, that approach begins to get cumbersome. A prepared material list can consist of a number of pages in a three-ring binder (or files on a computer) with pages for each drawing bill of material (could be a copy of the original drawing) and an index to help find the more important ones. The advantage here is that the information in the notebook, or computer, can be changed to reflect replacements and not have to alter the original drawings. When a valve is replaced, the material list can be edited to include the manufacturer and figure number of the valve that was put in. The figure number on the drawing may identify a valve that is no longer available or the original manufacturer could be out of business. All those documents should be prepared initially by the engineer and contractor who built the boiler plant. They all should be available and, if not, take the time to create them. Once they are available, all that is needed is to keep them current and add maintenance history.

STANDARD OPERATING PROCEDURES

Historically, there was always one "old timer" who knew a lot about a plant. It is so regrettable that many boiler plants have lost valuable knowledge and experience that was developed over the years of the plant's operation when that person was lost (retirement, accident, death, etc.). This type of behavior was often considered to be job security. Also, plant operators tended to be long-term employees with long years of experience. However, such behavior should not be characteristic of the wise operator. As mentioned earlier, the wise operator trains a replacement. The wise operator will help document the SOPs and keep them up to date.

SOPs are known, followed and disregarded, and changed and updated, but they are seldom written down. It is the lack of SOPs in written form that not only leads to potential accidents but also causes more trouble in the aftermath, when inspectors try to determine the root cause of the accident. Keep a written set of SOPs and keep them up to date. Be certain that they are complete enough to be followed properly. When a bad experience demonstrates that something was done the wrong way, the SOPs should be changed so that nobody else has to face that bad experience. A footnote should be added in the SOP that reads something like, "To avoid the failure experienced on (date)" so that new and future operators will be able to look up the history of that incident in the log should they question the SOP. Documenting the operation that works well is one way to ensure that the experience is normally a pleasant one and often avoids the unpleasant ones.

Boiler manufacturers make it a point of stating that no two boilers are exactly alike. No two plants and no two boilers will function exactly the same way. The only way to determine how to handle those variances is through the experience of the operators. The manufacturer's instructions for operating the equipment are almost always inadequate because they cannot (nor do they even try to) foresee the unique situations that surround their equipment when it is installed in a plant. Do not expect the chapters that follow to be complete either. General activity can be covered and some important things identified. However, each plant has its own set of equipment, its own location, and its own past experience that will influence how the plant should be operated. Only the plant's operators can produce a quality document of SOPs for a given plant. English major writing skills are not required to write up a procedure. Write the procedure in the same words that one would use to explain it to another operator. There is no real difference between saying it and writing it. What is important is the message and not how it is expressed. Some operators have been concerned with the appearance of their writing and used the services of someone with more language skills to help. Be cautious. Read out loud what has been written. These helpers may have good English skills, but they do not know much about operating a boiler plant. Language modification can change the intended meaning of the original sentences. The final text should be understandable to another operator. It is not necessary to have the SOPs typed, but print (rather than cursive) should be used if they are hand written. Too many people have trouble reading someone else's writing. If a computer is used to store the SOPs, be sure to check them afterward for the same reasons. There should be a computer in every boiler plant control room so that the operators can use it to record log data, analyze plant performance, plan maintenance, document maintenance activities, etc. Using a word processor on it to produce the SOPs is a good thing to do. It just is not important that it be very fancy. Some advantages include the ability to change a sentence or paragraph without having to type a whole new page, indexing, and all the other niceties of word processing. Just make sure to have backups and at least one up-to-date printed copy. To make sure of using the current document, the date of the last revision of each page should always be written on the bottom.

The SOPs should be reviewed by all of the operators. They should initial each page that is read to assure that the procedure is correct and the one that they use. The implication of the initials is that the page has been read and agreed to. If there is any disagreement, the procedure should be reviewed and corrective action should be taken. The SOPs should include all the operating activities in the boiler plant and in other areas of the facility that operators are responsible for. They can include related items, such as how shifts rotate, which shift is responsible for operating certain equipment, or, in certain areas of the plant, what equipment a particular shift is responsible for maintaining. The SOP should contain a description of each of the operating modes of the plant along with all the detail associated with operating each piece of equipment. Some may have to contain special provisions for specific pieces of equipment, such as modifying flow loops for different hot water boilers because the piping arrangement produces different situations at each of the boilers even though the boilers are thought to be identical.

The SOPs can sometimes modify the order of operations where it is more convenient to do so (provided that no new problem is introduced). An example would be where the order of opening valves is reversed (without consequence) because the operator would have to go from one level to another and then back again to open them in the normal order. They should recognize additional valves, drains, vents, switches, disconnects, and circuit breakers that are particular to the plant or added over time. It is a good idea to tag all of the valves for ease of identification in the SOP. A valve that was not identified in the SOP may have been left closed but needs to be open for operation. Again, such a situation emphasizes the need to review and update the SOPs for completeness.

SOPs can also include standard maintenance procedures which, even though they are maintenance and not operating activities, are performed by the operating staff and, when included in one document, show the extent of activities performed by the operators. In a large plant with separate maintenance staff, there should be another document for maintenance activities and someone should check for coordination of the two to ensure that all procedures are documented and there are no duplications and no conflicting procedures.

Once a set of SOPs exists, the difficult work begins. They have to be kept up to date. After initial preparation of the SOPs and for a week on each anniversary of their completion, each function that is performed should be reviewed. The actual procedure should be checked against the SOP. If there is a difference, then the SOPs must be brought up to date. Be very attentive to any construction going on in the plant because that work may change the SOPs or require the creation of some new ones. Do not make them and forget them. There will be nothing worse than giving an EPA inspector a set of SOPs that describe coal unloading, coal firing, ash handling, etc., when the plant had been converted to oil ten years ago and gas three years later. When projects involve such things as adding a new boiler, replacing the burners, replacing a pump, and adding new controls or technology such as variable speed drives (VSDs), changes in the SOPs are a foregone conclusion. Prepare an initial draft of the SOP for the operation prior to project completion. That provides an opportunity to think about how to operate that new or modified equipment. Look in the manufacturer's instructions for keys to successful operation and mentally rehearse the operation before it is time to do it. After startup, commissioning, and a few normal operations of the project, the SOP can be edited to account for things that were learned during the startup and operation. Finally, know and follow the SOPs. An inspector may select a procedure and ask the personnel to run through it, describing what they would do and comparing the given description with the copy of the written procedure. Prepare for such inspections. Pretend an unannounced inspection of the plant will occur every quarter and review the written procedures.

DISASTER PLANS

Preparing disaster plans has taken on increased significance since a number of major accidents or storms have caused considerable and costly damage. A risk management plan is a document that a manager prepares to foresee risks, estimate impacts, and define responses to those risks. It also contains a risk assessment matrix. A risk is an uncertain event or condition that, if it occurs, has a negative effect. The EPA and OSHA are coordinating their efforts to update the requirements under the Risk Management Program Development Rule. Elements include emergency response, incident reporting, third-party audits, record keeping, drills, and training/certification. Risk, or disaster, plans should be carried out in a similar manner to SOPs for new installations. Consider what would have to be done in the event of a disaster. Preparing for such events is a wise action.

First, the plans have to consider what to do if a disaster is happening and what can be done to limit the damage. Plans for fire are essential, especially if the facility does not have sprinklers. Even with sprinklers, consider what has to be done if they were not available, as in loss of all water. Pick spots at ten-foot spacing all over the plant. Imagine a fire starting at that point, and then decide how to fight it with and without water supply. Of course, there will be some duplicate situations. Simply refer to plans for those other locations. In some cases, consideration must be given to protecting a bigger potential loss (like fuel oil storage tanks) before fighting the actual fire. Look at the equipment in the vicinity and pay special attention to electrical conduits because it is possible for a small fire in one location to completely shut down the plant. Some plants have all the control wiring for the entire plant run through one spot. These are extremely vulnerable. Pay special attention to what will be needed for a fire in the control room or at the control panels. Once the plans for fires that start are developed, work on the plans for fires that get out of control. Finally, work on how to restore operations after a fire. This exercise typically leads to some decisions to reduce vulnerability to a fire by adding sprinklers, relocating systems (especially wiring), and duplicating some services to make a fire more survivable.

A good appendix to put together for a disaster plan manual is a list of every piece of equipment in the plant with a source for that equipment. In the case of critical parts that are known to break down regularly, there are probably parts in inventory. Refer the inventory in the manual. Other devices that are too expensive to keep as spares or are not likely to break down are the ones that need sources. Sources can be a rental company, stocking parts distributor, or the manufacturer. Include contact names, phone numbers, fax numbers, e-mail addresses, and travel directions for each potential supplier. This list has to be maintained and kept up to date. Do not neglect anything when preparing this list. It should include items such as transformers, transfer switches, distribution panels, fuel oil storage tanks, large valves, and pipe fittings that are not the standard stock item for the local suppliers.

Some disasters are not expected but still happen. Total loss of the plant is one possibility. Boiler rooms can be flattened by an explosion. All the boilers in the boiler room can have their casings blown off by a simultaneous combustibles explosion. The disaster plan for such an incident would include a list of suppliers of rental boilers that have capacity and pressure ratings to match the plant, contact names, and phone numbers, and two sets of prepared directions for the contractors on truck routes to deliver the boilers and set them up (two in case the primary site is unusable). In addition, a design for piping to connect the boilers to existing service connections, with alternates for each source and each service pipe should be prepared.

It is best to have plans broken down by area. Each plan should include an option for temporary water treatment facilities, deaerator, etc., if needed. Also include options for the ability to use some existing equipment in a plan that considers what to do if the entire plant is lost. The following is a list of disasters that can be addressed by preparing a disaster plan to follow in each event. Even with total disasters, there should be a plan for what to do when they happen. Try developing a plan for each of the following disasters where the conditions described relate to the plant:

- The plant is experiencing heavy rain. Flooding is occurring all over the place. The nearby stream is over its banks and threatening to enter the boiler room. The relief crew cannot get in. Oil delivery is out of the question. The natural gas supply line over that stream is starting to catch debris and back up the water. The roof drains are plugged with leaves, and so the roof is flooded and water is running down all the walls.

- All the weather conditions just described happened up the river, and all of a sudden, the water is pouring into the boiler room because the river overflowed.

- A tornado just swept through the plant. All the windows are blown out. The roof is gone and rain is coming in. The insulation was swept off several hundred feet of distribution piping supplying an area where steam supply is critical. The stack for the largest boiler was buckled over by the storm.

- It is an unusually hot summer. The temperatures in the upper levels of the boiler room are so high that motor starters located there are tripping as if the motor was overloaded. Some ventilation fans have been lost. The boiler room is so hot that the personnel cannot remain in the room for more than 10 minutes at a time. Insulation on the steam lines that were soaked by an oil leak is smoking. The control room air conditioning is failing so that operator sweat is all over the log book while trying to record all the systems that are shutting down from overheating.

- The plant is experiencing heavy snow, well beyond normal such that personnel are trapped in the plant. The relief crew cannot get in. Oil delivery trucks cannot reach there for a day or two. The roof of the boiler room is buckling under the weight of the snow. The atmospheric vents for gas systems and the oil tanks are buried in a snow drift. Combustion air openings are plugged or plugging with snow.

- Today is the third day of subzero weather and systems that were supposed to keep operating in the cold are beginning to freeze up. For those in the south, it only has to be the first day of subfreezing weather.

- The electrical power is out. The electric company states that it will be down for at least a day. Two subsidiary considerations are when it is below freezing and when it is extremely hot.

- Consider the loss of city water supply due to a city line rupture. The city states that it will be at least 24 hrs before water pressure can be expected. The plant has to keep the plant going and makeup water is needed.

- Boiler No. 1 (or the lowest number that is still around) just blew up shredding all piping and wiring within 6 feet of the boiler. Steam, water, chemicals, fuel gas, and/or fuel oil are spilling into the area. The noise from the blast and the pressure wave has caused temporary deafness. Repeat this consideration for each boiler in the plant.

- The plant is next to a chemical complex that makes a hazardous gas. The complex has an alarm system to indicate a gas release and it has been blowing for 5 minutes which is a fair indication that it is not a drill.

While these events are fairly rare, they happen somewhere in the world ever year. Prepare disaster plans and do not be afraid to imagine the almost incomprehensible.

LOGS

Recording data in a log has already been mentioned, but the maintenance of logs is so critical to operating wisely that it deserves a section of its own. Logs are tools. They contain information that allows the operator to make better decisions. In many cases, they are the only records of a plant's operation and the activity therein. By looking at the log, an operator can determine if a current condition of pressure or temperature is consistent with what existed at another time under similar conditions. Today, with faster computers, a data historian can record and store all of the plant measurements for at least two or three years. However, that is just recorded data. Examination of the data can lead to understanding. There are software firms that provide software tools for data analytics that can review the past few years of data and determine trends and variables of importance to the operation of the plant. The wise operator should take notice.

The wise operator knows the value of the log. By maintaining an adequate log, the wise operator is demonstrating a skill, protecting the interest of the employer, and developing a database as a resource for evaluating the performance of the plant which allows for the improvement of plant performance. There are many sources of information available to an operator today, but the one resource that continues to be a reliable source of information is the log.

Modern plants are equipped with computers, recorders, electronic devices called data loggers, and other means of recording data. Those devices do not always record everything. The electronic devices may not retain data information for long periods of time. Some only retain data for 24 hrs. Frequently, the traditional boiler plant log is abandoned in the mistaken belief that all that modern instrumentation eliminates the need for a log. All too frequently, those plants realize, after a serious incident, that such a belief was ill founded. A major, or even a minor, incident can destroy electronic data to leave the plant and operator with no historic data for reference or evidence (accidentally wipe the hard drive?).

The typical boiler plant should have a log "book," not a three-ring binder or loose pages. A bound book with consecutively numbered or dated pages is the best type of log book. Contrary to what one might believe, handwritten paper logs have survived many of the worst boiler plant incidents, being lost only when the entire plant was destroyed. Others have survived a plant burnt down although the edges of every page were burned back.

Most importantly, if ever required as evidence in court, the log book should survive scrutiny. A judge or jury will be confident that the document was not tampered with or altered, believing that the bound document is factual and representative of what the operator recorded. Loose pages and electronic data can be altered readily without evidence of that alteration and are not considered a legal record. When facing a law suit, it is too late to create a log book. In today's litigious society, it is foolish to think that one will never be sued.

On the other hand, the maintenance of a log could support an employer's claim against a contractor or manufacturer. A log is more than just a piece of paper that has to be filled out. It is every operator's responsibility to maintain one. The best log today is a combination of electronic data, printed records, and handwritten logs. The handwritten log can contain data that is not stored electronically or it can include that data as an original source that is subsequently entered into an electronic database by the operator. There is no need to put all data on a single piece of media.

As technology continues to develop, an electronic database may eventually eliminate the handwritten log. An electronic log that could eliminate the handwritten log should consist of a non-erasable media (such as a compact disk read only memory (CDR)) with provisions for the operator to record all pertinent data in concert with electronic data storage. The log should be duplicated in another location to preserve it and should also be on non-erasable media. One or more could store the electronic data normally captured by recorders and data loggers, while another could store data entered by the operator. Password control can provide the equivalent of the operator's signature. Unless the data are secure and duplicates exist at a location outside the plant where they are not exposed to the same opportunities for damage, do not abandon a paper log.

Types of logs

A boiler plant log can consist of many documents and devices that, as a group, constitute the log. A typical set of documents that form a log include the following:

Operator's log: A paper document that contains consecutive dated entries made by the plant operators to describe activity on their shift or watch. The log can contain a record of data readings recorded by the operator along with a narrative on activities undertaken by the operator, a record of visitors, contractors, and others who visited the plant, work performed by contractors, problems encountered, etc. Of all documents, this one must be arranged to survive as a legal document of what occurred in the boiler plant. It should not be alterable or altered without an operator's signature. If an operator decides to change what has been written in the log prescribed, procedures must be followed.

Water treatment log: A paper document that contains a record of water analysis and water chemical additions. This document could be part of the operator's log if desired but normally consists of forms prepared by the water treatment service organization.

Maintenance and repair log: Documents that constitute a record of maintenance and repair of everything in the plant. This log should be arranged to facilitate locating the information. There is more on this log in the documentation and maintenance sections.

Visitor's log: A paper document recording the signatures of visitors to the plant. It is normally unnecessary unless the plant has a great number of visitors on regular occasions which would clutter the operator's log.

Contractor's log: A paper document recording the signatures of contractors working in the plant. It is normally unnecessary unless the plant has a great number of contractors regularly working in the plant so that the information would clutter the operator's log.

Recorder charts: All charts from recorders are a part of the plant log. They provide a continuous record of pressures, temperatures, and levels that would normally be recorded at intervals in the operator's log. These are normally paper documents that show values for pressures, temperatures, and levels over a 24-hr period or a week. Some recorder charts span a month and strip charts can easily hold data for three months.

Modern recorders are provided that store the data on a thumb drive. These also have their limits and their survivability is questionable. See previous comment on digital data.

Creating the log

Many plants simply visit the nearest stationary store to purchase journal binders. These are fabric covered cardboard bound books with lined and numbered pages. All data are entered by the operator according to SOPs. That is the least expensive approach to producing a log but not necessarily the best method. Anything larger than a small heating plant should consider using a custom log book. Why a custom log book? There are basically five reasons. First, it saves an operator's time. Second, it provides a consistency not available with a journal, even with well-developed SOPs for log entries. Third, it ensures that data are recorded consistently over time. Fourth, it invites contributions of a professional to assist in the development of the log to ensure that all important information is recorded. Fifth, a custom log provides a sense of professionalism that is not associated with the journal type. A preprinted log can provide assigned spaces for entering much of the data and recording normal activities. Every log must have space for an operator's narrative. The operator's narrative is that written portion of the log normally referred to as notes. It contains a description of what happened in the plant in the operator's own words. Custom preprinted logs also incorporate the feature of a carbon copy. Every other page is perforated at the binder so that it can be removed and carbon paper is used over that page to produce a duplicate that can be moved every day to another location. That copy is also used by the manager to perform a more detailed analysis and note comments by the operators that require the manager take action to correct deficiencies or have work performed that is not within the purview of the operators. Bound paper operator's logs that are maintained by the individual operators can be combined with a computerized log which provides the electronic database for the plant. The contents of the paper operator's log are entered in the database. Thus, the best of both worlds are possible. There is an original document that is prepared in the operator's handwriting and an electronic database that the operator transfers this information to. It also allows some independence on the part of the operator and will reveal the lack of understanding of an unqualified operator.

What to Record, Why, and When

Despite the installation of recorders and computers, there are lots of important data in a boiler plant that are not recorded other than in the operator's log. The content also depends on the provision of the other logs. Data can be recorded in different binders that, combined, form the plant's log. The amount of data recorded is dependent on factors such as personnel responsibilities, the type of plant, and the importance of plant reliability and efficiency. For that reason, a full evaluation of the log by a professional or an in-depth review by a facility's operators and management personnel should be conducted to ensure that the log contains all the data necessary for the plant. Frequently, operators and management are not aware of the value of certain data. For that reason, the following recommended list is included with a rationale for why that data should be recorded.

When to record data depends on the type and size of plant. A small heating plant may have limited visits by operating personnel and choose to record data once a week. There is a dramatic exposure to additional expense for fuel and water and serious damage to equipment that is seldom considered with that timing. A household heater receives more attention than those plants because the residents note deviations in temperature or noise. A boiler installation in any building should be checked at least daily by someone who is competent in checking the plant and recording and interpreting data. Probably one of the most serious exposures for limited operator attendance is in a country's schools. It is not in the least unusual for parents to discover, only after asking the children, that the temperatures have been irregular in their school for several weeks or even an entire season. In a school, the attendant is usually the janitor who has no operator training. A qualified person should check the boiler plant and record readings twice a day while school is in session. That same rule applies to apartment and office buildings. Plants with boilers larger than 300 horsepower and supplying critical loads such as hospitals and nursing homes should have a qualified person check the boiler plant three times daily as a minimum. High pressure boiler plants are commonly required to have a licensed boiler operator in attendance, but that is not the case in every state. Many times, the presence of a boiler operator is a function of a union contract rather than state law. When an operator is in attendance, recording data hourly is a common practice. The actual written log, however, may only include a record of data by shift or on a 4- or 2-hr interval. There is little value to hourly data other than requiring the operator to be within the vicinity of each piece of equipment every hour. It is a matter of professionalism. Operators with a sense of being a professional enter data in the log every hour to demonstrate that they are watching the plant.

Suggested Matter and Data to Record

Following is an abbreviated list of things that should be documented in the boiler plant log along with some good reasons for maintaining a record of the values or information. It is arranged in alphabetical order. Some of these items will not apply to every plant and they would not be included in the log.

Air heater outlet air temperature: Monitoring the heated air temperature along with flue gas inlet and outlet temperatures provides an indication of fouling of the heat transfer surfaces, leakage past seals or through corroded tubes, and other performance problems with the air heater.

Annual inspection: The operator's narrative should record the annual (biannual or fifth year in certain jurisdictions and with certain types of pressure vessels) inspection of the boilers and pressure vessels in the plant. Inspections are required by law in every state. Documenting that it happened is imperative. Do not rely on the inspector, some of whom have been known to lose paperwork. The record should include the name of the National Board Certified Inspector and any findings that inspector relates to the operator.

Blowdown heat exchanger drain temperature: This data provides a means of calculating the cost of heat lost to blowdown. The temperature is an indicator of the performance of the heat recovery system and blowdown/makeup relationship. The drain also dumps to a sanitary sewer which, by Code and law, cannot be higher than about 140°F so that it is also a record of compliance.

Boiler inlet water temperature: For steam boilers, it is an indication of heat lost in the feed water piping or heat added by feed water heaters and economizers. For hot water boilers, it is an indicator of load and required for output calculations. The inlet temperature for fluid heaters and vaporizers serves the same purposes.

Boiler outlet water temperature: Hot water heating boilers are typically controlled to maintain this temperature. It is required for output calculations.

Boiler water flow: Hot water boilers, especially certain types of high temperature hot water (HTHW) boilers, require a controlled flow of water. The value is required for output calculations and should also be monitored for reliability because minimal flow should trip a limit switch.

Booster pump pressure: See condensate pump pressure.

Burner gas pressure: The gas pressure at the burner is indicative of input and should be monitored for consistency relative to load. Increases in gas pressure relative to load are indicative of plugging of or damage to the gas burner. Decreases are indicative of failure or damage to the gas burner.

Burner oil pressure: The oil pressure at the burner is indicative of input and should be monitored for consistency relative to load. Increases in oil pressure relative to load are indicative of plugging of or damage to the burner gun or the atomizing medium controls. Decreases are indicative of failure or damage to the burner gun or atomizing medium controls.

Continuous emissions monitoring systems (CEMS) data: Plants that are considered to be major sources for environmental emissions or a part of an emissions trading program must install CEMS. These systems continuously monitor and record emissions data taken from the stack. CEMS data must be reported to the EPA on a quarterly basis for compliance purposes.

City water temperature/pressure: See makeup water.

Combustibles: Monitoring the combustibles content of the flue gas can lead to early detection of burner problems and fuel air ratio control failure. Larger plants may actually control air to fuel ratio using combustibles, and monitoring that value is very important to them.

Combustion air temperature: Frequently, this is also the boiler room temperature. The combustion air temperature is the base for a boiler heat loss efficiency determination.

Contractor's activities: The operator's narrative should describe which contractors were present in the plant, when they were there, how many people were there, and what they were working on. It would not hurt to list the names of each of the contractors' employees. Less needs to be recorded if there is a contractor's log. Even if there is, the operator's log should note the presence of the contractors as well. Use a simpler record such as "XYZ Contractors on site at 8:20 a.m.—seven men."

Condensate pump pressure: Also called booster pumps, these lift condensate to the deaerator, and the discharge pressure relative to plant steam load and deaerator pressure is indicative of the condition of the spray valves in the deaerator. The discharge pressure of condensate return pumps, not necessarily in the boiler plant, can reveal steam blowing through traps connected to the same header.

Condensate tank temperature: The tank temperature is a first indicator of excessive trap failures. Once the temperature exceeds 200°, a trap inspection is warranted. When makeup and condensate are blended in the tank, the temperature can indicate the percentage of returns. An upward shift in temperature of those tanks indicates trap problems.

Deaerator pressure: Small variations in the deaerator pressure relative to feed water temperature or plant steam load can indicate problems with the deaerator.

Draft readings: The draft readings are seldom recorded by electronic equipment, but they are indicative of the internal conditions of a boiler. Variations in draft readings are frequently subtle, occur over extended periods of operation, and are load-related. Thus, the operator can miss a significant change. Variations relative to load can indicate fireside blockage, loose baffles, and loss of refractory baffles and seals.

Drum pressure: For high pressure steam boilers, the drum pressure is indicative of load because of the drop through the non-return valve and the superheater when equipped. The drum pressure also permits a more accurate calculation of blowdown losses.

Feed water pressure: Changes in heating plants with cycling feed pumps indicate problems with the pumps or piping. Changes in plants with feed water flow control valves are relative to boiler load.

Feed water temperature: The amount of steam a boiler can generate is dependent on feed water temperature. Lower temperature feed water will reduce the capacity of the boiler to generate steam. It has an effect on evaporation rate and overall plant performance. The temperature is also indicative of deaerator performance.

Flue gas recirculation (FGR): See recirculated flue gas.

Flame signal strength: Upsets in burner conditions and soot or moisture accumulations on the flame detectors are indicated by changes in the flame signal strength. Monitoring them can preclude a sudden unexpected boiler outage. Gradual degradation of the flame detector can be monitored for guidance in replacement beyond the normal one year.

Fuel oil meter reading: The totalizer should be read at the beginning or end of the shift to track how much fuel was burned during each shift. These data are essential for calculating evaporation rate and fuel inventory maintenance. A fuel oil meter reading should be taken for each boiler whenever possible to determine the boiler performance. If there is no meter, then fuel tank level readings have to be used to determine consumption.

Fuel oil supply temperature: Measured at the inlet of the pumps, this provides an indication of the temperature in the tank(s) for inventory management and detecting leaks in underground storage tanks (USTs). When burning heavy oil, the temperature after the heaters is monitored to confirm heater operation. Temperature to the burners is critical for proper atomization and can vary with oil deliveries because the viscosity of the delivered oil can change.

Fuel oil tank levels: Required for fuel oil inventory management and detecting UST leaks.

Gas fuel meter reading: The totalizer should be read at the beginning or end of the shift to track how much fuel was burned on that shift. These data are essential for calculating evaporation rate and comparing with the gas supplier's meter readings. A gas fuel meter reading should be taken for each boiler whenever possible to determine the boiler performance. If the only meter available is the gas supplier's meter, it should be read to monitor consumption relative to steam generated, heat output, degree days, or other measures of performance.

Gas supply pressure: The pressure of the gas supplied to the plant is monitored to confirm the gas supplier's delivery promise. Gas supply pressure should also be monitored for possible loss of supply. Gas pressure supplied to each boiler, after the boiler pressure regulator, must be maintained constant and at a prescribed value for accuracy of boiler gas flow meters and/or air fuel ratio. Changes in the gas pressure supply pressure to boilers with parallel positioning controls can alter the air fuel ratio and must be monitored to prevent unsafe operating conditions.

Happenings: Anything that happens which is not normal should be documented. An operator's note that he heard what sounded like a gunshot proved beneficial in a later court case. Happenings must be recorded consistently to support the credibility of a single incident report in court.

Header pressure: In high pressure steam plants, this is the pressure that is controlled. Changes indicate problems with controls, excessively large load changes, and inadequate boiler capacity.

Low water cutoff tests: See testing.

Makeup water meter reading: The principal source of contaminants in boiler water is the makeup water. If makeup is consistent and there is no leakage of untreated water into the system (such as a domestic hot water heating coil break), the water chemistry should be consistent. A sudden decrease in makeup is an indication of an external coil break that can be returning untreated water to the boilers. Monitoring makeup water permits extending time between water chemistry analyses. The quantity of makeup has a significant impact on energy consumption. Every gallon of cold, say 50°, makeup water that replaces 180° condensate requires more than 1000 Btu of additional heat input. This one is essential, even in the smallest of plants and regardless of whether they are steam or hot water.

Makeup water pressure: This is a value that is seldom monitored because operators take the continuous supply of city water for granted. Someday, the city may not deliver. Monitoring the pressure from wells is more critical.

Makeup water temperature: Determines heat required for makeup. See makeup water meter reading.

Oil supply pressure: Main oil supply and boiler oil supply pressures must both be monitored. Variation in oil supply pressure is indicative of problems with fuel oil pumps, tank levels, variations in oil viscosity, or quality. Changes in burner oil supply pressure can upset fuel air ratio.

Operating hours: Recording the amount of time a piece of equipment is operating can permit output and input calculations as well as a record of the amount of time the equipment has been running. Logging equipment start and stop times or operating hour meter readings are invaluable for plant performance analysis and maintenance scheduling.

Outdoor air temperature: Preferably, the high and low outdoor air temperature should be recorded. The outdoor air temperature is a prime indicator of heating and ventilation loads. Taking the high and low temperatures for a day permits calculating degree days for the facility location. Sophisticated recording devices can record the time the outdoor air temperature is within a given range to provide bin data when desired. Bin data are records of the number of hours the outdoor temperature was within a certain range, and they allow very accurate evaluation of heating plants. In larger plants, outdoor air temperature is often the inlet air temperature to the boiler plant.

Oxygen: Monitoring and maintaining a minimum oxygen content of the furnace gases is good practice for maintaining efficiency and required for larger units for emissions compliance. Usually, however, the analysis is made of the stack gases. Recording oxygen readings can reveal problems with air to fuel ratio controls, damage to boiler casings, or burner problems. When available, it should be recorded regularly.

Primary air temperature (coal firing): Too high a temperature will result in pulverizer fires. Too low a temperature will result in pulverizer plugging because the coal is not dried adequately. The temperature of the primary air (leaving the pulverizer) when compared to the air heater outlet temperature is indicative of coal condition, moisture content, and/or pulverizer condition.

Recirculated flue gas temperature: This temperature should be monitored for changes that indicate fan seal leakage and stratification in boiler outlet ducts.

Reheater steam flow and inlet and outlet pressures and temperatures: On boilers equipped with reheaters, these data are required to determine the heat absorbed by the steam. Reheater outlet temperature also has to be monitored like superheater outlet temperature.

Softened water pressure: Comparing the pressures at the inlet and outlet of the softener is a simple measure for determining the cleanliness and quality of the resin bed. Higher pressure drop through a softener can limit the capacity of the makeup water supply.

Stack gas oxygen: See oxygen.

Stack gas combustibles: See combustibles.

Stack temperature: This list is in alphabetical order, but stack temperature is undoubtedly one of the most important data points to record. Monitoring stack temperature is like monitoring the human body temperature. Stack temperature is the most important indicator of boiler health. It should be recorded as frequently as possible. Stack temperature varies slightly with load. Load-related temperatures should be monitored to indicate scale accumulation, fireside accumulation, baffle failures, improper air fuel ratio, and other problems.

Steam flow indication: If the plant load varies considerably during a shift, say more than 10% of operating boiler capacity, recording the indication of steam flow consistent with the other data readings is desirable to maintain a correct relationship for evaluation.

Superheater outlet pressure: This pressure should be recorded because, combined with the outlet temperature, it is used to determine the amount of heat added to the steam. Variations (relative to load) in superheater pressure drop can indicate superheater leaks or blockage that is otherwise undetectable.

Superheater outlet temperature: The damage associated with an excessive superheater outlet temperature requires constant monitoring of the superheater outlet temperature. The superheater outlet temperature combined with the outlet pressure is required to determine the amount of heat added to the steam.

Total dissolved solids (TDS): The TDS content of the makeup, condensate, boiler feed water, and boiler water should be monitored at a frequency adequate to detect problems and any time a problem with water chemistry is indicated. The conductivity of the water is related to the TDS and can be measured continuously.

Testing: Regular testing such as testing operation of the low water cutoffs on steam boilers should have a check box where, by checking the box, the operator indicates that the operational test has been performed. An initial box, where the operator's initials indicate who did it, is appropriate when more than one person is on the shift. Most other tests, conducted infrequently, such as quarterly lift testing of a steam boiler's safety valves can be included in the operator's narrative. Tests that should be recorded, and their frequency, include the following:

- Combustion analysis—frequency is subject to State Environmental Regulations but should be performed at least quarterly for boilers that operate continuously and any time the efficiency of combustion is questioned.

- Flame sensor tests—each month for gas and oil fired boilers.

- Hydrostatic tests—for boilers, annually. For unfired pressure vessels, biannually except for compressed air storage tanks which may only be tested every five years. Note that these are common time frames. The local jurisdiction may require a higher or lower frequency. For any pressure vessel or piping system, a test should be conducted after the vessel or piping is opened for inspection or repair.

- Low water cutoff tests—each day for steam boilers, each shift for high pressure steam boilers, and semi-annually for hot water boilers. Testing of the low water cutoff is imperative since fully one-third of boiler failures are due to low water.

- Safety valve lift tests—each quarter for steam boilers operating at less than 400 psig and annually for hot water boilers.

- Safety valve pop tests—each year for steam boilers and hot oil vapor boilers. Alternatively record replacement with rebuilt safety valves. The boiler inspector normally governs the performance of these tests because many boilers have more than one safety valve and the seals have to be broken (and replaced by the inspector) to test the second valve.

- Stack tests—as required for emissions compliance under the State's operating permit.

- Water analysis—depends on the plant. High pressure steam boilers with highly variable loads and makeup water requirements should have water analyzed every shift. Other high pressure plants may test water daily. For steam plants where makeup water is limited and consistent, condensate returns cannot be contaminated, and makeup water is metered, weekly analysis should do. For hot water boiler plants with limited leakage and when makeup water is metered, monthly analysis should be adequate. Monitoring the makeup is the key. Analysis should be checked immediately when makeup usage changes abruptly, either up or down.

- Water pressure/temperature—see makeup, boiler, and feed water.

- Visitors—unless frequent visitors suggest having a visitor's log, the operator's log should record all visitors to the plant.

Log Calculations

The logged record of a boiler plant's operation should include calculations of fuel consumed (absolute minimum), steam generated or MMBtu output, and percent makeup as a minimum. These are fundamental values that, if not monitored, can allow plant performance to decay until it becomes a serious problem. Other calculations that can be incorporated into a log include evaporation rate or heat rate, a degree day calculation, and steam generated or heat output per degree day or according to a degree day formula. Reconciliation of fuel oil inventory (including shrink or swell of oil in outdoor above ground storage tanks) to account for variations in inventory is recommended for oil burners. Reconciliation of boiler fuel flow meters with gas service meters is invaluable for monitoring the quality of the gas service instrumentation as well as in plant instruments. Calculation of the plant heat balance will permit determining how much steam was delivered to the facility.

Chapter 2

Boiler Plant Operations

Sports heroes include the football quarterback who throws the winning touchdown, the baseball player who hits the last inning home run, and the jockey who rides the leader over the finish line. Inside boiler plants around the country are other heroes. The boiler operator demonstrates skill and experience by flawlessly lighting off a boiler, bringing it up to pressure, and putting it on line. That is controlling thousands of horsepower with explosive energy that exceeds the imagination of most people. The operator moves swiftly to respond to a cacophony of alarms, swinging valve handles and pressing buttons in a long practiced dance to restore operations to normal and the noise to the low roar that is customary. If it were not for the experience, training, and skill of today's boiler operators, there could have been thousands of accidents and significant numbers of injuries and loss of life, which were normal over a century ago. They are operating equipment with a lower designed margin of safety and more complex limits on operation than their predecessors ever dreamed of.

Operations are covered in this chapter without discussion about the equipment. That is not true for operation of refrigeration systems, gas turbines, and heat recovery steam generators (HRSGs), where their operation is described along with the equipment in general and without the details covered here. This chapter does, however, contain guiding information that can apply to those other systems. It would be wise to read it again to see how it may relate to those other systems as well.

OPERATING MODES

There are many different modes of boiler plant operation. The one normally dealt with is "normal operation," when the plant is generating steam (vapor) or heating water (fluid) and all the operator needs to do is monitor it in the event something goes wrong. The other modes of operation require an operator to act to change the condition of the plant. No book can provide a specific set of instructions to perform those activities because every boiler plant is different. The following are guidelines to use for writing plant specific procedures if they do not exist and to check them in the event they do.

VALVE MANIPULATION

If it were not for the fact that most piping systems are normally built with generous safety factors, the operation of valves would be one of the most critical skills for a boiler operator. It is still a critical skill. However, the piping can usually take the punishment that some operators hand out. At some point in time, all the piping in the plant will be shaking around and making banging sounds that seem like it is going to blow apart at any minute. With experience, start checking out the plant to find out where the problem originated. If the noise can be stopped in a reasonable period of time, the piping will usually survive. Most of the time, those banging and shaking incidents are due to improper operation of a valve.

Sometimes the problem is not involved in operating the valve. It is because the valve did not work or was left in the wrong position. One such incident happened after starting up a new boiler plant. Steam was not needed at night. The boilers were shut and a header drain was opened. The following morning, as the boilers came up, the whole main steam system started banging and thrashing about. After everything quieted down again, which took a while, and what seemed like an awful lot of water was drained out of the header, it was finally realized that the drain valve that was opened the night before was plugged. A vacuum had built up in the system and drawn condensate into everything. After the drain piping was dismantled and cleaned, along with the valve, a procedure was put in place to make sure that one drain or vent was open to ensure a vacuum did not build up in any steam piping that was shut down.

When manipulating valves on steam piping, it is important to remember that a cold line is either full of air or water. It is rare for it to contain a vacuum. When shutting down a steam system, the space occupied by the steam has to be filled with something when the steam condenses, either air or water, unless the plant injects nitrogen into cooling steam piping. Water sitting in any piping system will descend to the lowest level if allowed. Air can compress in piping to preclude admission of steam or water. Steam at pressures less than 15 psig (pounds per square inch gauge) is lighter than air and steam at 15 psig (actually a tad lower than that) and above is heavier than air. That is one reason to keep a high pressure boiler vent open until the pressure is above 25 psig and vent low pressure boilers until they are carrying a load, counting on the flow of steam to sweep the heavier air out of the boiler.

Air can be trapped high or low in a steam system depending on the pressure. It can create pockets where piping is suddenly heated as the air is displaced. Some air is desirable in water systems to serve as a cushion to absorb the shock of sudden changes in flow. There is always a standing length of piping at the top of any water system. It is there to trap air for that purpose. In a residence, it will be in the wall behind the medicine cabinet. Modern plumbing systems use a special fitting with a seal so that the air cannot be absorbed in the water to lose the cushion. Plumbers used to know that the solution to a hammering sound in the customer's pipes every time a valve closed was to drain, and then refill, the system to restore that air cushion. Draining and refilling the water piping in a house is usually all that is needed to eliminate pipes banging every time a faucet is opened or closed.

Every time a system is filled or drained, a prescribed procedure that has been proven successful for the plant should be followed. For a new plant, the procedures will have to be developed. Think about how others have been done and apply that experience in producing a prescribed procedure for each piping system in the new plant. The first step in filling a system is opening vents and drains. Keep in mind that they are never empty. Usually, they are filled with air and it is necessary to get the air out. When shutting down a system, open the vents and drains so that the liquid can drain out and the air can fill the space left by condensing steam. Speaking of the latter, it is always important to open some vents first. A little steam escaping proves that the valve is open.

Once a main steam valve to a piping system is closed, the pressure will drop quickly and a vacuum could be generated before a vent or drain valve is opened. Open the vents first and let a little steam escape because it is safer. On large systems, it may take several vents and drains to admit air fast enough to prevent pulling a vacuum. Any system containing large pieces of equipment (deaerators, tanks, heat exchangers, etc.) should be monitored closely as they are shut down to ensure a vacuum does not happen. The equipment is not necessarily designed for a vacuum and atmospheric pressure can crush them. Failure to do so can cause a heat exchanger failure that costs several thousand dollars to replace.

Simply draining water without venting a system can also create damaging vacuums. Anytime the column of water in the piping gets over 35 feet, it can create as pure a vacuum as steam. Draining a water system without venting tanks on upper floors can result in all those tanks being crushed by atmospheric pressure because the water draining out left a vacuum.

It boils down to knowing the fluid in the system, what is in the piping, and what will happen when a valve is opened or closed. Filling any large system, be it with water or steam, should be done with a valve installed for that purpose. Normally, it is a small valve mounted on the side of the shut off valve (Figure 2-1). It can also be piped as a bypass or even consist of a simple drain and hose bib, where a hose from the supply can be connected to fill the water piping. The problem is that sometimes it is put in a bypass or fill valve that is so small that it will take hours to fill the system. Take the time to fill the system or, if it needs to be filled regularly, put in a larger bypass or fill valve (like the additional one in Figure 2-1). Don't leave the insulation off the valve and piping.

Some operators choose to crack the main isolating valve to speed up the filling process. Make sure that the terminology is clear. To crack a valve means to open it

Figure 2-1. Warm up bypass valves.

until the disc lifts off the seat (creating a small opening or crack for the fluid to flow through). A 10 inch steam header shut off valve should have something like a 3 inch globe bypassing it to allow warm up of the steam main. It is not possible to crack a valve that large without producing a significant surge in steam flow. One particular valve took two turns of the wheel to close it back off after it was cracked open and the resulting jump in steam flow lifted the boiler water level in the boiler to the point that it tripped on high level. Regardless, always crack any valve as the first stage of opening it. When the valve is larger than 2 inches, wait a moment or two to see what happens while preparing to spin it. Shut again if necessary. If the system starts hammering, the pressure in the system was changed too quickly. The important thing to remember here is that it will do the same thing the next time. Change the operating mode to eliminate that action thereafter. Always open and close valves slowly. It is never a good idea to spin valves. Someone else may be watching and follow suit anytime they are directed to operate one. That can be a costly mistake.

A valve should be installed in a piping system so that someone can shut it off when necessary. Anything higher than 4 feet off the floor is a pain to operate. Putting valves up in the air so that a ladder is needed to get to them can result in a dangerous situation. Operators expose themselves to potential harm by having to climb up to get at a valve, particularly in the event of an emergency. If a valve is not located in a convenient position, an extension should be requested. A chain wheel or extension rod is going to cost the owner something. But the cost of a workmen compensation claim in the event of an accident is much greater. The chain wheels should have the built-in hammers that help drive the valve open whenever the valve is larger than 3 inches. Use oversized chain wheels otherwise.

A lubricated plug valve is a unique device that uses a stiff grease to create a tight seal of the metal parts inside the valve. The funny looking knob that sticks out of the square is just a screw that, when turned, presses a small amount of grease into the valve. The grease is not soft flowing material. It is very thick and stiff. In order to replace the grease, turn that fitting all the way out and put in a stick of grease. Give that fitting a quarter turn every time the lubricated plug valve is operated, unless it is operated several times per shift. In that case, give it one turn per shift. A lubricated plug valve is the only valve that can be stopped from leaking in service. When the plug screw is turned, stiff grease is being forced in between the metal parts of the plug valve to seal it. Unless nobody has operated the valve for years so that the grease has hardened and does not flow uniformly into the valve, it will always seal. That is one reason Factory Mutual first chose the lubricated plug valve for fuel safety shut off service, which is commonly referred to as an "FM Cock." They should never leak if operated properly.

With the exception of those lubricated plug valves, all valves do leak. Some soft seated valves can last what seems like indefinitely. An operator should always be conscious of the fact that a valve can leak and should never, even with lubricated plug valves, rely on a valve holding right after it was closed. Sometimes, indications like pressure dropping can give false assurance that a valve is not leaking. Always wait until conditions have stabilized, cooled down, or heated up, as the case may be, before taking the position that a valve is closed tight. Also keep in mind that zero pressure measured by a gauge at the high point of a system (or a gauge with a water leg that is compensated for it) does not reveal the pressure at the low point of a system, which could have several feet of static fluid pressure on it. A system is not down and without pressure until all the vents and drains have been opened and, to be absolutely certain, the lowest drain valve passed some fluid when it was opened (to prove that it really was open and the connecting piping was not clogged) and, finally, no fluid is leaving it. If there is a possibility of gas lighter than air entering the system (like natural gas), test for it at the high point vent and a high point closest to the potential source of that gas before declaring a system isolated. Also, don't count on a valve holding if it held the last time. Sometimes, leaks are random (i.e., they hold sometimes, and sometimes they do not). Don't assume. Check to make sure.

When isolating systems (see more under lock out, tag out), it is always advisable to ensure that there is double protection in the event that one of the valves fails or leaks. If there is another valve in the line, close it. A vent or drain between the two valves will release any leakage to atmosphere instead of into the system that is isolated. Resilient seated valves (butterfly, ball, globe, and check) can seal initially and then leak later if upstream pressures increase.

An important consideration in valve operation is the use of a valve wrench. All plants should have valve wrenches. Hang them where they are convenient. Don't slap a pipe wrench on a valve handle to open or close the valve. The pipe wrench is designed to grip a pipe by cutting into it. Using one on a valve handle will create sharp slivers and grooves in the handle's metal which can tear through leather gloves and cut up the hand of the next person who tries to operate the valve.

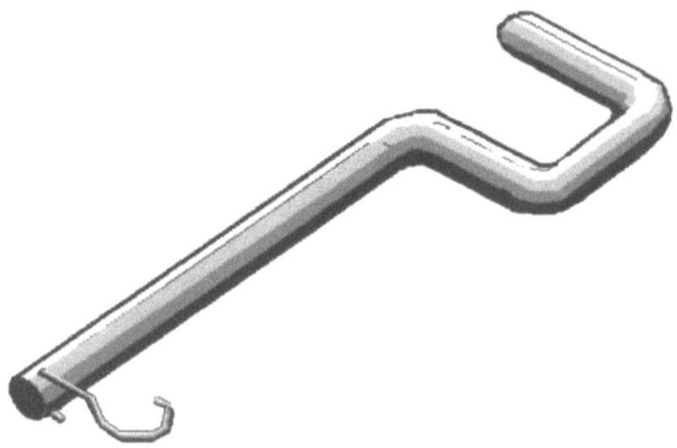

Figure 2-2. Valve wrench.

In order to make some valve wrenches, all that is needed is different sizes of round stock, a vise to bend it, and, for larger sizes, a torch to heat the metal so that it can be bent. Never put the portion that is to be gripped in the vise. Keep the grip smooth. The standard construction (Figure 2-2) includes drilling a hole for a hook for hanging the wrench near the valve for use when needed. Valve wrenches, by the way, are not for closing valves. They are only for opening them. One last comment on operating valves: It is a matter of courtesy that has almost been abandoned. When a valve is opened, always close the handle back down one-half, then back one-quarter, turn. That way, anyone coming along later will be able to tell immediately whether the valve is open. When they try to close the valve, it will make at least a quarter turn toward closed. If the valve was jammed open, someone can think that it is closed because it does not spin that quarter turn.

NEW STARTUP

There is a significant difference between starting a boiler plant that is new and one that has been in operation. Hundreds of wiring connections, pipe joints, and other work went into preparing the boiler and there are bound to be a few unforeseen problems as the first startup proceeds. These guidelines should help to achieve a smooth startup. They should also be used after any maintenance that resulted in opening a system. First, have a written procedure prepared, and not an outline. Each step should be described along with who is responsible for the action. In many cases, it will be the installing contractor's responsibility to produce this document. However, it should be checked for completeness and accuracy. Try to imagine all the things that could go wrong when preparing or checking a written procedure. The following should be addressed by the written procedure.

Preparing for Operation

Be certain that the safety shutdown push buttons, switches, valves, and other devices are in place and are ready to operate. Test each one, if possible, and refuse to continue the procedure if one is not present or not operational.

Check all electrical circuits for shorts and grounds before energizing them. Make sure all equipment and piping is electrically grounded before admitting fluids into the plant. Energize all electrical circuits before admitting fluids into the plant to ensure that they can be powered up. Test all electrical emergency trips and shutdown devices. De-energize circuits before admitting fluids. Prior to closing a boiler or pressure vessel, inspect it to ensure that there are no personnel, tools, or other things inside that should not be there. In Amsterdam in 1967, a boiler was almost closed with ten shipyard workers taking a nap in its furnace. In another incident, a worker left a screw driver inside a steam turbine. On the next startup, several rows of turbine blades were destroyed. Small boilers can come set up from the factory to reduce the chances of a problem on initial startup. It is rare that a boiler to be attended is factory tested. Even then, there is no certainty that the conditions in the plant are identical to the conditions in the factory. Thus, the initial startup of a boiler requires a careful approach to lighting the initial fire. Ensure air flow. Make certain it is linear on modulating boilers. Then establish safe light off conditions before thinking about starting to fire.

The codes require a minimum amount of building opening to admit fresh air for combustion. Sometimes, that is frequently overlooked. If the only boiler in the plant is about to start up, it is possible that there is no way for combustion air to enter that boiler room. If the boiler is an addition to an existing plant, the likelihood that someone paid attention to the requirements for combustion air is even more remote. A basic rule is two openings consisting of one square inch in each opening for every 1000 Btu/hr (British thermal unit per hour) of boiler input and a minimum of 100 square inches for small boilers. Larger installations allow 4000 Btu/hr per square inch. One opening should be high up in the building and the other near the floor. Prior to starting a new boiler, the availability of fresh air should be confirmed and the openings should be labeled "combustion air, do not cover." Sometimes, the air openings were blocked because the operators could feel a draft. Then they could not understand why their boiler was smoking. Once the fresh air source has been confirmed, make sure that there is linear air flow on any modulating boiler. Refer to the chapter on tune ups for establishing linearity. Make sure each fluid system is closed and ready to accept fluids before opening shut off valves. When preparing to admit liquids, identify vent valves and make certain they are open. The boiler will not accept much water if it is plugged full of air. If the fluid is admitted through a pressure reducing station, position a person to monitor the pressure in the system.

Position observers to detect leaks in the piping and equipment. Be certain that the observers are capable of seeing all drains leaving the plant to ensure that hazardous or toxic materials do not escape. Ensure that the person controlling the valve(s) admitting the fluid is in contact with all observers and can shut the valves immediately if a problem arises. Ensure personnel are positioned to close vent valves as the system is filled. Look at the instruction manuals. Know how much fluid is required to fill the system and estimate the filling time. It is another way to ensure that the fluid is going where it is supposed to. Wondering where all that fuel oil went several minutes after the tank should have been full is not a comfortable feeling. Whenever possible, have a means of detecting the level as the system fills.

Fill Systems

Fill the system slowly. Whenever possible, use bypass valves even though the filling may be slower than desired. The person attending to the valves controlling the fluid entering the systems should not leave that post and close the valves immediately upon instructions, or any sounds, from any observer. The valve operator should announce at regular intervals after closing a fill valve. In one instance, the startup team was standing around waiting for a boiler to fill for more than 3 hrs. When the team finally checked with the apprentice who was stationed on the valve, they found that it had been closed when someone shouted "hold it" and that was how it had been for 3 hrs.

Observe vent valves and close them as fluid reaches them. After the system has filled, operate the vent valves again to bleed off any air that may have been trapped and then migrated to the vents. When filling systems with compressible gas, use testers and bleed the system at the high or low points accordingly (high points for systems where the fluid is heavier than air, and low points for fluids lighter than air).

Allow the systems to reach supply pressure or controlled pressure slowly while diligently looking for leaks. Compressed gases (including air) will expand explosively if the container ruptures. Thus, the plan should provide for small increases in pressure, with hold points at regular intervals to check for leaks and any signs of distortion of the vessel or piping that could be caused by the pressure. A hold point, by the way, is when a certain time or condition in an operation has been reached where conditions will be held while checking that the procedure is happening as planned and all safety measures have been taken. In many cases, they are described in the standard operating procedure (SOP) as a hold point. In the case of very thick walled pressure parts, there may be a hold point designated by the manufacturer to allow the metal temperatures to equalize throughout the metal. Hydrostatically test each system after it is filled, following the procedures described for pressure testing. As with filling, there should be a person assigned to control the pump or valve that is pressurizing the system. Check electrical circuits that are connected to the systems during hydrostatic tests to ensure the liquid did not introduce an undesirable ground. Check them again after all test apparatuses have been removed and normal connections have been reinstated. Finally, make certain that all the tests performed are documented. A note in the log saying "tested Boiler 2" is not adequate. The documentation should contain values that demonstrate what was really done. The log should read "Tested Boiler 2 to 226 psig by the boiler gauge."

Start Makeup Systems

Once all pressure testing is completed, begin operation of the systems in an orderly manner. Water softeners, dealkalizers, etc., should be placed in service to condition the water to be fed to a boiler system. Provide a means to drain water until water suitably conditioned for the boiler is produced. If the installing contractor was sloppy, there could be mud, short pieces of welding rod, and other debris that will make flushing a major project. In one case, the pipe caps had to be cut off the bottom of the drip legs to get the large rocks out.

Establish Light off Conditions

The combustible range (see Combustion, Chapter 1) is so narrow that it really is difficult to establish conditions to create a fire in a furnace. Today's modern boilers that surround the fire with (relatively) cold surfaces do not provide heat or reflect it back to help maintain a fire, making firing difficult if conditions are not correct. On fixed fire boilers (no modulation), check the instructions for any measurements that will help to establish the proper air flow or conditions for the combustion air. On modulating boilers, set the air flow at a low fire (minimum fire) condition. If there is no other means of determining where to set the air flow, start at maximum

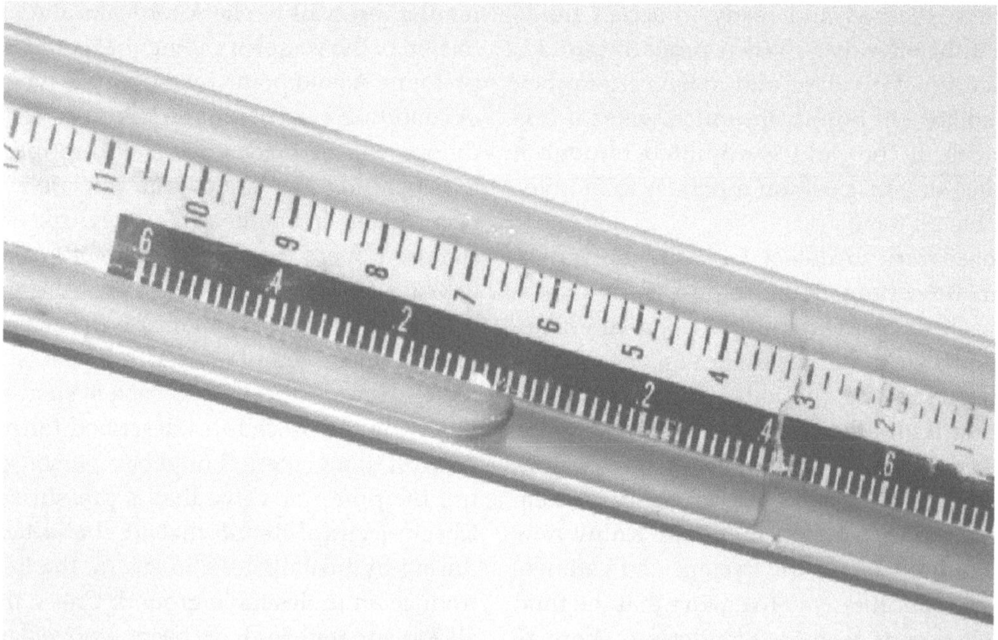

Figure 2-3. Manometer on slope.

on fixed fire units and 25% on modulating units. Maximum is easy to set. The 25% level is not that hard to determine. If a manometer is not available, make one by taping some clear tubing to a yardstick and leave a loop of tubing hanging off the low end to hold the water. All that is needed is a way to measure the air flow. Pressure drop across anything in the flow path is adequate. Set up the manometer on a ten to one slope (Figure 2-3) so that every inch on the ruler is a tenth of an inch in actual pressure. Position the end of the tubing at the inlet of the forced draft fan or air inlet. Then fill the manometer with water until the level is at zero. Run the fan to high fire (maximum) and record the reading on the manometer. Recall that pressure drop is proportional to the square of the flow. Thus, the measurement for when the flow is at 1/4 flow (25%) will be 1/16 of the reading at high fire. Run the modulating controls down to the bottom to see if the manometer reading is about 1/16 of the reading at high fire. If it is not, check that manual again. Some boilers are only rated for a two to one turndown. Thus, low fire is 50% and the differential pressure reading would be 1/4 of the high fire value.

If the linkage must be adjusted, and that is very possible, remember to check for any changes to the high fire reading after ensuring that the controls will stroke (go from high to low and back) without binding any of the linkage. Once light off combustion air flow is established, set up the fuel or fuels. Setting up fuel oil at low fire should be a snap. The only problems with it could be an improper piping design which, among other things, does not include any fuel oil return. Installation of a boiler without a fuel oil return line is setting it up for a furnace explosion! With a fuel oil return line, oil conditions can be set up without creating a fire. Before opening the oil valves, make sure the oil atomizer is not in the burner or open a joint at the hose or tubing to the burner to avoid dumping oil during this process. The safety shut off valves might be leaking. Once again, check the instruction manual to confirm the burner oil pressure at light off. That is either operating pressure for a fixed fire burner or a specific pressure for a modulating burner. Pressure atomizing burners will follow the rules for flow and pressure drop. Air and steam atomized burners do not. If the information is not in the manual, use the pressure that is half the range of the pressure gauge for fixed fire burners, 1/5 of that for modulating pressure atomizing burners, and 1/16 of it for steam or air atomized burners. Half the gauge is explained in the chapter on measurements.

Once the required oil pressure for light off is established, it can be set. Close the fuel oil recirculation control valve (a globe type valve in the fuel oil return line at the boiler). Position the controls at low fire on modulating boilers. That can be as simple as holding the "decrease" push button on a jackshaft controlled boiler to several adjustments on a pneumatic control valve. Slowly open the fuel oil supply valve while observing the burner supply pressure gauge. It is slowly opened because the flow of oil to another operating boiler could be interrupted and shut down the plant. The burner pressure should suddenly jump to oil supply pressure because the recirculating valve is closed and there is no flow through the piping. That explains why there should be no oil going to the atomizer. If the safety shut off valve is leaking, there will be oil dripping or spraying out of the burner yoke or the opening that was created. Needless to say, if the safety shut off leaks, the startup is halted. Assuming there are no leaks, the burner piping has been tested at operating pressure and the setup can continue. Crack the recirculating control valve, and then slowly open it until the light off pressure at the burner piping (after the firing rate control valve) is established. Then continue slowly opening the supply valve while adjusting the recirculating control as necessary to maintain the pressure.

Establishing a pressure for light off is not that simple. The main concern is flow. Establishing the pressure does not prove the flow. Use the oil flow indication on full metering systems or take two oil meter readings at a set interval to determine the gallons per minute (gpm) to determine the flow. It should be the design flow for fixed fire boilers and 20%–50% (depending on turndown capability) for modulating boilers. If there is no meter, count quarter turns of the recirculating control valve on another identical boiler, match that position, and establish light off pressure by adjusting the control valve. Barring any other means of setting it, listen to the recirculating control valve and set the low fire pressure while the squeal through the recirculating control valve sounds familiar. After having established a final position for the control valve, set the recirculating control valve to produce a pressure that matches operating pressure at low fire for a good smooth light off.

Since gas is not recirculated, the light off position cannot be guaranteed by measuring the flow. The pressure can be established. For fixed fire units, it is a matter of setting the pressure regulator. The pressure regulator on a modulating burner should be set for the design supply pressure. Light off pressure can be refined when the initial light off is performed. Light off pressure is not necessarily low fire but usually is. Some burners will operate at lower flows than that required for light off. The plant may have operating conditions where it is imperative to establish a low fire position independent of light off. If that is the case, the control provisions should include means of proving the light off conditions. Low fire is typically the light off condition on most boilers. It is imperative that the low fire conditions are fixed and reliable because many upsetting situations could produce unstable fires and explosive conditions otherwise. The fuel flow control valves should never shut. Their minimum position should be set mechanically so that something has to break before they shut. That way, any upset in the controls, including broken linkages, should establish a low fire condition.

It pays to look at the equipment to see how it will fail. If a linkage comes loose and can fail to open the fuel control valve, add weights so that it will close to minimum fire instead. The air flow controls should also rest on a mechanical stop at low fire so that the dampers never shut, unless they leak so much when closed that low fire air flow is still achieved.

Fill Boiler and Test Low Water Cutoffs

Before starting a fire in the boiler, fill it with water to a low level in the gauge glass, about an inch. Make sure the vent valve on the top of the boiler is open so that air can get out to let the water in. When the water is heated from cold to boiling, it will have swelled so much that the level will rise to over the middle of the glass. In unusual boilers, it is sometimes necessary to drain some water before the boiler reaches operating temperature because the boiler has a large volume of water compared to the room for expansion in the steam drum. Water will have to be drained to keep it visible in the glass. From this point on, it is necessary keep an eye on that water level. When the water level is visible in the gauge glass, it is time to test the low water cutoff. Proving a low water cutoff works on a new boiler is doubly important because there are so many ways to defeat those devices. The cutoffs should be tested without operating any bypass buttons or similar provisions to ensure they operate properly. Their failure is a primary reason for boiler failures. Be sure to test the low water cutoff properly. Simulate a loss of water due to evaporation by draining the water column or cutoff chamber slowly so that the water level drops gradually to the cutoff setting. If it does not shut the burner controls down, don't continue the startup until it is fixed.

Prove Combustion Air Flow

After the boiler is filled with water, it is time to start a burner cycle, which always begins with establishing and proving air flow through the burner and furnace. In very small boilers, like a home hot water heater, air flow is a function of combustion and is not proven. In most boilers, however, it amounts to starting a fan which will produce a measurable air flow that can be proven. The proof typically consists of a fan motor starter interlock contact and an air flow switch. The key words are air flow switch. On many systems, a simple pressure switch is used. Pressure does not prove that there is flow. Sometimes, a simple wind box pressure switch is used to prove combustion air flow. Its contacts will close when the fan runs and will open when the fan is shut down because a pressure switch simply compares pressure at the point of connection and atmospheric pressure. If one of those switches is giving problems, it can usually be made to function by closing the burner register. This should not be done. There is no air flow through the burner when the register is shut, but the switch has been made. Air flow should be proven by a means that is independent of such conditions. One of the best methods is using the differential pressure across a fixed (not adjustable) resistance somewhere in the air flow stream.

Purge the Boiler

Once air flow is proven, it is time to "purge" the boiler. A purge is a constant flow of air through the boiler that must occur long enough to ensure any combustible material is swept out the stack. That way, it cannot be ignited by the starting burner. On an initial startup, some math has to be done to determine the purge timing and the flow rate may have to be established. State law and, frequently, insurance company requirements dictate the flow rate and timing of a purge. These are the more common requirements: Single burner boilers can be purged at the maximum combustion air flow rate unless they are coal fired. The purge air flow requirements of multiple burner and coal fired boilers vary, but the basic rule is 25% of full load air flow. Single burner, fire tube boilers must purge for sufficient time to displace the volume of the setting four times. Single burner water tube boilers must purge for sufficient time to displace the volume of the setting eight times. Multiple burner and coal fired boilers must purge for sufficient time to displace the volume of the setting five times and for at least five minutes.

The purge air timing needs to be calculated. First, calculate the volume of the setting. The setting is everything from the point where combustion air enters enclosed spaces leading to the furnace to the exit of the stack. For all the fans, ductwork, air heaters, burner wind box, and similar parts, the inside is mostly air. The volume can be estimated by simply measuring the outside and multiplying length, width, and height to get the volume. Do the same thing for the boiler. The manufacturer's instruction manual will list the weight of the boiler empty and flooded. From this information, the volume of water, steam, steel, and refractory can be estimated and then subtracted from the total to get the volume of the gas space in the boiler. Divide the dry weight by 500, the approximate weight of a cubic foot of steel, to determine the steel volume and divide the difference between flooded and dry weight by 62.4 to determine the volume of water. Subtract the results from the outside volume of the boiler. The total gives the volume of the setting.

For single burner oil and gas fired boilers, use the required combustion air flow rate for full load air flow. If the boiler fans cannot be operated at full load air flow on a purge, determine the actual purge air flow rate (as a percent of full load) using the processes described for estimating the minimum air flow. For multiple burner and coal fired boilers, use 25% of the full load air flow as a purge rate. Given the volume in cubic feet and a rate of flow in cubic feet per minute, divide the volume by the flow rate and calculate the number of minutes it takes to displace the volume of the setting, which is one air change. Multiply that result by the required number of air changes (4, 5, or 8) to determine the purge timing. For an existing boiler, the exact same methodology can be used. It is wise to know the actual purge rate that is required. The required purge timing is often a lot longer than what the boiler was originally set up for. With the installation of microprocessor-based (programmable controller) systems, the operator is no longer able to reach up to the control panel to reset the timer (shorten the purge). Learn to live with the legally binding purge rate. It is much safer and reduces the risk of an explosion.

Once the correct purge time has been established, set the controls for it. Use the purge to clear the boiler every time before attempting ignition. It provides some valuable time to think about why or how the boiler tripped prior to the startup. A purge must be proven before the timing starts. The purge conditions must be proven during the entire purge period. Purge proving is one thing very few systems do well. Be sure that the system on a new burner really proves a purge air flow exists. The boiler operator should be the final authority on purge air flow and ensure that the automatic system's acceptance of the condition is correct. On small boilers, the typical proof is a fan running and the controls at high fire position. As the boiler size increases, a device to monitor flow should be provided. One that can measure air flow just like the one for combustion air flow is the best. A proven purge is imperative for a safe boiler operation. Many of the explosions and regular puffing experienced on boilers were the result of an inadequate or non-existent purge. Don't be satisfied with anything less and don't trust the sensors entirely.

Open Fuel Supply, Prove Light Off Conditions

When the purge system is working properly, open manual fuel shut off valves to bring fuel up to the safety shut off valves. Don't open the burner shut off valves yet. The piping should be checked to ensure the fuel is up to the safety shut off valves and there are no leaks before proceeding. Perform a leak test of the fuel safety shut off valves (see Maintenance, Chapter 6) to ensure that they are working properly before proceeding. Valves can leak, even new ones! After that final check, install the oil guns or gas guns that were intentionally left out so that no fuel could get into the boiler. It is a good habit to get into. Always remove the guns when the boiler is not to be fired, assuming that they are readily removable. That provides some degree of confidence that fuel cannot possibly get into the boiler. It is better for a leak to appear at the front where it can be seen or smelled than to quietly create an explosive condition inside the boiler. One of the most expensive boiler accidents to date, costing over a billion dollars, was the result of leaking fuel. If a fitting cannot be opened to show leakage at the front, then check an idle boiler regularly when there is (or could be) fuel in the burner piping.

Once a purge is complete, modulating boilers should be positioned for light off. Because most boilers light off at low fire, this is commonly referred to as the low fire position, and not the light off position. It should be noted that light off position and low fire do not have to be the same. Once a burner is operating, it can usually remain stable at firing rates lower than rates required to achieve a smooth light off. Where loads can require a boiler to operate at very low firing rates on occasion, and it is more desirable to keep a boiler going, separate minimum (low) fire and light off positions may be established. In those instances, the position switches have to prove the settings are high enough for ignition as well as low to minimize input during light off.

Low fire position switches have always proven to be difficult to maintain and set because the low fire is at the minimum stops described earlier. Understand that the position proving switch(es) do not have to be set right at the minimum position. Determine an acceptable upper limit for light off and adjust the low fire switch accordingly. The acceptable upper limit is determined by increasing the firing rate until light off gets rough. That is above the upper limit. Back it off a little to set it.

There is plenty of room for switch adjustment on a multiple burner boiler because the low fire setting has to be established by an independent means of control. The minimum stop on the main fuel flow control valve should be set so that the flow produces a pressure slightly less than desired at low fire with one burner in operation. The additional flow can then be provided by the minimum fire controls. Use minimum fire pressure regulators that bypass the main fuel control valve to maintain a certain minimum pressure in the burner header regardless of the number of burners in operation. Setting the main control valve with its minimum stop to produce almost enough flow for one burner helps make it possible to keep the boiler from losing all burners in the event the minimum fire pressure regulator fails. Some multiple burner boilers may be without minimum pressure regulators and will experience the difficulties of operating without them. All multiple burner boilers should have them, one for gas and one for oil, for more reliable operation.

A low fire position is not a certain solution to problems when lighting off a boiler. Fixed fire boilers light off at full fire. Thus, there is no switch or adjustment to be made. They can experience a rough light off. A rough lighting is due to creation of a fuel and air mixture that is outside the flammable range which finally lights when a proper mixture is established. When firing gas, it is usually because the mixture is fuel rich due to gas leaking past a regulator. A quantity of gas is trapped between the regulator and safety shut off valves at a higher pressure than normal. When the shut off valves open, the result is a flow of gas larger than normal for a few seconds until that buildup of gas bleeds off. On oil fired boilers, the gun can start empty with fuel mixing with the air in the gun to produce too lean a mixture. On the next try, if the gun is not purged, the mixture can be fuel rich. A rich or lean condition can be created depending on the operation of atomizing medium controls. If the boiler does not have smooth ignition, start looking for short-term surges or sags in fuel pressure and flow when compared to conditions after a stable fire is established.

There is usually a lot of room for low fire variations because most boilers have fan dampers that simply cannot close enough to produce a minimal excess air condition at low fire. Those dampers leak so badly that low fire is usually established with the dampers in what could be considered a closed position, and excess air is still 200%–300%. A good variable speed drive will provide lower excess air at low fire. However, the flow usually has to be controlled to overcome problems with changes in stack draft producing significant changes in the air flow. Remember, it is always important that low fire be a stable condition. With multiple burner boilers, where the code limits low fire air flow to 25% of full load air flow, that can be difficult because air fuel ratios can change with the number of burner registers open. Set procedures must be established to get the air flowing at the correct amount through the burners to be started. Those procedures must then be followed religiously.

Establish an Ignitor

With low fire (actually light-off) position determined, it is time to actually get a flame going in the burner, except for very small boilers that involves the operation of an ignitor. Most boilers will be equipped with a gas electric ignitor. Small boilers frequently use nothing more than an electric spark to light the fire. That is because their burner is of the size of an ignitor or smaller. There are also some with oil electric ignitors and a few with high energy electric ignitors and other unique methods. The bulk of boilers use a gas electric ignitor which will be covered here. Some may choose to call an ignitor a "pilot fire" or "pilot light," but ignitor is the more common term.

Since ignitors use an electric arc to start the gas or oil fire, it is appropriate to make certain the flame sensor does not think the electric arc is a fire. Begin by closing all the manual fuel shut off valves, including the ones that supply fuel to the ignitor. Next, go through several partial ignition cycles to see if the spark is detected. An "ignition cycle" is everything done from starting of the fans to igniting the main fuel on the first burner fired, including a purge. On multiple burner boilers, there is also a "burner ignition cycle" which includes waiting and then trying to light that burner. If the flame scanner "picks up" (the system indicates that the scanner recognizes a flame that is probably from another burner), the burner supplier has a problem. The boiler should not be operated until the problem is fixed. Sometimes, the problem can be corrected by re-sighting the scanner (adjust it so that it points in another direction). Even so, this check should be performed regularly to ensure the adjustment has not failed to prevent sensing a spark as a flame.

Flame discrimination is very important in multiple burner boilers. The flame scanner for a burner should not detect the fire of any other burner. If it does, it can improperly indicate the ignitor, or main flame, of its burner is on and allow the fuel valves to remain open when, in fact, there is no fire there. To prevent this, and any false indication of a fire, a Burner Management System (BMS) will normally lock out when a fire is detected that should not be there. That will be anytime a flame is detected, but the fuel safety shut off valves are not energized. This lock out should work even with a single burner boiler. Slip the scanner out of the burner assembly during the purge period and expose it to a flame. The burner should lock out. For almost all ignitors, the trial time is 10 seconds. This is the pilot trial for ignition (PTFI). It means that 10 seconds after the ignitor gas shut off valves open, the scanner must detect a flame to permit continued operation of the boiler. If the ignitor flame is not detected, the valves should shut and the BMS should "lock out." When the system has passed the spark test, check the timing and make sure that the system locks out.

Open the ignitor manual valves when the spark test and checking of PTFI is complete. Then determine if a proven ignitor flame can be established. Once the ignitor is proven, the BMS allows at least 10 seconds for main flame trial for ignition (MFTI) before shutting down. That period can be used to check the ignitor fire and do some other things. This is not necessary every time the burner is started but only after maintenance or adjustments have been made that could allow the scanner to see a spark as a flame. That includes a change in scanner alignment or simply removing and replacing the burner. There is no guarantee that equipment is back in exactly the same place.

The ignitor should light quickly (not just before the end of its 10 second trial) and burn with a clean and stable fire. If the ignitor is not stable, it cannot be expected to do a good job of lighting the main fire. It should be bright and ragged looking because there is a lot of excess air there. It should not be snapping and breaking up like fire from a machine gun where there are bursts of fire. An ignitor provides an opportunity to check operation of the boiler safeties and, during initial startup, maintain a minimum input into the furnace to slowly dry out any refractory. Drain the water from the low water cutoffs during the main flame trial period without pressing any bypass push buttons to be certain that the cutoffs shut the burner down. Some systems do an excellent job of alarming a low water cutoff but do not trip the burner. That is not desirable. Use the cycling of the fuel safety shut off valves to check fuel safeties as well. Every safety and limit switch should be operated to ensure that they will actually shut the burner down.

Start Refractory Dry Out

Over the years, the use of refractory in boilers has been minimized. One exception would be the circulating fluid bed (CFB) boiler. The life of refractory in a boiler is almost entirely dependent on how it was treated on the initial startup. By performing a controlled, slow warm up of the boiler, the life for the refractory can be maximized. Slam the fire to it and the refractory will need repair every time the unit is opened, until a complete refractory replacement is done. The ignitor provides an opportunity to begin a refractory dry out. It requires some temporary wiring and a relay in most cases (simply energize the ignitor fuel valves, not the spark, in place of the main fuel to keep the ignitor going). Operating on the ignitor will provide a very slow warming of the boiler. Only when it is apparent that the ignitor cannot bring the temperature up anymore, remove any temporary wiring to restore normal ignitor operation and allow main burner operation. A critical temperature during refractory dry out is 212°F because, at that point, any water in the refractor will start making steam. The steam, expanding rapidly, can erode or crack the refractory as it escapes out into the furnace. If the temperature is raised rapidly through that temperature, the steam generation can be so great that it creates pressure pockets in the refractory to force it apart, creating voids and cracks that will be repair items for years to come. That is why long-term operation on the ignitor can be beneficial to a new boiler, drying out any refractory so slowly that erosion, cracks, and voids are dramatically minimized.

For boilers with refractory that has already been fired, it is still a good idea to follow this procedure. There could be weather conditions that boiler was exposed to in traveling to the site that increased the water content of the refractory. Treat a new boiler as if the refractory was soaking wet. Run it to high fire right away and plan on a lifetime of refractory repairs. Once the limits of ignitor operation have been reached, it is time to establish a main flame and prepare for a combination of refractory dry out and boiler boil out. Repeat this operation to dry out any major repairs to the refractory as well. Refractory is one of those things that cannot be guaranteed because the manufacturer and installer have no way of knowing how the dry out was handled. Therefore, give it the tender loving care it deserves.

Establish Main Flame

Having spent a day or two on initially drying the refractory and testing ignitor operation, the unit is ready to light that main burner. This is not a time to be faint of heart or careless and quick. Although most small boilers come factory tested, there is no guarantee that it is set right for main flame ignition. On many boilers, the particular burner arrangement is being fired for the first time ever. Thus, nobody knows what the right settings are. Some operators slowly open the burner manual shut off valve after the automatic valves open as a burner starts. That is because they saw the technician do the same thing on the initial startup. Later on, that should not be done. But now, on initial startup, that is what has to be done. The goal is not to create a flammable mixture that does not light right away. If the manual valve is opened too slowly, very little fuel will be allowed, in that the mixture at the burner will always be too lean to burn.

However, the fuel can settle or rise and accumulate to create a mixture that is just right, waiting to finally get ignition. If the valve is opened too rapidly, fuel input will go past the point where the mixture is right and into a fuel rich condition that will not burn. That only happens if the controls admit too much fuel, and that could be the case on an initial startup. That fuel in its rich condition can mix with some air in the furnace to produce a flammable mixture and accumulate in preparation for an explosion, suddenly lighting when it is least expected. A plug valve or butterfly valve can be operated from closed to open in approximately 5 seconds. That provides enough time to stroke through all the potential mixture conditions within the trial for ignition period without going so fast that the proper point of ignition is missed. When lighting off on oil, there is usually use of a multiple turn valve that is really open enough for low fire in two turns. Practice getting it to two turns open in 5 seconds. Also practice to close the valve at the same speed.

Keep in mind when operating the valve that there is a delay involved. The fuel has to displace the air in the burner piping and burner parts before it can enter the furnace and start mixing with the air. An assistant watching the fuel gauge can read off pressures to get an idea of what is happening. Stop opening the valve the instant that the main fire lights and be prepared to close it a little or open it a little more depending on the perception of the fire. If the fire is bright and snappy, which is an indication that it is air rich, open the valve more. If the fire is lazy, rolling, and smoking to indicate it is fuel rich, close down on the valve. If no fire was obtained, then allow a full purge of the boiler. It is not uncommon to have several attempts at starting that first fire. Don't try to shorten the purge time. Every missed fire leaves an accumulation of fuel in the boiler that can produce a healthy explosion when it is lit by the next operation of the ignitor. Always do allow for a full purge. Further, if the fire was smoky, purge it twice to get all that fuel out of there before trying again.

Use the purge period to think about why a stable flame was not achieved. It might be because the gas piping was full of air and was not purged properly. It might be that the oil pump was not started (in which case, why did the low oil pressure switch not prevent an attempt at ignition?). Perhaps, the fuel or atomizing medium valve was not open. Maybe, there was too little or too much fuel. Adjust the controls accordingly. One problem with steam and air atomized burners is not enough fuel. It is often not apparent because the steam or air is breaking it up. Make some corrections. Then, after a purge, try again until a steady flame is achieved.

The next step is often glossed over. A pilot turndown test should be completed. It is a process to prove that, if the ignitor fire has decayed to the point where it cannot light the main burner, the scanner will prevent an attempt at ignition on a faulty pilot. Throttle the gas supply to the ignitor until a drop in the flame is observed. Make sure the ignitor can light the main flame. Continue dropping the pressure and checking to be sure the main flame ignites. If, during this process, the scanner fails to detect the ignitor flame and the BMS locks out, the test is complete. That seldom happens. What usually happens is that the ignitor fails to light the main fire. Now the flame scanner has to be adjusted so that it will not detect the ignitor flame when it is not adequate to light the main burner. Matching that scanner position, or orificing it so that it also allows reliable detection of the main flame, can frequently be a problem. It must be done or the system can be forced to repeatedly attempt light off of a main flame with an inadequate ignitor. The results have been very devastating in some installations.

Now that the main burner flame has been established, proper firing conditions must be set up so that they can be repeated for every light off. If the manual valve has been completely opened without changing the condition of the fire, then there is no need to balance the manual valve and controls. If not, then note the burner pressure and close down on the main fuel control valve a little (or adjust the minimum pressure regulator). Then open the manual valve a little to restore the pressure and repeat the process until the manual valve is wide open. Once the proper conditions for startup are known, the only reason for operating the manual valve is when the repeatability of the system is in question. A smooth light off should be achieved every time once the conditions have been set.

Now is the time to review the process. Open valves admitting fuel to the furnace only after purge and low fire position interlocks are proven. Open the valves in the main fuel only after a pilot (ignitor) flame is proven. Prove that the purge limits prevent completion of a purge cycle when combustion air is blocked by blocking it. Ensure that the purge requirements are not satisfied when the burner register(s) is (are) closed, when the fan inlet is blocked to the degree that the required air flow cannot achieve the specified flow rate, and when the boiler outlet is similarly blocked. Ensure that the burner startup cannot continue after purging until the low fire position is proven. Admit main fuel only after observing that a stable and adequate pilot flame exists and extinguishes at the end of the MFTI period (unless there are separate pilot and main flame sensors where the main flame sensor does not detect the pilot flame). Purge the boiler completely according to the code after each test or failure to produce a main flame. Don't alter flame trial timing of the control.

Boil Out and Complete Dry Out

This normally only applies to a new boiler. It will also be required after tube replacements or a complete dry out of some refractory repair. Follow the sequence when necessary. The entire process is skipped for normal operation of a boiler. Some boilers will have pipe caps or plugs in the casing drains where moisture can escape during dry out. They should be removed for this period of operation. Normally, the boiler is simply filled with treated makeup water or feed water before this stage. Once the process begins, that will have to change. Boil out chemicals should be as prescribed by the boiler water treatment supplier or the boiler manufacturer. Be certain not to have conflicting requirements. Handle those chemicals with extreme care, using all the required protective clothing and equipment. They are a lot tougher than normal chemicals. They should be added right before the start of the boil out and dry out and removed as soon as the boil out is done.

Burner operating time should be limited until the boiler is operational and refractory dry out and boiler boil out are complete. When it is possible to operate the boiler on main flame, make the first step a combined procedure of refractory dry out and boiler boil out. Neither function can be performed without having an effect of the outcome of the other. Procedures supplied by the boiler manufacturer should be followed or the selected procedure should be submitted to, and approved by, the manufacturer. The contractor may say "we always did it that way." However, that does not make it right. Insist on a written document. Be certain to remove brass, copper, or bronze parts exposed to the boiler water because the caustic water can damage them. In many instances, that includes the safety valves. Replace them with overflow lines run to a safe point of discharge where any liquid that passes through can be collected and treated. Be prepared to dispose of the boil out chemicals after the process is completed. Sometimes, it is necessary to interrupt the dry out procedure to dump the boil out chemicals, flush, and refill the boiler. Have a procedure in place for re-establishing the dry out. Be prepared to commence normal water treatment immediately after the boil out.

Don't rush these steps. Pushing activity along at this point can damage the boiler in a manner that will last its lifetime. Have adequate personnel on hand for the maximum period required because it is not unusual to start and stop the boiler frequently during the initial phase of a dry out. It is also possible for the procedure to take much more than an 8 hr shift. On any large boiler, it is common for it to take more than a day. Normally, the dry out and boil out are performed with controls in manual for minimal adjustments as necessary to obtain a clean burning fire. The dry out started before beginning the boil out. The boil out will most likely finish before the dry out is complete. That is because no steam pressure to speak of is produced while boiling out. The temperature is only a little over 212°F when the boil out is complete.

The boiler must cool a little before draining the boil out chemicals and refilling it. There are two arguments about dropping boil out water. One is that solids will stick to the metal and bake on. Allowing the water to cool is best. The other is that they will retain the solids while hot but drop them out if they are allowed to cool. Thus, dumping the water hot is best. Always look for recommendations of an appropriate temperature to drain the water from the boiler and chemical manufacturers. The boil out water is considerably more caustic than normal boiler blowdown. Provide for proper disposal of that water, neutralizing it before dumping it in the sanitary sewer or employing a licensed hauler to dispose of it. Once the boil out chemicals are drained, the boiler water must be treated. The boil out removed all the varnish and grease that was covering the inside of the boiler and protecting the metal from corrosion. It also removed that material so that it could not burn on to produce a permanent scale on the boiler heating surfaces. From completion of boil out on, those surfaces have to be protected by proper water treatment.

After boil out is complete, the safety valves and other materials that were removed for the boil out should be replaced. This can also produce an interruption in the dry out of the refractory and require a gentle reheating before continuing. Refractory dry out is complete when the temperature of the refractory at any point has gradually raised to something higher than atmospheric boiling temperature. That is usually 212°F but can be lower (203°F in Denver, CO, USA). Some people will accept termination of water flowing out of casing drains. Others are more elaborate. The minor expense of some thermocouples located at certain points in the refractory and monitoring them is the best way to determine a dry out is complete. The controlling temperature is the temperature of the refractory closest to the outer wall of the boiler, not the surface of the refractory in the furnace.

Boiler Control Adjustments

Tuning a boiler is a complicated process. The potential for an untrained individual to blow up the boiler is much higher for the first time. While it is far more dangerous with an older boiler, the potential still exists with a new one. Therefore, as a boiler operator, don't attempt burner adjustments without a significant amount of training and hands-on experience. The contractor's, or the boiler manufacturer's, service technician should tune up a new boiler on startup. That is because the boiler is the contractor's property until such time as the new owner accepts it. If any adjustments are made and the boiler blows up, the contractor or manufacturer is absolved. Therefore, leave them at it.

Transition to Automatic or Manual Control

Another requirement for a new boiler is establishing a smooth transition from light off to automatic operation. This is normally accomplished without any trouble on boilers with jackshaft type controls and is not a factor on fixed fire units. Making the transition with full metering controls is another matter. Normally, there is an interface between the combustion controls and the BMS which allows the BMS to control damper and valve positions to satisfy requirements for purge and light off (low fire) positions. At some point after a successful ignition of the main fuel, the interface lets the automatic controls take over. A stable, safe, and smooth transition between light off and automatic operation requires more than a simple switching from one to the other.

To begin with, a cold boiler with modulation should not be released to automatic control immediately. There is enough thermal shock for a boiler to experience going from relatively cold (even in a hot boiler room) to firing at low fire where the steel is less than a millimeter from hot flue gases over 1000°F. If the controls simply shift to automatic, that temperature difference will readily double. Limiting thermal shock as much as possible is important to extending boiler life. Thus, provisions to prevent the controls running to high fire right after ignition are important. The simplest approach is to set the controls in manual before the boiler starts and make sure that the manual signal is adjusted to low fire. Other approaches include low fire hold systems and ramping controls.

Low Fire Hold

A low fire hold consists of provisions to keep the burner at low fire until the boiler is near operating temperature. The normal arrangement is a pressure switch or temperature switch similar to the operating and high limit controls but with an electrical contact that is normally open. The pressure or temperature has to reach the switch setting before the contact closes to allow automatic operation. The switch has to be set lower than the normal pressure or temperature modulating controls so that the burner is not affected by the low fire hold system after the boiler is up to operating conditions. Sometimes during emergencies, the low fire hold controls will have to be bypassed or the boiler will not get hot until spring. Be certain that the unit can operate in manual to override low fire hold controls.

With the typical jackshaft control, the switch prevents an increase in firing rate above light off position until the pressure or temperature is reached. An automatic low fire hold is very important for modulating boilers that are controlled by a thermostat. A few warm days could prevent the boiler from operating until it is dead cold. The low fire hold will prevent the rapid heating of that boiler on high fire with severe thermal shock. When the outdoor temperature is swinging from warm to cold, the amount of time the boiler is held at low fire is almost proportional to the average heat load. It will be less as the average temperature drops and the delay before release to modulation will decrease. Unless an operator is always on hand to control the warm up of a boiler, there should be low fire hold controls. One final note: on some steam boilers where operating pressures are low, a temperature switch can be used for low fire hold because pressures can swing more significantly, generating control problems. It is generally not desirable to suddenly switch from light off position to modulating because the controls will simply run the burner right up to high fire when it is not necessary. If controlling the boiler manually, allow it to come on line while at low fire. Then, when it seems to have reached its limit, gradually increase the firing rate until the load has reached normal operating conditions. Then switch to automatic.

When the boiler is unattended, ramping controls function the same way and are recommended for high pressure steam boilers that start and stop automatically. They control the rate of change of the firing frequency. The unit gradually increases at a constant rate (like going up a ramp) until it is at high fire or, more normally, the set pressure is reached and the automatic controls take over. A ramping control should only function on the initial transition from light off to automatic or from low fire hold to automatic. The transition rate should be adjustable and should be set so that the rate is as slow as possible to minimize thermal shock. Pneumatic and microprocessor-based systems are described in the section on controls.

Test Safeties

Never forget that the safety valves are the last line of a defense against a boiler explosion. Test them as soon as possible. First, do a lift test on steam and high temperature hot water (HTHW) boilers when the pressure has exceeded 75% of the set pressure of the valves. Hot water boiler safeties can usually be tested before firing by applying city water pressure. As soon as possible in the startup of a new boiler, run a pop test of steam and HTHW boilers. A pop test is described later.

Boiler Warm Up

The boiler manufacturer should have indicated a warm up rate in the instruction manual. One problem is that, normally, there is no good way to determine if the unit is actually adhering to the required rate. If it were critical for temperatures to be below 212°F, then the boiler should be equipped with thermometers. Normally, it is a psi (pounds per square inch) per hour rate that can be tracked. On large boilers, it is not at all uncommon to have to stop and start the burners to limit that warm up rate. Most boilers that are smaller than a quarter of a million pounds of steam per hour can be allowed to warm up at the low fire rate. Fixed fire boilers are absorbing the maximum heat input every time the boiler is fired. They have to be started and stopped to reduce the warm up rate. If that is required, other than on initial startup, the manufacturer should provide automatic provisions for it.

Multiple burner boilers can be warmed up slowly by only operating one, or a portion, of the burners. The burners should be switched regularly, according to the manufacturer's instructions, or every 15 minutes to 1/2 hour so that the heating is more uniform. Always start another burner before extinguishing the one it replaces. Otherwise, a purge of the unit will be required. A purge is blowing cold air over the metal that was just heated to produce a sudden swing in its exposure to temperature. That could produce stress cracks in the metal that are undesirable. A boiler should be limited to the number of starts and full stops it is exposed to. When the manufacturer recommends limiting stops and starts, it is for high pressure boilers with very thick metal parts that are more susceptible to damage from stress due to temperature variations across their thickness (thick wall drums and headers or steam turbine casings).

Full Metering Switch to Automatic

Simply switching from light off position to firing rate control, whether it is manual or automatic, can be rough with a full metering control system. The fuel and air controls are pre-positioned by the interface with the BMS and may be lower or higher than the position that produces flow rates acceptable to the control system. This results in what is called a "bump," as the controls are suddenly allowed to react to the difference and make some rather abrupt, and usually excessive, changes in valve or damper positions in an effort to establish the required flows. On almost any pneumatic or electronic (not microprocessor based) controls, problems can be experienced with reset windup, where the controls detect an error and try to correct it, but cannot. Then the controller output continues to increase or decrease until it reaches zero or maximum possible output. The outputs are outside the control signal range (such as 3–15 psig where the signal can drop to zero or climb to 18 psig), the standard supply pressure. Similarly, a 1–5 volts range can be a negative voltage and go as high as 12. In either case, there is no response to controller action until the control signal winds back into the normal control range.

Modern microprocessor-based controls have anti-windup features and procedure-less and bump-less transfer (manual to auto and vice versa) features that eliminated the problems with earlier pneumatic and electronic controls. It is possible that the system designer did not properly configure those features and bumps can still be experienced on transfers. A fuel control valve should be positioned at a minimum (mechanical) stop where fuel flow after ignition is more than the controller's set point. If it is not, the controller would wind up to maximum output (and it has a lot of time to do it before a main flame starts). Then the fuel valve would suddenly swing open when the controls are released to automatic. If the flow is a little higher than the controller's set point, reset windup (in this case, it would wind down), there is simply a delay in the response. However, there may not be sufficient time between main flame ignition and transfer to automatic for the controls to wind down, and excessive fuel feed could still occur. If the controls do wind down before transfer, they will have to recover. Once the fuel valve starts to open, it swings open more than it should. To overcome those strange actions, the interface between the BMS and the combustion controls should actually adjust the set points to achieve purge and light off conditions so that the controls are controlling all the time. The ramping controls should help overcome that problem with reset wind up on light off. Bumps off low fire and maximum fire can occur during normal firing and are discussed in the section on controls.

Collect Performance Data

Among the many things that do not need a "break in period," a boiler should top the list. From the moment the first fire lights in the furnace, a boiler begins breaking down. Thus, collecting performance data on a new boiler as soon as possible is essential. It provides a record of what the boiler was capable of when it was new, completely clean, and not broken. Sometimes, there are problems with a new boiler that had to be corrected before it was accepted. Otherwise, a boiler is normally at its best when it is new. Should it happen that changes have to be made in a new boiler, this data should be collected once the boiler is accepted. This data provides a basis for comparison to actual operating conditions during the life of the boiler and is very helpful in detecting the source of problems, such as inefficient operation. Some of the data that should be collected once a new boiler is ready to go into service follows. The data should be collected as soon as possible and, when necessary, before performance testing. The following list is not necessarily complete. There may be certain things about the new boiler that are not indicated or documented and should be recorded as well. The more the data that is collected, the more likely the problems with the boiler detected during its life will be.

Data should be collected at each firing rate (each screw on a jack shaft cam as shown in Figure 2-4 or each 10% interval of control signal) and should include much more than readings of all boiler mounted instruments, pressure gauges, thermometers, etc. On steam boilers, the drum level using a signal level, a temporarily mounted ruler, or simply the number of nuts at the side of the gauge glass should be recorded. For hot water boilers, the water level in the expansion tank should be recorded, even if it is adjusted automatically. Some new boilers have been supplied without any pressure gauges on the fuel supply system. Despite that, find the required test ports that permit connection of a gauge to collect that data. This is especially important when there are multiple boilers, or other equipment, using a fuel gas supply which would prohibit taking fuel meter readings. If the boiler does not have a fuel meter, purchasing one is recommended. Install it right away so that data can be recorded and collected continually.

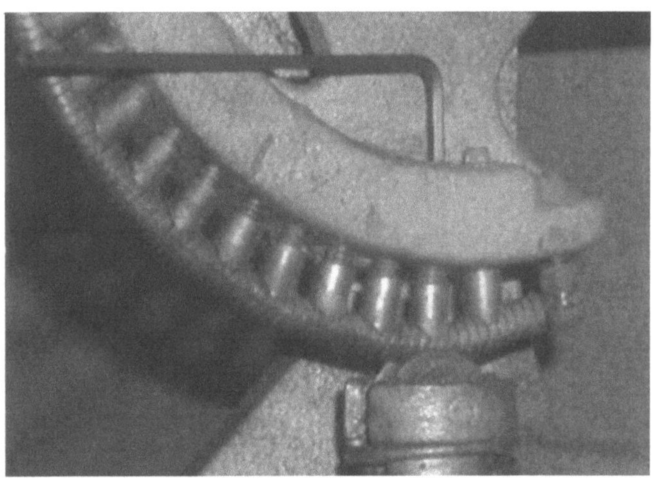

Figure 2-4. Adjustment screw on control valve.

Not all boilers are fitted with draft gauges. Hook up a manometer, measuring and recording the draft (sometimes a pressure reading) at the fan discharge, the burner wind box or burner head, in the furnace, at intermediate points within the convection bank (where accessible), the boiler outlet, economizer outlet (when equipped with an economizer), the air heater outlet (when equipped with an air heater), and the inlet to the stack.

It is also a good idea to take readings of shell or casing temperatures. The best way to do this today is with an infrared camera that is calibrated to accurately indicate temperatures. There are contractors who provide an infrared survey service that could be used. There are less expensive methods, including the use of an infrared thermometer or as simple as holding a thermometer against the casing or shell with a piece of insulation. That last method has to be described along with the recorded data because the insulation does restrict heat flow and produces a temperature higher than actual. In this way, the same method can be used for comparison at a later date. There can be considerable differences in the temperatures of casing surfaces on water tube package boilers. Take several measurements of those surfaces. New water tube boilers should be fitted with insulated drum head covers and their surface temperatures should be recorded as well. If the boiler (new or not) is not fitted with insulated drum covers, measurements of the drum head surfaces should suggest an early modification to the boiler to include them.

Photographs of the flame in the furnace for each fuel, as well as each load, can be used for comparison in the future. A piece of blue glass can be used to reduce the light intensity when photographing a fuel oil or coal fire. These flames have strong radiation characteristics and cameras will typically blur the picture when photographing flames of those fuels. If the boiler is fitted with automatic controls, the recorded data should include each controller's set point (whether manually set, fixed, or remote), process variable value and/or percent, and the controller's output signal, normally in percent. For plants with a data logger or data historian, be sure that all of the data that is desired is actually collected.

Acceptance Testing

The final step in startup of a new boiler should be the performance of an acceptance test. Data should be collected and recorded at the firing rate where the efficiency is guaranteed by the manufacturer. If the unit is a modulating boiler, collect and record data at no less than three other firing rates (a maximum of 75%, 50%, and 25% being common). All data collected should be carefully recorded and stored in a binder for future reference. If it is a new plant, the performance of all equipment should be documented at the various firing rates. Occasionally, a plant is started when there is no place to use the steam and no way to perform the test until other installations are in place. The installing contractor then requests a delay in testing until a load is available. When that occurs, collect data at firing rates which can be handled. Nearly identical readings at a later date will prove that the boiler was not abused while waiting for a load.

Acceptance tests vary. ASME PTC-4.1, the "Steam Generating Units" power test code, provides three means of testing a boiler for acceptance. However, a test in conformance with that test code is an expensive proposition, requiring continuous documented operation of the boiler for a period of 8–12 hrs. It is justified for a boiler designed to generate more than 60,000 pounds of steam per hour but not for a small 50-horsepower boiler that only generates 1700 pounds per hour. There are other simpler and acceptable means for testing those boilers. The important element is having established one acceptable to the owner and the manufacturer before the boiler is purchased. In the unlikely event that the boiler fails to perform, the manufacturer is then committed to make it right.

For small boilers, testing for 1 hour at each load point is recommended and, with that exception, testing using the "heat loss method" of PTC-4.1. That way, there is a formal acceptance test but not the expense of long runs. It should not require any overtime because the unit should be allowed to settle for half an hour before each test to establish test conditions. Six can be done in one day. On a boiler with two fuels, that would mean no less than four days just running acceptance tests.

A final acceptance test when a boiler is field erected is very important. A contractor can build, and many have, a boiler that does not meet the performance requirements. Even for the boiler that is factory tested, run an acceptance test of the final installation. The cost of a boiler is a fraction of the cost of the fuel that it will burn in its lifetime. A small difference in performance can represent a considerable sum. Estimate the cost of a 1% difference in efficiency for a particular installation and use that value, with the vendor's knowledge, in evaluating boiler offerings. Other startup activities that may be associated with a new plant are covered in the following descriptions.

DEAD PLANT STARTUP

All personnel working on the site within the boiler plant and equipment rooms should be notified of the plant startup so that they are not surprised and, more importantly, not injured as the system comes up to temperature. Confirmation from all those people indicating that they are prepared should be formally documented before the boiler is fired. Normally, a dead plant means dead cold. There is no heat in the boiler or any auxiliary equipment associated with normal operation. A dead plant startup is returning a dead plant that had been operating to operating condition. It is not uncommon to return a plant to service that was shut down for the summer or a protracted business slump. It is also occasionally necessary to return a plant to service after a loss of electric power or water supply that forced it to shut down. The operations mentioned here are assumed to occur after a plant was laid up according to the proper procedures. Some of these activities also apply to simply returning a plant to normal operation.

Remove sorbent from the boiler, deaerator, and other closed vessels. Install new gaskets and close manholes. Check to be sure that all personnel, tools, etc., are out before closing the vessels. Fill fluid systems as described in New Plant Startup. Everything from leaves to birds can find their way into air and gas openings to block them while a plant is shut down. Check to confirm stack clean outs, vent openings, and air inlets are clean. Confirm the vent valve on the boiler and the free blow drains are open. If the burner on the boiler was dismantled or repaired, the steps in New Plant Startup should be followed to ensure proper burner operation. As soon as possible, compare initial operating data with current operating conditions to ensure that there have been no significant changes in the boiler's performance. Record oil tank levels, fuel gas, steam, and water meter readings to establish values at startup. Leakage testing and other activities may have changed the meter readings from the shutdown or last recorded state.

A cold boiler should be returned to operating conditions slowly. When starting a boiler in a dead plant, it is advisable to bring the served facility up with the boiler. That increases the time it takes to raise pressure on the boiler and the facility to allow for gradual heating. Open all valves that lead to the facility only after confirming that all drains and vents in the facility have been closed or are manned by trained observers.

In steam plants, this process normally creates a flood of returned condensate as the pressure builds. Provisions for handling the condensate should be provided. Lower the operating level of the boiler feed tank or deaerator and condensate tank beforehand, if possible. If that is not possible, close the isolating valves for makeup to those tanks and manually maintain the lowest reasonable level until pressure in the facility is near normal. For hot water installations, the system should be flooded, the expansion tank level confirmed, and circulating pumps started to generate at least minimum flow in the system. This may require a walkthrough of all equipment rooms to ensure the systems are ready to circulate water. Any equipment still receiving maintenance should be adequately isolated using proper lock out and tag out procedures.

Lock controls in manual at low fire. Starting a dead plant or boiler should provide a very slow increase in temperature until the boiler's contents are above 220°F. That minimizes damage to the refractory from pockets of absorbed moisture. A sudden increase in volume as liquid changes to steam will build up pressure inside the refractory and rupture it. It is sometimes necessary to repeat an initial dry out because the refractory got wet or refractory repairs were performed while the plant was down.

Performing operational tests of the boiler's operating limits during the initial firing of the boiler will provide frequent interruptions to the heat. That will reduce problems with the refractory and provide early reassurance that the safety and operating limits are functioning properly. A wise operator will not only confirm limit operations but also record it in the log book.

As soon as steam is evident at the boiler vent, operate the vents in the facility to remove air from the steam distribution system. If the system has automatic air vents, it is a good idea to operate a few manual vents anyway to ensure the automatic vents are working. In high pressure steam plants, close the free blow drain valve only after steady steam flow is certain. The purpose is to prevent any condensate accumulation over the non-return valve that would slug over into the steam piping when an interrupted flow is re-established.

Close the boiler vent valve when the pressure is up to 10 psig on heating boilers or 25 psig on power boilers. Allowing a loss of steam until those pressures are reached helps ensure that all the air is removed from the boiler. If the boiler feed tank is fitted with a steam heating sparge line, it should be placed in operation after the boiler vent valve is closed. If it is a coil heater, it may be allowed to come up with the plant. Open the vent valves on a deaerator wide before admitting steam and gradually open the steam supply to the deaerator only after there is a constant flow of water to the boiler. Any sudden surges in water flow could rapidly produce a vacuum in the deaerator. Also avoid any rapid changes in facility steam consumption that could cause a drop in steam pressure. If a vacuum is formed, the deaerator and its storage tank could be damaged. Once the deaerator pressure is up to normal, open the isolating valves wide so that the steam pressure regulator can function and close the vent valve to its normal throttling position. Test the low water cutoff before reaching normal operating pressure and after the pressure is high enough for the boiler to return to firing. That is normally when the pressure exceeds 6 psig for heating boilers and 30 psig for power boilers. Lift test the safety valves when the pressure is above 75% of the safety valve set pressure. They could have corroded shut during the shutdown period.

At some point, low fire will not be adequate for the pressure to continue to rise. Increase the firing rate manually in small increments (less than 10%) and allow the pressure to stabilize before increasing it again. Initially, all the condensate will stay in a steam system because the pressure will be below atmospheric wherever automatic vents are not operating properly or do not exist. Condensate will not return until there is enough pressure differential to push it back to the boiler plant. At several points during the startup, the pressure differential will accelerate condensate returning. The slow steps will limit the rate at which that happens. Wait until the pressure is at, or slightly above, normal operating pressure to switch control to automatic.

After an hour or so of automatic boiler operation, the normal operating levels of the condensate tank and deaerator may be restored if they were lowered for the startup. Increase the level gradually to avoid any damage associated with a rush of cold inlet water. If the timing is right, there should not be an inrush because the vessels will be filled by the condensate stored in the system. Make a point of noting the amount of condensate returned to provide better guidance in an SOP for the next dead plant startup.

With steam generation stabilized, draw water for analysis and determine the setup of chemical feed and blowdown controls. Open cooling water valves to any quench system. Open valves to put the continuous blowdown heat recovery system into operation. Vent the flash tank until steam has been flowing out the vent for 10–15 minutes to avoid pushing air into the deaerator. Alternatively, leave the deaerator vent wide open until the blowdown system is in normal operation.

Record the startup activity in the log and begin monitoring the plant as required for normal operation. It is very important to note all problems that came up, any changes in operating procedures that were required to accomplish the startup or correct problems, and the conditions at various times during the process, with the times noted. That data can be used to compare with the original SOP for dead plant startup and modify it to improve the process (not shorten the process). Usually, when starting up a dead plant, there is time because many other operations will not even be contemplated until the steam or hot water is flowing normally. A slow startup ensures minimal stress from thermal shock and avoids the pitfalls of rushing to get the job done.

On the other hand, when the plant is being restored after an unscheduled interruption, one can take the shortest reasonable time based on experience with prior startups. If called upon to rush, know which boiler to select for it (the one that needs the most refractory repairs anyway). Selectively damaging the plant under emergency conditions, such as restoring heat to a hospital or nursing home where it is critical, is part of a well-prepared disaster plan.

NORMAL BOILER STARTUP

After that initial plant startup, don't relax and get too casual about a boiler startup. The equipment will deteriorate with age and use to the degree that something could go wrong. A certain amount of that should be addressed each year right after the annual inspection. Typically, the boiler has been shut down and opened up for inspection. Some items may have been repaired or replaced. Treat the next startup as if it were new. Keep in mind a few other good habits that take the wear and age into consideration.

Close circuit breakers as needed to apply power to the burner management control at least 24 hrs prior to starting a fire in the boiler. Flame sensors can deteriorate and provide false flame signals but may operate normally when they are first energized. The long warm up ensures that the sensors are properly checked by the BMS during startup. A normal boiler startup assumes that other boilers in the plant are operating and the boiler to be started has not had maintenance or other work performed on it. If there was work performed, review the recommendations for new plant and dead plant startup to determine if there is anything that should be checked or tested before proceeding. Make sure the vent valve is open. If the stop valve at the steam header is closed, the free blow drain valve should be open.

Set the firing rate controls to manual and low fire. Make one quick trip around the boiler to be certain that it is not open and all valves are in the proper positions before starting it. Open the fuel block valves slowly to ensure the fuel supply to operating boilers is not upset. When firing oil, check an oil burner assembly and then insert in the burner. If oil is steam atomized, open isolating valves to admit steam to the inlet of the burner steam shut off valve. If oil is air atomized, start the compressor and admit air to the inlet of the burner air shut off valve. Check to be certain that normal operating fuel supply pressures have been established. Blow down the gauge glass and water column while observing the water level in the glass to assure that the boiler contains water.

Turn the BMS control on to allow a burner to start. On multiple burner boilers and in older single burner plants, it may be necessary to initiate a purge and burner ignition. When the pilot flame is proven, gradually open the atomizing steam or atomizing air shut off valve at the burner. This ensures that any fuel oil that may be transported to the burner by the atomizing medium will be exposed immediately to the ignition energy of the ignitor and burned at nearly a normal rate. Opening the valves earlier can inject a slug of oil into the furnace that would subsequently vaporize to produce an explosive mixture in the furnace and ignite when the pilot comes on. Open the fuel shut off valve, if the atomizing medium did not produce a fire, to start the main burner. If the atomizing medium dumped in some fuel that produced a fire, it is best to repeat the purge. Sometimes, the opening of the steam or air valve is too slow. There is not enough time to get fuel on. That is okay. Wait until it has purged again. This is a normal startup and there should be no big hurry.

Shortly after the burner has started and is operating normally, close the burner manual valve. The BMS should detect a flame failure and initiate a boiler shutdown. Only if the boiler shuts down, reset the BMS for another start. Open the burner manual valve after the BMS indicates that an ignitor flame is proven. Having restored operation, check the low water cutoffs by blowing each one down and confirming the BMS shuts the boiler down. Don't use any bypass push button while testing the cutoffs at this time. A full operational test is needed. Once the boiler is up and operating, it may not shut down for months. This is the one and only, best and truest time to confirm that the flame failure and low water safety systems all work. Repeat the low water cutoff tests if it is necessary to shut the burner down to control the rate of heating of the boiler.

Close the boiler vent valves when the pressure is up on heating boilers or 25 psig on power boilers. If the non-return valve on a high pressure boiler is closed, open it so that steam will flow to the free blow drain. If the second steam stop valve was left open, open the free blow drain to drain the boiler header and leave the non-return closed. Allow the pressure to increase while observing it closely. The burner should shut down when the operating pressure or temperature control setting is reached. Once that operation is proven, test the high pressure or high temperature limit switch by temporarily installing a jumper on the terminals of the control switch. The high limit should shut the boiler down before the safety valves open or the temperature of a heating boiler exceeds 250°F. It should also lock out to prevent continued operation. Allow the pressure to fall until it is below the operating pressure. Then reset the controls so that the burner can be started again. Remove the jumper from the control switch terminals.

Once operation of the low water cutoff has been proven and the boiler pressure or temperature control and limit switches tested, successive tests of each combustion air and fuel limit switch can be made. Proving the operation of the low combustion air flow switch can produce a condition of flammable mixtures in the boiler. Be extra careful on that test. In some cases, the switch setting may have to be adjusted to simulate a condition. That is not the best of tests, but, at least, something will have been done to ensure that it operates.

- With the firing rate set at minimum fire, reduce combustion air flow by slowly sliding a blank over the inlet of the forced draft fan while someone watches the fire. The minimum air flow switch should trip before the fire gets smoky or unstable. Take care that the blank does not affect the switch sensing the air flow. Use another method of reducing air if it does.

- Increase gas pressure to the burner while watching the fire, again at minimum fire. The high gas pressure switch should trip before the fire gets smoky.

- Decrease gas pressure to the burner while watching the fire, again at minimum fire. The low gas pressure switch should trip before the fire becomes unstable.

- Decrease oil pressure to the burner while watching the fire, again at minimum fire. The low oil pressure switch should trip before the fire becomes unstable.

- If the oil is heated at the boiler, double check operation of high and low oil temperature switches (if present) by adjusting the oil temperature while observing the fire. This takes time due to the thermal inertia of the system. Be prepared for that. If the fuel is heated at a common supply point, the testing should only be done when it will not interrupt the operation of other boilers.

If the burner does not trip on low water cutoff, flame failure, or high pressure limit, secure it, note the failure in the log, and notify the supervisor that it is not working properly. A boiler with malfunctioning safety controls should not be placed in operation.

Open the boiler isolating valve on a heating boiler when the boiler pressure is reasonably close to the header pressure. It is best to open the second stop valve on high pressure boilers when the pressure in the boiler is within 20 pounds of the header pressure. The minimal difference in pressure limits the steam wire drawing of the second stop valve seats and makes it easier to open the valve. When there is a bypass built into, or around, the second stop valve, use it to pressurize the boiler header. The normal way is to open the non-return valve, when ready, to put the boiler on line to build up pressure in the boiler header. In either case, always be certain that the free blow drain valve is open and blowing steam to ensure that there is no condensate in the header that would suddenly enter the plant steam header.

After steam is flowing to the header, as indicated by a steam flow recorder or a drop in boiler pressure as the non-return valve lifts, close the free blow drain of a high pressure boiler. Once the boiler is "on line," which means it is delivering heat to the facility, record the fuel and steam or other output meter readings. In that way, the amount of fuel needed to bring that boiler up and onto the line can be calculated. If the plant changes boilers frequently, it may be a very important calculation because there is a considerable amount of fuel used to do that and an associated amount of energy lost when a boiler is taken off line and left to cool.

The final step in a normal boiler startup is to establish its manual firing rate or place it in automatic control. Since the unit should still be at low fire, this can require increasing the firing rate manually until the desired firing rate is reached. If the unit is intended to be placed on automatic control, increase the firing rate until it is about the same as the same sized boiler that is already on automatic before switching to auto. Simply throwing the switch to auto is not the appropriate way to do things because the boiler controls could swing for some time before they are stable again. On November 13, 2020, two steam workers at the VA Hospital in West Haven, CT, USA were killed while starting up a steam line. The preliminary report states that this was a "pressure event."

Especially critical from a safety standpoint is making sure that all water is eliminated from the steam system prior to and during startup. (It is also good practice to drain the system upon shutdown due to corrosion.) Failure to do so can likely result in water hammer events that can endanger personnel and damage system components. For this reason, it is imperative that facility managers provide proper startup procedures to their employees. Personnel should be held accountable for diligently and consistently adhering to these procedures. Including the following items in system startup procedures should be considered:

- Format procedures as a CHECKLIST
- Warm up to be slow and methodical
- Ensure any removed insulation has been replaced
- Use radio communication between participants
- Use warm up valves on any isolation valve > 3"
- Open drip leg blowdown valves
- Open strainer blowdown valves
- Confirm that all steam trap stations are in service
- Open test valves downstream of steam traps to ensure evacuation of condensate
- Once system is running, check test valves again to confirm proper steam trap operation
- Monitor expansion joints, blowdowns, steam quality, movement, and noise throughout
- Hold personnel accountable for diligent and consistent use of startup checklists

All drip leg steam trap stations should be evaluated regularly. At-a-glance saturated temperature indication, along with operator training, should be a standard specification.

EMERGENCY BOILER STARTUP

Emergencies come in two forms, instantaneous and impending. With well-maintained equipment, regardless of who maintains it, there should not be too many of either. Instantaneous emergencies involve an immediate shutdown of the plant or an operating boiler such that steam cannot be supplied to the facility served by the boiler plant. Impending emergencies are the ones where it is only a matter of time until that steam cannot be supplied to the facility. Impending emergencies involve things like the severe squeal of a fan belt or motor bearing on operating equipment that indicates that it is bound to fail very soon. It can also be clouds on the horizon and the sound of thunder when the power is likely to fail because the plant never seems to make it through a thunderstorm without a power failure. Natural occurrences from flood to excessive heat to deep snow and forest fires seldom come without a warning. They should be considered impending emergencies. With the disaster plan in mind, appropriate action can be taken, which is why there is a plan in the first place. If those disaster plans have been rehearsed, it will likely be a more comfortable situation. Steam may be such a precious commodity in the plant that a boiler may be placed on hot standby. In that case, start the standby boiler so that it can be brought on line. Once the problem with the boiler that tripped is resolved, it can be put on standby or restored to service.

Frequently, the reason for a boiler shutdown can be determined, and corrected, quickly so that it can be returned to service. Many times, it is resolved quickly and the boiler is returned to service even before anyone else notices that there is a problem. When that cannot happen, then it is time for an emergency boiler startup. Any emergency that results in the shutdown of a boiler should be attended with an instant evaluation of the condition of that boiler. If it cannot be returned to operation or the cause of the shutdown is unknown, the first step would be to start another boiler, if available, so that it is warming up. Starting the other boiler takes time from finding the problem with the unit in operation, but it also allows for a more gradual warm up of that boiler in the event that the one which was running cannot be restarted.

Of course, if the cause of the shutdown is known, such as a short power interruption, and will not prevent a restart, there is no reason to start that other unit. If there is water and steam pouring out of the boiler that shut down or large gaping holes in what used to be square casing, there is no hope for the boiler that went down, and all that can be done is to secure it.

An emergency boiler startup is one that requires operation of a boiler from a dead cold condition in as little time as possible. There are things that can be done to limit the damage to the boiler in that process and actually accelerate the startup time. Use these suggestions as guidelines to prepare a specific disaster plan that describes an emergency boiler startup for the plant in question.

Frequently, heating boilers are allowed to sit idle with their steam valves open. This frequently gives the operator an impression that the boiler is ready to go because there is pressure on it. Nothing could be further from the truth. Steam at the surface is at saturation and hot. The original boiler water and condensate below the surface can be much cooler. Even systems that drain the condensate from the bottom of the boiler do not correct for the fact that the majority of the water in the boiler is relatively cold. Ramping up that boiler will cause a significant temperature swing in the boiler shell right at the water line.

Power boilers will always be considerably colder than normal steam condition. The principle concern in an emergency start of a boiler is the development of stresses in the boiler metal associated with rapid heating of the boiler. Whether it is a low pressure fire tube boiler or a large water tube boiler does not matter much. Both have thick steel parts in contact with the boiler water, which have to be heated to normal saturation temperature. The time spent in doing that will determine the extent of damage by thermal overstress.

Rather than heating all the water in the boiler, bring in warm or hot water to help accelerate the warm up. That is especially true when the boiler is the only one available. Temporarily shutting off the makeup water and operating the boiler blow off valves to drop level so that the heated water from the boiler feed tank or deaerator displaces much of the cold water in the boiler will both add to the heating of the boiler and provide some movement of water to help transfer that heat to the thick parts of the boiler metal. Once a boiler feed tank has been nearly drained, the makeup water can be restored. Let a deaerator sit until steam is being produced. Then bring the makeup on very slowly.

In an emergency, the need is to push the envelope as much as possible without damaging the boiler. The disaster plan should have been developed after some testing that determines what firing rate provides the fastest warm up of the boiler within the limits recommended by the boiler manufacturer. That rate can immediately be set to get the fastest possible warm up.

If the normal procedure for warm up includes shutting the burner down, don't do it. That activity only needs to be done for refractory dry out. The operation of the burner followed by a purge produces dramatic swings in the metal's exposure to temperatures on the fire sides. Less damage to the boiler will be done by firing continuously, although at low fire, than by cycling the burner on and off. In multiple burner boilers, where only operating one or a portion of the burners during warm up, operation should consist of firing another burner before shutting one down, as explained in the first discussion on startup.

Then, of course, there is the matter of how serious the need for steam is. Loss of steam for blanketing chemical reactions may be more critical than damage to the boiler. In a hospital during a disaster where every operating room is handling emergency surgery, maintaining steam for sterilization is a must. In such situations, the disaster plan can call for ignoring the manufacturer's recommendations so as to bring a boiler up to operation as fast as possible.

At some point, how much time it will take to recover has been established and documented in the disaster plan. That information should be supplied to the facility served by the boiler plant so that they are aware of it when preparing their disaster plans. Some facilities may question the amount of time. It is a good idea to estimate the value of the steam production (amount of steam/hr times $/lb of steam) and provide that figure as a cost to be compared with probable failure of the boiler due to excess stress during a rapid startup.

With regard to refractory, if the boiler has been laid up properly, there is no reason to believe that serious damage to the refractory could occur during an emergency startup. Very old boilers and coal fired units may have sufficient thicknesses of refractory that can be a concern. In that case, the plan should address those conditions when they exist. Finally, log it all!

NORMAL OPERATION

What is the most sensitive, precise, and accurate sensor in a boiler plant? It is the operator's ear. Even before the pressure gauges drop or the alarm goes off, when something starts to go wrong, the operator can hear it. Today, there are sophisticated systems that analyze "big data" in an attempt to anticipate the maintenance needs or the operational needs of the plant. Still, in many situations, hearing something that is not quite right will be the indication that something needs to be looked at. Modern distributed control systems (DCS) can watch and monitor thousands of sensors in a plant. There can be so many alarms and warnings that the operator can suffer from sensory overload. The operator still has to hear and decide what is important and take appropriate action.

During any typical working day in a steam plant, a boiler operator will spend no less than 4 hrs plus 1 hr per operating boiler and 1/2 hr per idle boiler to do the following:

- Note the weather forecast for the next shift and predict the steam load to see if another boiler must be started or one stopped to accommodate that load. Transfer the number of local degree days to the log. Review communications from the prior shift, the chief engineer, and the plant engineer to see if facility operations will change the load and plan accordingly. In production facilities, review the production schedule for the same purpose. In some cases, today's operator checks the standing orders and production schedules on the plant's Intranet to determine the boiler load.

- Check each boiler in operation to note water level, steam pressure, feed water pressure, fuel pressure, fuel temperature, stack temperature, draft, casing color and temperature, firing rate, position of control linkage, security of control linkage connections, condition of air inlets, temperature of blower bearings, temperature of blower motor and its bearings, signs of vibration at blower or its motor, flame signal strength, flame appearance, flue gas appearance, and signs of leakage.

- Check each idle boiler to note water level, internal pressure, position of the vent valve, stack temperature, draft conditions, casing temperature, position of the control linkage, security of control linkage connections, condition of the air inlet, furnace and boiler pass conditions, and signs of leakage.

- Check auxiliary equipment and systems to note salt storage level, brine level, softener in service, other pretreatment equipment as applicable, condensate tank level, deaerator level, deaerator pressure, condensate temperature, feed water temperature, condition of deaerator vent gases, temperature of condensate pump bearings and the pump's motor and motor bearings, temperature and condition of the condensate pump seal and seal flushing flow, temperature of the boiler feed pump bearings and the pump's motor and motor bearings, temperature and condition of the feed pump seal, continuous blowdown discharge temperature, flash tank pressure, blowdown drain temperature, chemical feed tank levels, fuel oil supply pressure, fuel oil service pumps, motors and bearings when firing oil, fuel gas supply pressure, fuel tank levels, and signs of leakage.

- Draw representative samples of boiler water and test the water for partial alkalinity, total alkalinity, phosphate residual, sulfite residual, chlorides, iron, total dissolved solids (TDS), and other concentrations as dictated by the water treatment supplier. Draw representative samples of condensate and test for hardness, pH, iron, TDS, and other concentrations as dictated by the water treatment supplier. Draw multiple samples of condensate and test when necessary to isolate hardness leakage. Draw representative samples of the boiler feed water to test for pH, chlorides, TDS, and other concentrations as dictated by the water treatment supplier. Draw samples of raw water and test for hardness and TDS. Draw samples of softened makeup water and test for hardness, repeating frequently near ends of softener runs to detect breakthroughs.

- Record, in the boiler plant log, many of the levels, pressures, and temperatures described above, maintenance activities described below, unusual activities and events, and observations of conditions that are precursors to failures. Record water, fuel, and steam flow meter readings. Calculate and record evaporation rate and fuel consumption per degree day. Then evaluate the results to identify changes or upsets in system operation and quality of control adjustments. Calculate percentage of returns and compare with history to detect system leaks and upsets.

- Perform normal operating activities including: Test the low water cutoffs on each operating boiler during each shift for three-shift operation and at least twice each day. Calculate effect of changes in raw water hardness on softener capacity and adjust softener regeneration rates accordingly. Adjust the continuous blowdown rates at operating boilers to maintain dissolved solids concentrations, iron, alkalinity, or whatever is the controlling factor. Adjust the chemical feed pump rates to restore normal water chemistry for each concentration. Clean fuel oil filters when firing oil. Operate boiler soot blowers as required. Adjust firing rate controls to maintain normal operating pressures and/or cycling controls to maximize cycle time according to the load. When indicated, sample and test boiler flue gases to evaluate firing conditions. Then adjust fuel to air ratio accordingly.

- Provide escort for visitors, inspectors, and contractors. Note the work being performed by contractors and service providers and inspect their work where required. Receive shipments of fuel oil, water treatment chemicals, maintenance parts, and other materials. Document all visitors, contractors, deliveries, etc., in the log.

In addition to the daily activities described above, perform weekly activities including: Inspect air inlet louvers and screens for blockage and clean as necessary. Restore full levels to all lubricating oil reservoirs in pumps, blowers, fans, air compressors, etc., using the required lubricant. Check salt elutriation conditions and adjust brine feed accordingly. Draw representative samples of the boiler feed water and test for dissolved oxygen. Take direct level readings and check for water incursion in fuel oil storage tanks. Perform bottom blow off of operating boilers (this activity normally requires the presence of two operators).

In addition to the foregoing, perform monthly activities including: Lift test safety valves on all operating steam boilers. Conduct slow drain test of low water cut offs. Test flame detectors. Check along all fuel gas piping elements with leak tester. Check fuel gas regulator vents to detect diaphragm leaks and vent valve vents to detect leaking vent valves. Inspect all piping in plant for loss or dislodging of insulation. Inspect stack cleanout for accumulation of debris and clean as required. Changing and cleaning of filters is usually performed on a monthly basis, but each one is staggered to provide a level load of work as much as possible.

Annually, the operators should prepare each boiler for the internal annual inspection by the National Board Commissioned Inspector. During that process, the operators should inspect the boiler internals on the water side to assess their performance in maintaining water quality and on the fire side to detect any soot accumulation, refractory damage or dislodging, seal damage or loss, and other problems that might change the heat transfer rates in the boiler. At least, two people are needed for inspections to satisfy confined space requirements. Biannual, five-year, and ten-year inspection and maintenance cycles need to be considered as well. Programs for greasing motors and driven equipment can be scheduled in a manner that spreads this work out rather than doing it all at once.

Annual tests that should be performed by the boiler operators include: Leak testing of fuel oil safety shut off valves, regulators, and vent valves; calibration checks of gauges and thermometers; removal and replacing of safety valves where the insurance inspector requires rebuilding, normally on a five year per valve basis. All the above assumes a bare bones boiler plant. There is always additional equipment and systems that need to be monitored and maintained on a regular basis and service the facility and/or the boiler plant including (but not limited to) domestic hot water heaters, air compressors, cooling towers, chillers, air handling units, etc. Adding the monitoring, maintenance, and water conditioning for those systems can easily consume another operator's time for a normal day. Even though much of the readings and data are available on a computer screen, they still have to be checked and trends monitored to be sure that operations are within the normal ranges. Suppose the vibration tolerance on a fan shaft is up to 8 mils before a warning light goes on. The "normal" reading is 2 mils. If the trend is increasing from 2 to 3 and higher, that is an indication that there is a problem with the fan. There may be deposits on the fan blades or a bearing problem on the shaft. These are the kinds of things to be monitored and noted. On the next shut down, that fan can be inspected and possibly fixed before the next startup, without waiting for the high vibration alarm to go off, or worse.

SAFETY TESTING

Both the National Board of Boiler and Pressure Vessel Inspectors and the ASME have recommended that safety valves be lift tested monthly on boilers operating at pressures less than 400 psig and pop tested annually. For higher pressures, they recommend testing based on operating experience. Since there are more boilers operating at pressures lower than 400 psig, many should be testing the safety valves. Since those safety valves are the last line of defense to prevent a boiler explosion, checking them at a reasonable interval is very important. This section covers steam boilers here. Safety relief valves on hot water boilers are another story and will be addressed later.

As to the frequency, monthly lift testing may not be such a good idea. That is because plants will set a schedule for testing on, say, the first of each month or the first Monday of each month. With such a regular schedule, those that object to the noise from the test will get local officials to be on hand to hear all that dreadful noise. Testing quarterly, every three months, and at random hours on a random day will reduce the complaints because of the longer period and randomness.

Lift testing is raising the lever on the safety valve to open it, thereby proving that the mechanism will allow the disc in the safety valve to come off its seat to release the steam. If it does not lift, that means the valve is not operational and will not prevent a steam explosion due to high pressure. How it is done is rather important because of the potential to leak. If the test is done correctly, the valve will not leak. Also, it will not leak, provided there is nothing blocking its operation or that can clutter it up. The check round performed before lift testing should include a close inspection of the safety valve outlets, discharge piping, drip pans, and drains to ensure they are all clear. A safety valve pops open and pops closed with no feathering or weeping because it is either open or shut. The transition is so fast that there is no way wire drawing can occur. Wire drawing is a term used to describe the cutting of a valve seat, disc, or both by a small steam leak, where the high velocity of the leak erodes the metal.

According to the ASME, a safety valve should be lift tested only when the pressure in the boiler is higher than 75% of the valve setting. One reason is that the lifting lever can break upon lifting without the steam pressure to assist it. That is not necessarily a completely true statement. Another consideration is the load on the boiler. If there are only two safety valves, then lift testing when the load is higher than 50% is producing a temporary load greater than 100%. That means the steam and water separation in the drum will exceed what the manufacturer designed it for, and it could promote foaming and priming, which could carry boiler water droplets into the safety valve at high velocity to produce damage from impingement and possibly produce deposits of boiler chemicals on the valve disc and seat that dry immediately as the steam becomes superheated. That is how the valve can be made to leak. Thus, if there are two safety valves, they should not be lift tested unless the load on the boiler is less than 50%. With three valves, the load should be less than 66%. With four or more valves, the load should be less than 75%.

The problem with the lift test is that the lever and the valve are on top of the boiler. That way, if there is an incident, the vented steam goes up and away from any personnel. It is not convenient to get to and personnel should not be near a valve that automatically pops wide open. To circumvent that problem, get some lightweight welded chain, machine screws and nuts that fit through the chain, and two pulleys per valve that will carry the chain. Figure out where the chain will be located before purchasing the chain and then add 20 feet per boiler to have enough chain to do the job. Some clamps may be needed to attach to building steel, screw eyes for wood structures, or lead anchors and eye bolts to attach to a concrete ceiling to suspend the pulleys. Connect one end of the chain to the eye that is at the end of the safety valve lifting lever (that is what the eye is there for) and run the

Figure 2-5. Safety lift grip.

chain through the pulleys so that it drops to a reasonable height above the operating floor near a walkway. Finally, to prevent the chain from running out, a weight is needed at the other end to offset the weight of the chain between the two pulleys. One suggestion is to make a combination operating handle and weight that looks like the one in Figure 2-5. It looks sort of like a stop sign, and that is intentional. It should have a red background with white border and lettering to add to that effect. The idea is to stop some visitor to come wandering by and wondering what happens if the chain is pulled. Making it about 2.5 inches wide makes it relatively comfortable in the hand. Use at least 10 gauge steel with the edges ground smooth to make it very comfortable to operate the lifting lever. Finally, mount it so that it is above the top of the head of the tallest operator in the plant and low enough for the shortest one to reach it. A wire hook can be used to hang the end from a pipe off to the side of the walkway to store it and to keep it from knocking heads or catching on something being carried down that aisle.

To start the test, check the pressure and the boiler load, and hold the handle that is attached to the safety on the correct boiler. Pull on it and let go of it. Repeat for each safety on the boiler, allowing a few seconds between lifts to let things stabilize before doing the next one. What should happen is the steam will release out of the valve and then stop – pop and stop. The noise from the steam release can and should be heard. In the very unlikely situation (which normally occurs only when the valves have not been tested regularly) that the steam only weeps through the valve after a lift test, operate it again as soon as possible to blow whatever got under the seat out.

Pop testing of a safety valve is normally done as part of the semi-annual inspections on high pressure boilers and during annual inspection of another boiler in low pressure plants. That operation is described in the section on annual inspection.

The safety valves on hot water (HTHW) boilers can be tested using the same rules as those for steam boilers. Always make it a point to ensure everyone is clear of the vents on the boiler room roof before testing because the discharge will include water as well as steam. If problems occur when testing (like hot water splashing all over), the valve discharge piping is not arranged properly. Regular hydronic and hot water heating boiler safety relief valves should not be tested by lifting them. Normally, the system pressures are nowhere near 75% of the valve setting. Instead, the valves should be removed from the boiler during each annual or biannual inspection and examined by looking at the inlet and outlet. If there is scale buildup on either side, replace the valve. Hot water boilers are typically equipped with pressure and temperature relief valves (PTV). The temperature relief is accomplished by a small diameter cylinder that hangs out of the bottom of the valve and contains a fluid or wax that expands with temperature and will force the valve open at the set temperature. Unlike steam safety valves, they do not have to be installed vertically and must be installed so that cylinder is in the heated water. Those valves have to be installed according to the manufacturer's instructions and inspected regularly.

IDLE SYSTEMS

For some strange reason, people think a boiler plant that is shut down during the summer or an air conditioning system that is shut down during the winter does not need any attention. The contrary is true. They need more attention. It is during those periods when the equipment is not operating that they normally incur the most damage. Most of the rusting and corrosion in heating systems occurs during the summer when the boilers are shut down. A typical reason for catastrophic failure of a chilled water system is freezing when it is shut down. Idle equipment deserves just as much attention as operating equipment. Idle boilers should be warm (see the section on standby boilers) or laid up wet or dry. Concerns with warm boilers include checking to ensure that they are really warm. The temperature of the water at the bottom of the boiler should be the same as that of the water at the top of the boiler. Boilers that are not up to operating pressures and temperatures can weep enough to promote high rates of localized corrosion. Thus, casing drains should be checked daily to ensure there is no evidence of the boiler weeping excessively.

Idle boilers require more attention because an operating boiler is generating relatively inert gas. It is less likely to explode than an idle boiler. The fuel oil and gas supply shut off valves should be checked to ensure that they are closed and supply pressures after them are down to zero. Gas fired boilers should be checked by sniffing at an observation port or other sampling means to ensure there is no gas leaking into the boiler. A very expensive industrial accident occurred at the River Rouge Steel Mill in February of 1999. It was the result of gas igniting after leaking into an idle boiler. The result of that boiler explosion was six dead, several injured, and over a billion dollars in damage.

If the boiler is oil fired, the oil burner should be removed or the oil supply piping disconnected from the burner and plugged so that no oil can leak into the furnace. Separate ignitor gas supplies should also be isolated and checked. The ash pits, bunkers, and furnaces of coal and solid fuel fired boilers should be checked for accumulation of anything that could create problems, including water, trash, rodents, and sleeping contractor employees (it happens in some poor countries). Speaking of contractors, an idle boiler should be covered to prevent damage from contractor operations above and around it and panels and fan inlets should be sealed to keep construction dust from entering them. Leave the power on a BMS panel and control panels so that the indicating lights, transformers, and the like keep the enclosures dry. Alternatively, check for operation of panel heaters or temporary lights installed for that purpose. It is important to keep the panels dry. Open the panels once a week to check for condensation. Any rusting or discoloration indicates a need for heaters in them.

The boiler will have problems if it is full of holes when it is time to start it up in the fall. If the boilers are in wet layup, the water should be tested for sulfite content and pH weekly and corrected if the analysis shows the levels to be inadequate for proper storage. Boilers without stack caps should have the stacks covered if they are above the boiler. Stack base access doors should be opened to make sure rain is not entering the boiler and corroding it. Sometimes, that is not easy to do. It is more important to see to it that any rain that falls can dry out quickly by providing, and regularly confirming, good ventilation over the metal surfaces and up the stack.

During the winter, an idle boiler can freeze up if the plant is sealed so much that combustion air from operating equipment is drawn down the stack of the idle boiler. That is the reason why stack temperatures should always be recorded, even on idle boilers. Stagnant water piping and the like can also freeze if the cold outside air that is always drawn into a boiler plant for combustion happens to flow over that piping or equipment.

Chillers, cooling towers, and other air conditioning equipment, plus any equipment or piping system that contains water, should be drained completely when they are idle. If it is not possible to drain a system completely, then it should be filled with an anti-freeze solution that is guaranteed to prevent freezing at the lowest known temperature at the plant. If neither of those options is available, then there is a serious concern with freeze protection. Check every piece of idle equipment regularly during the winter months to be certain that it is not freezing. At one installation, the louvers for combustion air were located in the wall of a boiler room in such a way that the air drawn in traveled right over the chiller. Since it was inside the boiler room, it was supposed to be warm and the plant personnel failed to drain it. At the beginning of the cooling season, they got an expensive surprise. Remember that story. Don't forget that standing water in the boiler plant can freeze if cold air is drawn over it, including water in idle boilers.

The water supply piping is susceptible to freezing because the water is already cold. It will not take much more cold air to start freezing it. There has been more than one boiler plant shut down in the winter because cold drafts froze their city water line solid. Don't take an indicating light being on as proof that electric tracing is on. Touch the covering. If it is not warm, slip a thermometer under the lagging and, if necessary, push it through the insulation to the pipe (be careful with pointed thermometers and do not penetrate the tracing).

Salt storage tanks are usually idle, but they can overflow at any time. Brine can also freeze. An idle softener can freeze if exposed to a cold draft and can contribute to salt leaking into the effluent (another one of those engineering terms – it is the treated water leaving the softeners) if it is not checked while it is idle. Idle condensate and boiler feed pumps can also freeze up. That is why it is important to rotate them regularly. That means rotate, not bump. When a pump is bumped, simply push the electric motor's start and stop buttons one after the other so that the motor turns over. The problem with bumping any rotating equipment is that it tends to stop turning right where it stopped the last time. Any rotor suspended between bearings will tend to sag over time and, if left in, or returned to, the same position every time, the sagging increases. To rotate a pump, turn it by hand. Sometimes, that means temporarily removing a coupling guard or reaching under it. The final key is to turn it by 1° turns so that it is 90 degrees off its last position. Rotate it once a month and it will only be in the same position one-fourth of the year. All rotating equipment, anything run by an electric motor, gas or diesel engine, steam engine, or turbine, including the drives, should be rotated monthly. By maintaining a schedule of the rotating equipment and rotating once a day or once a week (depending on how many there are), all the equipment in a facility can be rotated on that monthly schedule.

Idle piping systems also deserve some attention. The first lesson of idle liquid piping systems should be to ensure that there is always one way for the liquid to expand out of the piping system. If a piping system is valved off to the extent that the liquid is trapped inside, the piping will be exposed to considerable swings in pressure as the liquid is heated and cooled. The liquids that enter a boiler plant are typically colder than the plant. It is very easy to isolate a cold liquid which will expand when heated. If that liquid is completely trapped, the only way it can expand is to stretch the pipe and it will.

Expanding liquid normally raises the pressure to the point of failure of a gasket or packing at a valve stem and operators will consider it a simple leak. If, however, all the leaks are fixed, the pressure will eventually split the pipe because expanding heated water can produce as much force as freezing ice. The best example for this is a typical run of hydronic piping where the temperature of the water ranges from an installation temperature of 70°F to a maximum operating temperature of 250°F. A run of 100 feet of that piping heated from 70 to 250°F will increase in length by 1.3 inches, but the water in the piping will extend its length by 67.44 inches. Since the water is not very compressible, it will try to stretch the pipe that far, with drastic results if that water does not have anywhere to go.

The simple solution for idle systems is to never isolate them completely. If necessary, then install provisions for expansion or a relief valve on them that discharges the liquid to a safe location. A favorite spot for this problem is the short length of piping between two fuel oil safety shut off valves. The engineering solution is a relief valve connected to that piping and discharging to the oil return line. If that does not exist, have a branch line with a small valve for leak testing closed with a nipple and pipe cap. Remove the cap and open the valve each time the boiler is shut down for an extended period. Then close it back up after a little air has gotten in. That little bit of air should not create a problem at the burner because it should pass through while the ignitor is still operating.

Fuel oil in idle piping exposed to the heat of a boiler room can gradually break down to form heavier hydrocarbons and gases that produce the equivalent of air pockets in piping. That does not necessarily create a problem for the piping, but pumping that fuel with its pocket of gas to a burner can create a flame out (there is not enough energy in the gas to keep the flame going) and subsequent re-ignition of the fuel oil to produce a furnace explosion. Always recirculate oil to eliminate any gases long before starting a burner on fuel oil. Fuel oil piping can also be a hazard if it is fully isolated.

Water piping can be reduced from a 4 inch internal diameter to less than a 3 inch internal diameter in a matter of months because it was sitting idle. Despite chlorination and other forms of water treatment, microbes manage to survive. Given stagnant water and a minimal source of nutrients (food to eat), they can thrive. Not only do those microbes construct rather solid homes on the inside of the pipes, they also generate waste that can be very acidic or caustic to corrode the piping. Microbe induced corrosion (MIC) can, in many cases, be comparable to oxygen pitting because the microbes concentrate under a little growth on the inside of the pipe and emit the acids and alkalis that attack locally.

Normally, the solution for idle cold water piping is simply opening a vent or drain valve to refresh the water in the idle piping once a week. Microbes cannot survive in water above 140°F and do not do well in water much warmer than 120°F. Water lines that are in the upper levels of a boiler room should not have a problem with microbial growth because of the heat, but would suffer from oxygen pitting if oxygen rich water was regularly added to them.

Oxygen is another problem in water piping, although not as persistent as in boilers. The cold city water usually warms up in idle pipes in the boiler plant. Raising the temperature of the water reduces its ability to absorb oxygen. Thus, some of it is released to produce the damage known as oxygen pitting (see Boilers and Dearators, Chapter 8). If the piping is to be idle for long periods of time, it should be drained and kept dry. That way, both microbes and water-borne oxygen cannot do damage to it. A dry line will develop a very thin coat of rust that will protect it. If the pipe cannot be kept dry, then adding chemicals to the water or filling the piping with nitrogen to inert it are options. A nitrogen inerting system, consisting of a regulator and safety valve on a portable cylinder, should maintain the inert status for several months. Only a few inches of water column as pressure are needed in that idle piping. Nitrogen can find some pretty small places to leak through, and maintaining high pressures will result in wasting a lot of nitrogen.

Vent and bleed lines for gas pressure regulators, gas pressure limit switches, and bleed of double block and bleed shut off valve systems are basically idle piping. The vent lines from a regulator or pressure switch are there to provide a direct connection for atmospheric pressure on the diaphragm of the control valve plus convey fuel gas to a safe location in the event that the diaphragm leaks. The bleed line is used intermittently to dump the gas trapped between the two safety shut off valves. Those lines should always be treated as gas lines, even though they may contain air most of the time. The condition of the terminations of gas system vents and bleeds, normally a screened fitting, should also be checked on a regular basis to ensure they are not blocked.

An ear to the line can detect a good sized gas leak. They should also be checked by stretching a rag over their outlet (or a union just inside the building when the outlet is inaccessible) and soaking it with soapy water. Bubbles indicate a leak. They should be checked whenever there is reason to believe they could be leaking or on annual inspection. In one case, the rubber disc of a bleed valve had been cut by the sharp seat of the valve and occasionally buckled to block the valve partially open while the boiler was operating, giving rise to an intermittent gas leak.

Fuel oil tanks that are not in use should be full except for one that may be filling. That will minimize the exposure of the metal in the tanks to air and its corrosive properties. It also limits the contact of air with the oil. Full, of course, does not mean up to the brim. Some freeboard (space between the liquid level and the top) is always needed to allow for expansion.

Propane and fuel oil storage facilities have a bad habit of becoming garbage dumps. In the fall, leaves accumulate in the diked areas around the oil tanks and on the ground around the supports of propane tanks. That is fuel for a fire from an inadvertent spark or cigarette that could produce a disastrous fire and possibly an explosion. Water can accumulate in diked areas or simply form ponds that stand on metal pipes, supports, and tanks to promote their corrosion. Every day shift should visit the fuel storage locations for the express purpose of identifying hazards and eliminating them. Raking leaves and mopping water may not be in the job description, but, typically, the operator has responsibility for those facilities and should take any action necessary to protect them.

A very important piece of equipment that is idle most of the time is an emergency generator. Many plants test them on a regular schedule. However, they deserve attention between tests to detect any problems that might arise. There are probably many items and systems in the plant that were not included in this discussion but deserve attention when they are idle because they are critically necessary when needed. Identify them and make certain that the SOPs include procedures to check on them.

SUPERHEATING

The deregulation of electricity generation has resulted in more superheated steam boilers to permit plants to generate their own electric power. That means knowing the important requirements of superheater operation. The first and foremost rule is the superheater has to have steam flowing through it to absorb the heat getting to the tubes or else the tubes will overheat and fail. Water in superheater tubes does not help. It can block flow in some tubes to permit their overheating or suddenly blow over at high velocities to create water hammer damage. The following guidelines should ensure proper startup of a superheated boiler. Note that some boilers, HRSGs in particular, can have special requirements. Be sure to read that instruction manual. When the boiler is equipped with a reheater, there should be valves to adjust to direct steam from the boiler through the reheater. Also, open the reheater vents and drains. When starting up a boiler with a superheater, make sure all vents and drains on the superheater are open. Similarly, check that all reheater vents and drains are open.

As soon as a reasonable flow of steam is evident at the boiler vent, close it to develop maximum flow through the superheater. When superheater drains appear to be blowing clear with no moisture present (a slight gap between the pipe and the cloud of water droplets), close down on the drain valves to increase flow through the whole superheater. Similarly, choke down on any intermediate vents. Constantly observe the superheater outlet temperature, paying close attention after any change in firing rate, number of burners or ignitors in service, and other activities that can change flue gas flow past the superheater. Close drains and vents except for the final superheater vent valve once the steam turbine has been rolled. Close the superheater vent valves after the steam turbine is carrying a load. Close the bypass valve to the reheater as well, confirming reheater flow from and to the steam turbine before closing the reheater vent valve.

During operation, note the superheater, and reheater if equipped, outlet temperature on a regular basis. There are many things that can go wrong to produce a problem with overheating the superheater or reheater that are not necessarily going to be associated with changes in sound. If the turbine trips, open the superheater vent valve before trying to reset the turbine trip valve. If the boiler has a reheater, establish flow through it as well.

Fooling around with a trip valve without superheater flow is dangerous. With no steam flow, the superheater outlet temperature indication will fall, even though the metal a few feet inside the boiler is overheating. It is very embarrassing and quite scary to see the superheater outlet indication peak well above the design temperature after the trip valve has been opened back up.

If the plant makes it a practice to lift check the safety valves, then do so with caution, waiting until the boiler has settled down after lifting each drum safely. Open the superheater vent first before starting to take a boiler off line. That means first before anything else. If other boilers are serving the load, any reheater will have to be set up to maintain steam flow as well. Whenever possible, keep serving the load after shutting off the fires to keep the flow up. Allow the turbine to drop off with the boiler so as to maintain maximum possible superheater steam flow. Don't open the other vents and drains until the boiler is down to 25 psig when it is time to open the drum vent. There are so many variables in superheater and reheater design today that it is not possible to ensure that these procedures are the best for every plant. Be certain to follow the manufacturer's instructions.

Some superheaters are equipped with gas bypass dampers inside the boiler as a means to control the superheat temperature to a degree. Others will have an intermediate desuperheater that injects feed water into piping, connecting two sections of a superheater to drop the temperature coming out of the first stage. There may also be desuperheaters on reheaters. Some of these devices can produce a false sense of security by producing safe superheat readings at the boiler outlets, but the temperatures upstream of the desuperheaters or in parts of a superheater that are not affected by the dampers go too high. Units that have tangential firing systems normally have a tilting mechanism for the wind box. This mechanism can be used to help adjust superheat and reheat temperatures by moving the entire fire ball up or down in the furnace of the boiler. In any kind of upset operating condition, check as many temperatures as possible and don't bet on the lowest reading being the right one. Always figure the highest reading is the right one.

Desuperheaters are used to increase the supply of desuperheated steam (the added water evaporates and becomes part of the steam). When the steam is used in heat exchangers and similar apparatus, desuperheating reduces the amount of heating surface required in the heat exchanger. They should always leave a little superheat in the steam to assure that there is no water racing down the piping looking for an elbow to run into. When operating a superheated steam plant, it is critical to know what the saturation conditions are for every service and what the maximum temperature ratings are for the equipment and piping.

SWITCHING FUELS

Any boiler plant of a reasonable size should be capable of burning more than one fuel. That capability provides the owner or user with an alternative fuel in the event that the supply of one is interrupted. It also provides a basis for negotiating price with the suppliers. Most boiler operators do not make the fuel supply or price decisions. However, they should be prepared to choose, and choose wisely, which fuel to burn. It is important to be aware of any permit restrictions associated with each fuel selection. For example, most gas fired units are only permitted for firing light oil as a backup fuel. In some cases, the number of hours on a backup fuel is also limited. The wise operator knows what is in the plant's operating permit.

In most northern states, the operator is informed by a phone call when to switch from natural gas to oil firing. Their natural gas is purchased in accordance with a special contract making the supply "interruptible." These contracts benefit the gas supplier and the consumer. The large pipelines that transport gas from the southern states, principally Texas and Louisiana, have a maximum capacity. This is also true for the so-called "fracked gas" from Pennsylvania. The pipeline owners want to optimize the use of those pipelines. They are limited by the pipeline capacity to the customers that are supplied "firm" gas. Those firm gas customers do not use much, if any, during the summer or when outdoor temperatures are mild. As a result, there is typically room in the pipelines for more gas to flow, except on very cold days. By selling interruptible gas, the pipelines make use of that extra room in the pipeline. The purchaser gets a discount, paying less for interruptible gas. The only compromise for the purchaser is a switch to an alternate fuel when notified by the supplier of an interruption. Since natural gas is considered a "clean fuel" and it has become more available due to the fracking of shale gas, state permitting authorities have become more stringent about the practice of switching to backup fuel. Fuel supply has to be truly not available. Price alone is not considered to be a valid reason to switch from a "clean fuel" to a "less clean" (i.e., "dirty") fuel. Again, knowing the plant permit requirements is crucial.

With experience, it is possible to know when to expect an interruption. On rare occasions, the supplier may have to take a pipeline out of service for maintenance or repair and will require an interruption. Most interruptions are due to load (see Know the Load, page 97). Most of the time, a weather forecast will predict the need to stop firing gas and change to an alternate fuel. The same process will indicate when to switch back to gas.

Here is an appropriate word of caution when considering a fuel transfer. There is no such thing as a "flick of the switch" fuel transfer. Most boilers have to shut down and go through a regular boiler startup to change from one fuel to the other. Some small plants have a system that automatically switches from gas to oil and vice versa. The system is called an "automatic interruptible gas service" and controlled by an automatic interruptible system (AIS). The AIS consists of a set of controls in a panel, normally sealed by the gas supplier, that sense outdoor air temperature and control the boilers to automatically switch fuel. These are typically small heating boiler plants, where only one boiler is required to carry the peak load and a short interruption in steam supply or a dip in steam pressure or water temperature is not considered a problem. At a prescribed cold temperature, the controls stop boiler operation and then automatically restart it on the alternate fuel. When the outside temperature rises to a higher value, the boiler is stopped and then restarted on natural gas.

Presuming the operating permit will allow switching fuels, it has become more important for the boiler operator to be involved in the decision because it relates to the fast paced financial situations of today. Some gas contracts do not set a fixed price for gas. The price varies according to any one or more sets of rules or price indices. Or the gas is purchased on the "spot market." A typical index is the "well head price," meaning the price of the gas where it is extracted from the ground (at the well head). Currently, that price is set for each month, but it could easily be set hourly in the future. The boiler operator may have to watch the Internet on a computer in the control room to be prepared to switch fuels when the gas price goes high enough. The winter of 2000–2001 produced some significant swings in natural gas pricing, with prices ranging from $2.97 to $10.81/MMBtu (million Btu's per hour). At the time, fuel oil cost was about $7.12/MMBtu. There were a few plant chiefs called upon to answer why they continued firing natural gas when it was cheaper to fire oil. Pricing is one reason for fuel switching. Loss of service is another. During an earthquake, buried gas piping is typically interrupted. Contractors can dig into the gas mains. Floods can cause gas piping breaks. Having an alternate oil supply is a way of recovering from those situations.

As with AIS, the simple way to switch fuels is to shut the boiler down and then restart it on the alternate fuel. One of the reasons AIS is seldom utilized today is that many people did not manage to get that right. There were several failures in the 1970s associated with systems created that simply switched fuel valves. The installer, or designer, did not understand that simple switch could result in a loss of flame with continued admission of fuel and subsequently in an explosion. There are specifically designed "on-the-fly" switching systems that will be described. Otherwise, shutting down and then starting on the second fuel is the only option.

One favored method of fuel switching is the "low fire changeover" method. The alternate fuel system (for the one presently not firing) is placed in service to bring the fuel supply up to the safety shut off valves. The operator also makes certain that the manual burner shut off valve for the alternate fuel is closed. The controls are switched to manual and the firing rate is reduced to minimum fire. The operator then begins the changeover by turning a selector switch on the control panel to "Dual" or "Changeover." The BMS will energize both sets of fuel safety shut off valves. The operator then throttles the manual burner shut off valve for the fuel being fired and slowly opens the manual burner shut off valve for the alternate fuel. When observation indicates the alternate fuel is firing, the operator spins the alternate fuel's manual burner shut off valve open while simultaneously closing the valve for the fuel that was firing. The selector switch is then turned to the alternate fuel position. The BMS will close the original fuel safety shut off valves. The controls are adjusted to bring firing rate back to slightly above the rate before the changeover until pressure or temperature in the boiler is near normal before switching back to automatic firing rate control.

The designers of BMSs incorporate additional logic in their systems to ensure that a low fire changeover is performed properly. That logic requires the low fire interlock be maintained while the selector switch is in the position to admit both fuels. They frequently add a timing sequence that limits the time when both fuels can be admitted. If the selector switch remains in the two fuel positions for more than a few minutes, the boiler is shut down. This can be a problem because it drives a sense of urgency in the operator that may cause the operator to make a mistake. Another approach is to allow the operator to initiate changeover by turning the selector switch. The control logic then knows that controls have to be in manual and at low fire. The logic switches controls to manual and low fire. The operator does not have to do it. Once the low fire position is established, the control energizes the ignitor and waits for 10 seconds for it to be established.

Gas is normally admitted at the perimeter of the burner, while oil enters at the center. Rather than accepting one will light the other, use the ignitor which is designed to light both. After 10 seconds, the normal PTFI timing, the alternate fuel safety shut off valves are opened. Then, after the normal MFTI timing, the ignitor and original fuel are shut down. Manual control of the fuel flows is not required, but the operator may do it. The controls should be set such that excess air at low fire is at least 150%–200%. During the period when both fuels are firing, the excess air would be 25%–50%. That does not guarantee complete combustion, but it will assure that a stable flame exists. Once the operator observes the stable firing of the alternate fuel and turns the selector switch to the alternate fuel, the controls are released back to automatic. Switching to manual and manual adjustment of the firing rate controls is optional. Ramping controls, mentioned earlier, can also be included. If the selector remains in the two fuel positions for more than a minute after both fuel valves are energized, the system shuts down the alternate fuel and returns to automatic. There is no reason to shut the boiler down.

Low pressure heating systems, and similar applications that do not have a critical steam pressure or water temperature requirement, can accept shutting down and restarting a boiler. The simple stop and restart method is satisfactory for them. The low fire changeover method manages to eliminate the loss of heat input during the purge period to reduce pressure or temperature loss. There will be some drop associated with holding operation at low fire. Generally, any facility that cannot afford a drop in pressure or temperature has two other means of switching fuels that will, unlike the previous methods, ensure a reasonably constant maintenance of pressure and temperature. Smaller plants will have a spare boiler that can be brought up on the alternate fuel and placed on line. Larger facilities normally do not have spare boilers. Therefore, a means of switching fuels on operating units while maintaining pressure or temperature is required.

Larger facilities will have full metering combustion control which allows dual fuel firing to maintain pressure or temperature. Dual fuel firing is simply operating with both fuels at once. When equipped with a full metering system, the two fuel flows are measured, their values are added, and the total fuel flow measurement is used by the controls to maintain a proper air to fuel ratio. The alternate fuel is started at low fire with its control in manual. The ignitor is brought on. Then the alternate fuel is brought on and the boiler simply fires both fuels. Once the operator observes a stable alternate fuel, the controls are adjusted to bring the alternate fuel up manually until the automatic control has reduced the original fuel firing rate to low fire. Once the original fuel is at low fire, the operator switches its control to manual and transfers control of the alternate fuel to automatic. Finally, the original fuel valves are de-energized to complete the transfer. This method has been successfully applied on multiple burner boilers with capacities of 250,000 pph (pounds per hour). When applied to multiple burners, the second fuel started one burner at a time to limit control upsets. An interlock requires all burners be firing on both fuels before the alternate fuel firing rate can be increased above low fire. Safety shut off valves for the original fuel are tripped in unison when at low fire. A sudden increase in excess air will not produce an abnormal furnace environment with a good control system.

There is a reason most boiler explosions occur on light off. An explosive mixture is created, which is then trying to burn instantly. When a boiler is operating, there is a fire. Thus, low fire changeovers or dual fuel firing do not involve that opportunity for an explosion. An inert gas is also being produced while the burners are firing. It means that any injection of fuel that is not burned is surrounded by inert gas instead of air. That mixture will not burn. There are reasons to be cautious about this when there are boilers with a common breeching.

The low fire changeover method requires significant quantities of excess air. There should be enough air for any introduced fuel to burn if it is not ignited immediately by the existing fire. That can be a bit of a problem because the existing fire is not very stable. All that excess air can make it even more unstable. Bringing on a second fuel during dual fuel firing with full metering controls results in the combustion air increasing as the fuel starts flowing to the furnace. The fire of the existing fuel is above minimum to produce more heat and is more stable than it would be at low fire. (Low fire position is normally determined to be when the fire is stable. Anything lower is unstable.)

The method that is available for switching fuels on the boiler should be documented by a detailed SOP for that operation. It is always possible for something to go wrong to produce an explosive condition. Finally, practice it. Before an operator is compelled to switch, that operator should have done it under supervision at least twice each way. It is also advisable to practice it in the early fall, before cold weather sets in, so that everyone has the memory of it refreshed.

STANDBY OPERATION

Firing a boiler to keep it on standby is typically inefficient, may be bad for the boiler, and nothing more than an indicator of an operator's lack of confidence in the equipment and/or the operator's skill. Generally speaking, boiler operation is highly reliable, perhaps more reliable than the electrical service, and should be treated that way. Boilers do shut down unexpectedly and loss of pressure or temperature will happen. Check the logs document to determine causes of the shutdowns. More than likely, especially for smaller gas fired units, the causes were primarily due to loss of electrical service. An unexpected boiler failure for these units is rare to non-existent. Of course, it is always possible to abuse a boiler. Poor water chemistry is probably the most common, followed by poor maintenance, and, then, poor operating procedures. A well-maintained plant where equipment is tested regularly and maintained properly will not have boiler failures and has no need for keeping a boiler on standby. The damage to the boiler and the fuel and electricity costs for keeping it hot normally outweigh any advantage of keeping it hot by regularly warming it up. On the other hand, the maintenance of pressure or temperature may be so critical that loss of a boiler is unacceptable. In one particular case, if the pressure dropped from 240 psig (normal operation) to 230 psig, it cost the plant a quarter of a million dollars. A standby boiler is not necessarily the solution in those cases. Rather, it is having a sufficient number of boilers on line such that the loss of any one will not prevent maintenance of pressure or temperature.

Firing a boiler regularly to keep it on hot standby consumes substantial amounts of fuel and tends to cycle the metal parts of the boiler. There are other means of maintaining a unit on hot standby if really necessary. They are (1) installation of convection heaters and (2) blowdown transfer. By installing a heating coil in the bottom drum of a boiler or installing a heat exchanger, circulator, and piping connecting the blow off and feed water to heat the boiler water using steam from operating units, a boiler can be kept hot enough that it can be brought on line as fast as one that has been fired to keep it warm. Blowdown transfer uses the continuous blowdown from operating boilers to keep an idle boiler hot. Depending on the amount of blowdown, it is possible to keep more than one boiler in hot standby without firing them. Either of these methods does not apply heat to the refractory. Thus, some minor refractory damage may occur if a standby has to be brought on line immediately. The pressure parts will be uniformly heated and the boiler will come on line quickly without the danger of stress cracking.

Some plants that deploy standby boilers also like to rotate them frequently. They are kept on standby to make it easier to rotate them. There is also a bit of confusion regarding the status of a boiler on standby that should be cleared up. It seems to happen frequently in plants with multiple heating boilers. Just because the pressure gauge shows the same pressure as operating boilers, it does not mean that the boiler is hot. Steam from the operating boilers will flow to an idle boiler. A power boiler with a leaking non-return valve can hold a head of steam. The problem is that the pressure and temperature are only above the water line. Everything below can be dead cold and, in one case, was actually freezing. For the same reasons that water circulates in a boiler when it is firing, it will stagnate when it is not. Boilers can show pressure while the bottom drum or a portion of the shell is cold to the touch. A boiler in that situation is not in hot standby. It is mostly thermally distorted steel. Any rapid changes in the water level can result in stress cracking of the drum or shell and tube sheets.

Systems that simply drain the condensate off at the surface of these boilers can maintain an artificial state that is dangerous. Those boilers should either be allowed to flood, so that they are all cold, with the condensate removed in a section of piping above the boiler, or isolated and put in lay up properly. It is not too expensive to replace a piece of piping compared to replacing a boiler.

ROTATING BOILERS

The act of rotating boilers, sometimes called alternating, is the operation of boilers in a manner that assures that all the boilers have the same amount of operating time. It has been common practice. Many facilities have alternating controls that ensure that every boiler takes its turn at operating. Why is it so important to make certain that all the boilers have an equal amount of use to improve the certainty that they all start having break downs at the same time? If three boilers are available, use the old rule of thirds (discussed later). Operate the plant so that one boiler has half the total operating hours and another has one-third of the total operating hours. The boiler with the most operating hours will experience problems, providing a good indication of when to maintain, rebuild, or replace parts. That, in turn, will ensure that the problems are not repeated on the other two. There will be two boilers with less wear than one and one with less wear than the other two. If only two boilers are available, one should have twice as many hours as the other.

Another inefficient practice is alternating systems that are constantly switching boilers, either each time a boiler cycles or every day. Heating up a boiler takes energy and switching to another results in all that energy being lost. If one boiler is too big for the load (it is cycling), why operate two to double radiation losses? Rotate the boilers on at least a quarterly schedule. In that way, they get at least three months of rest before they start up again. Startups always put a strain on a boiler. Don't strain them any more frequently than necessary. Also, lay up the boiler properly if it is not used regularly. Collect some data and do a little math, and it will become clear that it is costing the owner a considerable amount of money to keep two boilers running when one is adequate. Lay one up for a summer, or a year. The little bit of work it takes to do the job right will pay off in lower fuel bills.

BOTTOM BLOW OFF

The only purpose for bottom blow off is to remove sludge, scale, and sediment that collect in the bottom drum of the boiler. There is a prescribed procedure for it with some variations depending on the type of bottom blow off valves that are on the boiler. With excellent water pretreatment and chemistry methods, the accumulation of any sediment in that bottom drum should be minimal. The little bit that does collect is removed with each cleaning for annual inspection. Open the bottom blow off valves to drain the boiler for its annual internal inspection (biannual for some). However, draining the boiler is not the same as performing a bottom blow. Other reasons for opening those valves are simply not acceptable. The bottom blow off valves are not there to regulate the water level. If the water continuously runs high, then get the level controls fixed. The bottom blow off valves are not there to lower the concentration of solids in the boiler water. That is what the continuous blowdown system is for. Continuous blowdown removes water with the highest concentration of solids and, when diverted to a blowdown heat recovery system, wastes very little energy. They are definitely not for maintaining boiler operation. One operator was blowing the boiler down every 15 minutes. Enough cold water was added to prevent the boiler from cycling off. Fuel, water, and operator energy was being wasted to keep the boiler from doing something normal. Operating the bottom blow off valves without concern for operating conditions can interrupt boiler water circulation to result in an eventual failure. Use them only for their intended purpose.

The first and principal consideration for a bottom blow is to make certain to be in control of it. Preferably, it should be done at the change of shift so that two operators are there to do it. One operator can do it if, and only if, the operator can see the gauge glass while operating the valves. There are very few boilers set up that way. It is still a good thing to have another person on hand in case something goes wrong. Don't consider the blow off an option to test the low water cutoff. If there is no one there to keep an eye on the gauge glass, don't blow the boiler down until there is someone. Watch the glass every second until the entire process is complete. A bottom blow removes a considerable amount of water in a very short time and can change the natural circulation in the boiler. Unless the manufacturer's instructions specifically state that a bottom blow off can be performed below a certain load, never perform a bottom blow without shutting down the burners. Never blow a boiler with loads above the limit prescribed by the boiler manufacturer either. A bottom blow can temporarily stall flow in risers, resulting in high concentration of solids and scale formation in those tubes to promote subsequent failure. Water tube boilers are very susceptible to that form of damage. There should be written procedures in any plant for performing a bottom blow and they should be complied with.

Since the purpose of a bottom blow is to remove solids from that mud drum, there needs to be enough water flowing out to flush it well. The first step in preparing for a bottom blow off is to temporarily raise the drum level controller set point, use manual control, or bypass the feed water valve to raise the boiler water level up to within a couple of inches from the top of the glass. That provides the maximum reservoir of water for a good flush of the mud drum. Open the first valve. Then crack (see the "Valve Manipulation" section) the second valve to allow some water to slowly drain out of the boiler and heat up the blow off piping and flash tank or blow off tank. When the level in the gauge glass has dropped an inch, open the valve completely to provide full flow to flush the mud out of the boiler. Then, when the level is about two inches from the bottom of the glass, close the valves. Restore the set point or automatic control to establish normal water level.

Continue to monitor the level until it returns to normal and check it frequently for about an hour afterward. It is a good idea to blow down the water column a few times at 2 minute intervals after the bottom blow. If any mud was left in the boiler, it was loosened and will show as color in the fresh water in the gauge glass. If some color is observed, then the unit needs to blow down more frequently. Usually, when that happens, it is because the water supply or some other factor that would increase solids accumulation in the boiler had changed.

As to which valves to operate first, forget the arguments about the valve closest to the boiler. That is seldom the criteria. It depends on the valves. If the two valves are identical Y pattern globe valves, then the closest is opened first and closed last so that all the erosion is concentrated on the valve furthest from the boiler. However, such arrangements are unusual. There is a common mistake associated with the systems that have one slow opening valve and one quick opening valve. The operators tend to believe the slow opening valve should be opened first and closed last. However, the quick opening valve is there because it can be opened quickly without anything flowing. With the slow opening valve, sudden changes in flow are not produced. The valve can be cracked to warm up the piping slowly. A fast valve opening will immediately fill the cold blow off piping with hot boiler water. Eventually, there will be little puffs of steam where the cracks in the piping had formed from repeated shocks of that nature. It is bad enough to hit cold piping with 212°F water, let alone water well over 350°F.

Seatless blow off valves (Figure 2-6) must be operated in a manner based on their arrangement. The piston assembly in the valves creates a void as they are opened and closes one as they are closed. The valve closest to the boiler in this picture is opened last and closed first. The piston in the second valve in line creates a void in the blow off piping when it is opened, drawing back some air or water, and pushes it out as it is closed. If the second valve were closed first, the piston in the valve closest to the boiler would act to compress the water between the two valves as it is closed. Some operators then try to use a valve wrench to apply enough force to squeeze the water out of the packing or gasketed joint. This act will produce pressures so high that the gasket can blow out of the flange.

Figure 2-6. Seatless blow off valves.

Only use bottom blow off valves to remove sediment or drain the boiler. Operate them properly so as to avoid thermal shock on the valves and the piping. If they are used to drain the boiler, be sure to close them off once it is drained and the boiler is ready to be opened up. It is very embarrassing to blow a lot of dirty water into a boiler that was just drained because the valves were not closed. It is dangerous as well. The valve closest to the boiler in Figure 2-6 should be removed and the remaining valve locked and tagged closed before anyone enters the boiler.

CONTINUOUS (SURFACE) BLOWDOWN

The term surface blowdown is used mostly by old timers. Both continuous blowdown and surface blowdown are basically the same thing. In the early years of steam power, a lot of reciprocating steam engines were utilized. They required lubrication. As a result, lubricant ended up in the condensate and was returned to the boiler. Today, a reciprocating engine would use Teflon seals that can operate without oil lubricants. The surface blowdown was used to remove the oil and similar contaminants from the surface of the boiler water. Today, the proper term is continuous blowdown. The purpose of continuous blowdown is to remove concentrations of minerals dissolved in the boiler water. It should be noted that, with the advent of chelants and polymers, solids and suspended matter as well as dissolved minerals are also removed by continuous blowdown. When phosphates were used to remove dissolved minerals by converting them to a non-adherent sludge, that sludge had to be removed by bottom blow off. With current modern chemical water treatment, very little bottom blow off is required. Why is it connected to the boiler near the surface if it is not surface blowdown? It is because that is where the steam and water rising in the boiler are separated. The water at that point has the highest concentration of solids, more than anywhere else in the boiler.

Unlike bottom blow off, continuous blowdown should be continuous as its name implies. One reason is that there are limits to the concentration of TDS in the boiler water, normally based on operating pressure. If the blowdown is intermittent, then an average concentration of TDS has to be maintained that is lower than the limit. With continuous blowdown, the TDS concentration can be held near the limit. That reduces the amount of blowdown and the amount of energy removed from the boiler in the blowdown water.

Maintaining a continuous blowdown also makes it possible to use the heat in the water removed from the boiler to preheat the makeup water that has to be added to replace water removed by blowdown. That is accomplished with blowdown heat recovery systems. There are two types of blowdown heat recovery systems depending primarily on boiler operating pressure. One simply includes a heat exchanger arranged for counter flow, with the continuous blowdown on one side and makeup water on the other. These are only employed with low pressure steam systems and HTHW. For high pressure steam systems, the blowdown heat recovery equipment consists of a flash tank and a heat exchanger. The high pressure system blowdown heat recovery occurs in two stages. The high temperature boiler water enters the flash tank, which operates at a much lower pressure than the boiler. Some of the blowdown water flashes into steam. That steam is used in the deaerator to help preheat the boiler feed water. The remaining water then passes through a heat exchanger to preheat the makeup water.

Automatic continuous blowdown control systems come in two types. The least expensive and more commonly used has a sensor to detect the TDS in the boiler water and opens the blowdown control valve at set time intervals for a short period of time to allow the water in the boiler to reach the sensor. Then the valve is held open until the TDS in the boiler drops below the controller setting. This is basically an intermittent controller. The other type modulates the blowdown control valve to maintain the TDS at a set point. When the plant is fitted with a continuous blowdown heat recovery system, intermittent operation of the blowdown is undesirable. With the intermittent type control, the solution is to install a small control valve bypassed by a manually set valve. Then the operator can adjust the manual valve to handle the bulk of the blowdown and allow the automatic control to cycle the automatically controlled valve to maintain the TDS level. That produces a constant flow of blowdown with small intermittent increases in the flow to better match a constant flow of makeup. Increase or decrease the constant flow to keep the automatic valve open about 50% of the time.

Normally, the operator does not have to be concerned with the economics of blowdown heat recovery. Some feel that a blowdown heat recovery system cannot be justified economically. That is usually an incomplete evaluation. If there is no blowdown heat recovery system, then valuable resources are being wasted. With the exception of low temperature hot water heating boilers, boiler blowdown is discharged to a flash tank where some of the energy in the high temperature water is used to convert some of the water to steam that is vented to atmosphere. The remaining water, now at 212°F, is conveyed to a drain. The Plumbing Code clearly restricts the temperature entering a drain to 140°F. Therefore, high quality drinking water is used to cool the hot water by dumping more water down the drain. This is referred to as quench water and it is normally high quality drinking water that goes down the drain to sewer along with the blowdown. The typical blowdown heat recovery system cools the blowdown to about 90°F, in which case quenching with fresh water is not required. When the cost of the additional water is considered, the economics of the heat recovery system usually work out. Further, as time goes on, there will be more emphasis on overall efficiency as a means to reduce greenhouse gas emissions. Finally, potable water is becoming a prized resource that is not to be wasted. There is a Sustainability Accounting Standards Board (SASB) that prepares sustainability evaluations for companies. Wasting a prime resource such as potable water is not a good way to improve a sustainability score.

Don't disparage that heat recovery system. Make sure that it is operating properly and, if it is not, point out how important it is. The amount of blowdown is determined by dividing the average TDS of the boiler feed water by the average TDS level maintained in the boiler. The result multiplied by 100 is the percent blowdown. Determine how much is flashed into steam by dividing the enthalpy of the boiler water minus the enthalpy of saturated water at the pressure in the flash tank by the enthalpy of steam at the pressure in the flash tank. Multiply the result of that calculation by 100 to determine percent of flash steam. Then subtract that value from 100 to determine the percent going down the drain or through a heat exchanger. The heat lost to blowdown is determined by dividing the difference between the boiler water enthalpy and the feed water enthalpy by the difference between steam enthalpy and feed water enthalpy. That result times 100 is the percent heat in the blowdown. Finally, to determine the quantity of quench water, divide the difference between the temperature of the water after flashing and 140°F by the difference between 140°F and the average quench water temperature. The result times 100 is the percent of quench water. It is not unusual for it to be more than 100%.

To finish, some values are needed to justify blowdown heat recovery or the cost to repair it. The cost is determined by calculating the heat losses and their value and then adding the cost of the quench water. Since everything was calculated in percent, it is easy to figure out the cost of the heat in the blowdown. Multiply the annual fuel bill by the heat in the blowdown multiplied by the boiler efficiency. Divide by 100 each time a percentage is used. If there is a flash tank that recovers the steam, multiply that result by the percentage of water after flashing divided by 100. To determine the cost of quench water requires knowledge of how much steam was made. Use meter data or one of the estimating methods described in the section on instrumentation in the chapter on controls. Multiply the pounds of steam made per year by the percent blowdown divided by 100. Correct for flash steam used as just described. Then multiply by the percent of quench water to determine the pounds per year of quench water. Multiply that value by the cost per pound of water, the combined water and sewage charge.

ANNUAL INSPECTION

The annual inspection is a standard requirement except for some jurisdictions. Either the State or the insurance company will require that a National Board Commissioned Inspector inspect the boilers. The very limited number of incidents with boilers can be attributed to that one requirement more than any other. Normally, inspection is a maintenance activity. Every year, an inspector should stop by for an operating inspection. The inspector should visit to observe the boiler in operation and require demonstration of the operation of certain safety devices. At one time the inspector wanted to see those safety valves operate. Some still may. To make it possible to test the safety valves, temporarily jumper the high pressure safety switch or adjust it to a value above the safety valve settings. If other boilers are on line to carry the load, it is also possible to close the boiler's isolating valve(s) so that the other boilers and piping systems are not affected. The inspector will also require that his, or her, test gauge be connected to the connection adjacent to the boiler's pressure gauge. The inspector's gauge connection is required by code.

The boiler is then operated in manual control to raise the steam pressure until the safety valve lifts or the inspector refuses to let the pressure go higher. If the boiler is larger than 100 horsepower, it will have two safety valves and the inspector can ask to break the valve seals of the valve with the lower setting and gag it shut so that the higher set valve can be tested. After the higher set valve operates, remove the gag and the inspector replaces the valve seals. The code requires that the valves be open within a certain percentage of the pressure that their nameplate indicates. If one of the valves fails the test, the inspector will require it be sent out for repair or be replaced. To reduce their costs, some insurance companies have changed their requirements to reduce the amount of time an inspector is on site. It takes some time to set up the boiler, raise the pressure, and let it fall. In some cases, they will accept a lift test (see the Safety Testing Section) of the safety valves. Many insurance companies are now simply requiring the valves be sent out to an authorized shop for rebuilding at five-year intervals.

An authorized shop would be one that has received authorization from the National Board to use the "VR" (for valve repair) symbol stamp issued by the National Board. However, manufacturers who hold an ASME Certificate of Authorization "V," "HV," or "UV" (depending on the valve) Code Symbol Stamp can also rebuild safety valves. These changes may represent false economy. Rebuilding safety valves is not an inexpensive proposition and an owner typically ends up buying a spare set to switch out because the rebuild takes several days. Some owners simply buy new valves because they cost less than rebuilding. It is simply false economy again, save some time for an inspector and spend much more than the inspector's time on new safety valves and rebuilding. Pop tests of safety valves should be performed every year. There is no guarantee that they will pop when they should just because they can be lifted. After the safety valves are tested, remove the jumper or reset the high pressure switch. Then demonstrate that it opens at or near its setting and below the set pressure of the safety valves.

The inspector should also expect the demonstration of a functional test of the low water cutoff, either by an evaporation test or a "slow drain" test. The evaporation test consists of boiler operation with the boiler feed pump off or feed water control valve closed, in which case no water is fed to the boiler. As the water evaporates, the level drops until the low water cutoff shuts the burner off. A slow drain test is used when there is little or no steam demand. The blowdown valves are opened to drain the boiler slowly, until the low water cutoff functions.

Remember to keep an eye on the gauge glass when performing these tests (or have someone else watch it). Fully one-third of all boiler failures are due to low water condition according to National Board data. That means those low water cutoffs fail. That is why the inspector wants a functional test of each one. The inspector can also require a demonstration of the function of other safety interlocks. Specific tests are required by code, depending on the size of the boiler. State laws can include other requirements. ASME CSD-1 has a checklist requirement. NFPA-85 contains a list of mandatory tests. The National Board promoted adoption of those Standards in the mid-1990s and most jurisdictions have adopted them. Know what standards apply to the boiler in question. It is always possible that an inspector may not be up to speed on the latest requirements. In many cases, the inspector will simply require documented evidence that tests have been conducted.

Testing of safety valves and inspection of the boilers by inspectors are essential in reducing exposure to a boiler failure. Back before the boiler and pressure vessel codes were instituted, the conditions that existed led to millions of injuries and thousands of deaths from boiler failures. It is the benefit of a third party inspection (with no responsibility to the owner of the boiler) that makes the system as good as it is. Every boiler inspector is well trained and tested before receiving a commission as a National Board Inspector. Take advantage of their training and skills during every inspection, calling their attention to changes or conditions that are questionable. Never try to hide things from them. Remember, the operator is likely to be closest to that boiler if it does explode.

OPERATING DURING MAINTENANCE AND REPAIR

There are some additional duties when a contractor, or other employees, are working in the plant on maintenance or repair activities. Concerns are protecting the health and welfare of those workers, making certain they do not do damage to the plant, and making certain they do not disrupt normal operations inadvertently.

It may be necessary to start and secure boilers to provide access for the workmen to the equipment or parts of the plant. It can be as simple as operating to reduce temperatures where they are working above a boiler. It could also be as complicated as generating steam required for the contractor's operations. It is not uncommon to isolate sections of piping for work. Whatever the activity and regardless of who does the work, the operator should be the final authority for accessing any system. That should be made perfectly clear to anyone who enters the plant.

Frequently, the chief or manager of the boiler plant takes the attitude that an operator should have no authority over contractors working in the plant. If that happens, it is an indication of a lack of trust but can also be an indication that the chief cannot relinquish authority appropriately. In such instances, remind the chief of the safety concerns when others can do things without the operator's knowledge and consent. Make certain the chief understands that the concern is in the safe operation of the plant and that they should make certain that the contractor works with operator approval.

That is not an excuse to be dictatorial and unwavering. If the owner is paying the contractor to work on a time and material basis, the contractor will not complain a bit about waiting for operator approval. Every minute the contractor's employees stand around waiting for approval or to shut down a system, it simply means more time and more profit for the contractor. Treat every one of them as if they were working on a time and material basis. Probably the most difficult thing for the operator to remember during these periods is the requirement that everything done is recorded in the log. In the unlikely, but probable, revelation of problems later (either as a result of the workmen's activities or because they failed to do something), the log provides a documented history of the work for reference. There is frequently an air of distrust between boiler plant operators and contractors working in the plant. A log entry that reveals that distrust through nonspecific statements or general comments will not satisfy the requirements of a court. An owner whose operator made entries like "contractor XYZ is breaking everything" and "the stupid contractor broke it" could not get the jury to accept it. The jury could not get past the implication that the operator logged an opinion rather than fact. All log entries regarding a contractor's activities should be factual and devoid of comment. Log entries should indicate what was done, who did it, and when it was done, and nothing more. It is very important for the operator to do it because there may be nobody else to see it. Without a log entry, whatever happened may be nothing but the operator's imagination. In one case, a simple seven-word entry "Cliff working on Boiler 3 control panel" later proved to recover a rather expensive burner management chassis that Cliff had simply removed and taken with him.

Whenever possible, there should be checklists prepared for any repair or maintenance work in the plant. That is so that it can be consistent with normal operating procedures. Otherwise, what is a normal activity could be made unsafe. Many a contractor has decided a line has no pressure or contents and started working on it without realizing it could suddenly be filled with boiler water (bottom blow off). That also provokes the thought that operating procedures may have to be changed to accommodate work in the plant. Despite the fact that the blow off lines should be locked out and tagged out when working on them, people make mistakes or bad assumptions. A notice for the day, regarding operation of bottom blow off, should also be prepared by the chief or maintenance manager so that operators know the piping will be worked on. When contractors are working in the plant, maintain relatively constant observation of their activities. Be sure to enforce the owner's safety rules and regulations, informing the contractor when the rules are violated and reporting any refusal to comply. If a contractor's employee is injured as the result of a hazard addressed by the safety rules and that employee was not informed of the rules, the owner could be found liable for the person's injuries.

Make sure safety rules are complied with but don't help the contractor comply. The contractor should do confined space testing before the contractor's employees enter a confined space. The contractor should perform the lock out/tag out. The operator should have to add a lock when everything is proven out. The best projects for repair, retrofit, or maintenance in a plant by a contractor exist when the operator and the contractor work together. Preparing a schedule and working to it will help the contractor get done and get out of the plant as soon as possible. When several people are in a plant and their goals differ, that situation produces many opportunities for things to go wrong. If the contractor and the operator share a goal of limiting interference to plant operation and getting the work done readily and quickly, then there is less likelihood of problems cropping up. Keep the proper priorities in mind.

CODE REPAIRS

Even what are called "repairs of a routine nature," which appear to require little skill, are still required by Law (because the State, Commonwealth, or Province has adopted the National Board Inspection Code) that not only addresses the requirements for inspection of boilers but includes requirements for the repair of boilers. Any repair within the environment of a boiler or pressure vessel requires the work be performed by an organization (which can be a one man shop) that has Authorization to use the National Board "R" Symbol Stamp or ASME Authorization to use the applicable Code Symbol Stamp. Without those credentials, a person is not qualified to perform the repair, regardless of their particular skills. Be wary of contractors who may have convinced some purchasing agent that they can do the job cheaper but do not have the proper credentials. It costs a considerable amount of money to satisfy the National Board Inspectors who review firms to determine if they are qualified to do the work and that expense has to be included in the cost of the work. The firms have to repeat qualification every three years. Thus, it does not matter how long they have been in the business. If the contractor cannot produce the Code Symbol Stamp, or a copy of their authorization to use it, they should not be working on the equipment.

The skill of the welders and boilermakers that work on boilers and pressure vessels is impressive. Take a look at the result of a window weld to replace a blistered or bulged boiler tube in a water tube boiler. Proper repair should be convincing that they are the right people for the job. If, however, it looks sloppy, inconsistent with any prior observations, or just plain wrong, ask to see their Authorization or the Code Symbol Stamp and report them to the State, Province, or Commonwealth's Chief Boiler Inspector if they cannot produce it. That should get rid of any unqualified contractor. Don't accept shoddy or inadequate repairs.

PRESSURE TESTING

The most catastrophic incidents within a boiler plant are due to sudden releases of steam and water under pressure. To help ensure that the equipment, piping, etc., is capable of operating without rupture, regular pressure testing is performed. Pressure testing is normally limited to hydrostatic testing. That is not always possible. The procedures to be used should be consistent to ensure that the systems are safe for operation under pressure and not damaged while pressure testing. Hydrostatic testing will be covered first because it is common and preferred.

As with filling a boiler, there should be a person assigned to control the pump or valve that is pressurizing the system. Be as certain as possible that all air has been removed from the system. A system is usually air free if the water pressure increases rapidly once everything is closed. If the pressure does not jump to city or system pump pressure, there may be air in the system. Once the hydro pump has been started, look at the gauge. If the pressure is not jumping up with each cycle of the pump, then there is still air in the system. Get that air out. If the system ruptures with compressed air in it, the air and water will pass out through the point of failure with dramatic force. Hydrostatic tests should be conducted with water between 70°F and 120°F for reasons of safety. That temperature range is also required by code. Normally, hydrostatic test pressure is 150% of the maximum allowable pressure or the setting of the safety valves. Newer equipment is manufactured with steel at higher rated pressures. Because higher stresses are allowed, the requirements for a hydrostatic test pressure may be 125% of the design pressure instead of 150%. There is still a lot of equipment out there that were tested at 150%. It is also a good idea that it be tested at that pressure after any repair or for a preventative maintenance check.

Don't just apply 150% test pressure to a system without concern for what is attached to it. Many pressure switches, transmitters, etc., cannot withstand the hydrostatic test pressure. They will have to be disconnected. That includes some thermal wells and temperature switches and sensors. Be certain that they are okay for the pressure or remove them for the test. It is all that cumbersome removing stuff and putting it back that many contractors may wish to avoid by trying to get away with a lower test pressure.

Many times, even boiler inspectors permit testing at normal operating pressures. This approach could be dangerous because the system can fail and allow the pressure to reach the settings of the safety valves. Also, the valves can stick a little resulting in higher pressures. A large number of compressed air storage tanks were tested for an installation in the 1980s at the request of their inspectors. It was a good thing that it was done hydrostatically. Eleven of them failed: four at pressures below the safety setting and one just slightly above normal operating pressure. A hydrostatic test, done properly, will not result in injury if the containment fails. A little water will run out and the pressure will drop instantly. A boiler in operation does not fail that pleasantly. It is preferable to have a rupture (consisting of a leak of cool water) due to a hydrostatic test rather than or an explosion of steam and boiling hot water (or worse) when it is least expected. Testing at anything less than the standard test pressure is providing false hope that the containment will not fail in operation.

Pressure testing a vessel at 150% of its maximum allowable working pressure still does not mean it cannot fail at lower pressures. During operation, the temperatures experienced by boilers and many pressure vessels are substantially higher than the maximum hydrostatic test temperature. Those higher temperatures introduce additional stress levels into the vessel. These higher stress levels can contribute to the failure of a vessel that just passed a 150% hydro. It is even more likely to fail in service if it passed a hydro at normal operating pressures. One reason for always having somebody at the pump or valve controlling the application of pressure is to release it immediately if a problem is detected. Another is to make certain that the pressure does not exceed the chosen test pressure. If a manufacturer (who has to test at 150%) exceeds the test pressure by more than 6%, the engineering of the vessel must be repeated to ensure that it was not subjected to excessive stress during the hydro. There is no excuse for letting the test pressure run above the 150%. Make sure that it does not happen. Ensure that the pressure in the system never exceeds the specified test pressure by more than 6%. If it does, note it in the log and notify the manufacturer to determine if any damage was done by exceeding the test pressure.

Check electrical circuits that are connected to the systems during hydrostatic tests to ensure that the liquid did not introduce an undesirable ground. Check them again after all test apparatuses are removed and normal connections reinstated.

When testing is performed pneumatically (with air), the test pressure should not exceed 125% of the maximum allowable working pressure. Also, the pressure must be increased in steps with inspections for leaks at each step. The rapid expansion of the air, in the event that the vessel ruptures, could do serious damage. That is why flooding a vessel with water for a hydrostatic test is so important. The water pressure will drop instantly with a rupture. However, any air in the system will expand to push the water out with considerable force.

A sound test requires the source of the pressure be disconnected and the pressure observed for a period of time to ensure there are no leaks. Occasionally, the pressure will increase or decrease as the testing fluid heats or cools. If leaks are found, drain the system for repair. Repeat the test when the repairs are complete. Note that any air test requires precautions and should only be used when there is no other option.

A special test not normally performed is a boiler casing test. It ensures that there are no significant leaks of the products of combustion from the boiler into the boiler room. The test requires blocking the stack, preferably at a point outside the boiler room, and the burner opening(s) into the boiler. The actual test pressure should not exceed the manufacturer's rating for the casing or any ductwork connected to the boiler that is also included in the test. The best way to apply pressure is using the test setup shown in Figure 2-7 which, by its construction, serves as a gauge for the test and a way to prevent exceeding the test pressure. When some bubbles rise through the loop, the test pressure is achieved. Once the pressure is reached, the air supply is disconnected and the level drop observed. It should not drop more than 1 inch per minute after bubbles stop rising through the

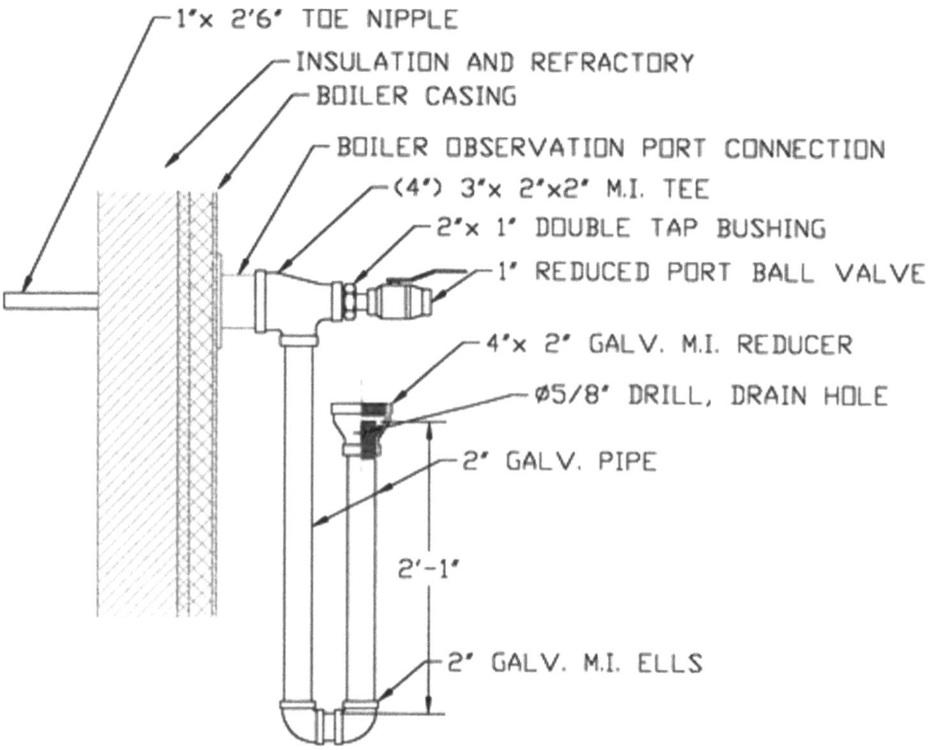

Figure 2-7. Casing test assembly.

column. If leaks are indicated, drop the pressure, insert a lit smoke bomb through the capped connection, and reinstate test pressure to locate the leak. Reduce the air line size to 1/2 inch and other piping to match the size of the observation port if it is less than 2 inches.

Note that the 25 inch water leg is selected for boilers designed for a maximum casing pressure of 25 inches water column. Many are only capable of 10 inches, in which case the leg should be shorter. The application of a smoke bomb is necessary to spot leaks in the casing. It is normally done for a replacement casing job. This test only applies to boilers with casings designed to operate under pressure. A person should remain, with a hand on air valve, at the test apparatus whenever the compressed air connection is open. Be certain to remove blanks and any combustible sealing material (caulking) when the test is completed.

LAY UP

When a boiler will not be used for an extended period of time (more than a week or so), it is important for operators to be certain that the boiler is maintained in such a manner to prevent corrosion or other damage while the boiler is inactive. The operating activities that prepare the boiler for an extended period of inactivity is called laying it up. There are two means of boiler layup: dry and wet. As the names imply, it depends on whether the boiler contains water or is drained.

Wet layup is the common method because it is used for short-term layup and does not require as much preparation to put the boiler into layup and restore it to operating condition. The first step in laying up a boiler is to shut it down and allow it to cool completely. During the cool down period, some circulation of boiler water occurs, and it is the best time to measure boiler chemistry and establish water conditions for layup. The sulfite concentration of the water (for oxygen scavenging) should be doubled compared to normal (60 vs. 30 ppm) and alkalinity raised to a pH of 10–11. In this manner, the boiler internals will be protected from corrosion. During a short-term layup, the only other provision that is made is raising the water level to the top of the drum to minimize the internal surfaces that are exposed to air.

For a longer term wet layup, the entire boiler drum should be protected from contact with air. It should be flooded. The installation of an expansion tank on the boiler to maintain a flooded condition is recommended. The expansion of the water can be determined from values in the steam tables, the difference between dry and flooded weight of the boiler, and the normal range of boiler plant temperatures (40°F to 135°F) in order to size the expansion tank. A tank with a capacity of 3% of the boiler (in gallons) should work in most situations. Best is a bladder tank connected to a branch connection off the boiler vent with another vent valve to bleed water when chemicals have to be added or the pressure adjusted. Starting with a tank drained of air until there is no pressure over the bladder will allow the pressure in the boiler to rise by 15 psig when the tank is half full. An alternative method is to install a bucket on a pipe nipple, set it up on the vent valve, and add or remove water to maintain the level in the bucket. Water should be added by introducing additional sulfite using the chemical pump and maintaining 60–120 ppm in the chemical pump's storage tank. Long-term wet layup requires addressing the condition of the fireside of the boiler. When it will be down for more than a month, it is advisable to seal the stack or block the boiler breeching at a point inside the boiler room. The daily swing in temperature and humidity can produce conditions that promote condensation of water in atmospheric air on the surfaces of the boiler because the water and steel are colder at some times. By restricting air flow, potential for condensation is reduced but not eliminated. Once the air in the boiler is confined, silica gel can be used to absorb moisture. Alternatively, a little heat with lights or a short length of tubing using condensate, blowdown water, or steam to raise the temperature of the air in the boiler to a couple of degrees above the water temperature will minimize condensing on the surfaces.

Dry layup, as the name implies, is achieved by draining the boiler. It is not that simple, however. If left exposed to air, as well as the varying temperature and humidity, around a boiler plant, there will be significant deterioration of the boiler's interior unless it is protected. To prevent corrosion, the boiler should be free of moisture. After the boiler is drained, all drain valves should be closed, the drum covers or inspection openings opened, and dry air blown through the boiler to remove any remaining moisture. Checking the exhaust air with a hygrometer until the humidity in the boiler is less than 10% or 5% above the humidity of the drying air is recommended. Then, insert a package of silica gel with a corrosion proof drain pan under it and close the boiler completely. The air will simply compress and expand in the boiler as it heats and cools. There is no reason to install an expansion tank. The silica gel must be checked twice a year to ensure it is active. Any moisture found in the drain pan should be removed.

Larger and higher temperature and pressure boilers may require nitrogen blanketing of the pressure parts. Nitrogen blanketing is used to provide an inert and corrosion-free environment. Nitrogen gas has an extremely low dew point. It is used routinely to purge oxygen from enclosed vessels. Corrosion cannot occur in an inert nitrogen environment. Under wet layup conditions, nitrogen may be connected to the steam vent to provide a low pressure nitrogen blanket to prevent oxygen ingress. Alternatively, nitrogen may be used during dry layup conditions to provide a positive nitrogen pressure (5 psig) in the closed boiler vessel to prevent oxygen and moisture intrusion. It may also be used to inert superheaters and reheaters and to provide a nitrogen blanket in deaerators and feed water heaters.

The fire sides of the boiler have to be considered for long-term layup. The connection to the stack and the combustion air inlets should be blocked off. The enclosed spaces should be dried and maintained with a silica gel dryer as described above. Normally, boiler control panels, motor starters, etc., can be maintained by simply leaving the power on the panels. The indicating lights in the panels should supply sufficient heat to lower the internal humidity and prevent corrosion from moisture. If the panels are exposed to the weather, the addition of some light bulbs inside the unit to lower the humidity is recommended. Wiring two 100 watt lights in series will produce about 25 watts of heat. The likelihood of one of the bulbs failing is very low. Add lights to panels that do not have any. Motors for combustion air fans, boiler water feed, or circulating pumps can be heated by applying reduced voltage to the windings or using heaters that are supplied for such a purpose.

Always refer to the manufacturer's instruction manuals for suggestions or requirements for layup. Regardless of how the boiler is laid up, its condition must be monitored on a regular basis, preferably weekly, to ensure that it is not deteriorating. Make sure the seals are intact (nobody opened it and left it) and there are no external signs of corrosion or other problems. Initially test the water during wet layup on a weekly basis to ensure that it has sufficient sulfite to remove any oxygen. When testing reveals consistent retention of the sulfite, extend the testing interval until a period with a drop of 10 ppm is experienced. Incorporate that testing interval in the SOPs. Check the silica gel inside the furnace on a monthly basis and inside the boiler on a semi-annual basis. Even if a boiler is no longer needed, don't abandon it to the ravages of weather, etc. At some point in the future, the only option for removing it will be to pay for its removal. Even if the equipment is not needed, preserve it. Someone may eventually need it. If it is in good shape, the owner will get enough for it to pay for its removal.

When the whole plant is put in layup, these guidelines can be extended to other equipment. Special consideration should be given to a long-term layup. Valves and pumps with packing should have the packing removed and replaced with fresh material, heavy in graphite. Packing that was in use and allowed to dry will harden and be almost impossible to remove later. Pumps that have mechanical seals cannot be reliably preserved. The best practice is to disassemble the seals, coating the sealing surfaces with a mineral oil, and reassemble them. Pumps containing oil and such materials that lubricate without freezing can simply be isolated after filling with liquid that is confirmed water free and not prone to form acids while stagnant. Pumps containing water should be drained completely. Close their supply and discharge valves. Then use the vent and drain connections to blow dry air through them and dry them completely before isolating.

TUNE UPS

On November 20, 2015, the US Environmental Protection Agency (EPA) published a final rule containing national standards for Hazardous Air Emissions from Industrial Boilers located at major sources. This rule is referred to as the Industrial Boiler MACT rule. MACT stands for maximum achievable control technology. A major source is one that emits 10 tons/year of any one hazardous air pollutant (HAP) or 25 tons/year of any combination of HAPs. The source is defined as the entire plant, and not just any one single boiler. Sources that emit less than these amounts are defined as area sources. Major sources must demonstrate compliance with either continuous emissions monitoring systems (CEMs) or performance tests. As gas fired units now dominate the industrial boiler market, the main requirements for the Industrial Boiler MACT center around CO emissions. In particular, the EPA uses the term "combustion optimization" for describing a tune up. If a plant does not have a CEM system, an annual tune up is required to show that an effort is being made to reduce CO emissions. Boilers smaller than 10 MMBtu/hr only require a tune up every two years. There is a section for combustion optimization in the chapter on maintenance because it is a normal maintenance function. It is included in the startup of a new boiler. Operation of the plant during a tune up is covered here.

It is rather uncommon for a boiler operator to be expected to perform the tune up of a boiler. A few plants do their own tune ups but use other personnel with labels like Instrument Technician to do the work. Rarely is it done by a licensed boiler operator. Tune ups should be performed on an annual basis and whenever there is reason to believe that the controls are out of tune. It is always the boiler operator's role to identify a problem with the controls that require a tune up.

Another factor in tune ups is the requirements of the local environmental office or whoever is responsible for enforcing the Clean Air Act. Many states now require a tune up be performed each year. However, performance of a tune up should be done as soon as evidence of misoperation exists. The tune up will pay for itself in as little as a couple of weeks. The larger the plant, the more fuel will be burned and the sooner a tune up will pay for itself. The important thing is that the operator monitors operations to determine when it is needed independent of regular intervals.

Sometimes, the evidence is rather apparent, such as smoke pouring out the stack or frequent flame failures. However, those are the extreme cases. The operator should detect problems long before it gets that bad. Generally, a plant should use a contractor for tune ups because the contractor's employees are doing the job at a higher frequency. Consequently, their equipment is maintained in calibration, their skill level is higher, and they are not distracted by other things going on in the boiler plant. A contractor can afford to invest in high tech equipment for tune ups when doing several in a month.

That same equipment is too expensive for a plant that only needs to use it once or twice a year. That does not mean that a contractor is always the best option. In some cases, the contractor considers the tune ups as fill-in jobs and pulls the employee regularly to handle other emergencies. Thus, the tune up loses the continuity that is required to ensure that it is done properly. The single biggest problem with operators doing tune ups is that they get pulled away to handle other situations. If the contractor's operation is the same, that is a disadvantage to using that particular contractor.

A boiler operator has to consider if a tune up is needed whenever plant operating conditions indicate it. Monitoring of evaporation rate or heat rate and other conditions can indicate that a tune up is necessary. Of course, the operator has to be aware of situations that can create a problem that could be wrongly attributed to controls (like blocking of plant air entrances) and correct them first. Something coming loose and shifting position from vibration or for other reasons should also be sought out before committing to a tune up. A number of things must be considered in association with a boiler tune up. Some of them are best accomplished by the operator. In order to tune a boiler, it is necessary to create stable firing conditions for at least a short period of time to enable the technician to collect data that are all relative to that firing rate. This can mean anything from operating the subject boiler in manual, while using another to handle load, to controlling steam dumped to atmosphere to produce a constant load.

An operator can be so involved in simply maintaining the firing condition that there is not enough time to collect the data. That is another reason for using a contractor. In many cases, there are problems creating the load conditions for tune ups because there is not enough load. Wasting steam may seem like a logical solution. However, if the plant normally operates with high condensate returns, wasting steam may be impossible because the water pre-treatment system cannot produce enough water to waste as steam. That is why, in some cases, boiler tune ups are restricted to the winter. Under the Industrial Boiler MACT rules, the testing must be done at the highest expected load on the worst fuel at the worst (from an emissions point of view) ambient conditions. If the boiler is not tested at the highest expected load, it will not be allowed to operate at any load higher than which it was tested.

When a boiler is tuned up in the summer, the data and adjustments at high fire may be made by temporarily running the firing rate up to grab readings, which is not the same as establishing a stable condition. Performance at those rates may be a lot different than the report indicates. A boiler plant log should always include a note to the effect that a tune up was achieved by grabbing readings so that the assumption that it was a normal load tune up is not made.

It is not necessarily important to fire a boiler at or near full load to tune it up with a full metering combustion control system. When properly configured, a full metering system can be set up with a few readings, preferably at loads to at least 50% of maximum firing rate, because the variables associated with load are corrected for by the system, with one single exception, assuring what the maximum firing rate according to air flow is.

A final note on tune ups: They are not a final fix. As the boiler continues to operate, the linkage, fan wheel, and everything else are subjected to friction and wear. With jackshaft type parallel positioning controls, everything in the plant can alter the burner's air to fuel ratio. It is always possible that something can slip, wear, or change in some manner during normal operation. The tuning process will likely be needed to restore efficient and clean firing before the year is up. When that happens, it is best to treat the time between tune ups as the required interval, unless a couple of repeat runs prove that one time was a fluke. Then go back to annual tune ups, or whatever interval the equipment normally needs.

AUXILIARY TURBINE OPERATION

Contrary to popular belief, auxiliary steam turbines are not there just in case of a loss of electric power. While it is true that an auxiliary turbine will operate without electricity, their more important function is to reduce operating cost while contributing to the heat balance of the plant. The auxiliary turbines are an optional source of power. The wise operator will make the best use of them because, if operated properly under the right conditions, they can reduce the cost of powering the auxiliary equipment by about 75%. It should also be noted that, if run under the wrong conditions, it can increase the cost of powering the equipment by 1000%.

It is important to understand exactly what a heat balance is. In its truest sense, a heat balance is the result of calculations that determine exactly where heat goes in a boiler plant. Balance means that the heat out equals the heat in at steady state. Every piece of equipment in the plant can have a heat balance. It is one of the more important tools for trouble shooting. A typical piece of equipment would be the deaerator.

If there is a sparge line in a boiler feed tank and the boiler feed water is heated by injecting steam into that line, that is something similar. Such a system seldom uses enough steam in that feed tank to effectively run a turbine. Maintaining a heat balance is operating a deaerator and auxiliary turbines to get the most efficient use out of the steam going to the deaerator. When steam flows through an auxiliary turbine, some energy is extracted from it to drive the pump, fan, or other auxiliary device. The exhaust steam then flows to the deaerator where it is used to preheat and deaerate the boiler feed water. That steam condenses as it mixes with the feed water, delivering virtually all the heat that is left in it to the feed water, which is then fed to the boiler.

For all practical purposes (by ignoring the little bit of heat lost from the piping and equipment through the insulation), all of the energy in deaerator steam is recovered and returned to the boiler. If it happens to flow through a steam turbine on its way to the deaerator and produce a little power, the cost of generating that power is only the little bit of the heat lost by the steam as it passes through the turbine.

When compared to the typical electrical utility plant, where 60% of the heat from fuel ends up lost, the auxiliary turbines are very efficient. Despite their economies of scale, burning cheap coal, etc., the utility cannot make power as inexpensively as can be done with auxiliary steam turbines. That is why a piece of auxiliary equipment can be powered with an auxiliary steam turbine for nearly one-fourth of the cost of doing it with an electric motor. The electric power plant burns fuel to liberate heat stored in the fuel and captures that heat as steam. That steam turns a steam turbine to convert the captured heat into mechanical motion (i.e., rotation of the turbine). That mechanical motion turns a generator to make the electric current that is sent down the line as electricity. That steam turbine is a heat engine that must reject some heat to a relatively low temperature heat sink. In a power plant, that heat sink is the condenser, which condenses the steam that can no longer do any more work in the turbine back to liquid water that can be pumped back to the boiler. The condensing steam loses that heat of vaporization to the cooling water used in the condenser. That heat is lost to the system. Combined with the heat that is lost with the products of combustion going up the stack, the total heat loss is about 60%. The electricity is then sent down the power lines to the plant, where it can be used to run that electric motor. Line losses account for another 10% of the energy. There is also a small loss in the electric motor itself. The auxiliary steam turbine works the same way. The big difference is that the exhaust steam from the auxiliary steam turbine is being fully used to heat the deaerator which contains the boiler feed water. Thus, the heat energy is being fully recovered instead of being thrown away in a condenser. The electric power plant uses a similar approach through the use of extraction steam. Some steam is taken (extracted) from the steam turbine and used to heat boiler feed water. That slightly reduces the power output of the steam turbine but recovers 100% of the energy in the steam. On an overall basis, that process improves the overall efficiency of the power plant. The use of extraction steam to heat feed water in the feed water train (a series of heat exchangers) is called a regenerative steam cycle. In summary, the auxiliary steam turbine generates a small amount of power at a relatively low cost compared to the purchase of electricity to do the same job because it avoids the heat losses experienced by the power plant, the line losses in the grid system, and the profit margins for the power company.

On the other hand, if too many auxiliary turbines are run so that steam is being dumped out the multiport (relief valve) to the atmosphere, all of the energy that should have gone to the deaerator is wasted. It now costs more than ten times as much as electricity. The trick is to operate the turbines so that as much steam as possible flows through the turbine without pushing any steam out of the multiport.

The best auxiliary turbines to use are boiler feed pump turbines. They require power proportional to the feed water requirements. Deaerator steam is also proportional to the feed water requirements. Forced and induced draft fans are the second best. Regrettably, turbines do not use steam proportional to their power output. They need a certain amount of steam to overcome friction and windage (like fighting the wind, it is the losses associated with the rotor of a turbine whirling in the steam). Thus, the steam consumption of an auxiliary turbine is not perfectly proportional to its power output.

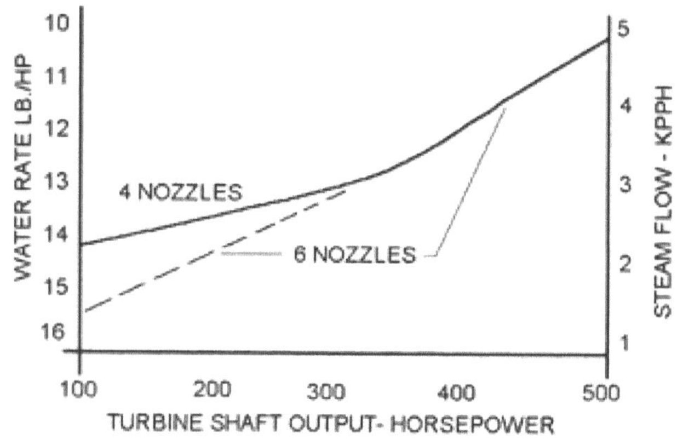

Figure 2-8. Willians line.

There is a reasonable degree of proportionality that is evident when observing the Willians line for a particular turbine. The Willians line is a line on a piece of graph paper that shows the relationship of steam consumption to turbine power output. It looks something like that shown in Figure 2-8. Since there is a fixed amount of energy needed just to keep the turbine spinning, there is some point where the turbine's steam requirement per gallon of boiler feed water pumped exceeds the requirement for heating steam at the deaerator. When operating a feed pump turbine below that point, some of the steam is wasted. When operating above that point, the deaerator needs more steam than the pump does. The basic task is to determine the boiler load closest to that point. Then operate one or more auxiliary turbines accordingly. Run the turbine whenever possible, without wasting steam. If there is more than one turbine driven feed pump, determine the boiler load above which two turbines can be run. If the turbine drives are of different sizes and there are some for other services (like condensate pumping or driving fans), learn how to juggle them for making the most use of the auxiliary steam going to the deaerator.

When there are many auxiliary turbines of different sizes, using the Willians lines in their instruction manuals will help to determine ways to mix them for maximum utilization. When there is an option of changing turbine nozzles (note the two lines in Figure 2-8), determine when the extra nozzles are needed by when the turbine seems to be inadequate to power the pump. Note the feed water flow or steam flow when that occurs to determine when to adjust turbine nozzles. Boiler feed pump turbines can help maintain the heat balance because they are equipped with controls. These vary from constant speed controllers, which will vary steam usage as the water flows change, to special control loops for maintaining a constant feed water pressure or constant differential between feed water and steam headers. As the boiler load increases, the pump horsepower has to increase to pump more water. The increased load will tend to slow down a speed regulated turbine. The controls open the steam valve more to restore the speed. Similarly, the steam supply to the turbine is increased to maintain feed water header pressure or water to steam differential as load increases.

Very large auxiliary turbines may actually have a control linkage that opens and closes the turbine nozzles. Those systems will open one nozzle control valve entirely before starting to open the next one. Thus, only a small quantity of steam is throttled. That increases the efficiency of the turbine and improves the ratio of feed water to turbine steam demand.

The steps in starting up and shutting down auxiliary turbines are all pretty much the same. The first task is deciding which one to start. Then set up its driven equipment the same way as that would be done in preparation for starting one powered by a motor. The turbine casing vents and drains should be open. Check to make sure that they are. Check oil levels in the turbine bearings or sump, any reduction gear, and on the driven equipment. If the turbine is fitted with an electric motor driven lubricating oil pump, start it to start oil circulating through the bearings. If it is possible to get at the shaft, rotate the shaft a quarter turn every 5 minutes while it is warming up to help ensure uniform heating.

Damage to auxiliary turbines is normally due to alignment problems associated with thermal imbalance. Take the time to ensure that the casing and rotor are uniformly heated. Large auxiliary turbines can have some very thick metal parts, especially around the nozzle blocks and shaft seals. Thus, the larger the turbine, the more the time that is needed to warm it up.

When a bypass is provided on the exhaust valve, crack it to start warming up the casing. Admit only enough to get steam at the vent. Then throttle down the vent so that the air is pushed out the drains. If there is no bypass, then crack the exhaust valve. Leave the vent open enough to dispel any air that is heated by the steam. Don't leave it wide open. With a wisp of steam coming out, there should be enough pressure to push air out of the drains. The steam from a typical 100–150 psig supply (or higher) is about half the density of air when dropped to atmospheric pressure. It is so light that some force is needed to push the air out of the turbine casing drains.

Since most auxiliary turbines operate with exhaust pressures of 15 psig or less, the steam will always be less dense than the air. Be certain that the entire casing is flooded with steam so that the rotor and casing are heated uniformly. As the casing warms, less steam will be used to heat it up. The drains will begin blowing more and more steam. Throttle the drain valves to limit steam waste, but be sure to keep them open enough to drain all the condensate.

When there is little to no condensate evident at the drain valves, open the exhaust valve. At this point, the steam has nowhere to go and is not condensing. The casing pressure should be close to exhaust line pressure. Open the drain valve above the steam supply shut off valve to drain any accumulated condensate above the isolating valve. Then throttle it until most of the condensate is draining. If there is a bypass on the steam supply valve, crack it to bring steam up to the trip valve. Once the supply line is dry, open the supply valve. If there is no drain at the trip valve body, don't open the supply until the turbine is ready to roll. While that supply piping is warming up, open outlet and then inlet valves of any turbine bearing coolers, throttling the inlets if the coolers are lacking temperature controls.

There have been concerns that heating up the casing will heat up the oil in the bearings. That is true and it should. If the cooling water to the bearings is opened first, the oil may still be colder than design operating temperature. When the turbine is rolled, there may not be sufficient lubrication because the oil is too cold. By using the casing heating to warm up the oil, it is possible to ensure that it is at the right temperature for operation before rolling the turbine. This kind of consideration should be included in the SOPs.

Once the casing and the steam supply piping is warm and dry and the oil is up to operating temperature, it is time to start rolling the turbine. Sometimes, the trip valve will have to be run down (turn it as if closing it) because someone tripped it earlier and did not reset it. If the valve does not seem to be opening, try that first. There is a spring loaded trip mechanism that shuts the valve by releasing the yoke screw. It is necessary to turn the valve as if to close it until the trip mechanism is reset. Open the supply shut off valve. Crack the trip valve and continue slowly opening it until the turbine starts to turn over. The minute the shaft starts moving, stop opening the valve and close it back down to maintain a slow rotation of the turbine.

If there is a loud screeching noise, shut the trip immediately and back up in the startup process. It is likely that the exhaust valve is closed or there is another valve in the exhaust piping that is closed or throttled. Auxiliary turbines are equipped with what is called a sentinel valve. It is expensive to put a full capacity relief valve on every turbine casing in case someone forgets to open the exhaust valves. That is why sentinel valves are used. They are like a relief valve, but they do not have much capacity. They just let enough steam out to make one loud squeal that is designed to shake up any operator who forgot to open all the exhaust valves.

All of this activity is prior to the most critical stage of auxiliary turbine operation. This is where things can go very wrong. It is important to take enough time and allow the turbine to gently roll over for a while. Steam has just started flowing in the exhaust piping. Any pockets of condensate should be slowly flushed out during this time. If there are known areas where the piping may have pockets of condensate and they are equipped with drain valves, those valves should still be open. If there is a reduction gear between the turbine and driven equipment, it needs time to warm up and get the oil properly distributed over the gears and bearings. Some will have heaters to keep the oil hot enough while the turbine is down. Some will have coolers. Some have nothing but a sump full of oil. Let the turbine roll slowly until all the temperatures are in the normal operating range. Make absolutely certain that there is no screeching, bumping, or grinding in the whole assembly. It does not hurt to use the screw driver at the casing with handle by the ear trick to listen for any unusual sounds while a turbine is slowly rolling over. Open the valves for cooling water to any reduction gear or oil coolers on the driven equipment.

The final step before bringing the turbine up to speed is checking the trip. Normally, there is some linkage between the turbine and the trip valve. All that is needed is to push gently on the lever closest to the turbine shaft to trip the valve. Some turbines will have a means to manually operate the trip. Make sure it works. Then reset and open it again to restore normal rolling. When it has been established that the turbine is rolling over without problems and the over speed trip should work, start bringing it up to speed. First, make certain that no one is in line with the rotor. If it flies apart and pieces penetrate the casing, there should be no one in the way. Open that trip valve very slowly. A fair amount of energy is required to overcome the inertia of the rotor, driven equipment, and any gears to get them moving. However, once they are moving, it does not take much to keep the speed up. If the turbine is brought up to speed too quickly, it will over speed. Sometimes, that trip just does not act fast enough. If the turbine has a tachometer, watch it and slow down the opening of the trip valve as normal speed is approached.

The turbine speed controls should eventually take over control of the steam flow. Once that happens, run the trip valve the rest of the way open. If the turbine is equipped with a process control (like feed water header pressure or feed water to steam pressure differential), that valve or controller should take over. Resist the temptation to bring a turbine up on one of those controllers, especially if they are in automatic. Neither the controller nor the manual signal output can control the steam flow as well as an operator with "hands on" of that trip valve.

If starting a centrifugal pump, it is time to open the pump discharge valve. Open it slowly so that the turbine controller has an opportunity to respond to the increased load. Once the turbine is up to speed and carrying load, close the vent and drain valves, provided no condensate is coming out. If the exhaust line from the turbine is routed up from the casing connection, then the casing should have a steam trap to continuously remove condensate. Make certain that such traps are really working by temporarily opening a manual casing drain about 5–10 minutes after it was closed. Nothing but steam should be vented.

Stop any electric driven oil pump and observe oil pressures to ensure that the turbine's pump is satisfactorily providing proper lubrication pressures. Some electric pumps will automatically stop as the turbine's oil pump generates a higher pressure. The potential for damage to an auxiliary turbine by rapid starting is very high. If the operation absolutely needs a turbine to be brought up quickly, then the turbine needs condensate traps on the casing and steam supply drains, an automatic air vent at the top of the casing, and means to rotate the turbine regularly, either automatically or by prescribed manual means, making it always ready. When starting one of these units, always check by opening a free blow drain to ensure the casing is dry before starting the turbine. They will make a lot of racket. Exhaust steam piping and the deaerator can get pretty rattled if the turbine is started with any accumulation of water in the casing.

On shut down and with a turbine that has an electric oil pump, make certain the pump is running. Begin to shut down the turbine by slowly closing down on the trip valve. The steam supply shut off valve should be open or shut, not throttled, to eliminate erosion or wire drawing to cause it to leak. Make certain that the load served by the driven equipment is handled by another turbine or motor driven device as the turbine being shut down starts slowing noticeably. When the turbine has slowed a little more, close the discharge valve of any pump powered by the turbine to ensure that a hung up check valve does not allow reverse flow to start driving everything backwards. Throttle down on the trip valve until the turbine is gently rolling over and allow it to continue rolling for 20 minutes to 1/2 hour. This slow rolling allows the turbine parts to cool from operating conditions to exhaust temperatures, and slow cooling is desirable for the heavy metal parts.

After that cool down period, close the trip valve and high pressure steam supply valve. Then immediately open all the drains a couple of turns. If the turbine will need to start back up again in a few hours, leave it under exhaust pressure. Otherwise, after the turbine stops rolling, close the exhaust valve, open the vent and drain valves completely, and stop any electric oil pump, once the rotor has stopped turning. It is a little complicated. It takes time. Small hand wheels in tight spaces around the turbine may have to be operated. Nevertheless, proper operation of auxiliary turbines can make a real difference in the overall operating cost of a boiler plant. Wise operators know that and operate them wisely.

POWER TURBINE OPERATION

Power turbines can operate in a manner similar to auxiliary turbines, in that their exhaust consists of steam under pressure that is used for other purposes than simply powering the turbine. These turbines are referred to most commonly as back pressure turbines. The only limitation on their exhaust pressure is the capability of the turbine casing to withstand the pressure. A common exhaust pressure for back pressure turbines is 150 psig. This back pressure is the key difference between these turbines and those used in a utility power plant. In the utility plant, the condenser creates a vacuum at the exhaust of the turbine in order to get the maximum mechanical work from the turbine. The back pressure turbine exhausts steam at a pressure that will be used somewhere else in the plant. These turbines are commonly used to generate electric power from superheated steam produced by the boilers and their exhaust used to service the facility. The amount of steam generated and passing through the turbines is typically controlled to maintain the facility service pressure by throttling the steam to the turbines. When both electricity (power) and thermal energy (steam) are generated together in this manner, the process is called cogeneration.

Unlike auxiliary turbines that may have simple ring lubricated bearings or a startup lubricating oil pump that is supplanted by a shaft driven oil pump, power turbines normally have a complete lubricating oil system that includes pumps to pump the oil to the bearings and gears of the turbine and the equipment it powers. On board a ship, never shut the main lubricating oil system down, other than for maintenance of that system. It has gravity tanks, a large reserve of oil in tanks installed much higher than the turbine, that use the static head to produce the pressure for feeding the oil to the turbine, and oil pumps that lift the oil from the sump of the turbine and gear casing back up to the gravity tanks. The pumps always deliver more oil than the turbine and gear require. There is an overflow line from the tanks that drops oil back to the sump. The piping has an illuminated observation window at the operating level. Use it to see that oil is flowing through that port back to the sump. It means that there is a reserve of oil that would keep the turbine and main gears lubricated for a period of time, even if electric power is lost. Shore side turbines may rely on emergency generators to operate the lubricating oil pumps, but some may have a similar arrangement to provide enough lubrication to allow the main turbine to come to a stop without losing lubrication in the event of loss of power. In some cases, with large power turbines, there is a backup battery followed by an emergency generator to provide power to the lubricating pumps in the event of a complete loss of power to the plant.

On board ships, to prevent distortion of the rotor shafts, always engage a "jacking gear" driven by a small electric motor that powers one of the turbine rotors and the gearing to drive the other rotor and the main shaft. That jacking gear constantly rotates everything taking about 10 minutes for rotating the main shaft and propeller over one revolution. Many power turbines are equipped with the jacking gear. These can be manually engaged or automatically engaged. Even if there is no vacuum or steam on the turbine, the jacking gear should be operated continually when the turbine is out of service. On board ship, always have a sign hanging on the two throttle control valves (ahead and astern) which indicates that jacking gear is engaged. Large steam turbines have a device called the turning gear. This device engages the turbine rotor on shut down when the speed of the rotor is less than 2 rpm (revolutions per minute). For large turbines, the casing takes a long time to cool down. The rotor will deform (sag) as the upper casing cools more slowly than the lower casing. The turning gear continues to rotate the shaft until the temperature in the upper and lower parts of the turbine are the same. Most steam turbine manufacturers will recommend a schedule for operating the turning gear during any length of shut down. Prior to startup, the turning gear will be engaged to make sure that there are no rubs or noises in the turbine for about 4 hrs. If there are any trips or problems during startup, the turbine is put back on the turning gear and the length of time increased. Follow the manufacturer's recommendations for startup and be sure to include them in the SOP.

Power turbines can be used to drive large pieces of equipment, everything from sewage treatment plant pumps to chillers. Every electric utility plant, whether burning hydrocarbon fuels or with nuclear energy, uses power turbines to generate electricity exclusively. In order to maximize power output and efficiency, those plants use condensers. Some plants with back pressure turbines will also use condensers. Condensers can be air cooled or water cooled. Most are water cooled. Water cooling is preferred because the temperature of the water is normally much lower than the air temperature. That, in turn, means that the temperature at which the steam condenses is lower. The lower the temperature, the lower the vapor pressure of water at that temperature. That lowers the exhaust pressure of the steam turbine. The water used for cooling the condensers can be drawn from a well, a city water supply, a river, a lake, or the sea. When using well water or city water, a closed-loop system utilizing cooling towers dramatically reduces that water consumption. For that reason, nearly all electric generating stations in the US now must use cooling towers.

Condensers for power turbines operate at a vacuum. The vacuum in the condenser is related directly to the temperature of the cooling air or cooling water. The lower the temperature, the higher the vacuum, and, therefore, the higher the pressure drop through the turbine, with correspondingly higher power output. Operating a turbine with a condenser introduces a number of pieces of other equipment to maintain the vacuum. Unlike auxiliary turbines, power turbines with condensers have low casing exhaust pressures and, therefore, lower casing temperatures. On ships, the vacuum in the main condenser was maintained at all times, exclusive of requirements for maintenance and long-term shutdowns of the turbine. This had very little effect on steam consumption and stabilized casing temperatures. Temperatures throughout the rotors (there can be a high pressure turbine rotor and another rotor that handled low pressures for propulsion and reversing) cannot be expected to be stable because the pressures along those rotors and the corresponding temperatures vary from the boiler outlet conditions to the saturated conditions in the condenser.

Shaft seals on a power turbine will consist of those seals exposed to pressure. They normally incorporate piping connecting to the shaft seals exposed to lower pressures or vacuum. This permits recovery of the steam that would leak through seals exposed to high pressures, while preventing air entering the seals exposed to a vacuum. Normally, the turbine will have a regulator installed to maintain a constant pressure in the seal piping. The first steps in putting a back pressure power turbine in operation are to ensure that the lube oil pumps and jacking gear are running, admit cooling water to the seal steam condenser, and open the bypass of the turbine exhaust valve to admit steam to the turbine casing to start warming it up. Start the fans on an air cooled condenser or the cooling water pumps for a water cooled condenser and admit makeup steam into the turbine seal system.

Once turbine seal steam pressure is established, the next step to take is to start the vacuum pumps or air ejectors. The typical vacuum pump uses water to seal it. As the pump rotor turns, it forces changes in the water level so that the water acts like a piston to pump the air and other non-condensable gases out of the condenser. Steam from the condenser is condensed into the water in the vacuum pumps, which typically makes up for any water that is lost in them. There is normally some automatic means of providing makeup water to the pumps to accommodate leaks and evaporation. The operator should make sure that the makeup isolating valves are open. The operator should also open isolating valves for any lube oil coolers on the vacuum pumps.

On board ships, steam powered air ejectors were used to initiate the vacuum in the condenser. Shore side plants can also use steam ejectors. These typically consist of two steam jets that pump the air by simply accelerating it with a flow of steam. Two jets are used in series to increase the vacuum in steps. The jet steam and any steam drawn from the condenser are recovered by condensing it. Condensate from the condenser is normally used to condense the steam in the air ejector condenser. The air ejector condenser contains two separate condensing chambers to serve the two steps in vacuum. The recovered condensate from the higher pressure condenser drains to the lower pressure condenser through a loop trap. The condensate from the lower pressure condenser flows to the main condenser through another loop trap. A loop trap is simply piping that loops down and back up so that the difference in pressures between the two chambers that the loop connects is offset by static head in the loop. Before starting a steam jet air ejector, its cooling water has to be established, which requires starting at least one main condensate pump. To ensure continuous flow, a recirculating valve is opened so that some condensate runs back to the main condenser. Then the steam jets can be placed in service by opening their steam valves. It should be noted that the steam jet ejectors do not set the pressure level in the condenser under normal operation. The pressure level is set by the cooling water being used in the condenser. The role of the ejectors, at that point, is to remove any non-condensable gases that might have found their way into the boiler water and then the steam turbine. These gases will increase the pressure in the condenser as well as the exhaust pressure of the turbine. That increased pressure will result in a reduction of power output, as the turbine output is related to the pressure ratio in the turbine. By removing those gases, the pressure will match up with the vapor pressure of water at the temperature of the cooling water. Of course, the turbine and condenser system must be designed to accommodate the cooling water temperature. North Sea water is always colder than Gulf Coast water. The colder cooling water allows the condenser to operate at a lower pressure, which makes those steam systems somewhat more efficient. The steam turbine and the condenser for those systems are designed to handle 40°F cooling water. On the Gulf Coast, the cooling water temperature will be higher. If the steam turbine and condenser on the Gulf Coast were designed for 80°F cooling water, they could not handle the 40°F cooling water. The lower temperature in the condenser would also lower the temperature in the back end of the steam turbine. Condensation would start inside the turbine. The water droplets that would be formed become erosive to the turbine blades spinning at 3600 rpm.

While the casing is pressurizing on a back pressure turbine or vacuum is building on a condensing turbine, steam piping upstream of the throttle valves can be pressurized. Drains in that piping must be opened to remove the condensate. Once the turbine is up and running, the superheated steam will not produce any condensate in the piping. Typically, there are no steam traps attached to that supply piping. The drain valves are always operated manually. Ensuring that there is no water collected over any one of the control valves or in the piping is necessary to prevent a slug of water banging on the turbine blades. A water induction incident is the worst type of accident for a steam turbine. In one incident on a ship, a boiler was being brought up at the same time that the main turbine was being prepared for operation. The isolating valve from the boiler had not been opened. The boiler room suddenly got a bell, indicating a need for full astern operation. A "bell" was the ringing of a bell on a large circular dial marked in different pie-shaped sections with "Finished with Engines, Stop, Slow Ahead, Slow Astern, Full Ahead, and Full Astern," providing instruction from the bridge as to how they wanted the engine (the main turbine) to be operated. The First Assistant Engineer (the one who normally operates the engine during maneuvers) spun open the astern throttle valve and things went awry. Sensing a drop in pressure in the main header, the controls increased the firing rate on the two boilers. However, since the stop valve from one was not open, only one provided steam. The pressure in the boiler with the closed valve climbed, as the one connected to the header dropped. Its water level dropped as the one serving the header went up. That stop valve was opened, but there was some water lying on top of it. The water raced toward the turbine, making considerable hammer noise. Luckily, there was no damage to the turbine. Make sure that all those drains are clear of condensate and all the right valves are open well in advance of turbine operation.

The main exhaust valve of a back pressure turbine can be opened as soon as the casing pressure matches the main exhaust pressure. It must be opened before admitting steam to the nozzles. Then, after the jacking gear is disengaged, steam can be admitted to either type of turbine to start rotation. It is not necessary to establish a complete vacuum before starting operation of the turbine. As long as there is cooling water flow and evidence that the vacuum systems are working, the turbine can be started. Before starting that turbine, always disengage the jacking gear. Admitting steam to the nozzles may automatically force the jacking gear to disengage. However, it is preferable to do it manually. Admitting steam with the gearing engaged will not turn the turbine. Increasing the steam flow in an effort to get the turbine turning can result in damage to the jacking gear.

As with auxiliary turbines, admit enough steam to achieve rotation and allow the turbine to slowly rotate and come up to operating temperature. The higher the temperature of supply steam, the longer it will take to bring the turbine up. As with many other elements of operation, always read the instruction manual for the turbine before trying to operate it. Pay special attention to this part of the operation. On board a ship, the turbine (and the propeller) was actually rotated forward and backward initially to evenly warm up both ahead and astern turbine components.

An over speed trip is provided on power turbines. It should be tested during startup, similar to the auxiliary turbines. These can fail with disastrous results. Be very careful in testing the trip to be certain that it will work when it is needed. Regardless, always bring the turbine up to speed slowly so as not to over speed. There is no load on the turbine to prevent it. At some point, the controls for the turbine should automatically take over to control the speed. Then spin the trip valve the rest of the way open.

Once an electric generator turbine is up to speed, it must be electrically connected to pick up a load. On older ships, direct current power was prevalent. It was rather easy to raise the speed until the voltage on the generator was above the main line and then throw the breaker to engage the generator. It is another story with alternating current generators. The alternating current frequency must be matched with the system frequency. There are panel mounted devices that inform the operator when the generator is synchronous with the line. It may be desirable to have the generator to be running just a little faster in order to pick up some load as soon as it is engaged. Then throw the breaker at the precise instant when the generator and the line are in synch. When connecting to the power grid, this action, called synchronizing the turbine, is very critical. The US power grid is much more massive and powerful than any one turbine. Any difference in frequency will cause a backward force on the turbine shaft. It is very easy to break the shaft by engaging the circuit breaker when the turbine is not in synch with the grid. This is one more place where reading and understanding all the instructions are very important. Modern turbines will likely have sensors and controls to automate this process. In any case, the plant's SOPs should always address a turbine startup.

Chapter 3

What the Wise Operator Knows

To know is to perceive or understand clearly and with certainty. Knowledge is based on training, experience, and the ability to use that training and experience to develop perceptions of outcomes that have not occurred. When in control of a facility that has the potential to level a city block under the worst of circumstances, that certainty becomes very important.

KNOW THE LOAD

The product generated by a boiler plant is steam, hot water, or similar products that deliver the heat to the facility served by the boiler plant. The load is the rate at which heat must be delivered to the facility served by the boiler plant. The normal concern (remember the priorities) is to maintain steam pressure or supply (return) water temperature. The load can vary depending upon the circumstances. Demand will vary with peak load, low load, weekend load, winter load, summer load, etc. The load must be known in order to operate the plant properly. If it is late Friday evening near the middle of October, and the weather forecast calls for a stiff cold front coming through before the end of the shift, it will be important to know whether or not to start another boiler. It is not always wise to rely on the chief leaving instructions either. The load must be known.

The heating load is one of the first things to know because the weather is fickle and changes without notice. Maybe the plant is simply a heating plant. The heating load is the most important load to know about. On the other hand, in a production facility, the weather may have a minimal effect on the total load. Regardless, it is the load to be aware of and be able to quantify.

The amount of heat needed to maintain temperatures in a facility is a function of the difference between the temperature in the facility and the outdoor air temperature. For more than half a century, degree days have been used as a measure of the heating load, normally on a month to month basis. Degree days are, as the units imply, degrees multiplied by days. They are calculated for a particular day by subtracting the average outdoor temperature during the day from 65°F. A typical example would be a day with a high temperature of 50°F and a low temperature of 40°F, where the average is 45°F. The degree days are 20 (65–45). The use of 65°F is based upon the assumption that it is not really necessary to turn the heat on until the temperature drops below 65°F. With that assumption, it is reasonable to say that the heating requirement for a 65°F day is zero. The numbers for each day are combined to provide the number of degree days for a period of time.

The numbers for all the days in a heating season (normally, October 15 to March 15) are added up to provide the number of degree days in a season. A geographical region can be thought of in terms of their seasonal degree days. Degree days can be compared for one heating season to an average that is based on a collection of data over more than a century. Some utilities now list the average temperature for the month, which may also be converted to degree days. The number of degree days is about equal to the number of days in the month multiplied by the difference between the average temperature and 65. Historical data for degree days is available on the internet. Start with the US Energy Information Agency (EIA).

Currently, it is common to preface degree days with the word "heating" in order to distinguish them from a comparable value for cooling degree days. In September and May, care must be taken to ensure that the data is for heating degree days. It could be a hot month that produced more cooling degree days. That may be the figure that has been reported. The problem is that degree days are reported after the fact. They are not available for predicting a boiler load. However, the same logic can be used to predict load. Whether the plant is strictly for heating or provides heat for other purposes as well, a heating load can be estimated based on outdoor air temperature. The 65°F value for zero load is known and there are published extreme temperatures. Data are provided in the appendix for locations throughout the United States and Canada that will allow the determination of what temperature matches full load or 100% heating load.

Local equipment vendors can provide the design low temperature in the local area. If the plant has been in operation for several years, there should be log book data to check back through to find the typical coldest temperature. Don't use one or even four extremes. They are so uncommon that there is little need to satisfy heating requirements for such temperatures. It is also unlikely that those temperatures will produce the predicted load because they are normally of short duration, only that cold for an hour or two, and the mass of the building will limit the effect on the load.

As an example, a home in Joppa, MD, USA reports an extremely low temperature of 5°F, one degree cooler than the Baltimore airport. The range of temperatures for heating that home is between 5 and 65°F, where the load is zero at 65°F and 100% at 5°F. To determine the percentage of load for a given outdoor temperature, divide the difference between 65°F and the current outside air temperature by 60. The heating load is 50% at an outside air temperature of 35°F.

All that is needed for a specific location is to determine the range by subtracting the extreme low for that location from 65. The current degree day value can be obtained by subtracting the outside air temperature from 65. The percent load is the degree day value divided by the range times 100. Remember that to convert a fraction to a percentage, multiply the fraction by 100. For an outdoor temperature of 42°F in Joppa, the load is calculated as 65 minus 42 divided by 60 to get 0.3833, which times 100 gives 38.33%. That is how to determine a common heating load. Simply checking the weather forecast by whatever convenient means will allow an estimate of what the load will be. Of course, the truth is that very few plants have a simple heating load. Boiler plant output is usually used for other purposes, a common one being hot water heating. Hospitals have sterilizers that run year round. Kitchens or cafeterias in the building can introduce substantial loads independent of outdoor temperatures too. However, they also require considerable ventilation. It means that much of that load is outdoor temperature related as well. The heat in the steam or hot water can be used for many things that are not related to outside air temperature.

In most systems used just for heating, the loads are rather consistent in the summer. That value can be considered a base load or summer load, to which the heating load can be added. Formulas can be generated for steam loads that are very consistent for apartment buildings, nursing homes, and similar loads. The formula becomes the base load plus a factor times the number of degree days. Each combination of base load and degree days should be for a specific period of time. Degree days should be for a specific time frame (hour, day, or month).

When generating a formula for heating load, it is important to realize that the actual steam load at any one time will seldom match the formula due to everything from people opening and closing doors to the kitchen starting up in the morning while everyone is getting up and taking a hot shower. It is likely that the actual load will swing by 25% of the maximum heating load in a typical heating plant. In general, the actual load should be equal to the formula value plus or minus 25%. Having a formula helps to explain and justify projections for upcoming load requirements.

To generate the formula, keep track of the plant load using steam flow or Btu (British thermal unit) meter readings, fuel meter readings, or tank soundings, preferably recorded each day. Also record the average temperature or number of degree days each day. Use a properly installed (in the shade and away from sources of heat) high/low thermometer and average those two readings to have an accurate value for the site. If there is a nearby airport, compare the site temperatures with those reported by the airport. If they are consistent, the airport information can be used, with an appropriate correction, instead of continuing to take the temperature readings. Then convert the average temperatures to degree days by subtracting them from 65. Any negative values should be converted to zero. Once there is sufficient data, start determining the value of the formula. Typically, it will take a year's worth of data to produce a reasonably accurate formula. Once enough data has been collected, begin by determining the base load. During the months of July and August, when it is never cold, it is safe to assume that there is no heating requirement. The average steam generation, Btu's, or fuel consumption is representative of the base load. If the plant is far enough north, take the average of those readings on days when the outdoor temperature never got below 65°F. With a spreadsheet program, determining the formula is rather easy. Lacking that, create a table of values using the recorded data. In the first column, enter all the degree day readings. Precede that one with values such as average outdoor temperature or the low and the high temperatures if those are the values that were recorded. Then use them to calculate the degree days. In the second column, record the steam generation, Btu's, or fuel use for that day. For the third column, calculate the heating load by subtracting the base load value from the value in the second column. If any of the results are negative, substitute a zero for that result. For the fourth column, calculate the heating ratio by dividing the heating load value of the third column by the number of degree days in the first column.

The values in that fourth column should all be close to each other. If there is one, or some, that seems to be significantly different and cannot be resolved, cross out that row of data. A common mistake is temperature recordings on weekdays being matched up against fuel use on weekends. Count the number of rows of good data (each daily set of readings) and write the number down at the bottom of the page. Add up all the values in the fourth column and divide by the count of good data rows to get an average of the values in the fourth column. The load formula can now be determined as being equal to the base value plus a factor times degree days. The factor is that average value of the valid fourth column data. To get an idea of how accurate it is, use it to calculate another value (enter it in the fifth column). Then compare that to the steam generation, Btu's, or fuel use in the second column. Monthly data should normally be within 5%. Daily data should fall within 10%. Hourly data are within 25% of the actual values. Continuing to record data and then adjusting the base and factor values will improve the accuracy.

Use the formula to compare the performance of a building at different times. Adjusting for the number of degree days corrects for variations in outdoor air temperature. It helps to detect when something goes wrong in a boiler plant. It is also useful to evaluate the degree of improvement in efficiency that a particular installation provided. Use the formula to predict loads and to detect problems with the plant.

There is another factor that changes the heating load and influences other uses of the heat that is generated. That is the people load. The use of the facility will determine most of the people load. A nursing home or prison will have a relatively constant people load because the people are always there and doing the same thing. Apartment buildings will have a more variable people load, which is one of the more difficult to determine. College dormitories are another story because nearly all of the students are on the same schedule. If the schedule is known, the loads are fairly predictable, despite the fact that they will vary considerably. Simply picture all the students rising at the same time to get ready for class, taking showers, and washing up. Then they vacate the building. They will create a short-term peak load during that time. If the building was equipped with night set back thermostats, the load swing will begin with the warm up and end with the students leaving for class.

When people are present, the loads will be higher. When they are absent, the loads will be lower. In an office building, for example, everyone but the cleaning staff goes home in the evening. The building does not need to be heated to a comfortable 75°F at night. In that case, all the thermostats may be set back to 55°F. Under those circumstances, the peak heating load is not based on 65°F. It will be based on 55°F. The difference between the thermostat set point of 75°F during the day and the 65°F base that is used for calculating degree days is covered by the people themselves (an office worker puts out about 550 Btu/hr of heat). Then there is the equipment that they are using (computers, etc.) and the lights. People have other effects on heat load depending on what they are doing. When everyone is arriving for work in the morning, they manage to pump a lot of the building heat out and the cold in when passing through doors. In one building, they set the lobby thermostat to 85°F about an hour before starting time in order to store some heat in the area to offset all the cold air that comes in with the arriving workers.

Remember that everything can store heat to one degree or another. The thermostat setting will need to be raised to 75°F in that office building well before the workers start arriving, or it will still be 55°F when they arrive. It takes time for the temperature to return to 75°F because the air in the room has to warm up the walls, floors, ceilings, furniture, etc., from 55 to 75°F. How fast it warms up depends on the weight of the materials and their specific heat, i.e., the amount of heat required to raise the temperature of the substance by 1°F. The appendix has a table of specific heats for various materials. When the outdoor temperature is mild, the materials in the building may never get to 55°F before the thermostats are reset in the morning. When it is very cold out, the temperature of walls and other surfaces exposed to the outdoors will drop quickly and may get cooler than 55°F. Because partitions, floors and ceilings, furniture, etc., cooled slower, they might still be warmer and help offset the effect of the colder walls. Warm up loads can be higher than heating loads if ventilation is not controlled. Unless the thermostat settings are timed to compensate for the variation in storage temperatures, there may be complaints in cold weather or heat wasted in milder weather.

Ventilation loads are primarily people loads. For all practical purposes, a facility has to introduce 20 cfm (cubic feet per minute) of fresh outside air for every person in the facility. There are more specific requirements that vary with the jurisdiction, but that is a good rule of thumb. Many older facilities may still be set for ventilation rates as low as 5 cfm per person. It pays to check the actual values before trying to determine the heating load they create. The amount of heat required for ventilation air is easy to determine. It is the total of ventilation air in cfm multiplied by a constant of 1.08 and the difference between the outdoor air temperature and room temperature. As an example, for 100 people, required ventilation air is 2000 cfm. If the outside air is 0°F, this requires 162,000 Btu/hr (2000 × 1.08 × (75 – 0). Using the quick approximation of 1000 Btu/lb of steam, that is equivalent to about 162 pph (pounds per hour) of steam. Note that 75°F was used and not 65°F. That is because, for ventilation, the heat from people, etc., cannot be counted to cover that portion of the load.

In areas containing a high concentration of people (movie theaters, stadiums, or office buildings), the ventilation load can be the largest single load of the facility. In the core of a building, in the middle where there are no outside walls, and floors and ceilings separate them from other occupied spaces, the ventilation air can produce a heating load that would not exist without it. If the facility has large changes in the number of people from day to night or over weekends, there should be swings in the load due to changes in the ventilation air.

Of course, many older buildings do not adjust ventilation air depending on building occupancy. Some just continue the full flow ventilation at night when there are only a few people, if any, in the building. If there is a way of closing that off at night (it will never be able to get zero ventilation), the building will save a lot on heating all that air unnecessarily. Modern facilities are using a combination of security and air conditioning controls to determine how many people are in the building and adjusting ventilation loads accordingly. Another method is measuring the carbon dioxide content of return air, which indicates how many people are in the building or a certain area of the building. The new technical name for that is demand controlled ventilation. Lacking the advantage of one of those specialized controls, there will probably be systems like time clocks that set the ventilation at a minimum when people are not supposed to be in the building and adjust them to a value for full occupancy the rest of the time.

Any of those controls should be set for minimum ventilation air during the period when the building is warming up in preparation for occupancy. That way, the ventilation load is avoided while handling the warm up load, which then limits the load on the boilers. It also makes no sense to heat up cold outside air to warm up walls. The ventilation should increase for a short period before people start entering the building to flush out the stale air.

Except for some process requirements, the hot water heating load is largely a function of people activities. People have a direct relationship with hot water needs for cooking, showers, and washing. Each of those hot water uses is sporadic, occurring at specific (sometimes inconsistent) times so that they are more on and off than a constant load. There are several means of producing hot water and satisfying the irregular loads. There is a section in this book devoted specifically to hot water heating. When the hot water is heated by many heat exchangers throughout the facility, there is little control of those loads. It will be necessary to monitor plant loads to determine their effect.

In an unusual case at a chemical production facility, an operator was suddenly rushing around trying to get the boiler operating. Once it was on line, the operator announced that it was about to rain. That plant experienced a 30,000 pph increase in boiler load every time it rained! Many district heating plants experience a delayed rain load, which is due to rain leaking into the manholes and tunnels containing the steam lines. It is a load that indicates inadequate or ineffective maintenance. It should not be as significant as the one at that plant. If there is one, it should not be difficult to identify it and get it fixed.

Finally, there are production loads. These are requirements for heat to warm raw materials for production, to convert the product to another form (like melting it), or steam actually injected into the product to alter it. They can include tank heating and heat tracing, where heat is used to keep the product in tanks and piping hot enough that it will flow or remain a liquid. Those heating requirements are independent of the actual production. Treat those requirements like heating loads with a higher base temperature.

An asphalt plant, for example, may operate at 500°F to keep the asphalt a liquid. That temperature is so high that swings in outdoor temperature between 0 and 100°F, an extreme winter to extreme summer outside air temperature, would produce a variation between 100% and 80% [(500 – 100) ÷ (500 – 0) = 80%]. If they are significant, they can be treated the same as heating loads by using the product temperature instead of 65°F. That is a way to determine production heating requirements, which will exist as a load independent of the amount of product made. Actual production loads can be related to production output. It is one reason that boiler operators should know how many widgets or pounds of product the plant makes and be informed of how many are planned for production during the next shift.

Some production facilities produce a negative load. These include plants with waste heat boilers that can generate steam or hot water from exothermic reactions (chemical reactions of the product that generate heat). A boiler operator can be called upon to control those boilers. For the most part, they conform to all the rules described for regular boilers. However, each one can have unique characteristics or operating features. The operator should make sure to fully understand all of the manufacturer's and process designer's instructions for their operation. Except for simple heating plants, the operator has to learn the contribution of each type of load and monitor loads to determine how much each one contributes to the total load. The simple mathematical relationships described here should help to explain some of the variations in loads that will be experienced to provide a way to determine what the load will be when plant operations change.

The operator should be able to tell how much change in load will be associated with a change in outdoor air temperature, a change in production rates, a shutdown of any particular part of the plant, and short-term swings associated with personnel activities. At the bare minimum, the operator should know what the maximum, minimum, weekday, weekend, holiday, and total plant shutdown loads are. Once the load and the plant are known, the operator can begin to operate wisely.

KNOW THE PLANT

The operator should know certain things about the plant and be able to respond to questions without hesitation. These are usually not tricky questions. If the pressure gauge reads somewhere between 120 and 125 psig (pounds per square inch gauge), a question about the pressure is reasonable because it could be either one of those values. Here is a quick list of common questions that the operator should be able to answer without looking them up:

1. What is the normal operating pressure/temperature?
2. What pressure/temperature are the safety/relief valves set at?
3. What is the capacity of each boiler?
4. What is the normal feed water/return temperature?
5. What fuels are fired?
6. What is the capacity of the fuel storage?
7. Where does the fuel come from? Are there alternate suppliers?
8. What is the turndown for each boiler?
9. What is the electrical power supply (208/230/460, 3 phase)?
10. How reliable is the electric power?
11. How many power interruptions are there in an average year, and what are their lengths?
12. What is the normal compressed air supply pressure?
13. What is the peak load? Peak day? Peak Hour?
14. What is the normal winter load?
15. What is the normal summer load?
16. What is the minimum load?
17. What is the water supply pressure?
18. What is the normal hardness of the water supply? Of alternate water supplies?
19. Where does the water come from?
20. Is there an alternate supply for water?
21. How many boilers are run in the summer?
22. How many boilers are run in the winter?
23. How frequently are boilers switched?
24. What is the condensate return system leakage/percentage?
25. What is the normal condensate temperature?
26. Is the condensate return pumped?
27. What does the blowdown drain to?

In addition to these questions, the plant piping should readily be identified as to its contents, source, and destination. The operator needs to know this in order to react quickly and responsibly if that piping fails. On board a ship, the engine room is always at the bottom and there are no windows. In the event that the electric generator tripped, the operator has to know how to get around in pitch dark. A working flashlight should be at hand at all times. Night shift at the plant will have the same requirements.

There are a lot more questions about the plant that do not necessarily have immediate answers. Typically, they are not asked frequently and are not immediately needed to operate. Still, there is a lot to know. It probably cannot all be committed to memory. The key is to know where to find the information. Know the location of historical documents, logs, and maintenance records and, basically, where all the paper and spare parts are stored. Know how to find something in that maze of paper or shelves of boxes. The next best thing to knowing an answer is knowing where to find it. If it is stored on a computer, know how to access it reasonably quickly. This knowledge will be especially helpful when an unannounced inspector visits the plant.

MATCHING EQUIPMENT TO THE LOAD

In the first chapter of this book, the last priority that was listed as the one that the operator would spend the most time on was operating the plant economically. Without a doubt, matching the equipment to the load is the easiest way to do that. It is really not very economical to operate with two boilers on line and with not enough load to keep one running constantly. As sustainability goals become more of a priority, the need to conserve energy will become paramount. Perhaps, 20% of the energy used in institutional heating plants in this country can be saved by simply matching loads.

Take the example of two low pressure heating boilers operating when one could carry the load easily. The load may be less than half the capacity of one boiler. Radiation losses, normally 2% of input (or less) at full load, account for 11.5% of the input at the lower load. Off cycle losses of the boiler that is not firing account for another 1.5%–2%, depending on the effective stack height. Purging losses are doubled. Demand charges for electricity increase when the two boilers just happen to be running at the same time. Then there is the additional time an operator spends attending to an operating boiler. All of these items add up to a considerable additional cost for operating two boilers where one could do the job. Further, that is ignoring the fact that cycling losses are doubled when the load is less than low fire capacity of one boiler.

Demand charges are calculated by the electric power company for some small and medium plants and all large installations. Maximum demand is determined by a separate meter that constantly measures the electrical load and keeps track of the maximum average electric load during a 15 minute period in each month. The utility bill includes a charge for that demand. At $12 per kW, that is equal to about $9 per horsepower. Any activity that produces a higher demand simply boosts that charge. Any temporary operating condition that produces that demand creates the charge for the entire month. In some areas, the utilities charge for the highest demand in the prior 6 months or even 12 months. Running two feed pumps when one will do is not only boosting the demand charge but also using electricity (the energy charge or the number of kilowatt-hours (kWhrs) consumed). Although it is not advisable to stop one feed pump before starting another, to avoid a bump in the demand charge, the air compressor can be stopped for a few minutes to switch over pumps. A drop in demand of 10 horsepower while the air compressor is down will reduce the demand while the 30 horsepower feed pump is switched over. That little bit of attention to the electrical demand could save the owner as much as $90 on the monthly electric bill.

When coal was the primary industrial fuel, conventional wisdom called for boilers to be of three sizes: one that could handle full load, one two-thirds of that size, and one a one-third of full load size. The two smaller units served as backup for the larger one. The variation in size ensured a closer match to steam load. Coal fired units did not provide the turndown that is possible with modern gas fired boilers. Cycling a coal fired boiler on and off left an operator very tired at the end of a day.

There are many plants with only one boiler, one feed pump, etc. Often, choices are limited or non-existent in operation. With today's interest in efficiency, many of these plants no longer have the steam load that the boiler was originally designed for. In that case, observe the operating conditions and estimate what could be saved by having another smaller boiler to carry the normal loads. The installation of a smaller boiler can usually be justified in any plant where the boiler cycles at the average winter outdoor air temperature. Cycling boilers are very inefficient, and, many times, a much smaller replacement produces fuel savings that pay for it in one heating season.

For multiple boilers, it should be possible to make a significant difference in the fuel and electricity consumption of the plant by matching the equipment to the load. It should be simply a matter of realizing that there is a difference and acting to reduce the costs. There may be a need to change old habits and rationale. The graph in Figure 3-1 is provided to show how to make wiser decisions about boiler operation, especially when the equipment capability is known. This plant started out with only one boiler. It then grew until two were required. After a few years, boiler three was added and, shortly thereafter, boiler four. The graphs show the evaporation rate (pounds of steam generated per therm of natural gas) for the boilers and a combination firing of boilers one and two. For all practical purposes, the evaporation rate is representative of boiler efficiency.

The curve for boiler one is typical for boilers manufactured in the middle of the 20th century, in that the operating point of highest efficiency is not at the maximum firing rate. A number of elements in the construction of those boilers combined to produce a curve similar to that shown in the figure. Boilers two, three, and four all exhibit properties consistent with later designs of package boilers. The curve for the combination of firing boilers one and two together is generated by adding the individual values for the two boilers with a specific trimming of the firing rate for boiler one so that the operation is at maximum efficiency. Using the bias adjustment of the boiler master for boiler one to limit its steam generation to 38,000 pph, a higher evaporation rate for loads over 76,000 pph is achieved. It is obvious that boiler one is the most efficient at loads less than 30,000 pph. Boiler two is the most efficient at loads between 30,000 and

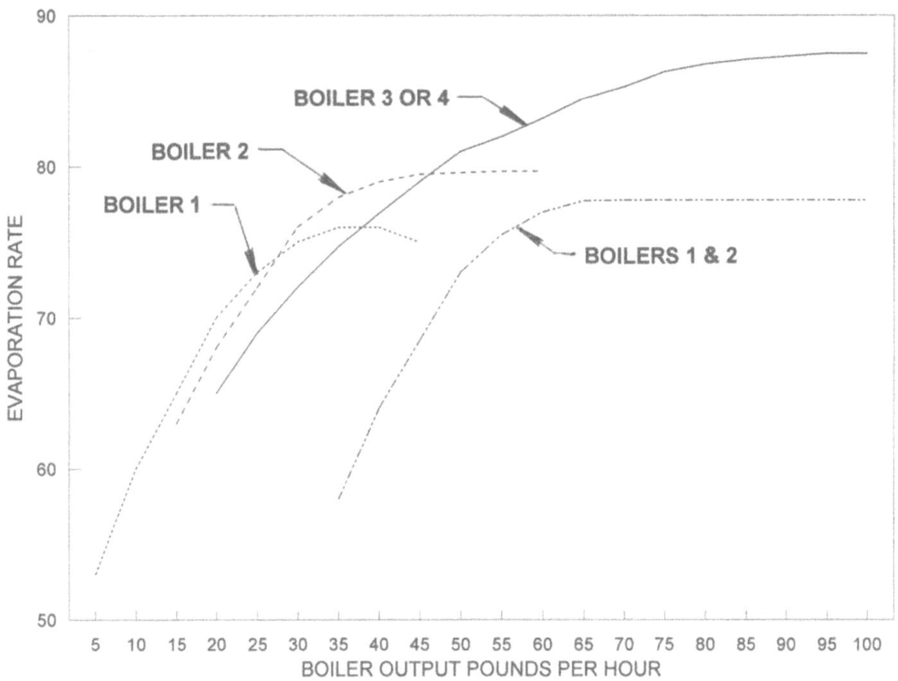

Figure 3-1. Evaporation rate curves.

48,000 pph. Boiler three or four is most efficient at higher loads. However, the plant was operated without this knowledge for several years. When the plant only had two boilers and the total load was less than the capacity of boiler one, the operators chose to fire them alternating every two months. Therefore, there were periods of operation when considerably more fuel was burned than was necessary. If it were not for the fact that the plant load was almost always over 15,000 pph, the operators should have elected to run boiler one at loads where it was the most efficient.

Once boiler three was installed, a necessity to match increased plant load, the operators chose to fire it alternating with a combination of boilers one and two. The significant discrepancy between those two operating setups was obvious to the plant engineers. An investigation was conducted to determine what to do about it. It was determined that the principal reason for the difference was that boiler three was equipped with an economizer. Due to the age of the first two boilers, a decision was made to purchase another new boiler with an economizer, rather than installing economizers on the older boilers. Part of the investigation included the generation of the graphs. Now the operators know that for certain periods of operation, including the Christmas shutdown, it is advisable to operate the oldest boiler because it is the most efficient for low steam loads. Evaluation of the plant's steam loads later led to the decision to abandon, dismantle, and scrap boiler two because plant operations almost never produced steam loads that were best served by that boiler and its maintenance was not economical.

This example should demonstrate that knowing the load and the equipment can make a significant difference in the cost of operation of the boiler plant. Graphing the data helps provide a picture to make it evident. Decisions to simply alternate firing boilers, without regard to load performance, cost that manufacturer hundreds of thousands of dollars. Knowing the plant and the load will help make decisions that reduce the impact on operating costs. Frequently, operators will decide to put another boiler on line whenever the load on one approaches 70%. That immediately converts operating conditions from one boiler operating at its maximum efficiency to two boilers operating near minimum efficiency at 35% load. Radiation losses are doubled with no change in the load. All losses associated with lower firing rates are encountered. Knowing the load, being able to forecast its changes, and knowing what the boiler can do will frequently prevent putting that other boiler on the line until the load exceeds 100% of what is already on line. Establish values based on experience and don't hesitate to experiment to see what the best matches are.

Matching equipment to the load is not restricted to the boilers. Many plants operate one boiler feed pump for each boiler on the line. Since feed pumps have to be capable of delivering water at the boiler safety valve pressure, it is not uncommon for them to have significant capacity relative to normal operating pressures. As a result, never associate the number of pumps in operation with the number of boilers. They deserve their own set of rules, established by experience and observation. Many operators do not realize that there is a lower limit to efficient operation of water softeners. Once the flow in a softener, or any ion exchange bed for that matter, drops below a set value (usually 3 gpm (gallons per minute) per square foot of flow area), they begin channeling. The water tends to bypass much of the resin and its capacity is not used. Operators can allow a lot of scale forming hardness to sneak through their softeners if they run too many of them in parallel.

If everyone in the plant is doing their best to conserve that valuable condensate, the demand for makeup water will be reduced. It may be reduced to the point that the softeners start channeling. Check the softeners more closely if it is the only one because it might start regenerating automatically. That will shut off the supply of makeup water. Some plants are constantly having trouble with condensate loss. It is either due to contamination indications or leaks. In those cases, it is better to have the technician who services those softeners modify the programming to limit the softeners on line, unless the pressure drop through them gets too high. It is a matter of adding a differential pressure switch so that another softener will come on line when needed. Also add a bypass switch that permits manual intervention to put a softener in service.

At one plant, the gas booster was running constantly when the gas supply pressure was more than adequate to serve the boiler load during the summer. The owner had his operators shut the booster down and bypass it. Of course, they had to check it when the temperatures got colder to determine when they might need it. A standard operating procedure (SOP) was developed to check it out by running it temporarily every fall so that they would be capable of putting it in service should they need it in the winter. It turned out that they did not really need it. That was not just a case of matching load. It was a case of recognizing that there was no load.

Don't confuse matching loads and reacting to changing loads. At one plant, they started up a boiler every morning to handle the warm up as the night set back thermostats switched back up. An hour or two later, the boiler was shut down until the next morning. First of all, that is rough on the boiler. It is literally shortening the boiler life. It is also wasting a certain amount of energy because the energy it took to heat the boiler up is lost up the stack before the next morning. If an operator is doing the job properly, of checking all the operating limits when a boiler is started, then that daily start up would be rough on the operator. Short-term operation for an intermittent peak load should not be considered, unless there are problems with the steam pressure or supply water temperature drop associated with that load. In other words, it is okay for the steam pressure or water temperature to drop a little when everything starts heating up in the morning. The drop will limit the heat flow to the load because there is a smaller temperature difference and everything will eventually recover. Don't hesitate to try it. Let the pressure or temperature drop. A slight dip in conditions on an operating boiler is much less damaging than running a boiler up from cold. If the pressure or temperature dips cannot be tolerated, learn what average night time temperature signals that limit so that more boiler capacity can be in operation when it is necessary.

It is not at all uncommon to find a two-boiler hot water plant where both boilers are always operating. In most of those plants, the boilers were each sized to carry the full load. The operators discovered that they could shut one down and never worry about having enough boiler capacity. The cost of fuel to simply keep a boiler hot can be considerable. By shutting one down, they also found that they saved the owner a lot of money. Of course, at least one valve must be shut when that hot water boiler is shut down. Otherwise, the hot water flowing through it will heat up a lot of air that is lost up the stack. It is not necessary to run a proportion of boilers to match the load. In some plants with four boilers, any one of which could carry the full load of the facility served, they will run one or two boilers in the summer and three in the winter, whether they need them or not. They are also usually the plants where the boilers are regularly switched so that they will all wear out at the same time.

EFFICIENCY

There are so many definitions of efficiency. Many operators (and most engineers) are confused as to which is which or simply assume they are all the same. For the most part, the ASME definitions used in their power test codes will be used here. The first point of confusion involves the definition of boiler efficiency. It can be officially defined as one hundred times the heat absorbed in the steam and water divided by the heat energy added by the fuel and other sources of energy. That is the definition established by the American Boiler Manufacturer's Association (ABMA) and the one that is most accepted as the true definition. Those other sources of energy include electric power supplied to the fans and pumps that are integral to the boiler. If all of those values considered to be inputs are accounted for, then a correct value of efficiency can be calculated. It is also the one that is used for the performance test to determine if the boiler actually achieved its contract performance. The energy added to the water and steam is the "output" of the boiler. There can be multiple outputs that have to be considered. If the boiler has a reheater, the energy added to the steam that flows through the reheater is an output in addition to the water that is evaporated to produce steam and the energy added in the superheater.

Soot blower operation to maintain boiler conditions is one of the reasons that a boiler efficiency test in accordance with ASME PTC 4 (Fired Steam Generators) is supposed to be run for a minimum of 8–12 hrs. The Test Code does account for the soot blower steam because it is required to keep the heat transfer surfaces clean. For boilers firing fuel with little or no ash content (i.e., natural gas), a shorter test period is allowed.

Several years ago, the ABMA agreed to guarantee boiler efficiency at only one firing rate and, unless otherwise specified by the customer, set it at full load. If there are some efficiencies listed at other firing rates in the boiler documentation, they are typically labeled "predicted performance" and only the full load is guaranteed. The problem with that situation is industrial boilers seldom, if ever, operate at full load. That is different from a very large utility boiler that was designed for base load operation. It might be a good idea to suggest that any new boiler that is purchased be guaranteed for performance at a load that is anticipated, say 50% or 75%. That does not violate the ABMA's rule. Today, some chiller manufacturers and some boiler manufacturers may guarantee the part load operating efficiency of their equipment. Occasionally, there will be a boiler efficiency guaranteed at somewhere around 50%–75% load. That is typically the load range where the small boiler operates at best efficiency. As the load and firing rate decreases, the volume of flue gas decreases. The heating surface, on the other hand, stays the same. Therefore, the flue gas spends more time in contact with a proportionally larger heating surface so that more heat is transferred. For the large utility boilers with both superheat and reheat surface in the boiler, control load is down to 60% load. Below that load, it is difficult to maintain both the superheat and reheat temperatures of the steam.

Take note that the stack temperature will drop as the firing rate is reduced from full load. At some point, the efficiency will start to drop off because the flue gas is channeling. Only a portion of the gas is contacting the heating surface. As the firing rate decreases, it becomes more difficult for the fuel and air to mix completely. Excess air must be increased to prevent CO formation and efficiency suffers further. The radiation losses also become more significant as the load decreases. All these factors influence the operating efficiency of the boiler to different extents at different loads. Make note of these figures in the log book and eventually create the curves shown in Figure 3-1.

Heat loss efficiency is determined by backing into the value. An efficiency is considered to be the output (basically, the steam and hot water) divided by the input (what was put into it) with the result of the division multiplied by 100.

$$\text{Efficiency} = \text{Output} \div \text{Input} \times 100$$

The loss (and in the case of a boiler, it is a loss of heat) is the difference between the input and output. Therefore, the output is equal to the input minus the heat losses. By substituting input minus losses for the output in the formula, a revised formula is derived that does not include the output at all.

$$\text{Efficiency} = (\text{Input} - \text{Losses}) \div \text{Input} \times 100$$

If the losses are measured, or calculated, as a percent of the input, then all that is needed is to subtract the losses from 100 to get the percent efficiency. Surprisingly, it is easier and far more accurate to determine some of the heat losses as a percent of the input. Determining efficiency using the heat loss method is the most widely accepted method. The increased accuracy of the heat loss method is derived from the fact that only the losses are measured. Since the boiler efficiency is typically in the range of 82%–90%, the losses represent only 10%–18%. Thus, a 1% error in the measurement of a loss will result in an error of around 0.1%–0.18% or less in the efficiency. In trying to measure the input and the output, a 1% error will translate directly to a 1% error in efficiency. PTC 4 lists the following 16 losses:

1. Dry gas
2. Water from fuel
3. Moisture in air
4. Unburned combustible
5. Sensible heat in residue
6. Hot Air Quality Control System (AQCS)
7. Air infiltration
8. NOx
9. Surface radiation
10. Additional moisture
11. Calcination
12. Water in sorbent
13. Wet ash pit loss
14. Recycled material
15. Cooling water
16. Heat internally supplied to air heater coils

Note that boiler blowdown and blow off are not listed. That is because these two losses are a function of how well the owner (and operator) takes care of the water chemistry for the boiler. Since this factor is beyond the control of the boiler manufacturer, it is not guaranteed by the manufacturer. Since the test code is used to verify the performance of the boiler prior to the owner acceptance, the boiler blowdown and blow off are not used in the boiler efficiency calculation. For those units firing natural gas, there are no residues, ash pits, sorbents, calcination losses, or AQCS. Normally, unburned combustible (mainly CO) and NOx formation are in the modest ppm range. Air infiltration in a well-maintained unit should be minimal. That leaves items 1–3 and 9 to be the main considerations.

The combustion calculations (see Chapter 1) allow the determination of the pounds of theoretical air required to burn a fuel of a given composition, either in lb/MMBtu (million Btu's) fired or in lb/lb of fuel fired. This information provides the ability to compute the majority of the heat losses for boiler efficiency on a percentage basis based on the fuel fired. For these calculations, a fuel analysis is required. In its broadest sense, a fuel analysis refers to the determination of all of the physical and chemical properties of a fuel. That includes the heating value of the fuel. A strip of narrow paper with a list of analysis values, temperatures, and a calculated boiler efficiency is not representative of a boiler test. Unless there is a fuel analysis, the test is simply flawed because the hydrogen to carbon ratio of fuels varies considerably. The modern flue gas analyzer contains programmed calculations based on an assumed fuel analysis. The odds that the fuel being fired and the values used by that program are identical are slim to none. The results are only representative and based on an assumed fuel. They are sufficiently accurate to determine relative efficiency over the load range and to compare the boiler performance to another boiler burning the same fuel. However, if those results are used to challenge the boiler manufacturer's higher prediction, that argument will be lost. Calculations in Appendix L permit the determination of boiler efficiency using the heat loss method and a fuel analysis for those purposes.

A common term in use today is referred to as the "combustion efficiency." This terminology, as it is applied to boiler efficiency, is incorrect. Strictly speaking, the combustion efficiency refers to the fraction of the fuel that is burned (i.e., the efficiency of combustion). This is particularly relevant for solid fuel fired boilers. Typically, there is always some unburned carbon. A coal fired unit with 99.8% combustion efficiency means that it has burned 99.8% of the fuel, while 0.2% of the fuel remains unburned and will likely show up as carbon in the ash (the residual solid left over from combustion of the fuel). When a technician visits the plant to perform the annual combustion optimization (typically required by the Environmental Protection Agency (EPA) or its equivalent in the State) or stack samples are drawn that allow a calculation of boiler efficiency, it is often called combustion efficiency. It is basically a heat loss efficiency that assumes a fuel analysis and determines the energy lost up the boiler stack. It is the one that is printed on that little strip of paper by the analyzer. Assuming the analyzer was properly calibrated, the value is a reasonable indication of the boiler efficiency when it is adjusted for radiation loss. That is because the stack loss is typically the largest single loss associated with boiler efficiency and the analyzer does a pretty good job in determining it. Nevertheless, it should be recognized that the term "combustion efficiency" is being used incorrectly in this case. What is really being calculated is a rough approximation of the boiler efficiency.

It is not much, but radiation loss has to be considered in addition to that combustion efficiency (i.e., boiler efficiency). The manufacturer will provide a value of radiation loss, equal to a percent of input at a prescribed boiler load. To determine its impact at the actual load, divide the manufacturer's predicted loss by the percent of boiler load. If the predicted loss is at a load other than 100%, multiply the result by the percent load for the prediction. In most cases, the manufacturer's prediction is at 100% load. Thus, only divide the predicted loss by the percent load. A few examples should suffice:

- A boiler with a predicted radiation loss of 3% at full load is tested and found to have a combustion efficiency of 79% at a 50% load. The radiation loss at that load is 6% (0.03 ÷ 0.5). Thus, the operating boiler efficiency is 73% (0.79 – 0.06). Note that this analysis is still incomplete as the other 14 losses were not considered. Note also that the term combustion efficiency, as used here, is an estimate of the boiler efficiency, and not the fraction of the fuel burned.

- A boiler with a predicted radiation loss of 2% at 80% firing rate is tested and found to have a combustion efficiency of 80% at full load. In this case, the operating boiler efficiency is 81.6% (0.8 + 0.02 ÷ 1 × 0.8).

- A boiler with a predicted radiation loss of 1.5% at 75% firing rate is tested and found to have a combustion efficiency of 78% at a 40% load. In this case, the operating boiler efficiency is 73% (0.82 + 0.015 ÷ 0.4 × 0.75).

Why bother with the radiation loss? To ignore it is to invite some crucial errors in operating decisions. Radiation losses are, for all practical purposes, constant regardless of firing rate. Thus, their proportional effect varies with load. An example is a plant with an old horizontal return tubular (HRT) boiler and a newer cast iron boiler. Since the HRT furnace was substantially hotter, it was easier to get low excess air with a newly installed burner than was possible with the cast iron boiler at the same loads. The predicted full load radiation loss for the HRT boiler was slightly more than 8%, while the cast iron boiler had a predicted full load radiation loss of 4%. At the normal load of 50%, the combustion efficiency of the HRT has to be 8% higher than the cast iron boiler to overcome the higher (16% versus 8% of actual input) radiation losses. The operators were firing the older boiler because combustion analysis indicated that it was 5% more efficient. Evaporation rate data later proved that they could not rely on their "combustion efficiency." Note the confusion here. The HRT furnace could operate at a lower excess air to burn the same amount of fuel (that is combustion efficiency). The combustion analysis indicated that the stack loss was lower since there was less hot gas going up the stack. The true combustion efficiency (fraction of fuel burned) was likely the same. The 5% improvement was due to the lower amount of excess air as measured at the stack. However, the radiation loss at normal load was 16%, compared to 8% for the newer boiler. The difference of 8% was greater than the 5% and resulted in a lower boiler efficiency as much more heat was just radiated away. Radiation losses are also more significant for smaller boilers since they have a much higher surface to volume ratio.

The Power Test Codes always relate boiler efficiency being relative to the higher heating value (HHV) of the fuel fired. The HHV accounts for the water that is produced by combustion of the fuel being condensed at ambient temperatures. The HHV is measured in a laboratory instrument called the bomb calorimeter, which is carried out at room temperature. When the water condenses, it gives up its latent heat of vaporization and that is measured by the calorimeter. That heat energy is available to be recovered if a suitable heat recovery device can be utilized. If the water vapor remains in the vapor state, the lower heating value (LHV) is measured. For solid fuels, the hydrogen content of the fuel is typically fairly low and there is only a small difference between the HHV and the LHV. For natural gas, the hydrogen content of the fuel is 25% or more. The difference between HHV and LHV is on the order of 10%–12%. The advent of combined cycle and cogeneration plants has resulted in the return of the use of LHV. An efficiency calculated using the LHV will always be significantly higher than an efficiency calculated using the HHV. In those applications where a condensing heat exchanger can be used in the exhaust gas, a calculated LHV efficiency could be greater than 100% because the system recovers heat, while the LHV does not acknowledge that heat as existing. LHV does not include the heat that could be extracted if the water in the flue gas was condensed. In this text, the heating value will always be considered to be the HHV, unless specifically noted. The gas turbine industry quotes their equipment efficiency using the LHV. Europe also uses the LHV, primarily because it results in a higher calculated efficiency.

Can a boiler efficiency be greater than 100%? Logic says the answer is no. However, by the definition of some efficiency labels, some of them can. At one project, the system used heated air from a process as combustion air. It contained a small amount of hydrocarbons with negligible heating value. However, when one particular process was operating, 360°F combustion air was produced. When supplied to the one boiler with an economizer and a stack temperature of 303°F, it produced results in the accepted definitions that exceed 100%. That, by the way, was efficiency at the HHV. If the true and full definition of boiler efficiency were used, the heat in that combustion air would be included as an input. However, the simple input–output efficiency calculations only include the heating value of the fuel. They are used to avoid measuring the energy added by fan motors and pump motors along with that hotter combustion air. "Combustion efficiency" calculations will show a negative loss because the temperature of the hotter air is subtracted from the temperature of the flue gas.

For some reason, everyone concentrates on boiler efficiency when it does not change very much and has little to do with the overall "plant efficiency," which the boiler operator should be attending to. This is a bigger problem when there is so much confusion over what boiler efficiency really is. Two identical boilers in different plants can have the same boiler efficiency and combustion efficiency. Yet, one will produce less usable energy than the other because it has a higher blowdown rate. The energy absorbed by the water and steam in the boiler (ASME definition) includes the heat added to the blowdown water. Two plants with identical boilers and loads can have different plant efficiencies simply because one plant does not have water softeners. Consequently, it must blow down more. Maybe they both have softeners, but one has very little condensate return. It must heat the makeup water to replace that condensate and blow down more. Those, and other variations, can produce plants with boilers, having an 80% efficiency, operating with a plant efficiency as low as 40%. Take a plant with a mismatch between equipment and load and that plant efficiency can be as low as 20%.

What is "Plant Efficiency?" It is the amount of heat delivered to the facility, the usable heat that is generated, divided by the energy used in the plant. What is delivered to the facility is the output of the boiler. Think of the energy in the steam or hot water going down the pipe to the plant less the energy in the condensate or return water. That way, the output is what the facility is using. The energy used in the plant includes electric power in addition to the fuel.

A kWhr is 3413 Btu. Multiply the kWhr on the electric bill by that number to know how many Btu's were added by electricity. If the fuel is natural gas, it is often sold in therms. Multiply the kWhr in the electric bill by 34.13 to convert the electricity use to therms. If the plant is larger and uses decatherms, or millions of Btu, multiply it by 3.413. With identical units, add the electrical and fuel energy inputs to the plant to get the total energy used. If steam is delivered to the facility and gets nothing back, this is a 100% makeup plant. The energy that is being delivered is all in the steam. Look for the enthalpy of the steam in the steam tables in the appendix. Then subtract the enthalpy of the water supplied to the plant and multiply by the number of pounds of steam produced to get the output in Btu. Divide by 100,000 to convert to therms and one million for decatherms, or million Btu.

If condensate is coming back, it will have to be metered. Alternatively, subtract makeup and blowdown from the steam output to determine the quantity of condensate return. Use the enthalpy in the steam tables for water at the condensate temperature. Multiply by the pounds of condensate returned to get Btu. Adjust that result to match the output units and subtract from the steam output to get the plant output. If electricity is also being generated, use the conversion and add that to the output.

For hot water plants, determine the water flow rate. Hopefully, it is constant. Convert gpm to pph. Then multiply by the number of hours in the day, week, or month being evaluated. One gpm is approximately 500 pph. Thus, multiplying gpm by 500 is close enough. The time period is determined by how the fuel usage is measured. If the gas billing is the basis, it is usually the month. Use 720 or 744 hrs depending on the month (except February which will be 672 or 696 hrs). With the number of pounds being pumped around, multiply it by the temperature difference of the water. Remember, the definition of a Btu is the amount of heat required to raise the temperature of one pound of water by 1°F.

An average temperature will have to be used for return water (or supply water if controlled on the return temperature) to calculate the output. Since the loads swing, a Btu meter, which constantly performs that calculation, should be an integral part of the plant so that the output can be measured. That is it. Plant efficiency is the output divided by input. It can be calculated regularly. Or use some of the rate measurements that will be covered. Why is this important? By measuring the plant efficiency, it is possible to develop a measure that will allow the determination, first and foremost, if the plant performance is consistent, increasing, or decreasing. The goal is to produce the highest efficiency or highest rate of output per unit of input that is possible. It is called burning less fuel and using less electricity while still satisfying the load. This will become more important as sustainability requirements become more stringent. First, measure it to determine the current status. Then, with operating experience, it will show that running one boiler instead of two makes a big difference. It will demonstrate that shutting down the continuous blowdown heat recovery system costs a lot more to operate without it. However, continuous blowdown saves more money in water than it does in fuel.

Now these considerations show how the wise operator can make some difference. All that attention to the tuning of the boiler to get optimum boiler efficiency is not as productive as making certain that the energy converted to steam and hot water is used efficiently. Plant efficiency deserves more attention because it is the sole purpose of the boiler plant. That is to deliver heat to the facility. Note that the "facility" means the buildings, production equipment, etc., that are served by the boiler plant. The facility itself is involved in the energy equation under these conditions because it can contribute to the performance of the boiler plant. It does so primarily by returning condensate and, in some cases, generating some of the steam or producing some of the heat.

A facility can also waste much of the heat energy produced in the boilers to increase fuel and electricity consumption. It may not be the operator's responsibility to reduce that waste. Nevertheless, it should be monitored and documented for the benefit of the owner so that it can be reduced. To identify the operator's overall performance, calculate the plant efficiency as defined. To get a measure of the facilities performance, compare fuel used to production quantities (production ratios), heating degree days, or a developed formula that accounts for the load variations. Keep track of the difference in energy returned by the facility. It can make a difference. If the third shift is assigned cleanup and discovered that the hot condensate did a better job in cleaning than the heated domestic water, it would be noticeable. After all, condensate is distilled water. It will dissolve a lot more than city water.

Which efficiency should be used? Plant efficiency is the one that should be monitored for overall plant performance. For comparing boilers, it is reasonable to use the boiler operating efficiency, which is basically the "combustion efficiency" with an accounting for radiation loss. For a performance test, use the ASME definition in PTC 4, as that will be the basis for the manufacturer's guarantee. Blow off and blowdown losses, as explained earlier, are functions of water treatment and operation and not boiler efficiency. They have to be accounted for in the plant efficiency because the heat lost to blow off and blowdown is not delivered to the facility. Steam generated that is used in the deaerator is not delivered to the facility nor is steam used to heat the plant.

For all practical purposes, every piece of equipment has an operating efficiency that is separate and distinct from predicted efficiency. Equipment is not frequently operated at its designed capacity. Thus, be aware of what its efficiency is at the actual operating conditions. When steam pressure is raised or lowered, the operating conditions for the boilers, economizers, boiler feed pumps, and system steam traps have all been changed. An increase or decrease in pressure will alter the pressure drop in steam mains to amplify the change at the steam utilization equipment. In some cases, there will be charts or graphs that will predict the efficiency at the new condition. Some, like pump curves, do so with an accuracy that can be used. Performance of other equipment may have to be measured to determine if the change is beneficial or detrimental.

In some cases, operating efficiencies are described using terms other than percent. Chillers, for example, will list the kilowatts (kW) per ton values at different loads. In those instances, the important thing to know is whether the ratio should be increased or decreased to increase efficiency. As operators, it is not so much the precise value as it is the direction (increase it or decrease it). In the case of kW per ton, the goal is to decrease it.

Another concern is "cycling efficiency." It is not addressed in much of the literature and is not given the attention it deserves. It is very important as many plants do not have a constant load. Whenever the load on a boiler is less than that boiler's output at low fire, the boiler has to cycle to serve the load. All the time that it sits there, it is radiating heat. The radiation loss that is only a few percent at the most at high fire may be 10% or more of the input when the unit is cycling. When the pressure or temperature control switch contacts close, the boiler starts, warms up, and serves the load until the pressure or temperature control switch contacts open. Every time it is off, the boiler loses heat to the load and the air drafting through it. When it starts, the boiler loses heat as the purge air cools it down. Those heat losses, purge air cooling, and off cycle cooling become very significant as a percentage of the input. Cycling efficiency accounts for all those losses. A boiler that is serving a load at 5% of capacity may be operating at a cycling efficiency of 30% or less. It means that it burns more than three times as much energy in fuel as it delivers to the facility. Now consider the fact that so many boilers are oversized. In that case, they are running at those low loads most of the time. Now, that cycling efficiency becomes meaningful. An oversized boiler is one that is no longer the right size for the facility. With added insulation, sealing up air leaks, adding double glazing, and other activities that have been taken to conserve energy, the plant load has been decreased so much that the boiler is now too big for the plant.

When a modulating heating boiler is cycling at temperatures that are halfway between the winter design low and 65°F, cycling efficiency has to be determined because that efficiency is so low that a replacement of that boiler with that of the right size will result in fuel savings that will pay for the new boiler in one or two heating seasons. Use that half the load and cycling determination to identify boilers that are cycling excessively and get an engineer to do an evaluation to determine if the boiler should be replaced. Perhaps, the management will not be willing to go to the expense of hiring an engineering firm to do the analysis. An alternative would be to contact an Energy Service Company (ESCO) and invite them to look into it. ESCOs install modifications to plants to reduce energy consumption and get their money back from the savings with no money layout by the owner.

PERFORMANCE MONITORING

Calculating boiler efficiency may not be considered part of the duties of a boiler operator. However, monitoring and optimizing plant performance is. To make it simple, use values that are less complicated to determine and easier to understand and work with. Of course, it is still necessary to understand how they are calculated and whether the desired results should be higher or lower to indicate an improvement in performance. Otherwise, they are a waste of time. To work in terms of efficiency, use the guidelines in the previous section. Don't be surprised if the resulting numbers seem out of place. Don't accept them as necessarily true either. It is simply unrealistic to believe something can operate at more than 100% efficiency, even if the calculations would imply that.

The best method for evaluating steam boilers is evaporation rate. Divide the quantity of steam generated by the gallons of oil or therms of gas burned to get it. Don't, as one plant did, simply enter 122 in the column on the log for evaporation rate because that is what it is. In that instance, and in many others, the operators enter a value in the log that the chief wanted so that everyone was happy. It was not anywhere near the actual value, which could be calculated from the other entries in the log. In the case of that plant, the math showed that the actual value was around 108 pounds of steam per gallon of oil. Further, two of the three operators managed to run the plant so that their value was 105, while one managed to maintain 114. Once the other two were clued in as to what they were doing wrong, and settled down, the average went to 114. There were sound reasons as to why the plant could not manage an evaporation rate of 122. However, the specification was for 122 and the operators put down what they thought was requested in the log book.

Evaporation rate can be used to compare boilers to each other and to performance at other loads and at other times. It is comparable to a boiler efficiency as far as variations are concerned. A change in evaporation rate should be relative to a change in "combustion efficiency." Of course, that does not come close to monitoring plant efficiency. For that, the delivery rate must be compared (the number of pounds of steam delivered to the facility divided by the amount of fuel burned in the same time frame). The actual value of the number itself is not important. The objective of calculating these rates is to see if they changed and, if so, whether they changed for the better. Whatever is used, it should be treated as a flexible number with a goal of increasing or decreasing that number depending on how it is calculated. The concept is exactly the same as monitoring the gas mileage on a car, where the miles per gallon dropping off indicates that there is something wrong. Or, perhaps, a lot of city driving was involved. Changes in the rate can be an indication of improved performance or changes in the load.

Evaporation rate provides a value very consistent with boiler operating efficiency and delivery rate is consistent with plant efficiency. They are good parameters to measure, log, and compare to monitor boiler performance and plant performance. Evaporation rate can indicate problems that cannot be determined by combustion analysis or other methods of monitoring boiler efficiency because the latter are instantaneous readings. Frequently, combustion analyses are performed while the boiler controls are in manual and the service technician has adjusted them to optimum. That can be a significantly different condition when compared to operating at varying loads in automatic.

Some steam plants have no steam flow meter. There are still ways of determining the amount of steam generated. A simple one in many plants is achieved by installing a 20 dollar operating hour meter on the boiler feed pump motor starter. This will work in all cases where the pumps are operated to control the boiler water level. The pump has a listed capacity in gpm which, when multiplied by 60, gives gallons per hour. Then multiply the gallons per hour by 8.33 (or the actual density) to get pph. Multiply differences in hour meter readings by the pump capacity, 60 minutes per hour, and density to determine how many pounds of steam were made. Then divide that by the amount of fuel burned to get the evaporation rate. If there is a lot of blowdown, then calculate its percentage, subtract that from 100, then divide the result by 100, and then multiply that result by the meter reading to get the steam generated.

A hot water plant is a little more difficult. If the water flow through the boiler is constant, a recorder for the water temperatures will provide an average temperature difference. Multiply that difference by the water flow to determine how many Btu's went into the water. If the boiler water flow varies, a Btu meter will be needed that calculates the heat added based on flow and temperature. Any decent sized plant will have a Btu meter that makes that calculation. Since a Btu is the amount of heat added to one pound of water to raise the water temperature by 1°F, the degree rise and number of pounds of water need to be figured out. Number of pounds times temperature rise gives the heat out. Dividing that by fuel used provides a heat rate. Since most hot water plants are heating plants, it may be possible to get along with a degree day ratio.

Plant efficiency can also have a relative parameter that is easy to calculate. If the plant is used solely for heating, then a degree day ratio can be used. Divide the quantity of fuel burned by the number of degree days in the same period. Typically, the ratio changes with load. Thus, always compare gallons per degree day or therms per degree day to periods with the same or a similar number of degree days. That value is the opposite of evaporation rate. The goal is to keep it as small as possible.

If the boilers are also used to heat hot water, the hot water use is reasonably consistent with variances that are insignificant compared to the heating load. In that case, treat it as a constant value. Refer back to that earlier discussion on knowing the load.

If the boiler is serving an industrial plant, there is the potential for a variety of plant efficiency comparisons. There are pounds of product per pound of steam, a very common measure, and complex calculations that vary depending on the industry, method of production, and product manufactured. Usually, these plants are large enough that process steam metering is justified. Then a Plant Rate, pounds of steam delivered to the plant divided by the quantity of fuel consumed, can be calculated.

There are not always fuel meters. If firing oil, then sound the tanks regularly and after every delivery. If firing gas, the gas company always has a meter that can be used. If firing coal, there has to be some way to get an idea of the weight burned. In plants that are so small that the price of a fuel meter is not justified, the boilers usually fire at a fixed rate. Another 20 dollar operating hour meter connected to the fuel safety shut off valves will provide a reading. While how many gallons or therms were burned can be determined, a formula as simple as hours of operation divided by degree days will provide a performance value that can be monitored. Put another operating hour meter on the feed pump. That will provide a comparison of fuel input to steam output. Don't bother with all the other math. Just divide the difference in readings of one meter by the difference in readings of the other.

Always make sure the ratios are quantities divided by quantities or flow rates divided by flow rates (that is, use the same units). A ratio has no dimensions.

Keep in mind that, unlike a car, the boiler plant is in operation 8760 hours a year. A little change in fuel consumption represents a significant change in the cost of operation. Monitoring the performance using one of the several ratios that are available will allow the wise operator to make those little differences in plant performance that can amount to significant reductions in operating cost.

MODERNIZING AND UPGRADING

There are two ways of looking at modernizing and upgrading. Either an operator arrives for work one day to find contractor's personnel swarming around the plant or the operator simply sits and dreams of what would be nice to have. Occasionally, there is some blend of the two, but, for the most part, operators only get to experience one or the other. There are ways to become more involved in any modernization or upgrading of the plant. Even if not involved, always respond to an upgrade professionally. Don't be close minded and insist that it will not work. An operator is in the position to make sure it will not work. Many engineers and contractors dismiss an operator's contention and put the project in anyway, figuring the operator will learn to live with it once it has been demonstrated that it does work. Most of the time, it does work but only until the engineers and contractors leave. The wise operator should be honest enough to admit a lack of understanding of the concept and show concern about how to make it work.

People buy the same make of car as a matter of comfort. They are familiar with that brand. Similarly, operators tend to want the replacement boiler to be just like the old one. They know how the old one worked and are comfortable with it. Many engineers and contractors are more than willing to give the operators what they want. It is easy for them to copy what is there. It does not take any imagination and it does not really require any engineering. As a result, millions of dollars of fuel go up the stacks of plants that were expanded, supposedly modernized, or upgraded, with no improvement in performance. Don't repeat that error. Another concern is that of being put out of a job. However, qualified, experienced boiler operators are becoming a rare commodity. No wise employer will get rid of a wise operator.

The typical employer is concerned first for the reliability of the plant and, second, for its cost of operation. Many of them do not realize how much it is costing them to run their plant. Many of them never think about the sum total of all the monthly fuel bills. Energy cost is not always the highest priority problem for an executive. However, sustainability and carbon foot print issues will get their attention. The wise operator will know how to relate energy use to these major issues.

Be aware of how the operation compares with others and what is available to improve the operation of the plant. That requires obtaining information on how other plants perform and what is available to improve the plant operation. The National Association of Power Engineers (NAPE) might be one source as it is an association for boiler plant operators. Attending local chapter meetings will provide an opportunity to talk to other operators and learn what they are doing. There are also a considerable number of publications that target decision makers in boiler plants and similar facilities. A lot of them provide the subscription at no cost. That association, and similar others, are a good resource for information. Use them to learn about the industry. That will be good preparation for any planning for modernization or upgrades in the plant.

There is always a part to play in the modernization or upgrade of a plant. The first and most important thing to do is listen. Listening is often hard. Yet, it is the best way to gain insight into what will happen. Right after listening comes reading. The wise operator is the one who reads the instruction manual. Every piece of equipment is unique and has its own unusual features. Those features should always be in the instruction manual. When it comes time to operate that new stuff, it will be good to be prepared.

Chapter 4

Special Systems

A working knowledge of steam systems makes it possible to understand the use of special and unique systems and heat exchange materials because all the rules of heat and flow do not change with a system or the fluids used as the heat exchange medium. This section provides a little insight into some of the special systems that a boiler operator can encounter and may be called upon to operate.

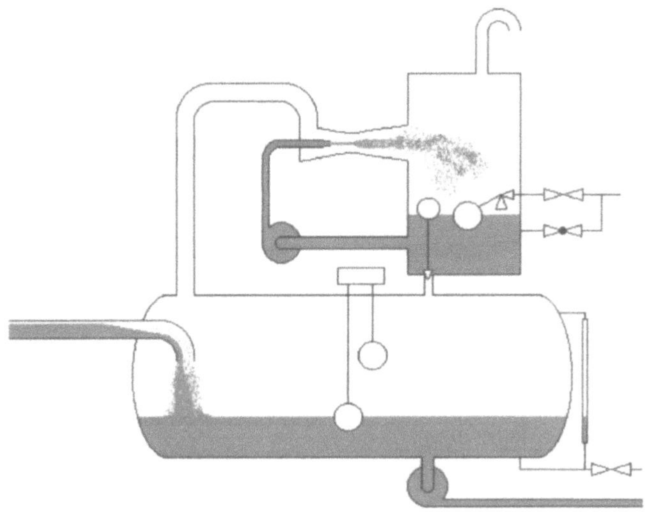

Figure 4-1. Vacuum pumps for condensate system.

SPECIAL SYSTEMS

Not all plants will have a special system. If there is one at the plant, always read the instruction manual. Some will be covered here to provide an idea of what is involved before opening the instruction manual.

VACUUM SYSTEMS

There are systems that are designed to operate with a vacuum. Vacuum pumps (Figure 4-1) intentionally produce a vacuum by removing air from the piping system, both the original air on startup and air that manages to leak in. Condensate flows to the vacuum system, which is operating as the lowest pressure in the system and is pumped out to the boiler feed tank or deaerator. The system shown in Figure 4-1 is a common one that produces a vacuum by pumping water through a water jet that acts as an ejector to pump the air out of the system. The vacuum system allows users of the heat to operate at lower temperatures, maybe a necessity in some situations where there is a concern for someone touching a radiator. The problem is solved by operating at 25 inches of mercury where the steam temperature would be 134°F.

There are not many of these vacuum systems. Many engineers and boiler operators consider them to be inoperable. A singular big problem with them is air leakage, which is impossible to locate during normal operation. Even when the system can be pressurized, the leaks do not show up because a drop of water or piece of scale can prevent water from leaking out but will allow air to leak in. Once air leaks start, they tend to get worse because the air dries out the joint sealing compounds. Technology could probably provide a joint compound that could maintain a seal in a vacuum system, but that horse has already escaped the barn.

Another problem with vacuum systems is that someone works on the system with no knowledge that there is a vacuum pump back at the boiler plant. They decide to put in a vent. Now that is an assured leak because someone created it and it looks perfectly normal. Such open vented condensate return units on vacuum systems are not uncommon. If someone does this, the simple solution is to connect the vent to the steam line instead of to atmosphere when the tank can take the steam pressure. Install a valve to allow service on the unit and put a liquid trap in the overflow line to block it. The water in the trap tends to dry out and will need a way to be refreshed as well. Use the manufacturer's instruction manual as a guide and other information in this book to get a better understanding of what is happening with these systems and how the standard operating procedures (SOPs), etc., should address them.

HYDRONIC HEATING

Much of this book addresses the steam generating boiler plant. While much of what is covered applies to water heating as well, there are many considerations in a water plant that are not a concern in a steam plant. Hydronic is just a word used to differentiate low pressure hot water heating systems from other types of boiler plants. Unlike a steam plant, a hydronic system can be shut down without admitting air to prevent a vacuum. For that one reason, hydronic systems should last at least twice as long as a steam system under otherwise equal operating conditions. How long is that? About 60 years.

It is the system of choice today for residential boiler applications and most commercial buildings because it does not require as much attention as a steam system. Being properly maintained, it will require a minimum of makeup water (almost nothing at all when new), and, therefore, it needs little attention to chemical treatment. With all that said, there is some reason to wonder why anyone even considers having an operator in a hydronic heating plant. It is not necessary to admit air to a hydronic system as is done with a steam system because the change in volume from operating to idle is not significant. That does not mean that changes in volume are of no concern for the operator. The problem with most hydronic systems is due to changes in volume that are not accounted for in various stages of operation. Close off a section of steam system and the steam will condense, leaving a vacuum that might permit atmospheric air to crush some thinner walled vessels attached to the system. That is all that will happen. Of course, one of those vessels could be a very expensive stainless steel heat exchanger. That has happened.

Hydronic systems will also produce a vacuum as the water cools. Therefore, expect air in that piping if it is isolated. Hot water and steam piping is usually strong enough that it can withstand the vacuum and nothing happens. Close off a section of chilled water piping in a building so that water is trapped, and that is another story. As the chilled water heats, it expands to build up pressure rapidly. It will rupture the piping if it cannot leak out somewhere. Unlike steam and air, water is not compressible. The best thing to do is close only enough valves to stop the flow but not so many that the system is completely isolated. When isolating for maintenance, open some vents as soon as the system is isolated.

Hydronic heating systems must have provisions for thermal expansion. When water is heated from a nominal building temperature of 65°F to an operating temperature of 180°F, each cubic foot of water in the system will swell by almost 3%. That is not a lot percentage wise. However, when the total volume of a heating system is considered, that can be several hundred gallons. A plant that is water logged (all elements full of water) can experience extreme swings in pressure associated with the expansion and contraction of the water. An expansion tank is provided in a hydronic heating system to reduce pressure swings to a tolerable range.

The tank can be an open type, located above the highest point in the system at a height adequate to maintain the desired system operating pressure. The top of the tank is open to atmosphere and the gauge pressure at any point in the system is a function of the height of the water. The tank has to be large enough to accept the expansion of the water in the system without a considerable change in level because the system pressure will change about 1 psi (pounds per square inch) for every 2.31 foot change in tank level. Sometimes, the tank is too small to handle full expansion and the water overflows from the tank as it expands. A float valve can be added to replenish the water when the system cools. Open tanks are used infrequently and normally only in systems using ethylene or propylene glycol and rust inhibitors for freeze and corrosion protection. The principle problem with an open tank is that it allows oxygen to get into the water with corrosion as the outcome.

Closed expansion tanks can be a simple pressure vessel or be fitted with a neoprene or Buna-N bladder that separates the water in the system from the air that provides the expansion cushion. Pressure maintenance in systems with closed expansion tanks is established by controlling the air pressure over the liquid and/or the amount of water in the system. Some systems use nitrogen instead of air to eliminate the oxygen as a source of corrosion of the tank and system. Tanks without bladders are usually epoxy coated internally. That is why they have those "do not weld" stencils that someone painted over several years ago. Most plants are served by an expansion tank that can take the full swing of expansion from an idle condition to design operating temperature. A few plants, however, either due to space or price limitations, or as a result of expansion of the building and adding boilers without changing the expansion tank, will not have enough room in the expansion tank. All systems are normally fitted with a makeup pressure regulator that admits city water to maintain a certain minimum pressure in the system and a relief valve that will drain off water when the pressure builds. Open and simple closed expansion tanks are fitted with a gauge glass to provide observation of the water level. Bladder type tanks do not provide any indication of level unless special instruments are provided. With a simple closed tank, in addition to knowing what is happening in the system by looking at the water level, a low water cutoff can be added to any tank mounted above the boiler for primary protection in the event of a loss of water. The tank low level cutoff cannot work alone because steam can be generated in the boiler to displace water in the tank. That will not provide a low water indication at the tank. That is why a low water cutoff is needed on the boiler and why a low system pressure alarm switch, shut down if the plant is not attended, is a necessity as well.

Unlike steam plants, the fluid in a hydronic heating system does not move around on its own. Some units use a glycol mixture, not just plain water, and hence the term "fluid." The glycol changes the boiling point of the fluid. That means another set of tables besides the steam tables are needed. Otherwise, they work the same. Steam will readily flow from one point to another with a very little difference in pressure. A hydronic heating system is full of water with the only pressure variation being the elevation at a particular point. There may be a little thermosyphoning going on, where lighter hot water is lifted up as heavier cold water drops down to displace it. However, it is never enough for heating any reasonably sized system. There might be a gravity system in a house, where the pipes are large enough to allow the liquid to move around. These are not likely to be found elsewhere. For most installations, there is no pressure differential to force the heated water out of the boiler and to the load.

That is why every hydronic heating system has circulators. Circulators are pumps that push the water around the hydronic heating system. They are not sized to fill the system or capable of pushing the water up to the highest level in a system. They are selected to overcome the resistance to flow through the system at the designed flow rate and that is all they do. If there is any large volume of air in the system, it will create differential pressures that can prevent or limit system flow (Figure 4-2) because the pump was not designed to overcome that differential. The pump in Figure 4-2 was designed to pump the water around the system. Once air accumulates in the radiator to produce a condition where the water drains to the boiler, the pump has to push the water up to the radiator and frequently does not have the ability to do it. Opening the vent on the radiator allows the pressure in the expansion tank to push the water up to displace the air. Air in water systems can create all sorts of problems. If there is not enough water in the system and a makeup regulator is not provided, water will have to be added to the expansion tank manually in order to restore the operating level.

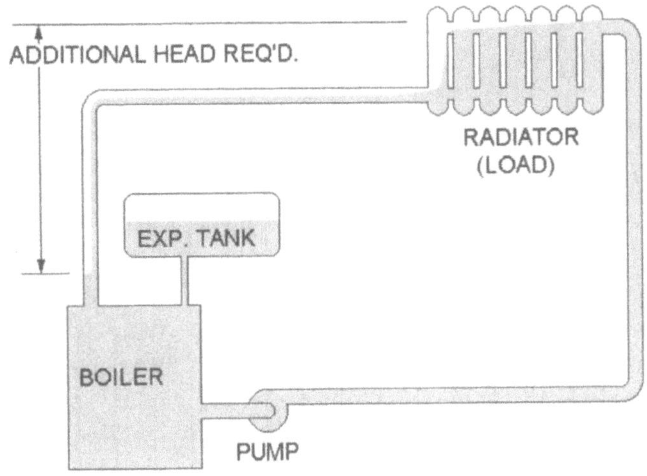

Figure 4-2. Differential produced by air in hydronic system.

One neat thing about hydronic systems is that they are easy to measure. Given the definition of a British thermal unit (Btu), all that is needed is the temperature in, temperature out, and the flow rate to know how many Btu's a boiler is putting out or how much a particular piece of equipment is using. That is true at any instant anyway. It is another story when the average or total readings are needed. The flow rate has to be close to the rating of the circulator. There are pressure drop curves (Figure 4-3) in the instruction manuals for most equipment. Read the pressure drop through a coil and read the flow off the curve. A differential gauge is preferable, but using the same gauge on both connections will give a fairly accurate differential. Just reading both installed gauges assumes that they are identically calibrated. They almost never are.

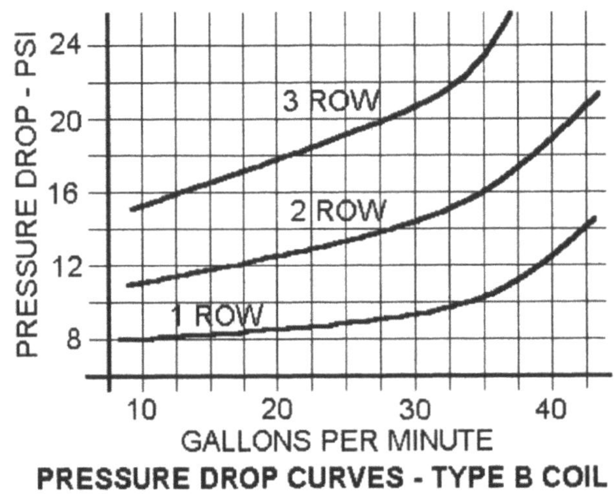

Figure 4-3. Pressure drop curves for heating coil.

The reading from a coil table is usually in gallons per minute (gpm). Multiply the gpm by 500 (to convert gpm to pounds per hour (pph)) and then the difference between the inlet and outlet temperatures to get the Btu/hr.

Hydronic systems in the US tend to have much higher flowing pressure requirements than systems in Europe. The Germans, in particular, look down on the US because these US systems introduce so much unnecessary differential. That wastes a lot of motor horsepower. That is a matter of initial design. In many systems, the operators throttle down on a valve here and there to resolve heating complaints until the whole system is operating at a fraction of its design flow, while in other situations, they adjust valves open enough that flow through some systems prevents flow in others. Building owners do not like to hear that their distribution system is totally upset and they have to bring in a balancing company to put everything back in order, a task that is very expensive relative to building size. If a small adjustment will solve a problem, then try it. Just count every turn or partial turn of that valve and log it so that it is possible to always return and put it back to where it was.

Sometimes, the flow control valves in hydronic systems, or piping loops themselves, accumulate mud and sludge because the flow is slow enough to allow the sediment to drop out. What should happen is the accumulation reduces the size of the flow stream. The velocity should increase until a balance is reached where no more material accumulates. In the initial years of a building system, that sediment accumulation can reduce the flow through the loop. Thus, it is necessary to open a throttling valve (TV) a little to return to the design flow. Use the measuring device (it may have to be rented) and the flow sensing taps on the valve and restore the design flow, which should be shown on the piping drawings. Also, check some of the other valves in the same area to be certain their flow rates were not altered, taking readings on them before and after the adjustment on the one is made. Often, it is just a matter of blowing the sediment out. One means to do this is to open a valve on each loop after noting its position. Then count the quarter turns and restore its position afterwards. The temporary jump in flow would flush out that particular loop and may return its operation to normal.

Hydronic systems need blowdown just like steam systems. There should not be a lot of sludge and sediment in a system. The problem is that there is always a little bit of it. Water contains dissolved solids. Chemicals are often added to treat water. There will always be some in the water. It will be swept along in the areas of the piping that have higher velocities and settle out in the areas that have the lowest velocities. Systems with sections designed for future expansion include piping larger than necessary for current operation. The velocity in those sections will be considerably lower than individual unit loops and other parts of the system. A unit loop is piping from the supply headers to the return headers that serve one piece of equipment that uses the heat. When there are future service connections, they are the ones that should be used to blow down occasionally to flush the mud and sediment out because that is where it will settle (in addition to the bottom of the boiler). If they are not cleared occasionally, the sludge will build until it can be swept up in chunks by the flowing water and jammed into a smaller distribution or unit loop. Then there will be a real problem to fix.

As for the frequency of blowdown for a hydronic system, it will depend on the quantity and quality of water that was added to the system. The installation of a meter on the makeup water supply is recommended for a plant because that will be the guide to how much water has been added. Then it is simply a matter of knowing the quality of the water to see how much mud, sludge, etc., was added along with that water. The mud and sludge, which is dirt that entered with the makeup water and sludge created by the water treatment to remove scale forming salts, does not leave with a water leak, unless the leak is a big one. Usually, the leak is in the form of steam. If water is heated to 220°F, a lot will flash off as it drops in pressure at a leak and flow out as pure steam. All the mud and sediment that was in that water stays in the system. It is one reason leaks are not as much of a problem. The remaining mud and sludge will plug the leak.

Typically, it is safe to blow down a new system once a month as long as makeup is minimal. Remember that blowing down removes water. Thus, makeup water will need to be added and more treatment chemicals with it to replace what was lost in the blowdown. Watching the first gush out of the drain valve will be the clue to frequency. Normally, a hydronic system should be tested for total dissolved solids (TDS) (see chemical water treatment) just like a steam system. The blowdown should be managed to keep the TDS below a prescribed value (usually 2500 ppm (parts per million)). However, if there is a slug of mud (the water will be discolored) for more than 10 seconds, the blowdown frequency is insufficient. Increase the frequency. If there is no sludge, decrease it. TDS is dissolved solids, not settled solids. There is a distinct difference. Unlike a steam system (where everything solid stays in the boiler because it cannot become a gas and leave with the steam), the settled solids tend to pick many points in the hydronic system to accumulate.

Don't believe that old lie that no water chemistry testing and maintenance needs to be done in a hydronic system. Even systems with zero leaks have problems with the water chemistry changing as it reacts with the metals in the systems and any air it comes in contact with. It is essential to maintain the proper pH of the system and a supply of nitrite or sulfite to prevent corrosion due to oxygen getting in (see Water Treatment, Chapter 8). If there are system leaks that must be replaced by makeup water, then that water has to be treated. As systems grow older, the number of leaks tends to increase, despite good maintenance practices. The water treatment program has to improve to handle the larger volumes of makeup water. Many hydronic systems are equipped with nothing to pretreat the water. Thus, more chemicals are required, and, in many cases, adding pretreating equipment is justified.

The major concern with hydronic boilers is preventing thermal shock. Be sure to read the section on why boilers fail in the chapter on thermal shock. It is particularly important when the plant has more than one boiler. Sending a slug of cold water from an idle boiler into a system operating on another boiler must be avoided. Also, avoid dumping hot water into a boiler that is cold. Most hydronic heating plants permit firing the boiler without any water flow through it so that the boiler can be warmed up without pumping its cold contents into the system piping. There might be situations and conditions where there are slugs of cold water in the piping even though the boiler is up to temperature. Careful manipulation of the boiler's isolating valves is required to warm up that piping. It is best to crack open one of the two valves (return or supply) connecting the boiler to the system before starting the boiler to maintain consistent pressures throughout the system. Leaving one valve open when a boiler is out of service but not isolated for repair or other purposes is not a bad idea. The selected valve should be in a position where thermosyphoning will not generate any thermal shock. Sometimes, warming the boiler up with a valve open allows thermosyphoning to warm up piping to avoid thermal shock. Since every plant is different, develop an SOP that allows starting and engaging a hydronic boiler with minimal thermal shock.

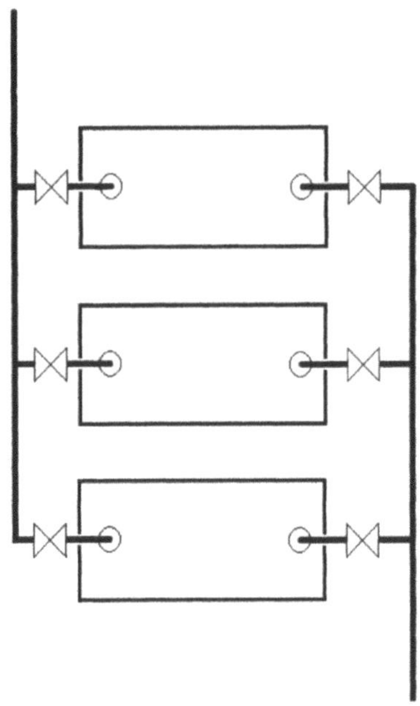

Figure 4-4. Hydronic boilers in parallel.

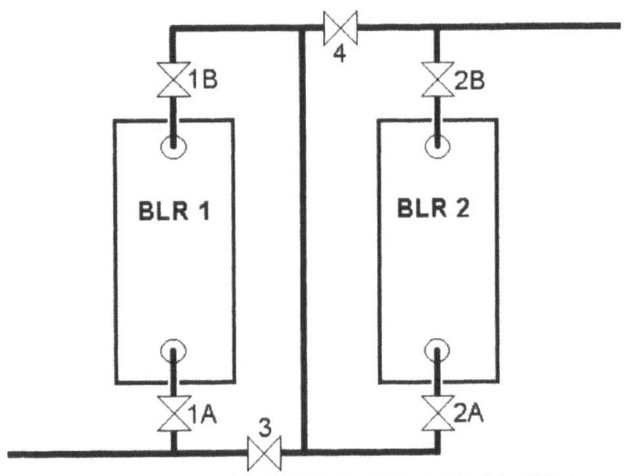

OPERATION VALVES	1	2	3	4
BOILER 1	O	C	C	O
BOILER 2	C	O	O	C
BOTH	O	O	C	C

Figure 4-5. Hydronic boilers in series.

Arrangements of hydronic boilers in multi-boiler plants come in two forms. Parallel installations (Figure 4-4) are most common and can be used with any number of boilers. Serial installations (Figure 4-5) are less common and the number of boilers is limited to two or three. In parallel installations, each boiler handles a portion of the system water and care is recommended to ensure that the water flows to each boiler uniformly. In some parallel installations, the system water is left flowing through each boiler. In that case, a boiler that is shut down acts as a radiator, wasting heat to the air that is drawn through it by stack effect to actually cool the system water. If there is no alternative to this type of arrangement, put a cardboard blank over the combustion air inlet to minimize the airflow due to draft. The hot boiler will still waste heat to the boiler room, as radiant losses and some thermosyphoning of the air will occur in the stack Thus, it is not the best solution. Closing one of the valves (supply or return) on an idle boiler will eliminate the heat losses. However, it will change system and boiler flows, and those effects have to be considered. Some boiler plants have a bypass line between the supply and the return headers which simulates the pressure drop of one boiler so that it can be opened after closing off a boiler to restore the flow rates in the operating boiler and system to normal.

When operating with less than the full complement of boilers on line and bypassing around or through others, be aware that the system supply temperature will be less than the boiler outlet temperature because it is mixed with the return water flowing through the idle boilers or bypass. Some plants use a header temperature control so that the idle boilers or bypasses do not change the hot water supply temperature. It will require higher temperatures in the operating boiler.

If there is a common header temperature control, it should be on the return. These systems usually have a proportional control so that the firing rate of the boiler will be proportional to the difference between return temperature and the set point (desired return temperature). The return temperature will be held near the set point, but the supply water temperature will vary depending on the blend of firing and the idle boilers or bypasses. It will not hold a constant return temperature either because there is a delay in response to changes in the boiler firing rates.

Checking the temperatures and a little math will enable the determination of what percentage of the water is flowing through the operating boiler. When waters of two different temperatures are mixed, the resulting temperature is dependent on the quantities of water at each temperature. The percentage of water flowing through a boiler will equal the difference between the mixed water temperature (Tm) and the return temperature (Tr) divided by the difference between the boiler outlet temperature (Tb) and the return temperature times 100. Boiler water flow as % of total = $(Tm - Tr) \div (Tb - Tr)$. This formula comes in handy when the flow rate of water in each part of a mixture is needed. The basic formula for energy can be used to determine how much heat is lost in an idle boiler. The temperature at the outlet will be lower than the temperature at the inlet. As in all cases where differences in gauge or thermometer readings are compared, it is a good idea, where possible, to switch the devices so that there is a different reading from the same instrument.

Series operation of hydronic plants requires that the piping arrangement allow for total flow through each boiler and a means for isolating the boiler, which requires three valves: two valves to isolate the boilers and one for bypass as shown in Figure 4-5. The water is heated first in one boiler. Then its temperature is raised further in the second boiler. These systems commonly use a header temperature controller to regulate the firing rate. Thus, the two boilers fire at the same rate. When the boilers are controlled independently, the modulating controller for the first boiler has to be set lower than the second one so that it does not take all of the load. Without the common controller, the controller set points (or firing one boiler on hand) will need constant adjustment to fire the two boilers evenly. An alternative to the common controls is using the position of the second boiler as a controller for the firing rate of the first boiler. Add another rheostat to the modulating motor of the first boiler and install a selector switch that will allow both single and two-boiler operations.

BOILER WATER CIRCULATING PUMPS

Boiler water circulating pumps are common on hot water heating boilers and consist of a small pump that delivers a fraction of the total flow through the boiler. The purpose of the pump is to help eliminate thermal shock. They pump enough water from the boiler outlet back to the inlet to produce a blend of return and circulated water to raise the inlet temperature well above the return water temperature. These should be started before the boiler and normally have proof of operation switches that prevent boiler operation if the pump does not operate. Some boilers were manufactured with the requirement that the circulation of water in the boiler had to be assisted by a boiler water circulating pump. The pump typically takes suction from a drum or header at the bottom of the boiler and pumps into a header a little higher. It is difficult to tell if the pump was provided as an element of the design of the boiler or added once it was revealed that natural circulation could not provide sufficient flow through the risers. On larger high pressure boilers, the difference in density between steam and water narrows, making natural circulation more difficult. One company featured large boiler water circulation pumps in the downcomer from the steam drum to force the circulation of the water through all of the tubes in the water walls of the boiler. These units were especially prevalent in boilers with steam pressures greater than 1800 psi.

Because boiler water circulating pumps are designed to pump hot boiler water, some units require that the pump should not be started until the boiler has been warmed up to a certain temperature or pressure. That is because the pump motor was sized for handling the hot boiler water with assistance in producing a differential from natural circulation. Thus, operating it with colder water would result in overloading the pump motor. The pump may also need cooling for the lubricating oil or direct cooling of the bearings, using an external source of cooling water because of high operating temperatures. There should also be cooling provided for the pump shaft seals and, of course, pump operation should require proof of cooling water flow before the pump motor can be started. There should be a differential pressure switch, used to prove that the pump is in operation, incorporated into the burner management system to safely shut down the boiler in the event of the failure of the boiler water circulating pump.

For the very large and high pressure boilers that use boiler water circulation pumps, the pumps are placed in the down take circuit to provide sufficient head to ensure adequate, positive upward circulation under all boiler operating conditions. Orifices are typically used in the inlets of the water wall circuits to assist in obtaining a predetermined, proportioned flow to the tubes of varying length and heat absorption. The steam drum is internally shrouded to provide uniform heating and cooling of the drum shell, for maximum maneuverability during start-up, load changes, and shut down. Downcomers installed on the steam drum carry the re-circulated boiler water, mixed with the feed water into the circulating pumps. The pumps are connected through a common suction manifold that insures flow through all downcomers regardless of the number or location of the circulation pumps in service. This feature minimizes the water level difference along the length of the steam drum for these long steam drums. The head developed by the circulating pumps is only that required to supplement the thermal head, which is typically 25 psi. Pumping water at or near saturated steam temperatures and high pressures requires specially designed pumps. All are vertical, single stage, centrifugal pumps with overhung impeller. Specially designed motor systems are used to minimize any problems with overheating and, typically, zero leakage. The use of these pumps gives the operator the capability to insure positive circulation for a wide variety of pre-operational procedures and part load cycling conditions. These pumps can be used to start water flow in the boilers prior to firing the fuel, thus minimizing the possibility of overheating the tubes.

HTHW BOILER PLANTS

High temperature hot water (HTHW) plants have all the characteristics, features, and problems associated with hydronic systems. The defined difference is that an HTHW plant operates with water temperatures higher than 250°F. The typical HTHW boiler plant has design conditions of 400 psig (pounds per square inch gauge) and 400°F. HTHW plants also have some other unique characteristics that are not found in the typical hydronic system. In most HTHW plants, the boilers are called HTHW generators. They differ considerably in construction and operation. The typical HTHW generator (Figure 4-6) is a once-through boiler. They are just unique boilers. They do not have steam drums and the headers are usually small. Those generators require water flow through them to operate. Because they do not store any hot water, their water volumes are very low. The controls will include low water flow switches that prevent burner operation and will shut the burner down if the water flow in the boiler is too low. The controls require Btu calculation with measurement of the return water in order to ensure that the outlet temperature is close to steady. Flow through the boiler consists of several parallel circuits and the tubes are frequently orificed at the headers to ensure proper distribution of water. It stands to reason that a tube designed for water flow on a once-through basis will have a real problem if steam is generated in it because the larger volume of steam will fill the tube. Once steaming starts in one of those boilers, failure due to overheating rapidly follows. Each HTHW generator is commonly fitted with its own circulating pump (standby circulating pumps are normally shared) to ensure adequate water flow. Normally, there are separate pumps used to circulate the HTHW through the system.

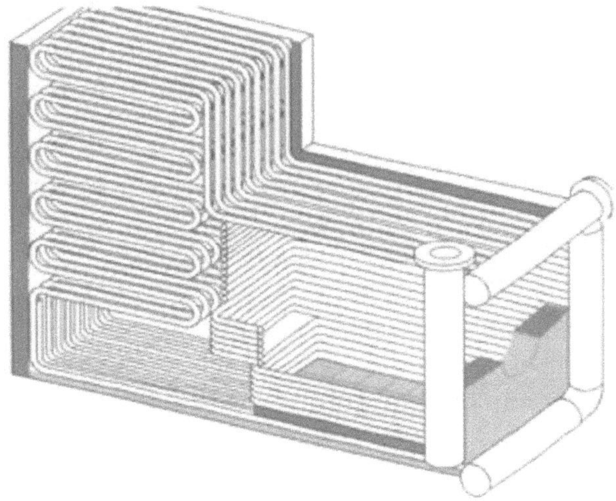

Figure 4-6. HTHW generator.

The circulators, or circulating pumps, have to pump water much hotter than the standard pump. Even though they are installed to pump the water into the boiler, like hydronic circulators, they are exposed to temperatures that are so high that the oil or grease in the pump bearings could be overheated. The pump seal or packing would also be exposed to those high temperatures and few can handle it. Any leakage of the hot water along the shaft would start flashing into steam. That could do serious damage to shaft and seal or packing.

To prevent problems with the seals or packing, the circulating pumps are normally fitted with sealing fluid systems. Where the seal or packing is exposed to the suction side of the pump, sealing fluid is commonly drawn off the pump discharge. Some may extract water using a pitot tube inside the discharge of the pump so that the velocity pressure is used to generate the differential to move the sealing fluid. In others, it may be necessary to have a seal pump draw water off the system and produce the differential necessary to force the water through the sealing fluid system. Newer pumps may be fitted with a special impeller on the shaft inside the seal housing that pumps liquid through the cooler and back to the seal.

Sealing fluid systems typically consist of two elements, a strainer to remove any particulate that might damage the pump seal, packing, or shaft, and a cooler to reduce the water temperature to values that the seal or packing can accommodate. After the sealing fluid passes through the strainer and the cooler, it is returned to the pump to flow over the seal and back into the pump. In the case of packing, it provides the little leakage that separates the packing and the shaft. In the case of packing, it is supplied to a lantern ring (see Pumps, Chapter 8). Proper control of the cooling of the sealing fluid is required to ensure the fluid is not overcooled to cause thermal shock.

The expansion tanks for HTHW plants are occasionally called accumulators. They can serve the typical expansion tank role but can also become a storage space for the hot water. To limit corrosion problems at the high temperatures, they are always pressurized with pure nitrogen instead of air. A true accumulator might be pressurized with steam and can contain electric heating coils to build up the steam pressure on a system startup and to maintain pressure when the system is shut down.

It is common for the low water cutoffs to be mounted on the accumulator because the generators do not have any point where a low water level can be detected. To avoid thermal shocks in the system, the makeup water is added to the accumulator where there is a considerable volume of water for it to mix with before it hits any metal. Preventing thermal shock is even more of a problem in HTHW boiler plants. Most HTHW plants have more than one boiler (unlike the hydronic plant that typically has only one) and the higher temperature operation requires careful management of the system when starting a boiler and putting it in service. The temperature differences between atmospheric and operating conditions are significant.

Care should be taken not to suddenly expose metal at 80°F to high temperature water at 390°F. In some circumstances, that is difficult to do, but operations that mix the two fluids (hot and cold) to gradually warm up a boiler, pump, or piping system can be managed. Steps in bringing a boiler on line and taking one off line can get very involved because the pumping and piping arrangements have to be reconfigured to ensure even distribution of the load on the boilers. Some plants have piping arrangements that restrict single boiler operation during periods of low load to a particular boiler because the system arrangement did not permit isolating the other boilers. In another plant, where the facility load had increased significantly, the design did not permit operating two boilers to carry the load because there was no way to arrange the piping to parallel the boilers. It is possible for HTHW boilers to operate in series, but it is uncommon. The piping arrangement has to provide for it. Unlike low pressure hydronic plants, HTHW boiler systems seldom have accumulators large enough to hold all the expansion of the system from atmospheric to operating conditions. A large pressure vessel designed to hold several hundred gallons of water is very expensive. They are occasionally reduced to a size that provides a cushion on the operation instead of allowing for complete expansion and contraction.

Those larger plants are equipped with provisions to fill the system as it cools from normal operating temperatures and tanks that allow steam to flash off and recover the remaining hot water as the system expands. In some cases, the requirements for expansion tanks to accommodate normal operating temperature swings is so great that even smaller smaller tanks with operating and standby provisions for fill and drain are installed instead. A lower pressure or open storage tank is used to prevent wasting the treated water as the system heats and cools..

Any HTHW system requires makeup water pumps to force the makeup water into the system. The pressure in a city water supply just is not adequate. Lack of electric power in these plants cannot be tolerated because the liquid in the system will cool and shrink to require makeup. A drop in pressure will result in steam flashing in some systems and driving water to others, with much noise and pipe rattling. The emergency electric generator is very important and some plants even have engine driven makeup pumps as a backup.

In general, HTHW plants are more dangerous than any other kind of boiler plant. The heated water contains a lot of energy. Any rupture of a piping system or a piece of equipment will result in a steam explosion. The rupture of an HTHW pipe will discharge almost 100 times as much steam as a steam pipe with steam at the same temperature. The number and location of exit doors from an HTHW boiler plant should greatly exceed those for a steam plant and any control room should have at least one exit that leads directly outdoors.

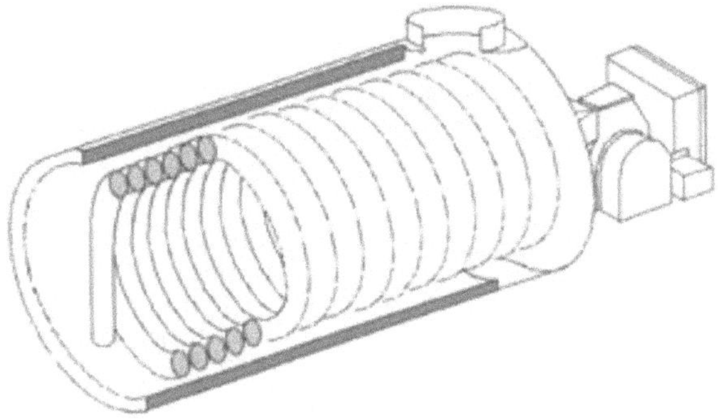

Figure 4-7. Fluid heater.

ORGANIC FLUID HEATERS AND VAPORIZERS

Organic fluids are basically like oil. These hydrocarbons are used as heat transfer fluids because they have much lower vapor pressures than water. What that means is that they can be heated to higher temperatures before they evaporate. Organic fluids are available that will remain a liquid and not evaporate at temperatures as high as 800°F at atmospheric pressure. By and large, these materials function the same as water and steam. They simply evaporate and pressurize at much higher temperatures. Organic fluids are used to produce high temperatures without the expense of handling high pressure. A system can be designed to operate at 500°F (a common maximum operating temperature) and pressures not exceeding 30 psig, where a steam or HTHW plant would have to operate at almost 900 psig. Both liquid and vapor systems are considered high pressure plants because the temperature is always higher than 250°F. The boiler is a power boiler, even if the operating temperature is below 15 psig. A fluid heater is basically the same as a hot water boiler and a vaporizer is very much like a steam boiler. The principal difference is the operating temperature.

The typical fluid heater (Figure 4-7) looks a lot like a common fire tube boiler from the outside. Many operators confuse them with a fire tube boiler. They are actually water tube boilers. What looks like an outer shell is a casing. The tubes form one continuous coil surrounding the furnace. In many cases, there are two coils to produce a secondary pass surrounding the furnace pass. The fluid to be heated flows inside the tubing. In a fire tube boiler, the hot combustion gases flow through the tubing, and hence the term "fire tube." Unlike a fire tube boiler, flow has to be proven in these units before the burner is started. The flow must be maintained or the burner should be tripped.

Other significant differences between steam and organic fluids include flammability, especially when they are heated to such high temperatures. If a water or steam boiler has a leak, the tendency is to put the fire out. If an organic heater or vaporizer has a leak, the tendency is to add to the fire. Almost any plant with organic heaters will also have a steam boiler that must be in operation in order for the organic device burners to function because the steam is used to quench any fire that might occur in the organic device. Normally, a thermocouple in the outlet or stack is monitored. Any rapid increase in temperature automatically results in burner shut down and opening of the steam quench valves. A few small units are fitted with compressed CO_2 extinguishing systems to avoid the provision of a steam plant. However, it takes a lot of CO_2 to put out an organic heater fire. Once it takes off, any leak adds enough fuel to melt more of the boiler metal to allow a bigger leak and a bigger fire.

The higher temperature fluids tend to have high pour points. That means they do not flow well, if at all, at normal atmospheric temperatures. Such a system will freeze up on shut down. Fluid systems for those high temperature fluids use steam tracing to warm up the organic fluid enough that it can be circulated in the system in order to get it started. While it is true that organic fluids do not need the attention of a water plant because the systems are designed to retain the fluids and vapors so that there is little to no makeup, the fluids do break down. Regular sampling and chemical analysis is still required.

Over a period of time, the fluid can break down and has to be replaced or reconditioned. Scale, as known in water-based systems, is not a problem. However, carbon can build up on the inside of tubes just like scale if the boiler is fired too hard, fluid flow is lost, or the fluid begins to break down. That can eventually result in a tube failure. A tube failure can result in the entire heater melting down. Thus, there is a concern for proper operation to prevent carbon formation, just like there are concerns for scale formation in a water boiler.

Monitoring the pressure drop across the liquid side of a fluid heater is critical to detecting a buildup of carbon in the tubes. Monitoring is not as simple as reading the gauges at the inlet and outlet and then subtracting the difference. Since viscosity changes with temperature, a record of the pressure drop at different average temperatures is needed. In that way, relative pressure drops can be used for comparison. Be as precise as possible with the measurements in order to catch the carbon formation the very instant it starts.

Even a very thin coating of carbon is so rough that it can produce a significantly rough surface on the inside of the tubes, causing the pressure drop to increase significantly. That is usually not a big problem because the circulating pumps are normally positive displacement types that will continue to force the designed flow of fluid through the heater. When carbon builds up, failure tends to be instantaneous because the increased pressure drop is handled until the pump motor is overloaded and trips out. Systems with centrifugal circulating pumps are uncommon because the viscosity variation with temperature has a significant effect on the flow in the system and the performance of the pump.

Any organic fluid system should be checked throughout its entire length at least once a shift, with special attention paid to any signs of leakage. The insulation is typically calcium silicate in order to handle the high temperatures. It is also very thick. A slow leak can penetrate a lot of insulation (store a lot of fuel) before it is detected. System leaks are dangerously close to becoming fires. They must be caught before they become a fire. There is no steam quenching on the piping like there is in a boiler.

Since most organic fluid systems are used in petrochemical and similar industrial production plants, immediate shutdown to repair a leak could result in thousands of dollars of production loss. It may be necessary to simply monitor a minor leak and be prepared to extinguish any fire that results until the entire facility can be economically shut down. It is one of those situations where the operator has to consider multiple risks and the cost of each. Any leak that cannot be made up, or becomes extensive to the degree that it is a dramatic hazard, requires a shut down.

Shutting down a fluid system takes time. Thus, the growth of a leak also becomes a factor to consider. The fluid has to be circulated long enough to allow the heater to cool until it will not carburize the fluid left standing in it. It also has to be cooled enough so that it will not spontaneously ignite when exposed to air. Then the fluid must be drained from the system back to storage until the level is below the point of the leak. Some facilities do not have sufficient storage to completely drain their systems and require a supplier's empty truck, on rental, to hold the fluid as it is drained.

Organic fluid heaters and the occasional vaporizer make some chemical processes possible only because they can produce high temperatures at low pressure. A common application is in the asphalt industry, where the product must be heated to high temperatures so that it can flow readily. All the rules for high pressure boilers apply. Every plant will have unique and special provisions that the operator should know. Among all plants, these are the ones where the SOPs must be memorized because lack of rapid and proper response to an upsetting condition can lead to hazardous conditions or long-term shutdown of the facility.

SERVICE WATER HEATING

Service water is the term currently used by the American Society of Heating, Refrigerating, and Air Conditioning Engineers (ASHRAE) to describe what is typically called domestic hot water heating. Heating of water for cooking, showers, baths, washing, etc., is not the same as heating water for closed hydronic building heating systems. The term "service water" is used to describe it. Service water heating systems are frequently ignored. Service water heaters do not enjoy the presence of chemically treated water to prevent scale and corrosion. Yet, most of them have such problems. In one area, the well water contained so much calcium sulfate that it would form heavy scale if the water temperature was increased by 6°F. There is nothing that can be done to the water to prevent scale formation or corrosion. Thus, the equipment has to be made for the service and will have to be operated and maintained properly to provide continued operation.

Service water heaters usually have much lower rates of heat transfer than steam and heating boilers in order to reduce scale formation. They are also fabricated for the application. Some of them are glass lined with glass coated heating surfaces. The water cannot be treated to make it non-corrosive. Thus, the heater must be protected from corrosion. The equipment sold in local area is usually suitable for service water heating of water used in that area. Small electric and gas or oil fired service water heaters require more attention in a commercial application than the ones used in the home because they get more use. There should be a schedule for blowing them down on a regular basis to remove any mud, scale, or other debris that may accumulate. Regular checking and recording of the stack temperature is also a must for the fired heaters because that can indicate problems with scaling. As scale forms, it insulates the heating surfaces, requiring higher flue gas temperatures to do the heating. There is also some checking and adjustment required for storage water heaters to keep everything working right.

Some installations use instantaneous hot water heaters. Instantaneous hot water heaters do just what they say they will do. They heat water quickly, primarily as it is used. Except for facilities where the instantaneous hot water heating load is less than about 25% of the lowest plant loads, those heaters can be a real problem for smooth and reliable boiler operation. It is also hard to believe an instantaneous heater is anywhere near efficient. They are capable of heating more water than is normally heated. As a result, they only operate a fraction of the time, allowing considerable off cycle losses. The amount of hot water used is a function of the activities of the occupants of the buildings. The curve in Figure 4-8 is based on ASHRAE data indicating the typical hot water consumption for a family over a 24-hr period. It is obvious that an instantaneous hot water heater has to be able to produce the quantity of hot water drawn between 7:00 and 8:00 in the morning but is required to produce a fraction of that load for the rest of the day.

The best system will always consist of a proper mix of water heater and storage that handles the load without excessive cycling of the water heater. See the discussion on cycling boilers for reasons why excessive cycling is a problem. When the hot water loads are large and variable, a modulating burner on an instantaneous hot water heater will reduce cycling or eliminate it. Instantaneous heaters with modulating burners can only eliminate cycling if the burner's turndown capability exceeds the variation in hot water usage. As can be seen from the figure, that service would require a burner with a turndown better than 20 to 1. Such burners are very expensive. Consequently, cycling is a normal condition. Steam powered instantaneous hot water heaters will produce load swings in the summer that prevent smooth and constant operation of the boilers. That means some type of storage is a necessity.

Figure 4-9 is a graphic of a boiler and storage tank system typical of that used in a large apartment building. Cold city water enters the system at the bottom center of the graphic, where it can either enter the circulating pump or the storage tank. Service water is drawn off the top of the tank. The arrow at the bottom right side of the tank represents flow of water circulated through the system to maintain hot water in the piping distribution system.

This combination of heater and storage will cycle, but it has the advantage of extended cycle operation and a fixed firing rate for the burner that makes it efficient but still simple to operate and maintain. A service water boiler deserves the same attention as a heating boiler on initial startup. Before the system is started, the owner, design engineer, or installing contractor (depending upon the requirements associated with installation) should contact the owner's insurance company or the authority having jurisdiction (normally the state, county, or municipality) to obtain a boiler certificate (or a document of similar title) which authorizes the owner to operate the boiler. There may be provisions in the jurisdiction to exempt certain equipment, but any requirements should be determined before placing the system in service. Normally, the boiler is subjected to a visual inspection by a National Board Certified Inspector before the certificate to operate is issued.

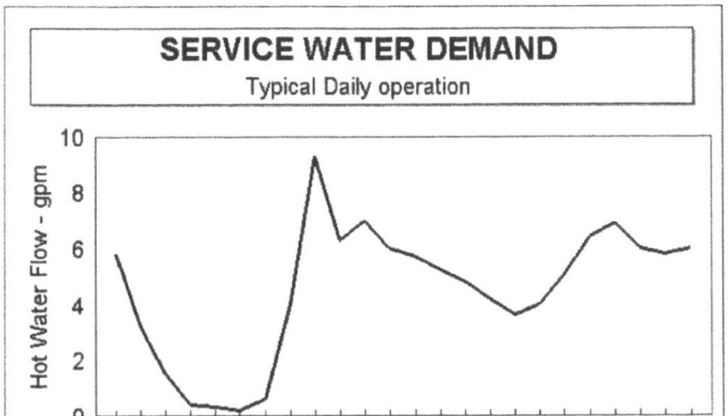

Figure 4-8. Daily hot water consumption curve.

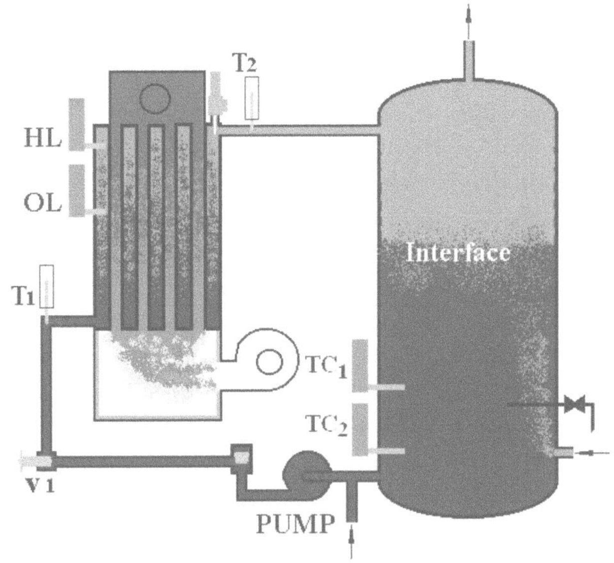

Figure 4-9. Service water heating system.

Initial operation of the burner should be achieved under the supervision of a technician trained in the proper setup of a fired piece of equipment. That technician should produce a "startup sheet," a document that includes, as a minimum, the following: the name, address, and phone number of the technician's employer, the technician's name and signature, and the date the initial startup was performed; a record of the actual settings of the operating limit (OL) and the high limit (HL) temperature switches, and an indication that their operation was confirmed; a record of the setting of the pressure and temperature relief valve (PTV) and a record that its operation was confirmed; a record of the burner performance while firing including, but not necessarily limited to, stack temperature, flame signal measurement, percent oxygen in flue gas, carbon monoxide level of flue gas, if measured, smoke spot test recording (oil only), if measured, gas consumption rate (gas firing), temperature of water at the boiler inlet during normal operation, temperature of water at the boiler outlet during normal operation, pressure at the inlet of the system, pressure at the discharge of the pump or other location between pump and boiler, and position of the TV. The startup sheet should be retained as a part of the original documentation for the system and referenced on each subsequent startup (after shutdowns for maintenance or other purposes) to ensure the conditions do not differ substantially from the original startup conditions.

All openings into the boiler and tank should be checked to ensure the system is closed and will not lose water unintentionally when placed in service. Before closing openings, the internals should be inspected to ensure that there are no loose parts, tools, personnel, or anything else inside the system that does not belong there. Valves and some spigots are opened to vent air and admit water until the system is flooded and at city water pressure. It is important to note that, if the city water supply to the inlet shown in the graphic is separated from the city water supply by a check valve or back flow preventer, an expansion tank or similar provision is required to prevent an increase in the system pressure when the water expands as it is heated. Disconnects, circuit breakers, and control switches are closed (in that order) to permit system operation. The circulating pump should start first, followed by the burner. The startup sheet should be checked as soon as operation stabilizes to ensure the conditions do not differ substantially from the original startup conditions.

The Codes in every state and province for all hot water boilers now require a low water cutoff. If the boiler is equipped with a low water cutoff, its operation should be confirmed during fill up and before the burner is fired. Since service water heaters are seldom fitted with a gauge glass, this test assumes a water level. Turn control power on before filling the boiler. The burner should not start. Then constantly monitor it to note that the burner starts to fire once the level reaches the low water cutoff and closes its contacts. Then, immediately shut off the water supply and open the drain. If the low water cutoff is operating properly, it should immediately stop burner operation. If it does not, secure the burner and see to it that it is fixed. Secure the drain. Turn the power off and fill the system, completely venting at all points before restoring power.

When stable operation is achieved, the TV should be adjusted to achieve the desired outlet temperature as indicated by the thermometer (T2) at the boiler outlet. Throttling of that valve is normally required to restrict the rate of water flow through the boiler to get the desired hot water temperature. If the valve is open too wide, the flow will exceed the design flow rate and the boiler outlet water temperature will be too low. If the valve is throttled too much, the boiler will heat the water excessively and the burner will start short cycling on the OL. If the heater is fired at a constant rate (most are), then there is a consistent output in Btu. Since the water flow is constant (the tank is a detour for any water that is not used in the system), the water temperature rise should be constant. Provided the demand for hot water does not exceed the capacity of the boiler, hot water will enter the tank faster than it flows to the building. Therefore, some of the water heated by the boiler remains in the tank, mixing with and displacing the cold water. Once the volume of the tank above the inlet pipe from the boiler is filled with hot water, an interface forms between the hot and cold water because the cold water is denser than the hot water.

Boiler operation continues and hot water displaces the cold water in the tank until the level of the interface drops to the level of the lower tank temperature control switch (TC2) to terminate the heating operation. The opening of contacts on the lower tank temperature control switch interrupts operation of the pump and burner to complete a heating cycle. During the period when the circulating pump and burner are shut down, the building is supplied by hot water from the tank. The weight of the check valve on the pump discharge provides sufficient differential pressure to prevent the flow of water through the boiler during this period. Sometimes, the valve is fitted with a spring rather than using weight. Don't put another type of valve in its place or it may not work.

As the hot water flows out the top of the tank, it is replaced by cold water entering the bottom of the tank. The interface level rises until it is above the level of the upper temperature control switch (TC1). Contacts on TC1 close to start the pump. Auxiliary contacts on the pump motor starter close to bypass the TC1 contacts so that the pump will not stop when the TC1 contacts open. The auxiliary contacts also permit burner operation. Whenever hot water demand does not exceed the capacity of the boiler, the system continuously repeats the operation described above. The pump and boiler start, heat a volume of water equal to the volume of the storage tank between TC1 and TC2, and then stop and wait until that volume of hot water is consumed.

When service water demand exceeds the capacity of the boiler, the difference between hot water demand and boiler capacity is made up by hot water flowing out of the storage tank and cold water entering the tank. The tank supplies all the hot water until the level is above TC1. Then the hot water from the boiler and the hot water stored in the tank combine to serve the hot water demand. Whenever the service water demand exceeds the capacity of the boiler, the elevation of the interface increases. Provided the high demand does not continue until the hot water stored in the tank is consumed, the boiler will continue to fire until the storage tank is once again filled with hot water down to the level of TC2, completing a boiler operating cycle.

Under unusual circumstances of sustained high demand for hot water, the reserve in the storage tank is consumed. Thereafter, the water leaving the system will be a mix of cold water passing up through the tank and hot water produced by the boiler. Hopefully, this will never be the case. If it frequently is, suggest a larger tank, larger boiler, or a combination because there is a hazard associated with it that is not desirable.

There are two temperature switches on the tank. The interface in a storage tank has a temperature gradient of 5–10 degrees per inch depending on turbulence. A system with a single temperature control would cycle on and off frequently as the interface rises and falls during each cycle. Each time the burner starts and stops, a purge is performed that, despite its purpose of safety, cools the boiler and the water with purge air. Provision of two temperature controls properly spaced (more on that will be discussed later) can significantly reduce losses and wear and tear associated with burner and circulating pump cycling.

City water temperature can vary significantly with the season, depending on the water source. If all the water is supplied from wells, the temperature varies less. When the water is stored in reservoirs or lakes and towers, the temperature can vary between 35 and 65°F. If the boiler operates at a fixed firing rate, as most do, the outlet temperature of the boiler will vary with the season. During burner operation, the operator should note the temperature on the outlet thermometer (T2) regularly and adjust the position of the TV to restore the desired tank temperature (±5°F) at least monthly. To increase the temperature, the valve is closed a little. To lower the temperature, the valve is opened further. Make the adjustment when the boiler operation has stabilized. Then wait a few minutes to see the results before adjusting the valve further. Normally, the boiler OL and HL do not function. However, when the boiler operates for extended times during periods of high demand, the OL could open its contacts because the temperature gradient in the boiler changes. The OL should not be adjusted to the point that it controls the boiler (starting and stopping it) during normal operation.

There is no provision to adjust the pressure in the system. It should follow the supply water pressure. The safety relief valve should not be tested to determine if it operates. Operating personnel wearing proper protective equipment should raise the lifting lever of the safety relief valve every three months to confirm that the valve mechanism is free and the water flow passages are not blocked. Testing of the safety relief valve should be recorded in the log. The purpose of the HL is to prevent overheating of the boiler in the event that the circulating pump fails or operating personnel inadvertently close a valve in the piping that prevents flow through the boiler. Its adjustment should be noted, lowered into the operating range to ensure it functions to interrupt burner operation, and then restored to the original setting on an annual basis. The test of the HL should be recorded in the log. The bacteria blamed for the deaths of several members of the American Legion in Philadelphia are frequently found in water supplies. When exposed to warm water in a confined environment, they can flourish. They are not the only type that can cause problems. The interface in the hot water storage tank always contains a level of water at the optimum temperature for that bacteria to grow and multiply. Sample the water from the interface for presence of Legionella at quarterly intervals after initial startup. If none is discovered, sample the water annually thereafter. Annual testing should coincide with heavy rains in the summer, where the bacteria are most likely to enter the system.

The process of checking for Legionella consists of drawing a sample and sending it to a laboratory for analysis. It requires a water sampling connection installed in the storage tank at the location indicated, which is just below the level of TC1. If the sample connection is above the return line inlet, it should penetrate the tank as shown to ensure a sample of the interface is drawn. To ensure the operating personnel are not exposed to the bacteria (in the event that it is there), they should wear protective equipment recommended for this operation. A sample bottle should be placed such that the sample piping extends into the bottle to the bottom in order to minimize splashing and generating aerosols while sampling. The sample should be drawn in the late afternoon or early evening, when demand is normally low and immediately after the pump and boiler start operating (when the interface is near the level of the sample line). If the laboratory test indicates Legionella is in the interface, it should be flushed from the storage tank. Connect a hose to the sample valve outlet and extend it into a drum containing sufficient sodium hypochlorite (Clorox™) to supertreat a drum full of water. Turn off the pump circuit breaker immediately after it starts to prevent pump and boiler operation temporarily. Then, after a few minutes of drawing hot water from the building system, open the sample valve and close the pump circuit breaker. When hot water is flowing to the drum, the sample valve can be closed because the complete interface was flushed to the drum. Repeat the procedure until a laboratory test of the interface does not show Legionella.

Even if Legionella does form in the storage tank interface, it should not contaminate the hot water delivered to the building unless the storage tank temperature is too low or hot water demands result in all the storage in the tank being consumed. In the latter case, the interface flows into the building's hot water distribution system. Operating the system to maintain hot water in storage at 180°F for more than one half hour should kill all bacteria except what is in the interface. Blending valves should be installed to provide the maximum 120°F water for hand washing, bathing, etc.

Thermal shock is observed by anyone pouring liquid into a glass of fresh ice. The ice cracks instantly, even when the liquid is very close to freezing. Iron, steel, and brass boiler parts are more malleable and slightly stronger than ice. Thus, the effect is not as dramatic, but it does happen. Boiler damage due to thermal shock is normally the result of repeated heating and cooling cycles. Damage occurs when the metal is overstressed because the surface is cooled or heated at a rate that exceeds the heat flow through it. As a result, one surface is at a different temperature than the one opposite to it. The differences in thermal expansion result in compressive stress at the hottest surface and tensile stress at the coldest surface. When the difference in stress reaches the breaking point of the metal, then tiny micro cracks form in the colder surface. Repeated exposure to the heating and cooling expands the cracks until leaks are evident. Thermal shock can also be associated with rapid changes in firing rate, but most service water heaters are designed to accommodate the changes associated with their on and off operations.

One would think that a hot water heater, with normal temperature differentials of 140°F, would be damaged regularly by thermal shock, if even smaller temperature differentials are a problem. They generally do not because the overall temperature differential is distributed along the length or height of the boiler. The boiler in Figure 4-9 would normally have 40°F water entering the bottom (at T1) and 180°F water leaving the outlet (at T2), with the temperature between those two levels varying almost linearly from top to bottom. The high temperature differentials, between the products of combustion and the water in the boiler, do not produce a significant temperature difference across the thickness of the metal because the heat flows through the metal much faster than through the thin film of flue gas between the metal and the products of combustion. The temperature differential across the metal is normally less than 30°F. Thermal shock occurs when a liquid in contact with the metal is quickly displaced by other liquid at a temperature significantly lower or higher than the original liquid. The direct contact with the metal parts and turbulence associated with the rapid replacement of the liquid heats or cools the metal surface rapidly, faster than the heat transfer through the metal itself.

What can cause thermal shock? In one case, the temperature control was different. Instead of installing a temperature switch that penetrates the storage tank at a level above the water inlet (as shown in Figure 4-9), the contractor provided a "strap-on" aquastat. That is a temperature switch with a bare thermal sensing bulb that is simply clamped to the outside of a tank or pipe to sense the temperature. In that case, the bulb was clamped to the pipe where the cold water enters the tank. Each time, the system filled the storage tank until hot water flowed out of the storage tank and into the piping and into the bottom of the boiler for a short period of time. Then the temperature controller finally responded to the change from cold to hot water. When the circulating pump started again, the hot water was immediately displaced by cold water. The thick metal at the bottom of the boiler was repeatedly subjected to swings between hot and cold water entering the boiler, which resulted in cracks around the bottom of the boiler shell.

Simply heating hot water is not as simple as it sounds. There is even an unusually different attitude about scale formation among people who maintain these devices. They manage to get away with a considerable amount of scale because the water temperatures are so low. It is a common practice to allow scale to build in one of these heaters (keep in mind that it cannot be treated because it has to be potable where someone could drink it) until the noise from the loose scale (lime deposits) rattling in the bottom of the heater, where steam is forming under the material and then collapsing as it contacts the colder water, can be heard. Since the water is not concentrated in a service water heater, it might not be expected to form scale, except under unusual conditions. Yet, it happens regularly. It is not uncommon for scale to form on the heat transfer surfaces to the point that the heater capacity becomes less than the demand. At that point, the demand for hot water cannot be met. At one location, the solids content of the water was so high that a mere 6°F increase in water temperature was all that was required for scale formation. The best solution for these applications is water softeners. That is not always accepted. Be prepared to clean a service water heater regularly as part of its maintenance when the calcium and/or magnesium content of the water is high (hard water). In particular, calcium salts exhibit what is known as inverse solubility. Most salts become more soluble as the temperature of the solution increases. However, calcium salts become less soluble as the temperature of the solution increases. Further, calcium bicarbonate is soluble in water. Upon heating, the bicarbonate breaks down and calcium carbonate is formed, which is very insoluble. These processes produce boiler scale. A water softener will substitute sodium ions for calcium ions, converting the dissolved salts into sodium carbonate, sodium bicarbonate, and sodium sulfate. All of these salts are much more soluble than the calcium salts and will remain in solution despite the water temperature being increased, which is the goal of the hot water heater. The downside is that the ion exchange resin in the water softener has to be either regenerated or replaced. Thus, there is a maintenance cost tradeoff between cleaning and softening that is dependent upon the calcium content (hardness) of the water source.

WASTE HEAT SERVICE

In these boilers, the cost of fuel, the single largest cost for any other kind of boiler plant, is essentially zero! Heat is being recovered from another heat stream and being used to make steam for plant use. That one great benefit also encourages to put up with some unique and, sometimes, hazardous flows that contain the heat that is extracted with the boiler. One of the most hazardous cases is a sulfur dioxide stream from firing pure sulfur to make sulfuric acid. Knowing about problems with sulfur in conventional fuels should provide for an appreciation of the special requirements for one of those boilers.

A waste heat boiler will always have a lot more heat exchange surface than a fired boiler because there is no radiant heat transfer. It is safe to assume that a waste heat boiler will have twice the heating surface of a conventional boiler for the same capacity. It is not uncommon to encounter a waste heat boiler with finned tubes to provide additional heating surface. This extended surface means the boilers will not have twice the number of tubes. Depending on the source of the heat, the boiler can incorporate an economizer section to preheat the feed water and can be of once-through design. The materials of construction may include materials that do not conform to the requirements of the Rules for Construction of Heating Boilers (Section IV of the ASME Boiler and Pressure Vessel Codes) or Rules for Construction of Power Boilers (Section I) because the liquids or gases that are the source of the heat would destroy those materials. In those cases, the boilers are constructed in accordance with the Rules for Construction of Pressure Vessels (Section VIII) as an "unfired boiler," which allows use of exotic materials including stainless steels, Inconel, and others. These boilers come in a variety of sizes and configurations that are so variable that there is no describing them all. Their operation varies significantly depending on the conditions of the fluid flow stream that the heat is coming from. Common examples include fume hoods for steel making and heat recovery steam generators (HRSGs) for gas turbine (GT) exhaust.

A low water cutoff is a required element for any boiler. They should always be provided on waste heat boilers unless the temperature of the fluid stream is less than about 750°F, where the metal will not overheat. In one system, the contractor installed a waste heat boiler connected directly to the exhaust of a steel annealing furnace. The exhaust heating stream was at about 1800°F. The new boiler was melted down two days after installation because the water source failed. If the temperature is high enough, there should always be a way of diverting the waste heat stream to prevent overheating the boiler. In some cases, there is no diversion of the waste heat stream. However, it may be possible to add air to dilute the hot stream until the boiler metal can withstand the temperature. With those exceptions, any waste heat boiler should be treated like a normal boiler.

ONCE-THROUGH BOILERS

Once-through boilers are more common in Europe. In the commercial to industrial range, there are a limited number of manufacturers. There are many electric utility boilers that have been designed for once-through operation. The boiler feed pump(s), or a boiler water circulating pump, is used to produce the differential pressure required to force the water and steam through the boiler. Similar to HTHW generators, these boilers can contain multiple circuits of tubing with inlet and outlet headers. Many have one single circuit to maintain a point of conversion from saturated to superheated steam. Feed water is forced into the tubes and steam, which may be saturated or superheated, leaves the tubes. All electric utility boilers of this design generate superheated steam. Needless to say, those boilers must use absolutely pure water because any dissolved solids would build up in the boiler tubes as scale.

Once-through boilers that generate saturated steam also incorporate a steam and water separator at the outlet so that only steam is delivered to the facility. The water that exits the boiler with the steam is separated and returned back to the inlet of the circulating pump. A safety feature on these boilers (that does not exist on other types of boilers) consists of a switch activated by the thermal expansion of one of the boiler tubes. If there is insufficient water flow through the tube, its coil expands as it is heated to trip the switch. Flow has to be proven before lighting a fire in one of these boilers. Treat the low water flow switch with the same concern and testing as a low water cutoff.

ENGINES AND EMERGENCY GENERATORS

The most common engine in a boiler plant is normally the emergency generator. Environmental regulations, developed under Environmental Protection Agency's (EPA) Tier 4 requirements, have changed the procedures for operating engines. Those requirements apply to all land-based equipment, whether stationary or mobile. Some operators are bound to be familiar with requirements for adding diesel exhaust fluid (DEF) into a separate reservoir of a diesel engine powered pickup truck. New emergency generators and stationary diesel engines will also have that requirement. They will be fitted with catalyzed diesel particulate filters (CDPFs) that capture the engine's particulate emissions, which are principally unburned hydrocarbons. Then they burn off the carbon while simultaneously reacting to break down nitrogen oxides using a catalyst alone or the DEF. To achieve the emission limits now established, engines will also have control features added to augment positive crankcase ventilation (PCV) and exhaust gas recirculation (EGR) valves. Because the implementation of Tier 4 regulations are progressing, there will likely be additional modifications to normal engines that will become normal in the coming years. Once again, read the instruction manual. Additional environmental requirements may be dictated by the US EPA regulations under the reciprocating internal combustion engine (RICE) maximum achievable control technology (MACT) rules. Be sure to understand all of the requirements, including reporting and documentation.

It is not necessary to describe putting an engine driven generator online or disconnecting it from the line because the operation is basically the same as it is for steam turbine (ST) powered generators. The only concern with an engine is that operators and automatic controls can connect an engine driven generator before it is warmed up. Some instability will result in fluctuating frequency and voltage. Except for actual emergency operation, it is advisable to allow the engine to be idle for a period of time so that it reaches normal operating temperatures before putting it online.

Engines can be used to power pumps, fans, blowers, and any number of pieces of mechanical equipment associated with a boiler plant, in addition to the more common emergency generator. Engines have hazards associated with them that make them almost as dangerous as a boiler. Blow by, which is the term used to describe the leakage of a fuel air mixture in an engine cylinder past the piston rings and into the crankcase, combined with the aerosol produced by the mechanical agitation of the lubricating oil, can be counted on to produce a combustible, hence explosive, mixture in the crankcase. Those spring loaded covers over the access openings for the crankcase have the dual purpose of first relieving "puffs" and then immediately closing to prevent the admission of air which could produce a larger puff or a true explosion, as air is added to the fuel. Recall that a puff is simply an explosion that did not do any damage. It is a very strong warning that a more extensive explosion could follow. Normally, the blow by consists principally of the products of combustion. Thus, the crankcase should contain a reasonably inert gas.

Of the two types of engines, diesels are the most common in boiler plants. Occasionally, there might be a gasoline engine that introduces additional concerns. Gasoline should not be stored in a house or attached garage. There may be a hot water heater in the garage. Note that it is installed on a platform. All this is because gasoline evaporates at normal atmospheric temperatures. The vapor is heavier than air. If gasoline vapors leak from a storage container, or an automobile, and mix with the air in the space, it will produce a combustible mixture. That mixture will gravitate to the lowest space in the garage. The solutions to these problems in a home consist of elevating the hot water heater and not storing gasoline inside, even if it is in an approved container. When it is necessary for the use of portable gasoline engines in a boiler plant, it is not always possible to keep the engine outdoors. It is, however, normally possible to take the engine powered equipment outside for filling and starting. During normal operation, when gasoline is drawn from the tank, a small orifice in the top of the tank admits air and no gasoline vapor is emitted from the tank. Operation of an engine inside a building, without providing additional openings for ventilation, may result in that engine running fuel rich. Its exhaust is discharging gasoline vapor into the building. Make sure there is adequate ventilation for those engines. Whenever possible, provide means to discharge the exhaust outdoors.

Procedures for starting an engine should always include checking the lubricating oil level first. The reason for checking it every time before starting is that someone else could have decided to change the oil, drained it, and failed to get around to refilling the engine. Yes, that has happened thousands of times. When lead acid batteries are used to power a starter for the engine, it is also a good procedure to check the battery water level as well. The typical emergency generator has radiator cooling. The level of the coolant in the radiator needs to be checked. Otherwise, the cooling systems for the engine must be in operation before starting. Then, whenever possible, after starting the engine, wait for it to warm up before connecting to the load. Closely monitor cooling water and lubricating oil temperatures after connecting to the load to ensure that they stabilize before leaving the engine. Frequently, engines are started with compressed air or hydraulic fluid pressurized by compressed air. Restoration of the pressure should be checked before leaving the engine.

During normal operation of an engine, check for proper lubricating oil level and pressure drop through air filters and fuel filters. Then log cooling and lubricating oil pressures and temperatures in every shift. Some engines do not permit checking the oil level while they are operating. They have to be shut down to check the oil level. Regardless, listen to detect any abnormal sounds and their meaning.

Remember that the air used for combustion must come from outdoors. Some air filter sensors (these usually provide a green or red indication) are simply connected to the air piping between the filter and engine. On large engines, it would be preferable to have the sensor detect the difference between outside air and the piping. In that way, failure of the automatic louvers, and the like, to operate properly would produce a red indication.

It is always a good idea, when shutting down an engine, to remove the load, if that is possible. Allow the engine to cool down from load conditions. Alternatively, reduce the load to cool the engine. Stopping it immediately could result in carburization of oil in areas where the oil is trapped, instead of flowing through. When the engine is equipped with an auxiliary oil pump, it should be started before shutting down the engine and stopped only after the oil temperatures entering and leaving the engine differ by less than 5°F. Then, treat the cooling system similarly. Don't shut it down until after the auxiliary oil pump is shut down.

Like boilers, idle engines deserve to be checked regularly. Potential problems include loss of starting power, freezing cooling water, and moisture condensing and accumulating in the crankcase to name a few. It is not unusual for emergency generators to be treated as peaking generators. In other words, the owner of the emergency generator has a contract with the local electric utility that pays the owner to operate the generators during periods of peak electric load. This arrangement is beneficial to both parties. The agreement does normally include provisions for penalty, should the owner fail to operate the equipment when it is requested by the utility. While someone could imagine that peak electric loads occur in the summertime, it is also possible in the winter. At one facility, on an extremely cold winter night, when temperatures dropped below 11°F, the utility called, but the engines would not start. It turns out that there were no heaters for the fuel oil or the lubricating oil. It took a combination of halogen lights and welding machines to heat the oil and get the engines running. Even if there is no agreement to deal with, consider the potential for an extremely cold start of the emergency generator and be assured that it will start and run and carry a load. Check the engine when it is not in use.

GAS TURBINES

There are essentially two types of GTs: aeroderivative and heavy duty. An aeroderivative GT is designed using a jet engine as the basis. The aeroderivative turbine assembly consists of a typical jet airplane engine with its exhaust discharging into a few stages of GT blading connected to the power output shaft. The heavy duty GT is designed specifically to be operated on the ground. This GT does not have the same weight considerations as a jet engine that has to be mounted on a jet aircraft. All GTs consist of a combination of compressor, burners, and turbine mounted on the same shaft as the compressor and connected to a generator or other piece of driven equipment. There are many models of GT. They are all different. Once again, the instruction manual becomes very important.

Unlike an internal combustion (IC) engine, most GTs are cooled by air flow and the lubricating oil. The lubricating oil is, in turn, cooled by air or water. Many of the aeroderivative plants have staged oil cooling, where the airplane engine's oil is cooled by another lubricating oil system. For air cooling, some of the air compressed by the compressor flows around the combustion chamber(s) for cooling and then mixes with the burner exhaust to reduce the temperatures entering the turbine section. More modern engines have cooling air from the compressor diverted through the turbine blades to cool them. The air leaves the blades through multiple orifices at the tip of the blade to flow over the external surface of the blade to help protect the blade from the hot flue gases.

It is not uncommon to require fuel gas compressors for natural gas fired turbines. Despite the large supply of natural gas, many GTs exist that are designed to fire fuel oil as well as natural gas. The oil nozzle may look a little strange (Figure 4-10).

GTs, like engines, need to be started. At minimum, they must be brought up to a speed that produces enough compression of the air to produce a differential pressure across the turbine that, when combined with burning of the fuel, can generate enough power at the turbine shaft to power the compressor and the rotating resistance of anything attached to the shaft of the GT. This is normally done with an electric motor, or the electric generator attached to a GT can function like a motor to get it started. Before everything starts turning, lubrication has to be established and proven.

Figure 4-10. Gas turbine oil nozzle.

Maintenance of lubricating oil temperatures during GT operation is very important because the oil can be too cool as well as too hot. Other temperatures that should be monitored closely are the compressor exhaust temperature, the burner exhaust temperature, and the turbine exhaust temperature. The first two must be within limits for continued operation. The last one should be as low as possible for maximum turbine efficiency.

During operation, the pressures, temperatures, and gas velocities through the GT vary as indicated in Figure 4-11. Atmospheric air accelerates toward the inlet of the compressor (1-2), accompanied by a drop in static pressure created as some of the static pressure of the air is converted to velocity pressure. In the compressor (2-3), the pressure is increased and the velocity decreases, as the air is compressed, because compression reduces the volume of the air. The temperature of the air rises from the work of compression, just as it does in diesel engines. Between the compressor and the burner (3-4), some pressure is lost as the air is directed through different passages with provisions to increase turbulence for mixing with the fuel. Velocity there can decrease when some of the compressed air is extracted to feed through the turbine blades for cooling them. Inside the burner, (4-5) pressure remains about the same and velocity and temperature increase. Mass flow also increases a small amount due to the addition of the fuel. Air to fuel ratios are considerably higher than those in a boiler because all the heat from combustion is retained in the flue gases. More air is required to absorb the heat and keep the temperature from exceeding the tolerances of the metals of construction. Older turbines use 400% excess air. More modern designs continue to reduce that value (about 200%) to reduce the energy used in the compressor for higher GT efficiency. With the lower levels of excess air, the temperatures entering the power turbine are increased, which also increases the overall efficiency (Ideal Gas Law). In the turbine (5-6), pressure and temperature drop as the gases convert energy to the turbine blades to rotate the shaft. Velocity increases as the gases expand through the turbine stages. A GT turbine only has reaction blading (see ST descriptions) because there are no nozzles. Finally, at the outlet of the GT (6-7), an evase (gradual expansion of ductwork) serves to convert some of the velocity pressure to static pressure. That allows

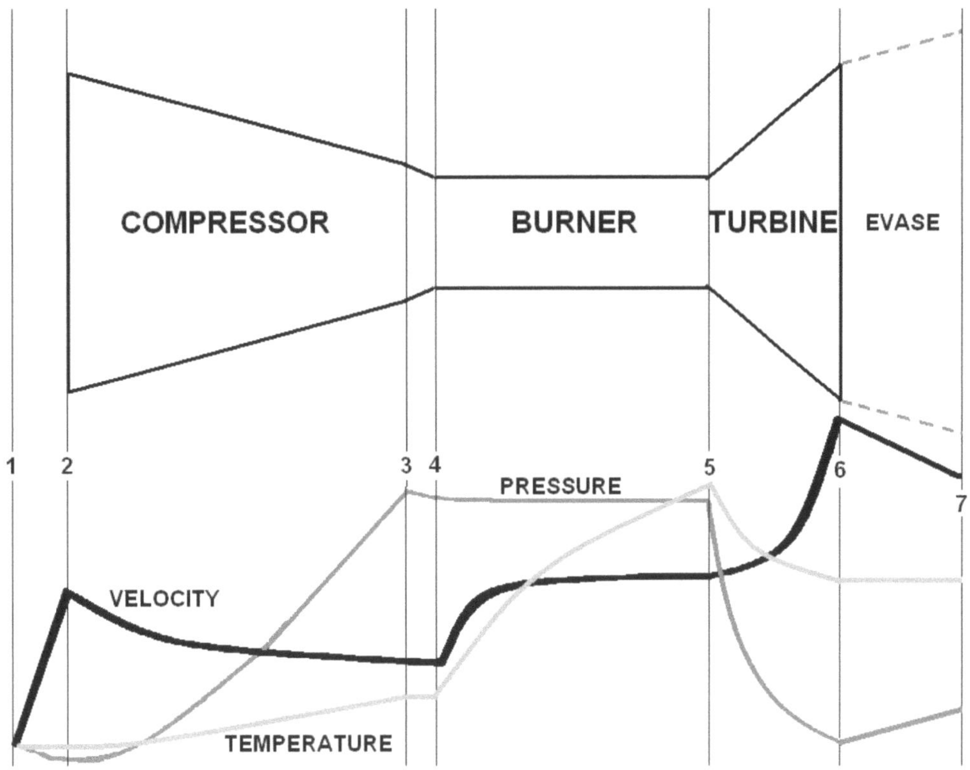

Figure 4-11. Gas turbine PTV curves.

for a slightly higher differential pressure across the turbine. The turbine outlet pressure reading can be lower than atmospheric for turbines without heat recovery and lower than the inlet pressure to heat recovery equipment.

A process used to boost power output of GT generators, when summer electrical loads are high, is misting. Immediately after the turbine inlet filters, spray nozzles inject water into the air stream entering the compressor. This cools the inlet air by evaporation of the spray water to produce increased mass flow of combustion air along with the additional moisture. The cooler air reduces the work load on the compressor. In turn, the GT can burn a little more fuel and produce more power. Overall, the performance improves as the compressor is essentially driven by the turbine. With lower compressor work, the power output increases.

HRSGs AND COMBINED CYCLE PLANTS

Heat recovery steam generators are referred to as HRSGs. HRSGs are connected to the discharge of a GT. It is not uncommon for people to use that abbreviation when discussing simple waste heat boilers. That usage avoids the term "waste." The term "waste" has a particular meaning at the US EPA. Treating "wastes" has particular requirements under EPA rules. The use of HRSG avoids any confusion. The typical HRSG captures the heat remaining in the exhaust of a GT generator and supplies superheated steam to feed a ST that is also connected to a generator. That is the typical combined cycle plant. A typical combined cycle plant consists of GT driven electric generators with an HRSG at the exhaust of each and an ST generator using steam from the HRSG. The steam cycle is very similar to the typical utility plant like the one shown in Figure 1-8. However, the utility sized combined cycle plant typically uses two or three GTs, each feeding an HRSG and then supplying one ST generator.

A single stage combined cycle plant is schematically represented in Figure 4-12. The ST and condenser, plus associated auxiliaries, are the same as any steam plant. Each GT drives an electric generator (GEN). The hot exhaust is converted to superheated steam in the HRSGs. The steam powers the ST to drive its generator. The ST exhaust is condensed in the condenser (COND) and is pumped back to the HRSGs to be converted to steam again. A cooling tower is normally provided to cool the condenser water. Condenser water can also come from an adjacent body of water. However, most regions of the US require a cooling tower. Multiple GTs and HRSGs permit turndown without sacrificing efficiency. Multiple

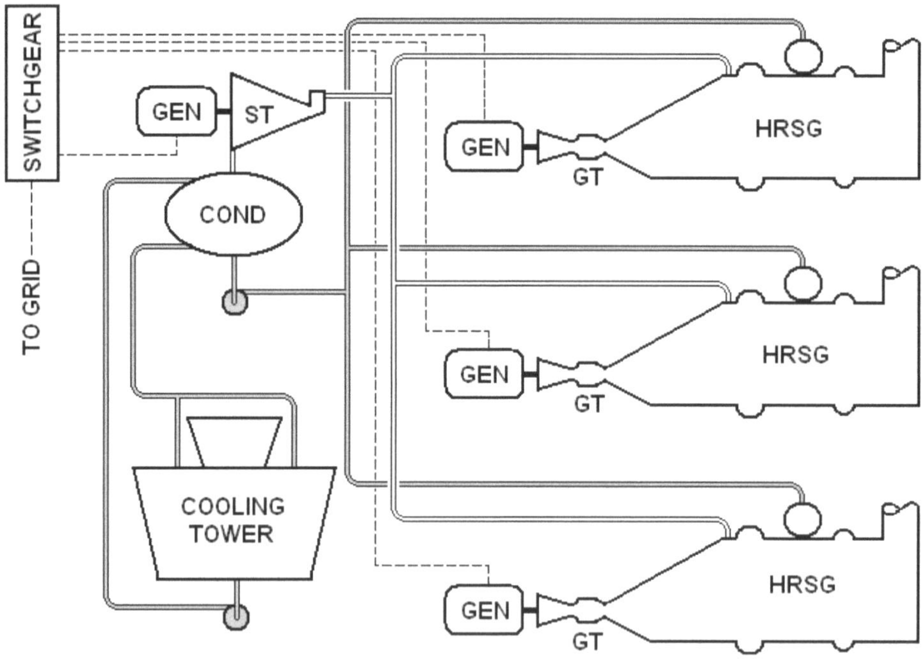

Figure 4-12. Combined cycle plant schematic.

nozzle blocks on the ST allow it to operate at varying loads with minimal loss in efficiency.

The HRSG is, for all practical purposes, a boiler plant combined in one package and is shown schematically in Figure 4-13. It consists of (1) a GT exhaust duct that conveys the hot turbine exhaust to, and distributes the exhaust into, the first stages of the boiler portion. A duct burner (2) can be used to provide additional heat to the boiler section. The superheater (3) raises the temperature of the steam, as described for utility plants, before continuing to the ST. An HRSG could also contain a reheater. The boiler (4) generates the steam that supplies the superheater from the feed water heated by the high pressure economizer (5). After leaving the economizer, the flue gases enter a lower pressure boiler (6) with its own economizer (7). That boiler generates steam for the deaerator (8), which is typically mounted integral to the HRSG and is confused by some as another steam drum. Before discharging up the stack (9), the flue gas could be exposed to a condensate heater that preheats a mixture of returned condensate and makeup water. Feed water is fed to the two boiler stages by independent feed water pumps. The main feed to the main boiler can also be heated by an external high pressure feed water heater using extraction steam from the ST.

Note that some of the boiler tubes and all economizer tubes are finned. The provisions of fins are dependent upon the operating flue gas temperatures (omitted where they would be burned off) and possibly eliminated on the last row of boiler tubes, which serve as downcomers. On large HRSGs for large size utility plants, even the high temperature surfaces are finned, using alloy steel for the fins. The use of fins reduces the number of tubes by about 20%, with commensurate cost savings. Since the fuels are primarily natural gas or light oils, the fins do not fill up with ash.

HRSGs are optimized in design to recover the heat from the GTs within the smallest possible footprint and lowest combined cycle cost (initial and operating costs combined over the life of the unit). The designer tries to extract as much heat as possible from each stage, which requires close attention to the pinch points. Pinch points are where the flue gas temperature approaches the temperature of the fluid that it is heating. If the two temperatures are the same, heat will not flow from the flue gas to the water or steam. A temperature difference is required to transfer the heat. The temperature profile of a typical HRSG is shown in Figure 4-14 and the pinch points are indicated. This also explains why there may be two or three pressure levels in an HRSG. In order to recover the most amount of heat energy from the GT exhaust, the gas temperature has to be reduced to typical stack temperatures (around 300°F). As the gas temperature drops, the temperature difference between the gas and boiler water decreases. At some point, it gets sufficiently close to the boiling point that heat transfer is limited. In order to further cool the gas, water can be boiled at a lower pressure, which translates to a lower temperature. Thus, multiple pressure levels can be used.

Since the HRSG is designed for these operating conditions, there are situations during startup and shut down that require special control, especially to prevent steaming in the economizers (which would result in water hammer in them along with potential damage to the economizers, feed water control valves, and the drum internals). At times, it is necessary to maintain flow through the economizers by returning some of the economizer outlet water to the deaerator. High pressure in the deaerator is normally prevented by a relief valve

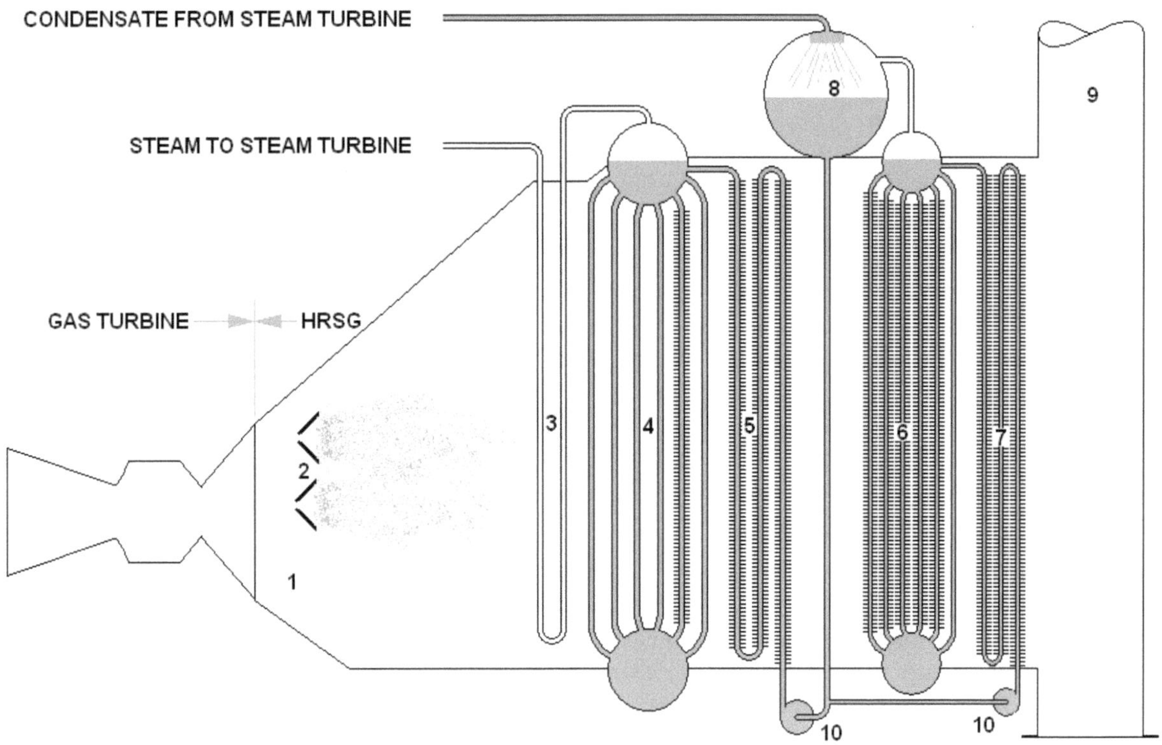

Figure 4-13. HRSG schematic.

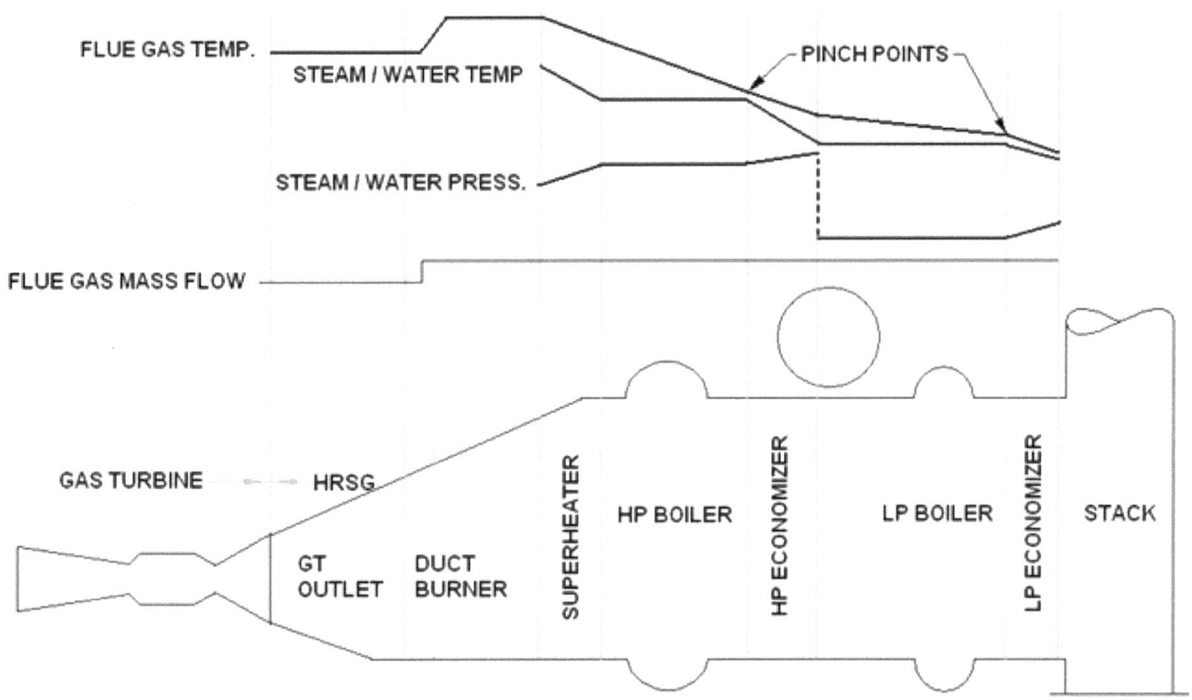

Figure 4-14. HRSG PTV profile.

and vacuum by a vacuum breaker. Return of some feed water to the deaerator from the low pressure economizer can also be used to help stabilize the deaerator pressure.

The duct burner is primarily to be used for cogeneration plants. Often times, the steam demand does not match the electrical demand. In order to provide more flexibility to the system, the duct burner allows for additional steam generation to match the steam demand. A duct burner is not normally used for electric generation. The fuel that is fired in the duct burner does not go through the combined cycle system. Thus, the efficiency of the overall plant is reduced. The only exception would be in the case where the power price on the grid is exceptionally high. In that case, the efficiency is less important than the ability to produce more power. Thus, some cogeneration plants that went out of business due to a plant closure, or the like, were sold to power companies. These plants were converted to combined cycle plants. The duct burners were retained. A somewhat larger ST was added to take advantage of the potential for extra steam generation. However, with the very low price of natural gas in the US, the exceptionally high price on the grid has been minimal. Since the duct burner is in a stream of hot gas with 200%–400% excess air, the only air added to it may be some for the ignitor. Being a radiant source in an otherwise simply convection flow, the duct burner principally increases the gas temperature to the HRSG. The mass flow through the HRSG is not altered as much by simple addition of fuel. The impact on steam generation is not as great as increasing the firing of a conventional boiler. In any case, the HRSG must be designed to accommodate the higher temperature gas and additional mass flow.

The literature (Power Magazine among others) is constantly revealing new techniques and features of combined cycle plant operation. Their use is increasing. The percentage of generation coming from natural gas has increased from less than 20% in the early 2000s to nearly 40% two decades later. It is not just the availability of natural gas. It is also the fact that these plants are more efficient, converting more of the energy in the fuel to electricity than conventional boiler plants. An advanced steam plant has a full load efficiency of about 40%. A combined cycle plant (clean and new) will have a full load efficiency in the range of 52%–54% on a higher heating value (HHV) basis at International Standards Organization (ISO) conditions (59°F, sea level, 1 atmosphere, 60% relative humidity). The GT industry quotes their performance at ISO conditions because the performance of the GT is much more sensitive to site conditions than a steam plant. GTs also come in specific designs (like most engines). The GT industry also uses the lower heating value (LHV) to quote their equipment performance. That is because a higher efficiency number can be calculated using LHV and ignoring the water vapor that is produced with the combustion of natural gas. Thus, industry advertisements will claim efficiency numbers of nearly 60%–61%. The wise operator will read the fine print. That figure is based on LHV and ISO conditions (clean and new). On a hot day, with a higher ambient temperature and higher relative humidity at even a modestly higher altitude, the performance of the GT will fall off significantly (6%–10% drop in efficiency).

In a combined cycle plant, the GT is started up first. The exhaust gases can be used to start warming up the boiler. When the GT is up to speed, it can be synchronized to the grid. At that point, the ST can be prepared and warmed up. Once the ST is up to speed, it can be synchronized to the grid. After that, the plant can be put in load following mode and controlled by demand from the grid. Because a gas fired combined cycle plant can swing load more readily than a large, base loaded coal or nuclear plant, the combined cycle plant is typically used for intermediated load following.

Chapter 5

Refrigeration and Air Conditioning

Many boiler plants have refrigeration equipment within the plant, attached, or in an adjacent room or building. Often, the boiler plant operator is expected to operate and help maintain that equipment as well. Several states and municipalities include questions on refrigeration systems and air conditioning in their boiler operator exams. If this comes as a surprise, it should not. That is because refrigeration is very similar to boiler operation. Refrigeration is simply transferring heat from one location to another. Advances in Supervisory Control and Data Acquisition (SCADA) systems and centralized monitoring of facility equipment have also served to drop responsibility for the operation of the systems in the lap of the boiler operator because the access terminals are in the boiler control room.

Refrigeration has many applications. Principal uses include food preservation and comfort cooling. Refrigeration can also be used in production processes. Of the latter, the most interesting one is making soap. One system was designed to operate at −25°F at Lever Brothers. At that time, Dove bars were made in a plant in Baltimore. The bars are formed in a press at a speed so fast that the heat of working the bars would melt the soap without the cooling. Air conditioning was used for a new Hystron fibers plant that made Trevira™, a synthetic silk. The synthetic fibers needed air cooling, as they were squeezed out of the plastics extruder quickly enough to prevent their sticking to each other as they were twisted into a single strand. A ton of refrigeration is a rate, not a quantity. The derivation of the value is the amount of heat that can be absorbed by one ton of ice melting in 24 hrs. It takes 80 Btu's (British thermal unit) to melt a pound of ice. Thus, a ton is $80 \times 2000 \div 24 = 12{,}000$ Btu/hr. Just as electric power is quantified in kilowatt-hours (Kwhrs), a quantity of refrigeration is ton hours.

REFRIGERANTS

What is a refrigerant? Why not just say Freon™? There are many refrigerants, including materials not normally thought of as refrigerants, such as water. The one to be used in a particular system is dependent upon a number of factors. There are also many regulations concerning refrigerants. The Montreal Protocol banned many traditional refrigerants as ozone depleting substances. These include chlorofluorocarbons (CFCs), halons, and hydrochlorofluorocarbons (HCFCs). It means that replacing a refrigerant in an older system could present a problem.

Of course, water can be a refrigerant. The first air conditioning system installed by Carrier himself was in a printing plant in Brooklyn, NY, USA, in 1902. It was a steam powered system using water as a refrigerant. All that is needed to get water cold enough for air conditioning is to expose it to a vacuum at 29.75 inches of mercury (Hg) and it will boil at 40°F (check the steam tables in the Appendix). Carrier created a vacuum over a tank of water using steam jets to remove the evaporated water plus any air and non-condensable gases. The water in the tank boiled until the heat absorbed by the evaporating water cooled the remaining water to 40°F. The cold water was pumped to coils that absorbed the heat in the air in the printing plant and simultaneously condensed excessive moisture in that air to lower the humidity in the printing plant. The slightly warmer water was then returned to the tank where the heat absorbed from the printing plant air was removed by evaporating some of the water. Why was a printing plant air conditioned first? It is simply because paper properties change with temperature and moisture content. Keeping it consistently spaced between printing rolls containing different colors of ink, a process called registration, required maintaining the temperature and humidity in the plant.

Many other fluids can be used as refrigerants but may have undesirable properties. Some known to cause cancer were used in early systems and others, such as propane, are not used because they are flammable or heavier than air. Along with water, another natural substance used as a refrigerant is ammonia. It is an excellent refrigerant, but it is also flammable, corrosive, and poisonous. If it leaks from the refrigeration system, it can catch on fire, ruin materials stored in the refrigerated spaces, and kill people. That is why it is only used in facilities that are not normally open to the public and monitored constantly for leaks.

Then, in the latter half of the 20th century, it was discovered that Freon (a trade name of the DuPont Company that became more or less a generic label for refrigerants) was not as safe as was thought. It was not flammable or poisonous. It was lighter than air, meaning it would not pool inside a basement like propane does. Instead, being lighter than air, it rose up in the atmosphere until it got to the stratosphere. There it disassociated, releasing chlorine atoms that break down ozone. Located about 30 miles above the surface of the earth, the stratosphere contains an ozone layer that works like a filter to screen out harmful ultraviolet (UV) radiation from the sun. By the time the chemistry of the action of most refrigerants (CFCs) on the ozone layer was recognized, the ozone depletion was so extensive that there was a "hole" in the ozone layer over the South Pole. International efforts to limit the release of refrigerants and development of new refrigerants that lack chlorine have reduced the damage to the ozone layer and it is now recovering. In 2016, HCFCs were added to the ban as they were determined to be greenhouse gases (GHGs).

A part of the solution is heavier than air refrigerants that will not reach the upper atmosphere. However, they can accumulate in spaces containing the refrigeration equipment, with the potential of suffocating people. If working in a plant that uses one of those refrigerants, find a self-contained breathing apparatus (SCBA) in a cabinet immediately inside the door. Don't fail to read the instructions and, preferably, try out the apparatus. Know how to operate and use it. Encourage other operators to do the same because they may have to use it to reach a colleague and remove them from the space in the event of an incident.

Figure 5-1 is a tabulation of some common refrigerants including replacements for refrigerants that are no longer allowed to be used in new equipment. Note that there are color codes for most refrigerants. Containers of the refrigerant are painted that color in an effort to ensure a system is not charged with the wrong refrigerant or refrigerants are not mixed. Most refrigerant systems will fail to operate properly when exposed to a mixed refrigerant. There are, however, some refrigerants which are intentional mixtures. These are known as binary refrigerants, a mix of two. R-134a has become the standard for all applications that used to use R-12. The orange containers of R-11 and containers of R-123 are actually drums because they are a liquid at normal atmospheric conditions. In use, the evaporator for those refrigerants operates in a vacuum. R-728 is liquid nitrogen. The current replacement for R-22 is R-416 A. Environmental Protection Agency (EPA) has banned a number of refrigerants starting in January 2021. Under its alternatives program (Significant New Alternatives Policy (SNAP)), it is listing R-452 B, R-454 A, R-454 B, R-454 C, and R-457 A. Check to make sure that an approved replacement is used when replacing a refrigerant. R-404a is one of the binary refrigerants that have a range of saturation temperatures between two conditions known as dew point and bubble point. Because it is a mixture of two refrigerants with different saturation points, one portion boils off at a slightly lower temperature than the other, creating bubbles of the lower temperature refrigerant. Then that vapor is superheated a few degrees until the other portion reaches its saturation temperature. The span between the dew point and the bubble point is referred to as "temperature glide." Handling of binary refrigerants requires special procedures. Once again, it is important to read that instruction manual. There are also ternary blends, three refrigerants blended together. Some blends are azeotropic. They have no temperature glide and act like a single refrigerant.

Venting of refrigerants, other than nitrogen and carbon dioxide, is illegal. Perpetrators can be subjected to a fine of up to $27,500.00. That is one of the many things to be learned about handling refrigerants. Take a course in that activity to obtain a certification as a technician for handling refrigerants to comply with Section

Application	Temperature Range	Refrigerant	Color	Uses	Replacement
High temperature	50°F to 90°F	R-22	Green	Comfort Cooling	?
Medium temperature	32°F to 50°F	R-12	White	Coolers	R-134a
Low temperature	0°F to 50°F	R-502	Purple	Freezers	R-507 & R-404a
Ultra-low temperature	-50°F to -250°F	R-13 & R-503		Cascade	R-403B
Cryogenics	-250° to -400°	R-728		Expendable	None
Low-pressure chillers	42°F to 55°F	R-11	Orange	Chilled water	R-123

Figure 5-1. Information on some common refrigerants.

608 of the Federal Clean Air Act of 1990. That class will teach about stratospheric ozone depletion, the Clean Air Act, refrigerants and oils, and recovering refrigerants. An examination with a passing grade is required to receive certification. This course is recommended for those operators who will handle or operate refrigeration equipment.

Oil for refrigeration equipment is special because it has to work in an environment where it is surrounded with refrigerant. The lubricating oils for refrigerants can be specific to a particular refrigerant or to a group of them. Extreme care is required in handling them. Almost all refrigerant oils are hygroscopic. They absorb water, and since water is not desirable in a refrigerant system, it is important that those oils are handled and stored in sealed containers to avoid absorbing water from the air.

THE REFRIGERATION CYCLE

As shown in Figure 5-2, the basic refrigeration cycle has four stages.

The four basic stages are evaporation, compression, condensation, and throttling.

Evaporation

Evaporation, as the title implies, is converting a liquid to a gas, just like making steam in a boiler. The refrigerant absorbs heat by evaporating (boiling). That is how the evaporator removes heat from another substance to cool it. The principal difference is that the refrigerant is exposed to pressures in the evaporator where the saturation temperature of the refrigerant is lower than the substance to be cooled. Heat flows from the substance to be cooled through the tubing or casing of the evaporator into the refrigerant, causing the refrigerant to boil. Despite some common descriptions, a refrigeration system does not push heat. It always flows in one direction, from a higher temperature to a lower temperature. By reducing the evaporator pressure, the boiling temperature of the refrigerant is lowered until it is colder than the substance to be cooled. Refrigeration is not only associated with changing the temperature of a substance, it can also change the state from gas to liquid and liquid to solid. Air is cooled and dehumidified. Water can be cooled and frozen.

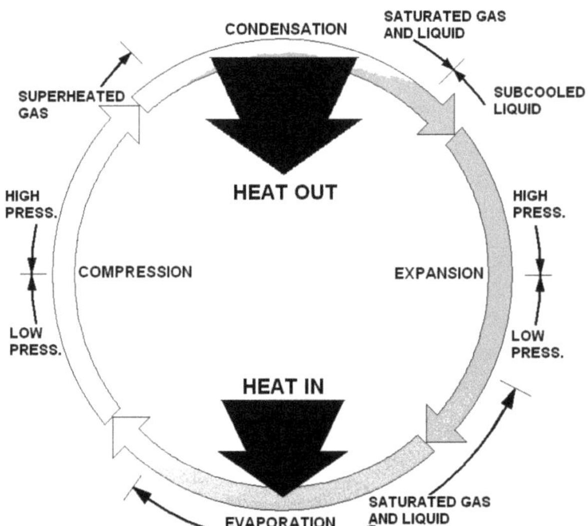

Figure 5-2. Basic refrigeration cycle.

Compression

It might be argued that the Carrier system did not compress the refrigerant. It pulled a vacuum. Another look at the steam tables should reveal that the volume of the pound of steam at 40°F is 2423.7 ft.^{3} per pound of steam. At 0 psig (pounds per square inch gauge), it is 26.8 ft.^{3} per pound. The steam jets compressed the steam removed from the tank at a compression ratio of 90 to 1. Other liquids that boil at lower temperatures under pressure can be used in order to avoid creating a vacuum. Absorption chillers, discussed later, do use water for a refrigerant. With any fluid, increases in pressure will increase the saturation temperature of the fluid. In order to eliminate the heat that was absorbed in the evaporator, the fluid is compressed to raise its saturation temperature to a value higher than the substance the heat is rejected to. The many forms of refrigerant compressors are covered later in this chapter. It takes power to compress a gas. The energy of compression is added to the gas. Therefore, heat is also added to the refrigerant in the compression stage.

Condensation

The compressed gas loses heat to a substance at a temperature lower than the saturated temperature of the gas when it condenses. The heat is transferred from the refrigerant through tubing or shell into the substance that absorbs the rejected heat. At this point, it should be evident that all that is happening is the transfer of heat. That is the same process used within the steam and water cycles. The construction of condensers is dependent upon the refrigerant and the substance accepting the rejected heat.

Throttling

Since condensers operate at a pressure higher than evaporators, a means of controlling the flow of refrigerant from the high pressure condenser to the low pressure evaporator is needed to ensure that the compressed hot gases do not enter the evaporator. It is also necessary to ensure that the liquid does not leave the evaporator to enter the compressor because, unlike a gas, a liquid is not very compressible. Therefore, a means of controlling refrigerant flow between the condenser and the evaporator is required. Many throttling systems utilize a means of sensing temperature to control the refrigerant flow, while others use liquid level. Two are simple orifices. When the fluid passes through the orifice, it expands and cools. As no work is performed and no heat is transferred, the gas molecules give up internal energy to move apart from each other at the lower pressure. This phenomenon is the Joule–Thompson effect, investigated in 1852 by British physicists Joule and Thompson. This process reduces the temperature of the substance and converts it back to a gas to cycle back to the evaporation step. The rapid evaporation is also called "flashing."

REFRIGERANT SUPERHEAT AND SUB-COOLING

Just as steam is superheated by adding heat after all the liquid has been boiled away, refrigerants are superheated. The heat added in the compression stage by the work done on the refrigerant to compress it increases the refrigerant superheat. In refrigeration, the temperatures are simply much lower. The refrigerant in an evaporator will continue to absorb heat by evaporating the liquid until it runs out of liquid. In most evaporators, the throttling controls limit the admission of liquid such that none exists in the evaporator. Because the gas temperature is still lower than the substance being cooled, the gas absorbs heat. That additional heat raises the temperature of the gas, superheating it. Terms need to be clarified so that there is no confusion. A superheated gas is a gas at a temperature higher than its saturation temperature at the pressure of the gas. Superheat is the difference in temperature between the temperature of the gas and its saturation temperature. A refrigerant at 83 psig, where the saturation temperature is 40°F, may be a superheated gas at 50°F. The gas has a superheat of 10°F. The same refrigerant compressed to 275 psig, where the saturation temperature is 120°F, can be a superheated gas at 130°F. Once again, it has 10°F of superheat. Normally, the superheat after the compressor is greater than 10°F.

While superheating occurs in the evaporator, sub-cooling occurs in the condenser. In the condenser, all of the gaseous refrigerant is condensed. The liquid is still exposed to temperatures lower than the saturation temperature. Heat leaves the liquid so that its temperature is lower than saturation. Liquid leaving a condenser is normally a sub-cooled liquid. The refrigerant described in the previous paragraph could be condensed at 275 psig and then cooled to a temperature of 110°F. In that case, the sub-cooling is 10°F.

It is important to understand these conditions because lack of superheat or too much superheat can result in damage to compressors. Inadequate sub-cooling can result in poor operation or damage to the throttling device. Inadequate sub-cooling also permits liquid flashing to vapor before the refrigerant reaches the evaporator, thus restricting the flow of the liquid refrigerant.

Sub-cooling is accomplished by removing heat from a liquid that has just condensed from a vapor to a temperature lower than the saturation temperature. Sub-cooling is normally accomplished in the condenser. Other provisions and equipment can also be utilized to sub-cool the liquid. A system can, for example, include a heat exchanger that uses the cool vapor coming out of the evaporator to help cool the liquid refrigerant. The cool gas is simply superheated a little more, as it absorbs the heat from the liquid. It is also possible to have an independent sub-cooler to cool the liquid refrigerant.

Sub-cooling is necessary to ensure that the liquid remains a liquid until it has exited the throttling device. If the liquid is not sub-cooled sufficiently, it can breach saturation conditions as the pressure in the liquid tubing drops due to friction or changes in elevation. When that happens, small bubbles of refrigerant vapor form in the liquid line. Because the vapor uses up more space than the liquid, this results in increased velocity in the liquid line for more pressure drop and generation of more vapor. The lower pressure reduces the performance of the throttling device.

Measuring Superheat and Sub-Cooling

Unlike a boiler plant, refrigeration systems do not always have pressure gauges and thermometers mounted in the piping. When there are gauges or thermometers, it is an indication that the reading should be monitored and logged regularly. Usually, a set of portable gauges, called a refrigeration gauge set, as shown in Figure 5-3, and a clip on thermometer are used to determine superheat and sub-cooling.

The gauge set in a facility will normally have the saturation temperatures printed on the gauge face along with the pressure indication. If the temperature is not indicated on the gauge set, then a refrigerant card, like the one shown in Figure 5-4, will be needed. Most of those cards will list several refrigerants to avoid having

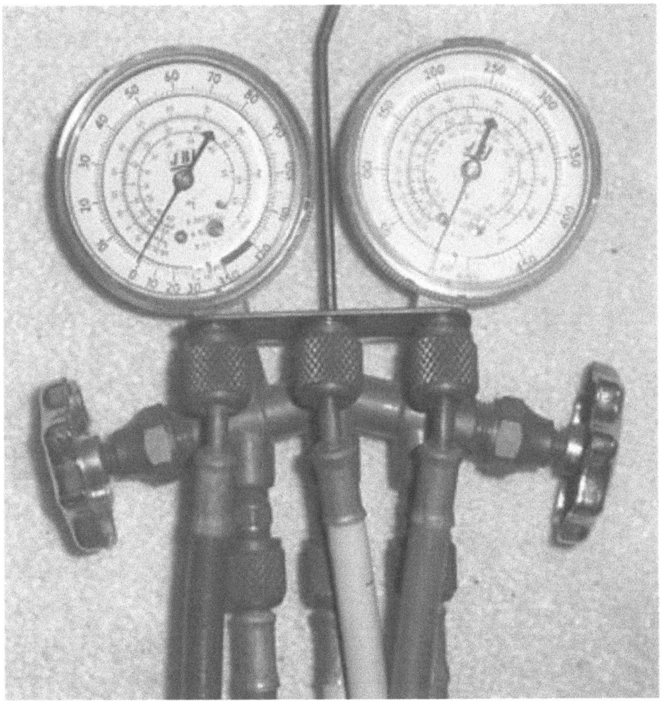

Figure 5-3. Gauge set.

to carry a lot of them. Remember to use the right column. The gauge set has three hoses connected and terminating in special fittings that match fittings provided for their connection on the refrigeration equipment. Sometimes, but not always, the fittings are keyed to avoid an improper connection. The color of the hoses is always a key as to how they should be connected. The blue hose fitting gets connected to the vapor line at the outlet of the evaporator. The red hose is connected to the hot line somewhere between the compressor and the condenser. Those hoses lead to the gauges to provide an indication of pressure and the corresponding saturation temperature of the refrigerant at the point where the hose fitting is connected. The third hose, typically yellow, the one in the middle, is separated from the other two by the valves and is typically used for a connection to a bottle of refrigerant. That hose is used to add refrigerant to, or remove refrigerant from, the system. The color of that hose can be keyed to the refrigerant. Gauge sets exist that can only be used for specific refrigerants. R134a is one. Their fittings will also be specific matches to prevent contamination of a refrigerant system by a different refrigerant. The clip on thermometer should be clipped onto the refrigerant piping as close as possible to the pressure gauge connection so that a true superheat, the difference between saturation and actual temperature, can be determined.

EVAPORATORS

Evaporators absorb heat by evaporating the refrigerant. Ignoring, for the moment, the pressure drop associated with the friction of flow through the evaporator, the heat is absorbed at a constant temperature, the saturation temperature of the refrigerant at the pressure in the evaporator. With the exception of chillers, all evaporators are intentionally operated to evaporate all the liquid before it leaves the evaporator. Once all the liquid is evaporated, the refrigerant, as a gas, absorbs additional heat to become superheated. The absorption to superheat is intentional in order to prevent liquid from entering the compressor and damaging it. The superheat is maintained automatically in most systems by the throttling device.

The evaporator most people can relate to is the one that is visible in the window air conditioning unit when the cover is removed and filtered in order to clean the filter. There are rows of tubing wrapped with aluminum that looks shredded or the tubing runs through sheets of aluminum. Those sheets are fins or extended surface. Heat transfer from a solid to a boiling liquid and through highly conductive metal is much faster than it is between a metal surface and air. Therefore, an evaporator to cool air is designed with a lot of surface area where the metal contacts the air. An evaporator in a window air conditioner cools the air and removes moisture by condensing it. The velocity of the airflow over the cooling surface has to be restricted to prevent blowing the droplets of moisture off the coil and into the circulating fan or the room. The additional surface compensates, to a degree, for the loss of turbulence that would be provided by higher airflow rates. The fins are also designed to accelerate the drainage of the condensed water off the heat transfer surfaces.

The typical water cooler in an office has an evaporator that is a coil of two tubes, one inside the other, with the refrigerant inside the inner tube and the water for drinking in the annular space between the two tubes. Heat transfer rates from metal to flowing water are higher than for a gas like air. Fins are not economically justified on something so small that only requires a little extra length of tubing to make up the difference between heat transfer to boiling liquid and heat transfer to flowing liquid.

Likewise, a walk-in freezer would probably not have fins on the cooling coils. That is because fins would hinder defrosting and maintaining cleanliness. A freezer's cooling coils not only condense moisture from the air but also freeze it on the coil surface. As that ice builds

SPORLAN — PRESSURE-TEMPERATURE CHART *at SEA LEVEL*

PSIG	TEMPERATURE °F						
	YELLOW	GREEN	GREEN	BLUE	PURPLE	TEAL	WHITE
	REFRIGERANT - (SPORLAN CODE)						
	12 (F)	22 (V)	124 (M)	134a (J)	502 (R)	AZ50 or 507 (P)	717 (A)
5*	-29	-48	3	-22	-57	-60	-34
4*	-28	-47	4	-21	-55	-58	-33
3*	-26	-45	6	-19	-54	-57	-32
2*	-25	-44	7	-18	-52	-55	-30
1*	-23	-43	9	-16	-51	-54	-29
0	-22	-41	10	-15	-50	-53	-28
1	-19	-39	13	-12	-47	-50	-26
2	-16	-37	16	-10	-45	-48	-23
3	-14	-34	18	-8	-42	-46	-21
4	-11	-32	21	-5	-40	-44	-19
5	-9	-30	23	-3	-38	-41	-17
6	-7	-28	26	-1	-36	-39	-15
7	-4	-26	28	1	-34	-38	-13
8	-2	-24	30	3	-32	-36	-12
9	0	-22	32	5	-30	-34	-10
10	2	-20	34	7	-29	-32	-8
11	4	-19	36	8	-27	-31	-7
12	5	-17	38	10	-25	-29	-5
13	7	-15	40	12	-24	-27	-4
14	9	-14	41	13	-22	-26	-2
15	11	-12	43	15	-20	-24	-1
16	12	-11	45	16	-19	-23	1
17	14	-9	46	18	-18	-21	2
18	15	-8	48	19	-16	-20	3
19	17	-7	49	21	-15	-19	4
20	18	-5	51	22	-13	-17	6
21	20	-4	52	24	-12	-16	7
22	21	-3	54	25	-11	-15	8
23	23	-1	55	26	-9	-14	9
24	24	0	57	27	-8	-12	11
25	25	1	58	29	-7	-11	12
26	27	2	59	30	-6	-10	13
27	28	4	61	31	-5	-9	14
28	29	5	62	32	-3	-8	15
29	31	6	63	33	-2	-7	16
30	32	7	65	35	-1	-6	17
31	33	8	66	36	0	-4	18
32	34	9	67	37	1	-3	19
33	35	10	68	38	2	-2	19
34	37	11	69	39	3	-1	20
35	38	12	71	40	4	0	21
36	39	13	72	41	5	1	22
37	40	14	73	42	6	2	23
38	41	15	74	43	7	3	24
39	42	16	75	44	8	4	25
40	43	17	76	45	9	4	26
42	45	19	78	47	11	6	28
44	47	21	80	49	13	8	29
46	49	23	82	51	15	10	31
48	51	24	84	52	16	11	32
50	53	26	86	54	18	13	34
52	55	28	88	56	20	15	35
54	57	29	90	57	21	16	37
56	58	31	91	59	23	18	38
58	60	32	93	60	24	19	40
60	62	34	95	62	26	21	41
62	64	35	97	64	27	22	42
64	65	37	98	65	29	24	44
66	67	38	100	66	30	25	45
68	68	40	101	68	32	26	46
70	70	41	103	69	33	28	47
72	71	42	104	71	34	29	49
74	73	44	106	72	36	30	50
76	74	45	107	73	37	32	51
78	76	46	109	75	38	33	52
80	77	48	110	76	40	34	53
85	81	51	114	79	43	37	56
90	84	54	117	82	46	40	58
95	87	56	120	85	49	43	61
100	90	59	123	88	51	45	63
105	93	62	126	90	54	48	66
110	96	64	129	93	57	51	68
115	99	67	132	96	59	53	70
120	102	69	135	98	62	55	73
125	104	72	138	100	64	58	75
130	107	74	140	103	67	60	77
135	109	76	143	105	69	62	79
140	112	78	145	107	71	64	81
145	114	81	148	109	73	66	82
150	117	83	150	112	75	68	84
155	119	85	152	114	77	70	86
160	121	87	154	116	80	72	88
165	123	89	157	118	82	74	90
170	126	91	159	120	83	76	91
175	128	92	161	122	85	78	93
180	130	94	163	123	87	80	95
185	132	96	165	125	89	82	96
190	134	98	167	127	91	83	98
195	136	100	169	129	93	85	99
200	138	101	171	131	95	87	101
205	140	103	173	132	96	88	102
210	142	105	175	134	98	90	104
220	145	108	178	137	101	93	107
230	149	111	182	140	105	96	109
240	152	114	185	143	108	99	112
250	156	117	188	146	111	102	115
260	159	120	192	149	114	105	117
275	163	124	196	153	118	109	121
290	168	128	201	157	122	113	124
305	172	132	205	161	126	117	128
320	177	136	209	165	130	120	131
335	181	139	213	169	133	124	134
350	185	143	217	172	137	127	137
365	188	146	221	176	141	129	140
380	192	150	224	179	144	132	143
400	197	154	229	183	148	136	147

* Inches mercury below one atmosphere

SPORLAN — PRESSURE-TEMPERATURE CHART *at SEA LEVEL*

PSIG	TEMPERATURE °F											
	PINK		SAND		ORANGE		LIGHT BROWN		REDDISH PURPLE		BROWN	
	REFRIGERANT - (SPORLAN CODE)											
	MP39 or 401A (X)		HP80 or 402A (L)		HP62 or 404A (S)		407C (N)		FX-10 or 408A (R)		FX-56 or 409A (F)	
	DEW POINT	BUBBLE POINT	DEW POINT	BUBBLE POINT	DEW POINT	BUBBLE POINT	DEW POINT	BUBBLE POINT	DEW POINT	BUBBLE POINT	DEW POINT	BUBBLE POINT
5*	-23		-59		-57		-48		-54		-22	
4*	-22		-58		-56		-45		-52		-20	
3*	-20		-56		-54		-42		-51		-19	
2*	-19		-55		-53		-39		-49		-17	
1*	-17		-54		-52		-36		-48		-16	
0	-16		-53		-50		-34		-47		-15	
1	-13		-50		-48		-31		-44		-12	
2	-11		-48		-46		-29		-42		-9	
3	-9		-45		-43		-27		-39		-7	
4	-6		-43		-41		-24		-37		-5	
5	-4		-41		-39		-22		-35		-2	
6	-2		-39		-37		-20		-33		0	
7	0		-37		-35		-18		-31		2	
8	2		-36		-33		-17		-29		4	
9	4		-34		-32		-15		-27		6	
10	6		-32		-30		-13		-26		8	
11	8		-30		-28		-12		-24		9	
12	9		-29		-27		-10		-22		11	
13	11		-27		-25		-8		-21		13	
14	13		-26		-23		-7		-19		14	
15	14		-24		-22		-5		-18		16	
16	16		-23		-20		-4		-16		17	
17	17		-21		-19		-3		-15		19	
18	19		-20		-18		-1		-13		20	
19	20		-19		-16		0		-12		22	
20	21		-17		-15		1		-11		23	
21	23		-16		-14		3		-9		25	
22	24		-15		-12		4		-8		26	
23	25		-14		-11		5		-7		27	
24	27		-12		-10		6		-5		29	
25	28		-11		-9		8		-4		30	
26	29		-10		-8		9		-3		31	
27	30		-9		-7		10		-2		32	
28	32		-8		-5		11		-1		34	
29	33		-7		-4		12		0		35	
30	34		-6		-3		13		1		36	
31	35		-5		-2		14		3		37	
32	36		-4		-1		15		4		38	
33	37		-2		0		16		5		39	
34	38		-1		1		17		6		40	
35	39		0		2		18		7		41	
36	40	30	0		3		19		8		43	
37	42	31	1		4		20		9		44	
38	43	32	2		5		21		10		45	30
39	44	33	3		6		22		11		46	31
40	45	34	4		7		23		12		47	32
42	46	36	6		9		25		13		48	34
44	48	38	8		10		26		15		50	36
46	50	40	10		12		28		17			38
48		42	11		14		30		19			39
50		44	13		15		31		20			41
52		45	14		17		33		22			43
54		47	16		19		34		24			45
56		49	18		20		36		25			46
58		50	19		22		37		27			48
60		52	20		23		39		28			50
62		53	22		25		40		30			51
64		55	23		26		42	30	31	30		53
66		56	25		27		43	32	32	32		54
68		58	26		29		44	33	34	33		56
70		59	27		30		46	34	35	34		57
72		61	29		31	30	47	36	37	36		58
74		62	30		33	32	48	37	38	37		60
76		64	31		34	33	49	38	39	38		61
78		65	32	30	35	34	51	39	40	40		63
80		66	34	31	36	35		41	42	41		64
85		69	37	34	39	38		44	45	44		67
90		73	40	37	42	41		46	48	47		70
95		76	42	40	45	44		49	50	50		73
100		78	45	43	48	47		52		52		76
105		81	48	45	50	50		54		55		79
110		84	50	48		52		57		57		82
115		87		50		55		59		60		84
120		89		53		57		62		62		87
125		92		55		59		64		65		89
130		94		57		61		66		67		92
135		96		60		64		69		69		94
140		99		62		66		71		71		96
145		101		64		68		73		73		99
150		103		66		70		75		76		101
155		105		68		72		77		78		103
160		108		70		74		79		80		105
165		110		72		76		81		81		107
170		112		74		78		82		83		109
175		114		75		80		84		85		111
180		116		77		81		86		87		113
185		117		79		83		88		89		115
190		119		81		85		90		91		117
195		121		82		87		91		92		119
200		123		84		88		93		94		121
205		125		86		90		95		96		123
210		127		87		92		96		97		124
220		130		91		95		99		100		128
230		133		94		98		102		104		131
240		136		97		101		105		107		134
250		140		99		104		108		109		137
260		143		102		107		111		112		141
275		147		106		111		115		116		145
290		151		110		114		119		120		149
305		155		114		118		123		124		153
320		159		118		122		126		128		157
335		163		121		125		130		131		161
350		167		125		129		133		135		165
365		170		128		132		137		138		169
380		174		131		135		140		141		172
400		178		135		139		144		145		177

* Inches mercury below one atmosphere

Figure 5-4. Saturation card.

up, it serves as an insulator, restricting the flow of heat. Timers, or devices monitoring either the superheat of the refrigerant or temperature of the fins on the coil, are used to initiate a defrosting cycle. Defrosting of freezer coils can be accomplished in two ways. One way is using an external heating source, normally found in frost-free refrigerators, where an electric heating element is used to melt the ice on the coils while the refrigeration compressor is temporarily shut down. If the freezer side of the refrigerator is opened during defrosting, a hissing sound of melted water dripping on the heating element can be heard. In larger commercial freezers, defrosting is achieved by first shutting off the supply of liquid refrigerant to the evaporator and then admitting hot refrigerant gas coming from the discharge of the compressor. Drain pans under those coils are designed to separate the ice that falls off the coils from the melted water so that the water is drained. Sometimes, maintenance is required to remove the ice. Fins on the cooling coils of a walk-in freezer would not permit effective defrosting because they would hold the ice in place. However, fins on the cooling coils of a refrigerator are effectively defrosted by the heater. In both the walk-in freezer and the refrigerator, flow of air over the coils while defrosting is reduced by shutting off air circulating fans.

It pays to check the condition of the drains that remove the water melted off the coils after a defrost cycle to ensure they are flowing clear. The drains in a typical household freezer drop the water into an external pan positioned at the condenser cooling air outlet where the heated air can evaporate the water. The tubing between the pan installed in the freezer to collect the melted ice water and the external pan or drain can get plugged with chunks of ice to block flow. That results in the pan inside the freezer overflowing to make the bottom of the freezer into an ice rink, a dangerous condition for a walk-in freezer.

Heat transfer between metal and flowing water occurs at a rate much faster than that between metal and flowing air. Therefore, fins are rarely used on tubing for cooling water or freezing water. Chillers frequently have a form of fin, which will be covered later. When making ice, the standing water and the ice itself restrict heat flow. That requires additional surface which has to be smooth to permit removal of the ice. Ice makers can use one of the two methods described for defrosting freezers to melt a thin layer of the ice at the surface of the freezer to allow removal of the ice.

There may be a P trap, often used in plumbing, in the piping of an evaporator or the gas piping between the evaporator and the compressor. It has a specific purpose when the compressor is higher than the evaporator. It accumulates coalesced oil until the trap is full and the gas pushes it up the piping to the compressor. In most systems, some lubricating oil circulates and those little traps make sure it does not get trapped in the system and always returns to the compressor.

FREEZING AND ICE STORAGE

The ability to make ice and freeze things is one of the major contributions to better health and longer living that are enjoyed today. On the other hand, freezing of water has contributed to considerable damage to refrigeration equipment, piping, and other equipment that contains water. Ice expands as it cools. That is different from most substances because they shrink as they cool and expand when heated. It is a good thing in actuality because, by expanding, the density of ice is less. Frozen water floats. If it were not for this phenomenon, the earth could be an ice planet because the sunlight could not get at the ice to melt it. An operator has to be conscious of the fact that ice expands when cooled in order to prevent damage to piping and equipment by preventing its temperature dropping below 32°F or by draining the equipment and piping.

When making ice, the operator also has to consider the fact that ice is a solid and heat transfer through ice is by conduction only. The rate of flow of heat through a solid, like ice, is proportional to the thickness of the solid. The thicker the solid, the slower the heat transfer. Freezing water requires the removal of 80 Btu from each pound of ice. That ice will absorb the same amount of heat when it is melted. Ice, therefore, increases the capacity of the system to absorb heat within 1¼% of the mass of water. Storing ice, principally for air conditioning systems, permits the installation of smaller equipment for cooling chilled water that is used in most facilities. Ice storage can also significantly reduce electricity cost for cooling when time-of-use electricity rates are in effect by operating the ice making equipment at night, when electricity rates are lower. Proper use of ice storage to limit peak electrical costs, and to absorb heat loads that exceed the capacity of the chilled water system, can significantly reduce the owners' operating costs, while ensuring the occupants of the building are always comfortable.

Ice storage systems do not work like the icemaker in a home refrigerator. In order to absorb the heat, the ice and chilled water have to be in contact and the rate of heat transfer between the two is proportional to the surface area where the water contacts the ice. In other words, one cannot simply create one huge block of ice and then melt it. One system that works well consists of a heat exchanger enclosed in a tank. The outer surface of the tubing of the exchanger is exposed to water that flows through the tank. Ice is formed on the outside of the tubes by cooling using a refrigerant, brine, or glycol water mixture inside the tubes. The ice builds up on the surfaces of the heat exchanger when making ice and is melted by water pumped through the tank that chills that water. To ensure an even buildup of ice on the surfaces of the heat exchanger, air is blown into the tank to bubble up through the water, thereby increasing turbulence to help improve heat transfer.

COMPRESSORS

The compressor takes in the refrigerant after it leaves the evaporator and pumps it up to a higher pressure to raise the saturation temperature to a point where the heat absorbed by the evaporator can be dumped to another substance in the condenser. A low pressure with a correspondingly low saturation temperature is maintained in the evaporator because the compressor removes the refrigerant vapor from the evaporator. Work is performed on the refrigerant by the compressor to squeeze it up to a higher pressure. The energy of the compression increases the heat in the refrigerant, thus increasing its superheat. Because many compressor motors are cooled by the refrigerant, friction and heat from the motor windings typically add to the superheat as well. High evaporator pressures or high temperatures at the outlet of the compressor are indicative of inefficient compression. A noisy compressor, one that is noisier than normal, may indicate liquid flooding through the evaporator to the inlet of the compressor. It could also indicate loss of lubrication in the compressor.

Because a compressor is a piece of mechanical equipment, it requires lubrication. Since the moving parts are in direct contact with the refrigerant, lubricating oils for compressors are specific to the refrigerant used. With the exception of centrifugal compressors, some of the lubricating oil is broken up into small droplets forming an aerosol that travels with the refrigerant through the system. The oil is cooled and occasionally heated by the refrigerant itself. That means oil coolers are not always required.

There is no dipstick in a refrigeration compressor. Other means are required to determine the level of the oil. Some compressors, like the one in a window air conditioner or a car, do not require a lot of oil. Because the systems are sealed, an adequate amount of oil is assumed. The wise operator always looks around and under compressors, evaporators, condensers, and associated piping to detect any accumulation of leaking oil. First of all, it is an indication of a possible refrigerant leak. Second, it can indicate a potential failure of the compressor. Larger compressors are usually fitted with a sight glass in the compressor crankcase that permits the observation of the oil level. An idle compressor can have a high oil level due to accumulation of oil drained from connecting piping as well as portions of the compressor itself. Check the oil level with the compressor running to see if it returns to normal shortly after the compressor was placed in operation. Typically, the oil fills the cavities in the compressor that have drained back to the bottom and some of the oil is circulated into the system. If the level returns to normal, there is no need to drain the oil.

While it may seem strange, every refrigerant compressor has a heater in it. The purpose of the heater is to boil off refrigerant that condenses in the crankcase or oil sump. When the compressor is operating, the heater is normally turned off because the operation of the compressor heats the oil. When the compressor is idle, or out of service, the refrigerant tends to migrate to the compressor crankcase, condense, and mix with the oil. This can happen if the heater does not work. The result will be higher crankcase oil level. This situation is problematic because as soon as a compressor starts operating, the lower crankcase pressure drops below the saturation pressure of the liquid refrigerant in the oil and it boils. The boiling refrigerant mixes with the lubricating oil to create foam, which results in slugs of lubricating oil entering the valves and cylinders of the compressor. Under the right conditions, liquid refrigerant can be carried into the cylinders as well. Always check to see if the oil heater is working before starting a compressor. Check that sight glass immediately after starting the compressor for foaming. A small amount of foaming that does not raise the oil level out of the glass would be acceptable. Compressor controls are typically fitted with a timer to prevent the motor from starting after power is interrupted for sufficient time to allow the heater to boil off any liquid refrigerant that is migrated to the crankcase. That is why a home air conditioner compressor will not start right away after the thermostat has been over-adjusted to stop it.

The description for evaporators mentioned the traps for accumulating and reinjecting oil into the flow of refrigerant gas. That has the potential of sending a slug of oil into the compressor that could be carried directly into one of the cylinders, resulting in damage to the compressor. The device shown in Figure 5-5 is called

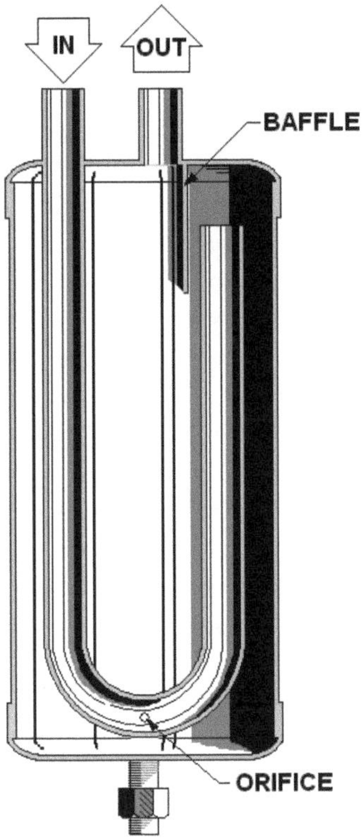

Figure 5-5. Suction accumulator.

a suction accumulator. It is installed in the vapor piping near the compressor inlet to protect the compressor from slugs of refrigerant or oil. A slug of oil, refrigerant, or a combination thereof is trapped in the accumulator. The small hole in the riser tubing allows a gradual reinjection of oil into the flow of refrigerant to return it to the compressor, and the superheat in the gas should vaporize any accumulation of liquid refrigerant. The operation of the accumulator means oil level can fluctuate. It is wise to observe the oil level for a while before reacting to a slightly higher or lower level by adding or removing oil.

Compressors are classified as open, hermetic, or semi-hermetic. Virtually, all small compressors are hermetic with the motor enclosed within the sealed compressor casing. Since the motor is enclosed within the compressor casing, the motor is cooled by the refrigerant flowing over it. A continuous flow of cool refrigerant while the compressor is operating is essential to prevent the motor from overheating. Small compressors like the one in a refrigerator, window unit, or residential air conditioners will not have means of monitoring conditions to detect a loss of refrigerant flow. However, another sense attributable only to a person can provide an indication of a problem: the sense of smell. If the motor is not cooled adequately, the higher temperature will result in vaporizing whatever managed to accumulate on the outside of the compressor casing, producing a hydrocarbon emission that can be recognized. Regular rounds by an experienced operator can usually result in detection of a problem and action to rectify it before the equipment fails even if there are no gauges or thermometers to read.

Medium sized and large compressors can be of any type, but, because most of them are constructed to permit maintenance, they are called semi-hermetic. Flanged and threaded connections permit opening the compressor enclosure for maintenance. Large compressors are either semi-hermetic or have the motor mounted outside of the compressor casing in which case they are referred to as open compressors. Because an automobile air conditioner is driven by a belt and pulley from the engine, it is an open compressor. An open compressor requires a shaft seal which can become a repeating maintenance problem. There are restrictions on the amount of refrigerant that can leak from a system, and shaft seals are normally the only source of leaks. If there is an open compressor with repeated seal failures, it is an indication that they are not installed properly, the compressor and motor are not in alignment, and/or the lubricating oil that lubricates and cools the seal has been interrupted.

Reciprocating Compressors

Reciprocating compressors are the most common type of compressors. The unit in a car, a window unit, a refrigerator, or a freezer typically contains two pistons and cylinders. Figure 5-6 is typical of reciprocating compressors used for home air conditioning units. There are compressors with as many as 16 cylinders, and units similar to the one in Figure 5-7(a) and (b), which has 8 cylinders. Note the head spring in the section which reveals that the head of each cylinder in these compressors is free to move within the cylinder, compressing that head spring whenever liquid might be drawn into the compressor cylinder. A louder "thump" (actually, it is more of a bang or rattle) is a sign that is happening to one of them. Reciprocating compressors function by the movement of the piston and cylinder to first increase the volume inside the cylinder, which allows the gas in the cylinder to expand until the pressure in that cylinder is lower than the pressure at the compressor inlet. Then, gas from the evaporator flows into the cylinder as the moving piston makes room for it. Gas enters the compressor cylinder through an intake valve or valves. A good description for the valves of a refrigeration compressor is that they look like popsicle sticks. They are guided within the casting of a cylinder head and are either a spring themselves or guided by springs that gently push them toward the seat. The flow of gas pushes the valve away from the opening and, when flow stops, the spring

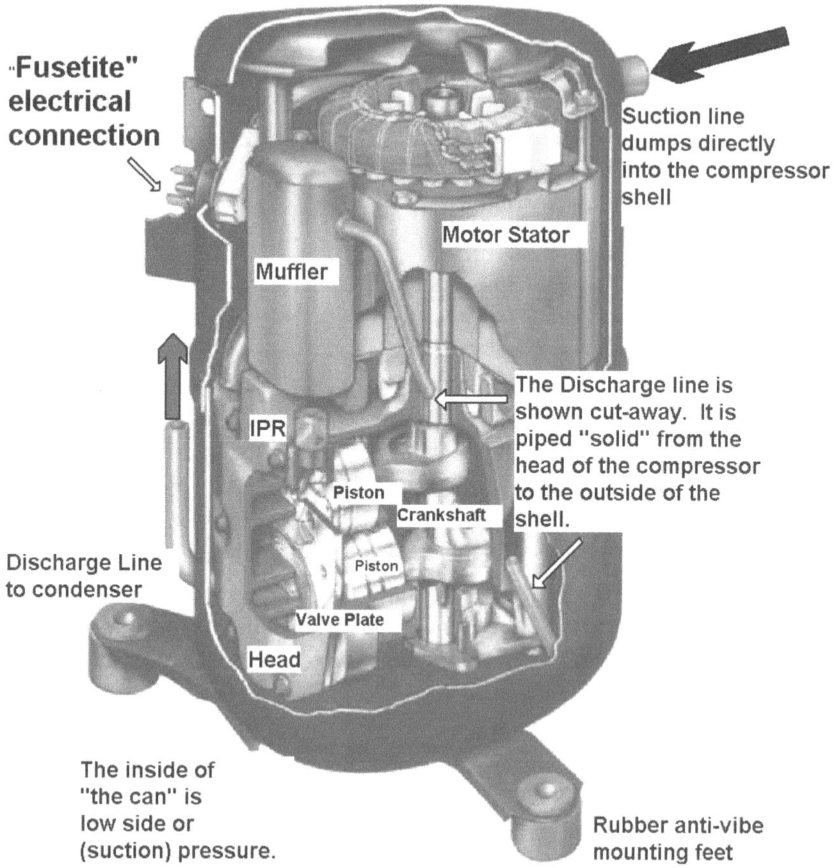

Figure 5-6. Small reciprocating compressor.

Figure 5-7(a). Eight-cylinder compressor.

force pushes the valve back onto its seat. As the piston approaches the bottom of its stroke, it slows down and then reverses due to the rotary motion of the piston connecting rod at the crankshaft. Then, as the piston rises in the cylinder, reducing the volume, the gas is compressed until it is at a pressure slightly higher than the pressure of the gas at the outlet to overcome the force of the spring on the valve and the discharge valve opens. For the remainder of the upward stroke, gas is pushed out of the cylinder into the discharge piping. The volume of gas delivered with each rotation of the shaft is not a function of the cylinder displacement because the gas has to expand and compress. The amount of gas pushed through the compressor is considerably less than the volume displaced by the piston in the cylinder.

Small compressors are started under load. To allow larger compressors to get up to speed before starting to pump refrigerant, unloading systems are used. Unloading a cylinder is accomplished with a pin that pushes up against each suction valve to hold it open. The pin is connected to a small cylinder containing a spring, as shown in Figure 5-8. When the compressor is shut down, the springs push the pins up to hold the valves open. After the compressor gets up to speed, the oil pump builds up pressure in the oil passages of the compressor and the small unloader cylinders to force the pins down and allow the suction valves to operate. When the unloaders operate, there will be a significant difference in the sound of the compressor as it comes up to speed. The unloaders allow a compressor to be powered by a standard duty motor instead of a high torque motor. Many modern compressors use a solenoid and spring to control the pins electronically.

In addition to reducing startup torque, unloaders allow multiple cylinder compressors to handle varying loads without starting and stopping. A controller senses suction pressure, loading and unloading cylinders as required to maintain a set suction pressure. With older hydraulic actuated unloaders, each one (sometimes pairs) can be set to operate at a different suction pressure allowing a wider range of evaporator temperatures or in a tighter range when temperature maintenance is more critical. If one is near the compressors regularly, one should be able to hear the unloaders cutting in and out. If one of two or more compressors is constantly running with fewer pulsing sounds, indicating the compressor is totally unloaded, check to see what happens if it is shut down to save energy and wear and tear on the compressor.

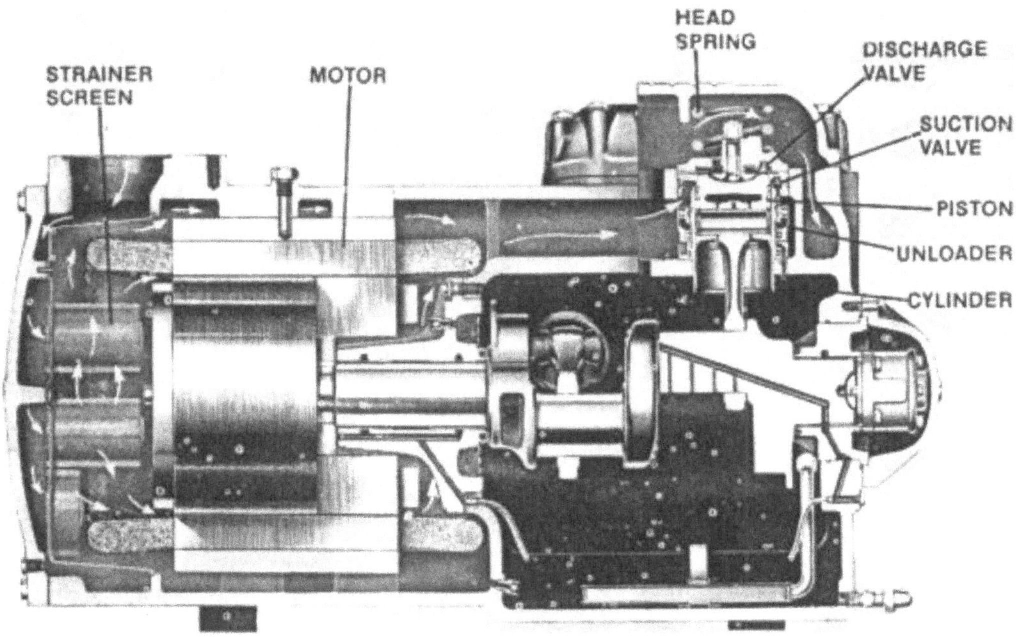

Figure 5-7(b). Eight-cylinder compressor section.

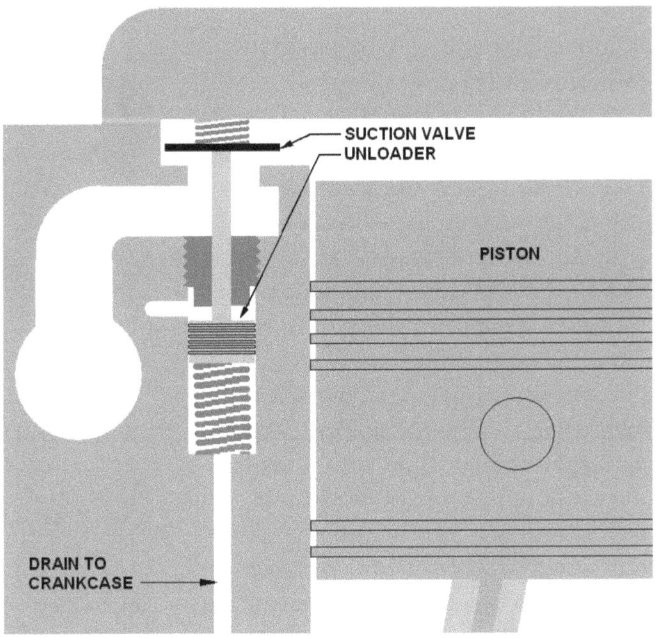

Figure 5-8. Unloader.

Vane Compressors

Vane compressors are used in automobile air conditioning and small systems with very low pressure differentials. The compressor shaft has vanes set in slots around its circumference that are spring loaded to press them against the compressor casing as shown in Figure 5-9. The eccentric setting of the shaft inside the casing changes the position of the vanes which slip in and out of the slots. The spaces between the vanes change volume as the shaft rotates. The space between each vane works almost like a cylinder in a reciprocating compressor.

Scroll Compressors

Scroll compressors are a relatively new technology dependent upon modern material technology. Compression is achieved in a scroll compressor by the two pieces shown in Figure 5-10, one of which is stationary, while the other wobbles eccentrically, creating crescent-shaped gaps between the walls that are moved along the scroll as the walls separate and then move together.

The result is pockets of refrigerant entering the scroll at the outside and being squeezed along the scroll to its discharge point in the center. The scroll pieces are mounted in the top of the compressor (Figure 5-11). Figure 5-12 shows a series of the changes in position of the two scroll pieces to demonstrate how the compressor grabs a volume of gas and gradually compresses it. Usually, a scroll compressor can be distinguished from a reciprocating compressor by shape, the scroll being taller and smaller in diameter for the same capacity. Scroll

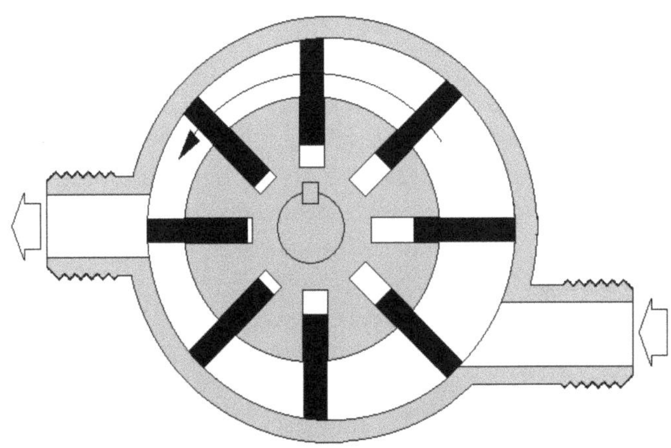

Figure 5-9. Vane compressor.

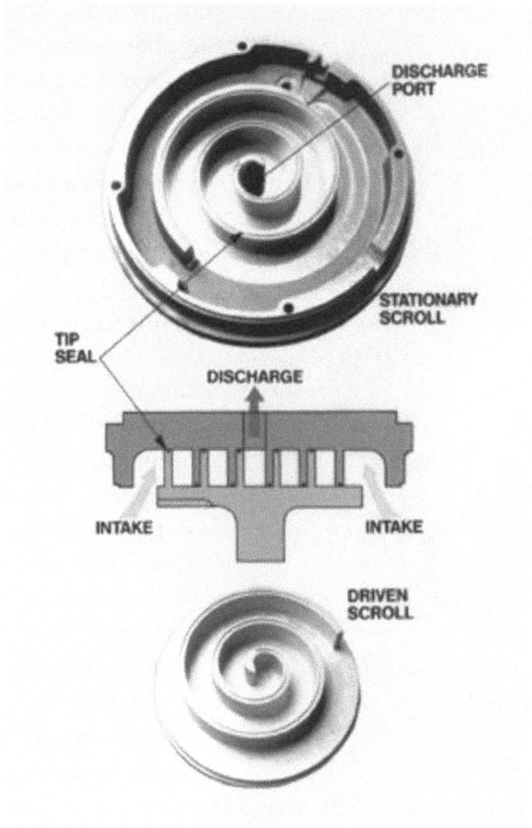

Figure 5-10. Scrolls.

Figure 5-11. Scroll compressor

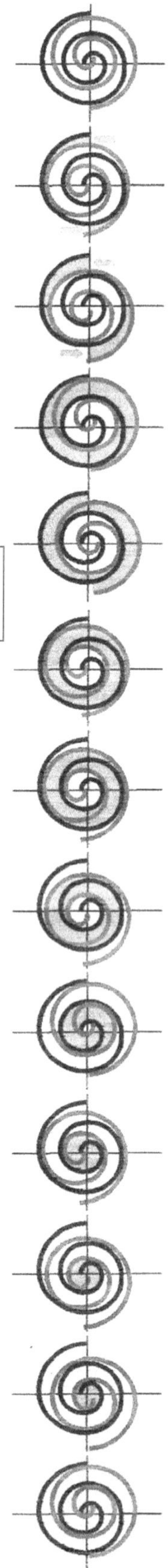

RIGHT: Figure 5-12. Scroll progression

compressors, at the time of writing this edition, are not made larger than 25 tons.

There is no way to unload a scroll compressor. Scroll compressors have an advantage of lower sound power levels because they have smoother operation. To a degree, scroll compressors can be turned down using variable speed drives or two speed motors for a better match to cooling loads. A home heat pump contains a two-speed compressor. In real cold periods, it is switching between high and low speed rather than starting and stopping.

Hot Gas Bypass and Defrosting

Many reciprocating and scroll compressors, especially the smaller ones, have on/off operation only. The typical refrigerator, freezer, icemaker, wine cooler, and window and auto air conditioners can be operated in this manner because swings in temperature can be tolerated. On occasion, the load can swing considerably and/or temperature deviations are unacceptable. A means is required to keep the compressor in operation constantly. That is because the delay time between starting and stopping is typically around 15 minutes for those designs of compressors to ensure the oil in the crankcase is free of liquid refrigerant. The solution to the problem is hot gas bypass.

Here is one situation where hot gas bypass to keep the compressor running was essential. An entryway for a plant laboratory at a plastic fibers plant was to be used to temporarily store pallet loads of production samples shipped to the lab for testing. The spools of fiber were not light. The laboratory technicians were not happy with a 15 minute wait time before cooling restarted. A pressure reducing valve and connecting tubing that delivered compressor discharge vapor to the inlet of the evaporator maintained the evaporator pressure to keep the compressor running. Hot gas bypass can be confused with hot gas defrosting because they are piped the same way. The difference in control operation will be the distinguishing factor

Screw Compressors

Screw compressors consist of two screw-shaped rotors that are

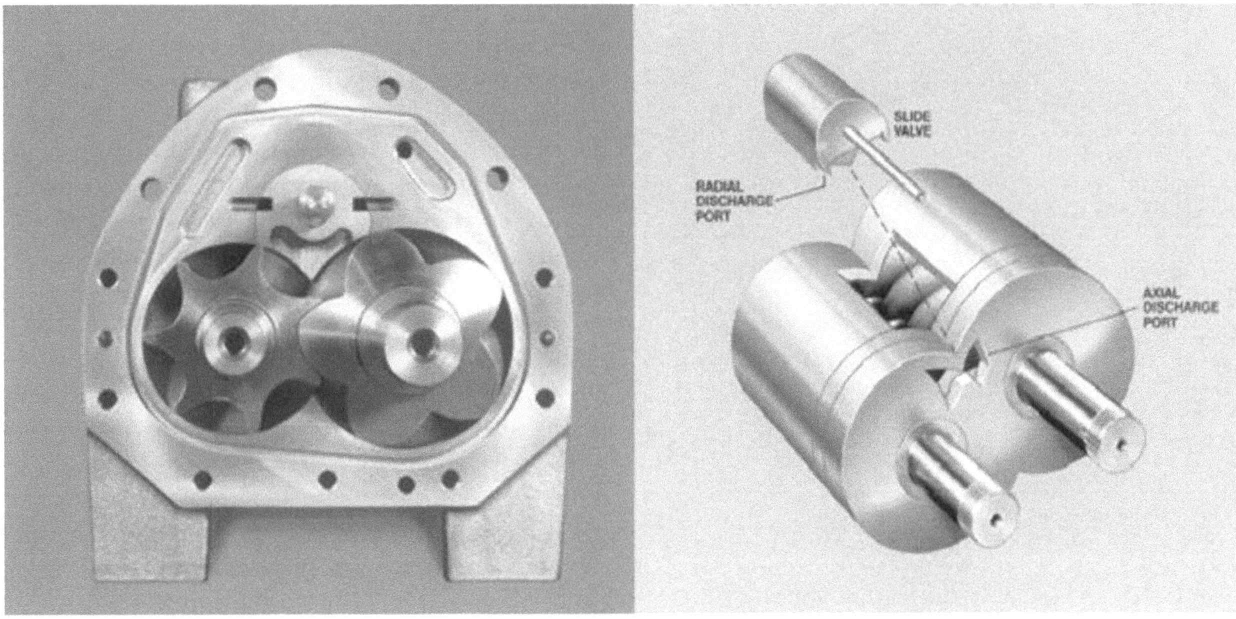

Figure 5-13. Screw compressor.

interlocked inside a casing. The two screws form cavities between the screws and the casing (Figure 5-13) that are sealed where the two screws meet. Compression is achieved by capturing a gulp of refrigerant vapor in a cavity and that cavity progressing from the inlet, where the pressure is low, to the outlet. Theoretically, the vapor would progress along the screws at suction pressure until the cavity is opened at the discharge, where high pressure vapor would flow into the cavity and, like a piston, compressing what was trapped.

However, leakage between and along the screws serves to raise the pressure in the cavity as it travels along so that the sudden compression when the pocket of gas reaches the discharge is not as violent as it could be. A certain degree of unloading is provided with screw compressors through the movement of a plug lying along the screws. That same plug also provides a means of partially unloading the compressor for startup.

Lubrication is very important for screw compressors because the oil helps seal the clearances between the screws and the casing to reduce refrigerant leakage at the same time as it creates a wedge of oil between the moving parts. A typical screw compressor will have provisions for removing oil from the compressor discharge vapor and injecting it back in at the inlet to the screws. Large screw compressors will also have gears mounted on the two screw shafts that maintain screw alignment without exposure to gas leakage. They must be lubricated as well. Screw compressors are inherently noisy. Never go near them without ear protection.

Centrifugal Compressors

A facility will typically utilize chilled water for cooling and air conditioning purposes instead of individual local refrigeration equipment because chillers, which will be discussed later, are very efficient. Refrigeration, using a centrifugal compressor, has energy requirements that are fractional compared to other types of compressors. With the advances in design, which include variable speed drives and magnetic bearings, a centrifugal compressor today has half the operating cost of one from 20 years ago. Still, those older machines produce cooling at half the power cost of local refrigeration equipment. A centrifugal compressor is a high tech piece of equipment and must be built in large sizes to be economical. A centrifugal compressor is very much like a blower (their forms are discussed in Chapter 10 under the section on fans and blowers). Considerable turndown is achieved with a centrifugal compressor using variable speed drives. However, an older one may have a fixed speed motor and inlet vanes (Figure 5-14). The inlet vanes are typically operated by a small motor mounted outside the compressor housing. They reduce flow through the compressor by closing off on the inlet while simultaneously creating a whirl in the flow of the gas in the direction of the impeller rotation to further reduce the motor horsepower required at the reduced flows. The inlet guide vanes on a gas turbine perform the same function.

Centrifugal compressors can be open or hermetic. They are normally hermetic when powered by an electric motor. Some may be powered by a steam turbine

Figure 5-14. Centrifugal inlet vanes.

or diesel engine. The principal use of centrifugal compressors is water chillers. Read about those later to learn more about the operation of centrifugal compressors.

CONDENSERS

Once the compressor has increased the pressure of the refrigerant high enough to produce a saturation temperature above the temperature of the substance that receives the heat (heat removed in the evaporator and added by the compressor), the condenser serves to transfer the heat from the refrigerant vapor to that substance. Operating pressures and temperatures in the condenser are dependent upon the substance used to cool the condenser and its temperature. The condenser has to remove the superheat, by cooling the refrigerant gas until it can start to condense, and then condense it.

Because heat transfer from a gas to a metal is a fraction of the rate of heat transfer from a condensing liquid to a metal, a fair portion of the condenser does nothing but desuperheat the gas. Once the gas temperature drops to the saturation temperature, the gas condenses to a liquid that drizzles over the heat exchange surfaces and drops to the bottom of the condenser. Many condensers are designed so that the hot superheated gas flows around and through the condensed droplets because a small portion of them will be evaporated in cooling the superheated gas and the rest will be heated a bit because the liquid in contact with the heat exchange surfaces is cooled below the saturation temperature of the refrigerant. The evaporative cooling of the gas by the droplets is more efficient than the cooling of the metal surfaces.

Finally, the condensed liquid must be sub-cooled in most systems so that it will remain a liquid until it reaches the throttling device. The amount of liquid in a system serves to balance the gas to liquid ratio in order to compensate for variations in load and operating conditions. A refrigerator, window unit, automobile, and house air conditioning systems hold varying amounts of liquid refrigerant in the bottom of the condenser depending on those factors. The amount of sub-cooling is an indication of how much liquid is retained in the condenser. A small amount of sub-cooling can indicate insufficient refrigerant and a large amount would indicate excessive refrigerant. A measure of sub-cooling is a good measure of the refrigerant in a system, when the liquid is stored in the condenser. The sub-cooling has to be measured at a consistent condenser pressure. Systems with receivers, discussed later, have other measures.

Free Cooling

When it is possible to obtain a source of fluid that is colder than what is to be cooled, the refrigerant cycle can function without the compression and expansion (throttling) stages. The refrigerant gas will migrate (a term used to describe the flow of the gas because it is a natural thing) to the coldest spot in a system because that is where it will condense if the pressure in the system is below the saturation pressure that matches that temperature. By assembling the equipment so that the condenser is above the evaporator, the gas will condense in the condenser, giving up heat to a substance and the liquid will drain to the evaporator where it can cool another substance by boiling. When equipment is arranged to do this, it is called free cooling because no energy is input to compress the gas.

Free cooling can also be accomplished with parallel systems that simply use the colder substance to remove heat from whatever has to be cooled. One example would be using cold river water to create chilled water.

To be truly effective, these systems have to allow warmer chilled water temperatures. That is perfectly acceptable because when the river water is cold, it is cold outside and the warmer chilled water will still do the job. Free cooling can also be utilized in more complex systems to precool chilled water or cooling tower water before mechanical equipment is used to achieve the necessary water temperature.

Heat Pipes

Devices that exchange heat between incoming ventilation air and indoor air exhausted to make room for the ventilation air can perform free cooling. Devices called heat pipes consist of a number of finned tubes, sealed at the ends, installed in a split casing, and sloped appropriately. They can automatically transfer heat from warm indoor exhaust air to hotter outside air and, with airflow reversed, heat cold outside air with warm indoor exhaust air. Each heat pipe is a refrigeration system unto itself. The half at the lower end of the slope absorbs the heat by evaporating the liquid refrigerant that drains down into it. The evaporated gas then travels up the pipe to the high end where it is condensed, thereby giving up its heat.

Air Cooled Condensers

There are many types of air cooled condensers. There is one in a car, a refrigerator, a freezer, and a home air conditioner, to name the principal ones. These have to be designed for operating temperatures of the cooling air at the maximum temperature one would expect to encounter. Most of these are designed to condense the refrigerant using air at 120°F. The auto manufacturers cannot take a chance on making vehicles that would not, on some day, pay a visit to the deep South or a visit to Death Valley. Heat transfer between metal and air is poor compared to metal and condensing vapor. Thus, most air cooled condensers have fins in tight contact with the condenser tubing to increase the surface for heat transfer between the metal and air. Provisions are also made to force the flow of air over the transfer surface to provide turbulence at the interface between the two substances.

A window air conditioner has a unique extra feature to help condense a refrigerant. The condensate from moisture removed from the air conditioned space is drained to a well in the casing, where it is picked up by a slinger mounted on the condenser fan and thrown onto the condenser tubes and fins. It is cold water and it absorbs over 1000 Btu per pound as it evaporates to help condense the refrigerant. It is why there is not always cold condensate dripping on the entry way of an establishment with a window unit above the door.

Some air cooled condensers may be called upon to operate in cold winter temperatures. Some examples are condensers for refrigeration equipment cooling the core of an office building, refrigerated food cases, or cooling a production process. In very cold weather, the colder air would result in low pressures in the condenser. That would reduce the power consumption of the compressor. It would also reduce the pressure differential across the throttling device and restrict the flow of the refrigerant for inadequate cooling. Simply shutting down the fan(s) can reduce heat transfer to raise condenser pressure to an acceptable value. Air cooled condensers in colder climates can also include dampers that can control airflow over the fins and coils to maintain a condenser pressure.

Condensing Units

This term is a label for a certain arrangement of refrigeration equipment, specifically a combination of a compressor and a condenser. It is normal to find condensing units outside residences, small office buildings, and grocery stores. Typically, these are reciprocating or scroll compressors in the same housing as an air cooled condenser. A typical air cooled condensing unit is shown in Figure 5-15. Liquid and suction lines connect the condensing unit to the evaporator and throttling device located inside the building. The condensing unit normally includes a receiver and suction accumulator and can contain multiple compressors with control switches set

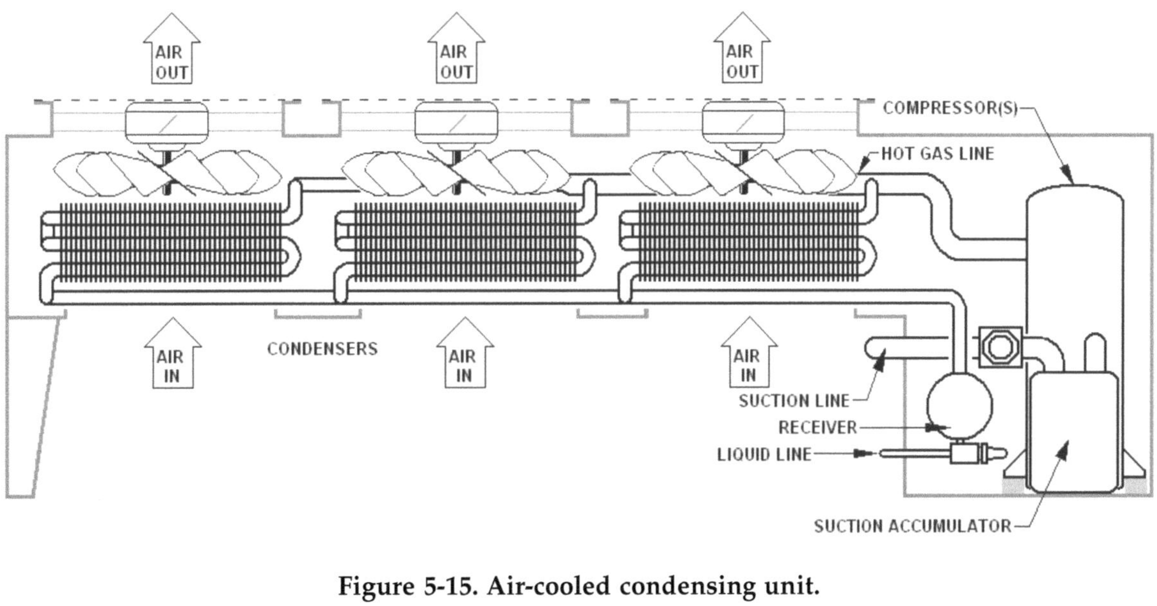

Figure 5-15. Air-cooled condensing unit.

for different pressures to bring them on and off in stages. When used for refrigeration equipment, they are also fitted with condenser pressure control switches that are staged to control the fans so that the air flow is regulated to maintain a reasonably constant pressure on the liquid line returning to the evaporator. If there were no pressure control, then the condenser could get so cold that there would not be enough pressure to push the liquid to the evaporator. On occasion in the winter, it may be necessary to cover two of the fan discharge screens to restrict airflow enough to maintain an adequate condenser pressure. Wax coated cardboard hinged on one side with some duct tape did a good job of it. The cardboard had to be flipped back over in cold weather if the fan of that section was operated.

Regular inspections of condensing units are required because so many things can happen to them. During each inspection of the condensing unit, always look for oily spots around the unit base and connecting piping. Those spots would be refrigerant oil, indicating a system leak. Sometimes, the units are mounted on the roofs of buildings, which require a roof hatch and/or ladder for access. It is not uncommon for the latter to be one that is carried on a truck. These are less susceptible to damage than units mounted at grade level. However, they are the ones that are frequently overlooked during normal inspection rounds. Problems with rooftop condensing units normally fall in the range of blockage of the coils and fins of the condensing unit by airborne contaminants. Cottonwood tree and oak tree pollen are common problems when the trees are blooming. Leaves can block the coils in the fall.

Newer condensing units mounted at ground level are exposed to other hazards. All new units are similar to a residential heat pump (Figure 5-31) because the three sided coil and vertical air discharge make them more efficient. A common problem is the accumulation of leaves, seeds, and small branches that fall onto the condensing unit and manage to drift past the fan to settle behind the condenser. The bottom tubes of the condenser then act as a receiver because tubes higher up have to serve as that portion of the heat exchanger that is dedicated to sub-cooling. The result is an apparent loss of refrigerant because more liquid is held in the condenser. Always look through the fan to check for accumulations that can cause this problem. It is not unusual for the discharge screen to be readily removed for access to remove any accumulations. However, always turn the unit off and disconnect the power to the condensing unit before reaching through the fan to remove any accumulation. Don't forget to put the discharge screen back on before restoring power and control. The inlet side of the condenser can also be plugged up with mown grass and debris drawn into the fins by the air flow, with the assistance of a lawn mower, weed eater, or wind. More frequent inspections in the spring, when the pollen and seeds are falling, and in the fall, when leaves are falling, should be made to ensure the condenser inlet is clean. Yard workers prove to be a hazard for most condensing units because the fins on the air cooled condensers are readily damaged by the plastic strings of weed eaters. At one problem site, the yard workers had elected to write their initials into the fins using a weed eater. Needless to say, the culprits were easy to identify. The grass and weed eater damage has the same effect as debris accumulation on the condenser outlet. Unless the damage is so extensive that the fins are torn, bent fins on an air cooled condenser can be repaired readily with a device called a fin comb. Even if none of these obvious conditions of condenser blockage are evident, a regular annual cleaning of the condenser coils is recommended.

Water Cooled Condensers

Every source of water for a water cooled condenser can be counted upon to be colder than air except in the winter. Even water cooled by a cooling tower is colder than the temperature of the air. Heat transfer between metal and flowing water is considerably higher than the heat transfer between metal and flowing air. Water cooled condensers are preferred because they permit lower saturation temperatures inside the condenser, which reduces the horsepower requirements of the compressor.

While most home air conditioners are not water cooled, those installed in the Southwest may be water cooled. They are called evaporative condensers (Figure 5-16). They can be found almost anywhere. The refrigerant is piped or connected with tubing to coils in a casing, where water drips on them to absorb heat as the water evaporates. The vapor is carried off by air drawn through the evaporative condenser. It requires a pump to circulate the water from the catch well at the bottom of the condenser to the top, where it is distributed over the tubes. A makeup water flow control valve is required to add water from a water supply to replace the water that is evaporated. A small amount of water must be removed by blowdown to limit the concentration of solids in the circulated water. The operation and maintenance of an evaporative condenser is very similar to cooling towers which will be discussed later.

Construction of a water cooled condenser is highly dependent on the quality of water used to absorb the rejected heat. The condensers on shipboard refrigeration

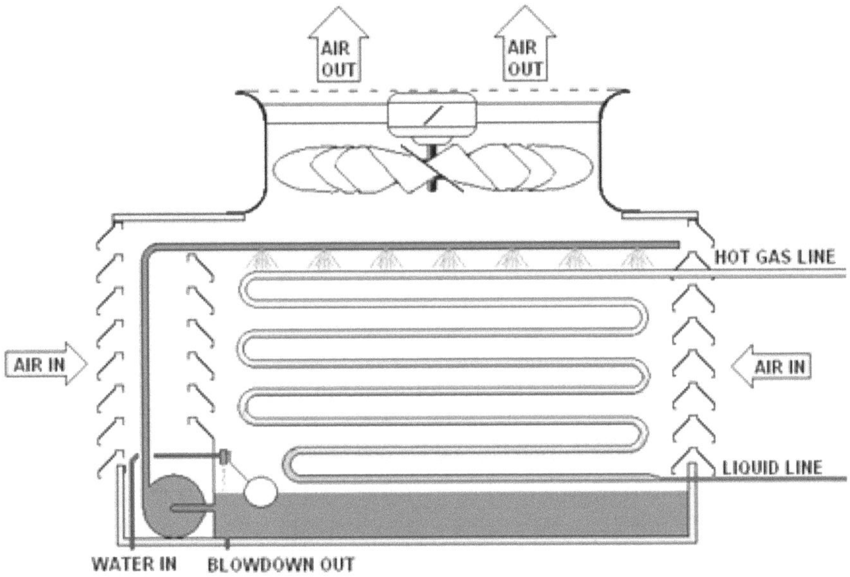

Figure 5-16. Evaporative condenser.

plants use seawater that is highly corrosive and contains all forms of plant and animal life. It had to be filtered before the section of the circulating pump that forced salt water from the ocean through the filter, the pump, condenser, and condenser water flow control valve and then back to the ocean. The shell, heads, and tubes of the condensers were made of high-quality brass and bronze. Sacrificial magnesium anodes electrically connected to the condensers provided additional protection against corrosion. Plants using seawater or brackish water would be constructed similarly. Large ones use electrically powered cathodic protection for corrosion protection. Growth of algae, bacteria, bivalves, etc., on the heat transfer surfaces are also concerns for those condensers. The pressure drop through and temperature differential of these condensers has to be monitored closely to identify conditions of organic growth hindering heat transfer.

Well water will have the highest dissolved solids content of any supply of condenser cooling water. Normally, when the well water is not circulated, it will not present a problem. However, it is possible to have sources that will form scale on heat transfer surfaces after a temperature rise of only a few degrees. Regular sampling and testing of the total dissolved solids (TDS) of the well water should allow the detection of conditions that could contribute to scaling. The important thing to do with high solids water is to maintain a minimum rate of water flow through the condenser and avoid concentrating solids content by blowdown or other means.

Water drawn from rivers, lakes, and reservoirs will normally have lower dissolved solids content. However, it can have a considerably higher organic materials content. Water circulating through a condenser and the cooling tower will also pick up organic materials. When those waters are used in condensers, they will require regular cleanings to remove the organics, algae, and bacterial growth. Monitoring of the pressure and temperature differentials under these conditions is imperative.

Condenser Pressure Control Valves

To maintain the pressure in a water cooled condenser above a minimum value, self-contained pressure control valves modulate the flow of the water leaving the condenser. A capillary containing a refrigerant charge is connected to bellows in the valve assembly (Figure 5-17). The pressure in the bellows is opposed by a spring that pushes on the valve stem to force the valve to close down and decrease water flow through the condenser. An increase in pressure on the

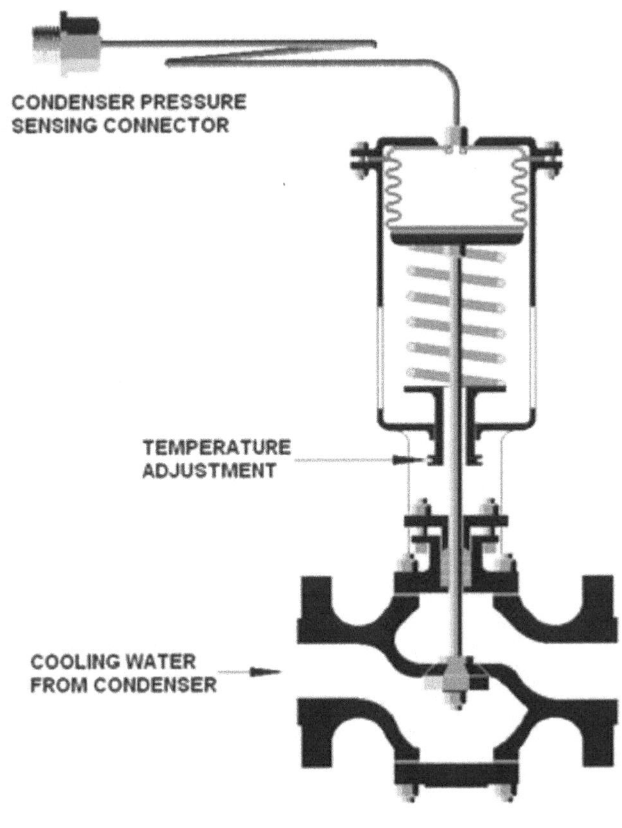

Figure 5-17. Condenser water control valve.

refrigerant side of the condenser results in an increase in the pressure in the bellows on the valve to force the valve to open, increasing the cooling water flow. Control of the condenser pressure is maintained within a narrow range of pressure, dependent upon the design of the spring. That range is adjustable up and down by adjusting the pressure on the spring.

Because this control valve is a proportional device, and water temperatures can change due to season, storms, etc., the pressure in the condenser will vary. Under extreme conditions, one might choose to adjust the setting of the valve to keep the range at the lowest possible values. That is because maintaining a low pressure in the condenser reduces the amount of effort on the part of the compressor, resulting in less cost to operate. Such a situation might occur when entering or leaving the Gulf Stream, where seawater temperatures could swing by as much as 50°F. It might be advisable to tweak the setting to increase the range between winter and summer. The control valve may completely block flow when the condenser is not in use and pressures inside it are low. It is always a good idea to have a small, valved bypass around it that can be operated to maintain a minimum flow, which helps discourage organic growth on the water sides. Organic growth can be a problem for any water cooled condenser. It is always a good idea to maintain some flow through them to discourage organic growth.

Why not just let the temperature of the condenser water drop in a refrigeration system? Primarily, it is because the condenser can get so cold that there is not enough pressure differential to force the liquid through the throttling device. Some modern chillers and refrigeration systems may take advantage of cold water (or cold air for that matter) to condense the gas coming from the evaporator without compressing it, by operating like a heat pipe. Those applications require a compressor bypass valve, a throttling device bypass, and, because the flow of refrigerant is due to convection currents, the condenser has to be mounted higher than the evaporator.

THROTTLING DEVICES

A throttling device is required to separate the high pressure of the condenser from the low pressure of the evaporator to ensure maintenance of the saturation conditions in each of those devices. The types of throttling devices vary considerably, from the very simple to quite complex, and can incorporate new high technology methodology. The throttling device on the existing piece of equipment was selected based on economic factors. Most of them will operate with a minimum of attention for years. That does not mean that they should be ignored. Constantly check for symptoms of failure of the throttling device.

Frequently, the throttling device is referred to as a metering device because it regulates the flow of refrigerant. The word "metering" refers to items that actually measure (or control) the flow. Avoid the use of that word. Another label is "expansion valve." The use of that title should be limited to those devices that are normally labeled with those words.

Failure of the throttling device to throttle sufficiently can result in liquid flooding the compressor, resulting in damage to valves, bearings, and piston rods, as in Figure 5-18. Occasionally, this flooding can result in crankcase rupture. On the other hand, excessive throttling will result in high superheat temperatures, making the compressor overheat. High superheat also burns up valves, burns up the motor, or melts parts as shown in Figure 5-19. If the throttling device is not working properly, insufficient flow will result in lower capacity cooling because heat transferred to the boiling liquid is much higher than heat transferred to a gas. If there is not enough liquid entering the evaporator to ensure most of its internal surface is exposed to boiling liquid, overall heat transfer will be reduced.

In some of the graphics that follow, it should be noted that bubbles can form within the stream of liquid leaving the throttling device. This always occurs because the temperature of the liquid at the outlet of the condenser is always higher than the saturation temperature maintained in the evaporator. Also, the ambient temperature is usually higher than the evaporator temperature. The liquid refrigerant cannot be cooled by air around the liquid piping. A very small portion of the

Figure 5-18. Broken compressor piston rods.

Figure 5-19. Overheated compressor parts.

liquid vaporizes absorbing heat from the rest of the liquid to lower its temperature to the saturation temperature in the evaporator.

Capillary

The throttling device for many small units is a simple capillary. An extended length of small diameter tubing between the outlet of the condenser and the inlet of the evaporator restricts the flow of refrigerant (Figure 5-20). As the pressure drops within the refrigerant liquid, it reaches a point of saturation and vapor starts to form. The bubbles of vapor take up more space within the narrow capillary. The velocity has to increase, which results in an increase in pressure drop. It only takes a small amount of vapor, just a few bubbles, to produce a significant difference in pressure drop. The result is a stable flow rate through the capillary. Capillary tubing diameters are tabulated for each refrigerant and capacity.

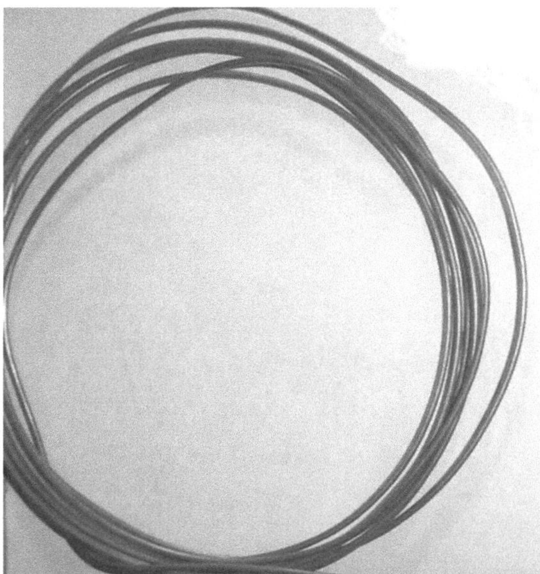

Figure 5-20. Capillary.

Adjustable Orifice

An adjustable orifice produces a pressure drop by restricting flow (Figure 5-21). Unlike a capillary, the pressure drop is generated right at the valve. As the name implies, an adjustable orifice is manually adjusted. It requires an operator to adjust it while observing evaporator pressure and outlet temperature. They should only be used when the load on the evaporator is very stable and consistent.

Automatic Expansion Valve

An automatic expansion valve simply maintains a constant pressure in the evaporator (Figure 5-22). In addition to the one shown, these can be manufactured with an external sensing line so that the pressure that is controlled is at a different location than the valve outlet. An automatic expansion valve can be used to control the refrigerant flow into an evaporator that has a means of separating the liquid and vapor so that only vapor leaves the evaporator. That is normally accomplished with a retention space for separating the liquid and vapor plus

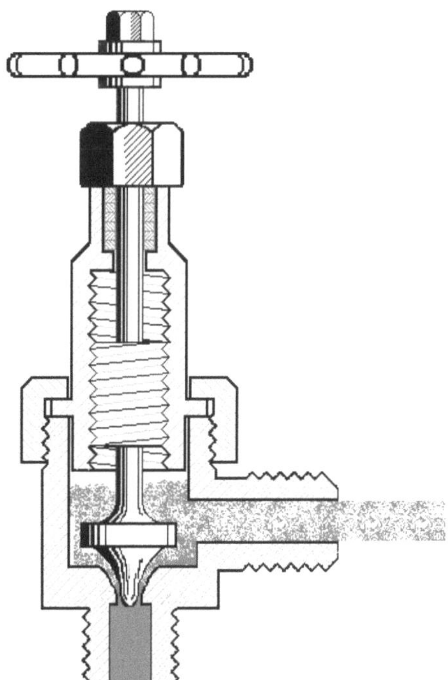

Figure 5-21. Adjustable orifice.

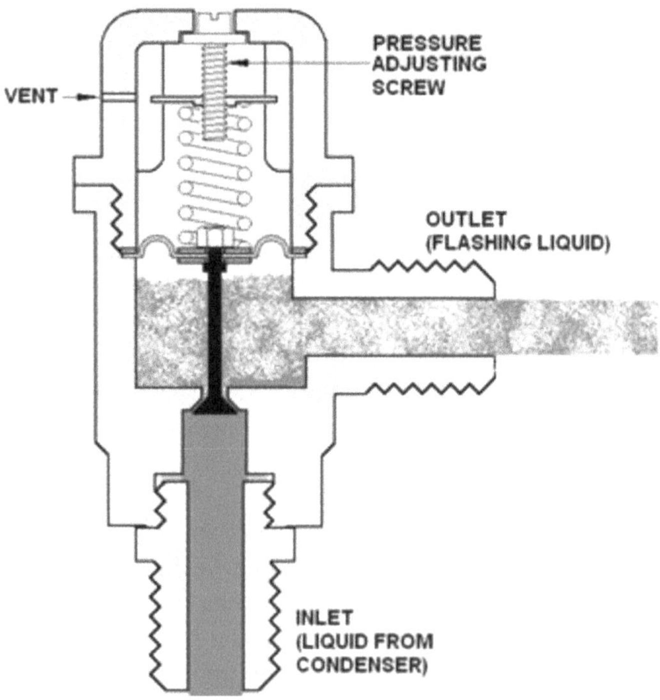

Figure 5-22. Automatic expansion valve.

enough room that the gas velocity does not carry any liquid droplets with it.

Float Valve

A float valve restricts the flow of refrigerant through the valve to liquid (Figure 5-23). Liquid, from the condenser, enters the float chamber and accumulates until the increased level increases the buoyancy forces on the float to overcome the force on the valve disc that is imposed by the pressure difference between the float chamber and the evaporator, over the area of the valve disc. The size of the float and the diameter of the orifice under the valve disc are selected to achieve a constant flow through the valve during normal operation. Similar to the automatic expansion valve, it can be used to control the refrigerant flow into an evaporator with liquid vapor separation. It simply drains the condenser so that there will be fluctuations in the level of the refrigerant in the evaporator. See the section on chillers for more on float valves.

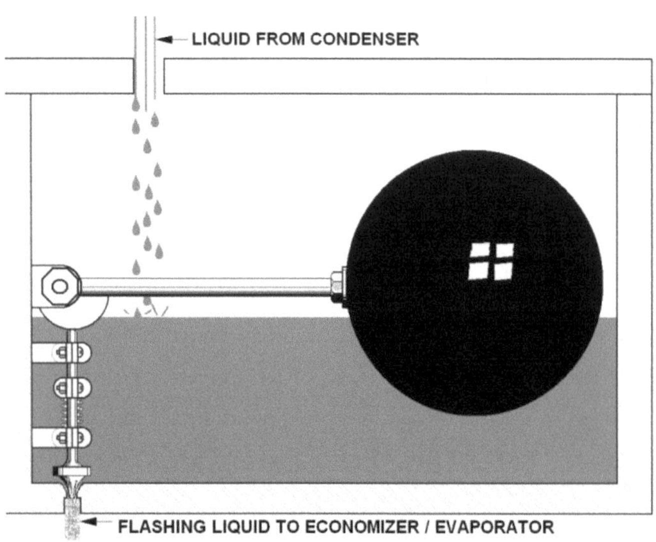

Figure 5-23. Float valve.

Thermostatic Expansion Valve

Thermostatic expansion valves are the most common means of controlling refrigerant flow. "Thermostatic" implies a combination of temperature and pressure control. The common abbreviation for the thermostatic expansion valve is TXV. The valve orifice is changed in size to vary the flow of refrigerant to maintain a constant value of superheat at the outlet of the evaporator. By ensuring a fixed value for superheat, liquid surging into the compressor or inadequate flow of liquid into the evaporator is prevented. Figure 5-24 shows a cutaway of a TXV. The thermal bulb and diaphragm at the top of the valve, connected by a capillary, produce a force within the valve assembly equal to the saturation pressure in the thermal bulb times the area of the diaphragm. That force is opposed by the actual pressure at the evaporator inlet or outlet. The two pressures correspond to two different saturation temperatures and the superheat is the

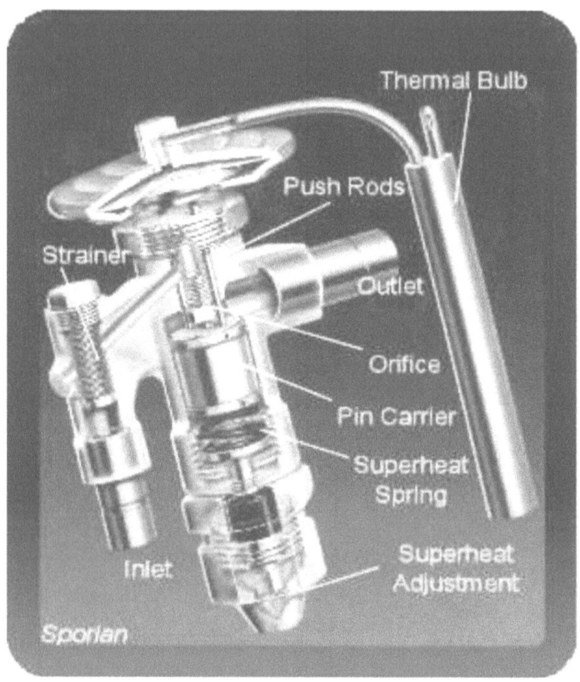

Figure 5-24. TXV.

difference in those two. The difference between the two forces on the diaphragm is opposed by a spring. When the difference increases, due to an increase in superheat, the spring is compressed to open the valve further and increase the flow of liquid into the evaporator. Conversely, a decrease in superheat will result in the valve closing down to reduce liquid flow. In some applications, where the pressure drop of the liquid and gas flowing through the evaporator is low, the evaporator pressure is sensed only at the valve outlet. For most installations, however, the lower diaphragm chamber is connected to the evaporator outlet with an external sensing line, commonly referred to as an equalizing tube, which is shown in Figure 5-25. The TXV outlet pressure sensing connection is a small orifice in the valve body. The flow through it is dumped off so fast, through the sensing line to the evaporator outlet, that the pressure on the bottom of the diaphragm is very close to the evaporator outlet pressure. This installation ensures that the TXV is controlling by the actual superheat at the outlet of the evaporator.

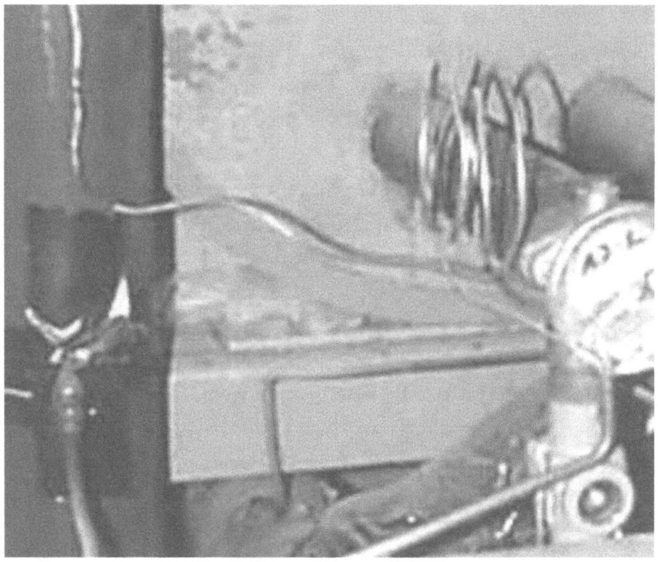

Figure 5-25. TXV installation.

The TXV has a strainer at the inlet that is accessible for cleaning. It cannot be removed for cleaning without removing the refrigerant from the system. Periodic cleaning is not required. The principal purpose of the strainer is to catch droplets of solder and contaminants in the liquid line upon startup of the system or products carried down the liquid line from breakdown of the filter dryer. The TXV also has a cap that can be removed to access an adjustment of the spring pressure that will change the setting of the superheat. This should be a "set and forget" adjustment. If there is a reason to believe that the valve is not working properly, there are many things to check before adjusting the superheat setting. Understanding the operation of the valve should aid in determining what the problems might be. A common problem is vibration loosening the mounting of the thermal bulb. Proper mounting of the thermal bulb is shown in Figure 5-26. The temperature of the bulb is maintained by heat transfer through contact of the bulb

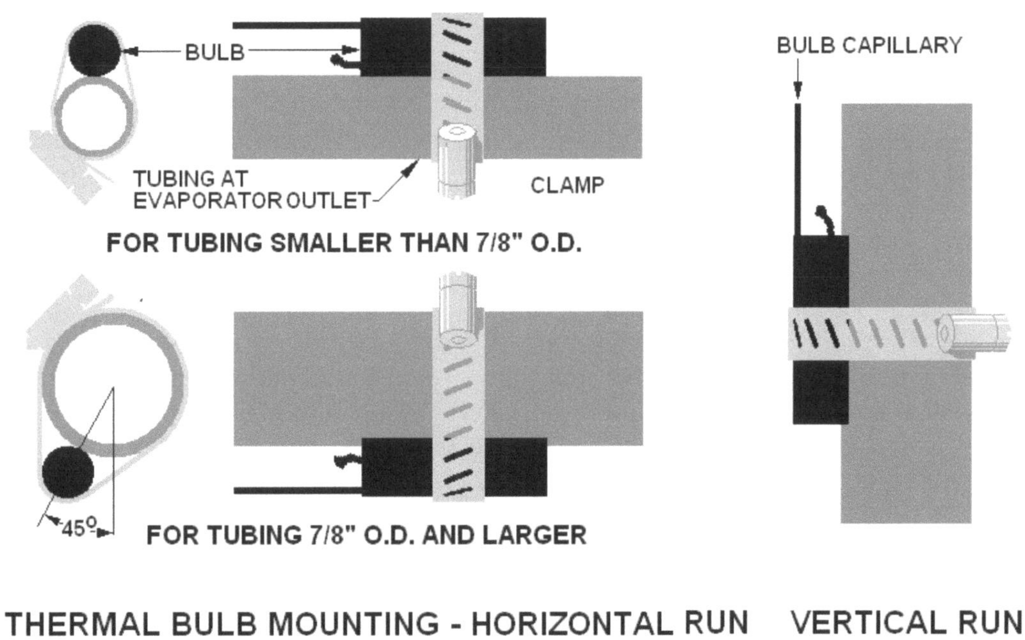

Figure 5-26. Mounting of thermal bulb.

and tube because heat transfer through the metal is faster than the heat transfer from the bulb to air. While it is not necessary, it is a good idea to insulate the bulb and suction line. On large tubing, the bulb is mounted at the bottom of a horizontal tube because boiling liquid will run along the bottom and superheated vapor could be at the top at low loads. The bulb is offset from the very bottom of the tube to eliminate the insulating effect of the thin film of oil that flows along with the refrigerant. The bulb is mounted with the capillary connection on top of vertical tubes to retain liquid in the bulb. If the connection were on the bottom, vapor would form in the top of the bulb to reduce heat transfer. Normally, the diaphragm and capillary contain some vapor because they are at a higher temperature than the bulb. The vapor in the capillary and diaphragm is usually superheated and does not materially affect the operation of the TXV.

Another possible problem with TXV operation is crimping of the capillary or the equalizing line. Crimping of the capillary can block the transfer of bulb pressure to the diaphragm or restrict the flow to increase the response time of the valve. Crimping of the equalizing line will result in a higher pressure on the underside of the diaphragm to reduce the superheat. If the evaporator has a high pressure drop, that could result in liquid surging to the compressor.

Technological advances and production methods have resulted in the introduction of electronic expansion valves. These are controlled by the electronic controls, which must sense evaporator outlet pressure and temperature to determine the positioning of the valve. The less expensive ones may use a solenoid that simply opens and closes (repeated clicking should be noted) to control the refrigerant flow.

What might appear to be an improper operation of a TXV may be attributable to problems with other pieces of the system. Evaporators and condensers can be fouled with dust and debris, changing system performance so much that the TXV cannot correct it. A worn out compressor may not produce enough differential. There are other elements of a typical system that could contribute to a problem including the filter dryer and other controls.

Regardless of the type of throttling device, it is important to remember what it is supposed to do, what harm can come to the equipment if it does not do what it is supposed to do, and, most importantly, how to monitor conditions to detect potential problems with the throttling device. For most systems, that means making regular checks of superheat and sub-cooling.

MISCELLANEOUS COMPONENTS OF A REFRIGERATION SYSTEM

A refrigeration system, consisting of an evaporator, compressor, condenser, and throttling device, contains everything necessary to transfer heat from a substance to be cooled to a substance that can accept the rejected heat. However, there are many additional components included in these systems for protection of the system, ease of maintenance, or diverse load service. There are also specific labels for certain parts of a refrigeration system. Among those labels are specific names for different sections of piping or tubing. The word "line" is used generically to identify piping or tubing. The line between the evaporator outlet and the inlet of the compressor is called the suction line. The line connecting the compressor discharge to the inlet of the condenser is called the hot gas line. The line connecting the outlet of the condenser to the inlet of the throttling device is called the liquid line. Since there is seldom any significant length of line between the throttling device and the inlet of the evaporator, there is no specific label for that portion. In many pieces of equipment, it is actually a sort of Christmas tree that feeds the evaporator with a lot of smaller lines.

Liquid Observation Port

In many older refrigeration systems, there will normally be a fitting on the liquid line before the throttling device that contains a glass window for observation into the line. Most of the time, there should be nothing to see because light passes straight through the liquid. If there is one, always check it because it only takes a glance to be sure no bubbles are forming in it. The presence of bubbles indicates that the liquid is not sufficiently sub-cooled and some of the liquid is vaporizing before it reaches the throttling device. Normally, bubbles indicate a loss of charge. They can also be an indication of inadequate sub-cooling in the condenser.

Filter Dryer

Practically, every system made includes a filter dryer. A filter dryer is always installed in the liquid line before the throttling device. The filter dryer (Figure 5-27) performs those two functions. It contains an element that consists of fabric or paper filtering media, combined with silica gel. Its purpose is to trap any rust, flaking, loose solder from installation or repair, and material from wear in the compressor to prevent plugging of the small orifices in the throttling device. Also, the desiccant, normally silica gel, absorbs water that gets into the refrigerant

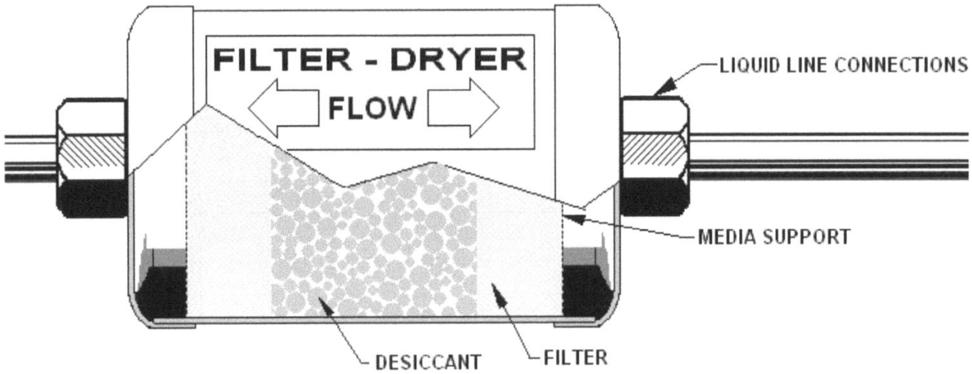

Figure 5-27. Filter Dryer

through leaks in the compressor seals or introduction of water from connected test equipment. Many filter dryers are designed for bidirectional flow (note the two arrows on the label). Others must be installed in the right flow direction. Occasionally, a filter dryer will be provided in the suction line before the inlet to the compressor. These have to be larger to handle the volume of vapor. Since most refrigeration systems operate under pressure, in-leakage of water is uncommon. If the gauge set is kept dry and filled with the refrigerant, there should not be any water introduction while testing. Therefore, replacement of the filter dryer is seldom required. Most of them are hermetically sealed. Replacement consists of replacing the entire unit. Most large systems will have a flanged cover to permit replacement of the element only.

Receiver

Refrigeration systems that serve a variety of loads with different temperature requirements in multiple evaporators require something like a surge tank to accommodate variations in system volume that occur. A receiver is a pressure vessel that contains liquid and vapor of sufficient volume to handle the fluctuation in the vapor/liquid ratio. Large receivers are normally ASME stamped pressure vessels and are subject to regular inspections by a National Board Commissioned Inspector at intervals of 2–5 years depending upon the State, Commonwealth, or Province in which they are located. Aboard ship, they are inspected by the Coast Guard. A common feature of a receiver is a glass port to permit observation of the liquid level. Some receivers have two glass ports. This allows a flashlight to illuminate one port and see into the other to get a better view of the liquid level. The purpose of a receiver is to absorb variations in the vapor/liquid ratio, where changes in the liquid level are common and normally do not have to be corrected.

Adding and removing refrigerant in an effort to maintain a constant liquid level is ignoring the purpose of the receiver and creating a lot of work for the operator. Changes in receiver level are associated with a change in load. Sometimes, turning on a load will result in a drop in refrigerant level and sometimes the opposite will occur. It all depends on things like elevation of the evaporator with respect to the receiver, the temperature to be maintained at the evaporator, and the length and run of the connecting piping. It is simply a matter of whether the liquid will end up in the receiver or be trapped in the evaporator and connecting piping.

Isolating Valves

Manual shutoff valves are provided on large refrigeration systems to permit isolation of compressors, condensers, and loads (evaporators) to place them in service or to remove them from service. Normally, the valves are only used to secure loads that do not need refrigeration and to isolate compressors and condensers for maintenance and inspection. Refrigeration isolating valves are designed to prevent leakage of refrigerant. That can be accomplished by valves with a bellows seal instead of packing, as shown in Figure 5-28. There are also packed valves, as shown in Figure 5-29. The handle of those valves also serves as a cap. In order to open or close the valve, remove the cap, flip it over, and position the square hole in the top of the cap over the square end of the valve stem. Then turn it to open or close the valve. After operating the valve, always return the cap, making sure the gasket is still inside it. Tighten it to prevent loss of refrigerant that could leak through the packing. The King Valve is a label identifying the valve located in the liquid line at the outlet of the receiver. Closing the King Valve prevents liquid flow from the receiver to all loads served by that system. Continued operation of a compressor then permits a "pump down" of the system,

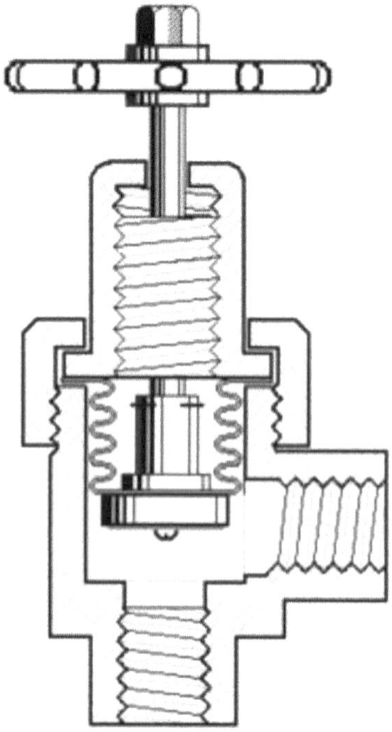

Figure 5-28. Isolating valve with bellows seal.

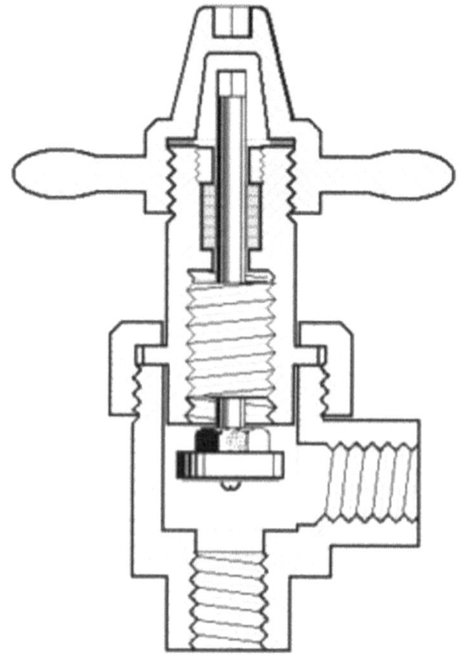

Figure 5-29. Isolating valve with packing.

transferring almost all of the refrigerant to storage in the receiver. Temporary overriding of the system controls can be used to lower the suction pressure and all elements between the King Valve and the compressor inlet, until it is almost at atmospheric, 0 psig. When pumping down, avoid pulling a vacuum which might result in ambient air containing moisture entering the system should it be leaking. A system can be tight under pressure, but leak under a vacuum, because a piece of rust or a defect in construction can produce the equivalent of a check valve at the leak site.

If the system is being pumped down for maintenance, that requires opening the system. A separate independent vacuum pump should be used to recover the remaining refrigerant and evacuate the system. A recovery system should also be used to remove refrigerant from a compressor, condenser, or receiver before opening them for inspection or maintenance of the refrigerant side of the system. Recall that evacuating a system for maintenance and opening a system will require the operator to be licensed by EPA to handle refrigerants.

Load Control Valves

When a refrigeration system serves multiple loads, control valves are installed in the liquid lines to start and stop the flow of liquid refrigerant to each load. These are normally solenoid controlled valves actuated by a thermostat. Aboard ship, there can be evaporators in many spaces. Some are used all the time such as ship stores, food freezers, refrigerators, and other equipment in the galley. Then, each deck in each hold could have coils for refrigeration or freezing of the cargo in that space. On a ship that carried bananas back from South America, all holds that contained bananas were refrigerated. Facilities ashore, such as food distribution centers, large supermarkets, and the like, can have central refrigeration condensing units serving freezers and refrigerated boxes.

Defrost Valves and Lines

In order to defrost evaporators used in freezing applications, some systems include a defrost valve and piping or tubing that transfers refrigerant from the hot gas lines to the inlet of the evaporator in order to melt ice accumulating on the evaporator. A defrost valve is identical to a load control valve, with the exception of size (larger because it handles vapor) and location. A defrost valve is always connected downstream of the throttling device. The defrost valve admits the hot gas into the evaporator. Thus, it has a higher temperature rating. When defrost controls call for defrosting, the load control valve is closed, any circulating fans within the freezer space are shutdown, and the defrost valve is opened to admit hot gas to the evaporator. Once the ice is melted off, normal operation is restored.

Backpressure Control Valves

If the system has various loads, backpressure control valves can be used to maintain a higher saturation pressure in evaporators that do not maintain freezing temperatures. That prevents the need for defrosting those evaporators. Backpressure control valves are also used in temperature sensitive applications, where the product being cooled is damaged by colder temperatures. Backpressure control valves simply restrict the flow of gas (vapor) from the evaporator to keep the pressure and, therefore, the related saturation temperature, above the preset minimum value. The throttling device controls the flow of refrigerant independent of the backpressure control valve. With changes in cargo, there may be a need to reset the valves several times.

Limit and Operating Controls

Loss of refrigerant, accidental crimping of tubing, a dirty condenser, an iced up evaporator, and loss of condenser air or cooling water flow can result in damage to the equipment, especially the compressor. Limit controls are provided on systems to prevent damage to equipment in the system by shutting down the compressor whenever operating pressures or temperatures exceed preset limits. These devices are either mounted at strategic locations in the system or are incorporated in remote sensors wired to the system controls. Think of each of them as a device operating a simple switch that turns a compressor on or off. Process refrigeration may include other devices not listed here. Always read the instruction manual to understand their purpose and operation. The most common limit device is a low pressure switch, which is connected near the outlet of the evaporator and opens an electrical circuit when the pressure in the evaporator falls below the switch setting. When evaporator loads are low, it works more like a control switch. A typical situation would be a windshield defrosting condition in an automobile in the winter time because the air conditioning system is operated even if the temperature inside the vehicle is below the setting of a thermostat. Of course, it assumes that there is a thermostat to control the automobile air conditioning. In larger systems, a second low pressure switch may be included, called a "low-low" or "limit" switch. That switch would require manually resetting it to restore system operation.

Operation of the low pressure switch when a system is not maintaining the set temperature is indicative of insufficient refrigerant. Check to see that the evaporator pressure remains low in the system while cycling on and off on the low pressure switch because there is inadequate refrigerant in the system. It is readily resolved by adding refrigerant. Doing so requires the EPA license. It may be necessary to bring in a licensed technician to do it. Checking for spots of oil that are indicative of a refrigerant leak before calling that technician can help to reduce the cost of the service because the extent of the work necessary will be known if a leak has been identified. A leak means a considerable amount of work has to be done to return the system to normal.

A high pressure switch is normally connected to the hot gas line near the compressor discharge. This is referred to as a high head pressure switch because it normally senses the pressure at the compressor head. In systems with multiple compressors, there will normally be a combination of high and low gas pressure switch mounted at each compressor, with connections inside each compressor's isolating valves for exclusive control of that compressor. One purpose of the high pressure switch is to shut down the compressor before the motor is overloaded. The high pressure switch should always be set at a value equal to or lower than the maximum design pressure of the compressor discharge, hot gas piping, condenser, and liquid line. A high pressure switch also prevents accidental operation continuing when a discharge isolating valve is closed. Regardless, make a habit of checking to ensure the valves are open before trying to start the compressor. It is a bit difficult with packed valves because the caps have to be removed. Nevertheless, it is always advisable to check, especially if someone else is also operating that equipment. The combination of high and low pressure switch also shuts down the compressor when someone forgets to open the suction isolating valve.

A high pressure switch may be provided on the condenser in systems with multiple compressors and condensers that can be isolated. In the event of a failure of a reversing valve, they are also used on heat pumps. There might be a high temperature switch that senses the temperature of the refrigerant in the compressor head and shuts down the compressor if that temperature gets too high. Low system pressure, loss of lubrication, or loss of refrigerant flow through the compressor could produce high head temperatures without a high head pressure. Low oil pressure switches are provided on almost all large compressors. They are occasionally incorporated with the combination of high and low refrigerant pressure switches into one assembly. Equipment such as reciprocating compressors may have an integral oil pump that builds up pressure after the compressor starts, just like an automobile. In that case, provision includes a contact that bypasses the low oil pressure switch for a few seconds during startup. If a compressor starts and then stops almost immediately, check the oil level again. Don't repeatedly try to start a compressor that will not continue to run because bearing damage can occur. Since refrigerant flow cools the motor and inrush current heats up the windings during starting, give it a chance to cool off. During annual inspection, simply lifting a wire from the bypass circuit should prove to prevent the compressor operating. If the compressor is fitted with an oil pressure gauge, make it a point to observe the switch operation and record the pressure at which the contacts close. Note also that oil pressure switches are actually differential pressure switches that compare the oil pressure to the suction pressure (typically in the compressor crankcase) because the flow of oil is determined by that differential.

Large compressors can also have high and low oil temperature switches. The oil temperature switches will prevent compressor operation if the oil temperature is too high or low. Don't confuse an oil temperature switch that controls the heater with a high oil temperature switch, especially if the switch is set by markings on the switch. Although the compressor could be completely unloaded, that unloading is never 100% complete. Either the compressor is set up with one or two of the cylinders always in operation or throttling devices like the adjustable plug of a screw compressor does not unload everything and the inlet vanes of a centrifugal compressor simply leak. Remember that most motors are cooled by refrigerant flow. Therefore, every compressor should be protected by high and low pressure switches and a flow switch would not hurt. Testing of high and low pressure switches should be performed on an annual basis. On individual compressors, throttling the suction and discharge valves while observing suction and discharge pressures on the gauge set should serve to confirm operation. As with other operating limits, record the actual values whenever possible in the log where the testing is described. Be certain to restored normal valve positions and any other alterations suggested for testing by the instruction manual before leaving the equipment, even if the switches are shown not to work.

Water flow switches are used to prevent operation of the compressor whenever water flow through the condenser or evaporator is not proven. These are typically differential pressure switches that measure the pressure at the inlet and outlet of the heat exchanger and open contacts to shut down the compressor whenever the water flow is so low that there is a minimal pressure drop. Their operation should be tested annually by closing down on the outlet valves of the condenser and/or evaporator water circuits and attempting to start the compressor. An established number of turns open for each valve (degrees, percent, or notches on a butterfly valve) should be determined and included in the standard operating procedures (SOPs) as starting points. Use a test light at the control panel terminals to detect switch operation when the equipment has both switches. Then gradually open the outlet valve and note when the first switch in the circuit closes. Record the inlet and outlet water pressures. Then do the same thing for the other. Using a test light is preferable to testing that requires frequent starts of a large compressor. Still, confirm that the compressor cannot start when one or both of the switches are actuated by low flow.

A low temperature switch is normally provided in any evaporator used to chill water to prevent freezing of the water. Even though water flow is maintained through the evaporator, a flow imbalance created by foreign substances or associated with the piping connections to the evaporator can reduce flow sufficiently that ice could form in one or more tubes. The result is similar to what is shown in Figure 5-30. The thermowell for the switch is typically inserted into the refrigerant space to sense the refrigerant temperature which, being at saturation, should be consistent throughout the evaporator. During each annual inspection of an evaporator containing water, the operation of the low temperature switch should be confirmed by removing the sensing element (normally, a bulb in a thermowell at the bottom) and inserting that bulb in a glass of ice water. After a few minutes in that condition, one should not be able to start the compressor. Don't forget to put the bulb back in its thermowell after testing it.

Note what appear to be grooves around the damaged chiller evaporator tube shown in Figure 5-30. They are not exactly fins, but they do increase the surface available for heat transfer.

When everything else fails, refrigerant systems are fitted with safety relief valves or rupture discs that will

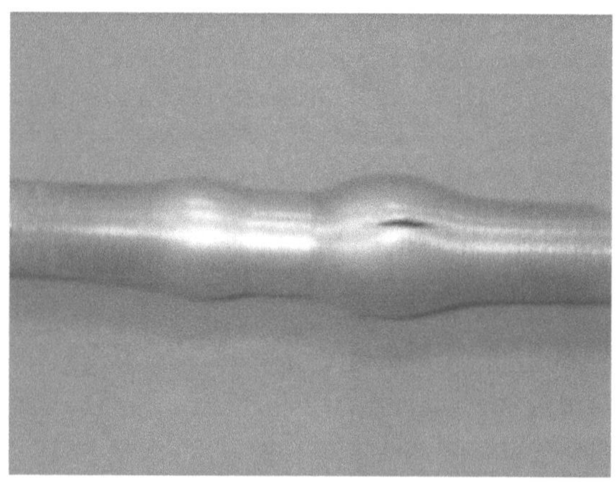

Figure 5-30. Chiller tube after freezing.

open and dump the refrigerant to atmosphere should the system pressure get too high, threatening an explosive rupture of some part of the system. A safety relief will vent refrigerant until the pressure drops below the setting of the valve. A rupture disc, once activated, will remain open, completely draining the system of refrigerant. With all modern refrigerants that are mostly heavier than air and, therefore, capable of displacing the air in the equipment room, a safety relief or a rupture disc should have connecting piping discharging to atmosphere high above the plant.

Heat Pumps

In truth, every refrigeration system is a heat pump. They all absorb heat at a lower temperature and pump it up to a higher temperature. However, certain pieces of equipment are reversible, while others are simply used to pump heat up. Heat pumps for household use are reversible. They can pump heat from the outside air into the house in the winter time as well as pump heat out of the house into the outside air in the summer. There are also heat pumps that operate only in the heating mode. The most common of those is a pump that extracts heat from the air or a water source and pumps that heat into a swimming pool. Systems such as the pool heater incorporate a simple refrigeration cycle.

Note the label of "pool heater." A piece of refrigeration equipment that is only used for heating is normally called a heater. The label of "heat pump" is primarily reserved for systems that can cool or heat. They are reversible. Reversible heat pumps operate to switch the condenser and evaporator operation. Figure 5-31 is a photograph of the most common one, a residential heat pump. The outdoor coil that forms almost three sides of the unit is a condenser in the summer and an evaporator in the winter. The enclosure also contains the compressor and the reversing valve. The solenoid-operated reversing valve switches the suction and discharge between the evaporator and condenser to control the route of the refrigerant. It fails to the heating mode because, normally, loss of heat can result in more damage to a facility than loss of cooling. Heat pumps can utilize a single capillary connecting the evaporator and condenser to handle the reversed flow. The unit shown uses two TXVs and check valves to control liquid refrigerant flow.

Figure 5-31. Residential heat pump.

Whenever the outdoor air temperatures are less than about 42°F, a heat pump will start forming ice on the outdoor (evaporator) coil. That ice can block air flow as well as insulate the coils to reduce heat transfer. The control systems for heat pumps that can be subjected to colder temperatures monitor the ice buildup by sensing air pressure drop across the coil or below freezing temperature at the coil surface to initiate a defrost cycle. The condenser fan is shut down, the reversing valve switches to cooling mode, and the compressor is operated to dump hot gas into the outdoor coil. For the comfort of the occupants, electric strip heaters in the indoor unit are powered to heat the air that is cooled during the defrost cycle.

Most residential heat pumps and air conditioners are arranged in the same way. It is difficult to tell if the outdoor unit is a heat pump or simply a condensing unit for air conditioning only. When a residence has a gas or oil fired heating system, a heat pump is only provided for operation at outdoor temperatures well above 42°F to produce heat more economically than firing fuel. It is important to note that residential style units, whether heat pumps or simple condensing units, usually have a vertical discharge. While this dramatically reduces the potential for recirculating the cooling air, it also provides a catch basin for pollen, leaves, and other debris. While it is not easy, it is a good idea to shut down and disconnect the unit, temporarily remove the discharge grill, and clean out the bottom of the unit twice a year. In that way, the accumulation of debris does not block the bottom two or three rows of coil. Also, inspect it about four times a year to catch any unusual accumulations.

Ground Source Heat Pumps

A ground source heat pump is a big investment. If sufficient space is available for installation, it would provide a good return on the investment. The reason is rather simple. Knowing the refrigeration cycle, a significant difference can be seen between the typical air to air heat pump in a residence and a ground source heat pump. Instead of heating the house with refrigerant boiling from heat in air at temperatures as low as 5°F, it is heated with refrigerant heated by the ground at an average temperature of 55°F. That defrost cycle is not required. As for cooling, instead of condensing refrigerant with air temperatures of 90°F or more, condensing heat is dumped to the ground at that same 55°F. Lower temperature differentials with heat pumps reduce the pressure differential that the compressor has to overcome, reducing power consumption and electricity costs. Most residential ground source heat pumps are also used to heat or at least preheat the domestic hot water.

There are office buildings, schools, and other facilities larger than a residence that are using ground source heat pumps. They take advantage of the ability of the ground under their parking lots to absorb and give up heat to either the refrigerant or, more commonly, water circulated from the condenser/absorber of the heat pump through wells in the ground under the adjacent parking lots. The systems are operated and maintained just like any other refrigeration system with the added duties of monitoring and recording the ground water circulating pump, flow, and temperatures.

Chillers

Hospitals, office buildings, retirement homes, schools, and similar buildings are commonly serviced with chillers for cooling and boilers for heating. Some hotels and motels are serviced as well. However, many of those opt for what is the equivalent of a window unit to provide heating and cooling of the guest rooms. That is because the units serving empty rooms can be turned off to reduce operating cost when the hotel or motel is not fully occupied. Chillers are used to cool chilled water that is circulated to air handling units (AHUs), convectors, beams, and other heat transfer devices in buildings that cool and dehumidify the air. A chiller can use any type of compressor. Centrifugal chillers, using a centrifugal compressor, are the most common. There are modifications of the typical arrangement that permit air cooled condensers and operation with a mix of water and antifreeze. That mix permits production of cold water at temperatures as low as −25°F. They see applications like ice storage systems, where the chilled water is cold enough to form the ice in the ice storage tanks. The normal chiller, however, produces chilled water at what has become a *de facto* standard of 42°F, with a return temperature of 52°F. It is a standard because all manufacturers make chillers with the same conditions to match users

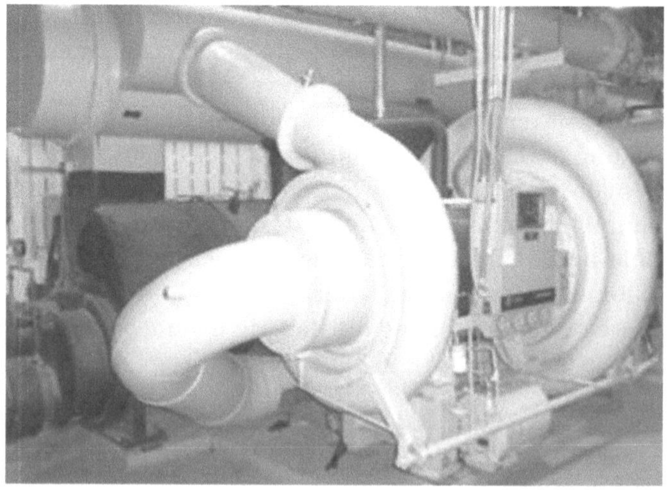

Figure 5-32. Centrifugal chiller with two-stage compressor.

of the chilled water with the same conditions. Needless to say, always monitor the temperature to ensure it is leaving the chilled water plant at a temperature of 42°F.

The typical centrifugal chiller (Figure 5-32) is designed to dump the heat to cooling tower water supplied at 85°F and returning it to the cooling tower at 95°F. Again, these are adopted standards so that equipment of different manufacturers can be used in a system. A typical centrifugal chiller also operates with a vacuum in the evaporator, 15–16 inches of Hg. Pressure in the condenser is typically 6–8 psig. The evaporator, commonly called a "chiller barrel," is a shell and tube heat exchanger, with the chilled water flowing through the tubes and liquid refrigerant surrounding the tubes inside the shell. The condenser is similar with (normally) cooling tower water flowing through the tubes. Unlike most refrigerant systems, the refrigerant is not superheated in the evaporator because all the heat is absorbed by boiling the liquid refrigerant. The evaporator is located below the inlet of the compressor. A demister in the top of the evaporator captures any droplets of liquid that are carried up from the surface of the boiling liquid, combining those droplets until they are large enough to drop back into the liquid refrigerant pool. The refrigerant is superheated as it is worked on by the compressor. Monitoring that superheat can detect problems with the compressor.

Despite the fact that the chiller barrel operates in a vacuum, it is possible for the refrigerant in an idle chiller to reach the setting of the pressure relief device. This is typically a rupture disc that looks like a large version of the lid on an old time canning jar. It is a convex sheet of metal that is bowed in toward the chiller barrel. When the pressure reaches the design of the rupture disc, it pops out (like the lid on that jar). In the process of doing that, it impinges on some sharp steel cutters that cut it so that it peels back and dumps all the refrigerant. Temperatures above 120°F can create a saturation pressure that exceeds the typical rupture disc design of 15 psig. A full charge of refrigerant in a typical chiller costs more than $25,000. Don't let those chillers get too warm. Since the centrifugal chiller operates in a vacuum, it is possible for ambient air to leak into it. There may be what looks like part of a window unit hanging around the condenser. Its purpose is to pull any air that leaks in off the top of the condenser and remove it, dumping it to atmosphere along with any moisture the air carried along.

Another device that can be applied to a centrifugal chiller is an oil separator, which recovers oil that leaked into the evaporator and returns it to the compressor sump. The oil can coat the tubes and block heat transfer. Thus, considerable effort is expended to ensure that it does not leak into the evaporator. The lubricating oil for a centrifugal chiller is not circulated with the refrigerant as it is with other compressors. The oil is sealed at the bearings and circulated in its own system, with a small portion of the chilled water used to cool the oil. If the seals leak, allowing oil to end up in the evaporator, it coats the tubes to reduce heat transfer and, therefore, the efficiency of the chiller. A separate electric oil pump is used to provide lubrication to the compressor bearings. It must be proven in operation before the compressor can be started. The oil sump is heated like other compressors and for the same reasons. The lubricating oil is typically sampled and tested regularly to determine the condition of the oil and the chiller itself. One concern is properly labeling the oil samples. Mixing up samples from several units can put the wrong oil in the wrong chiller, resulting in a rebuild of the wrong chiller.

Sometimes, people, experienced with the need for superheat to control the flow through the throttling device, will insist that the gas leaving the evaporator of a chiller has to be superheated. That is not the case with centrifugal chillers. The throttling device in a centrifugal chiller is a float valve (Figure 5-22) and normally consists of two float valves that separate two different pressures. The system of two float valves, and the chamber between them, is commonly called an economizer because it captures some of the vapor that flashes as the pressure is dropped and uses it to cool the hermetic motor of the chiller, entering the compressor between two stages. That system saves the compressor from pumping all of the gas from the evaporator pressure to the condenser pressure. Being a large and expensive machine, they are normally insured. The insurers typically require a breakdown inspection of a chiller every 15 years. Examinations include electrical insulation testing, leak testing, and eddy current testing of the tubes. There is a hint here. The typical chiller is expected to operate for 15 years without concern. It is a highly reliable piece of equipment.

A centrifugal chiller system can produce cooling for a fraction of the cost of the typical air-to-air cooling system. One chiller can produce chilled water at a lower first cost than self-contained units, all in a smaller space, for as little as half the operating cost. Centrifugal chillers are made with capacities, at the time of this writing, as high as 9000 tons (field assembled). That one would be a fraction of the size of 9000 ton window units stacked up.

Just like a boiler, the load for a chiller will change with outside air conditions. To accommodate reduced loads, many chillers use inlet vanes that throttle the flow of the evaporated refrigerant to the compressor, while causing it to swirl, thereby reducing the horsepower requirements of the compressor (Figure 5-14). More modern chillers control the evaporator temperature by varying the speed of the compressor. When the vanes, or reduced speed, cannot prevent the evaporator pressure from dropping to produce colder chilled water, a low pressure switch shuts down the compressor. The low pressure switch has a differential setting, which allows the evaporator pressure to rise over that range before restarting the compressor. A chiller, cycling on and off, is not much different than a boiler cycling. Whenever possible, a change to a smaller unit or shutting one down completely can reduce power consumption and demand charges.

Other considerations, when operating centrifugal chillers, include the monitoring of the temperature and corrosive/scaling condition of the cooling tower and chilled water, the level of the liquid over the tubes in the evaporator, the lubricating oil level, and the oil temperature. A centrifugal compressor takes some time to come to a stop. A problem can occur if all power is lost to the plant. The result will be the loss of lubricating oil pressure before the motor stops turning, resulting in wiping of the bearings. That means the moving metal parts are not separated by the oil. They rub stationary parts, resulting in rubbing (wiping) metal off the bearings. Some centrifugal compressors have surge tanks for oil that provide enough for lubrication on a power failure. Others require an uninterruptible power supply (UPS) to keep the oil pump running until the shaft stops rotating. Modern compressors with magnetic bearings will also require a UPS. The backup power in some facilities relies on the startup of emergency generators before the UPS batteries lose their charge. A very few units might even have a hand pump that can be operated to keep the chiller lubricated until it stops turning. Be sure that the SOPs and disaster plans include consideration for power loss to large chillers.

Figure 5-33. Absorption chiller.

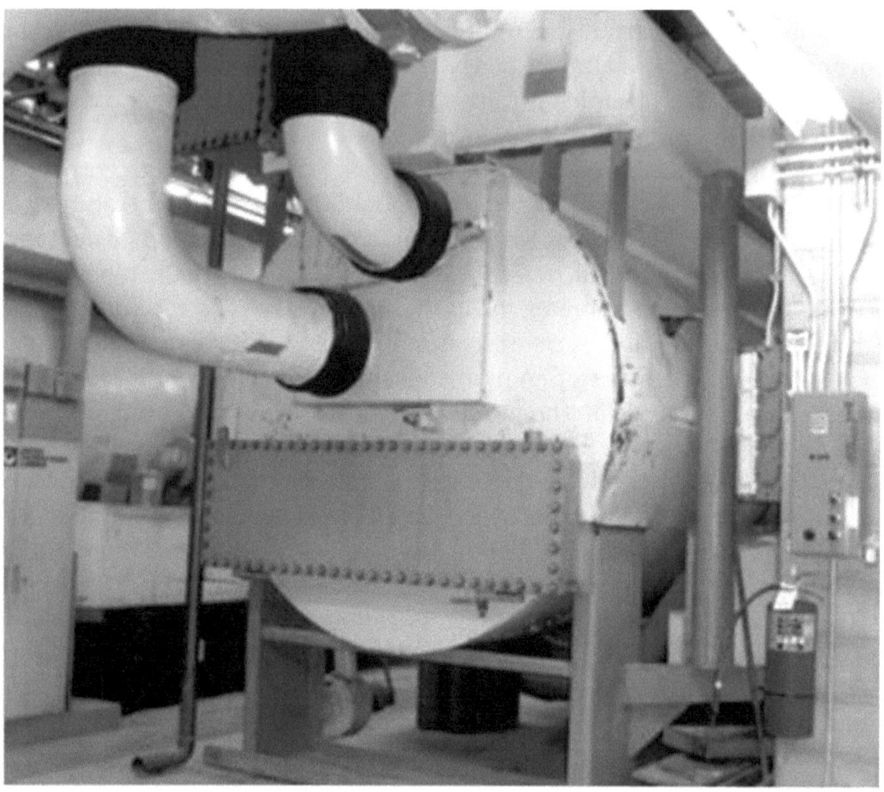

Figure 5-34. Absorption chiller rear view.

Absorption Chillers

The refrigerator in a typical recreational vehicle (RV) runs on propane. It boils a solution of ammonia and water to evaporate the ammonia, which is then condensed in coils on the back of the refrigerator and flows into the refrigerator, where it evaporates to absorb heat from the freezer and refrigerator sections. It uses the principle of partial pressures to achieve the cooling. The cooling is achieved because the liquid ammonia enters the evaporator sections that contain hydrogen gas, which provides a sensation of lower pressure so that the ammonia can evaporate. The evaporated ammonia is then absorbed in water that is cooled by another coil. The mixture is returned to the furnace section to be boiled again. The refrigerator thermostat controls the firing of the gas. It is not a particularly efficient cooling device. However, it works pretty much everywhere, as long as there is propane in the tank and there is a charged 12 volt battery to power the controls. There is not much difference between an RV refrigerator and an absorption chiller, in concept. Size, construction, only chilling water, and efficiency are the major differences. There are two systems for absorption refrigeration: water and ammonia like the RV refrigerator, where ammonia is the refrigerant and a solution of lithium bromide and water, where water is the refrigerant. The latter is more prevalent. The absorber is a large assembly of tubes, pumps, and piping, all as shown in Figures 5-33 and 5-34. Instead of using gravity like the RV refrigerator, the solution is pumped, spraying the solution on the coils for better heat transfer. Figure 5-35 is a diagram of a single stage absorption chiller.

Modern units are typically two-stage. The solution is pumped to the generator stage, where heat

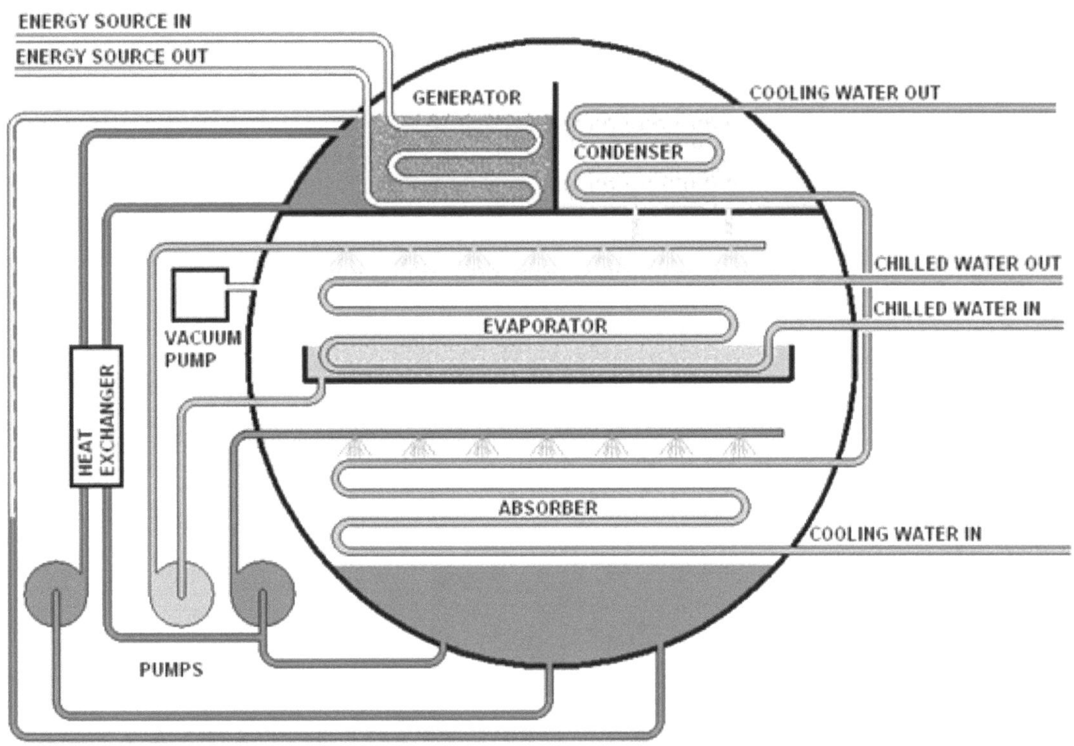

Figure 5-35. Absorption chiller schematic.

is added to vaporize the refrigerant, either the ammonia or water. The source of the heat can be a fire, steam, hot water, or any other fluid that is hot enough for the absorption chillers to take advantage of heat that would otherwise be wasted. The refrigerant is then condensed by a condenser, normally served by cooling tower water. Other cooling sources (ocean, lake, or river water) can also be used. The concentrated sorbent (water or the lithium bromide solution) drains back to the absorber. The liquid refrigerant drains from the condenser through a series of orifices back into the absorber section, over the chilled water coils, and into a pan connected to a pump that pumps it back up and, through sprays, over the chilled water coils. When the refrigerant evaporates to cool the chilled water, the vapor is absorbed in the sorbent, which is sprayed in a lower chamber to increase the rate of absorption.

The use of ammonia in an absorption chiller requires an additional feature not shown in the schematic. When evacuating the absorber, to produce the low pressure required for the ammonia to evaporate at chilled water temperatures, equipment is required to recover the ammonia, rather than just dumping it into the chiller room. The potential hazards with ammonia are what compel most designers to use lithium bromide. While much safer to use, lithium bromide is a salt. Under upset conditions, it is possible to concentrate the solution too much in the generator section. Then the salt crystallizes. The generator section subsequently plugs up with the salt.

Just like the RV refrigerator, an absorption chiller can be fired. After absorbing the vaporized refrigerant, the sorbent is pumped through a generator, which is just a boiler fired, by gas or oil, that boils off the refrigerant and concentrates the sorbent, which is recycled back to the evaporator. They can also be operated similar to a heat pump, where the sorbent is only heated and returned to make hot water, instead of chilled water, for a two-pipe air conditioning system. Absorption chillers have been in and out of favor, depending on both the price of electricity and fossil fuels and advances in efficiency of centrifugal and absorption chillers. Facilities that have both absorbers and centrifugal chillers can take advantage of the swings in the cost of energy to operate the chillers that are most economical. First, there is the opportunity to reduce demand charges by operating an absorber during peak load periods. Then there is the constantly changing difference in energy prices. An absorber can be cheaper to operate at certain times of the day. However, starting and stopping one takes time and energy. There will be a fair amount of calculating needed to determine how to operate, especially in regions where electricity prices can change every hour.

COOLING TOWERS

For refrigeration systems, cooling towers are used to reduce the pressure differential that a refrigerant compressor has to overcome in order to pump the refrigerant from the evaporator to the condenser. It is accomplished by cooling the water used in the water cooled condensers of refrigeration equipment, especially chillers. A cooling tower helps cool the water by evaporating some of it. A boiler operator knows that it takes about 1000 Btu to evaporate a pound of water. In addition to reducing the cooling tower water temperature by 10°F, the water is cooled by evaporation of some of the water. A good rule of thumb is that the cooling tower will evaporate 2 gpm (gallons per minute) of water for every one million Btu/hr removed from the cooling water. The tower can cool the water almost to the wet bulb temperature of the atmospheric air (see psychrometrics in the section on air conditioning). That is seldom higher than 85°F. The cooling tower operates by distributing the water over a fill which has atmospheric air flowing through it, normally forced by a fan. The fan can be on the outlet of the tower, the common location, or force air into the tower, where it is commonly referred to as a "blow-thru" cooling tower, as shown in Figure 5-36. Other terms for cooling towers are cross-flow and counter-flow. A "blow-thru" is commonly counter-flow, where the air is pushed up through the tower as the water drops down through. In a cross-flow tower, the air enters the sides or the ends, crossing the flow of flowing water and then rising to the fan inlet, as shown in Figure 5-37.

Normally, the water is delivered to perforated trays at the top of the cooling tower. Some use sprays, but the perforated tray serves to distribute the water evenly over the top of the fill. "Fill" is the word used to describe the material in the tower that the water runs over and around as it drops from the inlet trays or sprays to the sump at the bottom of the cooling tower. The fill is designed to convert the falling water to a film along the surface of the fill and thin sheets or droplets of water between parts of the fill. The air is drawn through the fill to contact the air over that extended surface to achieve the heat transfer and sweep vaporized water out of the tower. The fill can consist of redwood slats in several alternating layers (Figure 5-38). However, they present a fire hazard when idled during the winter. There are many variations and designs of plastic fill used today.

A part of the plume can be drift. Drift identifies droplets of water that are swept off the fill by the air and leave the tower without being vaporized. Those

Figure 5-36. Blow-thru cooling tower.

Figure 5-37. Cross-flow cooling towers.

droplets are usually much larger than fog and drop out on adjacent structures. Excessive drift is an indication of air velocities in the tower that are higher than design. Sometimes, it only occurs at high loads. It can also be a problem at lower loads, if the fill is distorted, damaged, covered with organic growth, or otherwise altered to reduce or block the air flow in some parts of the tower. Thus, the velocities have to be higher in other parts to compensate for the blockage. Towers can contain baffles to redirect water splashing off the fill and back into the tower. They can be bent or otherwise altered by high winds, falling tree limbs, and accidents to increase liquid water loss. Drift is simply a waste of water and should be addressed when it is detected. Consistent pools of water on adjacent roofs (evident in Figure 5-37) are indications of drift.

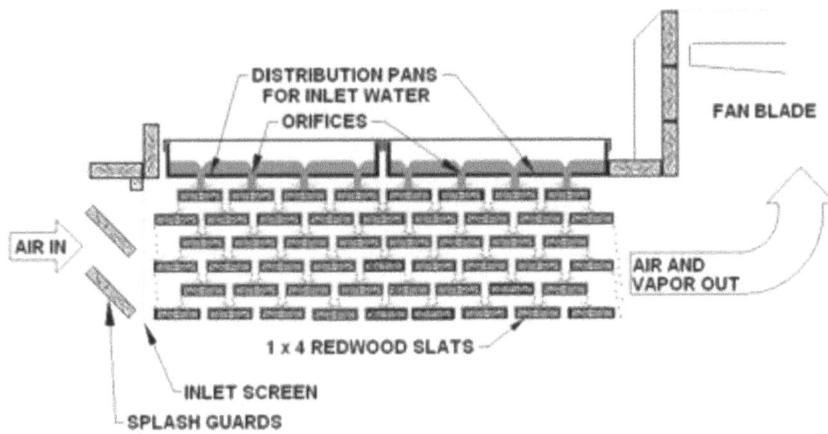

Figure 5-38. Cross section of redwood cooling tower

Figure 5-39. Hyperbolic Cooling Tower in a Power Plant

Cooling towers in electric power plants and large industrial plants can operate without a fan to force the flow of the air. Typical photographs of nuclear power plants show the tall hyperbolic cooling towers (Figure 5-39), which use the differential pressure produced by the difference in density of the atmospheric air and the air heated in the towers, which contains a larger volume of lighter steam. The plume that is associated with cooling towers is actually droplets of water that condense as the exhaust of the cooling tower is cooled by the slightly cooler atmospheric air. A better name for that plume would be fog. Neither the cooling tower itself nor the water vapor that is produced is radioactive. The cooling towers are just the largest pieces of equipment at the power plant and make for captivating photos.

In an electric power plant, colder condenser water, cooled by the cooling tower, increases the power that the turbine can produce. This is done by reducing the exhaust pressure of the steam emanating from the steam turbine. That pressure is set by the vapor pressure of the water that is condensing at the cooling water temperature. The cooling water is cooled as much as possible, stopping in the winter slightly above 32°F to prevent freezing of the water. Presuming the steam leaving the turbine is essentially pure, it will condense at the temperature of the cooling water. That sets the vacuum in the condenser.

Cooling tower water used in industry for cooling of production equipment or product can also be allowed to vary in temperature. Typically, the lower limit is set higher to minimize thermal shock and to ensure that the temperature control valves on the cooling equipment maintain control. For chilled water systems, a cooling tower is normally operated to maintain a leaving temperature of 85°F to match the design condition for most chillers. Maintaining the temperature maintains pressure to ensure adequate liquid flow through the throttling device. In circumstances of low loads, and where the atmospheric temperature is colder, fan speed is adjusted to maintain a minimum temperature, unlike the typical water cooled condenser, where water flow is controlled. Variable speed drives on cooling tower water pumps can reduce power costs but are limited to ensure good distribution of the water in the cooling tower.

The water vaporized in a cooling tower has to be replaced with makeup. In addition, because the vapor does not carry off the solids that were dissolved in the water, blowdown of the cooling tower is required to prevent scaling and deposits of mud on the fill. Small towers commonly have a float valve in the sump that adds makeup water to maintain the level. Larger towers can have level controls using an electric probe system that senses the water level to open and close a solenoid or motorized makeup water valve. Blowdown control can be manually set or automatic. Except for cooling the water, they work the same as a boiler's continuous blowdown. Any large system will also have meters on the makeup and blowdown water flows to produce a difference between the two that equals water lost to vaporization and drift. As a result, the owner does not pay for sewage treatment of water that was not dumped down the sewer. Just like a boiler, the cooling tower water chemistry has to be monitored and maintained to prevent damage to the tower, the condenser, the pumps, and the piping. In addition to scale and corrosion protection, chemicals are added to control the growth of bacteria and algae in the cooling tower water.

AIR CONDITIONING

Normally, air conditioning is thought of as cooling the air in a space, a room, or building. More appropriately, air conditioning should be considered to be the heating or cooling, plus adding or removing moisture, and removing airborne contaminants from the space to maintain conditions in that space that are comfortable for the occupants. That definition covers more than what is typically thought of. In many zoos around the country, there are spaces where the conditions are not necessarily comfortable for humans but are enjoyed by the occupants. With few exceptions, air conditioning is accomplished by removing air from the space, altering its conditions, and then returning it to the space, where it is mixed with the air in the space to produce the comfortable conditions for the occupants.

Another common term is HVAC, standing for heating, ventilating, and air conditioning. Strictly speaking, there are many facilities that only receive heating and ventilation. Since most people today expect complete air conditioning, it is not necessarily an applicable label anymore. Air conditioning is the best overall label.

The typical window unit (Figure 5-40) is a packaged air conditioning system that incorporates a complete refrigeration cycle within it to transfer heat from the room to the atmosphere outside the window. Air from the room (A) is drawn into the unit through a filter (B) that cleans the air. The air passes over an evaporator coil (C) with fins which removes the heat from the air and condenses some of the moisture in the air. Then a fan (D) blows the air back into the room (E) at a high velocity so that it mixes with the rest of the room air to produce the comfortable condition. To achieve all of the requirements for air conditioning, there is usually a small damper (F) in the barrier between the indoor and outdoor air that can be opened to admit outside air into the flow of room air to be conditioned to achieve ventilation. Condensate, which is moisture removed from the room air, drips (G) into a drain pan at the bottom of the unit and passes through a trap (H) formed in the bottom of the casing to the outdoor side of the unit. A slinging ring, attached to the outside of the condenser fan (I), picks up the condensate and hurls it at the condenser (J). Heat from cooling the air and condensing the moisture is absorbed by the refrigerant, with flow into the evaporator controlled by a metering device, in Figure 5-40, a capillary (K), and for the typical window unit with R-134a, into the evaporator (C) at a pressure of 35 psig, which corresponds to a saturation temperature of 40°F. The boiling refrigerant absorbs the heat from the room air and condenses some

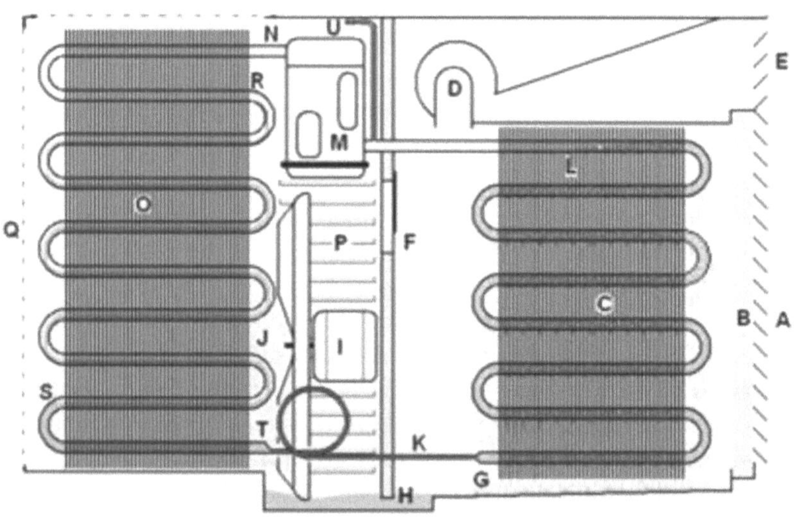

Figure 5-40. Window unit.

of the moisture from the room air to cool it from a design room temperature of 75°F to approximately 55°F. Once all the refrigerant has evaporated (L), the gas is heated to a leaving temperature of 50–60°F (10–20 degrees of superheat). The refrigerant passes through the insulated barrier between the room and outdoors to the inlet of the compressor (M). The gas is compressed by the compressor to a pressure of 100 psig, raising the temperature of the gas that enters the condenser to 200°F. The condenser (O) transfers the heat removed from the room, along with the heat added by the compressor, to outside air, drawn into the outside casing by the condenser fan (I) through louvers in the side (P), and passing over the condenser and out grills (Q) at the back and top of the unit. The grill at the top of the unit permits the admission of rain water that can assist in cooling, when the moisture contributed to the room air by the moisture from the rain penetrating the room increases the load on the unit. The first function of the condenser is to remove the superheat in the gas, reducing its temperature to the corresponding 123°F saturation temperature near (R). Then the refrigerant remains at the saturation temperature as it is condensed, until all the gas is converted to liquid near (S). Once all the gas is converted to liquid, the refrigerant is sub-cooled until it is about 105°F at the inlet of the capillary. As the pressure drops from the 100 psig in the condenser to 35 psig in the evaporator, some of the liquid is converted to a gas to absorb the heat required to cool the 105°F liquid to the 40°F saturation temperature in the evaporator. Making that window unit a heat pump simply requires a reversing of the refrigerant system. In doing so, it achieves all but one of the functions attributable to a proper air conditioning system.

- Clean the air normally by filtering it.
- Cool the air when the temperature in the room is above the set point.
- Heat the air when the temperature in the room is below the set point.
- Remove moisture from the air when cooling it.
- Add moisture to the air when heating it.
- Incorporate outside air for ventilation.

Additional provisions can include pressurizing the room to prevent contamination of the room air from external sources or reducing the pressure in the room to prevent contamination of other spaces by contaminants that are in that room. The typical window unit cannot perform the latter or either one of those with any degree of perfection and does not have means to add moisture to the air.

Introduction of outside air for ventilation may seem undesirable for facilities that lie in metropolitan areas, with plenty of vehicle and industrial exhaust contaminating that outside air. However, it is essential to add sufficient air for ventilation to restore the oxygen content of a room where the oxygen is consumed by the occupants and replaced with carbon dioxide. There are not many systems where there are enough plants within a space and enough sunlight to support photosynthesis to replace the carbon dioxide with fresh oxygen. The only spaces that have systems to do that, independent of nature, are on board Navy submarines.

A well-designed air conditioning system will draw outside air from a location that is unlikely to be contaminated. Nevertheless, there are still some systems that were designed when everyone thought the atmosphere was a giant sink that could absorb anything and remain fresh and clean. Be aware of potential problems with contaminated outside air. Be prepared to do something about it, even if it is only the poor, but essential, act of entering a regular comment about it in a log book.

The diagram in Figure 5-40 shows a feature common on small equipment, a service stub (U). That short length of tubing is provided to permit attachment of a tapping valve to permit the connection of a gauge set (Figure 5-3) to determine the unit suction pressure and add or remove refrigerant, if necessary. A service stub is common on window units, residential refrigerators and freezers, wine coolers, etc. Whenever a tapping valve is used, it should be removed by first crimping the tubing between the location of the tapping valve and the suction line. Then, after removing the valve, trim the stub end and solder it closed. If the valve is left in place, it is likely that it will vibrate loose, followed by loss of all the refrigerant. A single connection is all that is required for equipment with a capillary for the throttling device because the manufacturer always provides tables for determining the proper charge of refrigerant based on load data and superheat or sub-cooling. One of the keys to understanding the operation of an air conditioning system is learning what is comfortable for occupants and how to get the air in the space to meet those comfort requirements. In order to do that, an understanding of psychrometrics is required. Being able to use a psychrometric chart (Figure 5-41) is imperative.

Psychrometrics and Air Conditioning

The concepts and apparatus for air conditioning will be covered in a manner consistent with air flowing through an AHU. There are other labels for air conditioning equipment, but the generic term "apparatus" can be

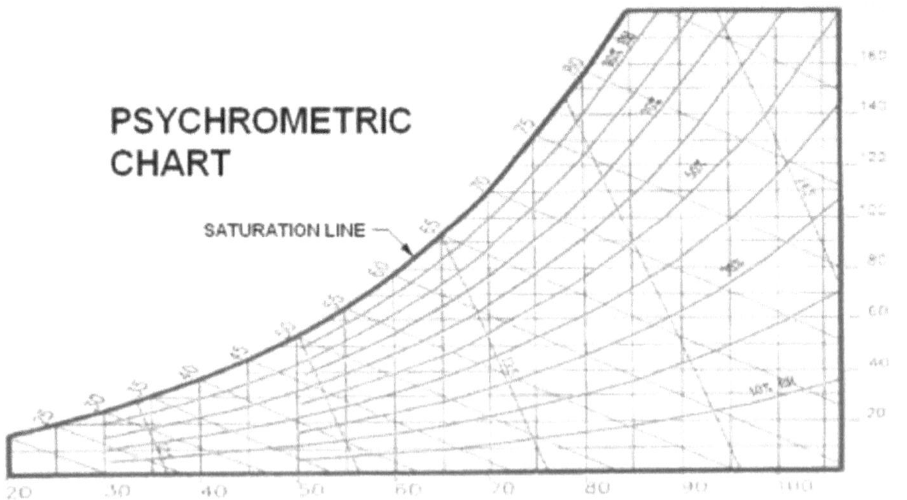

Figure 5-41. Psychrometric chart.

used to describe them all. Psychrometrics provides an understanding of the systems and processes, as the air flows from the conditioned space, through the ductwork and apparatus, and then back to the conditioned space. A psychrometric chart can be more detailed than the one in Figure 5-41. The one shown here is much simpler because it does not contain all the values and lines that are on a complete psychrometric chart. It should be adequate for the purposes of an operator. Normally, the operator does not need to know the volume of the water and air mixture, the enthalpy, and other values that are used by the engineers to design a system.

The concept of comfort is altered by being in a boiler room as opposed to being in an office. There will be times when it is not possible to isolate the reason why that lady keeps calling and complaining about the conditions in her office without using the chart. Plotting the conditions for her office and the equipment should provide a conclusion as to why she is uncomfortable and what can be done about it. The alternative is to receive constant calls to the boiler room and not being able to respond. The newer SCADA systems frequently allow the operator to get a readout of the temperature in the space in question. That does not mean that the humidity is under control and she will keep calling and complaining. Therefore, become an expert in simple psychrometrics and reduce those calls.

There is also the matter of productivity. Many studies have shown an increase in productivity in a space when the conditions were changed. Further analysis also showed that it declined for a while until the conditions were changed again. It does not mean that conditions should be constantly changing. It simply means that people are more comfortable when they believe the guy operating the air conditioner is concerned about their comfort. Ignore the psychology and simply give some attention to air conditioning complaints. There are multiple scales on the psychrometric chart. Some explanation is necessary. The horizontal scale read at the bottom of the chart is identified as the dry bulb (DB) temperature. The vertical lines from that scale are representative of that temperature. The vertical scale on the right side of the chart in Figure 5-41 is in grains of water per pound of dry air (a grain being one seven thousandth of a pound). Some charts will contain scales on the right side for pounds of moisture per pound of dry air as well. The horizontal lines are lines of constant moisture content, pounds or grains of water per pound of dry air. The scale on the curve of the chart is wet bulb temperature. The lines progressing down and to the right are lines of constant wet bulb temperature. Curves within the chart that are almost parallel to the terminating curve are values of percent humidity. Lines sloping steeply from left to right along the width of the chart indicate cubic feet of air per pound of the air and water mixture. Those are seldom used as well as the scale on the chart for enthalpy. The chart will also indicate that it is based on a barometric pressure of 29.92 inches of Hg. Deviations associated with changes in barometric pressure will not alter the basic understanding and use of the chart. Changes in barometric pressure are not significant unless the location is the visitor's shelter on Pike's Peak.

The first task upon getting the psychrometric chart is to mark it up for the comfort zones, copy it, and keep the copy handy. Because people typically change their clothes in response to outdoor conditions, two comfort zones have been established by the American Society of Heating, Refrigeration, and Air conditioning Engineers (ASHRAE). A significant number of samples for both men and women were used to determine those comfort zones for situations where people are sedentary or slightly active. Roughly, 80% of the people indicated that they were comfortable. It means that not everyone will be truly happy. When people are active, lower temperatures and humidity are required. Keep it close to the original design conditions (found on system drawings) and that should keep a lot more than 80% happy. The winter comfort zone is outlined by DB temperatures of 69 and 76°F at a moisture level of 31 grains per pound of dry air at the bottom and 68 and 74°F along a wet bulb of 64°F on top. Connect the left and right corners of the top and bottom lines to enclose the area. That is called the winter comfort zone. For the summer comfort zone, stick with the 31 grains and DB temperatures of 74 and 81°F at the bottom and plot the upper corners along the 68°F wet bulb line at 73 and 79°F DB temperatures. Now, just because the wet bulb and DB temperatures in that space indicate that the room is in the comfort zone, that is not an excuse to tell someone to get a sweater or take off their jacket. There are other factors that affect comfort.

A set of prints or drawing files on the computer for the air conditioning systems of responsibility will be needed. They contain information that can be used to isolate problems when they come up. Plotting the values from the drawings for the equipment on the psychrometric chart and copying them can help to prepare to tackle a problem. When an addition to the facility is under consideration or construction, it would help to find the design engineer and ask for copies of the charts for the systems and equipment to save some time.

In order to understand the function of a piece of air conditioning equipment, plot the values on a psychrometric chart. Comparing the design and existing conditions helps to locate the source of problems. Learning about psychrometrics is best done by using the design values. Do it before there are problems. Then have the design situation plotted out and ready to use when a problem occurs. Now, follow the design data to produce one as described in order to be familiar with the process and what is needed to plot a system on the chart.

The Air Conditioned Space (Room)

Engineers use the generic word "room" to describe the air conditioned space. Room conditions are indicated on the contract drawings and are normally expressed in DB and wet bulb temperature or DB and humidity values. Only two values are needed for a given condition of air to locate its point on a psychrometric chart. Locate the intersection of the DB and wet bulb or humidity values, mark it as a point, and then circle the point so that it is easy to find again. Noting where the point is within the comfort zone may be a key to the design engineer's understanding of the use of the spaces served by the equipment. A normal design point for office buildings is 75°F and 50% relative humidity (RH). A home for senior citizens may have higher values because seniors move around less and tend to prefer warmer temperatures. If the space is a gym, there might be cooler and dryer design conditions because the occupants are not sedentary. The same might be said for a plant's production line, where workers are always busy. Indoor swimming pools provide some unique considerations. Sometimes, the room conditions are specified for the manufactured product and not for the comfort of personnel. A good example would be a refrigerated space for food storage. On rare occasions, a room's pressure, relative to atmospheric, is specified as well.

Outdoor Conditions

Also, on the drawings will be the design outdoor temperature conditions. That will be the condition of the air pulled in to provide ventilation. Plot that point and circle it. Then draw a straight line between the room conditions and the outdoor air conditions. The room air and the outside air are usually mixed in the mixing box of the apparatus before conditioning of the air is continued. The equipment inlet conditions listed on the schedule on the drawings should fall on that line. Plot and circle it. When the two air streams are mixed, the conditions for the mixture should always fall on a straight line drawn between the two points for conditions of the air streams on a psychrometric chart. New regulations and design standards for building ventilation that are primarily concerned with the distribution of the ventilation air have led to the introduction of independent outside air supplies. Don't be surprised if some of the equipment does not have a mixing box with connected outside air ductwork.

New standards for ventilation produced by ASHRAE now require as much as 20 cfm (cubic feet per minute) per person. The design (essentially maximum) quantity for a given piece of air handling equipment is listed on the design drawings. Systems, today, use a control interface with personnel access controls, or the carbon dioxide content of the room air, to determine actual values. As with night set back temperature controls, the amount of outside air should be reduced when the building is not occupied or occupancy is very low. A round during night hours or weekends should include verification that the ventilation air is off or at minimums. It can cost a lot of money to heat or cool that air unnecessarily.

Economy Cooling

A feature of many air conditioning systems in the northern states is the provision for economy cooling (not to be confused with another concept called free cooling). When outside temperatures are low enough, air handling systems can be operated to take advantage of that colder outside air to cool spaces in the buildings, instead of cooling the air with refrigeration equipment. Core spaces, in the middle of the building, will need cooling even in the winter owing to the heat generated by people and lights alone. The savings in operating costs can be significant. The wise operator should ensure that the economy cooling is operating properly. When economy cooling is used, the outside air ducts and dampers are larger. As much as 100% of the conditioned air flow can be drawn from outdoors. Typically, the systems are fitted with a return air fan to draw the return air from the rooms and discharge it outdoors or into the mixing box. Room temperature controls are used to control the temperature of the supply air by mixing outside air and return air to produce the required temperature of supply air. Economy cooling systems need to be watched to ensure that the outdoor air is at minimum requirements during the summer. Only one of those large outdoor air dampers that hang up open can prevent adequate cooling and really pump up the electric bill.

In the deep winter, the moisture content of the conditioned air in an economy cooling system can get so low that the occupants encounter problems with static electricity. They get everything from shock, to clothing sticking to them, to hair issues. Means of adding moisture to the conditioned air to maintain a comfortable level of humidity are described later, but some systems will have overrides on the economy cooling to maintain the desired humidity in the conditioned spaces and refrigeration or chilled water cooling as needed for temperature control.

Apparatus Inlet

If the inlet conditions do not fall on that line drawn between the outside air and room conditions, it can be because the engineer has made allowances for heat gained between the room and the mixing box, which can come from heated piping, lighting heat, or sun on the roof of the space, any one of which heats the room air that returns to the unit as return air flowing between the ceiling and the floor or roof above the space. It could also be heat gained in ducts run outdoors to the equipment. To represent that added heat gain in the return air, erase the line from room air to outside air. Plot the point for the return air entering the equipment and draw a line from the outside air conditions through the inlet air conditions down to the line for the same moisture level as the room air. Draw a horizontal line that connects the room conditions with the end of the line just drawn. Unless it is an unusual air conditioning system, that should do it. When the air is simply heated, its DB temperature is increased without changes in moisture. Thus, the line is always horizontal and the percent humidity decreases. Confirm the condition by measuring the DB temperature of the return air immediately before it enters the mixing box. If it does not match, check wet bulb conditions. The return air must have picked up some moisture somehow. Any increase in moisture content between room air and return air entering the mixing box is potentially problematic, possibly caused by steam, water, or outside air leaks into the return air stream.

Filters

The equipment air flow, now a mix of outside and return air, passes through filters as it leaves the mixing box. The filters are there to remove contaminants from the air like dust or pollen brought in with the outside air (or produced by plants in the conditioned space) and any other solid airborne contaminants that might plug up the cooling and heating coils or bother the occupants. Since there is only one filter on a home air conditioner, it can also serve to remove bacteria, viruses, and even odors from the air depending on the grade of filter. Where those contaminants are a concern with industrial and institutional equipment, additional filters are normally installed at the outlet of the air handler to remove them.

Depending on the size of the facility, the replacement of those filters can become one of the duties of the boiler plant operator. Arguments go both ways in this regard. One benefit is that it places the eyes, ears, and incomparable senses of the boiler operator next to the air handling equipment on a regular basis to detect problems that can become very expensive. If that is the situation, be certain to use those senses to monitor the operation of the equipment. The argument that it takes away from attending to the equipment in the boiler room has its merits. However, properly spaced intervals of absence from the boiler room should have little, if any, effect on the quality of monitoring the boiler room. Large installations, with many air handlers, are normally maintained by contractors. Checking up on those contractors by reviewing the status of the equipment with a regular walk through or check of the SCADA system should also be a duty of the boiler plant operator.

One consideration of whoever is in charge of filters is the application of the right filter for the job. All too often, the filters are not specified carefully or the installing contractor put in the cheapest filter box and set of filters available. There are modern filters that not only do a better job and last longer than conventional ones but often cost less. When considering the cost of replacement, place a value on everyone's time as well. Cheap filters that have to be replaced more frequently may cost

Figure 5-42. Home air filter box and filter.

less. However, when the combined cost of replacement labor and additional filters is considered, more expensive filters can cost less. With the advent of Legionnaire's disease, SARS, MERS, and Covid-19, more and more attention is being paid to air handling equipment and filters. Don't hesitate to consider the replacement of the filter frames. The typical provision for a filter is a hole cut in the side of the duct and some channels that were pop riveted to the top and bottom of the duct as a guide. A good filter box, and filters (Figure 5-42), should be designed to filter every cubic centimeter of air that goes to that AHU. A proper, well-sealed, filter system will dramatically reduce cleaning and other maintenance chores on the air handling equipment. It will not eliminate it, but it can put the work off for years.

There have been some real disasters at some facilities due to the fact that filter replacement was ignored, abused, mishandled, or overdone. One can tell how well a facility is maintained by simply looking at a few filter banks. Another typical indicator is the condition of replacement filters in storage. If they are lying on a shelf or in an open box without protection from the environment and collecting dust before they are ever placed in the filter box, they can actually introduce dust and contaminants into the cooling and heating coils.

Cooling Coils

After the filters, the air normally enters the cooling coil, although the coil may have face and bypass dampers on it (Figure 5-43) that are used to control equipment outlet conditions. The air is not only cooled by the cooling coil but also by the moisture that is also removed from the air. What happens at that cooling coil is dependent on a lot of factors, including the inlet temperature and humidity of the mixed return and ventilation air, the number of people in the conditioned space, sunlight entering the conditioned space, outside air infiltrating the conditioned space through doors and leaks at windows, and a lot of things that occupants can do. The face and bypass dampers allow for varying supply air temperatures to match the cooling load. Combined with reheat coils, the face and bypass dampers also permit control of the moisture content of the supply air as well as the temperature.

Note that the damper blades do not operate in parallel. These opposed blade dampers provide a more linear position to air flow relationship for better control. The face and bypass dampers operate opposite each other. One closes as the other opens. The design provides for a pressure drop through the open bypass damper that equals that of the cooling coil and opens face dampers so that the air flow quantity is not altered by the operation of these dampers.

In constant air flow systems, two means are available for controlling the equipment outlet temperature when cooling. One is to cool all the air and then to reheat the air with heating coils (not necessarily in the air handler) to maintain conditioned air temperature. To control the apparatus outlet temperature, chilled water can be regulated by throttling the flow of water or bypassing

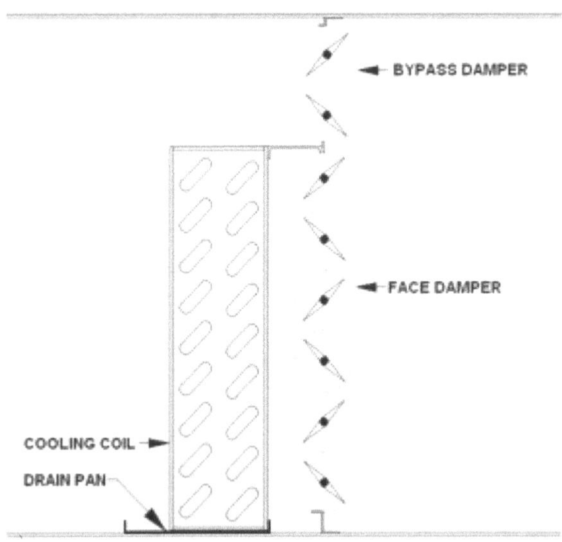

Figure 5-43. Cooling coil face and bypass dampers.

some of it. Refrigeration temperatures can be adjusted with compressor controls. Face and bypass dampers allow some of the air to pass through the apparatus without contacting the cooling coil and mix with the air passing through the cooling coil. The cooling coil not only removes heat, it also condenses moisture in the air to produce the equipment outlet conditions. The velocity of the air over the cooling coil is typically limited to 500 feet per minute so that the droplets of condensed moisture drizzle down the fins to the drain pan, instead of being carried off with the supply air. The condensed water dripping off the coils is collected in a drain pan connected to the drain piping.

Blockage of the pan or drain piping by condensate, contaminated with dust or other debris, is a common problem with air conditioning equipment. Elimination of the condensate can also be a concern. In hospitals, similar medical facilities, and production plants, the drain is piped to a sanitary sewer for obvious reasons. In some cases, the drain piping cannot slope down to discharge into the sewer or a suitable discharge point. In that case, small sumps and pumps are provided to lift the condensate into a suitable discharge area. The drains should be arranged in a manner that prevents sewer gases, or other contamination, from leaking into the air conditioning system. The normal solution for that is a "P" trap, similar to that used under a lavatory sink. A good remedy, for those plugging up, is to give them a fresh water rinse once a year. Regardless, they should be checked quarterly to ensure that water is not flooding the drain pan.

Inspection and Access Doors

Checking the drain pan would include checking the condition of the cooling coil, inlet and outlet, to detect any undesirable accumulations on the coil or in the drain pan. In order to check them, there has to be a means of looking into, and in large units, entering the space between the filters and the cooling coil. Inspection and access doors are normally provided for this purpose. Hopefully, access to them is not blocked by the piping and wiring on the outside of the unit. The covers, or doors, for these openings range from a thin piece of sheet metal to a rugged and insulated door with latching handles and hinges. They should not be left open, half hanging, or loosely mounted. The air pressure in the apparatus is negative after the filters. Any air leaking in bypasses the filters.

If, as in so many cases, the openings are simply covered with a piece of sheet metal, then it is good practice to replace the gasket that should be glued to the cover when it is accidentally damaged. Also, replace any screws that get lost. When replacing the cover, care should be taken to simply draw up the cover until the gasket is seated. Then stop turning the screw. Failure to stop will cause the metal cover to bow, leaving gaps between the screw holes. That will allow unfiltered air into the AHU. Whenever the cover screws simply spin, because the sheet metal of the AHU is stripped, look for larger screws that have plastic knobs on them to replace those old screws. The larger screws will engage in the stripped holes. The plastic knobs make it possible to remove and replace them without tools, while preventing application of too much torque that stripped the screw holes in the first place. Oh, by the way, get new screws that are just long enough to do the job to avoid wasting time removing and replacing long ones.

Hinged access doors need to have those hinges lubricated. Hear a squeak when opening one? That is the AHU indicating that it needs lubrication. Also for hinges that are worn down (cheap ones do that in a few years). They may need to be replaced to get the door to fit properly. First, try nylon washers. Remove the hinge pin, raise the door, slip the nylon washer in where the hinge has worn down, and replace the pin. Teflon washers would be better but are hard to find. Keep the access openings in good condition so that they seal out infiltration (air leaking in unfiltered). That will result in less trouble with the AHU.

Going back to the psychrometric chart, locate the point on the chart where the DB and wet bulb temperatures at the outlet of the cooling coil intersect. Then mark and circle it. Then, draw a line from the point of inlet conditions through the point of outlet conditions extended to the curve at the left side of the graph. The line between the inlet and outlet points indicates the changes in the air as it is cooled. The DB temperature decreases and the moisture content decreases. Extending that line to a point where the line intersects the curve on the left marks the apparatus dew point (ADP), which is the temperature of the cooling coil, either the refrigerant in it or close to the chilled water entering temperature. The difference between the ADP and the air temperature leaving the cooling coil is due to the fact that some of the air passes through the coil without contacting the coil but actually bypassing it. Dividing the difference between the coil outlet DB temperature and the ADP and then dividing the result by the difference between the coil inlet DB temperature and the ADP produces a value called a bypass factor. Multiplied by 100, it is the percent of the air that did not come in contact with the coil. A low bypass factor would be unusual and could be attributed to a replacement coil or change during construction, where

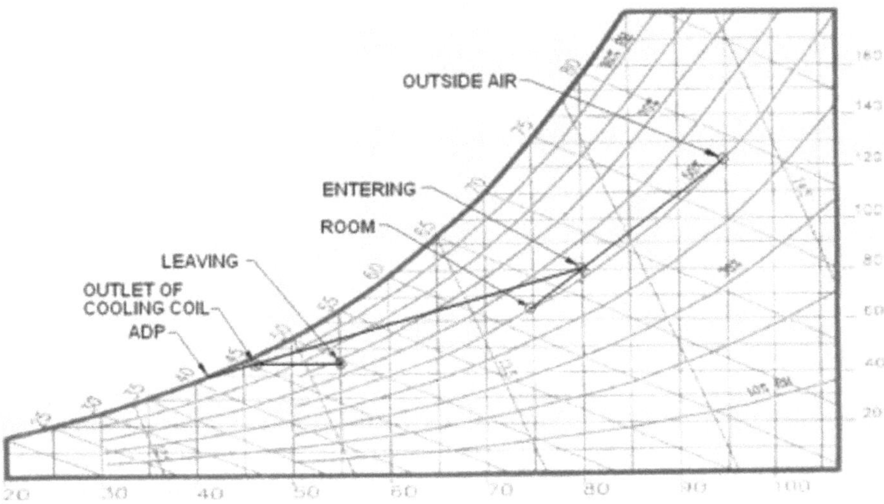

Figure 5-44. Psychrometric chart, reheat application.

a coil with a lower bypass factor was installed. A high bypass factor, one higher than that indicated by the original design drawings, or more than 20%, can indicate leakage around the coil due to changes in the air handler's structure, corrosion, and wasting away of metal baffles that prevent bypassing, structural problems with the construction of the AHU, or simply that the coil is dirty. Note that a dirty coil is detected by bypass factor and not by higher than normal outlet temperatures. A higher than design air pressure drop across the coil is also an indication. A dirty coil, like a dirty filter, changes the air flow, which normally results in cooler unit outlet temperatures because the coil has less air to cool. These points are shown on the chart for cooling and reheat in Figure 5-44.

Ultraviolet Lights

Some devices contain UV lights inside. They are there to kill bacteria and other growths that occur on the cooling coil. The cool and moist condition is an excellent breeding ground for all those nasty things. They are high intensity lights that can injure eyes (and exposed skin), which is why there should be signs on the access doors warning of exposure to them. Use a set of UV filtering eyeglasses to wear when inspecting the lights. It is necessary to open the access door and look in at them to see if they are working. When inspecting the cooling and heating coils, turn off the lights using the switch that should be nearby. When ready to close the unit up, turn the lights back on. Make sure they are all on before replacing the covers or closing the access doors. Check the upstream side of the bulbs while looking at the coils to detect any dust or debris that has impinged on them. Wipe that side with a white glove or cloth for best results. If the result is a dirty glove or cloth, there are likely problems with the filters.

Heating Coil

Some apparatus, but not all, will have a heating coil after the cooling coil. When cooling of the air is not required, heating may be. The coil is there to heat the supply air leaving the apparatus. It becomes necessary in many installations because the large, required, quantity of ventilating that air adds so much cold outside air that heating it is necessary. Even systems with reheat coils (heating coils in the ductwork supplying different portions of the air conditioned space) will have a heating coil in the main apparatus. The cold air in the ductwork would promote condensation that could damage the insulation and ductwork. The clue here is don't shut down that heating coil in the winter. Don't ignore its lack of operation, thinking the reheat coils will take care of things. The result may be wet and falling ceiling tiles. Heating can be performed with electric heating coils (very expensive to operate), steam or hot water heating coils, or a gas or oil fired furnace. Of course, the furnace is not a heating coil. It accomplishes the same purpose. When there is a heating coil, it can be used for partial reheat or simply heating. The apparatus can be operated for humidity control of the conditioned space by cooling the air more than necessary for temperature control in order to remove more moisture from the air and reheat it to provide the desired supply air temperature. When simply heating, the line representing the changing condition of the air on the psychrometric chart is represented by a line drawn horizontally from the apparatus entering conditions to the apparatus outlet conditions. In reheat applications, the line is drawn from the conditions at the outlet of the cooling coil to the apparatus outlet conditions, after the heating coil. A typical heating application is shown in Figure 5-45.

Humidifier

Whenever a considerable amount of outdoor air is used for ventilation, it has to be heated when the outside air is cold. Because the heating increases the DB temperature (as indicated by a horizontal line from left to right on the psychrometric chart), the humidity of that air drops. Unless moisture is added to the air, the personnel in the

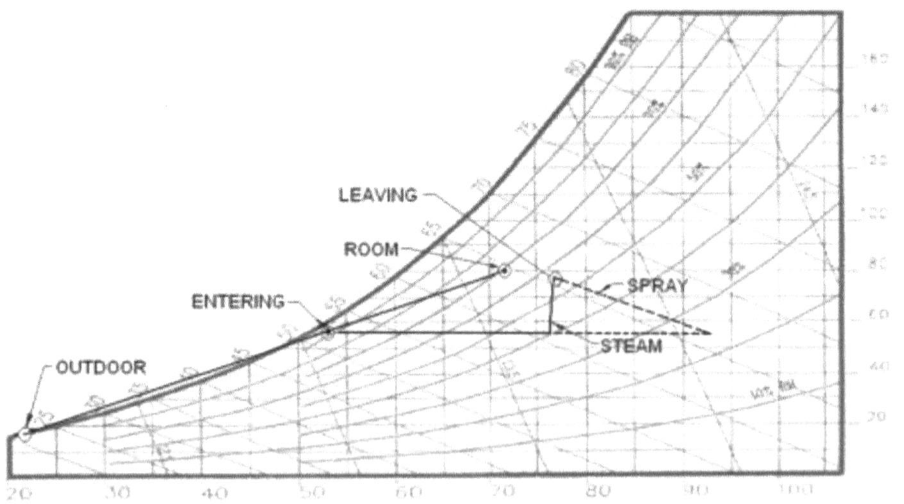

Figure 5-45. Psychrometric chart, heating and humidification.

conditioned space will experience electric shocks and other unpleasant conditions. Humidifiers add moisture to the air. Depending on their source of water, they can also heat or cool the air. The typical home humidifier consists of a high pressure spray that atomizes water to form fine droplets of it and injects them into the air stream coming off the heater. Others use a ceramic or fiber wick that is soaked with water and exposed to the air stream. Some of the heat in the air is used to evaporate the water. Being a boiler operator and knowing how much heat it takes to convert liquid water to a gas, it is easy to see how the line on a psychrometric chart for that service would be drawn from the conditions at the outlet of the heating coil up and to the left, both adding moisture to the air and cooling it some.

Larger applications can use similar means of humidification. They typically use steam humidifiers, which not only add moisture but also add heat to the air. The line on the psychrometric chart for that service would be drawn from the conditions at the outlet of the heating coil up and to the left, both adding moisture to the air and heating it a little. These two are shown on the chart in Figure 5-45. The dashed line shows that the air coming off the coil would have to be much hotter to vaporize the spray or wicked water because of the heat necessary to vaporize the water. The steam only adds about one degree to the air temperature, while cooling the steam which is already vaporized. One interesting element of the diagram on that chart is that the line for mixing the outside and return air indicates that if more outside air was in the mix, the conditions would be to the left of the saturation curve. That is when the air would be supersaturated with moisture and droplets of water forming, producing what is referred to as fog. The condition could occur as the return and outside air mix, resulting in wet spots in the mixing box or, worse yet, in the filters. One approach to eliminating the problem is a mixing box design that introduces sufficient turbulence to prevent droplets accumulating on the walls of the mixing box and ensuring good mixing before the combined air streams reach the filters. The other approach is to heat the outside air before it is mixed with the return air.

If a heating coil is used to boost the temperature of the outside air to prevent fogging at the apparatus inlet, it requires close attention when steam or hot water is used as the heating medium. It is also true when large quantities of outside air are required for ventilation (typical for movie theaters and similar venues). Then the cooling coil (if using chilled water) and the heating coil all require attention in order to protect from freezing. A poorly functioning steam trap is not all that is required for a steam coil to freeze up. Sometimes, the heating controls will throttle a steam valve so much that the pressure is inadequate to push the condensate out of the coil. Then, even the so-called freeze proof coils will freeze. Hot water coils have also frozen when the flow to them is throttled enough that localized freezing in the coil is initiated. That leads to blockage of part of the coil and freezing damage. An idle cooling coil, unless drained for the winter, can easily be frozen if the outdoor air heating coil fails. Today's SCADA systems typically monitor apparatus temperatures to warn of freezing conditions. Conduct an immediate inspection of the unit to determine if failure can be averted.

Furnaces

Gas fired and oil fired furnaces that heat the air directly can contribute to the discomfort and, under extreme conditions, injure the health of occupants. Even a small crack in a furnace section can introduce enough carbon monoxide (CO) to injure occupants. CO is not truly a poison. CO just combines with the blood much faster than oxygen. If a person breathes in enough CO, the person essentially suffocates. The process is reversible. Enough oxygen will drive off the CO attached to the blood. If there are operating carbon monoxide detectors, check their operation regularly. Don't disable one if it alarms repeatedly. It is probably indicting that there is a problem. Even a clean fire will produce CO under the right conditions. Unless shutdown introduces more serious hazards, such as freezing occupants, shut the unit down until the cause of the alarm is verified. Insist on having a portable analyzer if there are no fixed ones. Use it to check all spaces supplied by air from a furnace and any complaint of uncomfortable conditions. Notify the owner that all personnel should be evacuated from an area that contains CO. Anyone can call a utility or fire company that will come to the site. If they find CO in the building, shut off the fuel supply and order the place evacuated. CO kills should not be treated lightly.

Zone Dampers

When required to serve several zones (an independent group of spaces with different heating and cooling requirements), the AHU can be a push through design, with the fan mounted between the filter housing and the coils along with face and bypass dampers for each zone. Those dampers, installed after the coils, mix the outputs of the two coils to provide a supply air temperature to the respective zone controlled by that zone's thermostat. The zone dampers are like face and bypass dampers but simply direct air through the heating coil, the cooling coil, or a mix of the two to provide the desired supply air temperature to a specific zone.

Fan Housing

The normal arrangement of an air conditioning apparatus is to have the fan in line after the stuff already covered, making it what is called a draw-thru apparatus. If the fan or fans were located on the inlet, it would be referred to as a blow-thru apparatus. An advantage of the fan at the outlet is that it is in conditioned air. Blow-thru fans need regular attention to remove dust and dirt accumulating on the blades of the fan. That is a typical problem with return air fans as well. The fan can be powered by a motor contained within the casing of the apparatus or a motor mounted outside the apparatus and connected by belt and pulley to a shaft passing through the fan housing and supporting the fan wheels with separate bearings. A motor mounted inside the housing requires more space, plus room to access it for maintenance and is usually restricted to roof top units (RTUs), which will be covered later, or very large AHUs. The inefficiency of the motor (3%–7% of power consumed by the motor) is heat added to the air. Also, the fan simply working on the air adds heat that is normally considered a load on the supply air in the engineer's calculations.

The most common service for large units today is variable air volume (VAV), where the quantity of conditioned air supplied to a particular zone is controlled by a VAV box (described later) so that the amount of air passing through the apparatus is variable. To limit duct pressure and reduce operating costs, the fan is fitted with a variable speed controller that changes the fan speed to maintain a constant discharge pressure at the outlet of the apparatus or, when equipped, the high efficiency particulate air (HEPA) filters.

HEPA Filters

An HEPA filter is required to remove at least 99.97% of virus particles as small as 0.3 microns. HEPA filters are used in applications that require contamination control, such as the manufacturing of disk drives, medical devices, semiconductors, nuclear, food, and pharmaceutical products, as well as in hospitals, homes, and vehicles. With increased emphasis on protection from respiratory viruses, there will likely be more use of these filters in the future. They should be checked and changed at least every six months in commercial settings. In residential settings, they can be changed every two to three years. Failing to change an HEPA filter in a timely fashion will result in it putting stress on the machine or system and not removing particles from the air properly.

HEPA filters are normally installed on the outlet of the AHU. Since it is on the pressure side, it ensures that there will be no infiltration of air after the HEPA filter. The pressure drop required to achieve the filtration would be added to the negative pressure in the apparatus casing, which would make in-leakage of unconditioned air more of a problem, along with requiring a stronger casing. HEPA filters can actually be so fine that they can remove odors. It is another thing that requires attention to the instruction manual.

Heat Exchangers, Heat and Energy Wheels

Applications that require very high percentages of outside air are frequently equipped with some means of exchanging some energy between the outside air and exhaust air. Heat pipes were mentioned earlier. Dilution of pollution of the conditioned space by manufacturing processes and concern for cross contamination of work are typical situations where 100% of outside air is used. A simple heat exchanger that kept the two air streams isolated and ensures no cross contamination can be used.

Heat and energy wheels are what the name implies. The heat wheels contain a fine lattice of metal that absorbs heat from the hot air in summer and dumps it into the cooler exhaust air. The same wheel absorbs heat from the exhaust air in winter and heats the cold outside air. Energy wheels, sometimes called enthalpy wheels, use a desiccant coating that allows the wheel to absorb and discharge moisture as well. The wheel is installed in a casing with outside and exhaust air inlet and outlet connections. As it rotates, it picks up heat (and moisture) from one air stream and discharges it to the other. This application does not differentiate between heat and moisture and other things that can be in the air. Thus, it is possible that contamination from the conditioned space can be returned to the outside air drawn into the unit. A particular problem with the wheel units is maintaining the seals that separate the two air streams. Secondary problems include blockage due to filter leakage (some of them require filters on the exhaust air to protect the wheel). They are frequently equipped with a number of pressure sensors to detect problems of leakage and blockage.

Complete the Design Charts

The last line to draw on the psychrometric chart is from the equipment outlet condition to the room condition. That represents the heat and moisture added to the air as it enters the room and mixes with the room air to absorb the heat and moisture that is generated in the space. Once the charts for all the design conditions have been made, make copies of them. They can be used to compare with actual operation when problems arise. If there are actual measurements under full load conditions, save those charts as well. They are now "as-builts" and not design situations. They provide a better reference to use. The problem is that actual design conditions are seldom reached. Even then, often, there is no time to take the readings at exactly that moment. Don't forget that there are at least two design conditions, summer and winter. Now, instead of a steam cycle or refrigerant cycle, there is an air cycle. Label the chart with the equipment name or number, copy it, and put it away for the day when it will be needed. Do a couple of others, where the conditions are different. That provides a feel for what happens there. Gradually build a library of the design and normal operating conditions for each piece of air conditioning equipment. Then the needed information will be at hand, along with a full understanding of what is supposed to happen.

Air Handling Units

The typical AHU consists of a mixing box, where return and outside air are mixed before entering the filters, a filter box, face and bypass dampers (if included), an access space for maintenance of the cooling coil, cooling coil, another access space for both cooling and heating coils, heating coil, humidifier, and fan housing. Most AHUs are fitted with a fan driven by a shaft connecting to a sheave outside the unit, with the motor, sheave, and belts. These are typically prefabricated in sections for field assembly but can also be shipped as an assembled unit that is simply rigged into place. Most AHUs use chilled water in the cooling coil and steam or hot water in the heating coil. Valves controlling the flow of the heating and cooling mediums have to be added along with diffusers, supply, return, and outside air ductwork to complete the installation. Openings for access to inspect and maintain the filters, fan, and coils are provided in the casing. A drain pan is provided at the bottom of a cooling coil to collect and direct the condensate formed, as moisture is removed from the air

Ductwork

Every system, other than window units and the thru-the-wall units at hotels and motels, is fitted with ductwork to convey the conditioned supply air from the apparatus to the air conditioned spaces, return air from the conditioned spaces, and supply outside air to the inlet of the mixing box on the apparatus. As a general rule, air leaks in the ductwork were typically in the range of 20%–30%. Jointing methods for sheet metal duct were developed for low cost installation and, except where specified otherwise, subject to sloppy installation that produced lots of leaks. Even insulated ducts had heavy leakage. Newer standards have reduced the amount of duct leakage. Don't forget that it is still significant in anything designed or built before 1973 (the time of the first so-called energy crisis).

Ductwork is designed in one of two ways: constant friction or static regain. The first is simply choosing duct sizes so that the pressure drop is a constant (0.1 inches water column (W.C.) per 100 feet of duct). The required discharge pressure at the supply air fan could be calculated readily by multiplying the length of duct (in feet) to the furthest register with corrections for elbows and other fittings by 0.001. Static regain design allowed for determination of the recovery of static pressure as the air velocity decreases, providing additional force to convey the air to that furthest register. Regardless of the design, the operator only has to know that a boost in static pressure will occur when the air slows down, provided that it slows down in a manner that is not produced by a damper, or other restriction, where the regain is considerably less than 100%. The example in Figure 5-46 shows how

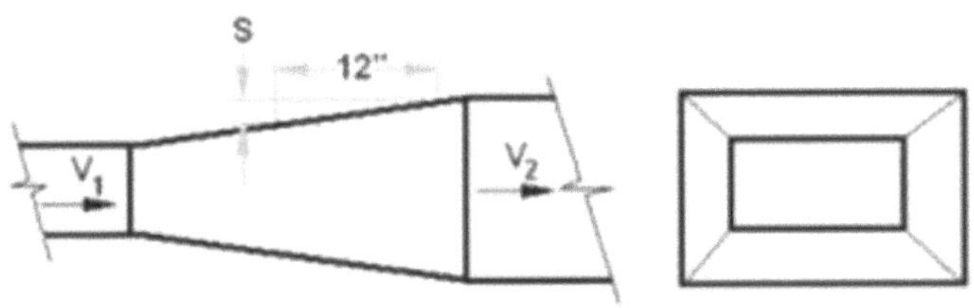

TABLE FOR DETERMINING THE VALUE OF 'R'						
$V_2 \div V_1$	SLOPE (S) INCHES PER FOOT (USE LARGEST)					
	1	2	3.25	4.5	7	10
0.20	0.83	0.74	0.68	0.62	0.52	0.45
0.40	0.89	0.83	0.78	0.74	0.68	0.64
0.60	0.93	0.87	0.84	0.82	0.79	0.77

VELOCITY PRESSURE (P_V) = ($V \div 4007$)
STATIC PRESSURE REGAIN = R x (P_{V2} - P_{V1})

Figure 5-46. Duct velocity and regain calculation.

static pressure may be recovered from velocity regain in a duct expansion. Different configurations provide other values for the regain factor (R). Simply looking at a configuration and how abrupt the changes are can give an idea of how much of the potential regain can be recovered.

Don't confuse a lack of air flow to a conditioned space with blockage of ductwork. There can be several problems with duct failures, many including that flexible duct which is easy to crush.

Reheat Coils

When apparatus serves spaces containing many rooms, offices, or cubicles, it is difficult to ensure occupant comfort because the loads can change. Some of the area can be empty with a large number of people gathered in one part of the space to produce a concentrated load in one section. To handle situations where that can occur, spaces served by a common piece of apparatus can be fitted with reheat coils that warm the air entering spaces where the cooling load is lower than normal (no people, sometimes no lights). Conference rooms and similar locations may also have time controls or occupancy sensors that actually reset the thermostatic control to allow the room to be cooler, or warmer, unless in use or scheduled to be in use.

Reheat coils are typically thermostatically controlled for the space in which they are located (which can be part of a number of spaces served by the same apparatus and reheat coil). The control system commonly uses a method of detecting the warmest of those spaces to control the temperature of the air leaving the apparatus and allows the remaining spaces to use their reheat coils to achieve comfort in them. It seems strange to cool the air and then heat it again. Still, it is one way of ensuring occupant comfort. At one major college, a considerable decrease in steam load was realized when the chiller plant was shut down for maintenance. To eliminate the energy waste associated with reheat, some new system designs, including VAV boxes, have been developed.

VAV Systems

Most modern systems have AHUs, or rooftop units, with a variable speed drive controlling the fan to supply conditioned air to VAV boxes. The space control thermostats are used to regulate the flow of conditioned air into the spaces to maintain the space DB temperature. Varying the air flow, however, can create difficulties with space ventilation. Passing of small clouds that block sunlight to one or two rooms of a zone served by the same VAV box can produce detectable temperature swings in those rooms, just like systems on reheat coils. A VAV box maintains temperature control in the room or zone by varying the quantity of conditioned air supplied to the room or zone. All the VAV boxes served by one piece of apparatus are supplied conditioned air to serve the largest load of the group. Theoretically, only one or two VAV box dampers will be wide open at any time. As long as the loads served by that AHU change uniformly, no room supplied by a VAV box will have its damper throttled so much that the air registers cannot mix the conditioned and room air enough to prevent poor ventilation and drafts. Despite that, many systems are now installed with separate ventilation air supplies.

Some VAV boxes are fan powered to provide the differential pressure required for proper distribution of conditioned air into the space. Electrical savings are realized when those fans are shut down during periods of no occupancy. The fans in the VAV boxes do not vary air flow since the airflow has to be maintained to provide proper mixing of supply and room air at the registers. The fan takes a suction on the inlet of the supply air after the VAV box damper and a return air connection from the conditioned space.

VAV boxes serving rooms or zones at the sides of the buildings, or under the roof, are typically fan powered and include a heating coil to absorb the load of heat losses to cold outside air. These are normally fitted with a filter in their return air connection to prevent accumulation of dust and airborne debris in the fins of the heating coil and on the fan (the conditioned air should be clean). Most of these maintain temperature control during the winter, and on cold days, by throttling the flow of the heating medium, usually hot water, through the heating coil. Since those filters are always in service, monitoring of their condition until an optimal replacement period is established is an added predictive maintenance requirement for a new system. Other than situations where the air in the zone can be contaminated inconsistently during regular activity, a filter replacement program can be established with longer periods between replacements than those recommended by the manufacturers.

Almost all VAV boxes are fitted with air flow sensors that normally permit the connection of a manometer to produce a differential measurement that indicates air flow. This permits a quick comparison of actual air flow to the design air flow, as shown on the drawings. As the cost of differential pressure sensors continues to drop, these will permit measurement of the air flow in any zone at a remote terminal. Some judgment will be needed in monitoring the air flow, as the loads served by the VAV box are seldom at design conditions. Most of the time, however, the problem will be during periods of maximum load, where the readings will be meaningful.

A VAV box can be a simple device, consisting of nothing more than an enclosure with a damper controlled by a modulating motor to open or close the damper depending on the temperature in the space served by the air passing through the VAV box. One VAV box can serve several offices along one exposure of one floor of an office building. Because of multiple exposures (sunlight being the major load in most air conditioning systems), corner offices typically require an independent VAV box. That is not always the case. When the box serves all offices along one building wall and, the executive gets the corner office, the thermostat will be in the corner office. When space utilization, or heating or cooling load, varies, such as interior versus exterior exposures, office space versus conference room, or operating room in a hospital compared to a patient's room, separate VAV boxes are normally provided. Changes in use of a space, installation of partitions, and the like can make space temperature control impossible. Thus, despite the advantages of VAV boxes, satisfying everyone is not always possible. The variable speed drive of the fan is controlled to maintain a constant pressure at one or more points in the distribution system. Cooling and fan motor horsepower are saved by these systems, while simultaneously providing more precise cooling and heating in the conditioned spaces.

Air Registers and Diffusers

Registers, typically in walls and floors, and diffusers, usually devices installed in the ceilings of the rooms, are designed to mix conditioned supply air with room air to absorb most of the rising heat load when cooling. The intent of the design is to mix the air within a couple of degrees of the room air temperature before it strikes a wall or occupants to produce sensations of draft. There is little need for concern with the delivery of heated air because very few occupants object to warmer temperatures of delivered air in the heating season. Temperature controlled registers that vary the delivery of supply air, by throttling the flow of air through the register, are available and a handy solution to isolated problems. Since the register has to draw in room air to mix it with the supply air, the temperature of the room can easily be detected at the register. Movement of the damper in the register is accomplished with a change in position of a bimetallic element in the room air stream. These registers are good for combating variations in cooling load in interior spaces, where heating is not required. They are only a trimming device, capable of making small corrections in the temperature of the room they are in. Sometimes, they are a solution for that person who is always cold in the summer. That person may have a blood circulation problem, but that is a subject for an entirely different book.

Almost every design drawing of an air conditioning system will show air quantities at each register. When troubleshooting problems, use a flow hood (Figure 5-47). The hood provides a reading of air flow through a diffuser, or register, and into the conditioned space to compare the existing condition with design. That will only determine what the air flow is and, not necessarily, provide a reason to start adjusting things, like the damper on the

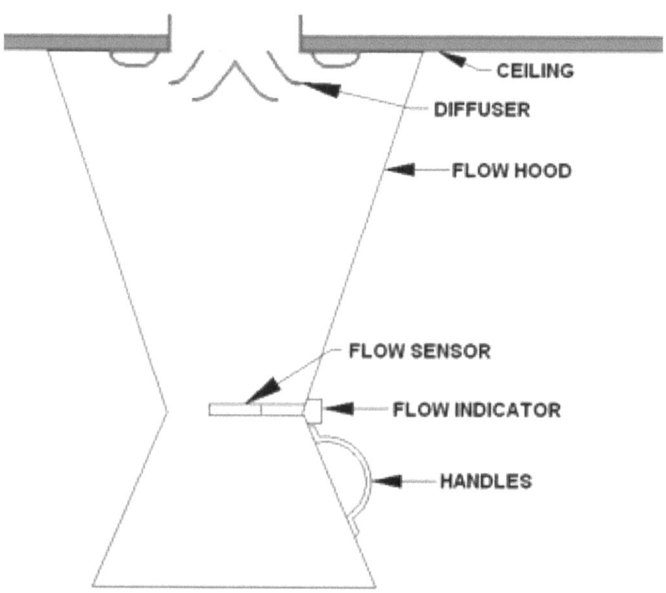

Figure 5-47. A flow hood.

register (if it has one). That can create more problems in other spaces by adjusting things to correct a problem in one space. Measure the actual air flow through a register or diffuser to compare it with the value on the design drawings. Then report problems with air flow that can be attended to by a testing, adjusting, and balancing (TAB) technician.

Roof Top Units

RTUs are applied in many commercial applications, schools, small office buildings, and other applications, where the first cost of a facility is the primary consideration. A unit that combines all the equipment and pipe and tubing necessary to heat or cool spaces in the building is prefabricated and rigged to the roof of the building to connect to ducts that distribute cool and dehumidified or heated air to the spaces served by that unit. In addition to steam or hot water, an RTU can be fitted with a natural gas or oil fired furnace for heating. Typical RTUs are shown in Figure 5-48.

RTUs can also be fitted with a variable speed drive and temperature controls to serve VAV systems as described for AHUs. RTUs tend to be a cheap substitute for a well-designed building and air conditioning system. Many of them are simply constructed to provide air conditioning at the lowest possible first cost. A typical rooftop unit will last about 10 years, whereas a typical AHU, installed indoors, will last more than 30 years. Self-contained refrigeration systems have twice the operating cost of systems using chilled water. Also, nearly every RTU has access opening covers that are simply pieces of sheet metal, along with the casing, attached with a minimal number of screws. Within a year or two, the covers are bent, lacking screws, and leaking outside air into the unit, disrupting its performance. It is not lack of maintenance. It is just cheap construction. They seldom use more efficient chilled water sources. Normally, they contain a complete refrigeration system with air cooled condensers and a DX coil. DX stands for direct expansion, which is an acronym that separates them from the separated systems that supply chilled water. That means the air is cooled directly by the refrigerant. Their operating cost is easily twice that of a unit using chilled water. They frequently are direct fired for heating, which would be more efficient than steam or hot water from a boiler if the burners and controls were not so cheap that they are less efficient than a steam or hot water operation.

Figure 5-48. RTUs.

Air Conditioning Systems Control

The normal air conditioning apparatus control consists of a thermostat that starts and stops the fan, compressor, and associated electrical equipment. That would be typical of those good old window units and the system in a home. While it is inexpensive at first cost, the operating costs are considerably higher than equipment that can operate continuously and modulate to control space temperatures. On a new heat pump, the new one was fitted with a two-speed, scroll compressor. It means that the fan runs more, but the compressor shifts back and forth between half and full speed (hi and lo on the thermostat) to keep it running longer, sometimes constantly at high heating and cooling loads. Every time a compressor shuts down, the refrigerant pressures balance out. The compressor has to reinstate the differential before the cooling or heating gets started again. It is not the best system, but the electric bills dropped to 38% of what they were with the older heat pump.

Instead of controlling the unit with a thermostat in the space, or sensing temperature in the return air duct, the outlet temperature of the air conditioning apparatus can be controlled by an independent system controller, according to the lowest air temperature required at any outlet in the air distribution system. All other outlets in the system control the temperature of supply air to the space by a thermostat in that space through use of a reheat coil or a VAV box.

New regulations and design standards for building ventilation that are primarily concerned with the distribution of the ventilation air have led to the introduction of independent outside air supplies. Don't be surprised if some equipment does not have a mixing box with connected outside air ductwork. Separate AHUs, frequently called ventilation air units, provide outside air distributed to the building through their own independent system of ducts and registers. Those units can be operated to deliver air at the room design conditions so that the ventilation air does not have any effect on the operation of the heating and cooling equipment.

One of the major considerations for load of an air conditioning system is sun light. A well-designed system will account for variations in solar heat gain at different exposures (North, South, East, and West) through the use of separate AHUs, face and bypass dampers, and/or reheat coils for each exposure. Of course, having corner offices with double exposures will need independent temperature controls or there will be periods when other occupants of the zone are uncomfortable.

The TAB Report

In any large installation, the construction project usually calls for air balancing. TAB specialists perform a series of tests and measurements to ensure that the building's design intent is met. Technicians will take readings and adjust air flows and temperatures, record them, and show the actual measured air flow at each apparatus, main distribution duct, VAV boxes, and registers. That document should also be kept handy in order to compare the setup conditions with actual. The report also summarizes the flaws and deficiencies of the building that can affect the overall performance of the system. It is crucial for every property manager and business owner to understand the TAB process, as it largely influences the future interior environmental conditions of the building. Moreover, prior knowledge about TAB reports can exempt business owners from unnecessary inspection charges.

Convectors

Convectors are usually floor mounted and usually below windows. They contain their own fan, or fans, to circulate room air over refrigerant or water coils, cooling and/or heating the room air to adjust the room air for comfort. Many use the thermostat control to start and stop the fans, while others will regulate the flow of the fluids through the coils. Ventilation is normally achieved by a small damper connecting to the outside so that the fans will draw air in when they are operating. Refrigerant, or chilled water, coils will remove moisture from the room air by condensing it. The drain is typically run out the building wall but can be piped indoors. Convectors can be supplied with or without filters. Filter replacement or regular cleaning of the unit is required. Convectors can also be two-pipe or four-pipe. Two-pipe convectors use the same coil for heating and cooling, using heated or chilled water supplied by the central system. These require the operator to monitor the weather, decide whether the building is to be heated or cooled, and set up the system to deliver water at the required temperature. Modern convectors sense the temperature of the water supplied and automatically switch the thermostat response to control the flow of the water or the operation of the fans. Four-pipe convectors have separate cooling and heating coils and can heat or cool independent of the weather.

Achieving Comfort for Occupants

If someone is not comfortable, there will likely be repeated complaints and, in some cases, direct orders to "do something" about it. Unlike being at home, the occupants are seldom able to walk over and adjust the thermostat. The thermostat is often locked up, or concealed, to avoid excessive adjustments that result in additional heating or cooling expense. The first step is to determine what the space conditions are. That involves using a sling psychrometer (Figure 5-49), a flow hood (Figure 5-47), a manometer set at a ten to one slope (Figure 2-3), a pitot tube (Figure 5-50), an infrared thermometer (Figure 5-51), and a scientific calculator. Sometimes, some of this equipment is left behind by the installer or TAB (testing, adjusting, and balancing) personnel. If so, put it away for future use. A combination of the manometer and the pitot tube is used to determine the air flow in a run of ductwork. It is great for checking for leakage in a long duct run. For best results, the test ports should be located

Figure 5-49. Sling psychrometer.

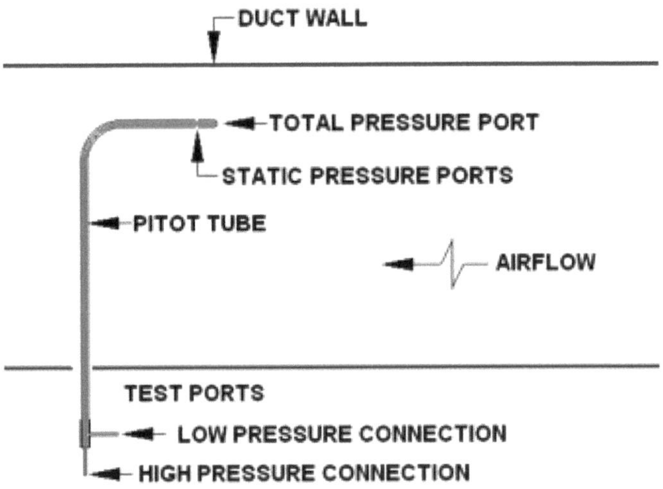

Figure 5-50. Pitot tube application.

Figure 5-51. Infrared thermometer.

20 diameters downstream and 4 diameters upstream of any change in the duct (elbows, transitions, etc.). For square ducts, use an equivalent round diameter, determined by a formula equal to 1.3 multiplied by the width times the depth of the square duct, multiplied and raised to the 0.425 power, and divided by the sum of width and depth raised to the 0.25 power. Width and depth are simply the two horizontal dimensions for vertical ducts. There should be a table for equivalent round duct sizes in one of the books available from the manufacturers of the air conditioning equipment. Test ports should be on two quarter points of round ducts and at least four on one side of a square duct. The more the readings taken, the more accurate the calculated result will be.

Connect the manometer to both ports on the pitot tube to get the velocity pressure, the difference between total pressure, and static pressure. Velocity (in feet per minute) at each measurement is determined by multiplying the square root of the velocity pressure (in inches of W.C.) by 4006. With round ducts, calculate the area of quarter points sampled in square feet and multiply by the velocity for the flow in that quarter for small ducts. For larger round ducts, determine the area in segments and multiply by the calculated velocity in that segment. For square ducts, it is much easier to simply select equal rectangles (that sometimes come out to be squares) centered on each measured point and multiply that area by the measured velocity.

Even when the air in a space is within the comfort zone, there may be other factors that affect the comfort of the occupants. These can be attributed to solar insolation, warm ceilings, cold floors, warm or cold walls, and drafts. Any one of these can result in more than 50% of the occupants being uncomfortable. That explains the need for all those instruments. Identification of all of the factors that have an effect on comfort will be required. It is not uncommon for complaints to disappear just because someone in the zone takes readings. People can get happy by just knowing that someone cares. There are many studies that indicate changes in the room temperature settings, above and below design, and at regular intervals, increased occupant satisfaction with the air conditioning system.

Problems can be local or area wide and should be approached with checking the room or checking the equipment serving the area depending upon the complaints. Complaints from one room or zone require a different approach than those from all areas served by one AHU or rooftop unit. Frequently, comfort problems in a room or zone are due to the location of the thermostat. It is a common problem. In hotel meeting rooms, some of the rooms are occupied. Others are not. All too often, the room with the thermostat was empty, with the lights off. On a few occasions, the other room containing the thermostat can have a much larger crowd, making the smaller meeting room feel cold. Those temperature controlled registers can help reduce those problems, along with zones containing several small office spaces. Hotels, conference centers, and similar venues might reduce those problems with multiple thermostats and thermostat selection based on occupancy sensors. There is nothing much an operator can do to reduce the problem, other than to inform the person scheduling use of the rooms to make sure the first or only room in use is the one containing the thermostat. Altering the setting of the thermostat to balance the loads will normally result in more complaints.

When there appears to be a problem with air conditioning equipment serving an area, collect the data on the system to see if it is operating as designed. Many factors can contribute to improper operation. Compare actual conditions to design to determine what may have changed. Take DB and wet bulb readings. The occupant discomfort can be due to humidity problems instead of temperature problems.

Action and Instrumentation

Listen to what are facts, even from the "dumbest" observer. It will provide a good foundation for the analysis of what is wrong. There are occasions when there is nothing wrong with the equipment, systems, or controls. Hopefully, all the foregoing explanation of equipment operation and things to check, along with data collection and analysis, will help identify the source of problems with occupant comfort and resolve them or provide the owner with the analysis and suggestions for what a contractor would be needed to do.

Data Collection and Analysis

Use of a psychrometer depends on the type available. Some prefer to use a good old sling psychrometer, as shown in Figure 5-49. All that is needed is to check that the wet bulb portion has water in the reservoir. There are electronic versions. Be sure to calibrate it every time before use, especially if it has been dropped. Since the gradations are on the thermometers, the sling is always right if it is not broken.

Progressively isolate the problem and determine if it is just one room, one zone, or affecting everything served by one piece of air conditioning equipment. Then check the obvious things. Is the equipment running? Does it have electric and control power? Has someone tampered with the thermostat or done something to block air flow? In checking a problem with a room or zone, collect and record room and supply air temperatures. Then use the flow hood (Figure 5-47) to measure the air flow through the registers. If the difference between room air and supply air temperature is equal to or exceeds design, air flow through the registers is equal to or exceeds design, and supply air wet bulb temperature is equal to design (or when cooling less, when heating more), then the system is working properly. That normally means the load in the room or zone is more than design and the equipment simply cannot handle it.

Unless it is a new facility, which should be under warranty, try to identify changes in load to see if they can be rectified. Look for any changes that increased the load and inform the owner of them. To check a problem with an AHU or RTU, collect and record the actual temperature and air flow data for that unit and mark a copy on the psychrometric chart with that unit's design data to see what might be off. Since the air carries the heat away, or delivers it to, the occupied spaces, a lack of adequate air flow should be addressed first. Problems with unit air flow are commonly dirty filters, followed by loose fan belts. For rooms and zones, VAV box filters should be checked, followed by isolating a loose splitter damper, control damper, dirty reheat coil, and failed or damaged ductwork. Note the changes in the space that might have produced the problem. In one case, a contractor had replaced all of the ceiling tiles in an office area, including replacing the return air grills that allowed room air to return to the AHU. The new ceiling tiles kept popping up. Other unique problems included one where a contractor adding sprinkler piping had removed the flex connections between the ductwork and registers to get the pipes into the ceiling space. That was revealed by return air temperatures considerably different than room air temperature. The air was flowing, not into the room, but bypassing it. Apply an understanding of the systems and what they are supposed to do in order to isolate problems with them.

Chapter 6

Maintenance

Operating a system is not as simple as starting and stopping equipment and opening and closing valves. An operator not only operates. The operator also ensures operability. That is the function of maintenance.

MAINTENANCE

Maintenance of the boiler plant is an operator's responsibility. An operator can be called upon to do everything from sweeping the floor to rebuilding a turbine, the simplest job to one of the most complex, and everything in between. In a small plant, with little equipment, the operator might be expected to do it all. As the size of the plant increases, those duties will increasingly be performed by others. The operator will still have a responsibility to make sure they do not interfere with the continuous and safe operation of the boiler plant.

The purpose of maintenance is reliability and cost control. Reliability of the equipment and systems in the boiler plant is ensured by limiting or preventing wear, vibration, erosion, corrosion, oxidation, and breakdown. Proper maintenance prevents failures of equipment that can result in significant repair costs. Maintenance includes many activities. The most important are monitoring and testing performed by the boiler operator.

There are many forms of maintenance. Contrary to many opinions, each one has its place. The choice of which form of maintenance to use depends upon the degree of reliability desired or is affordable. Maintenance methods fall into three general categories: breakdown maintenance, preventive maintenance (PM), and predictive maintenance. All three methods should be used to maintain the boiler plant. There are many items that do not need much attention until they fail. Then they will be replaced. That is breakdown maintenance. It applies to things like light bulbs, sump pumps, and other items that cost so little to replace and are so easy to obtain that any time spent maintaining them is futile. Some, like light bulbs, only allow breakdown maintenance.

Maintenance requirements vary but should represent a cost relative to the potential loss. It does not make sense to spend a considerable amount of time to check the lubrication of a little cooling fan motor (normally, they have permanent lubrication), when its replacement costs less than the labor to check it once. That is a situation where breakdown maintenance applies. On the other hand, lubrication of a steam turbine can include testing the oil and the operation of equipment that continuously cleans the oil because a failure would represent a significant cost.

A small 1/2 horsepower feed pump for a little heating boiler is not eligible for much more than breakdown maintenance. A 2000 horsepower feed pump for a super critical boiler plant will have vibration and temperature sensors at every bearing, speed sensor, suction and discharge pressure, and temperature sensors, and probably its own flow meter.

Between those two extremes are all sorts of variations on monitoring and maintenance. Most of them rely on the skill and dedication of the boiler operator. On each round of the boiler plant, look and listen to the feed pump, noting its condition, look for signs of vibration or shaft leakage, and possibly feel the motor and pump bearing housings to get a sense of their temperature. All of that is predictive maintenance. Adding oil or grease to bearings is PM.

Breakdown maintenance has the advantage of low cost. Nothing is really done to prevent a failure. Preventive and predictive maintenances require an expenditure of effort and materials, which represent an investment in reliability. There are varying degrees of effort expended in those activities depending on the cost of failure, the cost of maintenance, and the probability of failure. The only caution here is to remember that some equipment becomes obsolete. It pays to think about the condition of something that would normally only deserve breakdown maintenance but could be irreplaceable and force a major expense if it is not taken care of. An example would be a special bolt on a turbine speed control. The bolt might be easy to replace, if one could be found. However, its loss would produce hours of turbine down time.

PM is performed on a regular schedule to, as the name implies, prevent damage to equipment or systems.

Water treatment and lubrication are the two principle PM activities in a boiler plant. Those activities prevent failures by maintaining conditions that do not allow corrosion, scale, or friction to occur. Proper operation of some systems can also be called PM when they prevent erosion by ensuring that velocities do not get too high. Water treatment, properly performed, can prevent very expensive and catastrophic failure. The probability of such a failure, if water treatment is avoided or ignored, makes it the principle concern in all plants. It is so important that it deserves its own chapter in this book.

Predictive maintenance consists of monitoring, examinations, and testing to reveal problems that will, if allowed to continue, result in failure. Annual inspections of steam boilers and less frequent inspections of other pieces of equipment are conducted to detect formation of scale, corrosion, vibration, wear, cracks, overheating, and other problems that can be corrected to prevent eventual failure. The gas turbine and steam turbine manufacturers have taken advantage of the drop in cost of sensors and software and offer maintenance contracts that require these monitoring systems to send data back to the manufacturer. This is one use of data analytics to evaluate trends and provide warnings of incipient failure. They practice both preventive and predictive maintenances. The PM comes from experience with a particular model machine. This practice is a carryover from the airline industry. The hours of operation are monitored and tracked. At specific intervals, key parts are replaced. Inspections are dictated by the hours of operation (adjusted upward for stressful events). That is why the chance of a crash on a commercial jet liner is on the order of one in ten million. As the boiler itself has no moving parts, this advanced approach to maintenance has been slow to advance. Corrosion is usually localized. The inside of the boiler tubing is not visible. Thus, if oxygen pitting were to start on the inside of a tube, there is no good way to detect it in operation until the tiny hole comes through the wall. The hole may be 1/10th of an inch in diameter and there are miles of tubing in a reasonably sized boiler. Nevertheless, there are still good applications for data analytics in the plant. There are software programs that can review the historical data and provide insight into operating variables that impact efficiency and emissions.

Of course, there is that one instrument in the plant that is the best investment in predictive maintenance, the operator's ear. An operator can detect many problems indicating imminent failure and react to prevent the failure. An operator can detect changes in sound, vibration, and temperature (by simply resting a hand on the equipment) that would require a considerable investment in test and monitoring equipment. Constant attendance by a boiler plant operator is one investment in predictive maintenance that helps to ensure no surprises consisting of major equipment or system failures. It is normally the boiler operator who provides the principle maintenance of water treatment as well. Having a sound maintenance program is an essential part of the job. Repeating the importance of documentation, if the maintenance program is not documented, there is no proof that it was done and done correctly, including everything that is prudent and reasonable to prevent a failure. The oil may have been changed in that compressor the week before it failed. Without a document indicating that it was done, it will be very difficult to convince anyone of that fact. It is also very difficult to remember everything. A documented maintenance schedule serves as an excellent reminder of when something should be done. A schedule and a record of the work being done is the best evidence that proper maintenance has been carried out and a failure will not reflect on operator performance. With proper planning and execution of the maintenance plan, there should not be any failures.

Every piece of equipment that requires preventive or predictive maintenance should have that maintenance scheduled. Each plant is unique. That means the operator has to set up the maintenance plan for the plant. The best place to start working on that schedule is the operating and maintenance manuals. Do what the manufacturer recommends until there is a track record to find what has to be added and what requirements can be extended beyond the recommendations. Be certain that everything is covered. Failing to maintain something can be hazardous. One plant had three boiler explosions in as many months. It was determined that they had never bothered to replace the tubes in their ultraviolet flame scanners, despite the manufacturer's recommending they be replaced annually. Three boilers had extensive damage all because nobody replaced some three dollar electronic tubes.

CLEANING

If there is any distinct impression that one gets when walking into a boiler plant for the first time, it is the cleanliness of the plant, or lack thereof. In some plants, there are flowers in the control room and the floor looks clean enough to eat from. There are others that are so dirty that it is hard to see anything because the entire plant is black with soot. Which one gives the impression of better maintenance? Visitors and inspectors notice these things as well. Needless to say, cleanliness is not a sure sign of a quality plant. Lack of it, however, is almost always indicative of nothing but trouble. A boiler operator has the ability to make the difference in the appearance of the plant. It should be part of the PM program. Make time to keep the place clean. Like any other activity, it makes the shift seem shorter. Visitors, contractors, and inspectors do not necessarily make appointments to visit the plant. Keeping up a standard of cleanliness not only prepares for that unplanned visit but also demonstrates pride in the plant, which is noticeable to all.

Certain cleaning functions are, by their very nature, considered to be part of the operating function. That is because those devices are in operation. Only experienced and knowledgeable individuals (like a boiler operator) should be allowed to touch them because improper action could shut down the plant. These include cleaning burners, operating soot blowers, and cleaning oil strainers to name a few. The typical duplex oil strainer (Figure 6-1) is one of those devices that are in service when cleaned. If the wrong side is opened, the plant could be shut down. Another situation involves switching the strainer in service. It must be done carefully and slowly because it is always possible that the cover was not replaced properly and the strainer could leak.

In one shipboard incident, the new fireman was using a "helper" to change the strainer. A "helper" is a long piece of pipe stuck over the end of the handle. On the next watch, he reported that it was even tighter than the day before. A visit to the chief engineer's office later that day produced the instruction manual and revealed that there was a little jacking screw under the strainer that both lifted the plug valve so that the strainer could be changed and tightened back down. On the evening watch, the strainer was checked, and, sure enough, there was that little jacking screw. The fireman was amazed that once the jacking screw was operated, the strainer handle could be turned with one finger.

Figure 6-1. Duplex oil strainer.

INSTRUCTIONS AND SPECIFICATIONS

Read the manual first and every time before performing any maintenance. Then prepare a checklist that helps to ensure the instructions are followed. It is awfully easy to forget a step or get them out of sequence, with component failure being a result. Checklists are used in many areas, not the least in aircraft piloting. Before every flight, every engine start, every take off roll, every take off, and every landing, there is a checklist to follow. If a manual is not available, then contact the manufacturer to get one. They may charge an atrocious amount (consider their cost in producing one copy compared to several hundred during the period they manufactured and sold the original equipment). Yet, even as much as $300–500 can save ten times that amount in damage to the equipment.

A checklist will help ensure that all the steps are executed in the prescribed order and can save a lot of time. Just jumping in and doing it may seem faster, until the equipment has to be torn back down again because a part was left out or an adjustment was not made. It will be even longer if every step is being documented because there was a failure and the equipment was severely damaged. Check the instructions, even when highly skilled and knowledgeable.

Specifications define the requirements. Anything more complicated than a faucet or a toilet ballcock should be compared to the specifications to ensure that the right type and grade of material and parts are being used. That includes things supposedly simple, like bolts and nuts. There have been many situations where the wrong bolts or nuts were used, where, despite the drawings specifically listing the requirements, the steamfitters used the wrong bolts or nuts. Something that sounds good or looks right is not the answer. If a specification is not understood or the material may not comply with the specification, consult someone to be sure that the right material is being used.

Don't take the salesman's word for it. There will be no record of that after the catastrophe has occurred. Sometimes, the mistake is immediately evident. In one case, they started filling a piping system that took over a week for five men to install. Water was spurting from the longitudinal seam of every piece of pipe. Nobody checked the material. It was all "untested" pipes that were manufactured for structural use.

Sometimes, the problem shows up later. That is almost always the case when the material is not capable of withstanding the corrosive action of the liquids it contains. In one plant, a mild steel thermometer well was knowingly installed in a stainless steel piping system because the owner wanted the system up and running right away. There was no time to get a replacement well. The well was replaced with one of the right material a week later. There was not much left of that mild steel. Had the plant run for a few more days, the mild steel well would have corroded away, the thermometer would have blown out, and the highly corrosive liquid would have been spraying into the plant. There is one other thing about materials which needs to be addressed. There may be a modern material that does a better job, something like graphite gaskets for cast iron boilers instead of rubber ones. Refer to the section on replacements that follows.

LOCK OUT, TAG OUT

Lock out and tag out regulations and requirements are safety measures. Follow them religiously. They are there to protect personnel and keep them alive. Further, it is the operator's responsibility to ensure that all those regulations are followed and, more importantly, to be the person in charge of lock out, tag out.

Don't be too quick to allow that responsibility to reside in someone else. There will come a day when the contractor's crew closes and locks out the wrong valve (like on the plant's only water line) and then goes out to lunch! The operator is likely to be the only one in the plant who knows every valve that has to be closed to ensure that a system or vessel is really isolated. Another problem is that the owner of a plant is responsible for the safety of the contractors because any hazard in the plant involves the property of the owner. If the boss says "let the contractor do it," it might be pointed out that the contractor can do it wrong and sue the owner when someone is injured. The contractor will win!

The regulations for lock out, tag out are in Occupational Safety and Health Agency (OSHA) 29CFR part 1910. They are still changing and evolving. Obtain a copy of that document and be aware of any updates. Review it every time to prepare a system for maintenance. Right now, there are many methods for satisfying the requirements. One simple program seems to be a really clean and simple approach that satisfies the requirements with

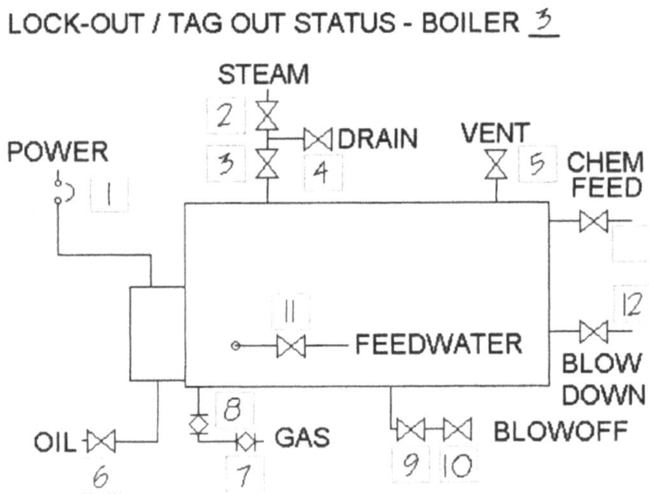

Figure 6-2. Lock out/tag out diagram.

a minimum of paper work and a great degree of understanding. It is demonstrated in Figure 6-2, which was prepared for work on a steam boiler.

A diagram or schematic of the system is prepared and laminated with plastic to serve as the key element of the program. It is mounted on a stiff board and hung near the equipment during maintenance so that it is easily seen and used. As each valve is closed, or opened, and locked, the number of the lock is marked on the diagram with a non-permanent marker. A quick look at the diagram will indicate if all the valves and disconnects are set and locked. All the keys for those locks are placed in one box, which has a lid secured by means of a latch that can accept multiple locks. As each worker places their lock on that lock box, his or her initials are added to the diagram to identify who is in there (or left their lock on) during the progress of the maintenance job. When they leave, they remove their lock and their initials. When all work is done and all workers' locks are removed, the keys can be removed from the lock box. Then remove the locks that ensure the equipment or system was isolated, erasing the lock numbers as the locks are removed.

In some cases, the job could have several operators removing locks and erasing the board as they are removed. This method ensures that they are all off. Now the board can be put away for use on the next turnaround. Note that it is simple and effective, while not producing a lot of paper. The locks can have tags permanently attached or the number on the lock can serve as the tag. The only time to cut a lock is when some worker leaves a lock on the lock box and goes home. Of course, there is a need to make certain that is what he or she did (i.e., went home).

It is always important to include venting, draining, and purging of systems as part of the procedures of lock out and tag out. That is very important when the system contains a hazardous substance, something corrosive or explosive. When dealing with certain substances, additional requirements should be followed. Any time a gas line is opened, it should be vented and purged. If the gas is considered hazardous to the environment, it should be purged through a flare or sorbent trap to prevent it from escaping untreated. Flammable gases should be purged with inert gas. Usually, that means a few bottles of nitrogen or carbon dioxide. Large and long lines could be purged with inert gas from a generator. Once the flammable gas is out (and proven), follow up by purging the inert gas with air. Just using air is only acceptable for very small lines (less than 3 inches) because flammable mixtures could be produced in the piping and ignited. Keep in mind that inert gas not only prevents combustion, it does not contain any oxygen and prevents breathing as well. Safety is an attitude. Acquire it. Lock out, tag out, purging, and environmental testing are things that should be automatic. Insist upon them happening before opening any equipment for maintenance.

LUBRICATION

Lubrication is probably the second most important element of PM. On larger pieces of equipment, drawing samples of the oil for testing is a predictive maintenance measure. It falls on the operator to ensure that every piece of moving equipment is properly lubricated. With the increased use of synthetic lubricants, that portion of the job is becoming more complex. Synthetic oils can save thousands of dollars in power cost for operating large pieces of equipment. On the other hand, adding the wrong oil to a crankcase can result in an instantaneous breakdown of the equipment because the two oils are incompatible and one oil causes the other to break down. Keeping an up-to-date lubrication chart that covers everything in the plant is important. Paying some attention to proper lubrication schedules can save time in the long run.

Most plants seem to have a program that consists of over-lubrication of some equipment and insufficient attention to the lubrication of other equipment. Many grease lubricated bearings need lubrication infrequently. Yet, they are lubricated regularly simply because the program does not provide for a proper schedule. That results in unnecessary lubrication and over-lubrication of that equipment. If the program does not allow for lubrication schedules over periods as long as five years, that will happen. Grease is not cheap. Nor is the labor that is required to move around the plant and lubricate equipment unnecessarily. Thus, developing a suitable program normally pays for itself.

Lubrication is a function of operating hours more than anything else. A program for scheduling it suggests installing monitors to record operating hours of the equipment in order to determine when lubrication is necessary. With the falling cost of today's sensors, install operating hour meters on everything. Tracking when equipment is in service in a log book is another way to determine operating hours.

Frequency of operation is also a factor. Equipment that is started and stopped frequently should be lubricated more often than those that run continuously because the constant heating and cooling of the bearing results in swell and shrinkage of the lubricant. That can easily result in air and moisture mixing with it to degrade the lubricant and rust the bearing. Systems that are oil lubricated also have a requirement for replacing the oil at frequencies that are based on the greater of operating hours or time. Grease is replaced with each lubrication. There is no additional scheduling to replace it.

It is that replacing of grease that many operators fail to consider. Just slapping a grease gun onto a fitting and pumping away does not necessarily provide grease to the right place. That frequently results in the bearing shaft seals failing because the grease forced them to upset (Figure 6-3), causing additional grease to be forced out around the shaft or into the equipment housing.

Combine that with the common over-lubrication associated with grease bearings and it promotes equipment failure. The grease eventually blocks cooling air flow passes within the equipment. Invariably, there is a plug or cap that can be removed to provide a passage for the old grease. That opening should be provided before

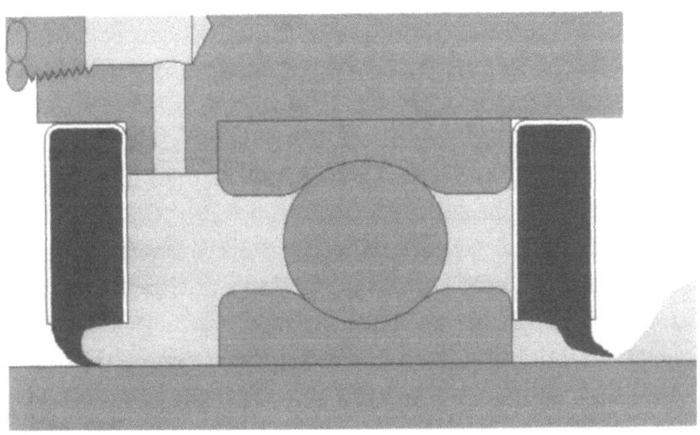

Figure 6-3. Grease seal upset by overpressure.

pressing new grease into the bearing. Don't forget to put the plug or cap back after the bearing is lubricated and, when the manufacturer recommends it, the equipment is operated to stabilize the volume of grease in the bearing.

Use of the proper grease is also important. Some facilities simply use the highest grade of grease required to simplify their activities, thinking that if they use the best in everything, they will not have a problem. There are two problems with that thinking. First, it is expensive. That high-quality grease is very expensive. Second, that high-priced grease may not work well in the bearings that can function with the less expensive material.

Grease requirements are a function of load on the bearing and speed. A grease designed for a high-speed, low-load, bearing will not adequately support the larger loads of a low speed bearing. A lubrication program that is designed to be simple, or make life easy for personnel, can result in shorter bearing and equipment life. Give up on the "one size fits all" concept. Lubricate the bearings in accordance with the manufacturer's instructions or the recommendations of a lubrication specialist. Painting a circle around each fitting with special colors to denote the grease to be used, and applying similar paint to the barrel of the grease guns and tip, will help to ensure that the proper lubricant is utilized.

Another problem is a failure to clean the grease fitting before attaching the grease gun. Use of a lint-free rag to wipe off the fitting is recommended. However, it will not always remove the paint and other materials that manage to find their way onto grease fittings over time. Using a plastic cap that prevents anything getting on that fitting between lubrications would be helpful. Even so, the fitting should be cleaned before attaching the grease gun. What if someone steps on the plastic cap or hits it with something and it comes off? Eliminating contamination of the bearing with contaminated grease in the tip of the grease gun is also important. Always carry an additional lint-free rag or small bucket to collect a small amount of grease from the gun before attaching it to the fitting. A quick shot into the rag, or bucket, will eliminate any dust or other debris that was picked up by the grease in the tip of the grease gun.

Sound like a lot more work? Grease lubrication requirements are so infrequent that a good grease lubrication program results in doing half the work. This is partly due to lubricating the equipment too frequently and partly due to over-lubrication when it is done. If there is a policy of greasing everything once a month, or more frequently, that is probably the case.

Oil, like grease, varies in its application. Be certain to use the proper oil for the equipment. A simple mistake involving oil can destroy a piece of equipment. One oil mixed with another can produce an incompatible mixture that loses all its lubricating properties. When that happens, the mixture tends to split into a light fraction that is too thin to support the load and a sludge that settles to the bottom of the sump or plugs up the pump and filters. Every piece of oil lubricated equipment should be marked to clearly indicate which oil is to be used in it. Refrigeration oils are addressed in Chapter 5. With all the changes in lubricants, it would be helpful if there was a standard color coding (like those for refrigerants) to make it a little more difficult to use the wrong lubricant.

Preventing contamination of the oil in the equipment by adding contaminated oil is fairly easy. Oil interacts with its environment more readily than grease. Always take every possible measure to protect oil in storage and on route from storage to the equipment. Keep it sealed. Many modern oils can absorb moisture and must be kept sealed until they are put to use. If the equipment has an oil heater, then the oil will probably absorb moisture right out of the air, contaminating itself if it is not kept in sealed containers.

Oil, unlike grease, can be cleaned and rehabilitated while still in the machine. In addition to oil strainers and filters, a lubricating system can contain water separators, magnetic separators, heaters, and coolers to maintain the oil at its optimum operating temperature and settling tanks to allow removal of solids and contaminants. The expensive oil is maintained by these systems to reduce the cost of regular replacements. It requires attention to the maintenance of the oil systems.

If there is no oil maintenance system, there may be the option of an oil maintenance service, a company that will pick up and refine the used oil and provide credit toward the purchase of new oil. Regular testing of the oil in those systems is essential to ensuring proper system operation and maintenance of the lubricating quality. Normally, the testing of oil (tribology) is performed by outside laboratories that have all the required equipment. The oil is tested for water, acidity, lubricating properties, and fine particles (microscopically). The examination by a skilled technician with a microscope can identify all the particles in the oil to reveal impending bearing failure or problems with gears or other parts of a machine.

Maintenance of oil lubricated equipment requires more attention than grease lubricated ones. Oil is exposed to the air in the plant. Grease systems are basically sealed so that air does not contaminate them. That is why some grease lubricated bearings can go 40,000 hrs, which is close to five years, without re-greasing. When equipment starts and stops, it breathes, as the oil and air heat up and then cool off to change volume. Air has to bleed out and then is drawn in. The grease changes volume, but it is normally such a small change that those seals expand and contract with it to prevent leakage of contaminants in or grease out. Systems with oil temperature control will also breathe with changes in load because the temperature of some of the oil increases and decreases, depending on the load. Therefore, equipment that is subjected to frequent stops and starts, or varying loads, requires more frequent checks of the oil than those that operate continuously. That is why there will frequently be an accumulation of oil around an oil sump vent. It is condensed vapors that were pushed out of the vent filter as the system breathes.

If there is an objection to the accumulation of oil around the vent, try putting an extension pipe on it, raising the vent at least 3 or 4 feet. For a more engineered design, calculate the change in volume of the air and oil in the system. Then put on enough pipes to provide that volume. Overhead clearances may prevent extending the pipe at its connection site. That does not prohibit the addition of a couple of reducers and larger pipe to the extension to get the volume needed. The concept of this solution is to create a vertical settling space, where the oil that would normally settle on something outside the vent settles in the piping to leave a volume of air substantially free of oil to flow out of the vent. A simpler solution is to carry a rag around and keep the area around the vent clean. Observation of the oil around that vent can provide an indication of a change in the condition of the oil in the equipment. That might be more helpful.

Oil has to be changed in any system that does not have its own conditioning equipment. Also, just like a car, there are rules of thumb that are wasteful. Most cars do not need an oil change every 3000 miles. Read the instruction manual to find out what it says to do. The instruction manual for the equipment will provide some guidance. Still, note the condition of the oil. A trained tribologist is not needed to tell that the oil needs changing more frequently when there are distinct changes in color or particles in the oil before it is due to be changed. A problem with water supply to the cooling system that resulted in a significant rise in oil temperature should be followed immediately by an oil change or testing to see if it needs changing. Other indications include the presence of a whitish, waxy substance that indicates water has contaminated the oil. The opposite is not necessarily true, however. Just because the oil looks good does not mean that it is okay. If the cost of the oil and labor to replace it is not significant (less than $100 per year), then change it according to manufacturer's recommendations. If the cost is significant, employ the services of a tribology lab to test the oil and make recommendations for changing it. Some systems have operated 100,000 hrs without an oil change. A manufacturer's recommendations are normally based on the most severe use and the wise operator makes every effort to ensure the equipment is not overloaded, or abused, so that the oil can last longer.

Replacing organic oils with synthetic ones can reduce wear and power requirements for equipment. In addition, the synthetics last much longer than the organic oils. There are balancing factors in the additional cost of the synthetic oil and reduced power and maintenance costs. If large volumes of oil in equipment are being changed on a regular basis (less than annually), a hard look at synthetic replacements is recommended.

Oil lubrication systems require maintenance of more than the oil. Filters have to be changed along with the oil, and more frequently in some systems. Coolers need to be cleaned on the water side to prevent fouling and maintain heat transfer. Temperature controls must be checked to ensure that they are operating properly and maintaining the right temperatures. Centrifugal separators, and the like, have to be maintained according to manufacturer's instructions. Anything that affects the temperature of a lubricating system is critical to continued safe and reliable operation. If a lubricant gets too hot, it will break down and lose its lubricating properties, allowing the metal surfaces in the equipment to rub, gall, and scrape with failure occurring rapidly. That is the reason to log an oil temperature that is always the same. The purpose is to notice when it suddenly does change so that something can be done about it.

Cleanliness is the next important factor. Clearances in bearings and gears are so small that a particle of dust that is almost invisible in the air can span the clearance to produce damage in the equipment. Any opening into a lubricating system should be fitted with a filter. Systems should not be opened unless provisions have been taken to prevent dust and dirt getting into them. A little contamination of a lubricating system can result in a total system failure costing a thousand times more than the oil.

INSULATION

Insulation is one of those items which, for whatever reason, never gets the attention it deserves. It is not uncommon for a plant to complain of high fuel bills, only to find that half the insulation has fallen off. Burning fuel unnecessarily because the insulation is not maintained is not what a wise operator does. Any discussion about insulation raises the concern for asbestos bearing insulation contaminating the air in the plant. While many facilities have spent the fortune it costs to remove asbestos bearing insulation, others have chosen to encapsulate it. If the plant is one of the latter, then maintenance of that encapsulation has a priority. Damage to the cover can occur as a result of normal operating and maintenance activities, from vibration that occurs during normal operation or a plant upset. A tour to check the integrity of encapsulation should be performed on a monthly basis.

When it becomes necessary to gain access to something covered by asbestos insulation, notify the employer so that the insulation can be removed and disposed of properly. The laws regarding asbestos bearing insulation do permit removal of small quantities without all the environmental controls required of a major material removal. Training courses are available for removal of these small amounts. Follow the rules that are taught in the class. If a contractor is used, make sure they are qualified (lots of dust blowing around is not to be accepted). Once the work is complete, make sure the asbestos that remains is encapsulated. Don't forget to mention its removal, and who did it, in the boiler plant log.

Whenever insulation is removed for maintenance or repair, make certain it is put back or replaced. Even small amounts of missing insulation can be a problem. Small areas tend to become bigger. After a while, the whole system is bald. Not only is it a waste of energy, it is hazardous, potentially causing severe burns. A qualified inspector will get a negative impression of the safety procedures at the plant. There is another safety reason. At one plant, the insulation had received no attention and was literally falling off the pipes. The hazard was associated with being hit on the head by falling insulation! Such instances are not uncommon. Never accept an insulation job that consists of nothing but stapling up All Service Jacket (ASJ, that white paper like material with the flap that comes on most insulation). It does not last. The staples eventually corrode and fail with the rest of what happens being most obvious.

At the very least, piping insulation should be secured with minimum 20-gauge galvanized wire wrapped around it, twisted, and bent back against the insulation (to prevent the sharp ends catching or cutting anything or anyone), twice on each section. For longevity, a light canvas wrap impregnated with a waterproof mastic will look better and will last even longer. Outdoors, and in areas where the insulation may be struck by people carrying objects such as ladders, the corrugated aluminum jacket with aluminum straps and fasteners is necessary to provide long life. Long runs of hot piping pose a special problem. The pipe expands, but the insulation does not expand anywhere near as much. The jacket, particularly outside in cold weather, can shrink from its original length. When restoring insulation on long runs, try to compress the existing insulation, as much as possible, without crushing it. Then compress the new material as much as possible when installing it. Jackets should have a minimum overlap of 3 inches outdoors. The longitudinal seam should always be on the side of the piping lapped down, to prevent rain entering the seam. On vertical runs of pipe, make certain any jacketing is lapped to shed water. Do it indoors as well. A leak can always spray water, or worse, all over the place.

Large flat surfaces require the installation of insulation studs. These studs are wire secured to the surface by stud welding or a special machine that shoots the wire into the surface. The studs hold insulation with special washers over the stud, thus pressing the insulation against the equipment surface. An impregnated canvas covering, or corrugated aluminum jacketing, is necessary to protect the surface of that insulation. Any repair job should return the insulation to a like-new condition using one of these methods.

What happens if some insulation gets wet? If it got so wet that it collapsed, it has to be replaced. Otherwise, let it dry. If it got wet while the pipe was out of service, and the line contains steam or hot water, warm the piping up very slowly to avoid generating steam under the insulation that will blow it off. Damaged or compressed insulation should be replaced as part of the annual cleanup operation. Where the damage is repeated, some consideration should be given to installation of better protection of the insulation. Consider replacing or covering the jacket with heavy galvanized sheet metal, thick enough to ward off the damage.

The argument that it does not make any difference if the piping is only used during the heating season and it heats the building anyway is a false one. The heat lost through the lack of insulation is almost never able to heat the space as intended. It is almost as weak an argument that it is only a little bit of heat. Little bits become lots when that attitude is taken. Being out of a particular thickness is not an acceptable excuse. Put something thicker on it or find a better replacement. The energy lost in the month or more that it takes for someone to get around to ordering the right thickness will pay for the additional thickness.

Speaking of various thicknesses, it usually does not pay to maintain an inventory of multiple thicknesses. Get pipe insulation in 1 inch increments, 1, 2, 3 (if 3 inches are needed), etc. Layer it for greater thicknesses. Limit the stock of 1 inch thickness to pipes 2 inches and smaller. For flat and large diameter surface insulation, 2 inch thicknesses should be adequate. Inventory should also be limited to the insulated pipe diameters that are actually available in the plant.

Be cautious with insulation on or near piping containing flammable liquids, such as fuel oil. The insulation can absorb it like a wick, to become a fire problem later. Insulation in the area of fuel oil pumps, strainers, burners, and such other places that could be splashed by a leak should have full aluminum jacketing over a mastic impregnated covering to prevent a leak or splash from soaking in.

Insulation for refrigerant piping, chilled water, and some ductwork has the added requirement that it be properly sealed. If there is an open path for water vapor to flow through the insulation to reach the cold surface of the equipment, piping, or duct, it can condense there, soaking the insulation to reduce its effectiveness, corrode the metal, and, occasionally, harbor mold growth. Cellular glass and foam insulations are inherently vapor proof. However, they only work if their seams are sealed tight. When there is damage to vapor tight insulation, try to take the system out of service for a while to allow any condensate to evaporate before restoring the vapor tight seal.

Re-evaluate the insulation once in a while. The old rule that says it should be insulated if it is too hot to touch still applies. The only place where insulation should not be added to is any part of a boiler casing. The wise operator maintains the insulation in the plant. The cost of material for repairing or even adding insulation is recovered in fuel cost in a couple of months.

REFRACTORY

Refractory is unique material in one regard. No manufacturer will absolutely guarantee that their material will remain intact. Materials exposed to the high temperatures of a furnace are also subject to components of the fuel that become very caustic or acidic at the high operating temperatures. Some components of fuels produce considerable damage, with vanadium being particularly offensive. Vanadium is common in many heavy fuel oils. It has a particular means to damage refractory. Vanadium pentoxide is molten at flame temperatures and as low as 1200°F. It remains molten at the refractory walls and soaks into the refractory during boiler operation. When the burner shuts down, the materials cool and the pentoxide solidifies. Being a metal oxide, it shrinks at a rate different from the refractory. The difference in thermal expansion, where the pentoxide soaked layer shrinks more than the regular refractory, creates a shear plane between the two materials, where they pull apart. The result is breaking off of a layer of the refractory from one-quarter to two inches thick, a process called spalling. The damage is very evident on inspection of the furnace. The pentoxide soaked layer has a glossy black appearance and is spotted with light tan areas where the pieces of refractory spalled off.

Refractory does expand and contract with changes in temperature. It is nowhere near as much as it is for metal, but it does grow and shrink. That must be accounted for. Repairing every crack that appears in the refractory in a boiler's furnace on each annual outage will, as a result, accelerate the damage. Smaller cracks will expand and contract. Use a number 2 pencil and put the sharpened point into the crack. If the yellow paint level is not reached, the crack should be left alone. Those are expansion cracks and will close up as the boiler heats up. Plugging larger cracks, as much as three quarters of an inch, with hard refractory materials is not recommended. Today, there are ceramic fibers rated at temperatures as high as 3200°F that should be used to fill those cracks. The ceramic fibers should not be packed into the crack to the extent that they are solid. Leave the fiber soft so that there is room for the major pieces of material to expand into the crack. In the past, asbestos was used for such repairs. There may be asbestos in joints and cracks of refractory in an older boiler. With good maintenance records, that fact should be identified. Lacking data, treat any fibrous material as asbestos until such time that it is proven that it is not.

One important location for providing thermal expansion is around the burner throat on oil and gas fired boilers. This also applies to pulverized coal, round burners. Tangentially fired coal boilers do not have a quarl that requires refractory protection. The throat material is usually rated for very high temperatures because the throat is closest to the fire and will be the hottest refractory in the furnace. Those firing gas know that the throat is glowing cherry red when the boiler is in operation. Actually, it is always red hot, regardless of the fuel. It is just very difficult to see the glow with pulverized coal or oil fires because the bright fire lights up the furnace.

Burner throats are either made up of pieces of a pre-fired refractory material called "tiles," or a plastic material. The word "plastic," in discussions of refractory, means a material that can be molded and shaped as desired until it is dried. Plastic refractory has the consistency of stiff clay and looks and feels like mud with lots of sand and fine gravel in it. Either of the throat materials will expand considerably during boiler operation. There should always be some form of expansion joint around the throat. There are some installations of plastic refractory where the throat and burner wall were monolithic (all one big piece). They do manage to stay intact for quite a while, despite the differences in temperature. However, a prepared joint provides a perimeter for expansion and, eventually, a repair.

Another problem is the sagging of a plastic refractory wall, which bears down on the burner throats to distort them. A "bull ring" is a circle of special pre-fired arch brick or tile around the burner throat that supports the wall. That prevents the weight of the wall from bearing down on the throat tile. The bull ring should be designed to provide a half inch gap between the inside diameter of the bull ring and the throat tile. Newer burners would be packed lightly with ceramic fiber.

At the next repair of the burner throat, give serious consideration to rebuilding the entire thing to get that flexibility. Burner throat repair and replacement is best left to the experts, men and women skilled in installing the materials. It is not easy to properly position throat tile to get a perfect circle or shape a refractory throat in perfect form along the sweep. A sweep is a special tool used to shape a burner throat out of plastic refractory. Normally, it is a piece of flat steel plate welded to a pipe that fits into the oil burner guide pipe and cut to produce the form of the burner throat (Figure 6-4). If the burner has a plastic throat, make certain the installers use that throat sweep and use it properly.

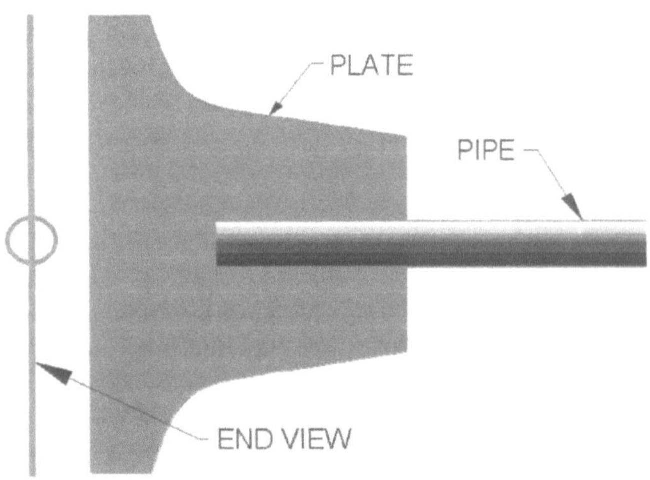

Figure 6-4. Throat sweep.

Don't be taken in by a refractory "maintenance coating." The so-called maintenance coatings do not indeed provide any real protection. They are usually harmful. Those materials are, in some instances, nothing more than a mud. Higher quality materials are seldom matched to the refractory in the boiler. It means that their thermal expansion rates are not matched. The result is that much of the spalling comes from the maintenance coating breaking away. It also fills the small cracks that provided for expansion to create stress on the face of the refractory.

Another regular problem with those materials is that they are applied carelessly. In many of situations, the openings in the gas ring were partially blocked with that maintenance coating. Instead of spending money on that, save it to pay for a complete replacement of the refractory some years in the future. If the refractory is suitable for the application, there will not be any serious degradation unless the unit is abused. The materials and installation methods have improved considerably in the past 40 years. If the boiler is over 40 years old, there may be some refractory issues. Modern boilers with mostly water cooled walls will have very few refractory problems. The only exception might be circulating fluid bed boilers. These units use a lot of refractory materials in the lower part of the furnace and the recirculating cyclones. Again, read the instruction manuals for proper care and maintenance of these refractory installations.

The one difficulty with modern boilers, especially the "A" and "O" type package boilers, is retention of the refractory seal where the tangent or finned tubes are offset or lacking fins next to the boiler drums. Those sections consist of very small pieces of refractory with very little to hold them in place. For those particular boilers, the grip has to overcome gravity. Their weight is a factor. The best way to repair those is to completely remove a section and replace it. The new material will not bond to old refractory at all. As the new material cures and dries, it shrinks and simply pulls away from the old material.

Any refractory repair that is not just for a short term should consist of complete replacement of a section with adequate provisions for expansion. That repair will last. Patches are exactly that. They do not last. Don't be afraid to improve on an installation either. If a repair is made because a furnace wall buckled into the furnace, improve the anchoring as well as provide for thermal expansion. Either lack of anchoring or buckling due to thermal expansion was the cause of the failure. Take appropriate measures to counter both problems. Any temporary patch has to be anchored or it will be more temporary than intended. It will fall out as soon as the boiler heats up. Since the repair material will shrink a little as it dries,

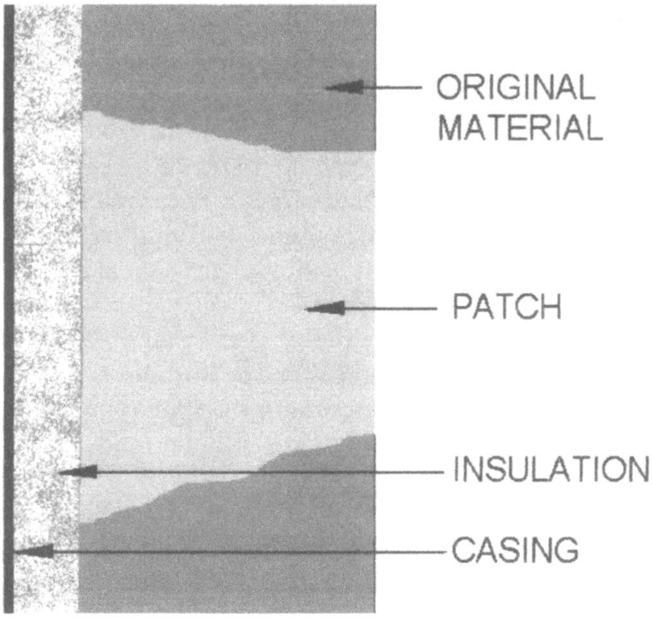

Figure 6-5. Undercut for refractory patch.

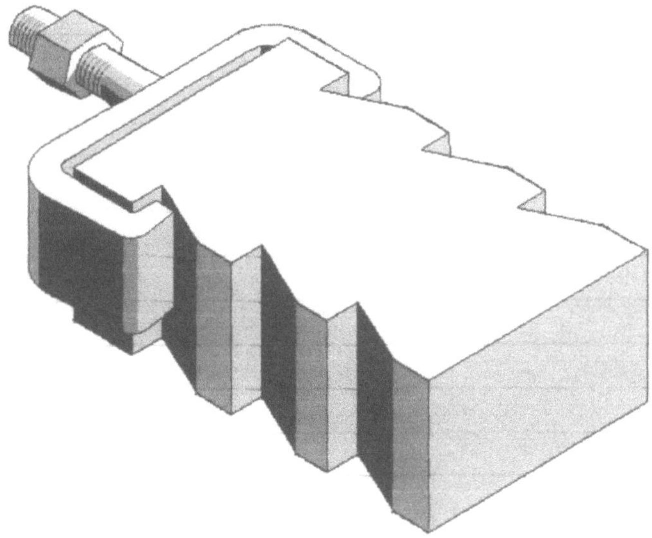

Figure 6-6. Refractory anchor.

it does not matter how hard the wet plastic refractory material is hammered on (or how thick any slurry of castable refractory is). It has to be anchored somehow. Castable refractory is a powder that is mixed with water to form a very dense, soupy mixture that can be poured into spaces surrounded by forms. Small areas, less than 16 inches in diameter, should be "keyed in" to the existing material. That is accomplished by undercutting the face of the existing material (Figure 6-5) so that the patch is wedged between the edges of the existing material and the casing insulation.

Larger patches should be anchored by installing a refractory anchor (Figure 6-6) secured to the casing or brick setting so that the patch is secured and will not tend to crack and buckle out as it is heated. Refractory anchors should be installed within 18–24 inches of each other if there is not a successful wall to compare to.

Almost any refractory repair requires a "dry-out," as described in the chapter on new startups. If the repair consists of brick or tile laid up dry, a common arrangement for sealing the furnace access opening on many boilers, then there is no need for a dry-out because there is no moisture imbedded in the refractory. Anything else will have to be dried out. When the patch is made with plastic refractory, the dry-out will be accelerated if vents in the material are provided. Provide vents by poking the material with a small welding rod to produce small round holes about two-thirds of the thickness of the wet material on 3–4 inch centers. Steam forming in the material will then have an escape route. If the repair is due to vanadium pentoxide damage, the venting is not recommended. The holes will provide places for the oxide to soak into the refractory.

Some refractory materials are labeled as air drying and some are heat drying. Most are combination air and heat drying. A heat drying material reacts to a small degree with the water that is in it to create another chemical that helps bond it together. When using heat drying material, it is important to avoid letting it air dry. Fire up the boiler to apply the heat in accordance with the manufacturer's instructions as soon as possible. The best option is to use a combination material. It is always important to treat all of them gently so that the repair is not destroyed in its first few hours of operation. Bring the boiler up to operating temperature as slowly as possible.

PACKING

A lot of modern designs and new materials are eliminating packing. Nevertheless, it will be a long time before a pump, a valve, or other device will come without packing. Packing is material pressed into a space between a metal housing and a metal shaft to provide a seal to prevent or control leakage of water, steam, or another fluid. The words "control leakage" are very important. Many times, an operator or maintenance technician who was thoroughly convinced that the packing on a pump should not leak destroyed the pump by tightening the packing to stop the leak. Unless a small amount of fluid leaks along a constantly moving shaft to lubricate the shaft, and protect it from rubbing, the packing will cut into the shaft. A pump shaft or sleeve reduced in diameter with gouges from the packing is the result.

Whether it is a pump, a valve, or a control float does not really matter. There is a standard arrangement for installing packing. Many leaky valves consist of a repair where the installer simply wrapped packing around the shaft in a spiral, cut it off, jammed it in, and expected it to seal. That does not work. Packing should be arranged in cut segments that barely fit around the shaft, stacked as shown in Figure 6-7. The stacking does not have to be precisely as shown. Just alternate the stacking by placing the open seams first 180 degrees out of phase and then 90 degrees to produce a complex path for any leakage to follow.

It is actually better to have the packing rings cut a little short than a little long. If the ends have to be jammed together to get the packing into the opening, it will create a hard bump that can bear all the pressure placed on the packing gland. Then the rest of the packing ring is not compressed and does not seal. If the ends are jammed when packing the gland on a gauge glass, the odds that the glass will break upon tightening the packing will increase.

Packing of pumps usually includes a lantern ring (Figure 6-8) that has to be properly positioned in the packing gland. Always count the number of pieces of packing taken out from under one. The lantern ring provides a space for distribution of leakage into or out of the packing gland. When the packing is sealing the high pressure side of a pump, the leakage into the space containing the lantern ring bleeds off to the pump suction, which is at a lower pressure. That recovers some of the fluid. The remaining packing, between the lantern ring and atmosphere, is only exposed to suction pressure. For cooling and lubricating, some flows between the packing and the shaft to the outside of the packing gland.

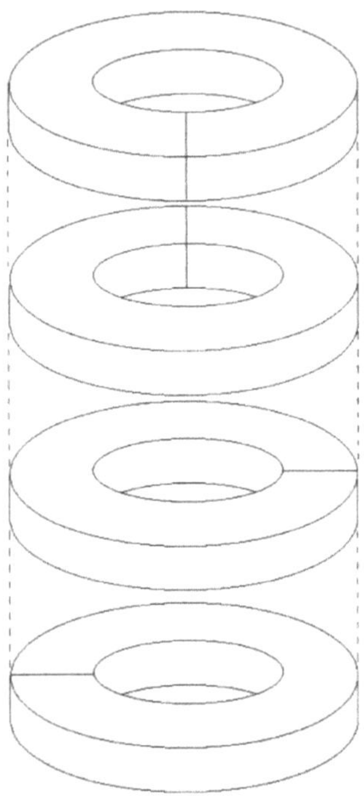

Figure 6-7. Packing segment stack.

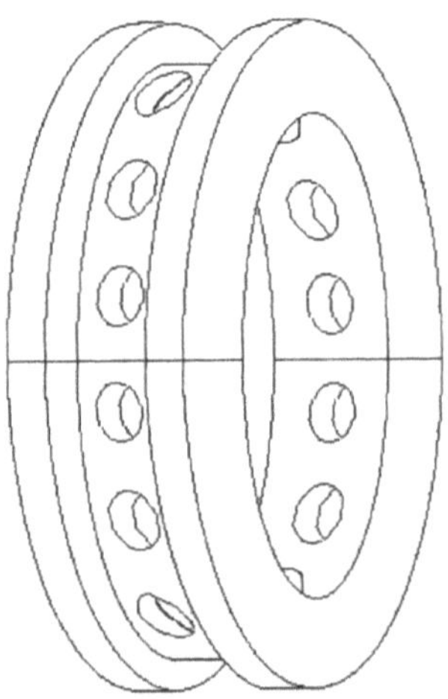

Figure 6-8. Lantern ring.

When the packing is on the suction side of a pump operating at pressures equal to or below atmospheric, the lantern ring space is piped to the pump discharge or an intermediate pump stage. The purpose here is to provide lubrication of the packing and shaft plus sealing the pump to prevent air leaking into the fluid. That is important for condensate pumps to keep oxygen out of the condensate. Flow in that case is into the lantern ring space. It then splits, with some flowing into the pump suction and the rest leaking out of the packing gland in the other direction.

Whenever repacking a pump, be aware that the gland could contain a lantern ring. In one instance, on a feed pump, the operators were not aware of the packing gland and had repeatedly pressed the packing down until the lantern ring was pressed past the location of the bleed connection. They could not stop excessive leaking because the entire packing set was exposed to the high pressure water and the erosion along the shaft was getting worse.

The split in the lantern ring should always be set 90 degrees from the split in any pump casing to provide a clear indication that it is a lantern ring and not the bottom of the packing gland. If there is a piping connection at the packing gland, open it up to look into the gland, repacking it to make certain that the lantern ring matches up to the opening. Sometimes, the count can be wrong when removing the packing because it comes out in pieces. It does not hurt to spend the extra time to make certain the lantern ring is positioned properly.

Packing of air actuators, compressors, etc., where there is no fluid for lubrication, will have grease fittings or piped oil connections to apply grease or oil to lubricate them. These usually incorporate a lantern ring to distribute the lubrication. Those packing glands use the lubricant as part of the seal. It is important to follow the manufacturer's instructions with that packing. Some have to be soaked in the oil or grease before installation in the packing gland, while others have to be installed dry and then "charged" with the lubricant before putting the equipment in service.

Valves, and a few other pieces of equipment, have very limited movement of the shaft through the packing. These generally do not need extensive lubrication. In most cases, the lubricant is part of the packing, typically graphite. There is no need for leakage of the fluid to lubricate the shaft. Pumps and other devices with moving shafts should allow some leakage to some degree. Valves and devices, like a Keckley float controller, should not leak. The most important maintenance practice for those packing glands is to tighten the packing as soon as it starts to leak.

Every time a valve is operated, check the packing gland afterward. Tighten it immediately if there is a leak. Quick response to a leak can prevent the need to completely repack the valve. If that leak is allowed to continue, it will cut through the packing, destroying it. Then it will be impossible to seal by simply tightening the gland. Since operators are the ones who open and close the valves, and since that is the only time the seal between packing and shaft is broken, there is no question that tightening valve packing is an operator's responsibility.

CONTROLS AND INSTRUMENTATION

Controls are the robots that do the boiler operator's bidding. Without them, the operator would have to make every little adjustment that the controls make automatically. The controls do not get tired or have a bad day. They are online 24/7. Once tuned, they perform consistently at all times. Instruments are an extension of the operator's eyes and ears to allow them to know what is going on in the process. It is important that the information they provide is correct. It makes sense to maintain them so that they keep doing their job. There is a separate section on the function and operation of controls and instruments in this book. This part is devoted only to their maintenance.

The modern microprocessor-based controls make the jobs as operators much easier. They are almost maintenance free. Make certain that cooling is maintained by keeping dust and dirt out of the slots and vents of devices and panels. Make sure they do not get wet. That nearly covers it. More control hardware is lost to water leaking into panels than for any other cause. Don't remove something from a panel and leave the opening for water to enter. Even small conduit covers can admit water that can find its way into a control panel or device. The wise operator looks for such things on every round and does something to restore enclosure integrity when a problem is spotted. Also, carry a clean rag to dust off cooling vents.

For pneumatic controls, the most important thing to keep up is the air compressor, storage tank, filters, and dryer. If the compressor fails, then the controls will not work. If the tank floods because the condensate was not drained, the controls get to try to work on water instead of air. If the filters get overloaded, then the compressor will not work or the controls get to try to work on oil. The oil coalescing filter and dryer are there to ensure that there is the clean dry air that the controls manufacturer specified. Without clean dry air, all that can be expected is one control problem after another. Refer to the previous section on lubrication and make sure to always check the oil level in the compressor. Keep the fins on any air cooler, and keep the ones on the compressor head clean so that they reject heat the way they are supposed to. It is better to replace a coalescing filter a little early than to put it off until it is too late. Once oil gets past that filter and into the system, it will take what seems like forever to get rid of the oil problems.

If the pneumatic controls do get gummed up with oil, someone will eventually have to clean them or replace them. That oil gets gummier as it dries and collects little particles of dust to really foul up the controls. If that problem is simply ignored, efficient operation becomes nearly impossible because the controls will always be hanging up.

If the problem is inherited (by taking over an old plant), try to convince the owner to replace the controls with microprocessor-based hardware to eliminate all the problems with the old pneumatics. Failing that, watch the systems for a while without changing anything. Some of the older pneumatic systems can work on oil or water. The old ratio totalizer seemed to be able to. Get a better understanding of how the oil affects everything. The expense of all the oil added to the compressor may help convince the owner to upgrade. Or perhaps, fresh oil flowing through the instruments will flush them and limit gumming up. Situations where the controls work anyway should probably be left alone. The only thing to do is keep good records of the costs associated with the problem to give the owner a justification for replacing those controls. Switching a system by adding coalescing filters or other oil removal devices could result in system failure because the oil remaining in the instruments will gum them up.

Keeping the control devices clean, free of dust, dirt, oil, and grease is the most important thing to do. Electrical and electronic, including microprocessor-based controls, are subject to dirty power supplies as well. That means power with harmonics, spikes, and all those other things that impact electrical equipment. Whenever a contractor tries to hook up a welding machine in the plant, make sure a connection designated for welding machines is used. Be certain any welding lead is not run over or around control cabinets or conduit containing control wiring. If an emergency generator is tested regularly, an uninterruptible power supply (UPS) may be needed for the controls to keep them from dropping out and doing unexpected things (some set up by the logic designer) like restarting everything in manual. Actually, a UPS on all electronic and microprocessor-based control supplies is beneficial because the UPS isolates the controls from the line and will protect the controls from surges and power line noise. It is like putting an oil-free compressor with a dryer on a pneumatic control air supply. Today, there are a lot of UPS systems designed for computers that can handle the normal control system load for a boiler. Putting one of those on the boiler control power supply will be well worth the little bit they cost.

Logging readings not only allows an evaluation of the continuing performance of the plant (See Logs in Chapter 1) but also provides indications of instruments and controls losing calibration or operating inconsistently. Maintaining local instruments, like pressure gauges and thermometers, provides a reference for the control and instrument indications and can be used to identify problems and schedule control and instrument tune ups.

The operator may be allowed to do the instrument calibration as well. With the proper training, tools, and by carefully following the manufacturer's instruction manuals, it is possible for an operator to maintain a calibration schedule during a normal shift. That is not only a big saving for the employer in contractor's costs, it will help keep fuel and power costs down as well.

Tuning firing rate controls is not always something an operator can do. There is a certain amount of skill and experience required to do it without blowing the boiler up. It can be done with hands-on training, under the watchful eye of an instructor. That instructor will know whether the operator has an aptitude for it or not. Tuning a boiler can be dangerous. Only the well trained and well skilled should be allowed to do it.

Pressure and draft gauges require maintenance to ensure their readings are accurate and reliable. All pressure and draft gauges in the plant should be checked for calibration every five years. If the gauge is observed to be constantly swinging (the needle is moving constantly) or it is subjected to frequent bumps (like the discharge gauge on an on–off boiler feed pump), it should be checked more frequently. The sensing lines of the gauges require more attention than the gauge itself. Lines to gauges (provided the gauge is protected by a siphon) should be blown down at least once a year. That blow should be long and large enough to fully flush out the piping.

Draft gauges should be checked for zero every time the boiler is shut down. There is little pressure available to blow them. Don't use compressed air. It has little effect and it is too easy to damage the gauges. Draft gauge lines are normally fitted with tees and crosses that permit cleaning them with a wire brush attached to special fiberglass extension rods. If they are dirty, that is the way to clean them.

Another important annual operation is to ensure that there is an air cushion in pressure sensing lines that are supposed to have it and no air in sensing lines that should not have it. Air in a sensing line can act like an accumulator, compressing when pressure is applied to the system to take on liquid and then expanding when the system is shut down to push the liquid back out. That is not a good thing for something like an oil burner gauge because the oil that is pushed back out will allow continued firing of the burner when it is not supposed to be. With heavy fuel oil, make sure the sensing lines are full of the separating fluid by pumping some through the sensing line during the startup after the annual inspection. Light fuel oil and other liquids that burn are best for this.

LIGHTING AND ELECTRICAL EQUIPMENT

In many plants, the operator has to change the light bulbs. Do it wisely. With modern lighting technology, there are more choices in lighting. Take advantage of them. Many of the modern lighting fixtures are energy efficient but will not pay for themselves in electrical savings because they cost so much more. A fluorescent bulb has an average life of about 10,000 hrs, five times that of an incandescent. Consider the value of the labor to replace one of those bulbs five times. That should be enough to justify the higher price.

Compact fluorescents, those curly bulbs, are becoming so common that their prices are dropping. Thus, they will pay for themselves in energy savings in less than a year, on top of the labor savings. Typically, a 60 watt bulb can be replaced with a 17 watt fluorescent. Use that ratio to get an idea of the right size. Light emitting diodes (LEDs) are another story. They are more expensive, but they have a life of about 100,000 hrs (over ten years of continuous operation). They are really invaluable for those applications where the reliability of the light is important. They take about one-quarter of the power of an incandescent bulb for comparable illumination and even less in applications that are not involved with illumination. Thus, with the extended life, they are fantastic for applications like control panel indicating lights. When it comes to a question of what is happening on the display because a light burnt out, the reliability of LED lights overshadows all the arguments about the little bit extra they cost. It is preferable to buy new LED light assemblies than to carry spare incandescent light bulbs.

Some operators are expected to perform normal checks and maintenance of electrical equipment in addition to maintaining the boiler plant. Changing light bulbs and performing the following maintenance functions can make the operator more valuable to the employer. Contrary to popular beliefs, electrical systems require maintenance. Ground fault interruption devices, called ground fault circuit interrupter (GFCI), all have a test push button on them. They are there to test the device on a regular basis. Instructions for the smaller units say to test them monthly. Don't forget to do it. Record the test in the log. Don't use those little stickers that come with the breakers. Insert a test light or some other device that is obviously using power to determine if the device passed the test for certain. When it is confirmed that everything powered by the circuit can be shut down, push the test button. The test light should go out and then come back on when the reset button is pushed or the circuit breaker reset.

GFCI circuit breakers trip without shifting the operating toggle all the way to the off position, just like a normal circuit breaker when it trips. That means it has to be turned off and back on. The GFCI has current detection devices in them to compare the current going out the hot conductor and the current coming back on the other conductor. If the two currents do not match precisely, it trips. Smaller GFCIs are also called personnel ground fault protectors. Their real purpose is to prevent anyone who accidentally touches a hot electric wire or any conductor (metal, wire, copper pipe, or whatever that will carry electricity) while in contact with a ground from becoming part of the circuit.

The concept of grounding needs some clarification. Grounds in electrical terms are conductors that are not supposed to carry electrical current. However, they can convey it to the ground. A concern in any installation is the lack of grounding. That situation refers to a conductor that is not supposed to carry electricity and is not connected to the ground. It is not grounded. The concern with ungrounded conductors is that they can become hot by coming in contact with a hot conductor. A hot conductor is anything in an electric circuit that is designed to carry electric current and there is a difference in voltage between it and the ground. If the ungrounded object is contacted by a person with their feet on the ground, they can close an electrical circuit between the hot conductor and the ground. Electricity will flow through them. If the current range is right, it will kill them instantly. If it is low voltage (less than 600 volts above ground), it should not kill them, but it can cause everything from a mild shock to severe burns.

Personnel GFCIs will sense the fact that the current is going to ground (because of the difference between the currents in the two conductors) and trip before the current reaches a value that could give more than a tickle. Regular testing of those devices helps to shift dust and debris that can settle in the mechanism and prevent its operation. Personnel GFCIs are very important in a boiler plant because there are a lot of grounds around. All receptacles in a plant should be fitted with personnel GFCIs because everything around is grounded (or should be). If an electric tool or trouble light being held has its hot conductor short to something, that device is needed to prevent a shock.

Larger GFCIs (in current carrying capability) are required because a current flowing through devices not intended to carry current can overheat them to the degree that they burn or explode. Look at the thickness of the metal in any large electrical panel compared to the size of the wiring supplying it. If the current were to suddenly start flowing from the wiring through that thin panel to ground, it would damage the thin metal in that panel. Those devices should be tested regularly by an electrician. Record the test in the log.

Operating circuit breakers has the same effect as GFCIs. It helps to ensure that they will function when necessary by keeping them loose. It is always a good idea to open the circuit breakers, in addition to disconnects, when servicing equipment. Add them to the lock out tag out procedures.

Maintaining grounds is a constant problem in many plants. A common way to ensure a good electrical connection between steel building structures and the ground is the installation of a grounding grid and bonding. A grounding grid is a pattern of copper rods laid out in the ground around and under a building to provide good contact with the earth. They are welded or mechanically attached to each other and to bonding jumpers that extend to the building structure. Bonding is the process of installing jumpers connecting one piece of metal to another to ensure that electrical current can flow from one to the other. If buildings were not grounded, lightning could create thousands of volts of potential between the building and ground, let alone the static electricity differences in a building from a cloud passing over it. If someone touched the building with their feet in contact with the ground, they would become the grounding conductor. Look around at the bases of steel columns. Some of them will have an occasional wire run up through the concrete to an attachment on the steel. That is a bonding jumper. The connections can be mechanical or the metals can be fused, using a thermite welding process. Thermite welding creates a puddle of hot molten metal that attaches itself to the steel and wiring. The bonding wires serve as the bonding jumpers. There is no guarantee that the anchor bolts, nuts, and column bases will maintain electrical continuity.

The problem with those connections is that they are exposed and can be broken loose by any number of methods. The operator should note any damage to one. Then repair it or have it repaired immediately. Caution is advisable because there could be a voltage difference between the two. Always make certain that there is no voltage difference before attempting to restore a connection. Be aware that any number of incidents in and around the facility could create a voltage difference, including a cloud passing over.

Grounds in and around boiler systems are of concern because there is so much steel and water, all good conductors of electricity (well water normally is). Lack of a ground invites problems with control operation. The deadly explosion of a boiler at the New York Telephone Company in 1963 was associated with ground paths bypassing some limit switches. The boiler continued to fire and build pressure until it exploded. To ensure that to not happen again, all control circuits must have one leg grounded. All final devices (control relays and fuel safety shut off valves) have one side connected to the grounded conductor. (A grounded conductor is a wire for carrying current that is connected to ground at one point to ensure that its electrical potential is the same as ground.) Any ground that forms in the control circuit should produce a fault that will trip the fuse or circuit breaker. If that does not happen, the ground should produce a short circuit between the fault and ground so that there is no voltage across the associated relays or safety shut off valves to keep them open. Of course, if the conduit, or other parts located where the wiring insulation fails, is not grounded properly, it becomes a point of high potential that can cause personnel injury. It is also a conductor that can bypass some of the limit switches on the boiler. To ensure there are no inadequately grounded metals around a boiler, an annual check should be made of their resistance to ground. Using a simple multi-meter set at the lowest resistance setting and one very long test lead, check the resistance between the grounded conductor in the burner management panel and every metal object (except wiring) on and around the boiler. The resistance should be less than 5 ohms everywhere. Usually, the resistance is less than 1 ohm, with 0.3 to 0.5 being common. The choice of 5 ohms is because a little more resistance can produce enough potential to keep a small control relay energized.

Along with checking motors for overheated bearings, check out the electrical panels and switchgear for loose connections that generate heat. The wiring can loosen, especially when the equipment is started and stopped frequently because the wire does heat up a little bit every time it runs. That results in expansion and contraction of the metal that can loosen the connections. Loose connections are very common with aluminum wiring because aluminum has a larger coefficient of expansion than copper. During a normal round, just laying a hand on the front of each panel and comparing to what is felt on previous rounds can give an indication of increased heat. With large panels, it is a good idea to sweep the hand over the front to note hot spots which are indicators of loose connections. If one is detected, plan to shut down that equipment to correct the problem, hopefully before the equipment picks its own time to go down. Prior to annual inspections, perform a detailed examination for hot spots at connections, opening panels whenever possible and scanning all connections with an infrared thermometer to find any hot spots. On a five-year interval, open all wiring boxes on motors to check the motor connections. Open rear covers on motor control centers to check the bus bars. Make that two years if they are aluminum. Regular annual tightening of aluminum conductors may be required.

High temperatures are the worst enemy of electrical systems. There is a rule of thumb that claims the life of electrical equipment is halved for every 10 degree increase in temperature. It is important to limit the temperature of the electrical equipment, even if not responsible for maintaining it. It is a simple matter of keeping cooling passages clean and unobstructed. Don't let painters lay their drop cloths over operating pumps or electrical enclosures so that they block the flow of cooling air. In one case, the fresh coat of paint actually froze a motor bearing on its shaft. Regular cleaning of vent screens, louvers, and the like will prevent blockages that could damage the equipment. Always use a vacuum to clean them. Blowing air and brushing will simply loosen the dirt and allow it to flow into the equipment, not keep it out. Use a damp rag for removing dust from the top of motors and electrical enclosures so that the dust is picked up, instead of brushed off and into the vents.

The concern in bathrooms and kitchens is that the water is there, contacting drain piping, etc., and is a ground which can be contacted at the same time as a hot conductor. That is why all new bathrooms have to have GFCIs. Electrical enclosures and motor housings should be grounded, not hot. Thus, a little scrubbing with a damp rag cannot cause a problem. Don't use a soaking wet rag that is squeezing water out and into the electrical appliance to become a conductor between hot and ground. A damp rag should not do that.

Transformers are frequently allowed to die for lack of maintenance. It is a shame that so many of them are neglected. They not only represent a significant repair or replacement cost but also the matter of the downtime associated with their failure. There is also the very large and very real additional cost of power that is wasted when the transformer is operating inefficiently. Whenever a transformer can be taken out of service, use the opportunity to maintain it. Open the enclosure and remove accumulated dust and dirt. Then inspect it for apparent hot spots. Tighten all the connections as a minimum.

Samples of oil, from oil filled transformers, should be drawn and sent to a qualified testing lab at least every five years. The lab should provide appropriate sampling kits. Refer to the manufacturer's instructions because there are a variety and forms of transformers with different requirements. Also be careful with some real old transformers that may still contain polychlorinated biphenyls (PCBs), a known carcinogen. During the operation of the transformers, a regular cleaning of any external fins should be scheduled based on an observed difference between metal temperature and ambient air. Also make sure to maintain the ventilation equipment for any electrical enclosure. It is a lot easier to replace a hundred dollar exhaust fan than several thousand dollars' worth of transformers. Even walking through the room containing a transformer and noting the temperatures will improve their reliability. Newer transformers can produce dramatic savings in energy cost because they are so much more efficient. Add to that, the problem with many transformers operating at very low loads (where the losses are more significant) to be aware that replacements should be considered on a regular basis.

VOLTAGE AND CURRENT IMBALANCE

Changes in electrical loads in facilities can create voltage and current imbalance that can generate serious problems with alternating current motors and other equipment. One significant contribution to changes in facilities includes the conversion to more efficient lighting. If that is not done with some attention to the share each phase of a three-phase power supply is altered, the plant will start having strange situations with motors and other electrical devices, including lighting. Prior to each annual shutdown, perform voltage balance checks and check all large motors for current imbalance.

Checking for voltage imbalance is relatively easy, is only applicable to three-phase power, and only takes a voltmeter and some protective equipment (PE, including face shield, leather jacket, and rubber gloves suitable for the voltage). The PE is required to protect from accidentally contacting hot conductors and arc flashes. Read the phase-to-phase voltage for the three phases at each three-phase motor starter. This requires opening the starter cabinet, which usually requires opening the disconnect. Then restore power by closing the disconnect with the door open. That usually requires manually overriding the interlock that proves the door is shut. After recording the voltages, calculate the average voltage by adding the three values and dividing the result by three. Then calculate the voltage imbalance by dividing the maximum difference between the average voltage and each of the three phases by the average voltage and multiply by 100 to get a voltage imbalance in percent. Do the same thing with the motor operating. Operating a motor with a voltage imbalance of more than 2%–3% is not recommended. If voltage readings vary considerably between the motor off and motor running values, the wiring to the motor should be checked. Voltage imbalance can be caused by a lack of balance of single-phase loads in the facility or, rarely, a problem with the utility feed.

Current imbalance in a three-phase motor can be caused by voltage imbalance (which should be corrected first), shorted motor windings, or a high resistance connection. For current imbalance, an amprobe, or multimeter, will be needed to measure the current on each leg of the three-phase motor. Current imbalance is calculated in the same manner as voltage imbalance. A motor should never be operated with a current imbalance of more than 10%. When a current imbalance of measurable value is detected, then it is advisable to check to determine if the problem is in the source or the motor. To determine the source of the problem, shut down the motor, isolate the power, and rotate the leads. Then take another set of readings. Note that rotating the leads is not the same as switching the leads. To change a three-phase motor's rotation, switch any two leads. Rotate the leads by connecting L1 to M2, L2 to M3, and L3 to M1 where L denotes the line and M denotes the motor lead. If the readings at the line match, then the problem is in the power supply. If the readings at the line show different values for each phase, the problem is with the motor.

EDDY CURRENT TESTING

Eddy current testing has become a standard practice for large chillers on their 15-year overhaul and is also used as a method for checking the condition of boiler tubes. It is a non-destructive test of the condition of tubing. A technician (because the interpretation of the readings requires someone skilled with it) has a helper insert a probe connected to a cable into each tube and pass it through the tube, while watching a monitor for indications. The technician may be willing to share an example of a similar tube, with holes drilled in it and sections where the metal is cut to be thinner, that must be detectable with the equipment at setup before beginning the testing of the tubes at the plant. The testing will not go well if the tubes are scaled up, especially if there is so much scale that the probe cannot be pushed through the tube.

MISCELLANEOUS

Painting is a maintenance activity that can create problems. In many plants, painting seems to be the only form of maintenance. If it is necessary to paint, then make sure nameplates, gauge faces, and other items that should not be painted are adequately masked before the painting process begins. Keep in mind that multiple layers of paint are insulation and can shorten the life of electrical equipment. Paint can block tiny openings that are required for proper operation of self-contained control valves and other equipment. Regular painting of screens and narrow louvers can reduce the free opening to reduce air flow, with possible hazardous or damaging consequences.

Instead of painting the plant, try cleaning it. Proper use of cleaners, soap, and water can restore the condition of a plant at a lower cost and with less harm than painting. It will look good when it is done and some people will think it was painted. The only things that should need regular painting are floors and handrails because they are exposed to wear.

ASME CSD-1 and the National Fire Prevention Association (NFPA) 85 Codes are adopted by law in many states and contain requirements for maintenance. Factory Mutual and other insurance underwriters also have their own requirements for testing of fire and explosion prevention devices to ensure their reliability. Be certain to incorporate all the applicable requirements in the maintenance program. A recommended program of testing safety devices is included in this book, but it may not contain every requirement that is legally or contractually required to perform. Keep in mind that those requirements are only safety related and concentrate on devices that were found to contribute to significant failures and warranted investigation due to their cost or loss of life. A system that is as safe as that of some insurers and that code writers would like is not necessarily reliable because it can shut down more frequently.

Maintenance of stored fuel oil is one item many operators forget about since they are primarily firing gas. Checking the inventory to be certain the tanks are not leaking and checking for water in the bottom of the tanks is critical to ensuring a reliable source of oil is available if it is needed. There are additives that can extend the life of fuel oil in storage and tests for the condition of the oil as well. Check with the oil supplier.

Whenever there is maintenance on a piece of equipment, replace the belt or coupling guard. Don't just set it there either. A loose coupling guard can vibrate around until it is caught by the coupling bolts and flung across the boiler room at someone. Always replace all the parts, especially the protective guards.

REPLACEMENTS

Failures in an aging boiler plant typically set the stage for a review. This normally results in a recommendation for a major replacement program because everything has been ignored and is so worn that it all needs replacement. Frequently, it is due to the plant being operated in a manner that ensures that everything wears out at the same time (see rotating boilers in the section on operating modes), a common practice that should be avoided.

Rotating equipment (fans and pumps) and similar devices, where movement promotes wear, top the list of equipment that must be replaced on a regular basis. Motorized valves, pressure and temperature switches, pressure gauges, and bi-metal thermometers all have moving parts that can wear, gall, and fail. They need to be replaced at regular intervals. Those devices can last for years when their use is infrequent and they are subjected to a limited number of operating cycles or changes in condition.

Scheduling replacements is not a simple process. There needs to be some reasonable degree of expectation of when the device is going to fail so as to not waste money by replacing them too frequently. That is one of the problems with a program that only considers PM.

If there is a schedule for operation of equipment that consists of an operating unit and a spare, the first failure provides a basis for determining the life of the other. Of course, if they were operated for equal periods of time, the probability is that the spare unit will now fail. By ensuring operating hours are proportional to the number of pieces of equipment, some time to operate the remaining piece or pieces is ensured before they will fail. Scheduled replacement of spares that have failed should not be questioned. There is also a reasonable basis for establishing a deadline for the replacement. The concept here is breakdown maintenance. It works well when there are one or two spares to deal with.

When there are no spares, the scheduling of replacement of devices is dependent on how critical their continued operation is. If possible, weigh the cost of the replacement of all the devices that have a greater than 50% probability of failing between now and the next maintenance period. Include the cost of labor to replace the devices and such contingent costs as disposal expense to establish a reasonable cost for replacement. The cost of a failure is dependent on the type of facility served by the boiler plant and can vary dramatically.

A hot water heater in a Boy Scout camp will have a minimal failure cost. They can use the time spent replacing the failed heater to train the scouts in providing their own hot water. On the other hand, failure of a hot water heater in a hospital borders on unacceptable because the lack of hot water prevents proper hygiene. The cost of canceled operations, bringing in food, and possibly relocating patients can all be reflected in the cost of failure of a steam boiler. Any production facility will normally have a high cost of failure. The costs could include damaged product and loss of sales that will destroy customer confidence. Then there is the high cost of paying employees when they are not making product and securing the facility, and then restoring it once the repairs are completed.

If there is no spare, there should be a contingency plan in the event of a failure. Consider the operation of a heating plant for an apartment complex that has only one heating boiler. In the event that boiler fails, there may be several options. However, lack of a plan will not only look unprepared but also could generate significant unnecessary costs. The wise operator will always have contingency plans for failure of each piece of equipment and service. That includes utility service. There needs to be a plan for the failure of every utility. Loss of electric power is a common occurrence. Don't get caught when the generator fails to start or shuts down shortly after the electricity is lost. Plans are needed that include procedures in the event that standby equipment fails, loss of the utility becomes long term, or conditions prevent delivery. Failure to plan is planning to fail.

When replacing small parts and items, make a concerted effort to ensure to replace with something of equal quality. A big problem with valves is that they cost less when furnished with reduced trim (a smaller opening). Motors with a service factor may be using it and a larger motor may be required. Modern technology has also provided better and lower cost alternatives, especially for motors and controls. They should be considered when replacing parts and equipment.

Boiler Tube Cleaning – Replacement

The ASME Boiler and Pressure Vessel Code (BPVC) posits a 30-year design life for boiler tubes, provided the water chemistry is maintained and no highly corrosive material is deposited on the tubes. They are designed to transfer heat rapidly and, as such, they are more likely to be coated with scale. They are relatively thin, also for heat transfer. Consequently, they will corrode through first. There are means for cleaning scaled tubes so that they do not have to be replaced. However, waterside cleaning occasionally penetrates the tube, making replacement necessary.

Fire side cleaning can be performed by wire brushing the tubes of fire tube boilers. A modern piece of equipment (Figure 6-9(a)) that connects to a vacuum

Figure 6-9(a). Fire tube cleaner.

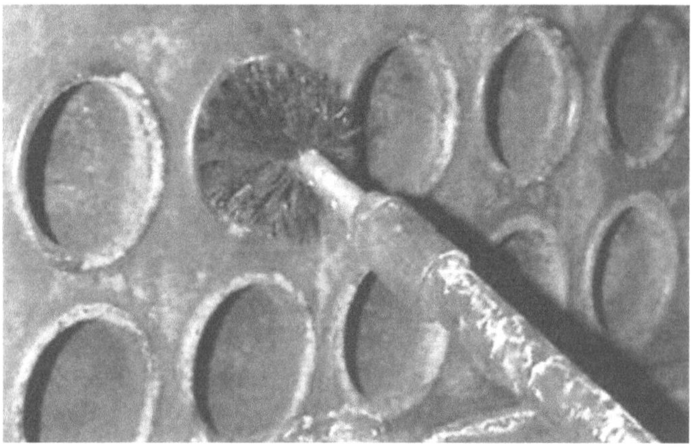

Figure 6-9(b). Brush with a pole.

to collect the removed soot, and a motor driven brush, makes the job relatively easy and a lot cleaner than using a brush on a pole (Figure 6-9(b)).

Fire side cleaning of water tube boilers is normally accomplished with the boiler in operation using soot blowers. Note that soot blowers should be used only when the boiler is firing. During boiler operation, the flue gas inside is essentially an inert gas. If soot blowers are operated with only the forced draft fan running, an explosive mixture of dust and air is created, with enough energy added by the steam to create a static spark. Of course, soot blowers have to be intact and installed correctly to do a good cleaning job. The sound of the blower should give an indication if they are working right. If the end of the soot blower has corroded, or burnt off, or the element is misaligned so that the steam, or air, jets are hitting the tubes (a good way to cut through the tubes), the sound should be noticeable. When soot blowing does not do the job and fuel additives do not do the job, then the boiler has to be cleaned with a high pressure water wash.

There are three methods for cleaning water side scale from boiler tubes. Under normal circumstances, with adequate pretreatment facilities and adequate boiler water chemical treatment, the tube should not need cleaning. Turbining is occasionally used as a general maintenance method in plants with very poor water pretreatment. Turbining of tubes is accomplished with a special water powered tool that rotates a set of small sharp gears around inside the tube (Figure 6-10).

The water not only powers the tool but also flushes the debris away. A tube cleaning turbine will remove most of the scale but leave small pieces unless it is repeatedly run up and down the tube, losing a lot of metal as well. They are not difficult to operate. Of course, they only work for removing waterside scale from inside the water tubes, that is, inside round water tubes. In a very old boiler, the tubes might be closer to square where they are bent. Turbines will jam in them. With modern water treatment methods, this picture should be the only time a tube turbine is seen.

High pressure washers are used to remove scale from the water side of fire tube boilers. Operating with nozzle pressures as high as 40,000 psig (pounds per square inch gauge), they blow the scale away and sometimes take some metal as well. These are best handled by contractors who are experienced with their operation. The application usually requires a vacuum system and truck to remove the scale from the boiler as it is washed off and separate it from the wash water to allow recycling of the wash water.

The third method for scale removal is acid washing. An inhibited hydrochloric acid is used to eat the scale off the tubes. The application requires care and regular testing to ensure the acid is removing scale and not boiler metal. The acid solution is heated and circulated. The entire boiler has to be flooded so that all the boiler metal is exposed to the acid. Any mistakes result in serious damage to the boiler. This method is also best left to contractors with the equipment and skill necessary to do the task. They also haul off the spent acid and dissolved scale when they are done.

When cleaning fails and so much energy is wasted by scale that something has to be done, plugging or replacement of the tubes is required. In a modern flexitube boiler, all that is needed is a wrench, a special tool, and a big hammer. They are designed to be replaced by individuals with a reasonable mechanical sense. Otherwise, the boiler tubes are installed by rolling or a combination of rolling and welding, processes that require more skill.

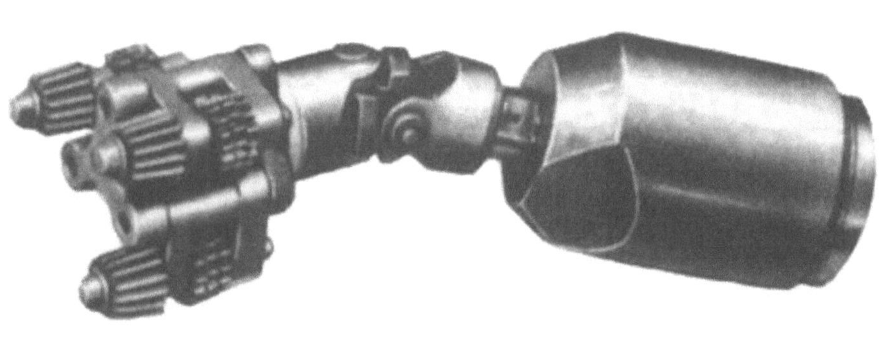

Figure 6-10. Tube turbine.

When only one or two tubes are defective, it is usually easier and more frugal to plug them than to replace them. Some tubes cannot be plugged because they serve purposes other than heat transfer. Tubes that form boiler walls or flue gas baffles cannot be plugged because they will melt down or burn off without water cooling and allow heat and flue gases through.

For water tube boilers, it is a little more than simply a matter of obtaining some machined steel plugs that fit into the ends of the tubes and inserting them. The first, and a very important, thing to do is to make sure to have located the leaking tube at both ends. Testing using rubber plugs and a water hose is recommended. To be certain the plugs do not blow out because steam is generated in the tube from water leakage, drill or chisel a hole in the tube so that any leakage is bled into the flue gas. Also, remove any scale from the end of the tube, making certain it is clean, round, and smooth to ensure that there is a good metal to metal fit between the plug and tube. Gently tap the plug into the tube, and the water pressure will hold it and hammering excessively can distort the drum or header.

Plugging of fire tubes requires not only a plug but also a means of holding them in because the water will get into the fireside of the tube and apply pressure to the plugs. A piece of cold rolled steel rod longer than the tube and threaded at the ends is required, along with nuts and plugs that are bored to accept the steel rod. At least with the rod, there is certainty about the right tube ends. The rod has to be large enough to overcome the force of the water pressure against the plug and produce enough force to seal the plug and the end of the tube. The tube end must be cleaned as described for water tube boilers. The plugs also have to have a means of sealing the space between the rod and the plug. An advantage of plugging a fire tube boiler is that the plug can be tightened while the boiler is under hydrostatic test to try to seal a leak. The boiler can also be plugged while it is under water pressure. However, for most operations, the plugging of a fire tube boiler is so involved that it is much easier to just replace the tube.

A common repair for many water tube boilers involves replacing a section of the boiler tube. Frequently, it is only a portion of the half of the tube that faces the furnace. When bulges or blisters form due to scale buildup, and sometimes rupture, the rest of the tube is still intact and of original thickness. The repair requires a skilled boilermaker welder. The tube is cut out around the failure, normally in an elliptical form. A piece cut from another tube is inserted in its place with the edges of the original tube and patch butt welded. Since the tube walls are so thin, the weld is normally made by Gas Tungsten Arc Weld (GTAW, also called tungsten inert gas (TIG)).

Entire sections of boiler tubes can be removed and replaced in a similar manner. The elliptical patch is used at either end so that the welder can reach through the opening provided for it to reach the butt joint at the back of the tube. The welder has to work on the inside of the tube at the back since there is no room to get to it from the back. Once the back is welded, the patch is set and welded to complete the repair. That method is referred to as using a "window weld." Replacing a boiler tube is best done by a boilermaker that has the skill and experience necessary to do the job right. If the tube is welded, check with the insurance company or state boiler inspector to be certain that it can be re-welded under the local law. Most states require all welded repairs to be performed by an authorized contractor that is approved by the State or holds a National Board Certificate of Authorization to repair boilers, called an "R" stamp. It really is a stamp. The authorized company actually has a steel stamp that is used to mark the boiler when the welded repair is done.

The first step in replacing a tube is removing the old one. Whenever it is possible, the tube should be cut off and removed, leaving the ends in the drum, header, or tube sheet. Replacement of some water tubes in bent tube water tube boilers requires removal of other tubes to gain access to the tube that is to be removed. It is possible that several good tubes may have to be removed to remove a defective one.

Removing a tube from a fire tube boiler is pretty much restricted to pulling it out of the hole that it is installed in. If the tube is heavily scaled, it may be necessary to remove it from the inside. That could require removal of several other tubes. A single tube replacement in a fire tube boiler is seldom located where the tube can be removed via a hand hole or manhole. The holes in the tubes sheet of a fire tube boiler are made a bit larger than the tube. Slight accumulations of scale will slip through the hole. In some cases, the scale is stripped from the tube as it is removed. In extreme situations, it is necessary to split and collapse the entire tube to get it out.

Removing the tube requires crushing or cutting away the tube end, where it is expanded into the drum, header, or tube sheet. Don't cut the tubes from the tube sheet with a torch. One contractor cut the tube sheet with a torch and put new tubes in without repairing the cuts. If a contractor uses a torch to cut the tubes, inspect every opening to ensure the tube holes are smooth and clean so that a new tube will seat properly in the hole when it is expanded. The best way to remove a tube end is to chisel it out, making certain the tube sheet, drum, or header is never touched with the chisel. It eliminates the risk of cutting the inside of the tube hole. It does take longer and, quite frankly, takes more skill. Cutting a shallow (about half the tube thickness) groove through the tube where it is expanded will produce the same effect as flame cutting. After the tube is cut, driving it to the center can collapse the tube into the middle, away from the tube hole, so that the end, or whole tube, can be removed.

Once the tube has been removed, "dress up" the hole, removing any tube metal stuck to it and any corrosion that would accompany a leak or defective rolled joint. Careful use of a file and sandpaper should produce a smooth surface. The edges of the holes should also be smoothed over to eliminate any sharp edges that will cut the new tube. The tube ends should also be dressed up to remove any corrosion for a tight metal to metal fit.

The new tube is expanded with a roller (Figure 6-11(a)) to compress the outside of the tube against the inside of the tube hole to seal the joint. The roller in Figure 6-11(a) expands the end of the tube inside the boiler, flaring it. The roller in Figure 6-11(b) has a beading attachment which forces the metal end of the tube out and back against the tube sheet to form the ends shown in Figure 6-12(b). As shown in Figure 6-12 of completed joints, a water tube (Figure 6-12(a)) is flared, but a fire tube end is beaded (Figure 6-12(b)) or restricted in protrusion to limit heating of the end of the tube. Typically, the inlet of the first pass of a four-pass fire tube boiler is welded (Figure 6-12(c)) to increase its ability to transfer heat to the water since the flue gases are much hotter in that first turn of a four-pass boiler. Once the tube replacement is complete, the boiler should be subjected to a full one and one-half times maximum allowable working pressure (MAWP) hydrostatic test. Many contractors and

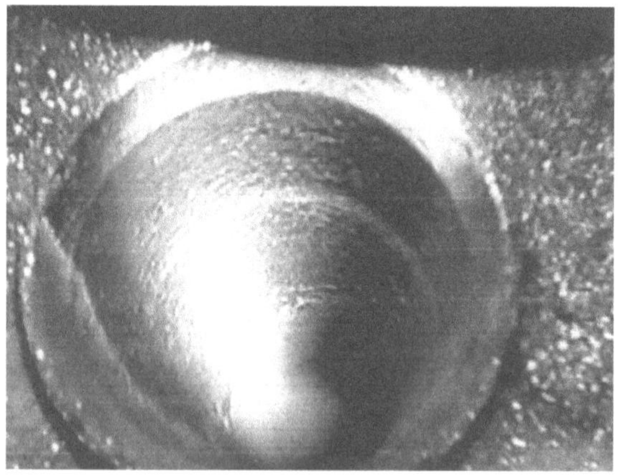

Figure 6-12(a). Rolled tube – flared.

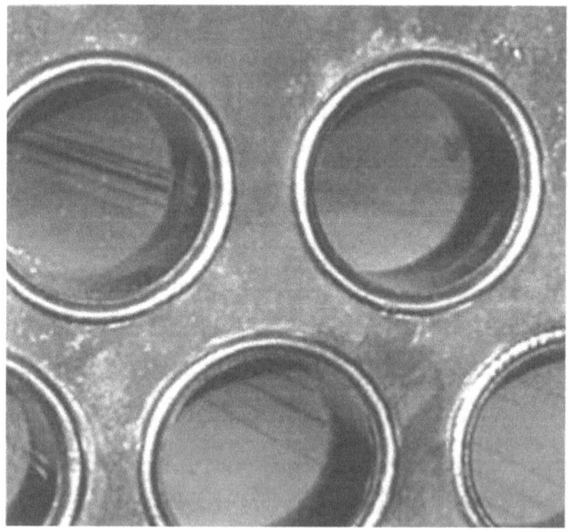

Figure 6-12(b). Rolled tubes – beaded.

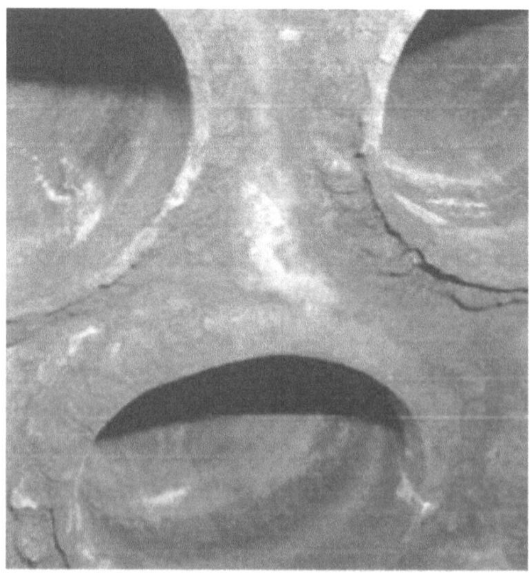

Figure 6-12(c). Rolled tubes – welded.

Figure 6-11(a). Tube roller with flare.

Figure 6-11(b). Tube roller beading attachment.

most inspectors will accept an operating pressure test, but why accept anything other than a test that proves the repair has returned the boiler to a like-new condition? Refer to the section on hydrostatic testing a new boiler. Testing a repaired boiler is done the same way.

For the larger, high temperature, high pressure boilers, the walls are typically made of boiler tubes welded together. They may be fin welded or fusion welded. The fin welded wall has a metal fin between the tubes making up the wall. The fusion welded wall has the tubes nearly touching and a weld bead in between all the way along the length of the tube. A 500 Mw, coal fired boiler might be 55 feet by 65 feet by 200 feet. A typical water wall might consist of 2 inch tubes on 3 inch centers. The typical tube wall thickness might be ¼ inch.

When replacing a length of damaged tubing, the cut and weld prep should be located at least 2 inches beyond the damaged area. If the new cut line places the new weld in the vicinity of an existing weld, the cut line should be moved so that the old weld falls within the removed section of tube. The minimum replacement tube length should be no less than 6 inches. For a water wall tube that absorbs heat and is in contact with water and steam, a backing ring cannot be used. A suitable welding process, capable of producing a defect-free root weld, such as a GTAW, must be used. Once the root weld has been completed, the remainder of the weld groove can be made with an acceptable welding process (usually shield metal arc weld (SMAW)).

Fit up of the weld joints is crucial to obtaining an acceptable final weld. Although it is difficult to make cuts on furnace tubes, it is important to get the existing tube ends squared and correctly chamfered. Cut the replacement tube to the correct length and allow for weld shrinkage. Remember that the weld and the parent metal are melted during the welding process and that molten metal shrinks when it solidifies. A butt weld in a tube will shorten the total tube length by approximately 1/16 inch. The required preheat should be applied to the weld joint before welding. A clamp or guide lug should be used to hold one end of the replacement tube in alignment while the first weld is made.

For a fin welded water wall, the fin must be split along both sides of the tube to be replaced. The preferred method for splitting the fin is flame cutting. Because this leaves a very rough edge on the fin, the cut should be biased toward the tube that is to be replaced. This will allow the cut edge to be ground back to the desired location and a weld prep bevel placed on the membrane. Once the fin has been split, the damaged tube can be cut at both ends. The replacement tube can be installed and welded according to the guidelines for the fusion welded wall.

MAINTAINING EFFICIENCY

An important part of maintaining the plant is maintaining efficiency. Since the cost of fuel is the largest single expense in a boiler plant activity, it is essential to prevent that cost from getting out of control. Efficiency maintenance relies on two activities: monitoring to detect any changes and tune ups when a problem arises. Monitoring is the boiler operator's responsibility. Tune ups are usually performed by outside contractors who have the necessary equipment and skills to perform that work. An operator should know enough about tuning to ensure the contractor is doing a proper job, and the sections on combustion and controls in this book are sufficient to impart that knowledge.

RECORDS

Schedules for maintenance are essential to ensure the longevity and reliability of most equipment. Whether it runs until it breaks or significant PM/predictive maintenance is performed, documentation is essential. For breakdown maintenance items, it provides information about when to order a spare device because the operating one is scheduled to fail. More importantly, the documents record what to buy, what oil to use, what grease to use, etc. That way, the maintenance is performed in a manner that keeps the equipment and systems running.

Maintenance is not complete until all of the documents are properly filed away (see the chapter on documentation). To anyone investigating the plant after an incident, a lack of maintenance records is an indication of a failure to see to it that the work was done. Make a claim that it was done and describe what was done, but without that documentation, there is no good way to prove that it was done. Further, for many types of equipment, there are legal requirements for documentation. The Environmental Protection Agency (EPA) hands out more fines for missing paperwork than for actual emission violations. When a check is listed as part of a standard operating procedure (SOP), then the entry into the log that the procedure was performed is documented proof that it was done. Be careful, however, that it is done consistently, or the entire log is questionable. The job is not finished until the paperwork is done.

Welding

Welding of boilers (including attached piping on high pressure boilers) and pressure vessels must be performed by a company that is authorized by ASME to do that work. Welding on duct work and other containments that operate at very low pressures are not covered by the ASME Code. All of the piping is. Piping for steam at pressures over 15 psig and hot water operating at pressures over 160 psig or 250°F must comply with the ASME B31.1 Power Piping Code. Piping for other services inside the facility must comply with the ASME B31.9 Building Services Piping Code. Both Codes require all welding to be performed in accordance with Section IX of the ASME BPVC, which describes the requirements for qualifying welding procedures and welding operators. The piping codes also contain requirements for examination of welds using methods and procedures covered by Section V of the BPVC. The following offers some advice on what to look for when a contractor, or other maintenance personnel, are welding piping in the plant. After reading the following, it should be possible to determine quickly if the welding is being done in accordance with the Code.

First there is the requirement for materials. Inside a boiler plant, every piece of pipe must conform to at least an American National Standards Institute (ANSI) specification. For all high pressure work over 125 psig MAWP, it must also comply with an ASME specification. The two specifications differ very little, with the requirement that the piping must be supplied with mill test certificates when conforming to the ASME specifications. Beware of false sales claims. Any pipe that goes into a boiler room should have a stencil running repeatedly along its length that begins with "ANSI" and for B31.1 piping also "ASME" followed by a specification number and other data including a heat number, which makes the material traceable to the piping manufacturer. Pipe fittings and valves must bear certain markings, but certificates are not required. They should, however, bear a marking that identifies the manufacturer and the appropriate ANSI specification. As for the welder, the only way to determine the quality without requiring radiation testing (RT, frequently called X-ray) is visual examination by someone who knows what to look for. Here is what to look for.

Fit up

The fit up consists of matching the joint between two pieces of pipe or a pipe and a fitting, or weld end valve, and attaching them with tack welds at no fewer than four points evenly spaced around the pipe. For a high pressure, water wall tube, a clamp, or guide lug should be used, not tack welds. The two ends should be ground clean to gray metal (no rust, paint, oil, or other coating) for at least one-half inch from the surfaces to be welded. The pipe should be aligned such that any differences in the inside edge of each pipe are less than 1/16 inch (less for very high pressures and temperatures). The joint should also be prepared in a manner that conforms to the welding procedure specification (WPS), which the welder should be able to produce on demand. Typically, the ends of the pipe or fitting are ground to a bevel as shown in Figure 6-13. That is the standard preparation for an open butt welded joint, where the two ends of the pipe are positioned such that a cross section of the fit up joint would look like that in Figure 6-14. Four tack welds are typically used and located at quarter points straddling the vertical and horizontal centerlines. A keyhole is formed at the end of each tack weld, providing a starting point for the rest of the root pass.

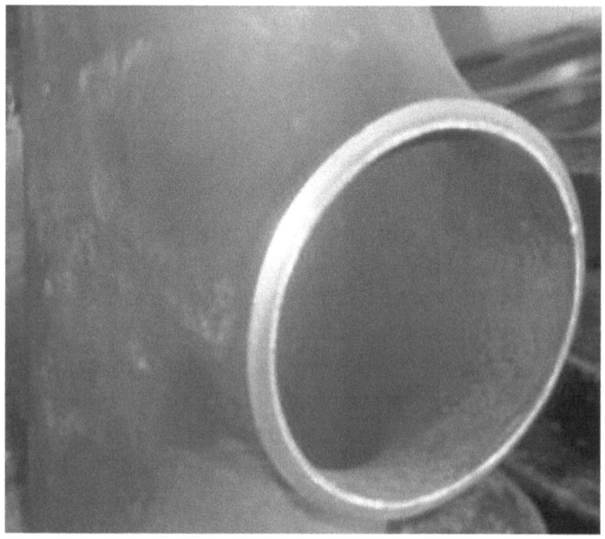

Figure 6-13. Tee showing beveled end prep.

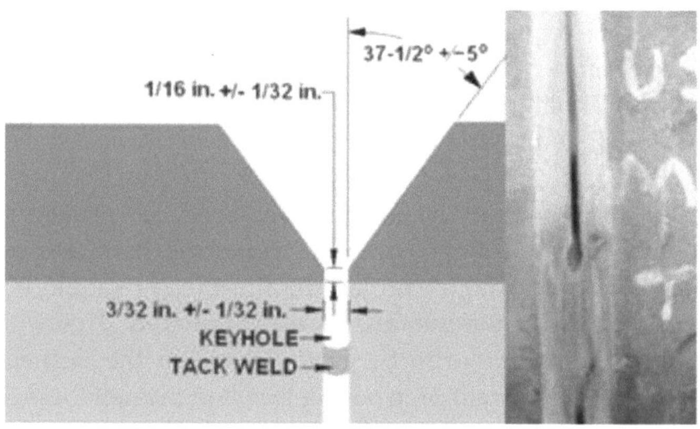

Figure 6-14. Cross section of fit up.

Figure 6-15(a). Welding inspection gauge.

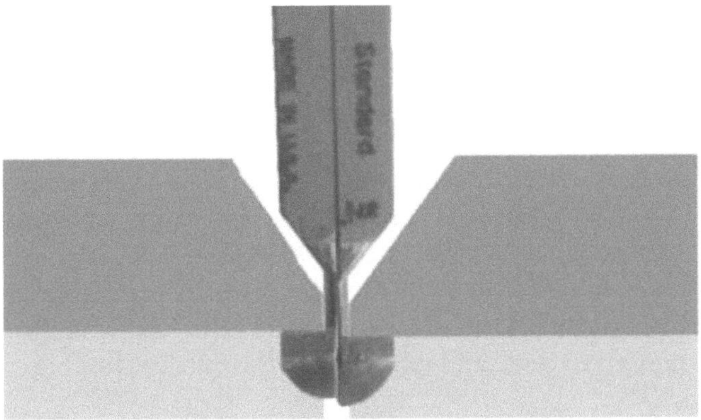

Figure 6-15(b). Checking fit up with gauge.

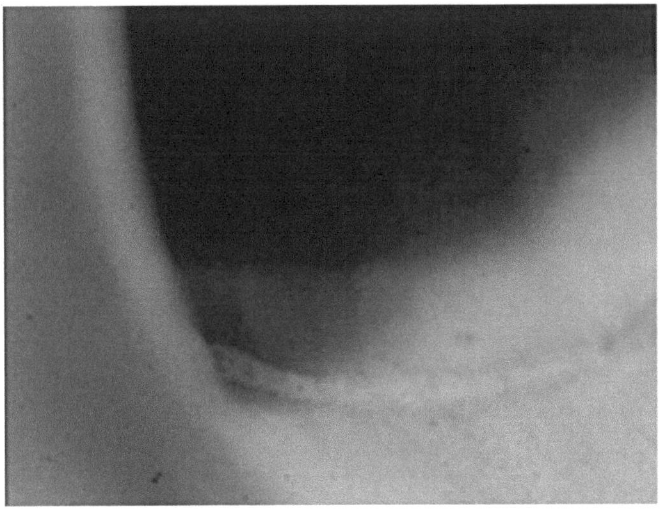

Figure 6-16. View inside of root weld.

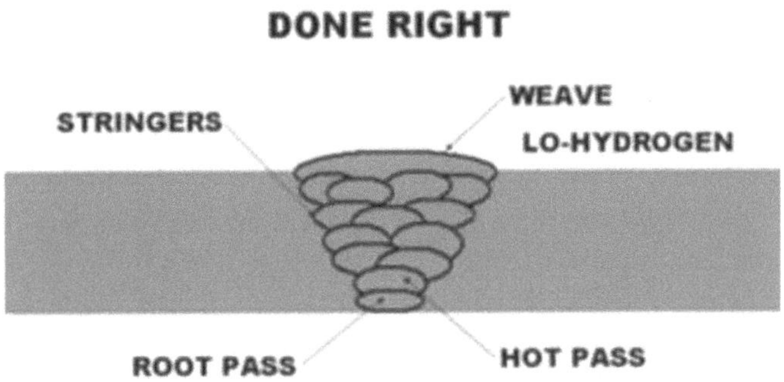

Figure 6-17. Weld cross section.

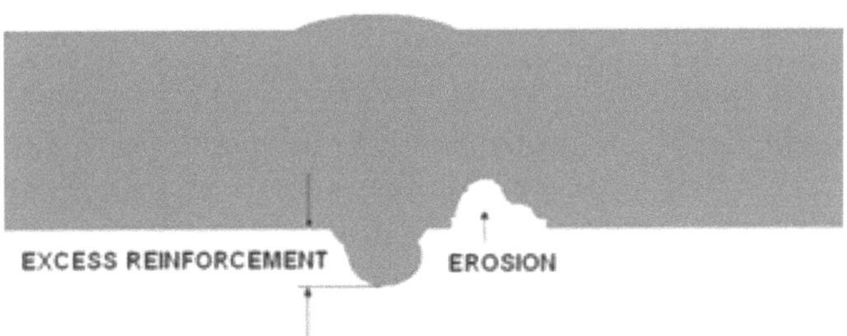

Figure 6-18. Excess internal weld reinforcement.

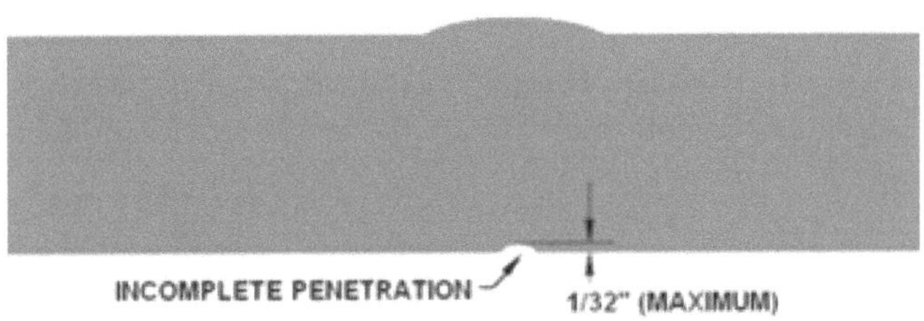

Figure 6-19. Incomplete penetration of a weld.

This arrangement permits the welder to fill in between the two ends to provide a complete weld like the cross section diagram in Figure 6-17. A section through a weld would not look like that because the black outlines shown in Figure 6-17 should not be visible. That is because the pipe and weld metal should be fused together so that they cannot be differentiated. The joint should be inspected by someone other than those preparing the joint and should incorporate the use of the gauge shown in Figure 6-15(a) to check for the appropriate bevel and internal alignment. Figure 6-15(b) shows use of the gauge to check for internal misalignment and a proper bevel.

Root Pass

The root pass, indicated in Figure 6-17, is the first full circumferential weld. The inspector should check it in stages to ensure that there is no excess internal reinforcement as shown in Figure 6-18. Welders normally refer to this as "grapes," and it can produce damaging eddy currents in the flowing fluid that will erode the piping downstream of the joint where indicated. Normally, internal reinforcement cannot exceed 1/16 inch. That value decreases with higher pressures and temperatures.

On the other hand, incomplete penetration, as shown in Figure 6-19, is not permitted in excess of 1/32

Figure 6-20. Stringers in progression of a pipe.

inch. The root and following passes should have a surface texture similar to what is difficult to see in Figure 6-20. It is produced by the welder swinging the rod in overlapping circles. It is difficult to see because the photo is one of a filler pass, commonly called a stringer, using low hydrogen electrodes. It is also hard to see in the photo of the inside of a root pass in Figure 6-16. A reasonably smooth root weld inside the pipe, and within the parameters described here, is always a sign of a good welder.

Hot Pass

The hot pass follows the root pass and is commonly checked for cracking. Once the hot pass is complete, the stresses associated with the cold pipe metal and hot weld during fit up are reduced by even heating. When welding carbon steel pipe, most welders use a procedure calling for low hydrogen electrodes for everything except the root pass. Low hydrogen electrodes have to be kept hot after the container is opened to prevent the absorption of water. Special ovens are usually used for that purpose. If those electrodes are left lying around, getting cold, and then used in a weld, that weld will have porosity (holes) in it. Low hydrogen electrodes produce a weld that flows and fills well with a slag layer (molten minerals that coat the electrodes that melt and form an airtight barrier over the weld to prevent oxygen from the air combining with the molten metal) that is readily removed.

Stringers

Additional passes that cover the hot pass can consist of several "strings" of weld to fill in the weld until nearly flushed with the surface. Careful cleaning of the slag from each of these passes is necessary to ensure it is not left in the edges, which dramatically weakens the joint. The stringer in Figure 6-20 is as wide as they are typically made, with additional ones running alongside each other.

Cover Pass

This is normally called a "weave" pass because it provides a smoother look to the welded joint, as shown in the photograph in Figure 6-21. On completion, it is checked with the gauge to measure the external reinforcement as shown in Figure 6-22. The external reinforcement should not exceed 3/32 of an inch (less as design pressure and temperature increase). Also, any grooves in the external joint should not exceed 1/32 of an inch, which is also called incomplete penetration.

The basic concern here is safety. When the weld section looks like the section in Figure 6-23, and such joints have failed, loss of life can occur. The photograph in Figure 6-24 shows a section of a failed joint. Even though a weld can look good on the outside, like Figure

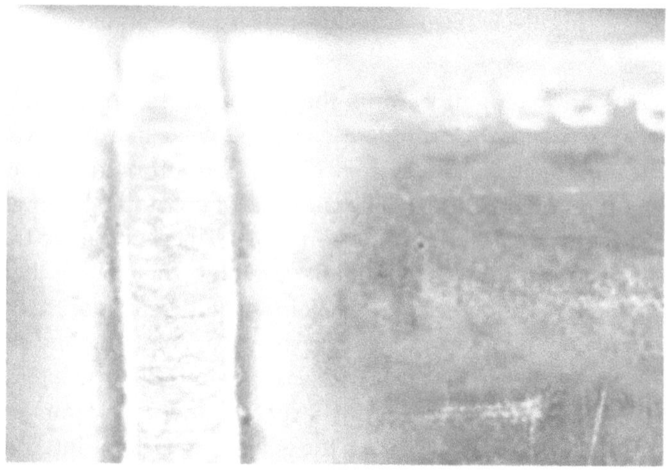

Figure 6-21. Cover pass of completed weld.

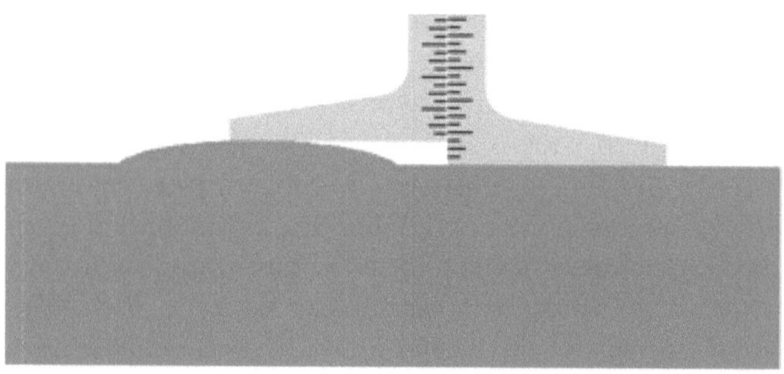

Figure 6-22. Check for excess external reinforcement.

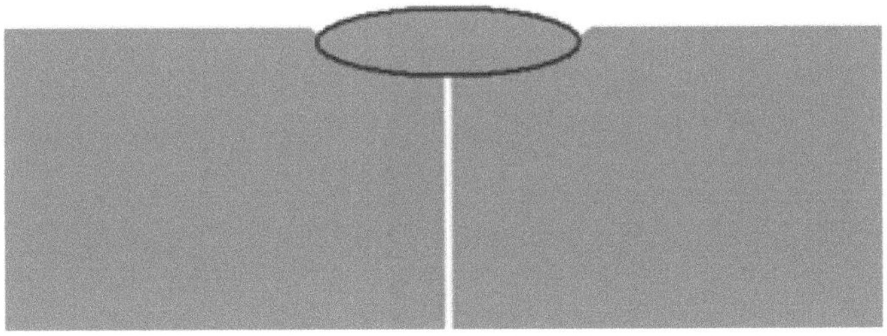

Figure 6-23. Incomplete weld section.

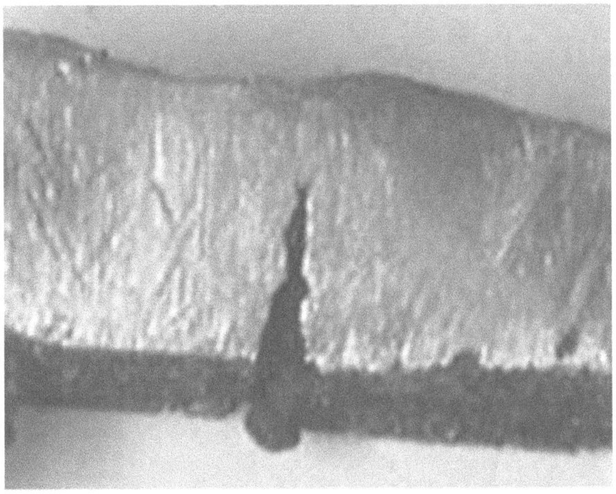

Figure 6-24. Section from failed incomplete weld.

6-25(a), when possible, take a look at the inside because it might look like Figure 6-25(b). When someone makes a weld like this, they are showing no concern for loss of life. Joints like those shown can fail dramatically. Low pressure steam in an 8 inch pipe produces an axial force of 750 pounds which is more than enough to do some serious damage if it gets loose. In the typical high pressure boiler plant, the force in the same 8 inch pipe is over 3 tons. In the typical utility plant at the throttle inlet, it is 30 tons. That is one of the reasons for having Codes and Standards. Don't allow welders in the plant to be cavalier about the welds they make.

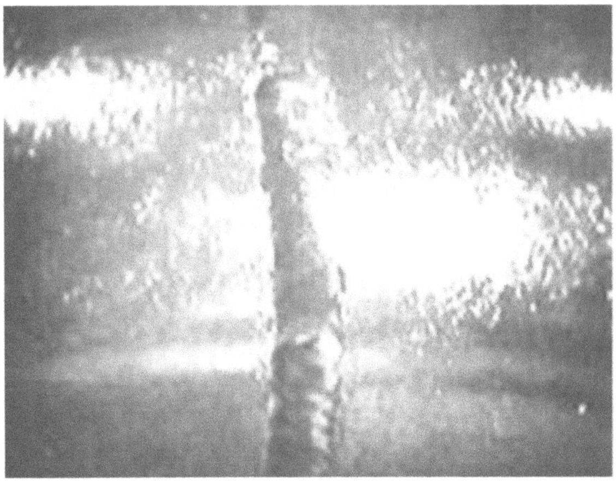

Figure 6-25(a). Apparently completed weld.

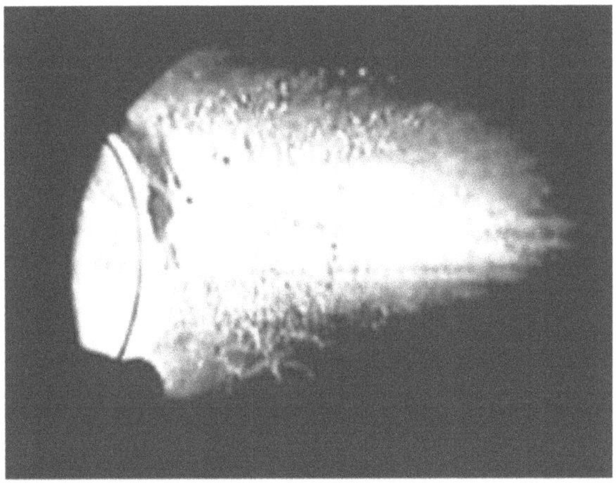

Figure 6-25(b). View of inside of weld shown in Figure 6-25(a).

Chapter 7

Consumables

Few people realize the value of consumables. The typical boiler plant consumes a million dollars' worth in each year. Boiler operators can have a significant impact on their consumption. There is significant trust placed in operating personnel to protect the income of their employer and, as a result, their fellow employees. The use, or abuse, of consumables is where the true value of operation is measured.

FUELS

The principle purpose of most boilers is to convert the chemical energy in a fuel to heat absorbed in water, steam, or another medium for use in the facility served by the boiler plant (there are also electric boilers). A wise operator should know as much as possible about the fuel that is burning, both to get it done safely and to get the most out of that fuel. This chapter will cover the most common fuels first and then touch on some of the others that might be encountered. The goal is to provide an understanding of what is required to burn any fuel, especially in the event that some unusual fuel is encountered.

Oil, gas, and coal are called "fossil" fuels. They are found in the ground where they were trapped as vegetable and animal matter hundreds of millions of years ago. As that matter decayed, they became the fuels available today. Wood, bagasse, corn, and similar fuels, all produced from living plants, are called "biomass" fuels.

The ultimate analysis of a fuel is a determination of the percentage of each chemical element in a fuel. An element is a material that consists entirely of one kind of atom. The determination is made in a laboratory using standard procedures which are included in the appendix. An ultimate analysis will normally list the amount of hydrogen, carbon, sulfur, oxygen, and nitrogen in the fuel along with any other element of significant quantity. For fuel oils and coal, water and ash are included. An analysis of fuel oil will also list "BS&W," which stands for bottom sediment and water. The laboratory will usually include the higher heating value (HHV) of the fuel as well. The HHV takes into account the water that is produced by the combustion of the fuel that is condensed in the bomb calorimeter used to determine heating value. Results are typically listed as pounds of an element per pound of fuel, a value that is readily converted to percent by multiplying by 100. Typically, the oxygen content of the fuel is not measured directly. A good laboratory will note in the report that the oxygen was determined by difference (100 minus the percentages of everything else). The values for the fuel are dependent on the fuel source and any treatment it endures before it is delivered to the plant to burn. When a fuel gas is analyzed, and an ultimate analysis is not requested, the laboratory will give a list of the gases in the fuel and their respective percentages by volume. Normally, methane is listed as the primary constituent of natural gas, with much smaller fractions of other gases.

It is a simple matter to convert a volumetric analysis (one that shows the percent by volume) to a gravimetric analysis (one that shows percent by weight) and to use those analyses. It is only essential for a boiler operator to know what the words mean and to be aware that the ratio of hydrogen to carbon in fuel will vary to affect boiler operation. The reason is clear when an efficiency calculation is performed (see the chapter on efficiency).

Sulfur in fuel contributes a small amount to the energy released in combustion. The problem with sulfur is its products of combustion, sulfur dioxide (SO_2) and sulfur trioxide (SO_3). These combine with the water in the flue gas and the atmosphere to produce sulfurous (H_2SO_3) and sulfuric (H_2SO_4) acids. When surface temperatures in the boiler and ductwork are so low that these acid gases can condense, the acids attack the metal and extreme damage due to corrosion is the result. The last half of the twentieth century saw a concerted effort to reduce the sulfur content of fuels to reduce the problems with acid rain caused by the burning of the sulfur in fuels. Since the late 1970s, SO_2 removal systems have been required on all new, large size units to reduce sulfur emissions to the atmosphere. As a result, SO_2 emissions have been decreased by 93.1% since 1970, and they continue to decline.

Liquid water in fuel can create all sorts of problems. It absorbs a lot of heat from combustion to convert it to a vapor (the hydrogen in fuel burns to a vapor, not a liquid) and it creates corrosive conditions that can damage the fuel handling and storage system. Water in coal is a major problem in the winter because it will freeze to convert a pile of coal to one solid chunk that cannot be fed to the boilers. Similarly, it can freeze in gas or oil systems to block valves and regulators, resulting in dangerous operating pressures.

When water separates from the oil in storage tanks, it settles to the bottom. It will eventually accumulate until, all of a sudden, the burner is trying to burn water. Water in fuel oil also provides a medium for corrosion of the fuel tank and piping. It is one of the reasons for leakage of underground storage tanks (USTs), with some serious consequences. Water can be emulsified (a process that mixes the fuel and water, distributing water throughout the oil), but it can still produce corrosion and will always require the addition of latent heat to vaporize it in the furnace. Small and controlled quantities of water emulsified in oil can help reduce soot formation, which can improve heat transfer to the degree that it compensates for the latent heat loss.

Water in fuel gas systems can be a considerable problem when the gas pressures are low because it can collect and produce blockages in the piping as well as promote corrosion. With wet fuel gas, there will be additional requirements for handling the liquids that settle in the piping because there can be liquid fuels as well as water. Water draining from a coal pile is highly corrosive and must be discharged to a sanitary sewer after it is neutralized.

The discussion in the chapter on combustion helps explain why firing conditions change when the fuel changes. Most of the time, the air/fuel ratio is close enough to ignore the variations. When a service technician uses a portable analyzer to calculate combustion efficiency, that analyzer contains a "typical" fuel analysis for the fuel. It determines efficiency based on that typical analysis. That brings up the question of the calibration of those analyzers. The carbon content of natural gas can vary from 20.3% to 23.5% between the east and west of the country. That amounts to a 15% variation in HHV of the fuel. The analyzer can be used to determine if a change has made an improvement using the same fuel. However, the absolute value of the reported efficiency is questionable.

The source of the fuel will make a difference. In the Baltimore area, natural gas supplies can come from Pennsylvania, Texas, or Louisiana as well as blends of gases from those sources. There is also a port for liquefied natural gas (LNG) that is imported from North Africa, which has an air/fuel ratio that is 10% higher than domestic natural gas. LNG is compressed and cooled until it becomes a liquid, is loaded into tanks aboard ships built exclusively for the purpose, and then transported across the Atlantic Ocean to special port facilities near Boston and Baltimore, among others. With the advent of "fracked gas," the US is now exporting LNG. LNG has slightly less higher hydrocarbons than pipeline natural gas. As a result, somewhat more air is needed to burn the fuel completely.

Being able to burn one of the liquefied petroleum gas (LPG) choices is one way to have a standby provision in the event that the gas supply fails. LPG is expensive. A storage facility capable of providing any extensive operation of a boiler plant is very expensive as well. Few plants use that option. Most of the time, fuel oil is used as a backup fuel in the event of a loss of the natural gas supply. Either LPG or fuel oil will be stored on site for interruptions to a natural gas supply regardless of the reason for the interruption. The Environmental Protection Agency (EPA) has regulations on the amount of time that back up fuels can be utilized during the course of a year.

Ash in the fuel, whether it is coal, oil, or biomass, can create problems with firing. The ash fusion point is the temperature at which the ash melts. If furnace conditions produce higher temperatures, the ash will melt and then solidify again when it cools, usually forming large accumulations of solidified ash that can block air or gas flow passages or grow in the upper portions of the furnace. They grow until they get too heavy to maintain their adhesion to the tubes or refractory and fall crashing to the bottom of the furnace, doing damage to tubes, grates, etc. When firing fuels with a low (less than 2000°F) ash fusion temperature, the operator has to monitor the furnace conditions, inspecting it and recording draft readings to detect hardened ash accumulations early. For a fuel like dust from a laminate sanding operation, where portions of the ash have very low fusion temperatures, the boiler can be operated at very high excess air levels, just to keep the furnace temperatures down to prevent the ash from melting and sticking to the tubes. Always know the vanadium content of the fuel. That material produces a lot of low melting point ash. Vanadium also produces corrosive ash when both vanadium and sulfur are present in the fuel. Even though the ash content of such fuel oils might be low, soot blowers will be needed to avoid ash build up as well as corrosion problems. In some cases, additives that neutralize the sulfur compounds can help, but at the cost of additional materials being added to the boiler.

FUEL GASES

Natural gas is mostly methane (CH_4), with portions of other flammable gases, oxygen, carbon dioxide, and nitrogen. A typical volumetric analysis is 96.53% methane, 2.38% ethane, 0.18% propane, 0.02% isobutane, 0.77% carbon dioxide, and 0.12% nitrogen. That is east coast gas. Gas constituents will vary depending on the well the gas came from. When a boiler is fired with oxygen trim controls to achieve very small quantities of excess air, those controls accommodate the varying air/fuel requirements of the gas supply. Domestic natural gas has a HHV of approximately 23,165 Btu (British thermal unit) per pound or approximately 1042 Btu per standard cubic foot. For combustion, it requires 11.48 standard cubic feet of air per standard cubic foot of gas or 185 standard cubic feet of air per minute per million Btu per hour.

LPGs are primarily butane or propane, with propane being the more common. They are transported as a liquid under pressure. They combine the clean burning properties of gas with the transportation properties of oil but at a premium in cost. In boiler plants where LPG is used, it is normally used as an alternate fuel for interruptible natural gas. Propane can be mixed with air in the proper proportions to produce a blend that will fire in natural gas burners without adjustment of the burners.

Propane has a HHV of approximately 21,523 Btu per pound or approximately 2573 Btu per standard cubic foot. It requires 28.78 standard cubic feet of air per standard cubic foot of gas or 186.45 cubic feet/min. of air per million Btu/hr. For hydrocarbon fuels, the air required per million Btu is about the same. All but very large LPG installations will absorb enough heat at the tank to convert the liquid to a vapor. Large installations require a vaporizer, a heater fired by vapor off the tank that provides the energy to evaporate a liquid stream for use in the boilers. Propane will condense at normal atmospheric temperature (70°F) at 109 psig (pounds per square inch gauge). Butane will condense at 17 psig. On a very cold day, butane will not vaporize and most installations will require a vaporizer. Butane has a HHV of approximately 21,441 Btu per pound or approximately 3392 Btu per standard cubic foot. It requires 37.57 standard cubic feet of air per standard cubic foot of gas or 184.64 cubic feet of air/min. per million Btu/hr.

Hydrogen is being proposed as a fuel that can be produced from renewable energy. There is now a color code that is being used to identify the "climate friendliness" of the hydrogen, depending upon its source. Green hydrogen is produced by using electricity generated by solar or wind power to electrolyze water to produce hydrogen and oxygen. Pink hydrogen is similar to green hydrogen, except that the power source is nuclear power. Blue hydrogen is made by traditional fuels but with the CO_2 that is produced along with the hydrogen being captured and sequestered. Gray, black, and brown hydrogen refers to hydrogen made from natural gas, coal, or lignite by conventional methods. Turquoise hydrogen is produced by the pyrolysis of fossil fuels. The pyrolysis process creates hydrogen and solid carbon, minimizing CO_2 emissions. Hydrogen is the principal fuel for fuel cells. Those devices are used on the space shuttle, space station, and space vehicles to generate electricity and water. A fuel cell does the reverse of electrolysis, taking on hydrogen and oxygen to produce water and energy, mostly as electricity. Fuel cells do produce some heat, but that is considered a minor byproduct in their application. Fuel cells are being used in vehicles as an alternative to all electric vehicles. Hydrogen burns with a very high flame temperature, roughly twice as high as methane, the prime component of natural gas. The HHV is 61,100 Btu per pound. As a result, combustion of hydrogen in boilers and gas turbines produces very high levels of NOx. Even when blended with other fuels and gases, NOx emissions tend to be high. Hydrogen is a component of refinery gases. It is being tested as a 15% blend with natural gas. Hydrogen is also a very small molecule. It tends to penetrate metals and leak through piping. Leaks can be very hazardous. Combustion systems for hydrogen are still being developed. This is one fuel that absolutely requires studying the instruction manuals.

Digester gas is actually natural gas, just very young natural gas. Like a young bourbon, it has a kick, and lots of things in it make it less desirable than natural gas, which had thousands of years to cure in the ground. Digester gas is a byproduct of waste water treatment, where the water is enclosed in the digester and anaerobic bacteria (bugs that do not like air) literally eat the waste and generate methane and carbon dioxide in the process. The principal difference between digester gas and natural gas from wells is that the digester gas contains a lot more carbon dioxide and usually has some other materials in it that carry over with the gas as it is generated. Some of the less desirable materials include water, hydrochloric acid, and solids. Some digester systems are fitted with filters, to reduce the solids, and separators, to remove most of the water and acid, before it gets to the boiler plant. The largest variable in digester gas is the amount of carbon dioxide. It is basically inert (the carbon and oxygen already combined). Thus, it dilutes the methane content of the gas to reduce its heating value to numbers in the 250–800 Btu per standard cubic foot range or 25%–80% of the energy normally found in natural gas.

Special considerations for firing digester gas include concern for the blockage of valves (especially safety shut offs), regulators, etc. All the piping should be fitted with drains, usually drain pots, where the collected moisture, etc., can be captured for return to the digester. The piping also has to be arranged so that it can be cleaned in the event of an upset in the digester, which could send over considerable quantities of water and solids to plug things up. Piping materials may be constructed of stainless and other alloys to prevent corrosion by the acids in the system. Precleaning can reduce the acids enough that normal steel can be used. Still, it is advisable to check its thickness regularly and after any severe plant upset.

The large fractions of carbon dioxide can dilute a digester gas so much that it will not burn with a stable fire. Special burners are required to pass the larger gas volumes required to get the fuel value needed for the boiler capacity. Many of them are fitted with standing pilots. Many applications include real natural gas as a support fuel to maintain ignition of the digester gas and to make up any additional energy requirements. Both fuels are fired simultaneously. The controls have to be able to cope with that.

Often the boiler operator on digester gas will have a responsibility to monitor the digester itself. If the gas is not fired in the boilers and allowed to escape to the atmosphere, there will be many complaints about the odor. When a boiler plant cannot burn all the digester gas, or the boiler plant is temporarily shut down for maintenance, the gas is usually burned off using a flare (Figure 7-1). A flare is a burner without a furnace and boiler around it.

Landfill gas is very much like digester gas. The anaerobic bacteria work on the garbage in the dump (a landfill is just a well-maintained garbage dump) to generate the gas. There are some potential problems with landfill gas that are not encountered with digester gas. The carbon dioxide content can vary more (over extended periods of time) and air can leak in through breaks in the cover of the landfill. The gas will also vary in the mix of fuel gases because the garbage in the landfill is not consistent. The US EPA has incentive programs to foster the use of landfill gas as a fuel.

Refineries produce a variety of gases with various blends which have different heating values and air fuel ratios. Flares are used to burn off those gases that are too dilute to support combustion on their own. They are primarily a safety device. Problems with hydrocarbon emissions from flares have led the EPA and others to discourage their use. The waste of energy, combined with modern technology that allows them to be burned efficiently, has reduced their numbers and use. Control systems that continuously measure the heating value and combustion air requirements of the gases can provide real-time information to a control system on a boiler to burn those gases. Here again, read the instruction manuals. Special training may be required for operating a boiler burning those gases.

Figure 7-1. Flare.

All too often, gas is taken for granted. It is just assumed that it will continue flowing out of the pipeline. The gas flow can stop if a line ruptures, a compressor station breaks down or has a fire or other emergency, or someone burning gas nearby has a failure. Interruptible gas service may cut off the gas supply. Some older plants had "gas holders." These were expandable tanks that used the tank weight to pressurize the gas in storage. Those gas holders provided a source of gas in case of an emergency. Utilities use mines, where they compress the gas for storage, and there are LNG storage facilities in a few spots in the country. Regardless of all these provisions, most plants have to be prepared for an interruption in the gas supply. The EPA has regulations about the use of fuel oil as a backup fuel. Price alone is not a justification to switch from natural gas to fuel oil, even light oil. Be sure to check the regulations in the region regarding the number of hours that can be spent operating on backup fuel. Also be sure that the plant has a permit for the backup up fuel to be used. Don't wait for the emergency to try to get permission from the state to burn propane if a permit is not already in place.

FUEL OIL

Fuel oils are identified by American Society for Testing and Materials (ASTM) specification D-396-62T, which replaced the Pacific Specifications (now obsolete) that originally identify the oils by a grade number. Number 1 is basically kerosene and is seldom used in boilers. The common fuel oils are grades 2, 4, and 6. The term "grade" was dropped. Now they are normally identified by the number alone. Number 2 is called "light fuel oil," which is not as dense as the others. Light fuel oil is basically the same as diesel engine fuel. It has a typical heating value of 141,000 Btu per gallon, weighs about 7.2 pounds per gallon, and has an air/fuel ratio requirement of 15.984 pounds of air per pound of fuel. That is approximately equal to 1534 cubic feet of air per gallon or 181.6 cubic feet of air per minute per million Btu/hr. It is a relatively clean burning fuel and has almost no ash. To reduce sulfur oxide emissions, transportation and off road equipment is now required to burn ultralow sulfur diesel oil (which does not materially differ from Number 2 oil), with a sulfur content of less than 15 ppm (parts per million). It stands to reason that restriction will shortly apply to heating fuel as well. It already does in most states. EPA has a proposed rulemaking on that level requirement. For ships, International Maritime Organization (IMO) 2020 (UN rule) limits the sulfur content for fuels used aboard ships to 0.5%, of the use of scrubbers to achieve the same result.

Grade 3 was dropped from consideration in 1948. Numbers 4 through 6 are referred to as "heavy fuel oil." They are dark in color, require some heating before they will burn, and exhibit varying degrees of soot formation and other problems with burning. Numbers 5 and 6 require heating to reduce the viscosity of the fuel so that it can be pumped. Number 6 fuel oil has to be heated so that it will flow. The viscosity (resistance to flowing) of these fuels varies considerably with temperature. The viscosity, not the temperature, has to be maintained at the value prescribed by the burner manufacturer. The operator has to set the oil temperature to achieve the required viscosity for proper atomization. The analysis of the fuel, provided by the fuel supplier, will indicate a viscosity at a standard temperature. Charts or graphs furnished by the fuel supplier, or the burner manufacturer, must be used to determine the required temperature for burning. If burning a heavy fuel oil, the fuel supplier should provide the appropriate temperature/viscosity charts and guidance in maintaining the proper viscosity.

That is just the starting point. An oil burner is designed to atomize the oil at a specific viscosity – most of them at 200 SSU (seconds Saybolt universal). That simply means that it takes 200 seconds for a 60 milliliter oil sample at 100°F to flow through an orifice in the Saybolt Viscometer. One test that can be run is to vary the viscosity by varying the temperature, a little each side of the specified value. Then see what it does for the boiler performance. If the performance improves or seems to be getting cleaner combustion at that viscosity, change it a little more. Eventually, the best viscosity for that burner will be determined. Then heat the oil to get that optimum viscosity. The result of that activity should be recorded in the maintenance log for that particular burner.

All oils are less dense than water (they float on water). Heavy oils are just heavier than lighter oils. One other confusing factor is the use of "gravity" to define an oil. The American Petroleum Institute (API) gravity of a fuel oil increases as the fuel gets lighter. API gravity is the ratio of a weight of oil of a specified volume compared to the weight of the same volume of water at the same temperature. To determine the specific gravity of an oil, add 131.5 to the API gravity and divide the result into 141.5. Multiply that result by 62.4 to determine the pounds per cubic foot. An oil with an API gravity of 10 will have the same weight as water. Higher numbers are lighter than water.

Number 4 oil has a typical heating value of 146,000 Btu per gallon, weighs about 7.7 pounds per gallon, and has an air/fuel ratio requirement of 14.14 pounds of air per pound of fuel. That is approximately equal to 14,528.2 cubic feet of air per gallon or 181.4 cubic feet/min. per million Btu/hr. Number 6 oil has a typical heating value of 150,000 Btu per gallon, weighs about 8.21 pounds per gallon, and has an air/fuel ratio requirement of 15.18 pounds of air per pound of fuel that is approximately equal to 1665.5 cubic feet of air per gallon or 185.3 cubic feet/min. per million Btu/hr.

Pour point is one of the important values the operator should monitor when firing heavy fuel oils, especially Number 6. Before acid rain was recognized as a problem, the pour point of fuel oils was fairly stable. When it became necessary to remove the 3%–5% sulfur in the oil to reduce emissions, the process changed the characteristics of the oils, introducing a problem with elevated pour points. The pour point is the temperature at which the oil will start to flow. Oil in a storage tank that is allowed to cool below its pour point will not flow to the heater to be heated and pumped out of the tank.

Heating the oil to a higher temperature ensures the oil will flow.

Desulfurized fuels have a tendency to develop elevated pour points. Once the oil cools below its pour point and sets up, it must be heated to a much higher temperature before it will flow again. Repeat the cooling and heating process for sufficient number of times and the oil becomes a solid mass that will not flow and cannot be pumped. The only solution to a gelled oil tank is to add chemicals and oil to dissolve the mass. Regrettably, it cannot be chopped up and burned as coal because once it gets in the furnace, it will melt, becoming a liquid again at the high furnace temperatures.

Flash point is another property of fuel oils that should be watched. Those Pacific Specifications required Number 2 fuel oil to have a flash point higher than 100°F. Heavier oils were listed for higher flash points, above 150°F. There are two methods for determining flash point, the common one being the open cup method, where the oil is heated and a technician passes a standard match over the top of the cup containing the oil. When the oil is so hot that it generates enough flammable vapor to be ignited by the match, the temperature of the oil is the flash point.

It is called the flash point because the flame starts and extinguishes rapidly, flashing rather than continuing to burn. When burning oil with a low flash point, any leak should be a concern. Temperatures in a boiler plant are frequently higher than 100°F, especially in the summer. Steam and hot water piping are so hot that they can generate flammable vapors if the oil leaks onto them. Natural gas has a comparable flash point at around 500°F. Gas requires more energy to ignite than oil and is not very hazardous. Of the major boiler explosions, the worse ones were generally light oil fired.

In addition to the normal grades of fuel oil, there are several sources of waste oils that can be burned in a boiler as fuel. A common one used in small installations is waste lubricating oil. If firing waste lubricating oils, more care is needed because it can be tainted by gasoline. At an army base, the waste lube oil was from helicopters and could contain a considerable fraction of jet fuel. Usually, waste oils are burned as a second fuel to limit the effect of their variable heating content and air requirements. Some systems use density meters to measure the waste oil flow to get a concept of air requirements and energy content according to its density. To date, there is not an economical means of obtaining instantaneous measurements of HHV and air requirements for waste oils.

Typical problems with waste oil firing include dirt and grit in the oil. There is also a concern for lead from bearings oxidizing in the furnace to produce high ground level concentrations of lead oxide around the plant. Any grade of fuel oil is a hazardous waste if it escapes the normal containers and piping to leak into the ground or sewers. Of particular concern is any floor drain in the plant. The wise operator should know where the floor drains in the plant discharge. Today, an oil leak can cost tens of thousands of dollars to clean up. Always seek to keep any leak contained.

Oil can be supplied directly to the plant via a pipeline. In such cases, the plant is relying on the supplier similar to a natural gas pipeline. Most plants could not justify a pipeline directly from a supplier. They have fuel oil delivered by truck and have to store the fuel on the plant site. Storage does not have to be in tanks, but potential hazards of leaks have eliminated the use of open pits, old mines, and similar measures. Tanks are generally one of the following three types: underground, above ground horizontal, and above ground vertical. Underground storage tanks are now labeled with the abbreviation "UST's" and are a lot different than 50 years ago. Above ground horizontal tanks are common for small plants and include the ones enclosed in concrete vaults for physical protection as well as fire safety. They are called horizontal because the tank is formed around a horizontal (parallel to the ground) centerline. Larger ones may exist, but the typical horizontal tank is limited to around 90,000 gallons capacity. Vertical tanks are formed around a vertical centerline and can range in size from a few hundred to hundreds of thousands of gallons.

USTs became a hassle when it was discovered how many of them were leaking. From tanks at gasoline filling stations to those at every boiler plant, more tanks were leaking than were intact. Much of it was due to a lack of understanding of how the tank and soil interact as the fuel was added and removed. For years, there was a standard procedure for installing an underground tank that consisted of pouring a concrete base and then resting the tank in the concrete. Only after several years was it discovered that the tanks changed shape, becoming more elliptical as they were filled and compressed the soil. The point between where the tank metal was trapped in concrete and bearing only on the soil provided a sharp corner that the tank was always bending around. That is where they cracked and leaked. There were other problems including corrosion due to electrolytic action in the soil. The initial solution to the UST leakage problem was their replacement with fiberglass tanks properly installed to allow flexing with the soil. Those tanks have also had problems. When replacing a UST with another one, get one with fiberglass resin encased steel to get the best of both worlds. All installations since the early 1990s are required to have means for testing the tank and connected buried piping for leaks. Most of the piping is also installed inside conduit so that a leak can be detected.

Any UST installed today has to be of a double shell construction, actually a tank within a tank. Monitoring of those tanks includes checking the "interstitial space" (the space between the two tanks) for signs of leakage. That is leakage either way, fuel oil out or ground water in. Any tanks of single shell design have to pass a "Petro-Tite" leak test regularly. Those tests are performed by a licensed third party.

An operator's responsibility, when it comes to USTs, is monitoring the existing tanks for leaks. That means keeping track of the oil. Know how much was stored, how much was delivered, how much was burned, and, therefore, how much should be in storage now. Storage equals previous quantity plus fuel delivered less fuel burned. Then check the tanks to determine how much fuel is in them and compare that to the calculations. Some modern microprocessor-based equipment is available that does all this, issuing an alarm when a leak is indicated. Regardless of that provision, always know if there is a leak of any significance.

Above ground tanks are not exempt from consideration. There have been many discoveries of leakage of above ground vertical tanks. Monitoring them and testing them on a regular basis is necessary. Above ground horizontal tanks are usually completely above the ground, making a leak readily apparent. That does not mean to stop keeping track of the fuel inventory. More than one above ground tank user has discovered mysterious disappearances of oil with no explanation. That is because some people know they can get away with burning No. 2 in their diesel vehicles if they are not too concerned for injector wear. Most of the heating oil that is not subjected to motor vehicle fuel taxes is now colored red. Anyone caught with red fuel in their car or truck faces serious fines. Any modern above ground tank installation should be properly isolated (fenced off) and surrounded by a berm. The berm wall will contain any spill. Sand should be used on the ground around the tank. If there is a spill, the sand tends to hold the oil and make a tar-like substance that can readily be collected and taken for disposal.

One special purpose label is a "day tank." That is a small fuel oil tank which is filled daily from the larger tanks in storage and used to supply the boilers. The initial purpose of a day tank was providing a supply of oil heated properly for pumping to the burners. It also eliminated double piping of oil suction and return to all the field tanks (the larger storage tanks). Oil in larger field tanks was allowed to be much cooler. A day tank requires means of filling it from field tanks and accepts the returned fuel oil from burners that are not operating and oil relieved from the fuel pump discharge. The day tank could be heated to supply oil at burning temperature or just heated enough to flow properly through the high pressure burner fuel oil supply pumps.

The oil is transferred from trucks to above ground tanks by fuel oil unloading pumps, "unloading pumps" for short. Those pumps are designed for high volume and low pressure to move the fuel from a typical delivery truck containing 8000 gallons to the storage tanks. Oil transfer pumps are used to move the oil from one tank to another and from field tanks to day tanks. An installation with USTs may have neither of these because the truck can drop the oil into the underground tanks and fuel is drawn from the tanks by the burner pumps. In some cases, fuel is drawn from storage tanks and transferred from tank to tank using the burner pumps.

The pumps used to deliver the stored fuel to the boiler burners are the only ones called fuel oil pumps, even though the others also pump oil. They are traditionally furnished in a package construction mounted on a steel base that supports the pumps and serves as a big drip pan underneath them to catch spills. When used for light fuel oil, the pumps and a suction strainer are mounted on the base. That is called a "pump set" or a "fuel oil pump set." Heavy oil fired installations include some heaters with the pumps to raise the temperature of the oil to a proper value for burning and another strainer, with smaller openings in the screen, to further clean the heated oil. The complete assembly with suction strainer, pumps, heaters, and discharge strainer is called a "pump and heater set."

Try to avoid oil pumps that are mounted on burners and fitted with a connecting shaft to the fan motor. That arrangement may be acceptable for small home heating furnaces. They can be problematical for anything larger. Those burners are arranged so that a short shaft with two coupling halves is inserted in the burner housing inside the fan wheel, where a matching half coupling receives one end. The other end of the shaft engages a coupling half on the oil pump, which is mounted on the outside of the fan housing, with its shaft through a hole in the housing. It is very difficult to access them, install that shaft properly, and tighten the set screws that secure the coupling halves. Do it wrong and the shaft flies off when the burner starts up, with subsequent damage to the fan wheel.

Heavy oil is not heated to a certain temperature so that the oil is hot enough to burn. It is heated so that it flows properly, viscosity providing an indication of its ability to flow. Storage tanks should be heated only enough to get the oil to flow to the day tank or the fuel oil heaters. Anything hotter is just a waste of heat. That is because most storage tanks are not insulated. Heating the oil to the right viscosity for burning should happen just before it goes to the burners.

It is necessary to run some of that heavy oil through the piping of an idle boiler to keep it flowing. That recirculation is essential for oils that could become solid in the piping and prevent the starting of the idle boiler. There is normally one globe valve in the piping that returns the oil to the pump suction or the tank (return oil piping). That valve is throttled for several reasons. If it is opened too far, it can return more oil than the pump is delivering, with a resulting drop in oil supply pressure. Carelessly open a recirculating valve too far and the entire plant could be shut down.

If not enough oil is recirculated, the heat losses in the piping will lower the temperature until the oil is too cold when it gets to the burner. Open the valve just enough to get the hot oil to the burner. On the other hand, the oil can return to the day tank to raise its temperature so high that the pumps cannot create enough pressure. The plant can shut down again. That happens because more oil slips back through the pump as the viscosity increases and, therefore, less oil is forced out of the piping to the heaters and burners.

In many plants, the operators are not trusted to do it right. The recirculating control valves (those globe valves in the return piping) are set and locked, or the hand wheels are removed, to prevent changing them. The best of both worlds is to throttle the recirculating control valve enough to keep oil flowing to the burners and back to the return line with only one boiler (the one that would start up if necessary, sometimes called the standby boiler) having enough recirculating flow to get the right temperature at the burner. That way, flow is assured, while not returning so much that the oil entering the pumps gets too hot.

Almost all fuel oil pumps are positive displacement pumps. Gear types and screw types for the most part, they are capable of raising the pressure of the oil considerably in order to deliver it to the burners at a pressure high enough for proper atomization. Since it is a positive displacement pump, it will deliver a relatively constant quantity of oil. The oil that is not burned is returned, sometimes to the pump suction, and other times to the fuel oil day tank or storage tank. To maintain pump discharge pressure and control the flow of oil to the tanks requires a relief valve, either pump mounted or piped onto the pump set. A relief valve is used, not a safety valve. Even when more precise pressure control is provided, there will normally be relief valves at the pump set.

The self-contained relief valve has to experience a change in pressure to change the flow of oil. In order to be stable in operation, a reasonable pressure drop of ten pounds minimum is required between conditions of no fire and full load on all boilers. All pressure reducing valves exhibit a drop in set point pressure (offset or droop) as flow through the valve increases. Offset is described by valve manufacturers as a % change in set point pressure as flow changes. Offset, or droop, is caused by a number of physical forces that act on the diaphragm, spring, and plug as flow changes. A relief valve might return all the oil to the tank at a pressure of 180 psig, and close off that port, so that all oil flows to the burners at 170 psig. If one is installed with a smaller droop, the flow and pressure will be unstable. It has to be that way. Don't expect the pressure relief valves at the pump set to maintain a constant oil supply pressure. It is the variation in supply pressure that makes for tiny variations in flow through the fuel oil flow control valves in certain burner systems. Additional provisions for pressure control are usually provided in an oil system. Since the pump set is usually remote from the burners, a second pressure adjustment is made closer to the burners by a back pressure regulator. It is a self-contained control valve that maintains a more constant pressure on its inlet by dumping some of the supplied oil into the fuel oil return line. It is really a relief valve but normally has a much larger diaphragm so that the swings in pressure are not as great as they are for the pump set relief valve. The two in combination produce a much lower droop.

For really precise oil supply pressure control, two measures are used. One is a pressure regulator at each boiler. The regulator has a large degree of droop. Since it is repeatable, the pressure at any particular firing rate is the same, regardless of the oil supply pressure. The other is installation of a more elaborate back pressure control system, from pilot operated valves to a complete control loop with transmitter, proportional-integral-derivative (PID) controller, and control valve.

Heating of the heavy fuel oil on small systems may consist of a simple temperature actuated control valve. Most of the systems use a temperature piloted pressure control valve. A valve that acts on temperature alone will allow large swings in oil temperature with swings in flow to the burner because the control valve does not know the oil flow has increased until the colder oil gets to it. By then, the lower steam pressure has allowed the metal of the heat exchanger to cool as well so that the temperature controller will have to over-react. A temperature piloted control valve simply uses the temperature of the oil to set a steam pressure to be maintained in the oil heater. When the oil flow increases, it will use more steam to heat it and the pressure in the heater will start to drop. The pressure controller opens the valve to compensate for the pressure drop, maintaining the pressure and a more precise output temperature.

Newer strategies include viscosity control. An instrument is installed in the piping to sense the viscosity of the oil. There are several methods for analysis, ranging from vibrating a heavy wire in the oil to trapping a sample and dropping a plunger through it. Whatever method of sensing is used, there is still the problem of response time. A viscosity controller should only be used to produce a set point for the steam pressure.

On many of the facilities that converted to light oil in the 1980s, it was suggested to the customers to retain the fuel oil heaters. Testing at that time indicated that a light oil burner would operate cleanly and more efficiently, with a little better turndown, if the oil was heated to 120°F. That was, of course, a temperature still below the flash point of oils supplied at that time. Now, with lower sulfur requirements, some light oils have flash points just barely above 100°F. In that case, heaters would not be used.

The most difficult activity regarding fuel oil handling for operators today is keeping the system in operating condition when it is rarely used. The standard operating procedures (SOPs) should include regular operation of the system to ensure that it is operational and a drill for switching to oil for each operator in the fall before the first winter interruption can be expected.

COAL

Coals are commonly identified by their source, sseither by the area or state in which they were found or a particular mine. There are three distinct classifications of coal: anthracite, bituminous, and lignite, which are principally related to the crushing strength of the coal, with anthracite being the hardest. Anthracite is typically referred to as metallurgical grade coal, as it has the strength to support a bed of coal, or coke derived from that coal in a steel making furnace. Lesser grades of coal, from bituminous down through lignite, are referred to as steam coals and are primarily used in boilers to make steam. Other criteria include size of the coal particles and characteristics that affect handling and burning. Coals that are fired on grates must be large enough that they do not fall through the holes in the grate and have a limited portion of fines that would fall through. Coals that are pulverized to something close to talcum powder so that they will burn in suspension (floating in the air very similar to an oil fire) are graded by how difficult it is to grind them. An operator in charge of a coal fired boiler plant should be aware of the specifications of the grate or burner manufacturer and boiler manufacturer and how variations in those specifications affect its firing.

Coal and oil require less air to burn than natural gas and LPG for the same heat output. That is because the gases have a higher hydrogen to carbon ratio. More hydrogen in the fuel produces more water, which increases stack losses. Some hydrogen in the fuel is always desirable, as it helps form volatile gases which burn far more readily than solid carbon. Thus, coal fired boilers will have a higher boiler efficiency than gas fired boilers. However, gas mixes with air much more readily than solid coal particles. As a result, the excess air levels that are typically needed in coal fired boilers will be higher than those for gas or oil fired boilers, offsetting some of the gains from reduced moisture loss. As solids are difficult to burn completely, there will usually be some unburned carbon left over, often referred to as soot. Coal also has a much higher ash content than either oil (less than 0.1% ash) or gas (nearly zero ash). Thus, coal firing will result in the ash depositing on boiler tubes and the need for soot blowers to clean the ash from the tubing.

Another form of coal that is being considered as a fuel is anthracite culm. That is the waste material removed from anthracite coal mines, which contains some coal but is mixed with dirt. The analogue for bituminous coal is called gob. There are several huge piles of culm and gob around the mines of this country. Some are big enough to supply a plant for several years. Modern fluidized bed boilers are capable of burning that material.

As a solid, coal requires different handling methods than oil or gas (both fluids which can be transported in pipelines). Coal is usually delivered by truck, barge, or railroad car. In either case, they can present a serious problem in the winter, if rained or snowed on, with subsequent freezing. Usually, a plant with railroad supply will have a melting shed, where the cars are heated to melt the ice so that the coal can be moved. The trucks or rail cars are dumped into a hopper, where a conveyor picks up small quantities of it and lifts it to the bunkers or a storage pile. A barge will require some kind of crane operation to bring the coal to the appropriate conveying equipment.

Storage piles are simply piles of coal stored for burning. Unlike a fuel oil storage tank, there is usually no enclosure. Thus, the coal is subject to degradation from the weather. They also have a bad tendency to ignite spontaneously if left sitting too long. Large coal piles need to be managed. They are typically moved and compacted to minimize air contact to avoid spontaneous combustion. When it comes time to burn, or move, the coal, another conveyor can be used. Or it might just be handled by operating a front loader or bulldozer. In any event, the coal is eventually transferred to the firing system.

Conveyors come in a wide variety of sizes and styles. Many use a belt, a wide, fabric reinforced rubber or synthetic rubber, riding on rollers that shape it into a trough that holds the coal. At some point in a belt conveyor system, the belt is pulled taught by a roller that is adjustable. The belt makes a full 180 degree turn over the roller. Belt conveyors are used mainly in large plants, where a constant movement of coal is required. Treat the belt with care. Sudden stopping and starting of large belt conveyors will tend to break the belt. One coal requires careful handling on conveyors. Coal from the Powder River Basin (PRB) in Wyoming and Montana is used throughout the country in pulverized coal fired boiler plants because of its lower sulfur content. PRB coal, however, is very friable (it breaks up easily) and it generates dust readily. After several explosions in conveyor systems, major rebuilds were required. When handling PRB coals, ventilate transfer sheds and other areas, thereby preventing flammable concentrations of airborne dust accumulation, and conveyors to reduce dust generation hot spots and sparking. Pay close attention to the special instructions normally received when a plant is switched to burn PRB coal.

The typical small coal fired plant will use a front loader to move coal from storage to a bucket elevator that lifts the coal from grade level to the bunkers. A bucket elevator can be a belt, with small containers (buckets) attached to it, or any number of unique arrangements of chains, connectors, and buckets that form a continuous and endless string of buckets to scoop up the coal and lift it to a higher level in the plant, where it is dumped into the bunker. In some medium sized plants, the bucket elevator will dump the coal onto a special belt conveyor that distributes the coal into the bunkers. The belt conveyor will have a special assembly consisting of a couple of rollers that flip the belt twice, all mounted on a set of rails so that it can be moved along the length of the belt. When the coal gets to the assembly, it is dumped as the belt turns at the first roller and is deflected past the second turn of the belt to fall into the bunker. Another special conveyor for coal is a Redler conveyor. It consists of a continuous chain with metal paddles that ride inside a rectangular metal tube. The top of the tube is eliminated at the in-feed hopper (where the coal is dumped or falls from a storage pile) so that the paddles can intercept the coal and start dragging it along. The tube is closed for lifting and transporting the coal horizontally past a series of gates. Each gate consists of a section of the tube where the bottom can be opened to allow the coal to fall out. The conveyor can then deliver coal to a large number of bunkers.

A bunker is sort of like a day tank for coal. A bunker can be a concrete room (for all practical purposes) or the more common catenary form of hopper. The shape was developed to hold coal without a lot of reinforcing and structural members. Steel plates were made long and literally slung, somewhat like a hammock, from the building framing, in the space above the burner fronts, also called the firing aisle. The result was something like a half ellipse in shape hanging down from above, with trap door openings that were used to release the coal for feeding to the burners.

To keep track of the coal and transport it from the bunker to individual burners, many plants have weigh lorries. These are mounted on tracks with wheels similar to those on a railroad car. The lorry can be moved from under the bunker horizontally along the tracks to a position above the coal hopper of the boiler being fired. The lorry incorporates a hopper to hold the coal dropped from the bunker and its own drop gate to empty the lorry hopper into the boiler hopper. The hopper on the lorry is suspended from the wheels and arranged like a scale so that the operator can weigh each load of coal. That can provide an idea of the coal firing rate in pounds per hour and track how much coal was burned on a shift. Smaller coal hoppers can be weighed continuously to provide a similar result. Weigh belt feeders consist of a short (up to 5 feet long) belt conveyor, with the belt assembly suspended so that its weight can be consistently monitored. For coal that is fed to pulverizers, a volumetric feeder is used to drop the coal into the top of the pulverizer. The pulverizer uses the primary air for combustion to convey the coal to the firing system. The feeder speed provides an indication of the rate at which fuel is being fed to the boiler.

For hard to burn coals, an indirect fired system can be used. In some countries, steam coal is not readily available. Sometimes, anthracite or meta-anthracite is available. These coals have very low amounts of volatile matter in the coal, which makes them difficult to support stable combustion. They can be fired in a specially designed furnace arrangement with down fired burners that are surrounded by refractory linings. In these systems, the coal is pulverized but delivered to a storage bin above the burners. The pulverized coal is essentially dropped into the burners for firing. The refractory lining retains heat and provides a stabilizing influence on the burner system. The flame goes down and turns back up to provide additional stability. This system was originally developed for the railroad industry and is called a Lopulco firing system, for locomotive pulverized coal firing. These systems are fairly rare and no longer used in the US.

When firing coal, whether on a stoker or with a pulverizer (see the section on burners), a continuous supply of coal to the hoppers or pulverizers is always a function of the operator. Usually, in a coal fired plant, there are "operators" because it takes more than one person to keep the coal moving and burning properly. There may also be operating equipment that changes the size of the coal particles or screens to actually separate out some of the coal to provide the size and form of fuel that is required for the firing system.

Coal units also require some handling after the coal is fired. A certain amount of ash and unburned fuel (frequently called LOI, for loss on ignition) collects in the bottom of the furnace, on top of the grate, on some of the tubes, in any hoppers in the duct work, and in the dust collector at the boiler outlet. It has to be handled back out of the plant to be dumped in a landfill or used in cement making operations. Over 50% of the ash generated in coal fired boilers is currently utilized in cement plants.

With the advent of climate change concerns, there is public and government pressure to stop using coal as a boiler fuel. Coal used to be the lowest cost fuel of the three major fossil fuels (coal, oil, and gas). However, with the advent of fracking technology and horizontal drilling technology, a lot more natural gas has been deemed to be economically recoverable, resulting in a significant reduction in price. Consequently, coal use in boiler plants has been reduced from over 1 billion tons/year to less than 600 million tons/year in the US. Nearly, 90% of the industrial boilers that used coal have been switched to natural gas. With the announced goal of being "carbon neutral" by 2050, it is expected that coal use will continue to decline.

OTHER SOLID FUELS

Biomass fuels can vary from firewood, the most common, to bedding, which contains some unpleasant animal waste but still burns. There are many varieties of wood and a considerable variation in other vegetation that can be burned. There are a number of ways to burn those fuels. These include stokers, bubbling fluid beds, circulating fluid beds, and pneumatically conveyed systems, similar to pulverized fuels. As with coal, the fuel has to be prepared to conform to the specifications of the firing system manufacturer in order to burn well. These fuels tend to have a higher hydrogen content. They tend to burn somewhat cleaner. The major problem with these fuels is their high moisture content. Liquid water in the fuel cools the fire in the furnace and the vapor produces high latent energy loss up the stack. The fuel's lower cost normally compensates for that. However, cost is typically influenced by the local availability of the fuel. Again, the higher moisture content of the fuel comes into play. It does not really pay to ship moisture around. If the fuel is not available locally, it is usually not economical to be used.

Wood can be fired in several forms, logs like on a campfire, chips as large as a playing card and about one-half inch thick, down to sizes rivaling sawdust and various sizes of dust from sawing (where the dust is more like a chip, sometimes as big as one-quarter inch square) to sanding. Some of the finer and lighter materials can be burned almost entirely in suspension (floating in the air) in a flame that is similar to an oil fire. Most of the chip is burned on a grate, although it is common to introduce the chips by tossing them in above the grate, where the finer dust in the fuel is burned in suspension. This type of firing system is called a spreader stoker.

Some wood burning systems are dealing with raw wood, which has a high moisture content (around 50%). Much of the energy in the fuel is used to vaporize that water. Others fire kiln dried wood, which has less than 10% moisture and is an excellent fuel. The construction of the boiler and the grates are designed for the fuel to be burned. Typically, it is usually difficult to handle a different material. A boiler designed for dry fuel will probably fail to reach capacity when burning wet fuel and may not maintain ignition. A boiler designed for wet fuel will probably have problems of burning up grates due to the higher flame temperatures of the dry fuel.

A principal problem with wood firing is sand. When the fuel is cut, hauled, and prepared for firing, a certain amount of dirt comes with it. Sand is primarily silica, which is just below diamonds on the hardness scale. Sand can erode boiler tubes quickly, as it is carried by the flue gas out of the furnace into the tube banks. Sander dust will always contain a certain amount of flint and other sharp sands that are very damaging to the boiler. When a boiler is designed to fire wood that is sand contaminated, the velocities through the tube banks are intentionally reduced to limit the erosive effects of the sand. Erosion is exponentially impacted by velocity. Operators should also avoid any action that produces high gas velocities (too much excess air and overfiring) to reduce erosion damage.

Leaves are another potential source of boiler fuel which are not used as much as they could be. A principal problem with leaves is that they are only available at certain times of the year. Firing problems with leaves include an ash content greater than wood. The major problem is that the fuel is tough to handle, can be messy if it gets wet, and can be contaminated with sand and dirt. There are some systems that convert dry loose leaves to compact fuel packages by extruding them.

Bagasse is sugar cane after all the sugar juice has been squeezed out. The long stringy material is tough to handle and burn. Cane stalks are sent through shredding machines. The disintegrated cane is crushed between rollers to squeeze out the juice containing the sugar. After this primary pressing, the issuing fiber, or bagasse, is sent through more roller mills for further extraction. It is often sprayed with water to help dissolve the sugar. On leaving the last mill, a typical bagasse might have 40% fiber, 55% moisture, 2.5% sugar, and 2.5% ash. The source of the ash is trash and dirt picked up during harvesting. A spreader is used to inject the fuel out over the grate. Overfire air (air introduced above or over the grate) is often used to help evaporate the water and increase the amount of combustion in suspension. The unburned portion falls to the grate to complete the combustion process. Other natural sources of biomass include hay, animal bedding (yes, it all burns), rice hulls, crop residues, cotton and wool products, woody plants, and corn cobs. Dried corn itself has been used for a fuel.

With the increased interest in climate change issues, biomass is being considered as one means of reducing CO_2 emissions. One result from the Kyoto Protocol was that biomass combustion was to be treated as "carbon neutral." This position avoided double counting when assigning a carbon penalty to cutting down trees, especially trees in the world's rain forests. The actual benefit is really dependent upon the type of biomass and its location. If a lot of processing and transportation is required to "grow" the fuel and get it to the combustion facility, it may not be as truly carbon neutral as it is given credit for. Nevertheless, fast growing crops such as switch grass, miscanthus, willow, and eucalyptus are being considered for sustainable energy production. These crops are grown, harvested, and replanted to constantly have the crop under cultivation. This process is called silviculture. The harvested crop can be burned for energy.

Waste paper, cardboard, and similar materials that have been contaminated cannot be recycled into more paper. These wastes are burned in trash burners. Some major government and industrial facilities that process a lot of paper may have boilers fired by those fuels, just to destroy the material for security purposes. Corrugated cardboard is one of those fuels that are very dangerous to burn. That is because it comes with its own air supply within the corrugations. When a corrugated cardboard is fed into a hot furnace, the heat will start boiling away the glues and wood to form gaseous hydrocarbons that mix with the air within the corrugations. When the mixture reaches its explosive range, it explodes!

One major issue for biomass is that it degrades in storage, particularly if it gets wet. Biomass can be pressed into pellets to make transportation and handling easier. Still there is a problem with most pellets that degrade and decay with time (rot and mold). That concern influences the handling and transportation of the fuel. Keeping the biomass dry and consuming the biomass relatively quickly once received can minimize this problem.

Trash burners, large boilers burning tons of garbage, are typically considered to be an air pollution hazard. Many localities chose to landfill their garbage rather than burn it. Today, the cost of landfill space, and the offset of better flue gas cleaning systems, has restored interest in trash burning plants. It is a unique boiler plant because the plant actually gets paid for burning the fuel, a far cry from paying for fuel. The costs of operating the plant, personnel, and the continuous repairs required (such as when an entire engine block gets into the plant) are covered by the value of the steam produced and the payments for processing the trash. There will be a considerable amount of ash left over, around 10%. That ash is usually returned to the county or city for placement in a landfill. What is basically accomplished is reducing the volume of trash to be dealt with. The EPA has special rules for air emissions from these plants.

Hospital waste is normally burned in an incinerator, with energy recovered by a waste heat boiler. These wastes are considered to be hazardous wastes and have additional rules concerning their emissions. In particular, the hospital, medical, and infectious waste incinerator (HMIWI) maximum achievable control technology (MACT) rules apply. The purpose of the separate incinerator is to ensure that all the material is exposed to the heat of the fire so that all the diseases and pathogens in the waste are destroyed. Unless the waste is mixed up to ensure that it is all exposed to the heat, there may be unburned, and sometimes untreated, fuel in the waste. The waste heat boilers must be designed for high ash loads and be capable of withstanding occasional acid attacks because of the acids produced while firing the waste.

There are many other forms of solid fuels, including such unique materials as laminate trimmings and plastic bags. Almost any organic material can be fired. The question is whether the source of the fuel is consistent in generation of quantity and quality as well as how much it costs to prepare, handle, and burn the fuel. The Council of Industrial Boiler Owners (CIBO) has identified over 200 different alternative fuels that are in use in industrial boilers. Often, these fuels are available at little extra cost. Even so, the cheap fuel has a lot of heating value and should be treated as if it were as expensive as any purchased fuel. If not burned efficiently, any deficit has to be made up with purchased fuels. Know the fuel. Know what the fire looks like when it is burning normally. Get very concerned when it is not normal. Keep in mind that how the fuel is stored and handled on its way to the burner can have an effect on plant safety.

WATER

As a consumable, water is not an unlimited resource. Continued growth of the human population will constantly expand the demand for fresh water. At some point, there will be a real water shortage, starting with drinking quality water. There are already parts of the country where water rights are causing serious issues.

A boiler plant has the potential to draw on, and waste, millions of gallons of water each year. Some plants consume and waste those millions in months or even weeks. Many jurisdictions are increasing costs for water, including costs for sewage treatment. Wise operators will address those costs and recognize their contribution to the preservation of this invaluable natural resource.

Major utility plants are doing something about it because water represents a significant cost to them. Where possible, they are using treated waste water (from sewage treatment plants) for makeup instead of fresh potable water. Despite the "yuck" factor, there is no reason to question the quality of that water after proper pretreatment (See Water Treatment in Chapter 8). The boilers do not care what it may have been as long as the water meets the required specifications. Public Service Electric & Gas's (PSEG) power plant in New Jersey saves 10 million gallons of precious drinkable water each month by using waste water as makeup, saving more than thirty thousand dollars a month in the process. As time goes on, more plants will be taking similar measures.

It is very important to understand what a gallon per minute (gpm) of water is worth. First, it is a good idea to know that a minute does not really give a fair measure of the cost. There are 525,600 minutes in a year, more than half a million. A 2 gpm leak that is allowed to continue represents more than one million gallons wasted every year. At the low range of water costs, a 1 gpm leak costs $1500. Consider the fact that half a million gallons of clear fresh water was converted to sewage. That leak becomes very expensive. A boiler water sample cooler should not be operating constantly. It does take a few minutes to clear lines and tune up a sample cooler each time a water sample is drawn and a little more time to close the valves when finished. It is also easy to argue that the boiler water would be removed by blowdown anyway. However, the typical sample cooler uses about 12 gpm to cool a boiler water sample. Leaving it running constantly wastes over six million gallons of water every year and costs at least $18,000 per year to convert good water into sewage. Don't do it.

Recycling the water in a boiler plant is becoming increasingly important. Some utilities are actually committed to zero discharge, where they do not put one gallon of waste water into the local municipal sewer or dump it otherwise. Part of that effort is to avoid the heavy cost of treating the plant's discharge of waste water, which is highly concentrated with solids and chemicals compared to water that is simply wasted to a drain. Understand that especially blowdown contains considerably more solids and chemicals than normal waste water. Minimizing blowdown is important.

Another consideration is the draining of a boiler and refilling it with fresh water during every annual outage. As the cost of treating sewage continues to rise, and concern for the treatment of caustic waters grows, there will come a time when dumping a boiler is restricted. If the plant does not have a connection to a sanitary sewer, and many do not, rent a tank trailer to store the boiler water while performing the annual inspections. That way, no alkaline water will be discharged into the environment unnecessarily. Also, there will be savings on the cost of the chemicals it contained (although loss of sulfite is expected) and the cost of treating fresh makeup water. Avoid the use of boiler bottom blow off as a means of water level control. Even if the controls malfunction, there is no reason to consistently waste water and boiler chemicals in order to maintain boiler water level. If the level tends to rise, it can be prevented by restricting feed water flow.

It is also not sensible to use bottom blow off as a means to reduce the total dissolved solids (TDS) content of the boiler water instead of using continuous blowdown. Removal of the boiler water to limit the TDS is best done with the continuous blowdown because it removes the most concentrated water in the boiler, the water that is left right after the steam is separated. Bottom blow off tends to be a blend of the boiler water and the feed water that just dropped to the bottom drum. It contains a much lower concentration of solids. None of the water or heat is recovered from blow off. Water and heat is recovered by a good blowdown heat recovery system.

Bottom blow off is used to remove sediment, mud, and sludge that either enter the boiler with the feed water or is created by the water treatment. With chelants and polymers, the scale forming salts are sequestered or trapped in solution and removed in the continuous blowdown. As a result, blow off is not needed anywhere near as much as it used to be. Especially after a switch to chelant or polymer treatment from 100% phosphate (maintaining some phosphate as an indicator of the operation of the others is not a bad idea), the chemical treatment representative should produce a blow off schedule that is much less frequent. If not, check it to be sure. Perform a bottom blow off before shutting down a boiler for inspection at a time before shutdown to match the scheduled interval for blow off plus a shift or day. Use a shift if the blow off is more frequent than every two or more days. Then, slowly drain the boiler after cooling and look at the bottom of the mud drum and headers right after the boiler is opened to see if there is any accumulated sludge or mud. If none is there, and the boiler had been operating normally (not in standby or other low load modes) before the shutdown, increase the interval between blow offs so that it matches the interval just established. After a period of time, there will be a small accumulation in the mud drum. That means the limit has been reached. Stick with the interval that works.

The EPA National Pollutant Discharge Elimination System (NPDES) permit program addresses water pollution by regulating point sources that discharge pollutants to waters of the United States. There have been a number of legal issues concerning the definition of waters of the US (WOTUS). However, nearly every lake, river, or stream is pretty much covered. It means that a plant must have a permit to discharge anything into a water way. Know and understand the conditions and requirements of that permit. Waste water may need to be treated before it can be discharged.

On steam systems, blowdown heat recovery systems capture much of the heat and a little bit of the water that is dumped by continuous blowdown. The blowdown is dumped into a flash tank, which operates at a pressure slightly above the deaerator pressure. Since the water is much hotter than the saturation temperature at that pressure, some of the water flashes into steam. The steam is separated by some internals and then flows to the deaerator, where it replaces some of the boiler output that would need to be used to heat the feed water. The remaining water then flows to the heat exchanger. Low pressure plants and small high pressure plants may not be able to justify the flash tanks. In that case, all the blowdown water flows to the heat exchanger. The heat exchanger transfers heat from the blowdown to the makeup water. In low pressure plants, the heat exchanger can be as simple as a barrel set above the boiler feed tank and arranged so that the makeup water is fed into the barrel and then overflows into the boiler feed tank. The blowdown is passed through a coil of tubing in the bottom of the tank. It then flows to the blowdown control valve (which can be manually set or an automatic one). In a plant with multiple low pressure boilers, each one could have its own coil. If the flow is not throttled after the heat exchanger, the boiler water will flash in the coil, making a lot of noise and eventually damaging the coil.

A heat exchanger in high pressure plants should be of high pressure construction and heat the makeup water before it goes to the deaerator. The control valve on the heat exchanger outlet is usually controlled by the level in the bottom of the flash tank. That way, the heat exchanger is always flooded and the blowdown is not flashing into steam, which can leave deposits that plug up the heat exchanger.

Blowdown heat recovery does save some energy and, with a flash tank, a little bit of water. The real savings, however, is in the water that would be used to cool the blowdown if there was no heat recovery system. Blowdown dumped through the blow off flash tank will dump some heat in the form of steam up the vent. That 212°F water has to be cooled to less than 140°F before it can be dumped in the sewer. The typical practice is to use good city water for that cooling. It will take a volume of cold water about equal to the blowdown to cool it before it can be dumped in the sewer. All that cold water is being wasted if there is no blowdown heat recovery system.

The best way to reduce water waste in a steam plant is to recover the condensate and use it as boiler feed water. There are many reasons for this in addition to saving water. Recovery of condensate recovers heat, eliminating the need to heat cold makeup water before it is fed to the boiler. Condensate is basically distilled water, converted to steam in the boiler and then condensed. It does not require all the pretreatment and chemical treatment needed for fresh makeup water. Recovery of condensate saves money that would be spent on additional fuel, boiler water treatment chemicals, and the additional water required for blowdown to remove the solids brought in by fresh city water.

All too frequently, the only consideration for recovering condensate is the value of the heat. Typically, the cost of heating the water is relatively minor. The cost of the water itself is more valuable than the heat and the cost of chemicals adds even more to that. Treat the condensate as a valuable resource. Recovery of condensate is the best way to minimize water waste from a boiler plant. However, there are times when recovery for use as boiler feed water is undesirable. Wasting of condensate is not unusual in a chemical or petroleum facility because the potential for contamination of that condensate is so high. In some cases, some of the energy can be recovered from it by using a heat exchanger. That does not preserve the water. Capability to monitor the water and filter it with carbon filters and other measures, including reverse osmosis, makes it possible to recover and use the condensate in those plants today.

In instances where the capital cost expenditure to recover condensate is so high that recovery cannot be justified, it is possible that the condensate can be used for other purposes, anything from makeup for cooling towers to use as sanitary water (where it has to be cooled). In chemical and petroleum facilities, there is considerable water used in scrubbers. Condensate makes a great replacement for scrubber makeup. In other words, if it has to be wasted, try wasting it in another system instead of fresh water.

Appropriate recovery of condensate is another matter. There have been plants that allow considerable waste of high temperature condensate by collecting it in open systems, where as much as 15% of the water and over 50% of the energy in that condensate is lost in flash steam. High pressure condensate should be recovered in a way that prevents flashing. The best way to recover high pressure condensate is to return it to the deaerator. Some of the condensate may flash off in the deaerator. That flash steam simply displaces some of the steam required for deaeration. There should be no reason to be concerned for oxygen in high pressure condensate. Typically, it is returned in a manner that allows some scrubbing of it to remove any oxygen that may be in it from startup and other operations.

Having explained that condensate that would flash off should be recovered in a manner that uses that steam, it only makes sense that any sign of steam escaping from a condensate tank vent line is a problem that requires an operator's attention. The normal reason that steam is leaking is leaking traps. Trap maintenance is very important in reducing water waste. Stop the leaks.

TREATMENT CHEMICALS

In the normal plant, the costs of treatment chemicals are about 2% of the total cost. The amount of chemicals used is a function of the amount of makeup water entering the plant. Thus, preserving water is the first important step in minimizing the cost of water treatment chemicals. The following section deals with water treatment because it is definitely one of the most important things that boiler operators have to do. Considerations of the chemicals as consumables are addressed here. The concentrated treatment chemicals are definitely hazardous waste if they escape their containers or treatment equipment. They are hazardous to handle and can cause severe burns. The wise operator uses all the protective gear. Not wearing that outfit is taking a chance on living with a serious injury. Wear it.

Frequently, people do not think of salt as a water treatment chemical. It is, and it is one of the cheapest and safest to handle. Be sure to make the best use of it first. Ensure that the water softeners are regenerated with adequate brine concentrations. Regenerate them before they are depleted to minimize consumption of phosphate or other scale treatments, which are a lot more expensive than salt. The water softener essentially substitutes the sodium in the salt for the calcium and magnesium that may be in the makeup water. Sodium salts are very soluble in water and become more soluble as the temperature of the water increases. The removal of the calcium and magnesium from the water eliminates the potential for those salts to form hard scale when the boiler water is heated. Proper care of the water softener is a worthwhile investment.

Take regular samples of the incoming makeup water to check for changes in hardness that will alter the capacity of the softeners and adjust the softener throughput accordingly. One plant was having problems with a new boiler. Blisters at the bottom of the boiler's water wall tubes were a sound indication of high degrees of hardness in the water. Plant personnel indicated that the softeners were regenerated every Wednesday, like they always did. It did not seem to matter to them that the steam demand on the plant, and makeup, had tripled in the last three years. The softeners ran out of sodium ions on Monday.

Applying the chemicals in a uniform manner, consistent with the rate of boiler water makeup, will

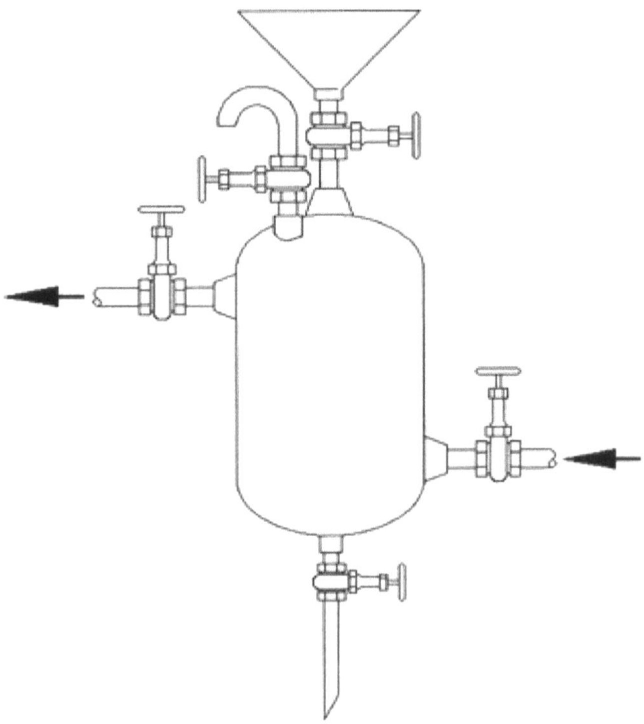

Figure 7-2. Shot feeder.

minimize their use by making them most effective. Some systems, such as low pressure hot water heating systems, require very little treatment because the system is closed and losses of water are very limited. Shot feeding of chemicals using a shot feeder (Figure 7-2) is capable of providing adequate treatment.

Those shot feeders do, however, often look much like a mess, where it is evident that the chemicals were spilled and wasted as opposed to injected into the system. Proper use of a shot feeder requires closing the isolating valves and proving them closed by slowly cracking the vent valve, while holding a bucket under the vent pipe to capture any discharge. It is possible for the shot feeder to accumulate some air or gas from the system. The contents could expand out dramatically when it is opened to atmosphere. It may require waiting several seconds, or even minutes, to allow the pressure to bleed off slowly before it is relieved. If only liquid flows out, that is an indication that one or both of the isolating valves are not shut. Be sure to wear the protective clothing because expanding gas can carry out slugs of water that could still contain concentrated chemicals.

Once the pressure is relieved, the shot feeder should be drained by opening the drain valve with a bucket under it to capture the contents. If the contents are system water, it is the best thing to use for mixing the new charge for chemicals. If the contents appear to be a concentrated mixture of chemicals, it means the feeder did not discharge its contents. In that case, close the drain, open the fill valve, pour it back in, and return the feeder to service to get the chemicals into the system. Be certain that the drain is closed, checking it by adding a cup or two of fresh water. Then open the fill valve and slowly pour in the new mixture of chemicals. To charge the chemicals, close the fill valve, close the vent valve down, crack it a little, and then crack the feeder outlet valve to fill the feeder pot. Hold a small container under the vent line to capture the first shot of water. Close the vent valve as soon as the water appears. Finally, open the feeder outlet valve and the feeder inlet valve to discharge the contents to the system. When the feeder is flushed by a high differential pressure (a typical arrangement is from the system pump discharge to the same pump's suction), it is advisable to limit opening the feeder inlet valve to limit thermal shock from any cold contents of the feeder. It also prevents sending a slug of chemicals into the system instead of a solution of them.

Failure to vent a pot feeder is a common problem. Always flood it before putting it in service. Failure to do so creates the risk of having a compressed gas burp blow concentrated chemicals around on the attempt to open it. It is also possible to send some air into the system to collect in some obscure spot and restrict system water flow. To prevent the loss of valuable sodium sulfite, keep the containers tightly closed. Sulfite is used as an oxygen scavenger. It reacts with dissolved oxygen that may have been missed by the deaerator to form sulfates, which are generally soluble. A sulfite mix tank for a chemical feed pump should have a floating top, or be otherwise sealed, to limit atmospheric oxygen getting at the contents to consume the sulfite before it even gets into the system water.

Be careful while mixing and handling caustic mixtures. Use appropriate equipment for the job. Aluminum and galvanized steel (actually the zinc in the galvanizing) react violently with caustic solutions. In one incident, someone used a galvanized bucket to mix some caustic solution. It literally boiled out of the bucket to create a hazardous spill and almost burned the individual seriously. Wear the protective gear.

MISCELLANEOUS

One consumable that a plant always seems to have troubles with is small tools. Small tools have a habit of going missing. Some people treat it as an acceptable practice, even implying a respect for the skill. It is still stealing. This is a matter of trust. If an operator cannot be trusted with a small tool that costs less than an hour's salary, how can that same operator be trusted with the operation of a plant worth millions? Some plants utilize a loan program. A tool can be signed out for a home project and returned. It is still a matter of trust. Treat other people's property with respect. Take pride in the plant having some tools that have been there for years. Another problem with small tools is breaking or damaging them. Wood chisels do not cut nails very well. Electric drills make lousy hammers. Again, treat the property with respect. Use the right tool for the job at hand.

Batteries are another commodity which are frequently converted to private use. Somehow, people get the idea that their alarm clock is required to get them to work and the batteries should be provided by their employer. Rags are another commodity that can be abused. These are not necessarily stolen. They are just wasted. Wise operators should always treat every little thing supplied by their employer as the employer's property and not something that somehow reverts to their possession. Paper, pads, pencils, erasers, scotch tape, and making copies, all can be abused. It does not amount to much and many employers say to use those resources without concern because it costs more to account for them than they are worth. If there is no policy for using the owner's property, don't take it.

Chapter 8

Water Treatment

Water is, unquestionably, a most unique substance. It is often said that, if it was not available, it would have to be invented. Its unique character is important to every form of life on earth. Controlling water's unique characteristics is one of the major occupations of the wise boiler operator.

WATER TREATMENT

While there are many water treatment companies, the trained boiler operator can have a major impact on the success of a plant water treatment program. Water treatment really is not a black art. The only problem with it is that it cannot be seen as to what is going on inside the tubes. Certain statements about it have to be accepted as fact, and decision making is based on them.

Water is sometimes called the "universal solvent" because it dissolves most things. It is such a great solvent that it even dissolves itself! The water molecule is lopsided. It contains two hydrogen atoms with an atomic weight of 1 (they are very light) and one atom of oxygen with an atomic weight of 16. The two hydrogen atoms hang around one side of the much larger oxygen atom. That lopsided condition results in a concentration of protons at the one side, where the hydrogen is hanging out, and nothing but electrons at the other side so that the molecule of water has an electronic polarity.

That is a reasonable explanation for how a microwave oven works. The microwaves are building up and then dumping a magnetic field in the food in the oven several times a second. The polarized water molecules keep twisting back and forth to align with the magnetic field. Other things, like plastics, do not have any polarity and are not affected. All those water molecules, twisting back and forth inside the food, rub the other molecules and heat everything up by friction. It is the water that gets hot first. Any other polar molecules would work the same way.

That polarity of the water is what makes it such a good solvent. It has a negative charge on one side and a positive one on the other. It can pull other molecules apart. It pulls molecules of H_2O apart, converting them to hydrogen ions (H+) and hydroxyl ions (OH−). That is how water dissolves itself.

Every solid material dissolved in water is present as an ion. Note the little plus sign and the little minus sign, which indicates that the atoms have something like an electric charge on them. It is what makes it possible for water to dissolve most things. In its pure form, where the only ions in the water are the hydrogen and hydroxyl ones, water is hungry. It looks for things to dissolve and will dissolve them until it has dissolved enough to satisfy its appetite for ions. Once it has dissolved a fair amount, it is not very aggressive. That is why it does not viciously attack the pipes, hot water heater, and other parts it contacts in homes. Everything that is done with water treatment is associated with what is dissolved in it, either ions or different types or gases.

One of the critical values in water treatment is the relative proportion of hydrogen ions in the water. Careful experiments have been developed to determine that there is one hydrogen ion in each million deciliters of pure water. That is 0.0000001 ions per deciliter. The normal range of hydrogen ion concentrations in water solutions runs from 0.01 ions per deciliter to 0.00000000000001 ions per deciliter. Since these numbers are a little cumbersome to work with, it was decided to relate the hydrogen ion concentration according to the number of decimal places, making the range of measurement easily described as 2 to 14 (the number of zeros after the decimal place plus one) and the number labeled "pH." The scientific definition is as follows:

$$pH = -\log [H^+]$$

The brackets denote the concentration of hydrogen ions in moles/liter. Neutral water has a pH of 7.

It really does represent the number of hydrogen ions in solution. Thinking in terms of decimal places, it gets smaller when there are more hydrogen ions. There are far more hydrogen ions in the solution when the pH is 2 than when the pH is 14. When dealing with pH, keep in mind that a change in value is a change in decimal places and not a proportional change. If chemicals are added to water in order to increase its pH from 7 to 8, it will take ten times as much chemicals to increase it from 8 to 9 and one hundred times as much to raise the pH from 9 to 10.

The value of pH provides a measure of the acidity or alkalinity of water. When the pH is less than 7, it is called acidic. When the pH is greater than 7, it is called alkaline. Acidic water is very corrosive. Highly alkaline water is also very reactive. Highly alkaline water will react violently with aluminum and generate some very toxic gases. Normal values of pH in a boiler plant are 7–8 for makeup and feed water, 8.5–10 for boiler water, and 6–8 for condensate. Water supply plants in the United States are required by law to maintain pH in the range of 7.6–8.5.

All the other things that dissolve in water use a scale that is a lot simpler than pH. The standard units of measure are parts per million (ppm), which is a ratio, the number of pounds of material that would be dissolved in a million pounds of water. Some operators find it easier to think in terms of pounds per million pounds of water. Of course, a million pounds of water are not needed to determine the ratio. Some water treatment departments will measure the concentrations of ions in solution in terms of micrograms per deciliter. That value is very close to ppm. Use it as such unless a critical evaluation of the water treatment facilities is needed.

Occasionally, there will be an analysis described as "ppm as $CaCO_3$" to describe a condition of water that includes a combination of materials dissolved. Since the materials have different weights, they are corrected so that the analysis can be expressed as an equivalent to calcium carbonate ($CaCO_3$). As calcium carbonate is the primary component of traditional boiler scale, that compound is chosen to represent the potential "hardness" of the water and its propensity to form scale. If the precise concentration of a substance dissolved in water is needed, there are tables of equivalents that provide a multiplier. Normally, that is not needed and best left to the water treatment specialists.

Most of the time, the exact number of pounds of a chemical in the water is not needed but only its proportion compared to the amount of water. Therefore, ppm is an easy way to measure the chemicals dissolved in the water. In those rare instances, when the quantity of chemicals in the water is needed, it is a simple calculation. Find out how much water is in the boiler (or whatever piece of equipment). If the value is in gallons, then multiply by 8.33 to convert to pounds. Divide the number of pounds of water by one million and then multiply by the ppm to get the pounds of chemical that is in the water. Normally, this only comes up when a system is being changed, such as filling the boiler and, in some cases, the piping, with water that should be properly treated. An initial fill of a hot water boiler system can be calculated by estimating the total length of pipe, multiplying the pounds of water per foot from the table in Appendix D and then adding that result to the number of pounds to fill the boiler. To establish the initial charge of sodium nitrite in the water (to achieve a content of 60 ppm), divide the weight of water by one million and then multiply by 60. The result is the number of pounds of nitrite to put in the chemical feed pot. Sodium nitrite is 63% nitrite. Therefore, multiply by 1.58 to determine how much of the actual chemical to add and then divide that result by the purity of what the chemical supplier provides.

Water is treated for two principal reasons: to prevent corrosion and to prevent scale formation. The most common form of corrosion is destruction of the metal by hydrogen ions. However, other chemicals dissolved in water can also attack the metal. Another form of corrosion is oxidation, where the oxygen in the air or water combines with the metal to form rust. A severe form of oxygen corrosion is oxygen pitting. The deaerator is the device that serves to remove dissolved oxygen from the boiler water.

Scale formation coats the heat exchange surfaces of the boiler to act like a heat insulator. The scale, being on the inner surface of the boiler tubes, separates the water and the metal so that the water cannot cool the metal. When enough scale builds up, the metal overheats and fails. The various water treatment processes serve to prevent corrosion and scale formation by pretreatment, which changes the corrosive and scale forming properties of the water before it gets into the boiler, and chemical treatment, which changes the properties of the feed water and boiler water.

WATER TESTING

Testing of water is required to learn what is in the water, what other people and other systems have done, and what results from the actions taken to maintain quality water for the system, be it boiler, chiller, or cooling water. The first requirement of water testing is to draw is a "representative sample." That means the sample of water taken to the test bench should be the same as the water in the system that the sample came from. If the sample is drawn from blowdown piping, it must come from a section that has almost the same pressure as in the boiler. If it is drawn after the water pressure drops, and some of it flashes to steam, there is no assurance that the sample is representative. It could be the water left after the steam flashes off and contain higher concentrations of solutes (the stuff dissolved in it, including the treatment chemicals) or it could be condensed flash steam and contain almost none of the solutes. If a sample is drawn off the blow off piping or any other volume where the water is stagnant, it is not a representative sample. The best point to draw a sample from is the continuous blowdown piping before the water passes through any orifice or throttling valve.

Samples of raw water, softened water, etc., can be collected by simply draining water from the systems and making certain the sampling piping is flushed so that the sample is fresh and representative of the water flowing through the system. Samples of boiler feed water, hydronic system water, boiler water, and most condensate require cooling to ensure getting a representative sample. Sample coolers can be as simple as a large coil of copper tubing to allow air cooling of a low pressure boiler water sample, to units designed for operating pressures up to 5000 psi (pounds per square inch). Read the instructions for the sample cooler and follow them. If they are not available, the following guidelines are suggested.

A sample cooler should be shut down, except when it is used to draw a sample. To ensure there is no vacuum created in it to draw air in and corrode it, and no way to over-pressure it through thermal expansion, leaving it under pressure is recommended. The preferred system is to connect a sample cooler so that there is only a cooling water supply valve. The cooling water outlet is piped to form a loop up above the cooler, vented, and then dropped to a drain. The loop keeps the cooler under static pressure so that air cannot get in. It will allow for expansion of the water and even generated steam to escape if someone opens the sample line first. If the cooler leaks, the boiler water will go to the drain, not back into the cooling water system. There is no way someone can close an outlet valve that is not there to force leaking boiler water into the cooling water system or to heat up the cooling water side of the cooler to blow it. It is also almost impossible to dilute a water sample with cooling water.

Close the water supply and sample outlet valves to shut down the cooler. When ready to draw a sample, first check the cooling water drain to be certain the cooler is not leaking. Then open the cooling water supply. Once cooling water is flowing, open the sample outlet valve to flush the sample piping and get a fresh sample up to

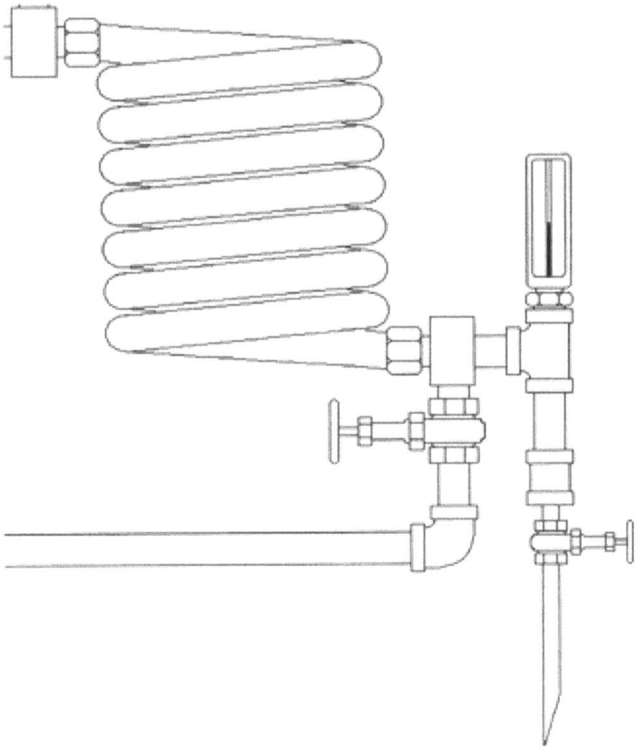

Figure 8-1. Water sample cooler.

the cooler. Boiler water and deaerated feed water should start flashing at the outlet, demonstrating that there is a fresh sample. Throttle the sampling outlet valve until a reasonable flow of cooled water is obtained.

To ensure there is no vapor vented off the sample, or condensate from the air getting in, the sample should be cooled to the same temperature as the air in that area. A thermometer sensing the water temperature leaving the cooler (Figure 8-1) works well. It has to be able to take the maximum possible temperature of the sample, the temperature in the boiler. The thermometer is also suggested to be certain to avoid skin burns when drawing the sample.

Once a sample is flowing, rinse all the apparatus that will be contacting the sample to avoid contamination from previous samples. If a sample must be drawn from a location away from the test bench, always draw enough to rinse the testing apparatus when back at the test bench. Note that the sample line should be long enough to submerge it in a sample bottle. That is necessary to provide a representative sample for testing sulfite content. Once the sample is exposed to air, some of the sulfite will start reacting with the oxygen in the air.

To minimize contamination of the water sample with air, insert the sample line to the bottom of the sample bottle, leaving it submerged as the bottle fills.

Allow the bottle to overflow for a couple of seconds to eliminate mixing of air with the sample and displace all the air from the sample bottle, flushing off the surface so that the sample was not in contact with air. If the sample is to be drawn from a remote sampling point, take another bottle for rinsing the apparatus. Unless the sample is tested for sulfite immediately, cap the flooded sample bottle. That is the right way to draw a sample even when not testing for sulfite. Always draw at least twice as much as needed. That small amount of sample is negligible compared to the cooling water that is wasted (see the section on water consumption).

Read the instructions for the test reagents. They degrade when exposed to light. A good bench will be closed up and dark. Also, the extra reagents and other test chemicals will be stored in their shipping containers in a dark area that has a reasonably constant temperature. Stacking them on shelves leaning against the sheet metal outside wall, that is cold in the winter nights and heated by the summer sun, is not the right place to put them. Don't order a 10-year supply. Be cognizant of the shelf life. If the expiration date is before the following week, throw it away and get a new one. The goal is to get the best results when testing the water.

Keep the test stand clean. Part of the cleanliness is associated with operating the test bench. Some reagents can damage or discolor paint if they are spilled. The automatic filling burettes will spray reagent out a little hole in the back if too much reagent is forced in. Those spatters on the back of the test cabinet are an indication of carelessness. If some reagent is accidentally pumped out, the discoloration will not happen if it is cleaned up right away. To make it easier to clean and limit breakage of glassware, many plants have rubber mats under the test equipment. The entire test stand should be white. It is a lot easier to see color changes and other things with a white background. Get white rubber mats as well.

A wise operator knows where the results should be. They add most of the reagent quickly to get to the point where it should be added drop by drop. That saves time in the process and has no effect on the outcome. Holding the sample container up so that its lip is above the reagent spout prevents spilling. That can be tiring if it takes too long to add the reagent until the color change is evident. If too much reagent is added, just measure up another sample and do it over. That is one reason for drawing a large sample to begin with.

When reading the liquid level in a burette or graduated cylinder, read the bottom of the meniscus. That is the level inside the glass (Figure 8-2), not the line at the edge of the glass, where the water tries to climb the sides. There is less than 99 milliliters in the cylinder of the figure, not 100. There is very little liquid in that edge. Don't read the level there.

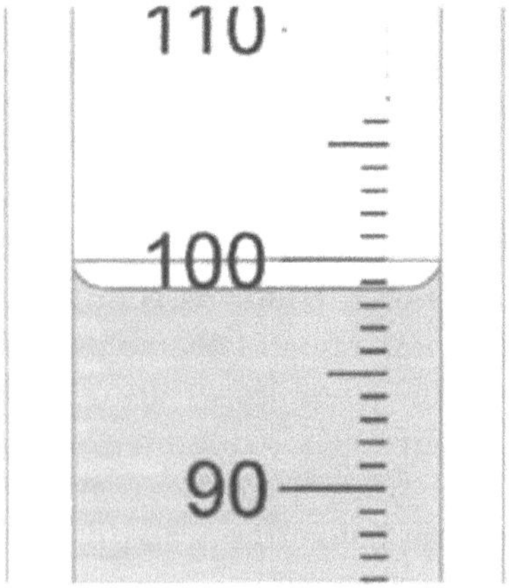

Figure 8-2. Meniscus.

Write down the result as soon as it is read. Make it a habit. Don't depend on memory. That will result in having to repeat the test to get the results right. Also, never assume that the results will be exactly the same. There will always be small differences, with emphasis on the word "small." Testing is one of the most important things to be done and should not be taken lightly.

Some operators are color blind. That is not a significant problem except for colorimetric testing. It is not something to be ashamed of. Just see to it that the chemical consultant provides a test method that can be used accurately. Some operators also have vision problems and have trouble reading the little numbers on the burettes. There are magnifying glasses for that. It is better to admit to having trouble reading those little numbers than to guess at what the reading is and potentially destroying a boiler. There are now more instrument methods for making some of these determinations. A pH meter can directly measure the pH of the sample without using an agent that turns color. There are flame spectrophotometers that take advantage of the fact that certain elements absorb light at particular wavelengths. A water sample can be aspirated into a flame, while a detector can look at specific wavelengths. The strength of the signal will be proportionate to the concentration of that particular element. Many of these can be used in place of the standard, wet chemistry methods. As always, read the instruction manuals before using these instruments.

The new conductivity meters provide a reading directly in ppm of total dissolved solids (TDS). Older meters gave the reading in "mho's" for conductance. Conductivity is measured in micromho, where a mho is "ohm," the label for resistance to electricity, spelled backwards. That required a table lookup to match the corresponding TDS level. It is a lot easier to have a meter that is simply labeled with values for TDS. TDS is a measure of the amount of solid material that is dissolved in the water. Those solids include what the water managed to dissolve as it hung around as droplets in a cloud, including gases from the atmosphere and fine particles of dust, what it picked up as a raindrop falling from that cloud, from the dirt and rocks it ran over going down the stream or river or as it trickled down through the earth to the well, and everything it managed to get from the piping until it entered the boiler plant, plus the chemicals that were added to it. TDS is measured in ppm. Steam boiler water should have the highest value of TDS and condensate the lowest, with makeup and boiler feed water falling in between. It is a value that is useful in determining percentage makeup and condensate as well as providing values for blowdown control (described later).

Modern titration or colorimetric methods for hardness testing are much easier to use and provide a better determination of the amount of hardness ions in water than the older, obscure method with so many drops of standard soap solution. The term "hardness" came from the fact that it was "hard" to generate any foam from soapy water that had relatively high levels of calcium and magnesium in it. Testing for acidity is a lot easier too. With the newer pH probes, just stick the instrument in the water sample and read the pH on the little screen on the instrument. Older probes were always a problem. The bulb of the probe was a semi-permeable membrane. With constant use, the reference solution was eventually lost. Solid state probes have resolved that problem.

Testing for alkalinity has not really changed much and still depends on titration testing with phenolphthalein for partial alkalinity. Acid is added to neutralize all the OH– ions from the caustic soda added to the water. Half the alkalinity is produced by carbonate dissolved in the water and one-third of the alkalinity is produced by phosphates dissolved in the water. The color of the solution changes from pink to clear at a pH of approximately 8.3. Alternatively, the pH meter could be used to record the acid addition to get to pH 8.3 on the meter. Testing for total alkalinity uses the same sample. Using methyl orange or methyl purple indicator, more acid is added until the color changes. The acid removes the remaining half of the alkalinity due to dissolved carbonates and the other two-thirds produced by dissolved phosphate. The color changes at a pH of approximately 4.3. Good results are another matter because the color change is very subjective. Add acid until the yellow turns pink or the green turns purple. Again, a pH meter can tell when the solution exactly turns to pH 4.3.

Since sodium carbonate is no longer used for water treatment, most of the alkalinity is due to the phosphate that was added to buffer the water. Some carbonate is dissolved in the makeup water, with the amount varying depending on the location of the plant and the source of the water. It is really not that important how much of each is in there, only to know that changes in the results of alkalinity testing can be due to the phosphate that was added to the boiler water. The main reason for looking for the difference between partial alkalinity and total alkalinity was the determination of how much scale treatment (carbonate or phosphate) was in the water. Keeping up the spread between partial and total alkalinity was, at one time, the only way to tell.

The chloride test uses silver nitrate solution, which stains skin, clothing, and almost anything else it touches. Care should be exercised to avoid spills, etc. Chloride tests are very handy. The chloride ion is not that corrosive in small quantities. It can be used as a tracer of sorts to determine the mixture of different waters. For example, the percent of makeup water can be determined by testing the makeup water and the boiler feed water for chlorides. The condensate should have zero chlorides in it (it is basically distilled water). Divide the feed water ppm by the makeup ppm and multiply by 100 to get the percent makeup. Of course, that does not work when there is some leakage into the condensate at hot water heaters and the like, which is best caught by testing for chlorides. Chloride concentration is used as a measure of dissolved solids on ships since the major source of contamination was salty sea water. In addition to checking for ratios of mixtures of water, chloride tests can indicate the performance of a dealkalizer, where chloride ions are exchanged for other anions (ions with a negative charge). It also allows a determination of the concentration of the boiler water, provided there is no carryover. Of course, they are used to check for carryover because, otherwise, there is no reason for them to be in the steam line condensate.

The test for phosphates involves mixing some boiler water with an indicator, filling another sample tube with plain boiler water, and then comparing the color. Those color comparator tests were always subjective. Similar tests are available for chelants. A quantitative method to determine the amount of phosphate present in samples, such as boiler feed water, is as follows. A measured amount of boiler water is poured into a mixing tube and ammonium heptamolybdate reagent is added. The tube is then stoppered and vigorously shaken. The next step is to add dilute stannous chloride reagent, which has been freshly prepared from concentrated stannous chloride reagent and distilled water, to the mixture in the tube. This will produce a blue color (due to the formation of molybdenum blue), and the depth of the blue color indicates the amount of phosphate in the boiler water. The absorbance of the blue solution can be measured with a colorimeter and the concentration of phosphate in the original solution can be calculated. Whatever the test method used is, read the instruction manual and carefully follow the instructions to get reliable, repeatable results. Repeatable means that the operator on the second shift should get the same reading as the one on the first shift and the third shift should concur. If everyone gets different results, someone is doing something wrong or the test is no good.

Any trainee should be allowed to test water with the operator repeating the test to see if the results are identical. If the results do not make sense, there is always a possibility that a step was missed or the sample was upset. The best thing to do is draw another sample and repeat the test. Sampling and testing water is the first step in a good water treatment program. Know how to measure the quality of the water and how to determine what is in it, both desirable and undesirable. That helps to ensure that the boiler plant remains intact. Keep in mind that carelessness and inattention to detail can result in major, sometimes catastrophic, failure of a boiler.

PRETREATMENT

Pretreatment is the conditioning of water to prepare it for use in the boilers. It is less expensive and easier to alter the conditions of the water before it gets into the boiler because it can be done at lower pressures and temperatures. Only the more common pretreatment methods are described in this book. There are other resources, with the best being the water treatment supplier, for descriptions of other methods.

Filtering is the most common form of pretreatment. Its use depends on the water source. If well water is used, filter it. City water is normally filtered by the city and is adequate for boiler makeup water. Filters vary from a simple cartridge filter to large sand filters that are tanks filled with sand that does the filtering. Sand filters are backwashed at regular intervals or when the pressure drop through them increases to a predetermined value. A backwash removes the accumulated material by pumping filtered water through the filter in the opposite direction of normal flow. The water used to backwash is sent to a sewer as waste and can, at least in the first few minutes of backwashing, contain a large amount of solid material. Backwashing also serves to fluff up the sand so that the water will flow through it at a lower pressure drop. Some other pretreatment equipment also does a certain amount of filtering.

The most common piece of pretreating equipment found in a boiler plant is a water softener. Softeners are just one of several types of ion exchange equipment. They are called softeners because they reduce the hardness of the water. Water is considered hard when it is difficult (hard) to make soap foam in the water. The original tests of a softener involved mixing a sample of the output water with a standard soap solution to see if it would foam. Water is soft when soap produces foam readily. The softener tanks contain resin that fills the tank one-third to half full. The resin just lays in the tank and is called a resin bed. The resin in a softener has an affinity for specific ions (ions with a positive charge), principally ($Na+$) sodium, ($Mg+$) magnesium, and ($Ca+$) calcium. The beads of the resin are selfish little things, always wanting what they do not have. They tend to collect ions until they are in balance with the solution surrounding them. The purpose of the softener is swapping the magnesium and calcium ions in the makeup water with sodium ions, exchanging one for the other. The reason for the exchange is that calcium and magnesium form scale in the boiler and sodium does not. The resin traps some of the dirt and large particles in the water, and, in that way, also acts as a filter.

The sodium ions for the softener come from salt. Salt is sodium chloride ($NaCl$), a common and very cheap material. It is dissolved in water by forming sodium (Na^+) cations and chlorine (Cl^-) anions. By using brine (concentrated salt solution) in the softener to remove hardness, the amount of expensive chemicals that are used in the boiler can be reduced. In very small plants, with very little makeup water, or where city water is fully softened or naturally soft, a softener is not justified. There are not many situations like that. The smallest plant can benefit from a softener if it does not use a more exotic form of ion exchanger or reverse osmosis (RO).

Operating modes of a softener include backwash, brine draw, fast rinse, slow rinse, and service. Backwashing removes dirt and "fluffs up" the resin. Water flow during a backwash is up through the bed. The space in the tank above the resin provides room for the resin to separate from the backwash water before the water leaves the tank. If the water flow rate is too high, then resin will be flushed out of the softener. It is a good idea to look at the water draining during a backwash to spot resin loss. That is best done with a flashlight pointed into the water. The resin will cause the light to sparkle.

An occasional piece of resin might leave because small pieces of resin break off.

The backwash also flushes out most of the dirt in the water that was filtered out by the resin bed. Under unusual and upset conditions, there can be a lot of dirt and mud collected by a softener. Try to take a look at the backwash water near the end of the cycle to ensure that it is clear. Sometimes storms, and at other times the water company crews flushing hydrants, can stir up mud and dirt to put a concentrated amount in the water for short periods. After the backwash is complete, brine is drawn into the softener. The brine solution has a high concentration of dissolved salt. Since salt is sodium chloride, brine is a solution of sodium and chlorine ions. The resin beads exchange ions to balance with the high concentration of brine in the softener, giving up the magnesium and calcium ions collected during the service mode and increasing the number of sodium ions they hold.

When the brine draw is complete, the softener is rinsed to remove the spent brine and the calcium and magnesium salts removed from the resin. A fast rinse flows down through the bed to quickly displace most of it. A slow rinse then follows to completely remove all the brine. A salt elutriation test is run occasionally to ensure the softener is operating properly, absorbing most of the brine.

Those previous modes of operation were all part of the regeneration cycle, which restores the softener's ability to remove calcium and magnesium ions from the water. They take from a few minutes up to 2 hrs depending on the size and capacity of the softener. Most of the time, the softener is in the service mode, where makeup water enters at the top and, as it flows to the bottom, calcium and magnesium ions are exchanged for the sodium ions on the resin beads. Since they are oversaturated with sodium from the brine draw operation, the resin beads readily give up those sodium ions when they can grab one of the calcium or magnesium ions from the water.

That explains those greedy little resin beads. They always grab what they do not have. They grab the calcium and magnesium ions when they are loaded up with sodium ions. Then they readily toss the calcium and magnesium ions when the water around them is full of sodium for them to grab. An important element of managing a water softener is knowing the hardness of the inlet water. A softener's capacity is normally listed in kilograins, thousands of grains. It depends on how much resin there is in the softener and how many sodium ions each particle of resin can exchange. Grains, by the way, are a measure of weight equal to one 7000th of a pound. The amount of water the softener can soften depends on its capacity and the hardness in the makeup water. Since the resin eventually degrades (chlorine is rough on it), some of it breaks up and is washed out. The hardness of makeup can vary. Check the operation by testing the water.

A condensate polisher is almost identical to a water softener. The differences are mainly due to the high temperature of the condensate. The resin beads and mechanical parts of a polisher are designed to take the higher temperatures. The resin also has an affinity for iron (Fe^{++}) in addition to calcium and magnesium to remove iron from the condensate. Products of corrosion, dissolved iron oxides, get removed by condensate polishers. Operation of a polisher is very similar to a softener, using brine to regenerate. High temperature and pressure boilers need extremely pure water. For those units, the condensate polisher is designed to take out nearly everything. Such resin-based systems are called "demineralizers." They remove both cations and anions and are regenerated with sodium hydroxide and sulfuric acid.

Dealkalizers are also similar to softeners and are regenerated with salt. The principal difference is dealkalizers contain anion exchange resin, accumulating a concentration of chlorine ions on the resin beads instead of sodium. Their principal purpose is exchanging the chlorine ions to replace the bicarbonate ions in makeup water. Salt water is not necessarily the best thing to put in a boiler. Yet, a combination of softener and dealkalizer will do exactly that. The reason is that salt, unlike many other chemicals, will stay dissolved in water as the water is heated up. The calcium, magnesium, and iron will not. They will drop out of solution as the water is heated to form scale. Some dealkalizers are also regenerated with a little caustic soda added to add hydroxyl ions for exchange instead of sodium. That helps to remove other anions while raising the pH of the water. As long as the concentration of chloride ions remains relatively low and the temperature and pressure are not too high, this approach is viable.

Demineralizers are combination ion exchange units that incorporate both cation and anion exchange resins. They can consist of trains of two tanks (one cation and one anion) in series or a "mixed bed" that contains both resins in one tank. Demineralizers differ from other ion exchangers because they actually remove dissolved materials from the water. The cation resins are regenerated with an acid to build up a concentration of hydrogen ions on the beads. The anion resins are regenerated with caustic soda to build up a concentration of hydroxyl ions on their beads. As the makeup water flows through the demineralizer, all the dissolved material is replaced with hydrogen and hydroxyl ions which combine to form water. The result is an output that is pure water, which is better than distilled water. As the temperature and pressure of the boiler increase, there is less and less tolerance for dissolved solids. The American Boiler Manufacturers Association (ABMA) and the ASME have recommended limits for boiler water depending upon drum pressure. These include limits for TDS, alkalinity, total suspended solids, conductivity, and silica. For the highest steam conditions (super critical boilers or once-through steam generators), the TDS limit is 50 ppb (parts per billion). The manufacturer will cite these limits in the instruction manuals. It pays to read the instruction manuals to know the water quality limits for the boiler in question.

One of the most important things an operator can do to maintain ion exchange equipment is to prevent condensation on them. The constant formation of moisture with access to air accelerates corrosion of the equipment and piping. Usually, good ventilation in the room containing the equipment is adequate. Sometimes, special coatings are required to act as insulators. Check the backwash water after any system maintenance and when water temperatures drop to ensure that the resin is not washing out. Colder water is denser than warm water and can carry out resin that warmer water could not. Another important thing to remember is that the ion exchange process is not perfect. A few ions manage to sneak through depending on the equipment design, loads, and operation. Demineralizers are almost perfect ion exchange devices. Softeners reduce hardness to 2–5 ppm and dealkalizers are about 80%–90% effective. All ion exchange devices have limited turndown and tend to "channel" at low flow rates, where the low flow of water takes the easiest route through the resin to consume the ions there and allows leakage of untreated water. Know the limitations of the equipment.

A good rule of thumb is to maintain a velocity of less than 2.5 gpm (gallons per minute) per square foot of resin surface to prevent channeling. When water demand is low, it is better to shut off the water flow through one or more of several ion exchangers in order to keep the flow rate up. To better understand this, imagine a small creek. When the water flow is filling the stream, all the rocks are wet. When the flow is down, many of the rocks will stick out of the water and be dry on top. The water will flow by in little channels around the rocks. When the water flow is low in an ion exchanger, some of the resin never sees water flow, while the resin in the channels sees it all. The result is that the resin in the channels is exhausted (all the ions it had to exchange are used up), while the rest is not used much at all. Since the control of the ion exchange units is done by measuring the throughput of the water, there is a good chance that untreated water will be passing through before the quantity of water that could be treated by the exchanger passes through.

An important part of an ion exchange operation is cleaning and replacement of the resin bed. The normal backwash does not remove all the sediment and particles that get imbedded in the resin beads during operation. Chemical cleaning with a resin cleaner that is pumped into the idle exchanger and then rinsed out is a normal function in many plants. A complete replacement of the resin every five years is common where the chlorine in the makeup is high.

RO is becoming more common as the cost of the membranes decreases. Rather than absorbing all the theory of osmosis, treat them as filters that will let water through but will not let the ions dissolved in the water get through. The pressure drop is high because the filter (the semi-permeable membrane) has very tiny holes in it. Also, some of the water has to be used to constantly carry the dissolved stuff away (sort of like blowdown). The filter membranes, depending on their make, can be susceptible to heat or certain chemicals in the water, chlorine being one. It means that the water may need to be treated before it gets to the RO unit. RO performance varies as well. Expect anything from 70% to 99% efficiency. Note that while they eliminate ions indiscriminately, they do not get them all. Thus, boiler water internal treatment is still needed. High-quality RO requires wasting a considerable amount of the water to carry off the contaminants, nominally about 20% of the water fed to the unit. The purified water is called "permeate" because it penetrated the membranes and is, therefore, about 80% of the makeup water supply. Lower waste rates usually accompany lower efficiency. However, some can be of low efficiency with high waste rates.

This is one piece of equipment that requires reading the instruction manual immediately. The membranes cannot be allowed to dry out. If they sit too long without water flow, there is danger of microbiological (very little bugs) growth. Don't shut it down for the summer and walk away. Feeding with a biocide (bug killer) during idle periods is required. They require some chemical treatment at their inlet to prevent chlorine damage. Cleaning at intervals as frequent as every month is necessary to keep the capacity up. Finally, the membrane cartridges have to be replaced about every five years. Current replacement cost is about $100 for every gpm capacity.

Some water sources, especially those in the middle of the country, have a high concentration of bicarbonate ions. The bicarbonate produces two problems for boiler operation. In the boiler, where the water is heated, the bicarbonate breaks down to form carbon dioxide gas and hydroxyl ions. That raises the pH and alkalinity of the boiler water enough that, frequently, the blowdown is based on alkalinity, not dissolved solids. The carbon dioxide that evaporates in the boiler flows with the steam to the steam users where it is absorbed in the condensate that forms. Each molecule of carbon dioxide dissolved into the water first produces carbonic acid. This compound further breaks down to produce a bicarbonate ion by combining with a hydroxyl ion. When it obtains the hydroxyl ion, another molecule of water is dissolved to replace the hydroxyl ion and increase the number of hydrogen ions in the water. The result is condensate with a very low pH and corrosion of the piping and other parts of the condensate system.

The best approach for high bicarbonates is to use a dealkalizer. Other equipment has been used, and is still used today, to remove the carbon dioxide before it ever gets to the boiler. These are caused by decarbonators or degasifiers and consist of a tank, usually wooden or fiberglass, with wood slats or pieces of plastic stacked inside to form what is called "fill." Treated water is dumped into the top and trickles down over the fill, while air is forced by a blower into the degasifier and up through the fill. The water has to be treated so that the carbon dioxide gas will separate from the bicarbonate ion. In some plants, the treatment simply consisted of adding acid, usually sulfuric, to the water to lower the pH. The acid causes the bicarbonate ions to break down. The other pretreatment consists of running some or all of the water through a cation unit. The hydrogen ions exchanged for others lower the pH of the water. In many demineralizers, the cation and anion units are separated by a degasifier. Then the bicarbonate is broken down, and removing it as carbon dioxide gas takes load off the anion units. The carbon dioxide, now a dissolved gas, is "stripped" from the water by the air flowing up through the degasifier so that it cannot recombine with a hydroxyl ion to form a bicarbonate ion again. A dealkalizer is an ion exchanger regenerated with salt, taking on chloride ions that are exchanged for the bicarbonate ions.

BOILER FEED TANKS AND DEAERATORS

Boiler feed tanks with heaters and deaerators are other common pieces of pretreating equipment. They have three principal functions: removing oxygen from the boiler feed water, heating, and storing boiler feed water. In the case of some deaerators, the three functions are served by separate tanks, a deaerator, and a separate storage tank. Both systems remove air from the water. There are variations in equipment construction and differences in how much air is removed. Neither removes oxygen completely. A boiler feed tank can only remove oxygen to small values. Deaerators, operated properly, will remove oxygen to minimal amounts.

Removal of the oxygen is achieved by raising the temperature of the water. As the water temperature approaches the boiling point, the amount of oxygen the water can hold decreases. Heating the water to 180°F reduces the maximum oxygen absorption to less than 2 ppm. Raising the temperature to boiling reduces that to 0.007 ppm. When the water is ready to boil, every molecule of water is prepared to change to steam. Thus, the water has very little ability to hold dissolved oxygen. The dissolved oxygen forms bubbles of gas in the water. Complete deaeration is not achieved until those bubbles are removed. It is getting the bubbles out that makes the difference in deaerators.

Boiler feed tanks come with two kinds of heaters. The water in the tank can be heated by a submerged heating coil or a sparge line. A sparge line simply injects steam directly into the tank. The steam heats the water, condenses, and becomes part of the feed water, while agitating the water. Agitation is important in that it helps remove the bubbles of oxygen from the water. Sparge lines are noisy. That should be considered when adopting a method of heating the water.

For all practical purposes, boiler feed tanks simply provide a place for storage of boiler feed water. They can return condensate with some capability of oxygen removal provided occasionally. They are normally fitted with a float controlled makeup valve to admit makeup water to maintain a constant level in the tank. The cold makeup water, being denser than the condensate, tends to simply drop to the bottom of the tank, mixing with the condensate as it enters the feed pump suction piping. Dripping or, better yet, spraying the makeup into the top of the tank will help reduce oxygen from it. Heaters and sparge lines seldom manage to effectively deaerate that water. Deaerators, on the other hand, are designed to remove air and the key is their operating pressure. Boiler plant deaerators are always operated so that pressure will force any removed air out of them.

Deaerators are provided in five types: vacuum, flash, spray, scrubber, and tray. A vacuum deaerator is typically a vessel filled with packing and operated under a vacuum. The packing is not like pump or valve packing.

It is like fill, loose pieces of ceramic or plastic materials stacked randomly that act like splash blocks. A lot of the water surface is exposed as it tumbles down through the packing. Producing a sufficient vacuum in a vacuum deaerator will bring the water to a saturated condition. For example, pulling a vacuum of 29″Hg (inches of mercury) produces a condition where 79°F water will boil. As long as the water is warmer than the saturation temperature that matches the pressure inside the deaerator, it will be at boiling and a little is actually vaporized. The air and non-condensable gases are removed from the deaerator by the vacuum pump or steam jet ejector, whichever is used. A steam jet ejector will normally discharge to a condenser that uses the remaining energy in the steam to preheat the water before it enters the deaerator. When a vacuum pump is used, provisions are made to heat the water and can include any type of heat. Vacuum deaerators are not normally used in boiler plants, as the water is heated to higher temperatures anyway.

By heating the water to a saturation temperature higher than 212°F, the pressure in the deaerator will be above atmospheric pressure. That higher pressure will push the air and non-condensables out to the atmosphere. That is typical of all boiler plant deaerators. The variations in the four types depend on how difficult it is to get the bubbles of air and non-condensables out. Non-condensables are gases other than air that can be released by bringing the water to boiling. They include chlorine gas, ammonia, and others that are not normally found in air but can be found, in very small quantities, in water.

Flash type deaerators use this concept to produce a pressure just slightly higher than atmospheric pressure to remove the gases. The makeup water is heated in an external heat exchanger to a temperature higher than 212°F. It is then passed through a spray valve into an open tank, where some of the water flashes into steam. Since all the water is above the saturation temperature, it cannot hold any oxygen. The oxygen should be removed with the flash steam which may, or may not, be recovered. There are a number of these devices in the field.

The best choices for deaerators for boiler plants are spray, scrubber, or tray types. The choice depends upon the normal temperature difference between the makeup water and the boiler feed water. They are all called DC heaters (for direct contact). The water is heated by mixing the steam with the water. The steam is condensed and becomes part of the feed water in the process. Heating the water to saturation only removes the oxygen and gases from solution. It does not get the little bubbles of air and gases out of the water. To do that, some form of agitation is needed. The means to get the agitation is determined by the temperature difference. All of these deaerators have spray nozzles that serve to break up the water as it enters the deaerator. The purpose of the water spray nozzles is to break the water up into small droplets so that they can be heated rapidly by the rising steam.

These deaerators also always have a vent condenser. A vent condenser can be an external heat exchanger or, as shown in the following figures, simply a length of tubing inside the water box above the water spray nozzles. The purpose of the vent condenser is to condense most of the steam that is carried out with the air and gases. The idea is to have only air and gases leaving the deaerator. Of course, the vent valve on a deaerator is always adjusted to produce a "wisp" of steam, just enough to be sure that all the air and gases are pushed out, because a little steam is coming out with them. Throttling the vent valve too much will recover all of the steam as condensate but can also trap air and gases in the top of the deaerator to prevent the steam from contacting the makeup as it enters through the sprays, which prevents proper deaeration. Opening the vent valve too much is just wasting steam. Operation of that vent valve is the key function of a boiler operator. The trouble is that most operators solve any control problem by simply leaving the valve so far open that steam is blowing out dramatically. That is a considerable waste of energy and water. The wise operator keeps that vent adjusted so that there is only that wisp of steam coming out.

On board ships, spray type deaerators (Figure 8-3) are fairly common since the water from the condenser is relatively cold and only heated slightly in the air ejector condenser and turbine bleed heat exchangers. Thus, there is a considerable difference between makeup and feed water temperatures. If the plant generates a lot of

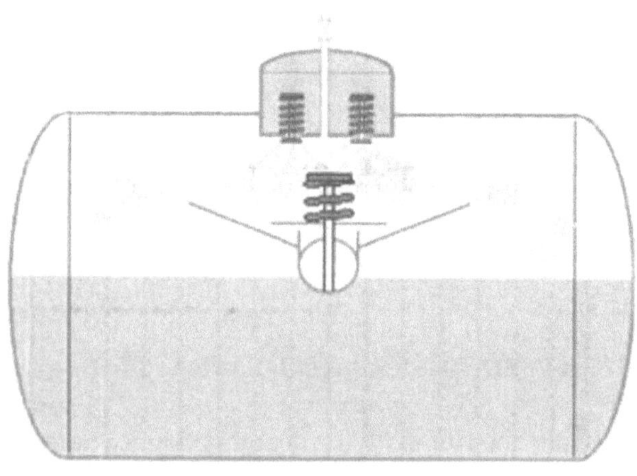

Figure 8-3. Spray type deaerator.

power by condensing turbines (a utility), then a spray type deaerator may be all that is needed. The large difference in temperature requires a lot of auxiliary steam to heat the water. The steam can be directed into the spray section, where it creates a violent mixing with the droplets of heated makeup water before it flows up to mix with and heat the water entering at the spray nozzles. It is the effect of all that steam rattling those water droplets around and breaking them up further, before they reach the outside and drop into the storage section, that removes the bubbles of air and gases.

Many people get confused with the term "spray" because all these deaerators have water spray nozzles. The terms "spray scrubber" and "spray tray" are often used to describe scrubber and tray type deaerators to avoid confusion. A spray scrubber uses a steam spray to provide the agitation to remove the bubbles of air and gas. There is no real reason to prefix the titles of the other two types with the word "spray."

Except for power generation plants, where the makeup is primarily colder water from a condenser, few plants can use a spray type scrubber. The combined condensate return and makeup water temperature is so high that the steam requirements are not enough to perform the agitation. When the temperature difference of the condensate and feed water can be consistently more than about 50°F, there is enough difference for a scrubber type of deaerator (Figure 8-4) to work well. The flow of steam along with the water up through the baffles of the scrubber provides enough energy to separate the bubbles. Some of the energy is achieved using the difference in density of the water and steam.

When the temperature difference between blended makeup and feed water is less than 50°F, always insist on a tray type deaerator. The trays (Figure 8-5) do not

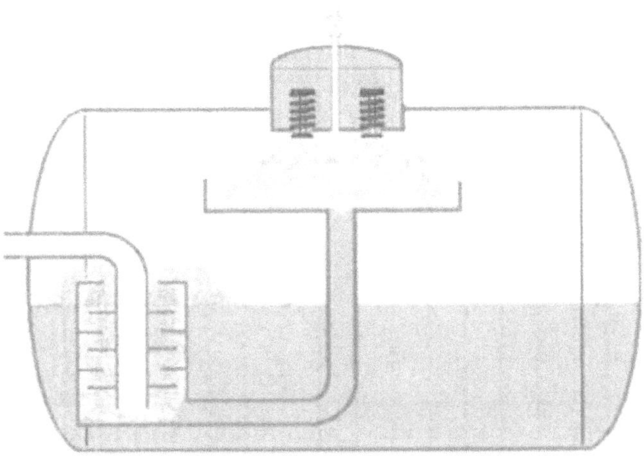

Figure 8-4. Scrubber type deaerator.

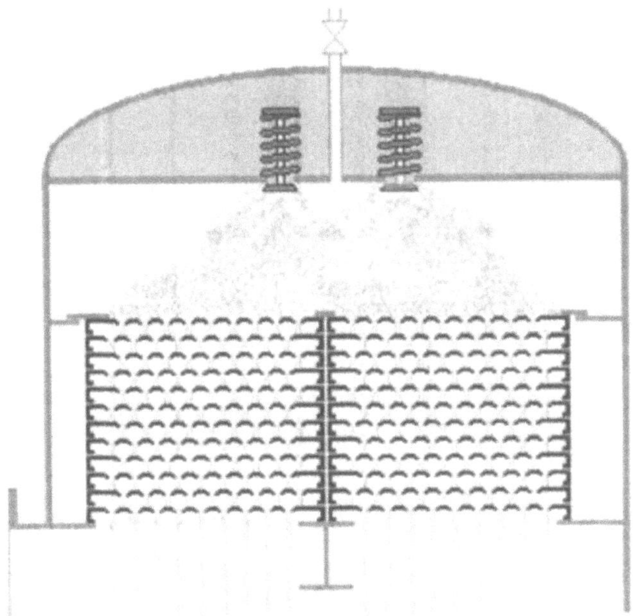

Figure 8-5. Tray type deaerator.

look like cafeteria trays. They are made up to produce hundreds or even thousands of little waterfalls. Distributing the water over the trays and producing thin little falls produces hundreds of square feet of exposed water surface for the bubbles to escape from. Some scrubbing of the falling water is achieved by the steam flowing up through the trays to the water sprays. However, most of the energy that is used to force the bubbles out of the water is provided by gravity. A tray type deaerator costs a lot more, but when compared to the added cost of sulfite and blowdown over the operating life of the boiler plant, the additional cost is justified.

In a power generating station, there are typically a number of feed water heaters that use steam taken from the steam turbine to heat the water from the condenser temperature up to the deaerator temperature. The steam taken from the steam turbine is called extraction steam. A feed water heater is a big heat exchanger. The feed water is inside the tubes and the steam cascades over the tubing to heat the water and condense the steam. The condensed steam is pumped back to the prior heat exchanger (or the condenser) and mixed with the feed water. It is not lost or wasted. The deaerator acts as a feed water heater. In a typical power plant, the deaerator operates at about 15 psig (pounds per square inch gauge) and 250°F. The boiler feed pump comes after the deaerator and puts the system pressure on the feed water. Additional feed water heaters are used to bring the feed water temperature up to the level that is appropriate for the economizer. Since extraction steam can be selected to be relatively close to the pressure required by the deaerator, most power plant deaerators are tray type.

Sometimes, a deaerator is not operating properly because the pressure control for the steam is at the control valve or senses the pressure in the steam line going to the deaerator. A proper installation, regardless of type, senses and controls the pressure on the top of the storage tank after the scrubber or trays of the deaerator to eliminate the pressure drop through the scrubber, or trays, and connecting piping. If there is a deaerator problem, check the location of the pressure sensor.

Modern tray type deaerators are normally furnished with tie bolts to hold down the trays because some people managed to dislodge all the trays. They only work properly when the trays are all stacked correctly and leveled so that the water flows uniformly over the entire bank of trays to interact with the steam.

Imagine what happens when someone shuts off the steam to a tray type deaerator. Colder makeup and condensate still enter through the sprays, but, now, there is no steam to heat the water. Below the deaeration section is a storage tank full of water at the original steam temperature. That water is going to start flashing off steam as the pressure falls. Thus, there is some steam provided for deaeration. Assuming the sudden flashing of all the feed water does not produce so much cavitation in the feed pumps that they trip (turbine driven ones normally do), the feed water in storage will boil as the colder makeup continues to enter the deaerator. Before they started bolting down the trays, the only sign an operator had that something was wrong was some clanging, as the flashing steam and water swelled up out of the storage section and lifted all the trays. Frequently, the insulation on the deaerator prevented the noise reaching the operator's ear. The next thing to notice was that all of the sulfite in the boilers just disappeared. Of course, by the time an operator got around to discovering that the sulfite was wiped out, the deaerator's trays were all lying in the bottom of the storage tank and not deaerating. A lot of oxygen had reached the boiler to corrode it.

Lower the operating pressure gradually to get down to atmospheric conditions to avoid shaking the deaerator. A deaerator should also have a vacuum breaker, normally a check valve installed backwards, connected to the steam space to admit air should there be a loss of steam pressure. It is possible to shut down the steam supplying a deaerator at full boiler load. The odds are that the check valve used for a vacuum breaker will not allow enough air in once vacuum starts to form. The storage tank could be crushed by atmospheric pressure.

Another problem that affects any unit is a water spray valve coming apart. When that happens, it is equivalent of a fire hose hitting the trays. There is no breakdown of the water initially. As a result, it is not heated. When the feed water temperature is lower than the saturation temperature matching the steam pressure, it is a good sign that there is a defective water spray valve, regardless of the deaerator design.

Except for vacuum deaerators, the feed water temperature has to be above 212°F (except in Denver where it has to be above 203°F) or the deaerator is not working. The saturation pressure has to be above atmospheric pressure or there is no pressure to push the air and gases out. Deaeration, getting the oxygen out of the water before it gets to the boiler, is principally done to reduce the cost of chemically treating the water to remove the oxygen. If the oxygen is not removed, it will create pits in the boiler metal, something that looks almost as if it were done with an electric drill. Oxygen pitting can destroy a boiler in short order. The sulfite is always added to remove the little bit of oxygen that slips past a deaerator, even when it is working fine. In order for the sulfite to be effective and remove the oxygen that gets past the deaerator before it gets to the boiler, the sulfite should be added to the deaerator storage section. Sulfite generates sulfate ions when it reacts with the oxygen in the water. Since sulfate salts form the hardest scale, don't put in any more than is absolutely necessary. Thus, maintaining proper operation of the deaerator is important.

If the sulfite is added before the deaerator, change that to provide the deaerator with something to do. The sulfite should be fed right below where the water drops from the deaerator (but below the low water line) so that it can start doing its job immediately. Sodium sulfite solution can be corrosive when concentrated and heated. It will produce a pattern of corrosion on the bottom of the deaerator storage section that looked like it was sandblasted unevenly, if the sulfite is not distributed properly. Having stainless steel sulfite distribution piping in the storage tank, with a long row of perforations to distribute the chemical just below the low level, should prevent that kind of damage.

Some plants cannot tolerate the additional dissolved solids that result from sulfite addition. These are primarily nuclear plants, heat recovery steam generators (HRSGs), and super critical (or once-through) boilers. These plants will use all volatile treatment (AVT). This treatment uses hydrazine as the oxygen scavenger. The hydrazine reacts with oxygen to form water and nitrogen gas. The water is desirable. The nitrogen must be removed first in the condenser and then in the deaerator.

Hydrazine is hazardous and must be handled with care. Wear that protective clothing.

Hot water boiler plants do not normally have the experience of constant makeup. Many of them are treated with sodium nitrite. The nitrite ion converts to nitrate, absorbing oxygen in the process. It is only usable at the low pressure hot water heating temperatures.

BLOWDOWN

Blowdown, continuous blowdown or surface blowdown on steam systems and low point blowdown on hydronic systems, is used to reduce the concentration of solids dissolved in the water. Even with demineralizers or distilled water for makeup, there will still be a growing concentration of solids in the water in the system. Some will come from corrosion of piping and other parts of the system by the condensate. Even in tight hydronic systems, there will be increasing solids from gradual dissolving of materials left in the system during construction and minor vapor leaks that are not always apparent. In steam boilers, all the solids remain in the boiler water, concentrating there as the water leaves the boiler as steam. If some of the solids carry over with water droplets in the steam, they are returned in the high pressure condensate. Thus, the boiler is where all the solids end up.

The amounts of solids and some liquids dissolved in water have an effect on the surface tension of the water. There are two sticky properties of fluids, cohesion and adhesion. Cohesion is a measure of how the material sticks to itself. Adhesion is a measure of how much the material sticks to something else. Water is high in both. Water actually climbs the sides of a glass because it adheres to the glass. High cohesion is evident at the surface of water where it sticks to itself. When separated from a large body of water, a small droplet becomes perfectly round because of the high cohesion at its surface, which is called surface tension. The combination of adhesion and cohesion contributes to the capillary action of water. It will literally pull itself up into narrow spaces after adhering to the surrounding walls and then reach out again.

As the quantity of dissolved solids increases, the physical characteristics of the water change, increasing the surface tension of the water until eventually the water starts to foam and carry over into the steam piping. While this is one way to get the solids out of the boiler, it does not do the steam piping a lot of good. Increasing solids can also result in saturation of the water with solids in the risers and some of the dissolved materials will drop out as scale.

A way to limit the concentration of solids in the boiler water to a value just below that point of carryover or scale formation is needed. Blowdown provides the answer. By removing some of the boiler water from where it contains the highest concentration of solids, some makeup water that contains very little solids can be added to the boiler and reduce the overall concentration of solids. In a steam boiler, that means removing the water right after it is separated from the steam in the steam drum. That is why the continuous blowdown piping is in the steam drum. The piping has the holes located where they are to accept the water to be removed as blowdown.

In hydronic systems, the blowdown is usually drawn from the boiler at the same place as that for steam boilers. Still, check the system for places where the solids are more concentrated. Usually, the return water will be more concentrated because the water shrank as it cooled but contains the same weight of solids. In that case, blowing down return water will waste less of it.

In boilers with multiple drum boilers, older sterling designs, the solids can concentrate in one section of the boiler but not the one where the continuous blowdown connection was located. Scale forming can occur despite the maintenance of low TDS at the point of blowdown. Regardless of the system, its operating pressure and temperature, and the quality of the makeup water, be aware that someone could have made a careless decision regarding the location of a blowdown connection. Any time there is scale formation or problems with carryover that are not related to pressure fluctuations, check the location of the testing and continuous blowdown connections. Be sure that they are located in the right place.

The amount of blowdown is determined by the TDS reading (described above in testing). The ABMA has set standards for the proper levels of solids concentration for boilers according to the operating pressure. The boiler's instruction manual should also contain that table or specific recommendations for what levels of solids concentration to run at. Note that it is a recommendation and not an absolute value. It may be possible that the boilers can operate with a considerably higher level of solids without forming scale or carryover. It depends on many factors including boiler load. Consider raising the settings for TDS levels gradually until some problem is detected or they get as high as 4000 ppm, either stopping at that value or backing down below the value where problems occurred. Values should be established for each boiler load since they can operate with higher solids content at lower loads. Usually, carryover is the limiting factor. Scale formation can also set the limit. Therefore, this test should be carried out each year one month before the annual internal inspection by raising the TDS limit by 50 ppm, while keeping a close watch on relative stack temperatures. Back off on any increase in stack temperature. That could indicate scale formation, which will reduce heat transfer. Since blowing down wastes energy and water, minimizing it is a wise operation. It is worthwhile to minimize blowdown.

Modern technology has produced instruments and equipment that provide a reasonable degree of automatic blowdown control. There are systems that provide continuous blowdown as intended, with continuous measurement of TDS and modulating of a control valve to vary the rate of blowdown to maintain a maximum level. The more common systems are intermittent in operation. The typical system incorporates a timer that opens the continuous blowdown control valve at fixed intervals. The valve then remains open until the TDS, measured at a probe in the blowdown piping, drops below the preset value. One potential problem with that type of blowdown control is the introduction of a surge of high solids water fed to the boiler by opening up a previously shutdown system. The solids will be high in the boiler until the valve opens again. A better method would use an automatic control with a high and low setting. Blowdown would be continuous through a manually set valve, with the automatic valve opening to dump additional blowdown when the high point is reached and closing when the low point is reached. It does not cost any more than the system with valve timing, constantly monitors solids, provides a continuous flow of water to any blowdown heat recovery system, and will react immediately when additional solids are introduced to the boiler. The only thing that is better is a modulating control valve. They are also rather expensive for small boilers.

Blow off is designed to remove solids that settled out of the boiler water. The sources include solids from makeup water, rust, and other solid particles returned with condensate, and the intentional production of sludge by chemical water treatment. It contributes to the reduction of dissolved solids but at a considerable expense in water and energy since bottom blow off is not recovered in any way. Use continuous blowdown to remove dissolved solids concentration and limit bottom blow off to its purpose of removing sediment, which will vary depending on the quality of the makeup water and the degree and type of chemical treatment. See the discussion on water as a consumable.

CHEMICAL TREATMENT

The first rule in understanding a chemical treatment program is to know what is in the container. There are not that many compounds for water treatment and they do the same thing, regardless of the name or number on the barrel. By knowing what is in the container, the purpose and the function of the chemical will be clear. Given the true title of the active chemical and the following paragraphs, there should be enough information to properly maintain chemical treatment of the plant's water. It is not a black art.

The only person who can effectively operate a water treatment program is the educated boiler operator. Chemical treatment consultants, vendors, and suppliers will have no idea of what has transpired between visits. They cannot possibly know that the boiler was shut down, drained and refilled, left sitting idle, or operated continuously. They may not know that there was an upset in the level controls and the operator used the bottom blow off to restore water level several times, dramatically reducing the chemical levels. Operators are able to communicate with each other so that they are aware of all the variables that affect the chemical concentrations in the boiler water and can make sound decisions about changes in the treatment program. Don't necessarily get rid of the consultant. It is like having a boiler inspector. It is always better to have some other, somewhat disinterested, party looking at the chemistry. It is really best if the consultant does not sell the chemicals.

A water treatment program only has two goals: prevent corrosion and prevent scale formation. The causes of corrosion and scale formation have to be understood to prevent them. Knowing how the chemicals prevent (or enhance) those conditions must be understood to maintain proper chemical water treatment. The process of obtaining representative water samples and properly testing them for chemical content has been covered. How to use that information to achieve the goals of the program is described in the paragraphs that follow.

Recording everything that happens, every test run and follow up actions, is important to understanding what is happening. Don't limit the record to the space provided on the log supplied by the chemical treatment supplier. The boiler, or water system, has boundaries and contains a certain volume of water. That volume, or weight, of water contents can be determined from the manufacturer's instruction manuals. Estimates, using actual measurements, and the data in the pipe tables in the appendix can be used to calculate the volume and determine the weights. Once an initial volume or weight has been determined, the weight of the water in the system or boiler can be calculated for when it is cold. At 70°F, water weighs 62.27 pounds per cubic foot. Freezing (32°F) water weighs 62.4 lbs/ft^3.

Once the system is up to operating temperature, the weight will be lower. Then adjust the data for the effect of thermal expansion. Determine the ratio of cold to operating by dividing the specific volume of water at 70°F (0.016025) by the specific volume of water at the operating temperature, using the data from the steam tables. Then multiply that ratio by the weight of the water when it is cold. That is the weight of the water in the system when it is operating. Move the decimal place of that result six places to the left or, if using a calculator, divide by one million. That provides how many million pounds of water are in the system. Unless it is a very big system, the number will be small. By knowing how many million pounds of water are in the system, the results of the chemical tests, in ppm, will have some meaning. That information can be used to estimate the effect of chemical additions. Don't forget about the complication with pH being steps of ten.

There are basically four sources for the chemicals that are in the boiler system's water: makeup, corrosion, leaks, and treatment. In order to effectively control the water treatment, the source of the chemicals must be determined. The principal source is the makeup water. It is a function of the quality of the water obtained from the well, river, city water main, or wherever it comes from. Test that water to know how much it is capable of adding to the chemical burden of the boiler water and how to treat it.

Testing that water for hardness provides an indication of the required frequency of regeneration of the water softeners. Tests for bicarbonates or TDS provide indications for other ion exchange equipment and bleed requirements for RO systems. When using well or river water, test the water for suspended solids to determine the loading of water filters.

Be aware of changes in the water source. Periods of drought or spring runoff can change the level of TDS in the water. Maintenance work on reservoirs may result in a change in source to a river. The TDS levels of the river are typically substantially higher than those of the reservoir water. Adjustments in softener throughput are essential to make sure all the hardness is removed. Also, blowdown has to be increased to compensate for the heavier solids loading. Regular daily testing of that raw city water is essential because they do not always notify the plant when they make the switch. The softener's capacity is based on the hardness removed. Testing the hardness and recording the meter provides a clue. If the hardness of the makeup is 50 ppm and the softener is set to regenerate after 20,000 gallons, the meter will need to be reset for 10,000 gallons when the hardness increases to 100 ppm. Stick with the ratios to avoid all those kilograin calculations. As the resin deteriorates, which is detected by noting some hardness increase at the end of the softener run, adjust the meter setting accordingly. Testing condensate can identify leaks into the system. A common source is steam heated service water heaters. That is always a concern because the water is not routed through the pretreatment equipment, such as softeners. Condensate will also contain iron, copper, and other metals from corrosion of the steam and condensate piping. It is also possible to receive water contaminated by some operation in the facility. A boiler plant operator has to know a little bit about the facility served to be aware of the potential for such contamination and to watch out for it. In one case, a boiler got contaminated with softener resin. It formed a hard, baked on coating over all the boiler tubes, wherever the resin hit the tubes and melted on. The operators found that one of the strainers in the softener had broken off, allowing the resin to leave with the treated water.

Water that is passed through a piece of pretreatment equipment has to be tested to ensure that the equipment is operating properly. Some of the tests are only significant at specific stages of the system operation. For example, testing of the output of a water softener near the end of the run is critical to make certain that the resin has not deteriorated to the degree that hardness is bleeding through. Some tests have to be combined with an analysis of chemical use. If more sulfite is being used than normal, it is an indication of problems with the deaerator.

RO and blowdown reduce the concentration of ions in the boiler water but do not eliminate them completely. Softeners and other ion exchange equipment, except hydrogen softeners and demineralizers, swap ions, replacing those that produce difficulties with ones that are not very damaging. They do not get every bad ion out. By maintaining a certain amount of special chemicals dissolved in the boiler water, it is possible to offset the nasty ions and remove any oxygen that may have managed to sneak past all of the pretreatment equipment. There is a "residual" of water treatment chemicals in the water. They reside there, waiting to pounce on any scale forming ions or oxygen that gets through before they can damage the boiler. Another reason to maintain a residual is that it can be measured. If it is there, it can be detected with a chemical water test, proving that it is there. For protection from corrosion due to oxygen in the water, normally maintain a residual of 30 ppm of sulfite. To stop hardness, a residual of 60 ppm of phosphate (less with chelant and polymer) is common.

There is one problem with sulfite use. When it has done the job, the sulfite ion becomes a sulfate ion. Sulfate ions can combine with calcium and magnesium to form the hardest and toughest scale there is. Low pressure hot water boiler systems and chilled water systems occasionally use sodium nitrite for oxygen removal. The mode of oxygen removal is the same as sulfite. Neither the nitrite nor the sulfite produces desirable elements in waste water. Other compounds are being considered. Chemicals cannot reduce the solids content of the boiler water. They actually increase it as they are added. Most of the water treatment chemicals are sodium based, consisting of sodium and other molecules that dissolve in the water. The sodium ions tend to remain dissolved. The other ions from the material are what are used to treat the problems of corrosion and scale formation. There is no need to test for sodium. The testing is for the ions that do something and TDS, which is a measure of all the ions in the water.

PREVENTING CORROSION

There are two basic ways that corrosion occurs in a boiler and an additional one for condensate systems and piping. As the number of hydrogen ions in water increases, the pH gets lower. The free hydrogen ions attack the metal in the boiler, changing places with the iron molecules in the steel. Preventing this kind of corrosion is solved by adding hydroxyl ions to the water to combine with the hydrogen ions, making water molecules. This greatly reduces the number of hydrogen ions and they cannot attack the iron. The chemical normally added to boiler water to raise the pH (which means fewer hydrogen ions) is sodium hydroxide (NaOH). Ammonia has also been used. It is easy to envision that chemical dissolving into sodium (Na+) and hydroxyl (OH–) ions in the water. Enough is added to keep the pH of the boiler water in the range of 9–10. Adding too much caustic soda will raise the pH so high that other problems, caustic embrittlement and caustic cracking, will occur. In some localities, the water is already alkaline. Additions of caustic soda are not required. Some of those actually require additional blowdown to prevent the pH from going too high. When there is a problem with caustic water or high pH, be very careful of leaks in the boiler. Evaporating water leaves a concentrated solution, where the pH is way too high. Severe damage to the boiler near the leak can result. The damage is said to be the result of caustic embrittlement.

The other cause of corrosion in a boiler is dissolved oxygen. Oxygen in a boiler will produce what is called "pitting." It looks as if some strange worm tried to eat a hole straight out through the metal. Oxygen pitting is usually easy to identify because it happens where water is heated to free the oxygen from solution and the oxygen comes in contact with the metal. Heating of boiler feed tanks and deaerators removes most of the oxygen. Some chemical treatment is needed to get the little bit that leaks through. If there are no heated feed tanks or deaerators, a lot of chemicals will be needed to make certain the oxygen does not eat away the boiler. The standard chemical for steam plants and lots of hot water plants is sodium sulfite ($NaSO_3$) which dissolves to free sulfite ions that remove oxygen. It takes two sulfite ions to remove a molecule of oxygen gas ($2SO_3$= + O_2 => $2SO_4$=). It takes a while for two of them to get around to ganging up on that oxygen to remove it from the water. That is one reason to feed the sulfite back at the boiler feed tank or deaerator, allowing the sulfite time to work.

Other reasons for feeding the sulfite there include protecting the feed system, storage tank, pumps, and piping, along with any economizer in the boiler. The ions move around and the sulfite ions will contact the oxygen in proportion to the temperature of the water (molecules and ions move around faster as they are heated) and the mixing of the water. If the recommended installation of the feeder is in place, the sulfite has lots of time to find and interact with the oxygen molecules.

The condensate line can be susceptible to corrosion from carbonic acid, dissolved CO_2. This is usually evidenced by a groove at the bottom of the pipe. The problem is associated with carbon dioxide gas coming from bicarbonate ions in the water. The testing for alkalinity, using the methyl orange or methyl purple test, showed either phosphate or carbonate alkalinity. Bicarbonate ions (HCO_3–) in the water break down when the water is heated in the boiler to form hydroxyl ions and carbon dioxide gas ($HCO_3^- \Rightarrow OH^- + CO_2$). The gas leaves the boiler and travels with the steam. Decarbonators and degasifiers help remove the bicarbonate, but, like other pretreatment processes, they do not always get it all. When the condensate forms, the carbon dioxide is dissolved in the condensate to return to bicarbonate, leaving a hydrogen ion in the process ($CO_2 + H_2O \Rightarrow H^+ + HCO_3^-$). It is those hydrogen ions that do the corroding of condensate lines after the carbon dioxide is dissolved again. The only effective treatment is to put something in the water to raise the pH (just like in the boiler). However, it is not a simple matter of adding caustic soda. If caustic soda were to be added, it would have to be put in at every little condensate trap in the system and then try to come up with a way of controlling it. It cannot be put in the steam because it would be a dry chemical and plug up the steam lines.

Special chemicals called "amines," and cyclohexylamine in particular, will flow with the steam as a vapor and then dissolve in the condensate along with the carbon dioxide. It will act to raise the pH of the condensate to prevent the acidic corrosion. Cyclohexylamine is questionable for cancer. Limit its use to what is necessary. Hydrazine (N_2H_4) combines with oxygen to produce water and gaseous nitrogen. It also formed ammonia, which flowed with the steam to dissolve in the condensate and raise the pH of the condensate. Hydrazine is a hazardous chemical and care should be taken in its use. Due to this concern, it is primarily used in boilers and steam generators that require AVT. Since the hydrazine breaks down to water and volatile gases, it qualifies for AVT. Nuclear plants, HRSGs, and super critical boilers require AVT.

Power plants typically use filters, demineralizers, and degasifiers to remove as many of these materials as possible before the water ever gets to the boiler. Thus, carbonate and bicarbonate ions are removed in those processes. Further, the power plant condenser has a steam jet ejector to remove any non-condensable gases from the vacuum in the condenser. This feature is desirable since these gases will raise the back pressure on the steam turbine. The steam jet ejector is not the source of the vacuum. That is caused by the temperature of the cooling water. At 80°F, the vapor pressure of water is 0.5 psia (pounds per square inch absolute), which is well below atmospheric pressure (14.7 psia). The cooling water causes the steam to condense at that temperature, producing the vacuum. Any non-condensable gases would add to that pressure, reducing the performance of the steam turbine. The steam jet ejector removes those gases, keeping them out of the condensate as well. The combination of enhanced water treatment, the deaerator, and the ejector in the condenser serves to provide high-quality condensate to return to the boiler.

Preventing corrosion is simply a matter of maintaining the pH and removing oxygen. It is also a matter of keeping oxygen out. Phosphates have been mentioned for preventing hardness issues. They also provide a chemical buffer for the pH of the water. Phosphates help to maintain the pH between 8.5 and 9.5 by absorbing additional hydrogen ions or hydroxyl ions that may enter the water. This feature avoids both acid attack and caustic embrittlement.

PREVENTING SCALE FORMATION

Scale is the result of all the rocks that the water dissolved as it traveled from the rain cloud to the makeup water piping in the plant. Once the water leaves the boiler as steam, it leaves all those dissolved rocks behind. Frequently, the water has so much dissolved in it that it is not a matter of converting it to steam. Just heat it up to get scale formation. In one application, well water at 57°F formed scale in a heat exchanger that only raised the temperature by 6°F. Water with that kind of scale forming property is going to plug up service water heaters with scale, let alone a boiler.

Softeners, and other forms of pretreatment, can reduce the amount of scale forming ions in the water by, with the exception of demineralizers, swapping them with ions that normally do not form scale (sodium). That does not necessarily eliminate the potential for scale. Some of the scale forming ions always manage to sneak past all that pretreatment. Chemicals are added to the water to either convert the scale forming salts to sludge or "sequester" (the word means to surround and isolate) them to prevent them from forming a scale. Both methods work fine as long as some water remains to hold the sludge or sequestered ions in solution. If all the water is boiled away to steam, then the dissolved solids that remain will appear as scale no matter what. After all, when salt water is evaporated, there will be crystalline salt left. It will be called scale if it is on the boiler tubes.

There are several chemicals that will combine with the magnesium and calcium ions that tend to form scale and convert them to a sludge. The idea is that the sludge is not going to stick to the heating surfaces of the boiler. Instead, it will settle out in the mud drum (where it can be removed by bottom blow off) to eliminate the scale forming salts from the water. Sources of treatment that accomplished this ranged from potato peels (a source of tannin which is the actual chemical) to the many blends of phosphates that are in use today.

An advantage of the sludge forming treatments is that they combine with the salts to form a solid, thereby reducing the TDS of the boiler water. They do not contribute to the dissolved solids content. The disadvantages of sludge forming treatments include problems handling the sludge and problems in certain boilers where there is not enough room in the mud drum to reduce the water velocity to the point where the sludge can settle out. If the sludge does not settle out, it can be swept around by the water and, eventually, reach a concentration where, despite treatment, the sludge sticks to a heating surface and becomes scale. If the boiler contains scale, and tests of it indicate a high percentage of phosphate, it is an indication of that problem. Sludge handling problems include plugging of blow off piping and valves, usually resolved by more frequent bottom blows. Problems with sludge remaining in suspension in the water are attacked with other chemicals called "sludge conditioners" that are designed to reduce the tendency of the sludge to stick and increase the density of the sludge so that it will settle out.

The conventional system for treating boiler water is called "soda-phosphate." Caustic soda is added to raise pH and alkalinity. Phosphate is added to remove scale forming salts by combining with them to produce removable sludge. The performance of the phosphates is dependent on the maintenance of alkalinity. The amount of phosphate is reduced as the temperature and pressure of the boiler is increased. For low pressure boilers, the pH should be maintained between 10 and 11. For boilers operating at pressures over 1500 psia, the phosphate level should be reduced and the pH maintained between 8.5 and 9.5. The general corrosion mechanism for iron (steel) and water is limited by a corrosion layer of iron oxide in the form of magnetite (Fe_3O_4) on the metal surface. The rate limiting step in the reactions to form the magnetite is the reaction with water to form iron hydroxide. By maintaining a degree of alkalinity in the boiler water, this reaction is minimized. The minimum occurs at about pH 9. The range for generally safe operation is from pH 8.5 to 11.

To be certain that there is sufficient phosphate for any calcium or magnesium ions that manage to find their way into a boiler, maintain a residual of 60 ppm of phosphate. Sometimes, that is a little tricky to do. Phosphate concentration can vary with boiler load. This phenomenon is known as "phosphate hideout." Phosphate hideout is primarily a result of a temperature-dependent interaction of sodium phosphate compounds with iron oxide, which creates a low solubility reaction product that precipitates in high heat flux areas or under high load conditions. At lower loads and lower heat fluxes, the precipitates return to solution. Reducing the phosphate level in the boiler water greatly reduces the variability with load changes and may eliminate the hideout phenomenon in many boilers. Since phosphate hideout is a phenomenon associated with precipitation of phosphate/metal solids in areas of deposits on heat transfer surfaces, it follows that adequate oxygen control and passivation of the metal surfaces are important in the prevention of phosphate hideout. Oxygen control efforts must be maintained for many reasons, but in the case of hideout, it is important to control oxygen in the condensate system so that corrosion products are not deposited in the boiler. The deposits can then interact with phosphate during high heat flux conditions to form sodium iron phosphates.

A better, and more complicated, method of controlling scale emerged in the late 1960s. The treatment is generally called "chelant" and it comes in many patented forms. Phosphates are used to remove scale forming salts from the water. Chelants simply build a barrier around them that prevents their combining with other ions to form scale. Keeping the scale forming ions in suspension allows their removal in the continuous blowdown, where the energy and some water are recovered, thereby reducing the losses associated with bottom blow off. Chelants also attack scale that is already formed, returning it to solution, where it can be removed with the blowdown. Being used properly, a chelant treatment program can remove scale formed on a boiler as the result of an upset condition.

There are two hazards associated with the use of chelant treatment. First, if used to remove the existing scale, it has to be performed in a manner that does not result in fast removal of the scale. The chelant tends to break the bond of the scale to the iron first. Any rapid attack on the scale will result in large pieces of scale releasing into the boiler water and transporting to points of restriction, where it can plug tubes, resulting in overheating and failure. Even when that extreme is not reached, it can produce so much loose scale accumulating in the bottom of the boiler that there will be problems with blow off valves and piping plugging up. The second hazard has to do with the fact that iron is related to magnesium and calcium. Chelant insists on having something to sequester. If it runs out of calcium and magnesium, then it will grab iron, which is the stuff the boiler is made of. That requires careful and closely controlled use of chelant.

To ensure the scale forming ions are sequestered before they get near the boiler heating surface, chelant is normally introduced into the boiler feed water. The typical means is to introduce it using a "quill," which is best described as a thermometer well with a hole drilled in the side at the tip. By injecting the chelant into the water through the quill into the center of the feed piping, it will encounter the scale forming ions in the water before it reaches the iron in the pipe. The quill should always be installed upstream of a long straight run, where it can uniformly mix. Any elbows, valves, or pipe fittings downstream of the injection point should be inspected one year after beginning treatment and at five-year intervals thereafter to ensure they are not corroded away by the chelant.

To ensure the chelant does not attack the iron, it must be fed at the same rate that the scale forming salts enter the system. A chemical feed pump capable of varying the feed rate automatically is required to feed proportional to the feed water flow. Testing of the hardness of the feed water before the chelant feed, to make adjustments of the proportions of chemical feed to water flow, must be made regularly. Testing of the water for any residual should be frequent when boiler loads or feed water blends vary to ensure that a residual does not build up that would result in attacking the boiler metal.

The typical commercial or industrial boiler plant today uses a combination of phosphate and chelant, which is introduced into a boiler the same way as phosphate. The phosphate residual reacts with any ions entering the boiler and the chelant works on the scale that has formed in the boiler because the phosphate residual beats them to the ions. Because of the thermal decomposition of chelants at saturation temperatures associated with pressures higher than 1000 psia, these substances are generally not used in high pressure, high temperature boilers. Instead, demineralizers are used to minimize the level of TDS, while coordinated phosphate treatment is used to control scale and pH.

One relatively new phenomenon has been experienced as a result of the increased use of natural gas fired combined cycle plants. The gas turbine utilizes an air compressor to raise the pressure of the combustion air to the level required by the combustor and the turbine. As a result, the air temperature is also increased. The exhaust gas from the gas turbine goes to a HRSG to produce steam to run a steam turbine. In order to achieve a high level of overall efficiency, the gas temperature leaving the HRSG must be reduced to conventional levels. In a boiler plant, this is accomplished with an air heater. The combustion air is heat exchanged with the flue gas leaving. In the process, the air is heated and the flue gas is cooled. In the combined cycle plant, the air is already heated. A conventional air heater cannot be used. Instead, feed water is used to cool the flue gas, eliminating some of the feed water heaters. This means that the economizers in the HRSG are operating at lower temperatures. As it turns out, at water temperatures of about 300°F, flow assisted corrosion can result. Bends and restrictions cause turbulence in the water flow, which greatly assists the corrosion mechanism. For such units, a higher pH level has been helpful in reducing the potential for flow assisted corrosion. For the higher pressure units, pH level of 10 is recommended, rather than the conventional range of pH 8.5–9.5. These units also use AVT for oxygen control.

Chapter 9

Strength of Materials

An understanding of the strength of the materials used in the construction of the boiler plant is essential. No element of a plant is designed to operate anywhere close to its breaking point for reasons of safety. Maintenance of that margin of safety protects the life of the operator and others.

STRENGTH OF MATERIALS

If an operator takes some action, it is important to understand a little bit about the strength of materials in order to make sound operating and maintenance decisions. The goal is not to break things. While there is a lot of technical jargon, the idea is to provide an adequate understanding of what is involved and what some of the buzzwords mean.

STRESS

Stress in materials is very much like pressure. It is measured in pounds per square inch (psi). It is basically force (in pounds) applied over a surface area, measured in square inches. Stress can be determined as applied to a material and, by testing, know how much it takes. Since most of the materials involved in a boiler plant are metal, that material will be used to explain the application of stress and the strength of the material.

Start with tensile stress because it is the most common. A material is subjected to tension when it is pulled apart. A common material is exposed to tensile stress on a regular basis. That material is not steel. It is rubber. Call it a rubber band. It is a little hard to imagine stretching a metal, but it can be done. Clamp one end of a piece of lightweight wire in a vise, lead it out about 20 feet, and grip it with a pair of pliers. Set a stepladder, or something else, next to the wire to get a reference point. Then pull on the wire and ease off. The wire will act like a rubber band. It can stretch and will shorten when the pulling force is reduced. That is the elastic action where a stress is applied, and the material resists it. If the wire is pulled a little too much, the wire will suddenly give and will not shorten to its original length when released. The wire has been overstressed.

That operation is a simple form of the tensile tests that are performed on materials to determine their strength. In the case of the wire, it was pulled with a variable force that could be measured in pounds. That load was applied to the cross-sectional area of the wire, which is the area of a circle with a diameter equal to the diameter of the wire. Since any wire that was stretched would be very thin, the area would be very small. The more the cross-sectional area of the material, the more the force that will be needed to stretch it. It is easy to stretch a rubber band. A rubber hose is another story.

Tensile tests on a material use a sample that is a little larger than a piece of wire to get an average value. The typical tensile test specimen consists of a piece of metal about 6 inches long, with the center 3 inches machined uniformly to a thickness of about one-quarter of an inch and a width of three quarters of an inch to produce a cross-sectional area of three-sixteenths of an inch (1/4 times 3/4 equals 3/16). Thus, the cross-sectional area is 0.1875 square inches. The two ends of the specimen are clamped in a machine that pulls them apart. For standard metal testing, the sample is also marked with a center punch about 1 inch from the center on each side. That way, the machine can sense the location of the punch marks and measure very precisely the distance between them. The stress–strain diagram (Figure 9-1) shows a common graph produced by the machine as the material is tested. The stress, which is the applied force per square inch of material, is indicated on the left of the diagram. The strain, which is the amount the material is stretched, is indicated on the bottom. As the machine pulls on the material, the force or pull on the material is recorded. That value is converted to stress by dividing by the cross-sectional area of the specimen.

Modern machines allow the operator to enter the area on a keyboard to allow the machine to calculate the stress (pounds pull divided by the area in square inches) and to imprint it on the diagram. The machine measures the change in distance between the two center punch marks

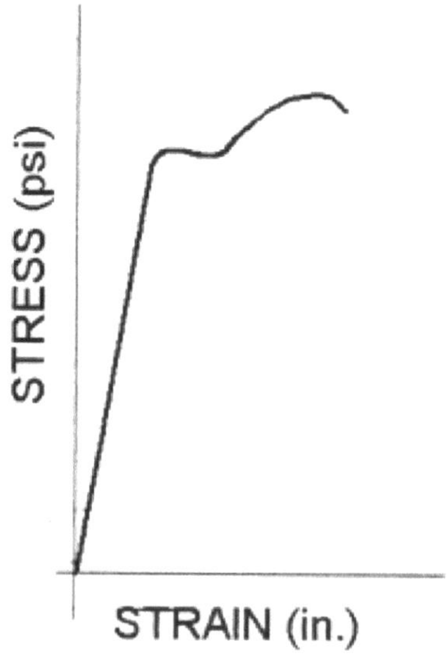

Figure 9-1. Stress–strain diagram.

to determine the strain. The stress–strain diagram shows what is normally called the proportional range, where, from zero stress, the stress and strain are proportional. If the machine were stopped while the metal was in the proportional range and the force removed, the metal would return to its original length. Metal in that range acts the same as the rubber band, always returning to its original shape. At the end of that straight line is the proportional limit, where the metal's properties change and it will not return to its original size when the force is removed. It is the same situation as when pulling on the wire.

Application of a little more force creates a stress where the metal simply stretches out without adding resistance (the slope of the line is horizontal). The point where that starts is called the yield point. When metal reaches its yield point, it deforms. That action is similar to "cold working" the metal, which hardens most steels, making them stronger. In general, cold worked metal is stronger than hot worked metal. The sudden cold working of the metal increases its strength and, despite the cross-sectional area being reduced a tiny bit, it can handle more stress.

The metal continues to resist force, but it stretches dramatically until the ultimate strength is reached, where the stress does not go any higher. That is where the coupon is deforming so much that its cross-sectional area is reducing. Even though the stress in the coupon increases, the force that it can withstand decreases because the area is decreasing. Shortly after the ultimate strength is reached, the material ruptures. If the coupon is not too deformed, the cross-sectional area can be measured at the rupture to determine the actual stress when it ruptured. That is how metal is tested. Most large boilers are hung from the structural steel that supports the boiler. Gravity provides a force pulling the metal parts downward. When the boiler heats up, the metal parts expand. The force of gravity pulls the parts downward. The length of the boiler increases. When the boiler cools down, it returns back to its original length, provided that it is not overstressed.

Cast iron, and similar materials, including concrete, that are not extremely strong in tension but very strong in compression are tested differently. The test method helps describe what compressive stress is all about. A metal sample is machined to prescribed dimensions over its entire length to form a test coupon. All those short round chunks of concrete that are lying around at any construction site are test coupons that were poured. The coupon is placed in a machine with a firm bottom plate. Pressure is applied to the top of the coupon (Figure 9-2). The force applied by the machine is divided by the cross-sectional area of the coupon to determine the stress. Some materials, like cast iron and concrete, withstand considerable stress until they fail. They fail quickly when their yield strength is reached. They produce a failure that is closer to shear than compression because it goes across the coupon at an angle. Since most metals would swell (increasing the cross-sectional area and strength of the coupon) when their yield point is reached, the test is not run past the yield point. The slope of the curve is usually the same for metals under tensile stress. The compressive

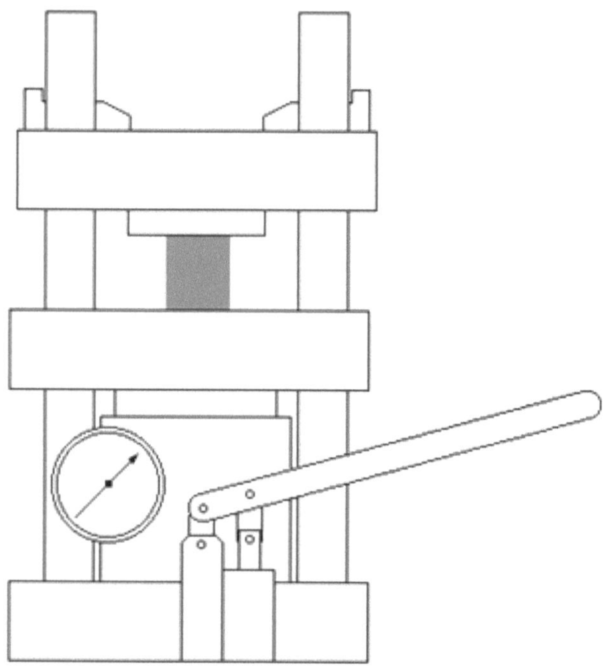

Figure 9-2. Compression stress coupon in machine.

stress–strain diagram matches the tensile stress–strain diagram in the proportional range.

Shear stress, as its name implies, is the resistance to being cut. It is considered primarily for fabrication activities, where the material is cut by shears. Unlike tensile and compressive stress, where the force is applied through the cross-sectional area in tensile stress, it is applied parallel to the cross-sectional area. It is seldom a consideration in boiler design. Shear stress comes into play in bending stress. Large boilers may have support lugs or hangers that may experience shear stress.

Bending stress is not really a special kind of stress. It is a function of compressive, tensile, and shear strength. The following example will demonstrate and can easily be reproduced. Take several pieces of 1 by 4 (that is lumber which is really about 3/4 of an inch thick by 3-1/2 inches wide) and stack them up on the floor between two bricks. Then stand on them. The result is something like that shown in Figure 9-3. The layers of lumber cannot support the weight of a person. Note, however, that the lumber ends are not flush like they were when originally laid out. Gluing all the layers of lumber together (or even securing them to each other with several nails or screws) prevents the equivalent of shearing stress from occurring in the material. They will now support the weight when stood upon. The force of a person's weight is countered by tension on the bottom layers of the material and compression on the top layers with shear stress applied to the individual layers.

Once the layers are glued (or fastened) together, they will still bend a little but will hold the weight. Just like the rubber band, the material length changes when force is applied to it. The bottom layers get longer and the top layers get shorter to compensate for the applied force of the weight. Since the layer at the middle neither shortens nor lengthens, it does not do anything to counter the applied force. The stress in the material increases from zero at the center to maximum at the extreme outer fibers or edges. That is why all the steel beams are typically made in the form of the alphabet I. By putting most of the material at the outer layer (where the maximum stress is), the strongest beam is produced.

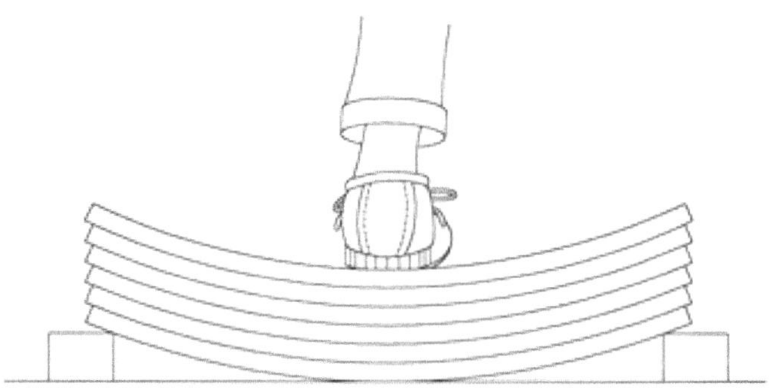

Figure 9-3. Layered board sample of bending stresses.

The foregoing covered the actual measured strength of the material. The next concept is the "allowable," or "design," stress. For everything that is boiler and pressure vessel related, those values are listed in the ASME Code in Section II, which is called "Material Specifications." Section II is broken down into three parts. Part A is for ferrous (engineer's and scientist's word for iron) metals. Part B is for nonferrous metals (like brass and copper). Part C is for welding materials (welding rod). Those sections define the quality of a material and how it must be made and tested. For the most part, the Section II contents are identical to the material specifications prepared by The American Society of Testing and Materials (ASTM) and differ primarily in the certification requirements. A boiler or pressure vessel manufacturer has to buy material that is certified by the manufacturer to conform to the specifications in Section II. Part D is called "Properties" and it lists the allowable stress for each of the metals described in the other three parts. Close inspection of Part D shows that the ASME has different values for allowable stress, depending on the use of the material and the maximum or minimum operating temperature. Allowable stresses vary for use as boilers (Boiler and Pressure Vessel Code (BPVC) Sections I and IV) and pressure vessels.

To relate to that yield strength determined by testing a coupon, take the minimum yield values for a material in the applicable part (A, B, or C) and compare them to the allowable stress in Part D. Typically, for ferrous metals, the allowable stress is one-fifth to one-fourth of the yield stress. That means the boiler is constructed of metal that should not fail (by deforming) until the pressure is four or five times higher than the maximum allowable pressure. It is a safety factor of 4 or 5. It is one thing that helps protect people from injury due to a material failure. The safety factor is there to account for small changes that might occur during the life of the material. The minimum design life for boiler materials is 30 years. Boilers can and do last longer. Provided that the water chemistry is handled appropriately and the boiler operation is not abused, a boiler will last a long time.

CYLINDERS UNDER INTERNAL PRESSURE

The basic calculations for determining the required thickness of a cylinder under internal pressure (like a boiler tube or drum or shell or piping) is best explained by looking at a cross section of the cylinder like that

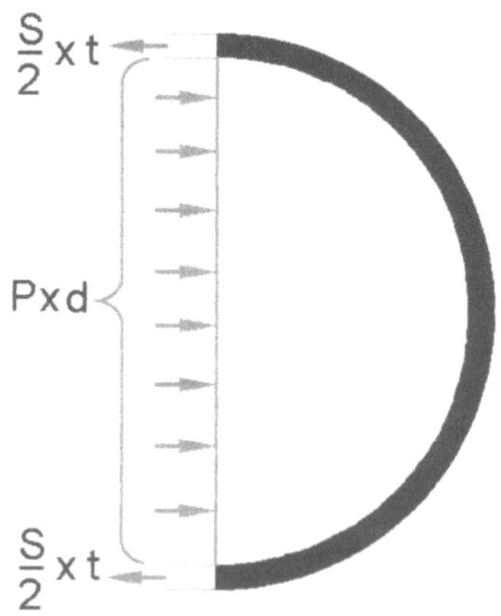

Figure 9-4. Cylinder analyzed for pressure stress.

in Figure 9-4. The figure shows half the cylinder, with arrows beside where the wall of the cylinder was sliced. In evaluating that view, the section over a unit length of the cylinder, normally 1 inch, is used. Imagine that the dark line is a piece of metal that is one inch deep into the page. Any inch along the length of a cylinder would be the same. The gray arrows, pointing toward the piece of metal, show the direction of the forces that are applied.

The pressure is inside the cylinder trying to get out and is pushing against the area that is equal to the inside diameter (I.D.) of the cylinder times the length. The area equals the diameter because the length is unity (1 inch). The pressure times the diameter equals the force produced by the internal pressure (P × D). The pressure, in psi, is applied against an area measured in square inches. The overall force can be measured in pounds. That force has to be balanced. The balance is the force produced by the metal cylinder. If the force were not balanced, the cylinder would rupture. The area of the metal in the cylinder is equal to twice the metal thickness (two walls). The stress can be determined in a known thickness of metal. Alternatively, the minimum thickness of the metal for a given stress can be calculated because the forces have to be equal. The force from pressure equals the pressure (P) times the diameter (D). It must be equaled by the force on the two thicknesses of metal (2T) and the stress in the metal (S). The mathematical formula for a cylinder under pressure is P × D = 2T × S. Substitute known values for any three of the letters and the fourth can be calculated.

- To determine the stress on the metal, multiply pressure times the diameter and divide that result by twice the metal thickness. S = (P × D)/(2 × T).
- To determine the minimum thickness of the metal, multiply pressure times the diameter, divide that result by the allowable stress, and divide that result by 2. T = ((P × D)/S)/2.
- To determine the maximum diameter for a cylinder of a given thickness at a selected operating pressure, multiply the thickness and the allowable stress. That result is multiplied by 2 and then divided by the pressure. D = (T × S × 2)/P.
- To determine the maximum pressure for a cylinder of a given thickness, diameter, and material, multiply the thickness and the allowable stress. That result is multiplied by 2. Then divide by the diameter. P = (T × S × 2)/D.

The ASME Code is not quite very simple. That is because the overall length of the material around the cylinder gets larger as the thickness increases. The code formula is

$$T = (P \times D)/(2 \times S \times E+2 \times Y \times P) + C$$

to determine the thickness and

$$P = (2 \times S \times E) \times (T - C)/(D - 2 \times Y) \times (T - C)$$

to determine the maximum allowable pressure for a given thickness

There are values in addition to those in the more simple explanation above, represented by C for corrosion allowance, E for a factor that depends on the method of welding (sometimes called weld efficiency), and Y, which is a coefficient that depends on the maximum operating temperature and the type of steel. These formulas are for power boilers. The ones for heating boilers and pressure vessels are a little different. For most purposes, the simple formulas should be fine. Just be aware that there is a little difference between them and the actual code formulas. The operator is not expected to design the boiler. However, having an understanding of how the design is determined will be helpful to the wise operator.

Calculating the stresses and the required thickness of a pipe or boiler shell is rather simple. Complications enter the equations when there are openings in the pipe or shell, for example, all the holes in a water tube boiler drum. In those cases, allowance for the holes is based on the required thickness of a cylinder without holes and how much metal has to be added where the cylinder is complete to make up for the holes in other locations. A steam drum, where the tube holes are 2 inches in diameter and installed on 4 inch centers, has to be about twice as thick as one without the holes. Normally, the code does not require any special consideration for an occasional opening for a connection smaller than 2 inch nominal pipe size. Larger openings may have enough extra material in the cylinder (because the standard steel plate thickness, greater than what was required by the code formulas, provided it). It may be included in the structure of the opening (like manhole rings) or a doubler (additional steel plate surrounding the opening) that is added to provide the required material. To know more about boiler design and construction requirements, there are courses provided by the National Board.

Cylinders under internal pressure are relatively easy to understand. The calculations are rather straightforward. There are other situations that are complex. Cylinders under external pressure are one of the situations. All the tubes of a fire tube boiler and its furnace are cylinders with the pressure on the outside of the tube. There is a difference in the amount of pressure a cylinder can withstand depending on whether the pressure is on the inside or the outside.

CYLINDERS UNDER EXTERNAL PRESSURE

The stresses that are applied to anything exposed to external pressure produce both compressive and bending stresses. Usually, the bending stresses produce the failure. Cylinders, and other parts exposed to external pressure, and flat parts of vessels exposed to internal pressure are thicker than they would have to be for the same pressure applied internally. Or they are made with stiffening rings, bars, etc., to help them resist the bending forces. The corrugated steel furnace of a fire tube boiler (called a Morrison tube after the man who determined it would be stronger) handles external pressure better than a simple cylindrical furnace of the same thickness and diameter because the corrugations stiffen the cylinder.

Calculating stresses becomes a lot more complex when making a valve, flange, or other pressure retaining structure. Many standard arrangements have been developed. In most cases, they have been tested to failure to determine their strength. That was a much easier proposition in the days before computers and all their capabilities. There are standards based on a maximum operating pressure. The ones normally encountered are 125, 150, 250, 300, 400, 600, 900, 1500, 2000, and 3000. An important thing to know about those standards is that they have secondary ratings. Take a valve with "500 WOG" (water, oil, or gas) cast onto it. The other side has "250," meaning it is a 250 psig (pounds per square inch gauge) valve. The 500 WOG means the valve is also rated for operation at a maximum allowable pressure of 500 psig if it is used for WOG at normal atmospheric temperatures. The 250 psig steam is at 400°F and the valve's strength is less at that temperature. The secondary ratings are an American National Standard.

The secondary ratings allow for differences in the maximum temperature of the system and permit using less expensive, but perfectly fine, materials for some processes. There will typically be 600 psi rated steel valves and flanges on feed water piping for boilers with a maximum operating pressure of 600 psig, even when the feed water pressure is as high as 800 psig. The secondary rating of 600 psi standard valves and flanges is 900 psig with 250°F feed water. An abbreviated copy of secondary ratings is in the Appendix.

PIPING FLEXIBILITY

Tension, compression, and bending stresses are all involved in determining the flexibility of boiler plant piping. The words "piping flexibility" mean the stresses in the piping and the stress and forces applied to boilers, pumps, turbines, building structures, and other things the piping is connected to, where those forces and stresses are produced by the thermal expansion or contraction of the piping.

In one example, there was a problem in the warehouse section of a plant where a wall had been damaged. The wall was at the south end of a large warehouse. It was made out of concrete block and it had a very large hole in it, right around a piece of insulated pipe. In the shipping area, opposite the wall, was a pile of broken concrete block. There were the remnants of a thin steel plate, that was welded around the pipe to seal the opening in the wall (required for a fire wall construction), still hanging on the pipe. According to the drawings, a similar plate was in the north wall of the warehouse. In between those two plates was 84 feet of 4 inch steam piping, a straight 84 feet of pipe. The operating pressure was 150 psig. The pipe was installed when the outdoor temperature was about 70°F. Using the tables and procedures in the Appendix, the length of expansion can be determined. The pipe would lengthen by about 2 inches when heated. Since the pipe had no place to go but straight south, it broke the wall. Later a "U" bend was installed in the pipe inside the warehouse and the wall repaired. Unlike the stiff, straight piece of pipe, the pipe with the U bend was sufficiently flexible that the pipe bent slightly. That way, the seal plates and walls were able to withstand the forces applied to them.

Keep in mind that pipes get stiffer as they get larger. Note the sag in different sizes of pipe when they are picked up in the middle. Bigger is stiffer. Smaller is more flexible. Valves and other devices in the piping make it stiffer. When stiff piping is heated, it tends to grow in length and diameter. Getting a little larger in diameter is not much of a problem to handle but the added length is.

Sometimes the pipe can do the same thing that railroad tracks do. They can spring sideways just a little to convert the straight line of pipe to a shallow S. That does not cover much of a change in length. Fortunately, railroad tracks do not get that hot. Other examples include roads. During one hot summer, a lane of the Baltimore Beltway got so hot that the pavement buckled up at a joint, producing the equivalent of a 2 foot high speed bump. The compression stress gets so high that any little change in cross section (the roadway joint) permits translation of some of that compression stress to bending stress, and, in that event, the roadway bends. Respect the forces associated with thermal expansion.

Usually, there are not many problems with piping flexibility in a boiler plant because the designers are aware of it. That does not always mean the designer did it right. There are times when the installing contractor changes the piping arrangement and it produces excessive stresses. If there are problems maintaining joints in the piping or the piping supports, there may be some problems with overstress. Welded steel elbows are buckled because an adjacent packed type expansion joint (Figure 9-5) froze up. This form of expansion joint allows the pipe to expand into the space between the flanged connection and a bare end. They have to have anchors somewhere else to take the axial pressure forces of the pipe or the joint will come apart. If the plant has some of these joints, be very certain that the anchors are not corroded.

The forces produced by a packed expansion joint, or bellows joint, on the anchors is readily determined by multiplying the maximum allowable working pressure (MAWP) of the fluid in the system by the cross-sectional area of the pipe (including the metal area) for packed joints, like the one in Figure 9-5, or the bellows area of bellows joints. The forces produced on these joints are substantial.

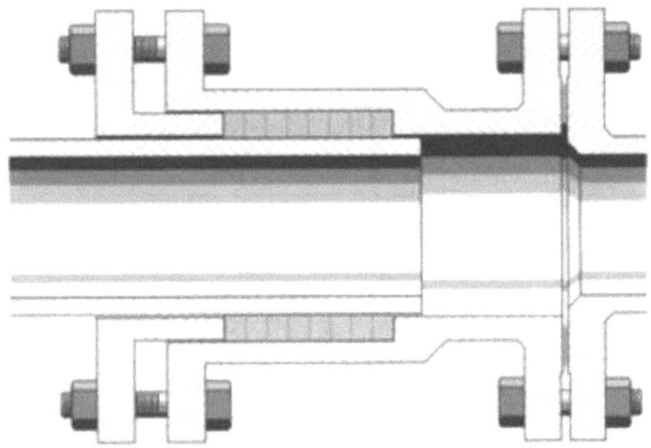

Figure 9-5. Packed type expansion joint.

Valves can leak because the piping stresses applied to them were too high. It happens frequently, where large stiff piping is reduced at a control valve, making the valve, and its flanged joint, the weakest point in the piping, and the one that bends or breaks. The largest problem with stiff piping is its effect on pumps, blowers, and turbines. Where piping is attached to boilers and pressure vessels, the vessel wall is normally more flexible, while remaining strong, and can take the stress. The shell simply changes its shape a little. Pumps, blowers, and turbines, on the other hand, go out of alignment when they are overstressed. A piping system that has very little stress in the piping can still overstress a pump or turbine to the degree that the impellers hit the casing and shaft seals are rapidly worn and fail.

The first thing to look at with a pump problem is the connecting piping. On one job, the stress got so high that it broke the concrete pad, under the pump, away from the floor and moved it almost by 2 inches. The maximum allowable forces on pump connections are described in API-610. Inspection of those allowable forces will indicate that some are so low that enough pipe to get from where it is attached to the pump to overhead will, along with the weight of fluid it holds, weigh enough to exceed the standard's limits. Sometimes, a plant modified, or had a contractor modify, the piping without careful analysis of the flexibility. They suffered the consequences. The wise operator realizes that the piping has to remain flexible and will not attach stuff to it or impair its movement to reduce its flexibility.

Chapter 10

Plants and Equipment

It would take volumes of books to adequately describe all of the variations of design and construction in boilers since Hero first produced steam under (a little) pressure in 62 AD. That is only boilers. It has nothing to do with all the other plant equipment and systems. This section has been limited to what might normally be encountered. If an older design is encountered, it should be described in one of the references listed in the bibliography.

TYPES OF BOILER PLANTS

In order to get a definition right, go to the source. In the case of boilers, the source is ASME, the American Society of Mechanical Engineers, which produced and maintains its Boiler and Pressure Vessel Code (BPVC). The code is the accepted rule for construction of boilers and pressure vessels in the United States, Canada, and much of the world. According to the code, a boiler is a vessel in which a liquid is heated, or a vapor is generated, under pressure by the application of heat from the products or combustion or another source. Vessel is the code word for an enclosed container under pressure. The BPVC, in its various documents, defines high pressure and low pressure but never addresses the term "medium pressure."

High pressure boilers are defined by ASME in the first document prepared to address the construction of boilers and pressure vessels, which is now known as Section I of the BPVC. It is simply titled "Rules for Construction of Power Boilers." That is a Roman numeral one, not a capital letter "I." All sections of the code are numbered using Roman numerals. Within Section I, a high pressure boiler is defined as a steam boiler that operates at a pressure higher than 15 psig (pounds per square inch gauge) or a hot water boiler that operates at a water temperature greater than 250°F or a pressure greater than 160 psig.

Low pressure boilers are defined by ASME in Section IV of the BPVC "Rules for Construction of Heating Boilers." It defines a low pressure boiler as a steam boiler that operates at a pressure no greater than 15 psig or a hot water boiler that operates at temperatures not greater than 250°F and pressures not exceeding 160 psig.

Under these definitions, there simply is no definition of a medium pressure boiler. If the boiler makes steam, it is low pressure until 15 psig and high pressure at any pressure higher than 15 psig. Hot water boilers are not quite as clearly defined. The temperature is normally the clue. Almost any hot water boiler operating at temperatures less than 250°F is a low pressure boiler.

The titles of the code sections are one key. A high pressure boiler is also called a "power" boiler. Low pressure boilers are called "heating" boilers. The definitions apply to those titles as well. A boiler plant that is only used for heating but operates at steam pressures above 15 psig or heats water to a temperature greater than 250°F is a high pressure plant with power boilers. A low pressure boiler could be used to power a steam engine to generate electricity. However, it is still called a low pressure boiler or heating boiler. The use has no bearing on the definition of the boiler.

There are similar difficulties with the US Environmental Protection Agency (EPA) definitions. For example, take an electric generating unit (EGU). Fossil fuel fired electric utility steam generating units are boilers that are capable of combusting over 250 MMBtu/hr (million Btu's per hour) of fossil fuel and that were constructed for the purpose of supplying more than one-third of their potential electric output capacity and more than 25 MW electrical output to any utility power distribution system for sale. Thus, a 5 MW boiler and turbine system that generates electric power at a university is not an EGU according to the EPA. Nor is a 100 MW boiler and turbine system that generates electricity for an industrial plant that does not sell any power. Those plants fall under EPA regulations for industrial boilers. Similarly, a 250 MW plant that burns wood is not an EGU. Wood, or biomass, fuels are not considered to be fossil fuels. Therefore, such a unit does not fire a fossil fuel and is not an EGU. Another issue concerns the combustion of a waste. If a unit burns any waste, it is considered to be an incinerator and falls under the EPA regulations for incinerators. The EPA definition of a waste is anything that is discarded or thrown away. Thus, a tire that is put out with the trash or sent to a dump is a waste. That same tire can be a fuel if it was collected by an approved recycle program and sent directly to a plant to burn it. The wise operator must know the requirements for the operating permit for the plant.

A superheated steam boiler is any boiler that raises the temperature of the steam above saturation temperature. It is possible that low pressure steam could be superheated, but, virtually, all superheated steam boilers are power boilers. On rare occasions, a plant will have a separately fired superheater, which is also a power boiler by definition in Section I of the code. A super critical steam plant has steam conditions that are above the critical point of steam and water. At that point, there is no longer any density difference between steam and water. These steam conditions are 705°F and 3200 psia (pounds per square inch absolute). Plants that have lower steam conditions are called subcritical boilers.

One other definition that is not clearly defined in the code, but is commonly used, is "high temperature hot water" (HTHW). These plants are typically referred to by the initials rather than the words. An HTHW boiler is simply a power, or high pressure boiler, that heats water rather than generating steam.

With the adoption of the label of HTHW, any low pressure, hot water, heating boiler plant is simply called a "hot water" plant, with the understanding that it complies with the code definition of a low pressure hot water heating boiler. With water heating plants labeled as such, it is understood that a low pressure or high pressure label means a steam generating plant. Don't ever be afraid to ask what somebody means. Requirements for licensing of operators frequently depend on whether a boiler is a power boiler or heating boiler. It is important to get it right.

BOILERS

Boilers do not have to have a burner. All of these types can generate hot water or steam by absorbing heat from another fluid. That other fluid can be steam and create steam or hot water. It can be HTHW and generate steam, or it can be a hot liquid or gas from some chemical process that is hot enough to do the job. One of the physically largest low pressure steam boilers was built in the late 1960s. It generated steam by oxidizing a liquid. The heat source was a large volume of oil in which air was forced through to oxidize the liquid, similar to combustion, but at a low temperature and nowhere near complete combustion. Twenty-four feet in diameter and 90 feet tall, with thousands of square feet of heating surface, it made about 25,000 pph (pounds per hour). Other projects included a hot water boiler using 500°F air from a steelmaking operation, rated at 100 MMBtu/hr. Operating that type of equipment to get the most steam out of it is wise because of the fuel saved that would have to be used to generate that steam. These boilers can be constructed as unfired pressure vessels in accordance with Section VIII of the ASME Code, "Rules for Construction of Pressure Vessels."

Boilers that are fired must be built to Section I or Section IV. Their construction is limited to materials that can handle the high rates of heat transfer required for direct fired equipment. Boilers using waste heat can require materials of construction that cannot handle direct firing but are essential to prevent corrosion in the waste heat application. In simpler words, a fired boiler cannot be built in stainless steel. An unfired boiler can be.

Since there is a fixed relationship between pressure and temperature for steam and water, pressure has to increase. When heat is needed for products, or other materials, at high temperatures, the pressures can get very high. To obtain temperatures greater than about 500°F, which would require steam or water pressure over 666 psig, another fluid might be used. There are several liquids, mostly hydrocarbons, that can be heated to temperatures as high as 1000°F without operating at such high pressures. The liquids are identified by the trade name given by their manufacturer and include Dowtherm™ and Paracymene™ as the more common names. They are supplied in different materials according to the temperatures required. The common label for boilers that heat these liquids is "hot oil" or hot oil boilers. The Appendix contains tables, similar to the steam tables, for the more common of those hot oils.

Some of those liquids can be vaporized, just like converting water to steam. A common name for them could be oil vaporizers. It is far more common for the label to use the trade name of the fluid and add the word vaporizer. Since all these plants operate at temperatures higher than 250°F, they require power boilers built in accordance with Section I. A plant could be operating one of these boilers, in addition to the steam plant, because steam is usually required to quench the fire in the event that the hot oil leaks into the furnace to feed the fire.

Equipment that heats water in an open container, or very small one, is not a boiler. A teapot does not have to be constructed in accordance with the code. It is too small. The hot water heater in a home is not considered a boiler, unless it holds more than 120 gallons. Another limit on the size of a boiler is an internal diameter of 6 inches or less. The exceptions found in the code are occasionally stretched to create boilers that, by definition, are not. Fired air heaters are not boilers, unless the air is under pressure. Any application that heats air or any other gas for that matter that does not contain the heated fluid in an enclosed vessel is normally called a furnace. If the fluid is air, or another gas, and it is under pressure, then it does meet the definition of a boiler.

There are many boilers unique to their respective industry. There are asphalt heaters, flux heaters (a raw material that becomes asphalt), many forms of waste heat boilers, and equipment like recovery boilers (used in the paper industry), which convert product by burning it. This book is limited to the more common types of boilers to provide a basic understanding of them. The principles discussed here will apply to those unique boilers. However, by virtue of their uniqueness, they are best understood by reading the operating and maintenance instruction manuals for them. This section contains general descriptions of the basic elements of a boiler plant to provide a basic understanding of the systems and equipment. Hopefully, an operator can append this information with the contents of the instruction manuals to develop a full working knowledge of any boiler plant.

HEAT TRANSFER IN BOILERS

An understanding of heat transfer is a fundamental requirement for a boiler operator. A lack of understanding of heat transfer can result in an operator's death. It is that simple. The energy transferred in a little 100 horsepower boiler is about eight times the amount it takes to power an automobile at 60 miles per hour. Screw that up to get that energy going in the wrong direction and there can potentially be an accident that can only be compared to eight or more cars crashing, all at the same instant.

There are three ways that heat is transferred: conduction, radiation, and convection. All three means occur in a boiler. Conductive heat transfer is the flow of heat through a substance, molecule by molecule. A molecule is the smallest piece of a substance that can be obtained without destroying its identity. The heat is absorbed by one molecule, which passes it onto the next one, and so on. A good example of conductive heat transfer is toasting marshmallows. Think of the metal fork without the wooden handle. As the marshmallow was toasting, the metal got hot. The metal over the fire was heated and that heat was conducted up the metal of the fork to the handle.

The sun is a good example of radiant heat transfer. It is almost 93 million miles away with mostly space (nothing) between it and the earth, but the heat is getting here. Radiant heat transfer is the flow of heat energy by light waves that can penetrate empty space and the atmosphere but is absorbed by solid and liquid in its path.

The last means, convective heat transfer, uses a transport mechanism to get the heat from one spot to another. In a house, the furnace or boiler heats air or water, which is then moved (blown or pumped) to the rooms in the house and heats the air in the rooms. There are two types of convection heating: natural and forced. Forced convection is the result of a fan, pump, or blower forcing the movement of the fluid over a heated surface, where it picks up the heat and then moves on to another surface where it gives up that heat. In a house with a radiator next to the wall, that radiator is heating the air around it. That air gets lighter (less dense) as it expands from the heating. It rises up in the room like a lighter-than-air balloon. When it reaches the ceiling, it starts to cool because it is giving up heat to the ceiling. It is pushed aside by hot air following it. When the air reaches a cooler outside wall, it gives up more heat, shrinks to become denser, and drops to the floor. It then travels back to the radiator. That is natural convection heat transfer. All these methods of heat transfer occur in a boiler.

The modern boiler, with its water cooled walls, absorbs about 60% of the heat from the burning of the fuel using radiant energy. That heat travels in the form of light waves from the glowing hot fire directly to the boiler tubes, in water tube boilers, or the furnace tube in fire tube boilers. The reason that so much heat is transferred is due to the low resistance to the radiant heat flow from the fire to the tubes. Though not quite as hot as the sun, a fire is an awful lot closer. Thus, there is a lot of heat flowing there. Open up an observation port to look in and the heat can be felt immediately. Once it hits the fire side of the tube, the heat is transferred by conduction to the water side of the tube and by convection to form hot water and steam.

Conductive heat transfer to the boiler water and steam is limited to the heat flow through the boiler metal itself. The steel parts of a boiler are selected for their ability to transfer heat with as little temperature difference as possible. The outside of a water cooled tube is no more than 60 or 70 degrees hotter than the inside. That is because the heat is passed through the tube easily and because the heat is drawn off the tube by the water and steam rapidly.

Other parts of a boiler count on poor conductive heat transfer to protect them from the heat of the fire. Refractory material not only can withstand high temperatures, it is a poor conductor of heat. When it is backed up with some insulation, the outer surface of the boiler's metal casing is less than 140°F, which is the maximum temperature that should be allowed (anything hotter will give someone a serious burn in a matter of seconds).

Now is a good time to point out that heat flows from points of higher temperature to points of lower temperature. If there is no temperature difference, there will be no heat flow. The converse is almost true. If there is no heat flow, there cannot be any temperature difference. This factor is a thermodynamic property. Consider a glass of ice water sitting on a table that is at room temperature. Over time, the ice will melt and the water will increase in temperature. The room will be slightly cooled. The room does not get hotter and the ice water will not freeze up solid. The heat flows from the room, at a higher temperature, to the glass of ice water, at a lower temperature, until the two temperatures become equal to each other. Then heat no longer flows. If a layer of insulation with a super high resistance to heat flow was added on the outside of the boiler, the refractory, insulation, and casing would get almost as hot as the inside of the furnace. That is why insulation is never added to a boiler casing that is not water cooled. It will overheat.

If the boiler tubes are coated with fireside deposits, they will get hotter and reflect heat back to the fire to reduce heat transfer to the water and steam. If the boiler tubes are coated with scale on the water side, then the tube wall will get very hot because the scale acts like insulation to block the flow of heat from the tube metal to the water. Other mechanisms are involved when the scale on the water side accumulates. It provides an early indication of potential failure. If the metal gets too hot, it will lose its strength and begin to bulge under the force of the boiler pressure. Usually found on the top of fire tubes and in the bottom of water tubes where exposed to the furnace, bulges are evidence of excessive water side scale formation.

When the tube metal bulges, the hard scale is released, breaking away from the metal that is stretched to form the bulge. Once the scale is broken away, the metal is exposed to water again, cooling it to stop the growth of the bulge. Repeated incidents of bulge formation can occur, with some of the metal stretched until it is very thin. Its chemical composition changes and the surface becomes rough oxidized metal, something called a blister. Sometimes, the bulges or blisters can be left in place if the processes that promoted scale formation are eliminated. However, blisters should eventually be replaced because the metal is thinner than permitted by code. Slight bulges, where the tube metal is not distended or deformed beyond its own thickness, can be left in place. See Code repairs for replacing bulges and blisters. Changes in heat conductivity of materials in the path of conductive heat transfer can create conditions that are inconsistent with the original boiler design and result in failure. Hopefully, the boiler will be operated and maintained in a manner that does not interfere with the design heat flow.

As for the radiant energy that hits the refractory wall, it is reflected right back to the flame or is reflected off toward some of the heat transfer surface. Actually, it could be argued that very little heat is transferred because the face of the wall and the hot gases are very close in temperature. The truth is that it is radiated back almost as fast as it is received once the system is at steady state. Everything radiates energy. Typically, boiler loads are a little higher on clear nights because of the black sky effect. Heat radiates from the earth and everything else right out into space on a clear night. Thus, it takes more heat to keep the buildings warm. On a cloudy night, the clouds act like a mirror reflecting the radiant heat back toward the surface of the earth, and the temperatures stay warmer. An important factor in radiant heat transfer is the emissivity of a substance. It has more to do with the color and finish of a surface than the actual material of construction. White and mirrored objects have a higher emissivity than black and rough surfaces. They tend to emit more radiant energy than the black and rough surface even though they are at the same temperature. Keeping those white rubber roofs clean in the summer, and letting them get dirty in the winter, will actually help maintain desirable building temperatures.

As the flue gases leave the furnace, they carry the remaining heat into what is called the convection section of the boiler. That is where the convective heat transfer takes place. In water tube boilers, it is also called the convection bank (a bank being a group of boiler tubes that serve a common purpose). Heat transfer in the convection section is driven by much lower temperature differences (typically, the flue gas leaves the furnace at less than 2000°F). For smaller boilers, 1400–1600°F is a normal range, which is much less than the 3400°F plus flame temperature. For larger boilers, the gas temperature leaving the furnace section may be closer to 2200°F. The reason is that the volume of gas compared to the surface area of the walls (surface to volume ratio) is greater for the larger boilers. With less surface available, the percentage of heat transferred to the walls is less. The temperature difference in the convective section drops from, perhaps, 1000°F to a typical leaving differential of 75–150°F due the fact that superheat and reheat tubing usually constitutes the first sections of tubes and the economizer tubing comes last, with the lowest metal temperatures. With these smaller temperature differences, a lot more heat transfer surface is needed in the convection section of a boiler to get the most out of the 40% of the heat energy that was not transferred by radiant energy in the furnace. As the gas was flowing in the furnace zone, there was a very small percentage of the heat transferred by convection. The dominant mode was radiation. There is even less conductive heat transfer from the gas to the wall, even though some gas molecules do come in contact with the wall.

Flame impingement can create a condition when there is conductive heat transfer from the flame to the boiler tubes. It is also called flame gouging because the tube metal is melted and swept away when flame impingement really happens. A cutting torch uses flame impingement. If there is flame impingement, the damage can be seen during an internal inspection. The truth is that flame impingement problems are relatively rare in a boiler, despite many people arguing that they have it. It generally does not happen because the flame is cooled so much by radiant heat transfer that it is normally quenched (below ignition temperature) before it gets to the tube. The flame may be bouncing off the tubes or the furnace wall that appears to be flame impingement. Even then, an examination of the boiler will find no damage at all on the tubes. Bulges and blisters (mentioned earlier) are not due to flame impingement. They are due to scale formation. If the flame seems to be rolling along the tubes or passing along them so close that they must be touching, it is called "brushing" the tubes and it does not do any significant damage.

The same thing that helps prevent true damage from flame impingement also makes it difficult to transfer heat by convection. The molecules of air and flue gas that are in contact with the tubes stick to the tube and each other to form what is called a "film." It is a very thin layer of gas that acts like insulation, separating the hot flue gases from the tubes. In the course of heat flow from the flue gases to the water and steam, it contributes the most resistance to heat flow. That film is mainly what protects the tubes in a furnace from the hot flue gases in the fire. Otherwise, the metal temperature would be so high that it would melt. The typical boiler steel will melt around 2800°F. It begins to weaken at temperatures above 650°F. (It actually gets a little stronger as it is heated up to 650°F.)

A film forms on most gas to metal or liquid to metal surfaces to resist heat transfer. Water really sticks to other surfaces. Its adhesion is greater than its cohesion, as evidenced by the meniscus (See Water Analysis, Chapter 8). In fluid dynamics, this adhesion is called the "no slip" condition at the wall. Essentially, the velocity at the wall is zero. The film is referred to as the boundary layer. In order to minimize the thickness of the film, the overall velocity must be increased. To improve convective heat transfer, the fluid flowing past the heat transfer surface is made turbulent (all mixed up and swirling around) to sweep against that film and transfer the heat from the fluid through the film to the metal. As velocities in a boiler drop, a point is reached where the flue gases can no longer disturb the film. It gets thicker and the heat transfer drops off dramatically. When flow is so low that the flue gases simply meander along, like congested traffic where the vehicles in the middle cannot get to the sides of the road, a lot of the gas leaves without contacting the tubes. It cannot give up its heat and becomes hotter, carrying that valuable energy out of the boiler and up the stack.

Something unique happens to that film on the water side when steam is generated. Heat transfer from metal to boiling water is a lot greater than heat transfer to water or steam. When heat is transferred from the tube to the water to make steam, a bubble of steam forms and it grows to several times the volume of the water it came from (in the typical heating boiler operating at 10 psig, the steam expands to 981 times the volume of the water). As a result, there is a dramatic movement of the steam and water interface. The steam bubble then breaks away from the metal (steam is nowhere near as cohesive as water) and water rushes in to fill the void. All that activity makes steam generation much easier than simply heating water or superheating steam. It requires less heat transfer surface to get the heat through. Similarly, when getting heat from steam, the steam forms condensate at almost one thousandth of the volume. More steam rushes in to fill that void while the condensate drizzles down the heat transfer surface, effectively scrubbing it clean. The range of heat transmittance (U) for steam condensers is 50–200 Btu/hr/ft^2/°F (British thermal units per hour per square foot per degree Fahrenheit) compared to water to water heaters at 25–60 Btu/hr/ft^2/°F, and superheaters that have values of 2.6–6 Btu/hr/ft^2/°F. Radiation heat transfer to the boiler water walls is 60–200 Btu/hr/ft^2/°F. Also see the comparison of equivalent direct radiation (E.D.R.) in Chapter 1. No wonder steam is an excellent heat transfer medium.

CIRCULATION

In addition to heat transfer, a boiler operator has to have a sound understanding of the circulation of steam and water in a boiler to operate it without damaging it.

If circulation is interrupted for more than a few seconds, all the water will boil away in areas of high heat transfer. With only steam inside the tube, metal temperatures will shoot up and the boiler will fail.

Look at a pot of boiler water on a stove. Note how rapidly the steam bubbles and water move in that pot. At a nominal one atmosphere, where water boils at 212°F, the volume of steam is 1603 times greater than the volume of the same weight of water. The weight of the steam is about six ten thousandths of the weight of an equal volume of water. Try to push a balloon full of air down into a bucket of water to get an idea of the force created by the difference in density. The steam forming in that pot of boiling water would blow all the water out of the pot, if it were not for the fact that it rises to the surface of the water and breaks out rapidly. The steam bubbles have to move fast to get out of the water without displacing it completely. If the pot gets boiling too fast, the level will rise and the water will spill over the top. That is despite the fact that some of it is converting to steam. Thus, there will always be less water in the pot than when at the start. The water in the pot is circulating. Water and steam bubbles rise up. The steam separates and goes into the air. The water that came up with the steam returns to the bottom of the pot, usually in the middle but not always and not consistently. Being much heavier than the steam, the water manages to find its way down. It will tend to go where the velocity of rising steam bubbles and water is the lowest. As a bubble of steam rises, it pushes water out of the way. The water slides around the bubble and fills in behind it.

The water in a boiler has to move around, or circulate, just like it does in the pot on the stove in order to let the steam out of the boiler. Enough water has to flow with the steam to carry the solids dissolved in the remaining water and keep them dissolved. Otherwise, they will drop out on the heat exchange surfaces to form scale. Luckily, water is highly cohesive (it sticks to itself) and tries to hold itself together around those steam bubbles. There are several pounds of water circulating up to the water surface along with each pound of steam that is formed. The steam forms round bubbles. A sphere (bubble) occupies 52.36% of a cube that would have sides equal to the diameter of the sphere. Even if every steam bubble was touching another one, only slightly more than half of the volume of the rising steam and water mixture would be steam. In the pot on the stove, the steam occupies 26.8 cubic feet per pound and water occupies 0.01672 cubic feet per pound (see steam tables in the Appendix). If the volume of the pot was one cubic foot, the weights of steam and water could be calculated

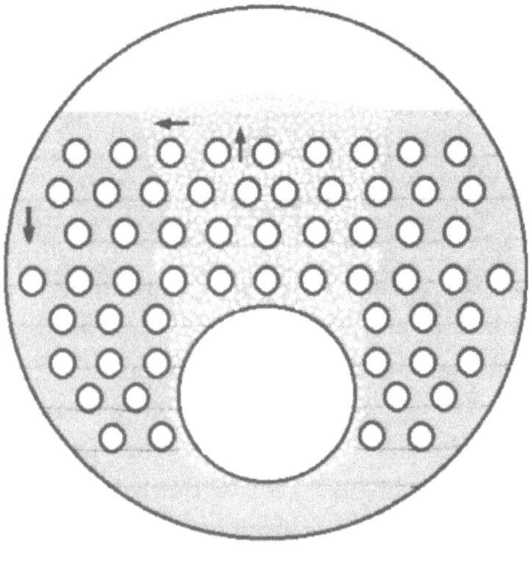

Figure 10-1. Steam flow pattern in fire tube boiler.

if all the bubbles were touching each other. The steam would weigh 0.01954 pounds (0.5236 $ft^3 \div 26.8\ ft^3/lb$) and the water would weigh 28.498 pounds (1 – 0.5235} $ft^3 \div 0.01672\ ft^3/lb$). The weight ratio of water to steam would be 1458 pounds of water per pound of steam (28.498 ÷ 0.01954). As the pressure increases, this ratio decreases until the critical point (3200 psia). Then, the density of steam and water are equal. There are no longer any bubbles. There is just fluid (just like air). This ratio is called the circulation ratio. It is typically quite high for low pressure boilers. Likewise, for high pressure boilers, the circulation ratio will decrease. It is desirable to maintain a circulation ratio of at least 2 in high pressure boilers in order to maintain that bubbly steam condition. This circulation, due to the density difference between steam and water, is called natural circulation.

A fire tube boiler might have a pattern like that in Figure 10-1. It is more complicated than that because the amount of heat transfer changes from the front of the boiler to the rear. In the typical Scotch marine boiler, the water rises around the furnace over the entire length and drops at the sides to varying degrees and considerably against the front tube sheet. The outer shell holds the water to be boiled. The hot gases from the combustion chamber move through the tubes, transferring heat to the water in the shell. The steam is collected at the top for use. The water remains in the shell.

Water tube boilers have circulation patterns that vary considerably with the boiler design and the firing rate. The typical example shown for circulation in a water tube boiler is that shown in Figure 10-2(a). The water and steam rises in the tubes that receive the greatest amount of heat because more steam bubbles are in that water. Water, along with a little steam that is generated, drops

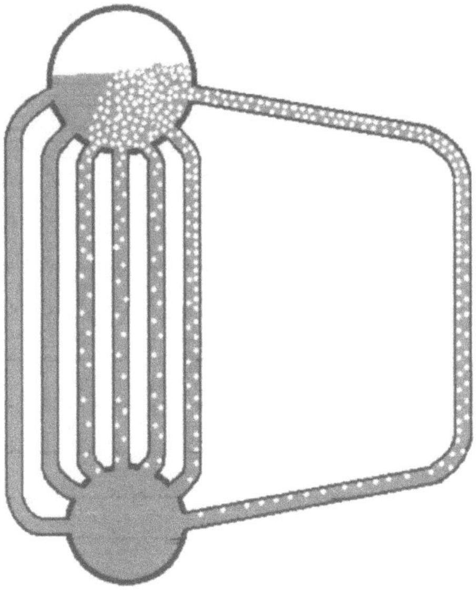

Figure 10-2(a). Steam flow pattern in water tube boiler.

in the tubes that receive less heat. The tubes where water and steam flow up toward the steam drum are called "risers." The ones where the water drops are called "downcomers." It stands to reason that all the tubes that face the furnace of a boiler must be risers. Remember that 60% of heat is transferred by radiation. When the boiler is operating at low loads, only a few of the tubes, those along the sides of the boiler that are heated on one side only (and do not face the furnace) will be downcomers. As the boiler load increases, even the downcomers will have some steam bubbles forming in them because they are absorbing heat. More tubes will have to become downcomers in order to move all the water that has to circulate in the boiler. Some tubes will always be risers. Some will always be downcomers. Some of them switch back and forth depending upon the load.

Some water tube boiler designs encountered problems with the translation from risers to downcomers. The water flow tended to be so low in those tubes that scale formed in them. In these boilers, there are certain steaming rates to avoid to prevent scaling problems in portions of the boiler. A number of designs were modified to include "unheated downcomers," tubes or pipes installed between the top and bottom drums (or headers) on the boiler to provide an unheated path for the water to circulate through.

In hot water boilers, there is some steam generation to force circulation. There has to be. Maybe there is none at low loads. However, the differences in density of heated water are not enough to produce the rapid flow of water needed to carry the heat away from the heat transfer surfaces. The steam that is generated condenses again when the bubbles separate from the heat transfer surface and find their way to colder (by a few degrees) water in the boiler. There are some hot water boilers, HTHW generators for example, that are designed to force the water along and absorb the heat fast enough to prevent steam formation.

Keep in mind that circulation is absolutely necessary to prevent scale formation and blocking of the tubes to the degree that they overheat and fail. If bottom blows are not adequately removing the accumulating sludge in a boiler, the normal circulation can sweep some of that sludge into some risers with almost instantaneous failure being a certainty. Growth of scale on tubes will restrict flow in the boiler and accelerate the scale formation as a result. If there is scale in the boiler, its demise is only a question of timing. Loose drum internals, which will break loose when exposed to the rapid movement of water and steam, can block flow resulting in the loss of circulation and boiler failure. Don't let those broken bolts and supports go. Get them fixed.

There are water tube boilers that are designed in a manner that overrides natural circulation. These units are equipped with boiler water circulation pumps that are designed to operate at the high pressure and temperature of a steam boiler and produce the differential that is required for the boiler to operate properly. This type of circulation is called forced circulation. It can be found on boilers operating at pressures higher than 1800 psig (Figure 10-2(b)).

Forced circulation ensures sufficient circulation at high subcritical pressures under all operating conditions. With forced circulation, the water flow to each tube can be engineered, providing for more uniform metal temperatures throughout the water wall. Also, water flow can be

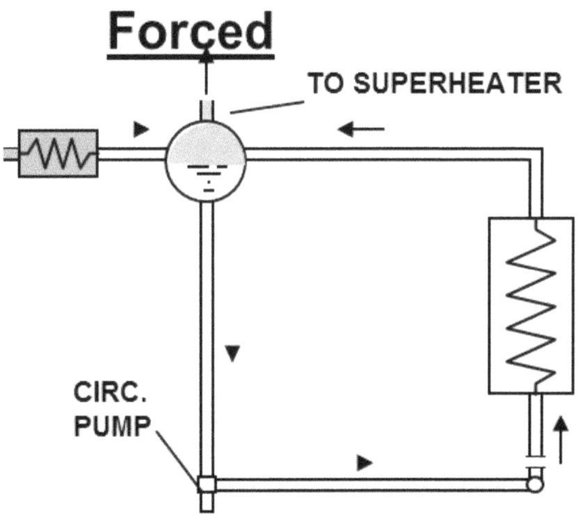

Figure 10-2(b). Forced circulation diagram.

established prior to firing any fuel. The circulating pump provides a convenient connection for chemical cleaning.

When the steam conditions exceed the critical point, there is no longer any difference between steam and water. There is just a fluid. That fluid flows through the tubes and is heated. There is no separation of steam and water. Therefore, the flow is referred to as "once-through." Special design features are required for such boilers. The drive for higher temperature and pressure in boilers is primarily due to the potential improvement in overall efficiency of an electric generating plant. These plants use heat engines to drive an electric generator. Heat engines convert heat energy into mechanical energy (usually of rotation). Heat engines include steam turbines, gas turbines, automobile engines, and diesel engines. These engines are governed by thermodynamics (the movement of heat). In particular, the maximum theoretical efficiency of a heat engine is determined by the so called "second law" of thermodynamics. This law basically states that there is no such thing as a perpetual motion machine. The consequences of this law are that heat always flows from the high temperature to the low temperature, that a heat engine must work between two temperature levels labeled T_{hot} and T_{cold}, that a heat engine must reject some heat at that lower temperature, and that the maximum theoretical efficiency is governed by the following equation:

$$\text{Maximum efficiency} = (T_{hot} - T_{cold})/T_{hot}.$$

These temperatures are given in absolute temperature (degrees Rankine). To convert from degrees Fahrenheit to degrees Rankine, add 460 to the degrees Fahrenheit.

Examination of this equation shows that as T_{hot} increases, the efficiency increases. Thus, increasing the steam temperature will increase the maximum efficiency that can be obtained by the plant. Since a steam turbine operates by taking steam at a high pressure and expanding it to a lower pressure, an increase in pressure is needed to go along with the increase in temperature. Of course, at these high temperatures and pressures, special materials are needed. Indeed, it is the development of these special materials that limits their use in advanced super critical boilers and steam turbines.

BOILER CONSTRUCTION

The construction of a boiler can be attributed to many things. The principle ones are code compliance and cost. The manufacturer has to build a boiler that complies with the applicable section of the ASME BPVC while, at the same time, to minimize cost and still do the job. Low price can be as simple as first cost but should be based on life cycle cost, where the selected boiler should provide the required steam or hot water with the lowest combined price, installation, fuel, and maintenance cost over its expected life. There is always an ongoing effort to design a better boiler. There have been many changes over the years and more can be expected in the future. There are many books that show the extent of construction variations. The goal here is to provide an idea of the development of the designs and why they are made that way.

Not only is a teapot a simple boiler, it is representative of many of the earliest designs of boilers. They were nothing more than an enclosed pressure vessel, full of water, suspended above a fire, with some piping leading off to the user of the steam. Some, like the early Roman baths, were even simpler, separating the fire from the water by a simple row of mud bricks, the earliest refractory.

Many fired boilers use some refractory. A refractory material is one that can withstand the heat right next to a fire. Looking like cement, or regular brick, it contains chemicals to bind it that will not melt under normal furnace conditions. There are very few that can stand to be right next to a fire and none can tolerate the highest possible flame temperatures. Refractory materials come in different grades based principally on the temperature they can reach without melting or failing. They range from 1200°F stuff on the low end to 3200°F material. Normally, the higher grade materials are used closest to the fire and lower grades are used where the temperature will be lower. Upsets in flame shape, openings in baffle walls, and other problems in a furnace can direct hot burning gases against refractory that cannot tolerate the higher temperature, resulting in early, and sometimes quick, failure of the boiler.

There are basically three types of refractory: brick or tile, plastic, and castable. Brick or tile are preformed and fired at the factory. A burner throat is normally made up of tile. Plastic is moldable, usually applied by positioning chunks of it and then beating it into position with a hammer. Castable is mixed and poured into forms like cement. In any large wall of refractory, special "anchors" are furnished with steel or alloy material that penetrates any backup insulation and attaches to the setting, casing, or buckstays for support. Some anchors are made up of a combination of metal and a piece of tile (Figure 6-6) to provide better attachment to the refractory. Setting is the name used for a boiler and furnace enclosure that consists of brick stacked up like walls to enclose the boiler and furnace.

Casing is the name used to describe the outside of the boiler enclosure when it is typically made up of steel plate. It is not the same as lagging. Lagging can range from steel plate to painted canvas but is normally thin sheet metal covers used to protect insulation applied to a boiler. Buckstays are structural steel components that stiffen the casing of a boiler or provide attachments for panels of water tubes. For the larger boilers, the buckstay system maintains the boiler shape as the unit expands and contracts from startup to operation to shut down. There was a time when all boilers were enclosed in a setting or casing, insulation, and refractory. The typical form was a box and could consist of a mixture of materials. Boilers were constructed with bottom support, top, and intermediate support. Top supported boilers require inverted thinking because they grow down as the boiler heats up. Intermediate supported units grow both ways. Top supported boilers required an external structural steel frame to hang from. Sometimes, they are made part of the building. Other times, they are independent of the building.

Boilers will grow upon heating. A list of materials and the amount their length changes is given in the Appendix. Since a boiler is made mostly of steel, it will grow around 0.6% for each one degree change in temperature. The steel in a boiler will always be very close to the temperature of the steam or water (saturation condition for a steam boiler, average temperature for hot water). If the boiler is supported at the top, basically hanging from the structural steel, it will grow down. If it is supported at the bottom, it will grow up. The boiler is not attached to the building structure. The tendency of boilers to grow as they are heated prevents it. There are some platforms that are supported off the boiler steel. Be aware that they will move!

There are three basic types of boiler construction: cast iron, fire tube, and water tube. Cast iron forms produce spaces for water, the fire, and products of combustion. A fire tube boiler contains the fire and products of combustion inside the tubes and the water and steam are outside the tube. A water tube boiler has the water inside the tubes and the fire and flue gas are on the outside of the tubes. There are also tubeless boilers (which can be classified as fire tube) that, like the whistling teapot on the stove, are small and inefficient. They are also cheap to build and are more than adequate for some small operations.

Cast Iron and Tubeless Boilers

Cast iron boilers are made up of cast pressure parts bolted together or connected by piping. There are arrangements of castings that form a furnace as part of the boiler (Figure 10-3) and others that require additional setting (Figure 10-4) and lagging. Cast iron boilers are restricted to heating boiler service. The maximum pressure rating is 60 psig. The corrosion resistance of cast iron makes the cast iron boiler very durable. Many of them have been in hot water service for more than 50 years. Their largest problem is that durability. They get ignored and they fail.

The tubeless boiler (Figure 10-5) uses the outside of its shell as part of the heat exchange surface. The gases

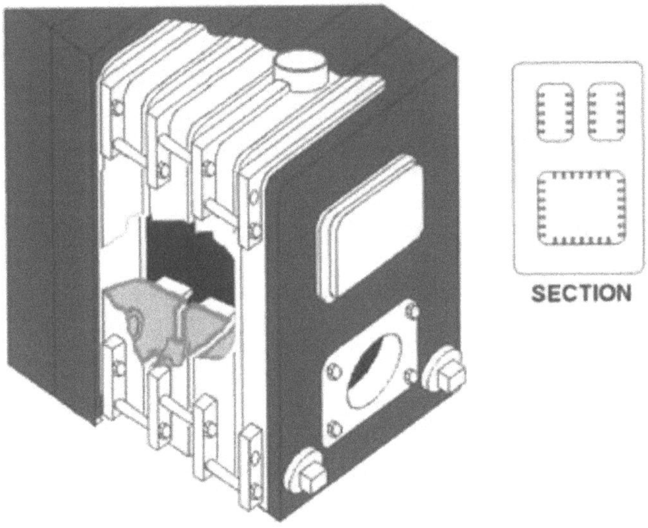

Figure 10-3. Cast iron boiler, integral furnace.

Figure 10-4. Cast iron boiler, pork chop sections.

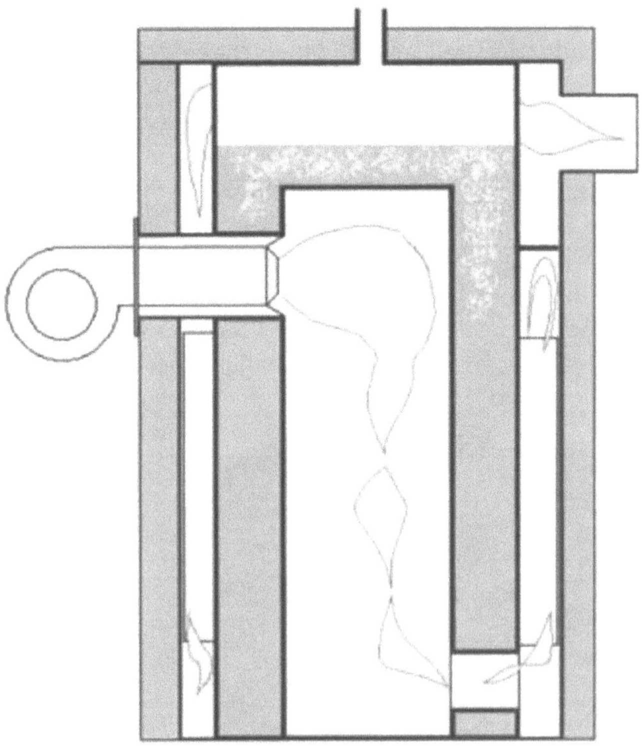

Figure 10-5. Tubeless boiler.

exit the furnace through a nozzle that connects the furnace and shell and then makes a couple of passes along the shell between fins formed by welding steel flat bar to the shell before exiting the stack. One manufacturer adds another pass around a boiler feed tank attached to the boiler shell and forming part of the assembly. Many of these boilers are sold to restaurants and other facilities for the sole purpose of steaming crabs. Since the crabs are exposed to the steam, there is no condensate return. These boilers do not last very long using 100% makeup. Their low price and vertical construction allows for relatively inexpensive replacement.

FIRE TUBE BOILERS

The fire tube boiler requires a "shell" to enclose the water and steam to complete the pressure vessel portion of the boiler. That shell is the principal limit on the size of a fire tube boiler. To understand why the shell is the limiting factor, review the chapter about the strength of materials and how to determine the required thickness of the shell, tubes, and other parts of a boiler. The required thickness of the shell of a boiler, or a boiler tube, is a function of the radius. As the tubes get larger, the thickness has to increase to hold the same pressure. Since the outer shell of a fire tube boiler is very large, it has to be quite thick. Thicker materials require more elaborate construction practices, in addition to more weight. Thus, the price of a fire tube boiler increases proportional to its diameter, with sudden large steps in price associated with different construction rules, depending on the thickness and temperature.

A big break point for higher pressure boilers comes at 1/2 inch thick and 650°F. The increasing thickness has imposed a normal limit on fire tube boilers of 250 psig maximum allowable working pressure (MAWP). It is possible to get a fire tube boiler for a higher pressure, but it is not a common one. The other practical limit on the size of a fire tube boiler is its diameter. Anything larger than 8 feet 6 inches in diameter will require special permits for transporting it on the nation's highways. Shipping a fire tube boiler without trim and panels on the sides (but with insulation and lagging) and without special roadway permits and escort vehicles limits the diameter to 8 feet.

To allow shipment with control panels mounted, the normal fire tube boiler is limited to shell diameters of 7 feet. There is also a limit on length which is around 20 feet (to fit inside a low boy trailer). Longer units have been made. Twice the length of the boiler is needed to permit replacing the tubes. A 12 foot long boiler would require 24 feet of space. That is the nominal distance between building columns in average construction. Many units are built backed up to rollup doors so that the tubes can be pulled outdoors.

All those factors place a reasonable limit on fire tube boilers at about 500 horsepower for a normal unit rated 5 square feet of heating surface per boiler horsepower, 600 horsepower if all the trim is removed or the boiler is rated at 4 square feet of heating surface per boiler horsepower, and about 800 boiler horsepower if roadway problems are not too expensive and the customer can handle a permit load or delivery by rail. That does not mean that a fire tube boiler cannot be larger. A 1400 horsepower fire tube boiler was some 10 feet in diameter and almost 40 feet long. The lower cost of manufacturing fire tube boilers has also increased the manufacturer's offerings to 1000 boiler horsepower. Sometimes, they do it by simply increasing the size of a burner on an 800 horsepower boiler. With the advent of low cost natural gas and the development of a condensing heat exchanger (CHX) for the flue gas, sizes up to 2500 boiler horsepower are now being offered. The boiler efficiency is increased with the water condensation and recovery from the flue gas and the cost of natural gas is relatively low. Good burner technology has reduced both CO and NOx emissions. Further, natural gas firing

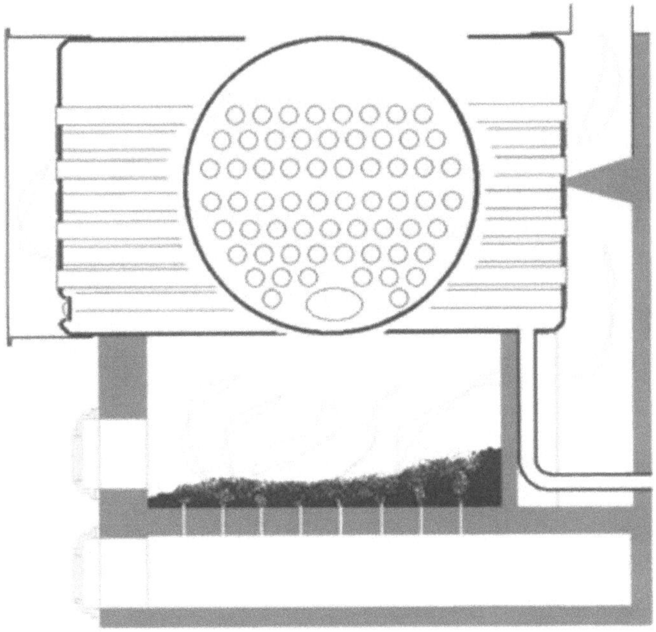

Figure 10-6. HRT boiler.

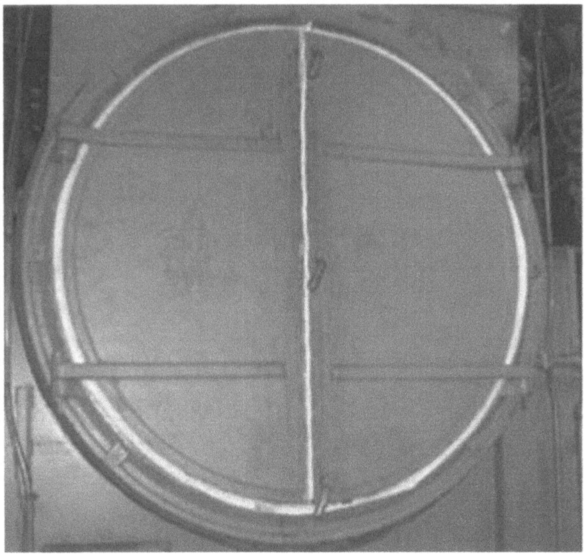

Figure 10-7. Cast doors on HRT boiler.

produces less CO_2 per MMBtu than other fossil fuels. All that translates into lower costs, as no additional air emission control equipment is required.

Fire tube boilers come in several configurations and arrangements. Basically, they are cylindrical in shape (Figure 10-6) and are further defined by position and modifications to the general form. The arrangement in Figure 10-6 is typical of an HRT boiler (the letters stand for horizontal return tubular), which is an early design of boiler that has survived to modern times. Return in the label indicates that the flue gases flow down some of the boiler tubes from one end to the other and then return through the remaining tubes.

A cross section is shown in the middle of the figure that shows the tubes, how they are arranged to permit the baffle at the rear, and the location of an access door for scraping off the bottom. Typically, the shell of the boiler is extended at the end where the gas makes the turn to form a "turning box," which is closed by large cast iron doors (Figure 10-7). The doors could be at the front or rear of the boiler, depending on how it is constructed relative to the furnace.

Most of these boilers were assembled without welding. The joints in the shell, the tube sheet to shell joint, and the piping connections were all made using rivets. The furnace was typically a brick walled enclosure constructed below the boiler. Many were built with the brick serving as a base to support the boiler. Few of those remain because a furnace explosion that dislodges the bricks would result in the boiler collapsing into the furnace. More modern HRT boilers are constructed with steel bases that support the boiler, or a steel frame straddling the boiler, and supporting it with suspension rods.

A constant problem with HRT boilers is maintenance of protection for the bottom blow off piping. In many cases, that pipe drops vertically through one end of the furnace and has to be protected by refractory because it would absorb so much heat that steam could not escape it fast enough to allow water in. They go dry, overheat, and rupture. The other concern with HRT boilers is the bottom, where radiant heat from the furnace is absorbed by the shell. Any accumulation of mud in the bottom of the boiler tends to prevent cooling of the shell with resultant failure. The only service one of these boilers is purchased for today is in firing solid fuel, normally small biomass applications. Those applications require a large furnace and have low radiant energy emissions compared to oil and gas fired boilers.

Take the standard form of fire tube boiler and turn it on its end to get a vertical fire tube boiler. These are seldom used for steam service because the top tube sheet is exposed to steam instead of water. The tube sheets to tube joints are exposed to considerable heat. They are commonly used for service water heating (Figure 4-9) and may find occasional use for hydronic heating and in waste heat service.

A locomotive boiler (Figure 10-8) is a good example of a fire tube boiler modified to provide some water cooling of the furnace. The increased cost of the boiler to create a water jacket around the furnace was justified for locomotive service because the steel and water were considerably lighter than the refractory that would

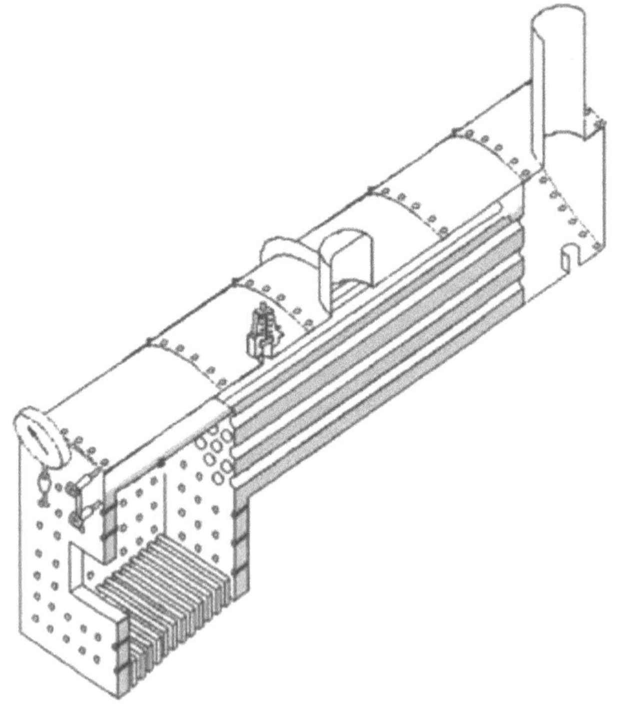

Figure 10-8. Locomotive boiler.

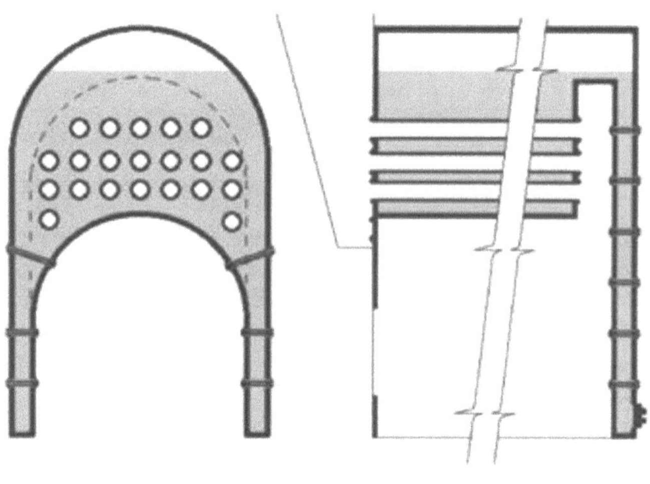

Figure 10-9. Firebox boiler.

be required, while providing more heating surface to make the locomotive more powerful. Staybolts are used to hold the flat surfaces against the internal pressure. Their failure was one reason that many of these boilers are no longer around. The techniques developed in the railroad industry were translated to stationary boilers to create the firebox boiler (Figure 10-9). The firebox boiler was the first potential "package" boiler. It only required construction of an insulated base in the field, with all other parts assembled in the factory. A partial form of the boiler was also built to provide comparable performance at lower construction and shipping costs by requiring

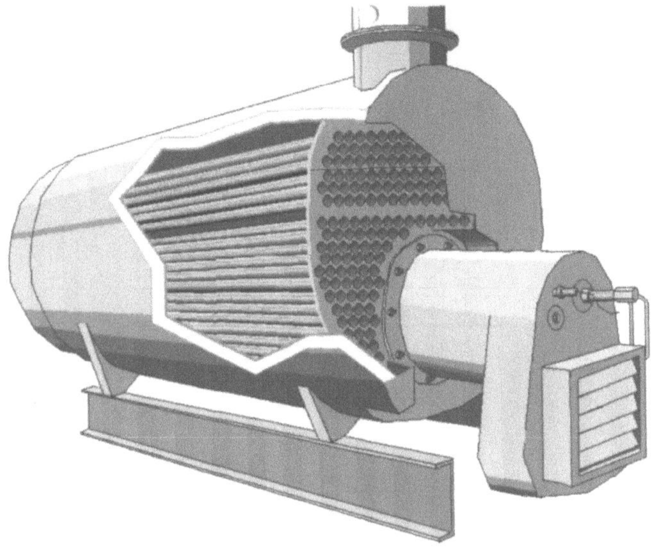

Figure 10-10. Scotch marine boiler.

construction of part of the furnace as a brickwork base and then setting the boiler on top of that base. It included some of the cast iron boilers shown previously. The terms "low set" and "high set" refer to these boilers. A high set firebox boiler incorporated all of the furnace. The burner was set high in the firebox. A low set firebox boiler normally requires that the burner be installed in the brickwork base.

Finally, there is the construction that is typical of all of the modern fire tube boilers. They are called Scotch marine boilers, although it will not likely be found on a ship and there is no proof that they were a Scottish design. This construction incorporates the insertion of a large furnace tube in the boiler (Figure 10-10), eliminating the requirements for an external furnace and providing a furnace that is almost completely water cooled.

Many of the original boilers of this design, the ones that were used on ships, were coal fired. They required multiple furnaces to provide enough furnace volume and grate surface. The furnace tube diameters range from 2 feet to 4 feet and are welded to the tube sheets. The tube sheet to shell joint is also welded. The Scotch marine design comes in two general arrangements. The most common is a dry back design, where the turning chambers at either ends of the boiler are formed by an extension of the shell and/or a door that forms the turning chambers. In either case, both ends of the boiler are fitted with doors to gain access to the tube ends. The doors can be full size, covering the entire end of the boiler or they can be multiple, with separate doors providing access to various portions of the tube ends and furnace. In almost every case, the door covering the end of the

Figure 10-11. Baffled rear door of four-pass fire tube boiler.

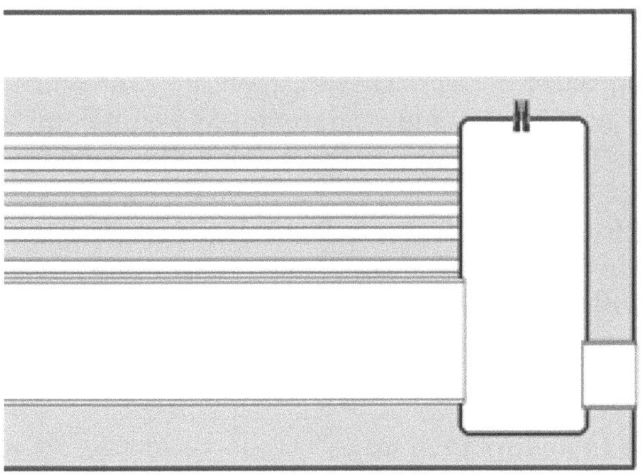

Figure 10-12. Wet back Scotch marine boiler.

boiler and furnace tube is refractory lined because the temperatures of flue gas leaving the furnace can be over 1200°F. Some doors contain integral baffles (Figure 10-11) to divert the flow of flue gas back into other tubes in the boilers. The baffle arrangement varies with the boiler design principally to separate the passes. The wet back arrangement (Figure 10-12) is a more efficient boiler, with less refractory to maintain. However, the higher cost and limited tube removal (front only) has resulted in a decline of its use.

Something common to firebox and wet back boilers, and possible to find on others, is a fusible plug. It is shown in Figure 10-12, where it belongs, at the top of the turning box in the middle of the flat top, which is a surface that would be exposed to high temperatures if the water level dropped enough that the only thing to cool it was steam. The plug is filled with a low melting point metal. It would (theoretically) put the fire out with steam if the top of the turning box was overheated. Of course, if it had to work, the boiler had to be shut down, cooled down, drained, and the plug replaced before returning to operation.

The locomotive boiler (Figure 10-8) is a basic single pass design. The flue gases enter the boiler proper and flow through all the tubes to the outlet of the boiler. The HRT design provided improved heat transfer by providing two passes. The flue gases are turned and return down a portion of the tubes on their way to the stack. Note that a pass consists of a path for flue gas to travel from one extreme end of the flue gas containing parts of the boiler to another. Neither of these designs required a baffle to direct the flow of flue gas. Scotch marine designs can have two, three, or four passes. A two-pass Scotch marine boiler requires no baffles, other than means to separate the burner from the returning flue gas. Three-pass Scotch marine construction requires one baffle in the rear of the boiler to separate the first and second pass turning box from the third pass outlet. Four-pass boilers require a baffle there, plus one at the front to separate the second and third pass turning box from the fourth pass outlet (Figure 10-13).

Four-pass fire tube boilers have a construction unique to them. The tubes at the inlet of the second pass are normally welded to the tube sheet. That is because the flue gases in the first to second pass turning box are much hotter in those boilers. The welding provides a better course for heat to pass from the metal to the water to prevent overheating those tube ends (See Why They Fail, Chapter 11, for a discussion of problems with four-pass boilers). Whether fire tube or water tube, the normal means of connecting the tubes in a boiler is by

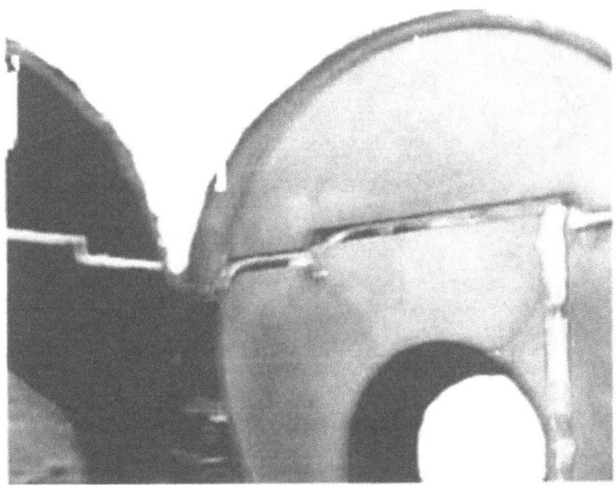

Figure 10-13. Front baffle of four-pass boiler.

rolling. It is a mechanical method of attachment that is strong, watertight, and reliable. It is also relatively easy to break, allowing the tubes to be removed. Refer to the section on maintenance for a description of installing a tube by rolling.

The furnace tube is normally connected by welding to the tube sheets. That is because it is large and thick, making it difficult, if not impossible, to install by rolling. Sometimes, furnace tubes are called Morrison tubes. Some furnace tubes are not Morrison tubes. They are the ones that are basically a simple cylinder. Morrison realized that the furnace tube could be made thinner and could still withstand the external pressure without collapsing if it was corrugated (Figure 10-14). Look closely at Figure 10-11. That boiler has a Morrison tube. If it is corrugated, it is a Morrison tube. If it is not corrugated, it is just a furnace tube.

The section through a fire tube boiler in Figure 10-14 also reveals another important element of their construction, staybolts. The tube sheet is not supported by the boiler tubes in the top of the boiler (what is called the steam space). Staybolts are required to keep that portion of the tube sheet from buckling out. Part of a boiler internal inspection is checking the fillet welds attaching the staybolts to the top of the boiler shell, and the staybolts themselves, for corrosion. The staybolts normally penetrate the tube sheet. Their welds should be checked on the outside as well as the inside.

There is another classification of fire tube boiler. They are called "oil field boilers." They are designed for that specific application. Boilers used in oil fields get little care. They normally run on raw water or worse.

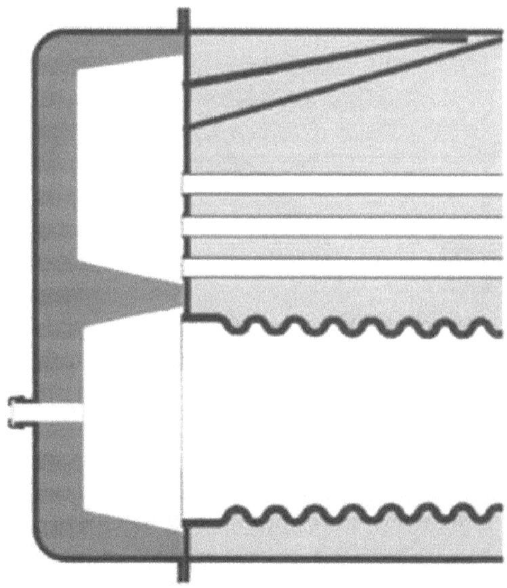

Figure 10-14. Morrison tube.

They do not get the quality treatment provided by a wise boiler operator. Consequently, they are designed for the abuse. They have thicker shells, thicker tubes, and lower heat transfer rates. Saturated steam is injected into the oil field to reduce the viscosity of the oil and improve the percentage recovery. This process is called enhanced oil recovery (EOR). An EOR boiler is a "once-through" unit with no superheating. Heat transfer tubes are arranged horizontally with water cooled supports in high exhaust gas temperature areas. Steam produced ranges from 60% to 80% quality at pressures from 880 to 2200 psig. Water quality requirements are much lower than that in a typical boiler because there are no downstream requirements that a steam turbine requires and many of the solids are carried out with water phase with the saturated steam. Additionally, these units are designed to be mechanically cleaned on the tube side. These boilers can be fired, typically with oil, or can use exhaust gas from another source.

There are many advantages to a Scotch marine fire tube boiler, which includes simplicity in design. They are relatively easy to clean completely on the fire side, once those heavy are doors off. They can be packaged in most of the sizes. They contain minimal refractory. Tube replacement is less expensive because all the tubes are straight. They also hold a larger volume of water compared to a water tube boiler, allowing them to absorb load swings a little better.

WATER TUBE BOILERS

Whether tubes are straight or bent is probably the first distinguishing characteristic for the multitude of designs of water tube boilers. Riveted boilers are no longer built. Most state laws require replacement of any riveted boiler that has a failure after a certain age. Those laws have effectively eliminated riveted boilers. Riveted boilers had issues with leaks. Joints had to be "caulked." To caulk a joint in a riveted boiler, a special chisel and a good heavy hammer was used to deform the metal at the joint working the two together. Blacksmiths still weld metal by heating the material until it is soft and then beating two pieces together to form one piece. Most of the time, the joints in a riveted boiler were sealed by caulking them cold. The real problem with riveted boilers was not leaks. It was cracks forming between the rivets. The crack formation was eventually identified as a byproduct of tiny leaks that left water concentrated in the metal to metal joint, resulting in caustic corrosion cracking (See Water Treatment, Chapter 8). A lack of skilled riveters and caulkers and the development of gas and electric arc welding, which formed a stronger and cheaper joint, produced the change from riveted boiler construction to welded construction.

Just as fire tube boilers need a shell to contain the water and steam, most water tube boilers require drums or headers to close off the ends of the tubes. This arrangement provides a path for the water and steam to flow into and out of the tubes and a place for steam and water separation. A drum is typically a long cylinder with a large enough diameter to provide some kind of separation of two fluids (i.e., steam drum, mud drum, knock out drum, etc.). A header is a long cylinder that collects a fluid and connects to the tubes, either at the inlet or the outlet. The steam drum separates the steam from the circulating water. The mud drum has the connecting piping for blow off but may no longer be the lowest point in the boiler. The inlet headers to the water wall tubing will likely be at the lowest point in the boiler.

As boilers got larger, the area of the furnace walls increased to the point that they represented a considerable waste of heat. At that time, fuel was so inexpensive that it was not the primary consideration. However, keeping the boiler room cool and limiting the cost of refractory was. Refractory walls were getting so high that they could no longer be self-supporting. Expensive structural steel was required to hold them up. To solve many of those problems, boiler manufacturers started making water walls, which are rows of tubes that help protect the refractory or actually replace it. The water walls on large utility boilers actually occupy more space than the boiler itself. Most of them are tangent tube walls (described later) and constructed in "panels" that are subsequently welded together to form water walls, some over 200 feet tall. Water walls consist of tubes that may be bent to connect to a steam or mud drum or connect to a header that is connected to one of the drums with more tubes. The water walls and boiler are all parts of the same pressure vessel.

A cross drum, sectional header boiler (Figure 10-15) is one where all the tubes are straight. That makes it a straight tube boiler. They are no longer made. However, it is a good one for explaining some of the unique characteristics and requirements of water tube boilers.

Note first that this is a three-pass boiler. The flue gases traverse the furnace from the burners to the rear, but that is not counted as a pass. The gases turn up at the back of the boiler and pass up through the superheater and boiler tubes until they reach the top (first pass). Then they drop down through the middle of the tubes (second pass) and finally up through the tubes at the front of the boiler and out the stack. The baffles are made out of

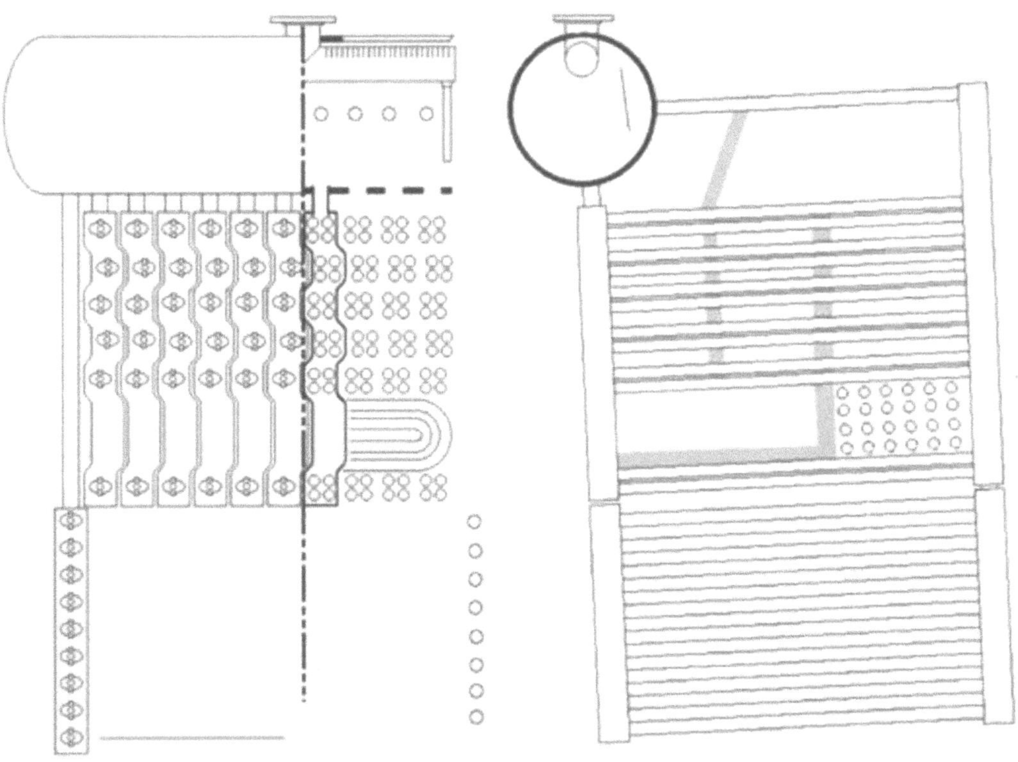

Figure 10-15. Cross drum sectional header boiler.

refractory and include tile laid on top of the screen tubes to form the bottom of the second and third passes.

The bottom two rows of tubes are called screen tubes because they form a screen that blocks the radiant energy from the furnace to the superheater (more on superheaters later). They also protect the baffle. The sectional header part of this boiler involved the forged square headers, shown in detail, which were connected to the steam drum and bottom header by tube nipples (short lengths of tube) and contained hand holes on the side to gain access to the tube ends so that they could be rolled. The headers were forged in a semi-square shape to provide a uniform surface for rolling the tubes. Drums are normally of sufficient diameter that there is no problem rolling a tube in them.

To gain access to the tube ends to roll them, and for other parts, the drums have man holes, usually a 12 inch by 16 inch oval opening. Hand holes are simple openings in the drum or header that are closed by a cast cover (Figure 10-16(a)), which is inserted inside the boiler and bears on the inner surface of the shell, drum, or header, usually against a gasket, so that the internal pressure of the boiler helps hold the cover in place. To keep them in place when the boiler is not under pressure, the bolt, nut, and dog are applied. Key caps (Figure 10-16(b)) are similar but tapered cast plugs that are wedged into the header or drum openings to form a metal-to-metal fit. A special "puller" was required to seat the key caps so that they would not leak as the boiler was filled.

That old sectional header boiler provides a simple look at the complex conditions surrounding circulation in water tube boilers. Water, separated from the steam, and boiler feed water mixes in the steam drum (a common arrangement) and then drops down the front headers (which are exposed to the coolest flue gas) and rises up the sloped tubes, going from the front of the boiler to the rear. In those tubes, the water is heated to the point of saturation and starts boiling, changing from water to steam. The steam forms small bubbles in the water, displacing the heavier water and reducing the density of the steam and water mixture as it travels along the tube. By the time the mixture reaches the rear headers, it is significantly lighter than the water. The weight of the water in the front header is just like a piston pushing down to force the water and steam mixture up the rear headers and back the return tubes to the steam drum. There is only a little difference in pressure between the water in the front header and the mixture at the rear header, perhaps half the height of the boiler (inches water column). That is enough to force the water and steam to flow around, with the flow rate of the steam and water mixture through the top tubes at least five times the rate of the steam going out the nozzle, perhaps more. In the case of this boiler, all tubes are risers. The front headers are the downcomers.

Another form of straight tube boiler was the box header boiler, which used fabricated boxes containing stud bolts (see the discussion on fire box boilers) and hand holes opposite the tube ends, in an arrangement very similar to the sectional header boiler. The straight tube boiler, with its headers, limited boiler size (it was difficult to support the tubes as they got longer) and included multiple sources for leaks (all those hand holes). In 1890, a man named Sterling came up with a better concept for constructing boilers to eliminate a lot of those problems: the use of bent tubes. There are particular designs of

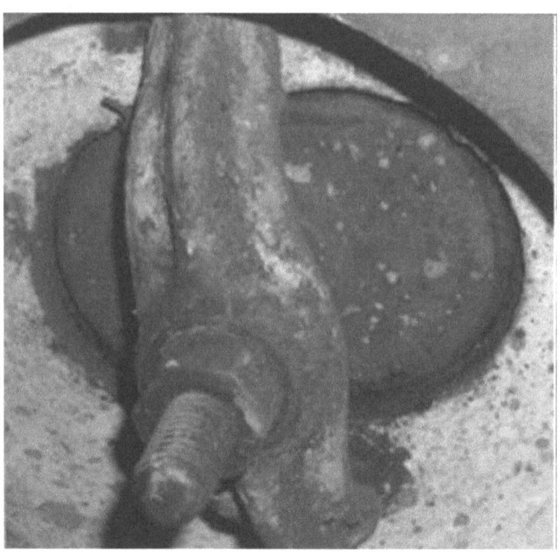

Figure 10-16(a). Hand hole and cover.

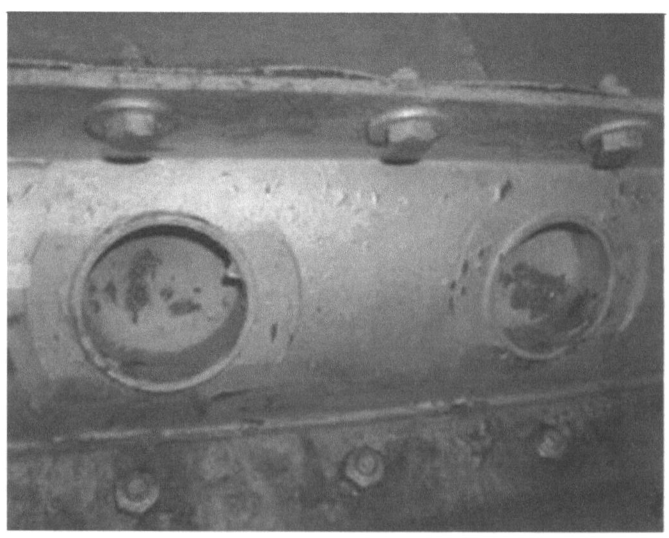

Figure 10-16(b). Key caps.

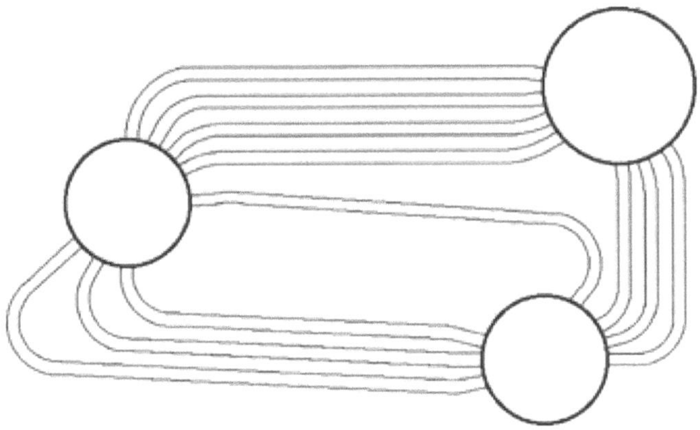

Figure 10-17. Sterling boiler.

boilers (Figure 10-17) that are identified as Sterling boilers, but for all practical purposes, all bent tube boilers can be identified as Sterling. Bent tubes added flexibility to the design of boilers to permit hundreds of designs and arrangements. With the evolution of the bent tube concept, water tube boilers consisted of many arrangements of the Sterling design, far more variations than can be covered here. Keys to sounding intelligent about them include how they are supported (top, bottom, or intermediate), drum position relative to movement of the fire (cross drum if the fire moves perpendicular to the centerline of the drum), and the pressure on the flue gas side (forced draft (FD), balanced draft, or induced draft (ID)). Fire tube and package water tube type boilers are mostly FD design. Most forms of Sterling boilers built today are balanced draft design.

A FD boiler has a fan blowing air into it. The pressure produced by that fan is used to force the air and the products of combustion all the way through the setting and out the stack. An ID boiler can use stack effect to produce the differential pressure necessary to get the air and flue gases through the setting, or the boiler can be fitted with an ID fan that creates a negative pressure at the boiler outlet and forces the flue gases up the stack. ID methods basically create a lower pressure at the outlet of the boiler, allowing atmospheric air pressure to force the air and gases into the burner, furnace, etc. The first boilers were primarily ID designs because motors and fans were more expensive to buy and run than building a tall stack. The stack effect is also a lot more reliable. However, tall stacks today are considered an eyesore, not the indication of prosperity that was welcomed in the 1930s and 1940s. A tall stack today is used to disperse emissions, not to create a draft for ID boiler operation.

As industry flourished, the cost of fans and electricity dropped. The pressure drop across the boiler heating surfaces increased to the point that a stack alone was not sufficient. ID fans were developed to save on the cost of a tall stack and a low pressure drop boiler. Almost all of those boilers were coal fired and had brick settings. The use of FD fans was not desirable because the pressure would force the flue gases out of little cracks in the setting and into the boiler room. As boilers got larger, the low furnace pressures required to draw the combustion air into the boiler and mix it with the fuel also increased the admission of tramp air which lowered the boiler efficiency. Tramp air leaks in after the burners. On large units, it required additional structure to overcome the force of atmospheric pressure on the furnace wall. To reduce the low furnace pressures, balanced draft boilers were developed, where the ID fan, or stack, produced a slightly negative pressure in the furnace and provided the force to move the flue gases out of the boiler, while a FD fan delivered the combustion air to the furnace.

Modern fossil fuel fired electric power generating boilers are all balanced draft and have significant pressure drops on the flue gas side to overcome draft losses in the environmental controls as well as the heat transfer elements. Some operate with ID fans capable of generating over 50 inches of water column differential. That is high enough that, if conditions were not controlled, it would implode the boiler. They create so much differential that atmospheric pressure would push in the casing around the furnace of the boiler. The walls are not necessarily designed to operate with that large a differential. Should the controls on those boilers fail, it is possible to get an "implosion," where the furnace walls collapse in. The problem occurs on a fuel trip. When the fuel feed ceases, air is still coming into the boiler. Since there is no fuel to burn, the temperature of the gases inside the furnace walls is reduced. That causes the volume of the gases to be reduced, resulting in a reduced pressure inside the furnace. The control system has to sense the fuel trip and maintain the combustion air flow from the FD fan, while the ID fan needs to turn down the flow leaving the system. In this manner, the amount of gas inside the furnace can be controlled to minimize the drop in internal pressure.

Package Water Tube Boilers

An interest in other water tube boiler designs can be satisfied by looking up a copy of "Steam" or "Clean Combustion Technologies." Most of the water tube boilers encountered in industrial applications today are what are loosely termed "package types" that come in one of four basic arrangements: A, D, O, or Flexitube. These designs provide the current optimum in cost and performance,

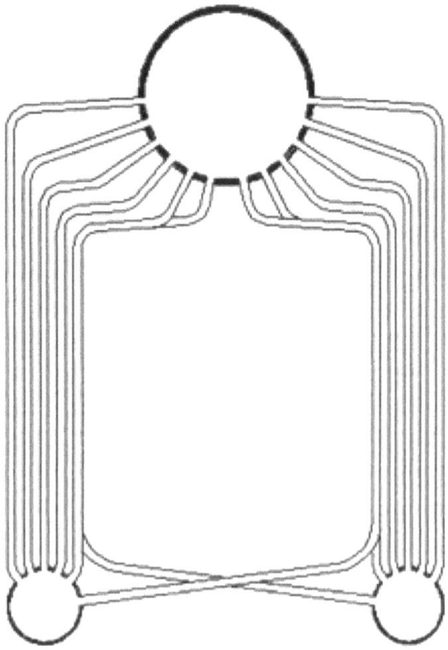

Figure 10-18. "A" type boiler.

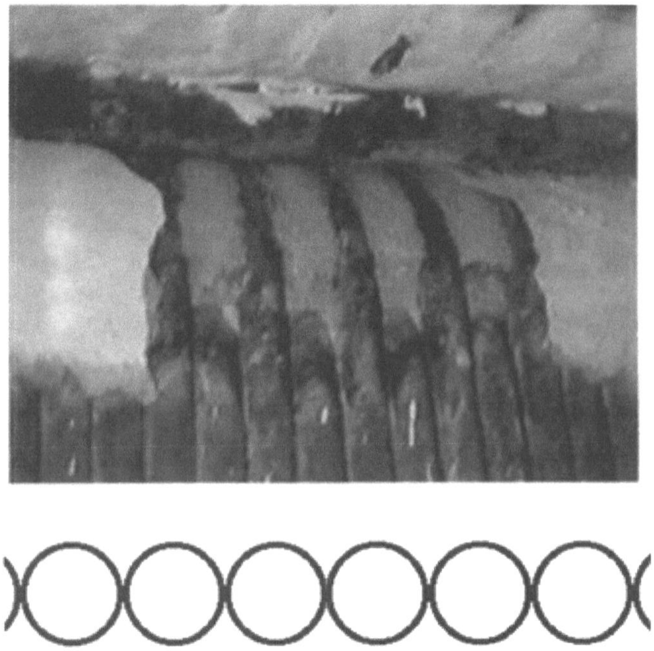

Figure 10-19. Tangent tube construction.

some better than others, and represent the heart of the packaged water tube boiler industry. A good understanding of their construction and operation will provide an understanding of other water tube boilers as well.

The A type (Figure 10-18) was originally developed by the Saginaw Boiler Works in Michigan and subsequently purchased by Combustion Engineering (CE). Other manufacturers produce comparable designs. The A shape is attributed to the single steam drum at the center top and the two mud drums, commonly called headers, at the bottom. They require a second blowdown line and more soot blowers. They provided features like a water cooled furnace from one end to the other and balanced construction, which makes them easy to transport as package boilers.

The tubes inside that form the furnace have alternating shapes. One will drop from the steam drum around the furnace and down into the bottom header while the next tube turns above the bottom header and crosses the bottom of the furnace to enter the side of the opposite bottom header. Shifting the tube arrangement by one sets up the crossing pattern, with a tangent tube wall construction (Figure 10-19) in most of the roof and sides of the furnace. The furnace floor tubes (the tubes at the bottom) have a maximum spacing of one tube width.

Normally, the bottom tubes are covered with refractory tile to limit heat absorption on the top of the tubes. The floor tubes are sloped at a 7 degree angle to allow any steam that might be formed to travel upwards. The tangent tube walls and installation of sealing refractory in the "crotch" under the steam drum close the furnace so that all the flame and flue gases are restricted to the center of the boiler. Four to eight rows of tubes from the back of the boiler are installed without the drop to the bottom header, forming tube gaps that allow the flue gases to turn and proceed down the convection bank tubes back toward the front of the boiler.

Most of these boilers have the flue gas outlet at the top front. Some were made with the convection bank terminated part way down the boiler to create a larger furnace. In that case, the side wall tubes are also the furnace wall tubes. One serious problem with the A type boiler is that the crotch refractory falls out on occasion, forcing an outage of the boiler because a lot of capacity is lost and there is concern for damage to the steam drum. They are also a pain to maintain because all the trim is above the burner. Fans and ductwork connected to the burner at that point makes access to the front drum man hole almost impossible. The front wall of all these boiler designs is normally a simple 13-1/2 inch thickness consisting of 9 inches of plastic refractory over 4-1/2 inches of insulating brick, with a 1/4 or 3/8 inch thick steel front wall plate. There are variations in thickness and materials of construction, including the use of ceramic wool, insulation instead of brick, and precast fired tile instead of the plastic refractory. They all perform the basic function of closing the front wall. A few use additional tubes bent to spread over the front wall to help protect the refractory.

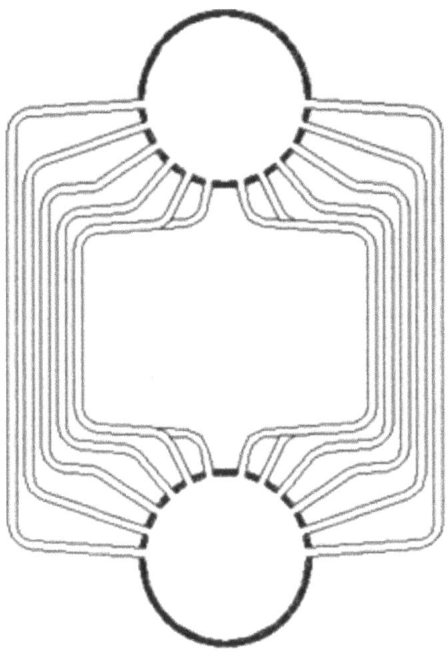

Figure 10-20. "O" type boiler.

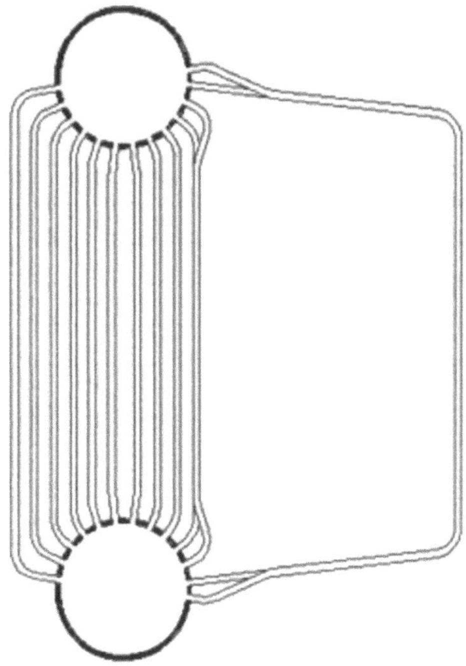

Figure 10-21. "D" type boiler.

The rear wall, on the other hand, is usually fitted with bent tubes spread out to cover it. The wall is typically much lighter in construction than the front wall, an allowance partially provided by the tubes and the distance from the heat of the flame. Frequently, the rear wall is called the target wall, as the flame is shooting straight at it. The tubes against the rear wall are called target tubes. The tubes form a framework of steel that helps to hold the rear wall in place, especially during shipment of the boiler. That is a major consideration in the wall thickness.

The O type boiler (Figure 10-20) is similar to the A, while eliminating one header by providing a drum in the bottom center just like the top. The headers required many hand holes for rolling the tubes in an A type boiler. The single drum eliminated that expense but produced a boiler with a smaller furnace cross section.

The single bottom drum saved one longitudinal weld as well. All the longitudinal welds in modern boilers are X-rayed, making them more expensive to form. Some of the same difficulties experienced with the A boiler are associated with the O design. This boiler is not a good candidate for firing solid fuels or heavy fuel oil because it is almost impossible to remove the soot and ash from the bottom of the boiler. It does work well on gas.

The predominant design is the D type (Figure 10-21), which has only one drawback. That is the problem with transporting and supporting something with most of the weight on one side. The D tubes extend out of the drum to form the roof of the furnace, drop to form the furnace side wall, and return under the furnace to the mud drum. It has one convection bank of tubes centered between the drums to limit soot blower requirements. This construction makes it possible for the flue gas to leave the boiler via the front or side. A more detailed diagram (Figure 10-22) will help identify some of the standard features of this construction.

There are many modifications to this design, with different manufacturers featuring different details. D type boilers are also manufactured in shop fabricated form, where the furnace portion is shipped as an independent assembly from the convection bank with the two drums. Another arrangement is the D tubes and casing being shipped loose, for installation in the field. These may still be referred to as "packaged" boilers, despite final field assembly. Shipping the furnace, or its components, separately allows for larger capacity boilers without the restraints of shipping clearances, while still retaining most of the advantages of a package boiler. Unlike the Scotch marine fire tube and other smaller boilers, "package" does not clearly describe the assembly for water tube boilers. A package boiler can be shipped without any burner or connecting piping. Almost any package water tube boiler with a capacity over 25,000 pph is not ready for connecting pipe, wire, and startup. There are always different degrees of assembly. When specifying a package water tube boiler, an engineer has to explain very carefully what is in the package. There

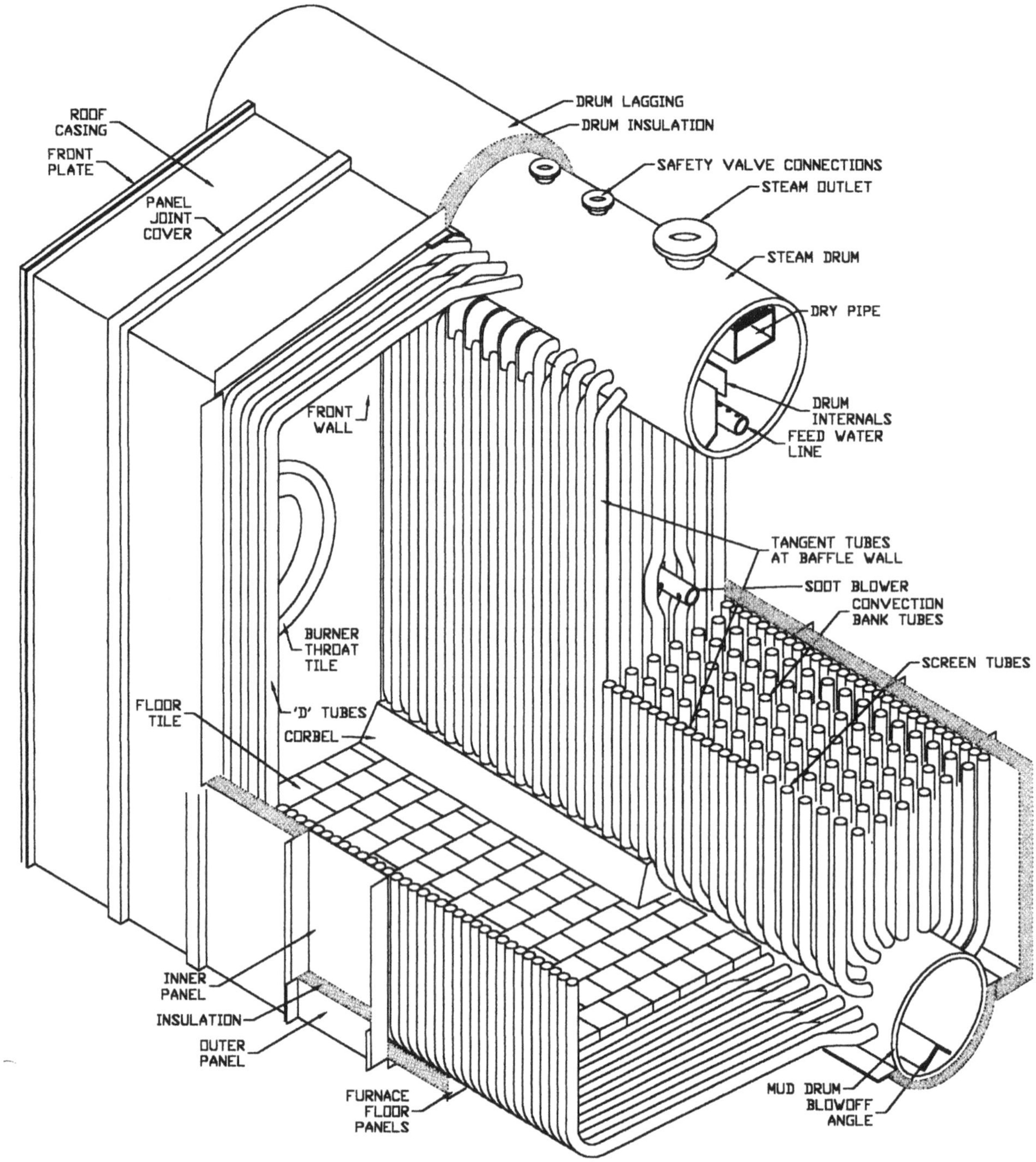

Figure 10-22. "D" type boiler details.

are also a lot of package boilers setting around that were not built in a factory. They were field erected. Problems of shipping clearances, where a bridge or a tunnel near a plant prevented delivery of a factory packaged boiler, or clearances into a building where the owner wanted the boiler installed, resulted in field erection of those boilers.

The boiler in Figure 10-22 has tangent tube walls at the side of the furnace, the side of the convection bank, and the baffle wall between the furnace and the convection bank (except for the short section of screen tubes). Other manufacturers provide finned tube walls (Figure 10-23), where bars are welded between the tubes to form a heat absorbing fin. This feature eliminates the special bending of alternate tubes near the drum, which is required to get a tangent tube wall.

Babcock and Wilcox (B&W) provides an integral finned tube (Figure 10-24), which provides the equivalent of a tangent tube construction without the need

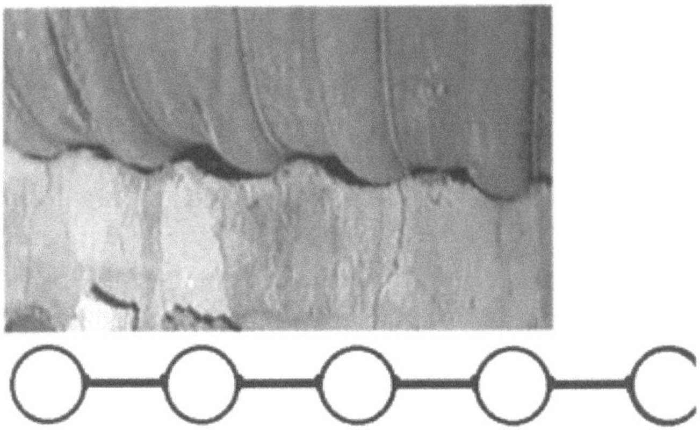

Figure 10-23. Finned wall construction.

Figure 10-24. Integral fin wall construction.

to weld the tubes. The finned tube provides a gas tight envelope around the furnace (with the exception of a gap where the tubes enter the drum). Tangent and integral fin tubes are easier to replace.

CE produced several boilers with swaged tubes to simplify construction of the boiler. Each D tube, outer wall tube, and baffle tube was swaged (mechanically formed to reduce the diameter, Figure 10-25) from 4 inch to 2 inch so that the tangent tubes could be installed in one row of holes. CE also built several boilers where the D tubes are made progressively shorter, top and bottom, so that the rear wall of the boiler could be formed of tangent tubes.

Looking at the construction of the A, O, and D type boilers gives the impression that they are only two-pass boilers. Many of them are, with flue gas traveling down the furnace to the back and then back to the front and out. A lot of D type boilers are not a simple two-pass design because they are fitted with baffles, consisting of steel plates set between the tubes near the outlet of the boiler. Those baffles redirect the horizontal flow of the flue gas to an up and down flow path to introduce additional

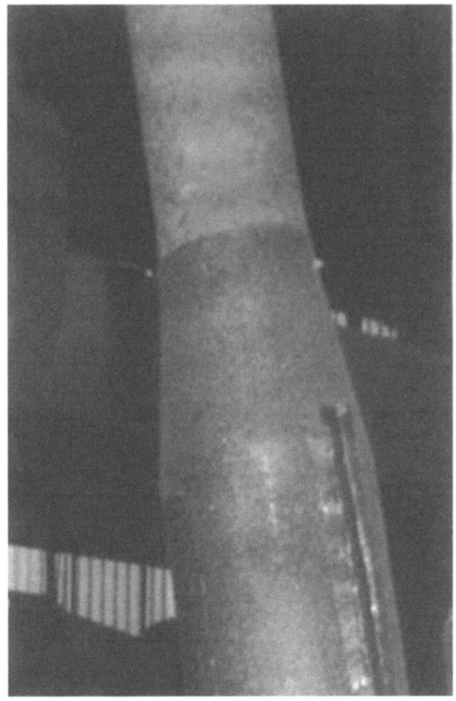

Figure 10-25. Swaged tube.

passes, usually making them a four-pass design when the switching of directions is accounted for. The boilers without baffling have higher velocities through the screen tubes and the initial portion of the convection bank, with attendant higher pressure drop on the gas side and higher furnace pressures to provide a balance of heat transfer comparable to a multi-pass boiler.

While most water tube boilers require drums or headers, a boiler that consists of continuous tube does not. Many hot oil heaters, and some steam and hot water boilers, consist of one coil of tube or two coils to produce a furnace and convection pass. A boiler consisting of one continuous tube or several tubes connected in parallel is called a once-through boiler. If they generate steam, the water used must be ultra-pure or some water leaves the boiler with the steam and is separated from it to remove the solids and impurities. Such boilers have no controllable steam and water line. Other means are necessary to ensure they are not dry fired. Some are fitted with temperature sensors that can identify conditions by superheat. One uses the coil of tube itself. When the tube gets hotter than saturation temperature, its thermal expansion trips a limit switch. When one of those boilers is in the plant, the best thing to do, once again, is to read the instruction manual.

A newer design is the "flexitube" boiler (Figure 10-26 being one example), which has taken advantage of the bent tube construction to produce a boiler that

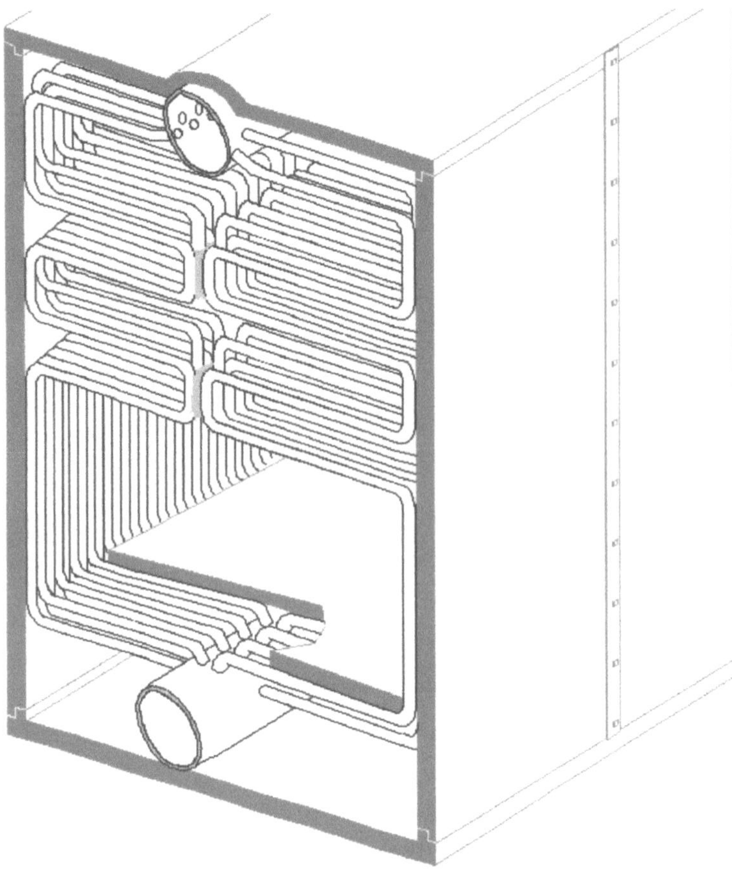

Figure 10-26. Flexitube boiler.

is lighter, easy to repair, easy to field erect, and highly efficient. The only disadvantage of these boilers is their very low water content. Tubes in these boilers are bent to very small radii to achieve the form that allows them to use the tubes as baffles and produce a five-pass boiler. In order to comply with code restrictions on the bending of tubes (which makes the wall at the outside of the bend thinner), they are constructed using 3/4 or 1 inch tubes, compared to the typical water tube boiler that is principally 2 inch tubes.

An additional feature of the flexitube design includes a new way of connecting the tubes to the drums or headers. That construction is shown in Figure 10-27(a). The ferrule is a forged, tapered plug that is bored to accept the tube. The tube is rolled into the ferrule instead of into the drum. They can also be welded together. To install the tube, the ferrule is driven into a correspondingly tapered hole, punched or drilled, and reamed into the drum or header. Precise machining of the ferrule and drum provides a tight fit. The dog is used to clamp it in position for added security.

This approach makes field erection of low pressure boilers much simpler. There are some questions about the long-term operation of these boilers because thermal cycling could loosen the ferrules. Also, movement could wipe out the ceramic fiber insulation used to seal the ends of the passes. However, when weighted against the ease of removing and replacing tubes, those questions are a little moot. These units have a wider range of thermal cycling under normal operation due to the small volumes of water. Reliability can be a concern. There can also be construction issues. It is possible to misalign the tubes where they form the baffles that separate the passes. That is what one contractor did. The leakage of flue gases, from the furnace into the second pass before combustion was completed, resulted in very noisy operation and regular explosions.

Superheaters

Most commercial and industrial boilers produce saturated steam only. Superheaters, associated with electric power generation and driving large equipment, will become more prevalent as more cogeneration facilities are built. A boiler plant will eventually become a power generator as well as a steam generator, unless it is a very small boiler plant or has a very inconsistent load. Since steam can only be superheated when there is no water left around to evaporate, any superheated steam boiler takes dry steam at the boiler outlet (usually the steam drum) to a superheater. The steam flows through a connecting pipe to a header where it is distributed through a number of parallel tubes exposed to the furnace (radiant superheater) or the flue gases after they pass through the screen tubes

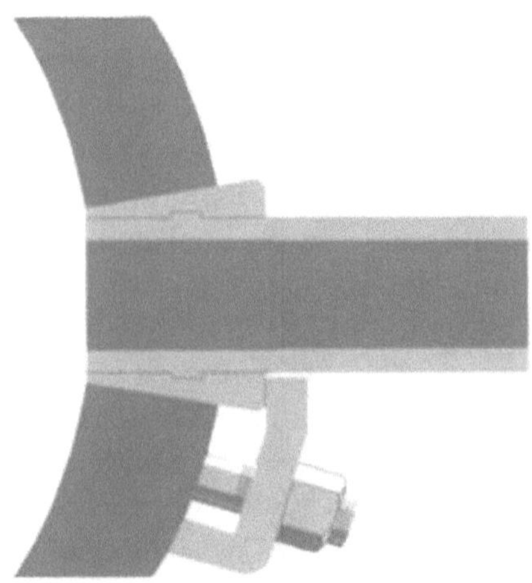

Figure 10-27(a). Flexitube tube to driving joint.

(convection superheater). There, the steam temperature is increased as it absorbs heat from the flue gas.

Since the heat transfer rate is not as efficient as boiling water, the steam velocity is rather high in the superheater tubing to ensure turbulent flow for the best possible cooling of the metal tubes. The full load pressure drop in a superheater is typically 10 psi (pounds per square inch). It takes a lot of pressure drop to create the turbulence for good heat transfer. The thin gas film that makes a conventional boiler tube much cooler than midway between the furnace gas and boiling water temperature when boiling water is repeated on the inside of a superheater. Thus, the tube metal in a superheater is considerably hotter. The superheater materials of construction are designed for those higher temperatures. Many of them use tube metal that is not as malleable (easy to mold or bend) as normal boiler tubes. In fact, they are so brittle that they cannot be rolled. A short piece of malleable tube that is rolled into the header is frequently provided as a stub end to be welded to the more brittle superheater tubes. Those stub ends are protected from the heat by baffles or refractory coatings.

To prevent problems with water depositing in them, many superheaters are designed to drain completely by installing the headers at the bottom, with the tubes extending up from the headers. They are called "drainable" superheaters. Boilers in most utility plants are of a construction that does not drain. The tubes hang down from the headers into the furnace or flue gas passages. They are called "pendant" type superheaters. Some superheaters are separately fired. Boilers on ships of the Navy usually have two furnaces, one before the superheater and one after it. That way, the superheat temperature can be controlled. In shore side applications, there is frequently a requirement for small quantities of superheated steam. A separately fired superheater can be installed to boost the temperature of that steam.

Large power generating boilers can also have reheaters. They are similar to a superheater in construction, but the steam passing through a reheater will be steam that has passed through part of the steam turbine after leaving the boiler outlet. That steam is brought back to the boiler and heated back up to the higher steam temperature. To ensure that the steam remains superheated in the lower pressure stages of the steam turbine, it is reheated in the reheater. Construction is about the same as a superheater. The provision of reheat in the overall steam cycle increases the overall efficiency in the generation of electricity. Boilers with superheaters will always have a safety valve at the outlet of the superheater and a valved vent line to atmosphere for ensuring flow through the superheater during startup and upset conditions. Another pressure gauge and a thermometer are also standard trim items.

Field Erected Industrial Boilers

Once the steam requirements get above 300,000 lb/hr of steam flow, it is difficult to design, manufacture, and ship a package boiler. These larger units will have to be field erected. They will also tend to look more like the larger boilers used for electric generation. Steam capacities in the range of 350 kpph (thousand pounds of steam per hour) to nearly 1 million lb/hr are still relatively small boilers compared to the average sized electric generation boilers. A 500 Mw boiler for power generation will use 3.5 million lb/hr of steam at high temperature and pressure. Industrial sizes tend to be in the equivalent to 35–100 Mw. The largest boiler size in service is 1300 Mw.

These larger industrial boilers are often called upon to fire a fair variety of fuels, depending upon the industry they are in. There are also special boilers for bark burners and chemical recovery boilers in the pulp and paper industry as well as units specifically designed to burn municipal solid waste (trash burners) or refuse derived fuel (RDF, trash that has been segregated into a more combustible fraction).

Some industrial units are called upon to provide both steam for thermal use and electricity. This process is referred to as cogeneration or combined heat and power (CHP). A separate section will cover that in more detail. These boilers often have steam conditions in the range of 900–1000°F and 1200–1800 psig. They utilize a back pressure turbine which takes the high pressure steam and makes electricity and then exhausts the steam at the lower temperature and pressure required for thermal use. An example of a large, field erected industrial boiler is shown in Figure 10-27(b).

This particular boiler can fire a solid fuel such as coal or wood as well as oil or gas. There is a grate for the solid fuel and burners for the oil or gas. The steam drum and the mud drum are located near the gas exit of the boiler. The tubing connecting those two drums is used for boiling water to generate more steam. That bank of tubing is referred to as a boiler bank. Boilers that have a boiler bank are classified as industrial boilers, regardless of their size or application. The original reasoning for this classification was that industrial boilers did not normally require the higher steam pressures and temperatures. At the relatively lower pressures, more evaporative surface was required. The boiler bank provided that additional surface.

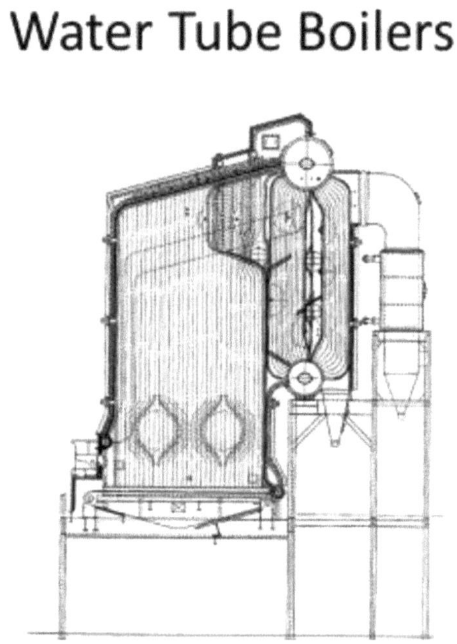

Figure 10-27(b). Large, field erected industrial boiler.

Large Utility Boilers

As the demand for electricity continued to grow, the size of the boilers needed to generate that electricity increased. After World War II, oil and gas were considered to be expensive fuels, while coal was relatively inexpensive. Hence, coal tended to dominate the utility industry until the availability of low priced natural gas became extensive, after about 2010. Up to that time, roughly 50% of the electric generation was fueled by coal. With the low price of natural gas, the cost of environmental controls for coal firing, and the concerns about climate change, the use of coal has decreased by about 40%. Coal now generates only 19% of the power, while natural gas provides 41%. The rest of the world is not as fortunate to have considerable amounts of low priced gas. Some parts of the world are still increasing the amount of coal firing, in spite of the climate concerns.

A side elevation of a large coal fired boiler is shown in Figure 10-27(c). Note that there is only a single steam drum. There is no boiler bank. The unit fires pulverized coal. A pulverizer is shown in the lower left hand corner. There is an air heater in the lower right hand corner. Air is drawn in through the air heater and warmed up by heat exchange with the flue gas leaving the boiler. The warm air is divided into two streams. One stream is directed to the pulverizer. This air stream is called the primary air. It enters the pulverizer and picks up the coal to be burned. In doing so, it dries the coal, removing most of the surface moisture. This air and coal mixture is directed up to the furnace. In this particular design, the coal is introduced into the furnace from the four corners. In this manner, the swirl that mixes the burning fuel and air is established in the center of the furnace. The other air stream is called secondary air. That air is also directed to the corners of the furnace and is introduced around the primary air flow that carries the coal. This type of firing system is called tangential firing. The alternative method of firing is called wall firing, where the coal burners are located in one of the walls of the furnace. If the burners are located across the furnace on two walls, it is called opposed wall firing.

The resulting fire ball radiates heat to the water walls. The typical water wall construction for these units is the fusion welded wall. In this construction, the tubes are welded together with a weld bead between the two tubes, rather than a metal fin. The weld runs all the way up the wall, creating an air tight enclosure. The tubes are first made in pairs. Then the pairs are welded together to make four tubes. Then those panels are welded together to make an eight-tube panel. The process is continued until the shipping width is obtained. Panels can be made 40 feet long and 12 feet wide. These panels are shipped to the field to be assembled into the water walls. A 500 Mw unit will be 55 feet by 65 feet by 200 feet tall. In this unit, the superheater panels are hung over the furnace. The lower bends will see radiation heat transfer from the fire ball. They also experience convection heat transfer from the hot gases rising up to the upper furnace zone. The gases then make a 90 degree turn into the convection pass. There is additional convection surface for the superheater and the reheater in this zone. The gas then turns down by 90 degrees into the economizer. The economizer accepts water from the feed water train and warms it up within 50 degrees of the boiling point. That feed water goes to the drum. The gases leave the economizer and go into the air heater, which heats the air and cools the flue gas. The water walls contain the boiling mixture of water and steam, which is directed to the steam drum for separation. The water mixes with the feed water and is directed to a downcomer external to the wall. In this design, there is a circulating pump for forced circulation. The steam leaves the top of the steam drum and is directed to the superheater. The hot superheated steam is then directed to the steam turbine. The steam goes through the high pressure turbine and is then sent back to the boiler to be reheated. The hot reheat steam then goes back to the steam turbine. After the steam leaves the steam turbine, it goes to the condenser and the feed water train. It then gets pumped back to the boiler to complete the cycle.

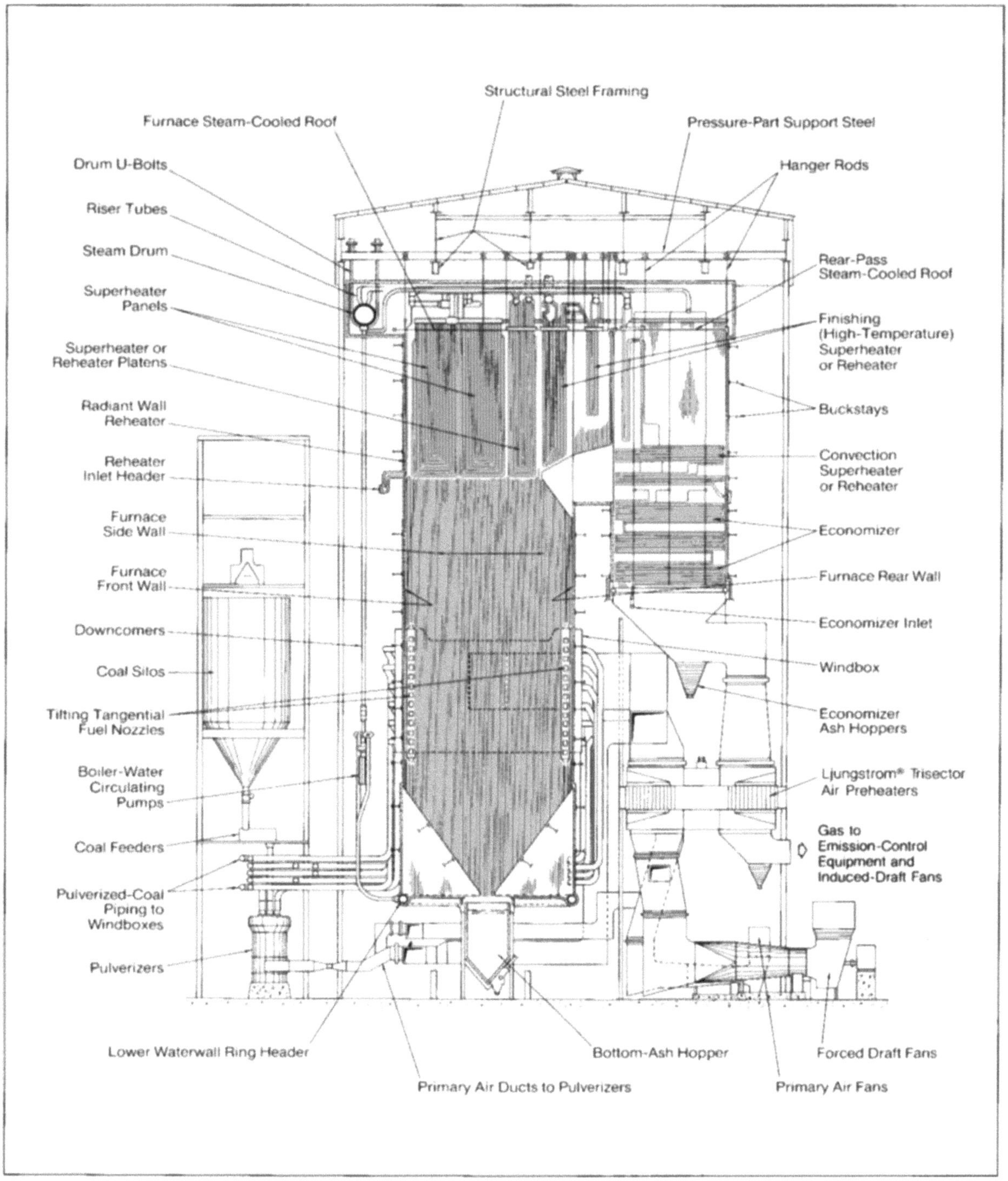

Figure 10-27(c). Large coal fired utility boiler.

Steam Drum Internals

All that steam and water entering the drum needs to be separated so that the steam can go out the steam nozzle and the water can drop down the downcomers. To aid in separating the steam and water, parts are installed in the steam drum. Everything that is installed inside the drum is described as "internals." That includes steam and water separating devices. Most steam drum internals are something like the details shown in Figure 10-28. Baffles deflect the steam and water mixture entering the drum to prevent water splashing up to the outlet. They spread the water and steam out over the surface to promote separation by gravity (heavier water falls and lighter steam rises).

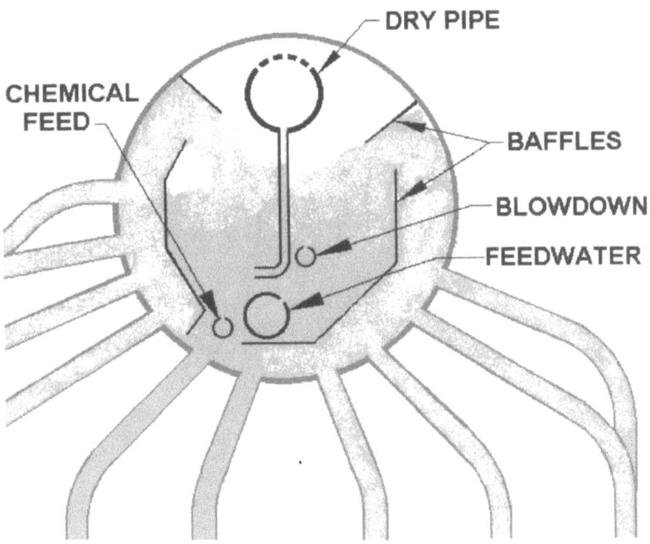

Figure 10-28. Steam drum internals.

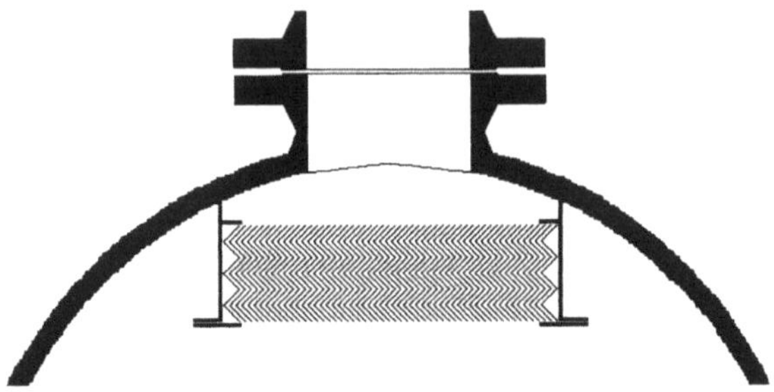

Figure 10-29. Chevron separator.

The steam then has to go up over the top of the dry pipe and down through the holes in it to get inside the dry pipe, which is connected by a tee to the boiler steam outlet. Small pipes, connected to each end of the dry pipe, extend into the water to drain any water that does carry over into the dry pipe and settles out before leaving via the steam nozzle. The other common form of steam and water separation device at the steam outlet is a chevron separator (Figure 10-29), which provides a tortuous path for the steam to travel on its way to the outlet, with several changes in direction that tend to throw entrained water droplets against the chevron elements, where they accumulate and then drain by sliding down the surface of the chevron to the bottom, forming large drops that fall off. Some modern boilers will have more complex baffling arrangements for separating the steam and water. A dry pipe or chevron separator will usually do the job for low to moderate pressure steam conditions.

The baffles are bolted to steel bars welded to the side of the drum to support them and keep them in position during operation. Since they have to be removed to allow for each internal boiler inspection, they are frequently broken. They should be replaced when broken. The movement of the water inside the drum is sufficiently violent that the lack of one connection could allow a baffle to break away and disrupt circulation to cause a boiler failure. A piece of metal flopping around inside the drum can cover some downcomer tubes and starve sections of the boiler tubing for water flow, causing them to overheat.

Another common internal for a steam drum is the boiler feed line. To prevent thermal shock, the boiler feed water piping enters the drum through a special arrangement (Figure 10-30(a)) that diminishes thermal stresses on the thick steam drum by isolating it from the feed water (which may be considerably colder than the steam and water mixture). The feed pipe extends into the drum, sometimes going the full length, and is capped off at the end. Holes are drilled in the feed pipe, normally in the top, to distribute the feed water over the length of the pipe. At least one hole is drilled in the bottom of the feed pipe to ensure it will drain. Occasionally, there are baffles added to the boiler to further distribute the feed water. There are always supports for the pipe attached to the drum and the pipe to prevent it from moving. A flanged, threaded, or slip joint is provided just inside the drum penetration so that the feed pipe can be removed to gain access to the tube ends.

In addition to that boiler feed pipe, drum internals commonly include a chemical feed line and a continuous (surface) blowdown line, which are installed similar to the feed piping. The continuous blowdown line does not require the tempering fitting used for feed water. A chemical feed line normally does. They are located in the drum in positions best suited for their purpose. The chemical feed is installed so that the chemicals can mix as thoroughly as possible with the water before it starts its trip down the downcomers. The continuous blowdown

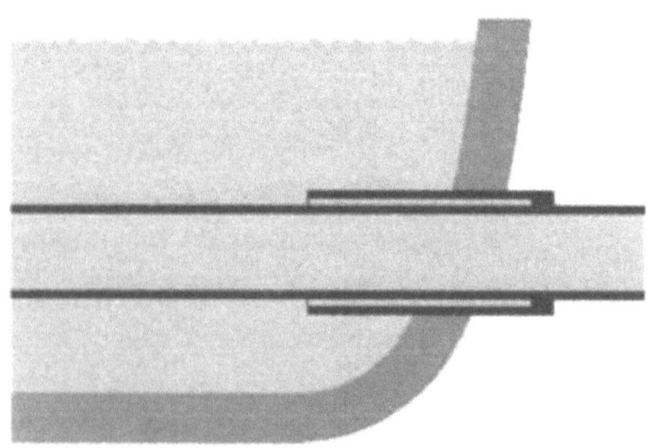

Figure 10-30(a). Feed water line entrance.

piping is located near the surface but not so close that it would draw off steam. The location should be as close as possible to the water that just separated from the steam, as that water will contain the highest concentration of solids. Occasionally, a mud drum will have one internal, an angle set in the bottom to spread out the flow of water when blowing off the boiler. There are more elaborate boiler internals, especially for higher temperature and pressure boilers found in electric generating plants.

As the temperature and pressure of the steam (and water) is increased, the density difference between steam and water diminishes. Thus, it becomes more difficult to separate the steam from the water. Water is still more dense than the steam, but the difference becomes smaller and smaller. Under these conditions, more types of internals will be needed inside the drum to promote the separation of the steam and the water.

A typical drum arrangement for high temperatures and pressures is shown in Figure 10-30(b).

Examination of this figure shows a major downcomer in the bottom center of the drum. Water coming into the drum from the water walls, along with the heated feed water from the boiler economizer, enters around the edge of the drum against a ring baffle that directs the water downward. The water turns upward, along with the steam, and enters the first set of separators. These separate the bulk of the water from the steam. In this particular drum, the can-like devices have swirlers that cause the steam flow to rotate, throwing any water droplets against the walls of the cans, where they can coalesce and run down into the bottom of the drum. In the top of the cans, screens, or chevrons, provide another tortuous path, causing any water droplets to further contact metal parts and drain down. At the top of the drum, there are final dryers. These may be fine screens or plates enclosed in the outlet section. Steam flows into these sections prior to exiting through multiple tubes in the top of the drum. Steam exiting the top provides the maximum chance for gravity to assist with the separation. Any water droplets that are still traveling with the steam have to find their way through the screens and plates. A molecule of steam is, perhaps, 10 Angstroms in diameter. A droplet of water might be 10 microns or 10,000 times larger. Essentially, the droplet of water cannot turn as easily as a molecule of steam and will have great difficulty finding its way through the final dryers. At the bottom of these final sections, there are drain tubes to drain out any water that is collected and bring it back to the bottom of the drum. In this way, only dry steam leaves the drum to pass on to the superheater.

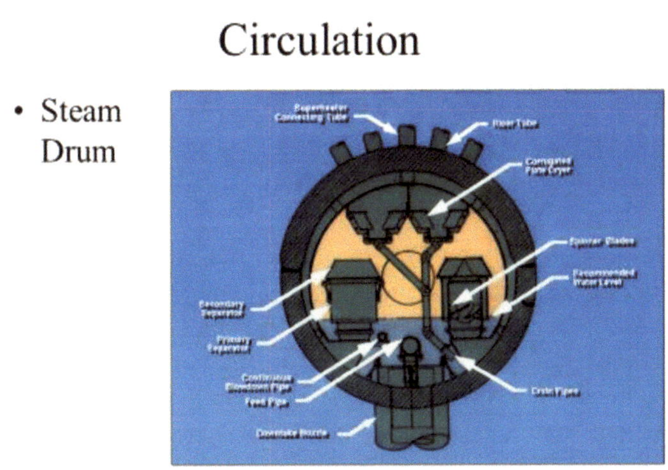

Figure 10-30(b). Steam drum internals.

TRIM

Boilers have a multitude of objects hanging on them called "trim." There is no concrete definition for what is included in boiler trim. Typically, it includes all devices normally attached to the boiler, including anything within the jurisdiction of the ASME Boiler and Pressure Vessel Construction Code, and anything that is not attached to something else. Since the code for construction of power boilers usually extends to the far side of the second steam valve from the boiler, those valves, and connecting piping, are part of the boiler trim. Sometimes, the blow off and feed water piping and valves may be included but not the steam piping and valves. Some manufacturers provide covers, or enclosures, around all or part of the trim to change the appearance of their product. Most of the trim is always there and some of it is essential.

Safety Valves

The correct title for safety valves is "safety relief valves," not to be confused with "relief valves" or safety shutoff valves. "Safety valve" will be used because boiler operators know that they are safety relief valves. Safety valves are the most important part of the boiler trim. They are the final defense against a real disaster, a boiler explosion. A safety valve may look simple, but it is the most refined device in the world. The ASME code contains extensive requirements for construction, testing, certification, and labeling them. A safety valve manufacturer has to be qualified to use one or more of the various stamps that ASME issues that authorize the manufacturer to make those valves. There are also rules and procedures for repairing safety valves.

Safety valves have to have a nameplate or stamp on them that includes the appropriate ASME Code Symbol

Stamp for the application. The stamps (see Appendix) identify valves that have met all the requirements of the code. Note that they are application specific. A safety valve for a pressure vessel (UV stamped, see Appendix S) should not be used for a boiler. If the valve does not have a label or stamping, use of that valve is like driving a car without any brakes. The ASME valve is an assurance that the valve will work when it has to. To operate without it is foolhardy and not the actions of a wise operator.

The valve nameplate should also bear the set pressure and capacity of the valve. The valve has to be large enough to dump all the steam (or heat) that the boiler can generate or the maximum fluid input to a pressure vessel. Never replace a valve with less capacity than the valve that is in place. Never add piping between the safety valve and the boiler. Under no circumstances should a valve, or a blind, be installed between the safety and the boiler. There are times, when testing the boiler, and for other maintenance activities, that a blank or plug will be installed in place of the safety valves, but never operate the boiler without them. Safety valves must be installed with their stems vertical. Adding an elbow to turn the valve so that the boiler will fit under some obstruction is unacceptable.

Steam safety valves have a special arrangement in their construction that makes the valve open completely. Sometimes, operators call them "pop valves" because they pop open. When the valve is closed, the disc of the valve is exposed to the pressure in the boiler over the area that is inside the seat, as shown in Figure 10-31. As soon as the valve starts to open, the pressure in the boiler is exposed to the full surface area of the disc (the larger circle). There is more force on the valve and it pops open. The pressure has to drop to a value lower than the set pressure of the valve before it will close. The difference is called "blowdown" (which has nothing to do with boiler blowdown). When operating too close to the set pressure of the safety valves, the operating pressure will have to be dropped to get the valve to reseat.

Service water heaters (for domestic hot water heating) have an added feature on their valves. They are called pressure and temperature relief valves (PTVs). They are essential for preventing the explosion of a service water heater. The hot water heater in a home has one. The problem with domestic water heaters is that the pressure is not provided by the source of heat. A typical valve setting is 125 psig. It will not lift to dump water with the normal variations in the water supply pressure. About the only time a PTV will operate on pressure is when the water is trapped by a check valve or backflow preventer (See Water Heating, Chapter 4) and the pressure is increased by the water expanding as it is heated.

Figure 10-31. Safety valve seat exposed to pressure.

If the controls fail to shut down the burner or electric element or steam supplying a service water heater, the pressure usually does not increase because the pressure is dependent on the water supply. Expansion of the heated water simply pushes the cold water back down the line out of the heater. Herein lies the problem. When the heat continues, the water eventually gets so hot that it starts to turn to steam. The steam takes up a lot more room than the water and pushes the hot water back to the cold water line until the heating element, or the bottom of the heater, is exposed to steam instead of water. Now, the steam picks up some heat as it is superheated. However, it cannot provide all the cooling that evaporating water does. The temperature of the heating element, or the bottom of the heater, rises until they get so hot that they fail. For an electric hot water heater, the element shorts out, or burns open, to stop adding heat. With a piece of fired equipment, the outcome is not so pleasant. The weakened surface of the heater ruptures. The steam expands and the hot water flashes to form more steam, resulting in an explosion. Hot water heaters commonly rocket their way up through as many floors as are above them and have flattened many houses.

The temperature element of a PTV is a small cylindrical tube that extends from the inlet of the valve. The valve must be installed so that element will be immersed in the hot water. Mounting the valve on connecting piping will not work because the element is not exposed to the heat. Since the element must be in contact with the heated water, PTVs can be installed horizontally or, when labeled for it, even upside down. Don't make the mistake of one contractor who decided the PTVs were installed wrong (the stems were not vertical). He went out to the local hardware store and got some street ells (piping elbows with a male thread at one end and a female thread at the other) to add and turn the PTVs. The worker who was assigned the job of changing the valves had a problem with the little pencil-like things hanging out of the bottom of the valves (they prevented installation on the street ell). He broke them off. Lacking the thermal element, the PTVs did not work when other controls failed. The heaters exploded. Six children and one adult were killed and 42 others were injured. It was an 80 gallon water heater.

Boilers larger than 100 horsepower must have two safety valves. That is a code requirement. Also, boilers with superheaters have to have a safety valve at the outlet of the superheater, which is set lower than the safety valves on the steam drum. It is essential that the superheater safety valve opens first to maintain a flow of steam through the superheater to prevent it from overheating. In addition to monthly and annual testing of safety valves (See Normal Operating Procedures, Chapter 2), they may be required to be sent out to be replaced or rebuilt. That is normally a requirement of the insurance company that does not want their inspectors to spend time observing the pop testing of safety valves. It is less expensive to simply replace a small valve. Valve prices increase with size and set pressure. At some point, they can be rebuilt at a much lower cost. A contractor who rebuilds safety valves should have ASME or National Board authorization to do that work. Replace a valve, or send the valve out for rebuilding, if it starts weeping or leaking. The steam condensing on the spring and stem will accelerate rusting in the top works of the safety valve, which can prevent it from operating. Continuously operating a boiler with a leaking safety valve is hazardous.

If a safety valve is leaking, check the vent piping immediately. That is the most common reason for a safety valve leaking. The boiler always grows (normally it expands upward) as it heats up. The conventional high pressure package boiler will grow at least three-eighths of an inch from cold to operating pressure and a little more before reaching the set pressure. Unless the vent piping allows the safety valve to move up with the boiler, a considerable amount of stress is applied to the valve to spring the vent piping. That stress can deform the valve so that it leaks. To prevent any stress on the safety valve, install a drip pan ell (Figure 10-32), which allows the safety valve to move with the boiler without any restraint.

Figure 10-32. Drip pan ell.

For top supported units, the boiler grows down. High pressure and temperature units may grow as much as 6 inches. However, the safety valves are usually close to the top support steel and undergo less movement. However, high pressure safety valves are more difficult to test. A hydraulic jack method is available for setting these high pressure safety valves. All personnel should keep a safe distance from the valve.

Buildings do settle as they age. There are times when the structure (which supports the vent pipe) shifts independently of the boiler. The foundation may change the relative position of the safety valve discharge stub and the vent pipe. The settling can shift the structure so that the vent pipe is not centered around the stub but pressing against it for another way to stress the safety valve. Annually, preferably right before doing the pop tests, check that the vent pipe is centered around the stub and that there is a 1-1/2-inch gap between the vent pipe and drip pan.

Water Column

In the list of trim, the water column and gauge glass come right after the safety valves in order of importance.

The water column is a surge chamber that provides a stable water level independent of the splashing and bubbling inside the boiler so that the level in the attached gauge glass is a true representation of the water level in the boiler. The water column is usually fitted with other trim items, such as a low water cutoff, or cutoff and pump controller combination. It can incorporate probes for remote water level indications. Usually, the controlling and high steam pressure switches are mounted on the piping connecting the water column to the boiler. There was a time when the code required petcocks on the column to provide a means of checking the water level if the gauge glass was damaged or its indication questioned. Many manufacturers still provide them. They are always a good idea for the original reason. If there is steam at the level of the petcock, then a second it is opened, nothing will be visible between the end of the petcock discharge and the cloud of condensate that forms. Steam is invisible. If there is water there, it will be seen coming out of the petcock.

A water column is always equipped with a drain valve. That permits blowing down the column to ensure that the connections between the boiler and water column are open. Refer to checking the low water cutout in the chapter on normal operation to learn more about blowing down water columns. Water columns can be separated from the boiler by valves, provided they are rising stem gate valves. Typically, they are seldom valved off. If they are, make it a habit of ensuring the valves are open (stems are sticking up). Keep in mind that the discs can come off the stem of a gate valve. The only time those valves should be closed is when the boiler is shut down to allow for maintenance of gauge glasses and other water column parts, while the boiler is still hot or under pressure.

Piping connecting the water column and the boiler must be installed so that it can be inspected and cleaned. That normally results in the installation of crosses in the piping. The opposite end of those crosses should be closed with nipples and pipe caps. It provides two possible joints that will break to gain access to inspect the piping. It is a lot easier to remove a pipe nipple than a pipe plug. No matter how good the water treatment is, there is always a potential for those pipes to plug. They must be inspected annually.

Another important consideration with the piping is connections. Nothing more than operating pressure switches should be connected to the water column piping. In one plant, someone had decided to connect the atomizing steam line to the column piping. All they had to do was remove the pipe cap and hook up to it. The pressure drop of the steam flowing from the inside of the drum to the cross immediately outside it was 12 inches of water column when the atomizing steam was on. Luckily, the boiler had a separately piped low water cutoff. The level at the gauge glass and water column read a false 12 inches higher than it actually was in the boiler. Never accept a leak in that water column piping for the same reason. Any small flow of steam out of a valve packing, or leaking pipe joint, can change the indicated level of the water. Another important factor with the column piping is that it must be installed so that it stays in position relative to the boiler. Any maintenance activity that involves removing the water column, or part of its piping, should be preceded by measuring the height of the column relative to the steam drum or above the boiler room floor. That way, its proper reinstallation position can be confirmed later. Most columns will have a mark in the casting that is the normal water line. That mark can be used as a reference.

Gauge Glass

The gauge glass is normally mounted on the water column and can be isolated with special shutoff valves. The valves are designed to shut off in about one-quarter of a turn. They are fitted with a T type handle so that they can be closed by pulling a chain hanging from the ends of the handle. For more effective shutoff, a chain link or small triangle shaped piece of metal is attached to the bottom valve handle and connected to balance the force of the pull chain between the two valve handles (Figure 10-33) for a positive shutoff.

The purpose of that valve arrangement is to permit an operator to close them when (not if) the gauge glass breaks. An added safety idea is to make two different tabs for the ends of the chains. One, that was a miniature copy of a stop sign, had "shut" instead of "stop." The other, looking like a yellow yield sign, had "open" instead of "yield" painted on it. Make sure that, even with the valves open, the shut tab hung a little lower than the open tab to make it easy to grab and pull when the glass broke. A couple of trips under the spray of hot water from a broken glass trying to grab the right chain to close the valve should be convincing that the arrangement will pay in the long run. The idea is to be prepared to avoid making several passes at those chains.

Locomotive boilers, and a few others, are fitted with gauge glasses independent of water columns. They normally have a liquid line that penetrates the boiler, with a few holes in it to restrict surging flow so that the glass level is stable.

Figure 10-33. Gauge glass shutoff chains.

Gauge glasses come in many forms, but they all perform the same function. The water level inside the boiler (or the drum) is repeated in the gauge glass. The water in the glass is usually very clear since it is all condensate. Steam is constantly condensing in the water column, connecting piping, and the gauge glass, then draining to the bottom of the column and gauge glass, and returning to the boiler through the connecting piping. Occasionally, when the water level is fluctuating, as the boiler water is surging into and out of the water column, it will mix with that condensate and any color in the boiler water will appear in the bottom of the glass.

Since the steam and water are both clear, it is difficult to identify the actual water level in some gauge glasses. The most common one is a simple glass tube, like the one in Figure 10-33, 12–20 inches tall, with some paint applied along one side. The paint is applied to form a thin red line along the length of the glass with a wider white line applied over that. That is the minimum to have. Don't try to save a few bucks and buy plain glass tubes instead of the red line tubes. In one, they also bought a new boiler because the operators made a wrong decision about water level. With a plain glass, it is difficult to tell if it is full of water or completely empty, when the level is beyond the limits of the glass. The red line glass utilizes the natural diffraction of light through steam and water to help to determine what is water and what is steam. When the level is within the limits of the glass, take up a position opposite the side with the red line. The narrow red line will be above the water level. A much wider red line will appear below the water level. It works because the light is bent at the intersection of the glass and water. It is not at the intersection of glass and steam. One important consideration is to install the glass with the lines painted on it away from the normal position when viewing it.

Tubular glasses should be fitted with an additional glass enclosure, usually wire reinforced, to protect personnel in case the glass breaks. It may not be necessary when the glass is 10 feet in the air and basically inaccessible. However, the glasses are used on vessels where people can be right beside them. Those should be guarded. Tubular glasses cannot handle pressure above 150 psig. Higher pressure boilers have other products that permit viewing the water level. Prismatic gauge glasses are heavy steel frames with a groove cut in them to form a tube between the steam and water connections and a special glass bolted to one side. The glass is thick and narrow to eliminate the stress associated with the difference in temperature between the water and air sides. A tubular glass tends to expand more on the inside, where it is hot, and the colder outside of the glass restrains that expansion, resulting in stress that will eventually result in the outer layer cracking, being pulled apart by the tensile stress. Since the prismatic glass is narrow, the stress is minimized. The glass to steel frame joint is sealed by a gasket. The glass is pressed against the gasket and frame by dogs, which are held against the glass and frame by bolts (Figure 10-34). The notches, cast into the glass that produce the saw tooth appearance, use the diffraction principle to differentiate between water and steam. Part of the installation of a prismatic glass requires a light shining on it from the side in order to illuminate the notches. They appear bright, almost white. Diffraction in the water shifts the

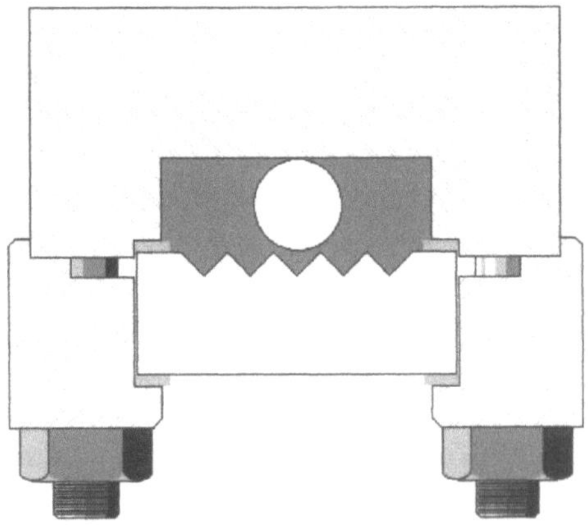

Figure 10-34. Section through refractive gauge glass.

light below the water line and that portion of the glass containing water looks dark.

When pressures get higher than 250 psig, the glass cannot withstand the heat of the steam. Flat glass is used with a thin sheet of mica (a mineral that forms natural transparent sheets) installed between the gasket and glass. As pressures increase, the problems with differential expansion prevent the use of full length glass. The gauge glass is converted to several small round flat glasses stacked one over the other on a steel frame. These have areas between each glass where the level is not visible. To allow differentiation of water and steam, the gauge glass is doubled up with another round flat glass behind, installed at a slight angle to the other one. Lights shine through red and green lenses and through the gauge glass. Diffraction, in this case, determines which color is seen: red if the glass contains steam and green if it is under water. Make it a point to carefully read and understand the manufacturer's instructions for the gauge glass. It will pay by reducing the number of times it has to be changed. Another problem with gauge glasses is regular packing leaks. Read the section on pumps to get some guidance on how to install packing properly to avoid having leaks right after they are packed. Another technological advance is graphite tape, which can be wrapped around the glass to form a packing ring that will do an excellent job of sealing a gauge glass.

Low Water Cutoff

LWCO is "Low Water Cut Off". Frequently integral with the water column, occasionally (on hot water boilers) built into the boiler, and regularly mounted as an external device, a low water cutoff is the primary protective device to save the boiler in the event that the water level goes too low. The cutoff must be installed in a manner that keeps it in a position relative to the boiler so that thermal expansion does not shift it relative to the lowest safe water level. A low water cutoff normally has a mark in the casting that indicates its operating point. That level has to be higher than the lowest safe operating level established by the boiler manufacturer. If there is no indication of that level in the documentation or on the boiler, the bottom of the gauge glass is a good place to set it. Be aware that a low water cutoff can be installed improperly. Any cutoff located significantly lower than the bottom of the gauge glass is a potential problem.

Any steam boiler should have two low water cutoffs (see why they fail at the end of the book). They should be piped to independent connections on the boiler. That way, if one connection gets plugged, the other cutoff can still work. Many are installed with a common steam connection because they are less likely to plug with two water leg connections. If there are valves located between the low water cutoffs and the boiler, they should be full ported valves to reduce the potential for plugging. They must be rising stem type, or quarter turn valves, that indicate their position at a glance.

The drain valve for any low water cutoff should always be a globe valve. Gate valves and quarter turn valves do not throttle flow adequately to permit the operator to drop the water level slowly.

The cutoff must be installed in such a manner that it will drain back into the boiler. A major university lost a brand new, field erected boiler because the erector installed the cutoff in a trap (Figure 10-35). Note that the

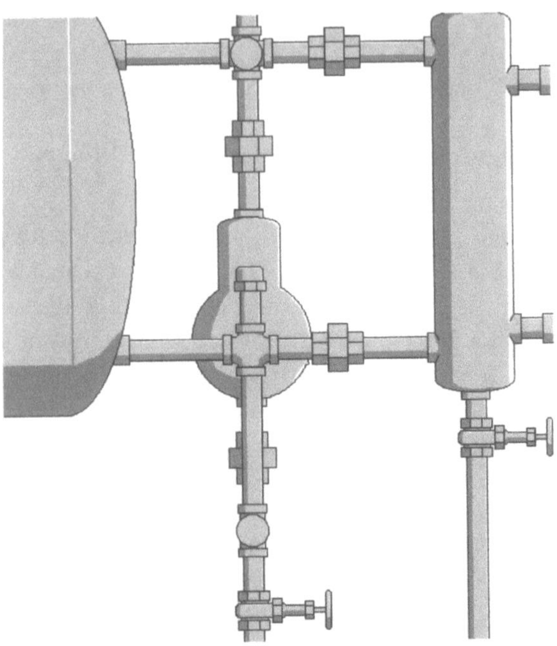

Figure 10-35. LWCO piped into a trap.

figure shows tees and crosses in the piping closed with nipples and caps. That is to allow access to the piping to inspect it and, if necessary, clean it.

If the boiler has low water cutoffs at the front and rear of the boiler, don't be surprised if they are not at the same level. Since the fire is concentrated in the front of the boiler, a slope in the surface of the water in the boiler from front to rear is not unusual. Depending on the distribution of the flue gas and tube arrangement, the level in the back of the boiler can be higher or lower than the front and there are some boilers where the level in the back shifts relative to the front with load changes.

Float actuated cutoffs require some means of sealing the part which connects the float rod to the electrical switches to prevent steam or water leaking into the portion of the switch that contains the electrical contacts. The most common method of sealing is to use a bellows (Figure 10-36), which allows the float shaft to transmit the float motion to the switches. The bellows provides a water (and steam) proof seal, which is flexible to allow movement. Another common method is to use a magnetic coupling, where a magnet connected to the float shaft is followed by external magnets connected to the switches (Figure 10-37). They work well in very clean environments. Another method is to transfer float motion using a transverse shaft (Figure 10-38) with packing. However, these are prone to leakage.

Any type can fail if the float is banged around by improper testing or fluctuating water level to create a crack so that the float fills and sinks. That is a fail-safe mode because the boiler should shut down. The cutoff is listed here, after the water column and gauge glass, because it is the operator watching the level in that gauge glass that is more reliable than the low water cutoff. If low water cutoffs were as reliable as desired, there would be almost 30% fewer boiler failures. Recall the low water cutoff testing in operations and read the comments in why they fail later in this book.

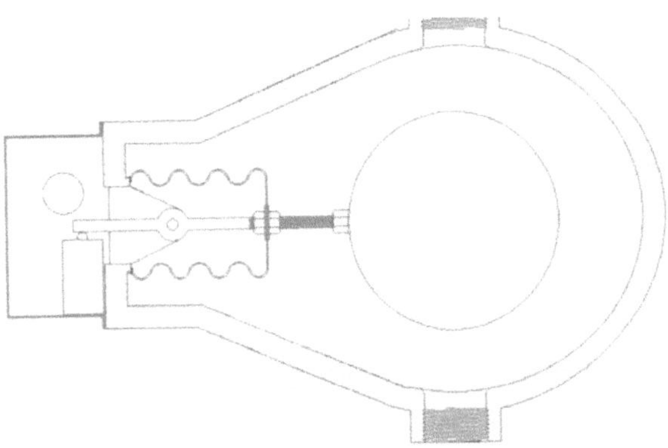

Figure 10-36. Bellows on float switch.

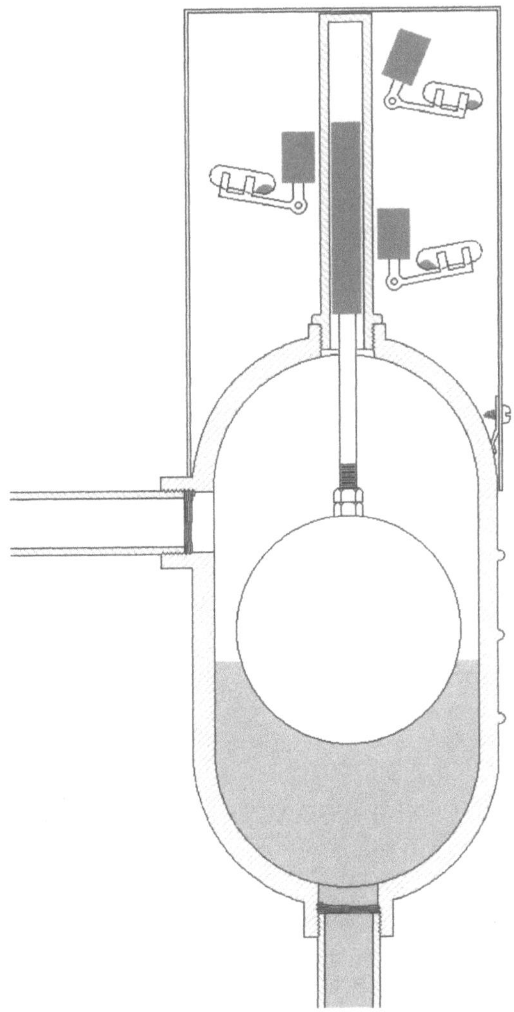

Figure 10-37. Magnet actuated level switch.

Pressure Gauge

A pressure gauge is a required piece of trim on a boiler. It is obvious that a pressure gauge is needed to ensure that the controls are doing what they are supposed to. A plant without a pressure gauge on the boiler is bound to have a lot of other problems. The pressure gauge is required by code. Its size and scale are also subject to requirements of the code. A favorite violation in many plants is replacing the gauge with a much smaller one. The owner thought to save some money but ended up buying two gauges because a smaller one does not meet the code requirements. There is no specific size required by code. The interpretation of the code requirement that the gauge be "easily readable" is interpreted to mean nothing smaller than what the manufacturer installed originally. For low pressure boilers, the size

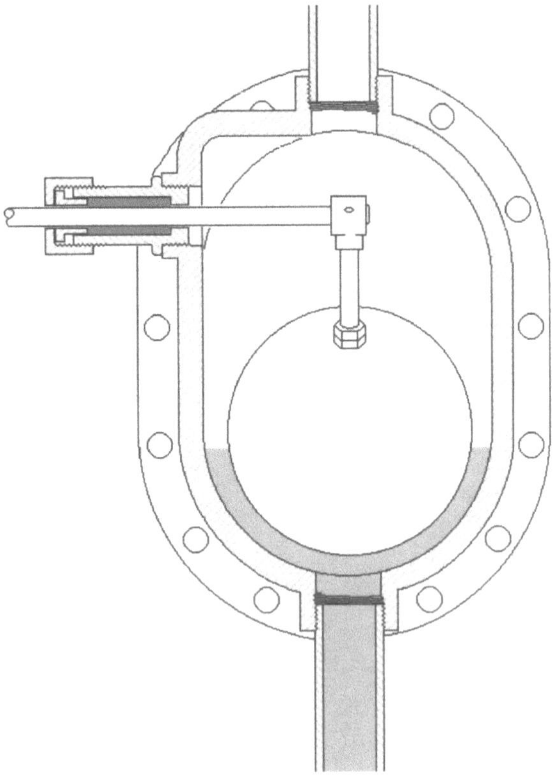

Figure 10-38. Packed transverse shaft for level switch.

is dictated by the travel of the pointer, which must be at least 3 inches for pressure swings from 0 to 30 psi. Manufacturers do not put on larger gauges to make the boiler look better. They put that large gauge on because the National Board Inspector monitoring that boiler's construction considered it as small as it could be and be easily readable.

A pressure gauge is normally selected such that at normal operating pressure, the needle on the gauge is pointing straight up. That makes it easy for the operator to determine if the controls are operating properly. The normal hydrostatic test pressure for a boiler is 1-1/2 times the MAWP. The gauge must always have a scale range that extends to that value. Hot water boilers must also have a thermometer that indicates the highest temperature in the boiler. Code rules for size and scale also apply to them. The piping connecting the pressure gauge to the boiler cannot have any other connections, except a drain connection that is open to the atmosphere and an extra valved connection for the inspector to attach a gauge. A valve in the piping must be a quarter turn valve, with the handle in line with the piping when the valve is open on low pressure boilers, and a rising stem valve locked open during operation on high pressure boilers.

Code requirements do not include provision of crosses and tees in the piping to permit cleaning it. However, it is a good idea to have them, as there have been instances where the boiler's pressure gauge piping was plugged with mud that managed to get into the sensing line over the years. The piping should be opened and inspected at the connection to the boiler every year. The rest of the piping should be inspected when there is any reason to believe it may contain some sludge or mud. The piping should also be blown down every year, right after bringing it up to pressure and before picking up the load. The piping should include a siphon, or pigtail, a curl of pipe or tubing, to ensure water is trapped between the gauge and the boiler to ensure the heat of the steam never gets to the working parts of the gauge. Sometimes, the gauge is connected to a section of piping that traps water for that purpose. Refer to the section on controls and instrumentation for other important considerations for application of pressure gauges.

On many larger boilers, the gauge can be considerably lower than the steam drum so that it is visible at the normal operating level. Those gauges have to be calibrated for the installation. They will have several feet of condensate standing in the connecting piping. The head of that water adds to the gauge pressure. If the gauge is 23 feet below the drum connection, it will read 10 psi higher, unless it is adjusted to compensate for that static head.

Pressure and Temperature Limit Switches

A boiler that vaporizes a fluid should always have a high pressure switch to stop the burner, or isolate the source of heat, in the event that the pressure in the boiler gets too high. If the boiler simply heats a fluid, it should have a high temperature switch. In some instances, fluid heating boilers are served by a common expansion tank, which can be isolated from the boiler. A high pressure switch is also provided. A hot water or fluid system boiler can also have a low pressure switch to prevent operation when the system pressure is so low that vapor would form out in the system (at the high point) to block liquid flow and possibly cause the equivalent of water hammer to damage the piping or heat utilization equipment.

All high limit switches require a manual reset. In some jurisdictions, this is interpreted as a switch which will not close, once it is opened by a high pressure or temperature, until the operating personnel push a button located on the switch. Preferably, the system should require the operator to reset it at the control panel or boiler front, rather than on the switch. In most cases, that switch is above the reach of a boiler operator. It is seldom mounted where the operator can conveniently get at it to push that reset button. Picture being alone in the plant at two in the morning and trying to climb up to the switch to push the button. Keep in mind that first priority.

In addition to the high limit switch, a boiler can have a pressure or temperature control switch, which provides for on/off control of the boiler. These switches are all directly connected to ensure that they sense the actual pressure or temperature in the boiler. Location of temperature switches is important. See the discussion on boiler construction. Pressure switches will not have any valves separating them from the boiler, unless they are rising stem valves and are locked in the open position when the boiler is operating normally (a provision on a boiler whose continued operation is critical). Pressure switch sensing piping can plug up, just like the pressure gauge sensing lines, although a pressure switch is normally mounted close to the boiler connection (it does not have to be seen all the time like a gauge). It is always important that the pressure switch has a siphon or piping arrangement which traps and holds some liquid in the switch and immediate connecting piping to protect the switch from the high temperature of the vapor.

Another concern with pressure switches that use mercury switches is the mounting of the switch. If mounted on a siphon, the switch can be tilted as pressure builds in the boiler. The siphon tends to straighten just like the Bourdon tube in a pressure gauge. That would alter the switch operating point. If there is a mercury bulb switch mounted on a siphon, make sure the travel of the mercury switch is perpendicular to the siphon.

Temperature limit switches are normally installed with a thermal bulb penetrating the boiler and the switch assembly right on the end of the bulb. When they are mounted separately, to keep the wiring and switch isolated from the high temperatures in the boiler, it is common for the assembly to include a capillary between the bulb and the switch bellows, or diaphragm, so that the fluid expands in the bulb as it is heated. Some of the fluid is pressed into the capillary, which displaces fluid in the capillary, pushing it into the bellows, or diaphragm chamber, to expand it and actuate the switch. If the capillary is crimped, by bending or by physical abuse, then the flat ends of the bulb tend to bulge out, making more room for the expanding fluid because the crimping of the capillary restricts the movement of the fluid in the assembly. The bulging of the flat ends of the bulb can act like a spring, maintaining pressure on the fluid such that it eventually bleeds through the small restriction and acts on the switch. After the boiler cools, the bulging is restored first and it may even reverse, caving in at the end, as the liquid shrinks to produce a pressure differential that forces the fluid back through the small restriction and the switch resets. The liquid slowly bleeding through the restriction results in the switch operating after a delay. Any significant delay in the response of a limit, or operating temperature control, is probably due to damage to the capillary. The only solution is to replace the switch. If the restriction is due to a repeated situation (like a plant where the operator climbed to reach a valve and repeatedly stepped on a capillary draped over a support), the capillary can be shut off entirely and the switch will not work.

Since the switches are normally mounted on the boiler, or the steam drum, it is not at all unusual for them to be located where they do not get regular attention. The heat that radiates from the boiler and leaks through the casing or lagging creates swirling air currents around the boiler and its trim. The air currents can be warm one minute and cold the next. The air around pressure and temperature switches promotes breathing of the switch housing that sucks dust into the housing. Dust settles inside the housing and can eventually block its operation. That is assuming that the cover is on the housing. All too often, the cover is dangling or removed. They are usually quite full of dust. They have to be cleaned.

Valves, Steam

The boiler codes do not have any requirements for steam valves for low pressure boilers. However, follow the discussion on high pressure boiler piping. The reasons for valve arrangements could apply to a low pressure plant. When a boiler plant has more than one boiler, and they are connected to a common header, two valves have to be installed on the steam outlet of each boiler, with a manhole. The piping between them has to be fitted with a free blow drain valve. The primary purpose of that arrangement is to protect anyone who is inside the boiler by providing a vented section of piping between the two valves to isolate them from steam generated by other boilers, or HTHW. A "free blow drain" is actually a combination of piping and a valve, or valves, to connect a piece of process piping to atmosphere, with a constant decrease in elevation of the piping so that it drains. In normal parlance, the label refers to the valve that is in that combination. The purpose is to provide an uninterrupted path from the process piping to atmosphere at a low point, where any liquid in the process will drain out. Since it is open to the atmosphere, any fluid in the process piping will also flow out until there is no pressure (or at least an insignificant pressure) in the process piping. A free blow drain ensures that the process line is not at pressure and does not contain any process fluids under pressure. A free blow drain should be provided in all piping systems that contain a process fluid that can be safely drained without contaminating the atmosphere, specifically to allow personnel to confirm lack of fluid under pressure in that piping. If the boilers have manholes, use the double valve and drain provision for safely working in them. It does not matter whether they are high pressure or low. Steam is deadly at any pressure above atmospheric and can be dangerous at any pressure and temperature.

The code does not require a non-return valve on all high pressure boilers. Non-returns are recommended in multiple boiler plants but are not required. Don't try to save a few bucks by using regular valves. Non-return valves just make operation of a multiple boiler plant a lot easier for the operator. The investment in a more expensive valve saves in operating headaches. A less tangible reason is that they prevent high thermal variations in the boilers (when operators do not isolate the boiler and there is cold water under the hot upper blanket of steam) and flooding (as the steam condensate accumulates in a cold boiler), which can result in early equipment failure. Since a non-return valve acts as a combination globe valve and lift check valve, it is treated by operators as an automatic shutoff valve for idle boilers (the check function isolates the boiler) which automatically opens when the boiler starts making steam. With non-return valves, the operator has to make a trip to the top of the boiler only for isolating it for annual internal inspection. It is important to use the free blow drain to remove any condensate from the piping between the two isolating valves to prevent a slug of condensate rushing down the piping with the first flow of steam.

Don't use a non-return valve on a low pressure boiler. The piston type disc in a non-return valve is heavy. It takes a lot to lift it, resulting in a 2–10 psi pressure drop across a non-return valve. Since low pressure boilers typically operate at around 10 psi, a non-return could prevent any steam getting to the facility. Instead, the same performance can be obtained by installing a low pressure drop check valve on the boiler outlet. Just remember that it has to have a low pressure drop. If the check valve is used to prevent steam entering and condensing in an idle low pressure boiler, then it should be soft seated. When that soft seated check valve is added to the steam outlet, also add another one as a vacuum breaker to a branch off the vent connection so that a vacuum will not crush the boiler. When there is no non-return or check valve in the steam piping, the valves have to be operated with each startup and shutdown of the boiler. Access to those valves should be as simple and convenient as possible. Either chain wheel operators or properly located platforms with safe ladders should be provided so that the operator can get at them. Operation of steam valves is scheduled by the boiler and the load more than anything else. Thus, the operators have to get at them quickly.

Valves, Feed Water

On low pressure boilers, the code requires a stop valve and a check valve on the boiler feed piping. The pipe itself is not controlled by the code. The code has specific requirements for the feed water piping on a high pressure boiler out to the second stop valve and also requires a check valve. That arrangement normally means that the bypass valve and isolating valve for the feed water control valve are both within the limits of the boiler external piping. The shutoff valve is there to allow isolation of the boiler from the feed water supply when it is shut down. More importantly, it isolates the feed water system from the pressure in the boiler. The check valve is there to help prevent draining water from the boiler, in the event that there is a failure of the feed water supply. More importantly, it prevents boiler water from leaking out to produce a steam explosion if there is a failure of the piping. Unlike the steam valves and piping, there is no code requirement for a free blow drain connected between the two isolating valves on a high pressure feed water line. There should be one installed, and for the same reasons.

Valves, Blowdown, and Blow Off

The valve for manual control of continuous blowdown (surface blow off) should be provided with an indicator that shows the position of the valve to allow an operator to restore a particular valve position. Some valves are fitted with indicators and tapered throttling guides as part of the disc so that the flow rate through the valve is proportional to the indicated valve position. The ability to closely regulate the flow of blowdown (independent of automatic blowdown controls) permits the operator to closely control the concentration of solids in the boiler.

Although not essential and not required by the code, the installation of a check valve between the continuous blowdown control valve and the boiler is strongly recommended. If the continuous blowdown valve is not closed, the check valve will prevent water from another boiler entering the idle boiler. It is also like using a non-return valve. Use of that check valve allows the blowdown control valve to remain open for short outages. An automatic blowdown control system can be isolated because the blowdown control valve failed. On most small boilers, these are quarter turn, motor actuated valves, or solenoid valves, which are not designed to handle flashing steam. There is supposed to be an orifice or manually adjusted throttling valve to take the pressure drop located in the piping close to, and farther from, the boiler than that automatic valve. If not, or the orifice is removed, or the throttling valve opened wide, the automatic valve will most certainly fail.

Bottom blow off valves come in a variety of forms. The most important part of their construction is that they do not have any pockets or cavities that sludge can settle into and plug up. That is the idea anyway. Installing the valves backwards allowed all the mud to settle on top of the discs, preventing opening the valves. There are two things that are stressed by these points. Use the proper valves (ones designed for bottom blow off applications) and make sure they are installed in the right direction. See the section on normal operation of blow off valves in Chapter 2.

The code for high pressure boilers requires two valves for bottom blow off designed for the service. The piping connecting them and the boiler must be at least schedule 80 (extra heavy) materials, certified to comply with ASME codes and all welded. Piping inside the second valve must be fabricated by a manufacturer or contractor certified by ASME to do that work (see the section on ASME Code construction). All other blow off and blowdown connections only require one valve. The code piping requirements are limited to the portion between valve and boiler. There may be two valves in other lines because the owner or contractor elected to have an accessible valve to use, with the code required valve and piping located close to the boiler, thereby limiting the extent of the code piping. Low pressure boilers, and some high pressure boilers, are equipped with quick opening valves. This type of valve works something like a gate valve but has a steel plate with a hole in it that is positioned in line with the pipe. Thus, there is no way for mud to plug the valve. The code permits one of the two valves required on high pressure boilers to be a quick opening valve. There are rules for operating those valves (see the section on normal operation). They should be installed in a manner that makes it easy to use them.

The seatless blow off valve (Figure 10-39) is a commonly used bottom blow off valve that is often operated improperly. It is easily repaired. Unlike other types of valves, it does not require skill or special tools to repair or even adjust to restore its shutoff capability. The manufacturer's instruction manual is a very important read before working on these valves.

Figure 10-39. Seatless blow off valve.

Boiler External Piping

The jurisdiction of the code for construction of power boilers extends to the far side of the second shutoff valve from the boiler on steam, blow off, and feed water piping. The jurisdiction extends to the far side of the first shutoff valve on all other connecting piping. All boiler external piping must be made of materials certified to comply with ASME codes. All welded piping must be fabricated by a manufacturer or contractor certified by ASME to do that work. A piece of welded piping should be stamped or fitted with a securely attached nameplate containing the stamping required by the ASME code. The stamping should include the "S," "A," or "PP" stamp. There is also the National Board "R" stamp, which indicates a contractor has repaired the piping. Be aware that the boiler inspector could look for those stampings at any time. They had better be there or the boiler will not be allowed to operate until a complying section of piping is installed.

Threaded pipe and fittings can be assembled by anyone. Damaged piping can be replaced, provided that the materials comply with the code. Replacing flanges and fittings is usually simple since they are marked. All that is needed is to find materials with matching marks. Pipe, on the other hand, can vary from code quality to what is called "untested" pipe, with some different grades in between. Usually, there will not be any markings on the damaged pipe to get a clue as to what material is required. There are many different grades of quality of material that can be provided in compliance with the ASME code. There are many ASME material specifications for a material that is not satisfactory for boiler external piping. The insurance inspector should be able to specify what material is required when replacing boiler external piping. When it is purchased, request the Mill Test Certificates and check the stamping (grade and heat number) on the pipe against the data on the certificate. Record in the maintenance log that the material and paperwork match or return both if they do not. If the pipe is welded, it can only be repaired or replaced by someone certified to do so by the ASME or the National Board. Unless plant maintains several boilers with a need for regular repair of boiler external piping, it does not pay to obtain that certification. It is much less expensive to locate a contractor who is certified and have them do the work for an occasional repair.

Adding a connection to boiler external piping can only be done by a certified contractor. Many an owner has had to employ a certified contractor to remove and replace connections that were installed by the operators, the maintenance personnel, or an unqualified contractor. How would the inspector know if a connection was added? All that the inspector has to do is ask for the ASME P-4 and any National Board R-1 forms available. They describe the initial construction and all repairs. If a connection is not described on those forms, it is not in compliance. Keep all those forms, which are actually certifications, for the boiler external piping, along with the forms for the boiler itself.

HEAT TRAPS

There is a general use of the term "heat trap" to refer to anything that is added to a boiler to absorb heat remaining in the flue gas. They normally return that heat directly to the boiler. Conventional heat traps are economizers and air preheaters. CHXs can be used as either an economizer or an air preheater but are commonly used to heat water for other purposes.

Economizers

An economizer traps heat by transferring energy from the flue gas to the boiler feed water. That heat does not leave the boiler but is returned to the system. Economizers are normally found in higher pressure steam plants. On most low pressure or any of the HTHW plants, the gas temperature leaving the boiler is already fairly low. The surface area of an economizer would be quite large (small temperature difference), making the device quite costly to recover very little additional heat. Some higher pressure plants do not benefit from the addition of an economizer as well. An economizer can work in a low pressure steam plant that has no condensate returns. There, the feed water temperature would be much lower than steam temperature. If there is a low pressure plant, with little condensate returns, such that the feed water temperature (before heating in a feed tank) would be around 100°F lower than the steam temperature, an economizer could be used to trap some of the energy lost up the stack.

When the boiler feed water is colder than the steam and water in the boiler, it can extract more heat from the flue gas. Fluids colder than what is in the boiler can also be used in an economizer to recover the heat. An economizer on a high pressure boiler plant makes it as efficient as low pressure boilers. The feed water supplied to the economizer inlet is about the same temperature as steam and water in a low pressure boiler. It is important to be certain the feed water flows through the economizer in the opposite direction of the flue gas so that it sees hotter flue gas as it heats up and the coldest water is exposed to the gas just before it leaves the economizer. Economizers can heat feed water to a higher temperature than the flue gas leaving the economizer because of this counterflow arrangement. For the very large electric generating stations, the feed water is first heated by extraction steam from the steam turbine in the feed water train. It is then sent to the economizer at a temperature in the range of 450–490°F. In the economizer, the feed water temperature is increased to within 50°F of the boiling point at the pressure of the boiler. The feed water is then sent to the steam drum where it mixes with the water that is circulated in the boiler. A typical high pressure boiler would be boiling water at temperatures in the range of 650–690°F. The steam is separated from the circulating water in the steam drum. The dry steam is sent to a superheater, where its temperature is raised to more than 1000°F.

At low loads, there are some concerns with economizer operation, which can restrict the turndown capability of the boiler. When the economizer is mounted in the stack or on top of the boiler, the water has to flow down through the economizer. The natural tendency of heated water is to rise up through colder water since it is lighter (the thermal-siphoning effect). The water flow through the economizer can become unstable at low loads. Combine that with the fact that the heating surface does not change. Proportionally, more heat transfer occurs at lower loads. That provides an opportunity for generating steam in the economizer. Generating steam in the economizer will promote scaling of the water sides of the economizer and potential damage from water hammer as flows change. In the large utility style boilers, the water flow in the economizer is always up flow to allow any steam bubbles that might form to rise up to be collected and sent to the steam drum.

When the feed water control valve is located between the economizer and boiler, the probability of steaming is reduced, as the economizer operates at a higher pressure. The control valve will take a beating, as some of the water flashes to steam as it goes through it. The feed water piping in the boiler drum will also be exposed to water hammering and erosion from the flashing steam. There are such things as steaming economizers, but they are designed to do it. A normal economizer is not designed to generate steam at any load. That is one reason that the economizers for large utility boilers are designed to approach the boiling point, not reach it. If there are wide variations in load, the economizer of each boiler should be fitted with a return line to dump some of the feed water back to the deaerator. By adjusting a globe valve in that line, the outlet temperature can be controlled at low loads. Bypass and isolating valves can be used for low load operation on lower pressure units. If the economizer has problems, draining it and bypassing it will not damage it because the flue gas temperatures will not be hot enough to hurt it.

An economizer is typically constructed of tubes just like boiler tubes, with those tubes rolled or welded into headers. The tubes can be bare for coal fired units. They are usually fitted with fins to increase the heat transfer surface (Figure 10-40) on oil and gas fired units. There are two standard arrangements of construction: square, where the tubes are straight and connected to each other by bends; circular, where the tubes form a coil between the two headers. The circular economizers are less expensive initially but almost impossible to repair.

Since economizers can be subjected to corrosive conditions more frequently than the boiler, the materials of construction may be special to resist corrosion. CE developed cast iron muffs, cast pieces that look like finned tubes pressed over the steel tubes, which provided a corrosion resistant covering for excellent performance on coal and heavy fuel oil fired boilers. The iron conducted heat well in addition to resisting corrosion. Modern metallurgy has created materials that permit construction of economizers that can withstand very corrosive conditions, permitting closer operation to the flue gas acid dew points without concern for serious damage.

In addition to corrosion, economizers can have problems with soot accumulation, occasional plugging with unburned fuel, and unique situations (Figure 10-41) with waste fuels. The tube in the top of the picture is the soot blower. The coated fins can be seen in the bottom of the picture. Soot and dirt manage to build up between the fins of finned economizers to almost completely block heat transfer. Even if there is no reason to believe there will be problems of blockage, an economizer should be supplied with a means to clean it or provisions to install them in the future. The common in-service method of cleaning is using soot blowers. They will be ineffective when the deposits form a hard mass. There should also be means to gain access to the economizer to clean it with water washing. Some economizer applications (like the one in Figure 10-41) require regular cleaning, a tough

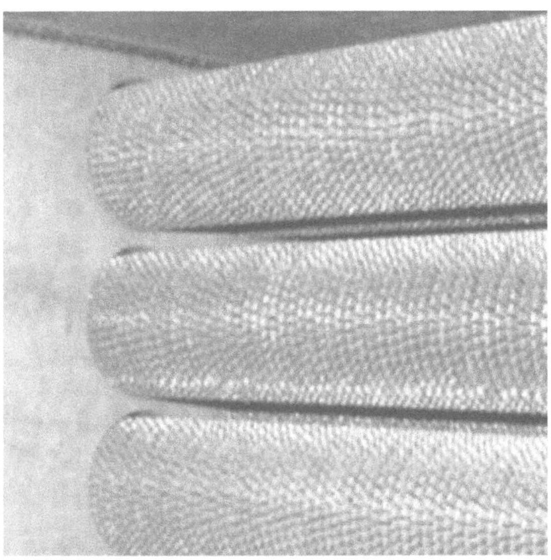

Figure 10-40. Finned tube economizer.

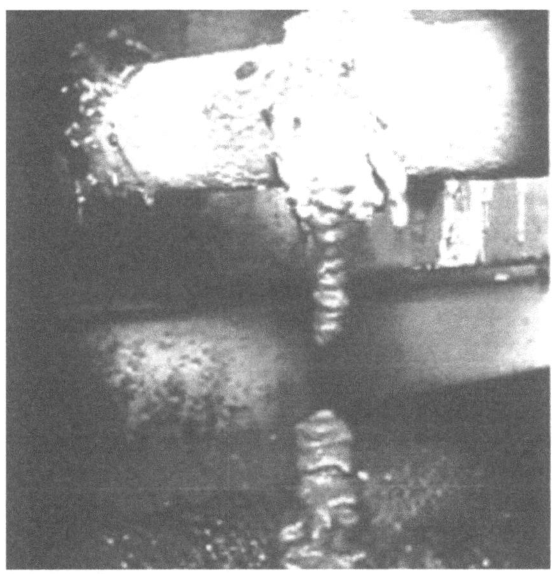

Figure 10-41. Plugged economizer, firing waste fuel.

and dirty job for the boiler operators. The savings in fuel makes the effort worth it.

Gas fired operations produce flue gas with very low acid dew points. The deaerator can be operated at lower temperatures, supplying economizers or low pressure boilers at lower feed water temperatures, when firing gas than when firing oil or other fuels with higher carbon and sulfur content. If gas is the primary fuel, adjust (slowly) the deaerator pressures to raise the feed water temperature when firing fuels with a higher acid dew point and lower it for firing gas. An alternative to that, found in plants with auxiliary turbines designed for low exhaust pressures, is to use a steam heated feed water heater between the deaerator and economizers to raise the feed water temperature to the point that corrosion will not occur when firing high sulfur fuels. Power generating plants normally use feed water heaters to condense some of the turbine steam and raise the feed water temperature. It raises the feed water temperature to reduce the potential for corrosion in the economizer and reduces the required size of later stages of the turbine for overall cost savings. This extraction steam also bypasses the condenser. The condenser discharges the latent heat of vaporization to the cooling water, which is lost to the cycle. By using some of that steam to heat the feed water, that energy is recovered, which improves the overall cycle efficiency in making electricity.

The best economizer arrangement (and also the most expensive) is where the flue gases flow down through the economizer. For counterflow, the feed water flows up through the economizer. That prevents problems with stratification and thermal siphoning. A turning box under the economizer can also serve as a drain pan for wash water used to clean the economizer. However, those installations introduce a hazard when the boiler is idle. Any natural gas, or other fuel vapors, which are lighter than air and can get into the setting, can accumulate as the boiler and economizer arrangement forms a trap to hold them. An access door in the top of the ductwork between the economizer and boiler to vent it will avoid this problem and should be used prior to entering the setting for inspection or maintenance.

In addition to capturing heat that would be lost, an economizer provides additional heating surface for transferring the energy that is in the fuel into the steam and water. Adding an economizer to an existing high pressure boiler installation can also increase the capacity of the plant. Heat that was absorbed through the boiler surface to raise the feed water temperature is now used to generate more steam. Capacities can increase by as much as 8%. Of course, the fan has to be replaced to overcome the pressure drop through the economizer or that added capacity is lost.

Economizers require some attention when starting a boiler and at low loads (to avoid steaming, enough water has to keep going through it). If there is a feed water recirculating loop, use that to maintain temperatures. Otherwise, during startups, add water to the boiler more frequently to provide some consistency to cooling of the economizer. Additional blowdown may be needed to provide enough water flow (that is a typical operation with heat recovery steam generators (HRSGs) with integral economizers). That little bit of extra work is worth the savings in fuel over the operating life of the equipment.

Air Preheaters

Use some caution with this term. Normally, a heat trap is meant when the term preheater is used. Some manufacturers will call a steam coil installed in an air supply an air preheater since it does do what the name implies. Within the trade, such devices are called "steam air heaters" or "steam coil air heaters" to differentiate them from the traditional heat traps. An air preheater uses energy left in the flue gas that leaves the boiler to preheat the combustion air. This makes the air preheater the only true heat trap. It does trap the heat without adding any surface to the boiler. The way the air preheater increases the efficiency of the boiler is by raising the temperature of the combustion air using the stack heat instead of fuel. There is also a slight increase in heat flow through the boiler heating surface due to higher furnace temperatures. An advantage of air heaters is higher temperature differentials. Instead of using 200°F plus feed water to cool the flue gas, the combustion air entering at 80°F is used. There is potential for lower flue gas outlet temperatures, which means higher boiler efficiency. However, corrosion of metal parts of the preheaters and ductwork to and including the stack must be given consideration. The steam coil air heater is used to warm up the cold air experienced in the winter to a temperature sufficient to protect the air preheater and the associated duct work. Since it uses steam, it detracts from the overall plant efficiency.

There are two basic designs of air preheater: tubular air preheaters, which consist of a box and tube heat exchanger to transfer heat from flue gas to the combustion air, and regenerative air preheaters. Tubular air preheaters are normally arranged with the flue gas passing through the center of the tubes and combustion air surrounding the tubes. Corrosion during startup and low load operation is reduced by bypassing the air around

Figure 10-42. Ljungstrom air preheater.

the heat exchanger, allowing the flue gases to keep it hot. Modulating the bypass damper to allow partial air flow does not work very well because the cold surfaces where the air first enters will promote condensation anyway. This is referred to as the "cold end" or the "cold corner." The bypass should be either open or closed.

Regenerative air preheaters use a rotating element to transfer the heat. A shaft rotates an assembly of "baskets" from the air side to the flue gas side and back. The metal baskets absorb heat from the flue gas and then give it up to the combustion air. The major manufacturer of regenerative air heaters makes a "Ljungstrom" (for its designer) air preheater (Figure 10-42). The regenerative air heater occupies less space than a tubular heat exchanger and can prevent corrosive conditions by simply stopping the rotation.

To accommodate varying combustion air supply temperatures, air preheaters are frequently fitted with steam air heaters to prevent acid condensation. There is a loss of efficiency associated with the steam use. That steam energy is recovered in added energy from the flue gas which could not be absorbed without damage to the preheater. Some of those heaters are adequate to permit startup and low load operation without starting and stopping or bypassing the air heater.

The Ljungstrom air preheater is designed with a single vertical shaft and oriented such that the flue gas flows downward and the air flows upward. The rotor is supported through a lower bearing at the cold end and guided through a guide bearing assembly located at the hot end, or top. The rotor can have between 12 and 48 radial members attached to the center post. The rotor compartments are closed with seal plates. The rotor sealing system consists of simple leaf type seals bolted to the rotor radial members at both the hot end and the cold end. The radial seals sweep across radial plates, again located at both the hot end and the cold end. To complete the system, axial seals are positioned at the peripheral end of the radial members of the rotor. These are also leaf type, labyrinth seals used with axial sealing plates. This system effectively separates the air system from the flue gas side. Nevertheless, there will be some distortion of the seals as the device is heated up. Some small amount of air will leak across to the flue gas side and escape up the stack.

For coal firing, some of the air is needed to be sent to the pulverizer system to dry the coal and carry it up to the boiler. This is the primary air. The air preheater can be divided into three sectors, rather than just two. One sector provides the primary air for the pulverizers. One sector provides the secondary air for the firing system. One sector carries the flue gas, which is cooled to an acceptable temperature. For coal firing, that temperature is generally in the neighborhood of 300°F, which is usually above the acid dew point. For natural gas firing, the exit temperature can be reduced to about 275°F. Specially designed air heaters have been attempted that can lower the gas temperature further. These CHXs have had limited success on coal fired boilers.

Air heaters are a little easier to operate than economizers. They can be left offline until the boiler is carrying a load. Then the bypass damper can be closed, or the rotor motor started, to put them into service. By simply not running the rotor motor during boiler warmup, the flue gas side is heated to prevent corrosion. The rotor should be run while purging the boiler to ensure that all the baskets are purged. There are small pie piece shaped sections that are sealed between the gas and air sides while the rotor is stopped. Regenerative air heaters require additional maintenance because of the moving parts and seals to separate the flue gas and air sections. Performance is usually more consistent than tubular air heaters. They can be cleaned in service, whereas tubular air heaters are usually bypassed for water washing or require a full boiler shutdown to clean them.

There were several attempts to use heat pipes for air preheaters in the 1980s. Heat pipes contain a fluid that can be vaporized at one end and condensed at the other end. The pipes are sloped so that the liquid flows down the pipe to the end exposed to flue gas, where the hot flue gases heat the liquid until it boils. The vapor then flows up the sloped pipe to the end exposed to the colder combustion air, where the liquid condenses, giving up its heat to the combustion air. The tubes are sealed in a casing where they penetrate a wall separating the hot flue gas and cold combustion air. These heat pipes have been used successfully on the Alaskan pipeline to keep the permafrost frozen. There were one or two applications on boilers. These were potentially for units with a high pressure differential across the air preheater or for district heating units that did not get much in the way of condensate returns. They could be designed with special coatings to avoid corrosion on the flue gas side. However, the acid condensation posed a water disposal problem in place of a corrosion problem. The regenerative air preheater is still the main workhorse of the industry.

Condensing Heat Exchangers

A CHX could be an economizer or an air preheater as well as other devices. What makes a CHX a CHX is the use of materials of construction that are corrosion resistant, allowing the heat exchanger to operate at temperatures below the acid dew point. Condensation of acidic flue gas components is expected and accounted for. There is a definite difference between a CHX and the other heat traps. The others are not designed to recover the latent heat in the flue gas. When the hydrogen in the fuel burns to form water, it normally leaves the boiler as steam. With natural gas firing, the energy that could be recovered by condensing that steam amounts to about 11% of the total heat input. A CHX is designed to condense as much of that water as possible to recover an additional 970 Btu per pound of water condensed.

The additional heat that can be recovered by a CHX helps pay for the exotic materials of construction. However, many of the materials that can withstand the corrosive acids cannot tolerate the high temperature of the flue gas. The current common application is a CHX used for preheating boiler water makeup and service water independent of the boilers. Flue gas is drawn from the boiler stacks by an ID fan downstream of CHX. By using city water, high temperature differentials (city water is normally between 40 and 70°F) can be achieved. The poor heat transfer capability of the corrosion resisting materials can be overcome. The typical applications use high grades of stainless steel for gas fired applications and Teflon coated copper for more acidic flue gases. Fire tube boilers have been developed for natural gas firing that use CHX in the final pass. The flue gas temperatures can be lower as there is very little sulfur in natural gas.

To withstand the corrosive properties of the flue gas after passing through the CHX, the exhaust ductwork is constructed of corrosion resistant materials, usually fiberglass reinforced plastic (FRP) piping. Those materials are not common to boiler plants, despite the fact that they have been used in some cooling tower operations and in SO_2 scrubbers for many years. They are not difficult to deal with in operation or maintenance. They just require learning about them. It is best to read the instruction manuals for the materials the plant may have because there are considerable variations in capability and handling.

The condensate from a CHX has a low pH because the condensed water will absorb the CO_2 and SO_2 in the flue gas to form acids. The drain piping should be FRP to a point where the condensate can be neutralized. Mixing the acid condensate of a CHX with the alkaline blowdown from the boiler can produce a mixture that may meet the local jurisdiction's requirements for sanitary sewage. If it does not, add caustic soda to neutralize the mix before it is dumped to the sewer.

A final consideration for heat traps is that they do not have to be used just on boilers or trap the heat from boilers. These devices can be used to recover heat lost up the stack from process operations. A device like an economizer, for all practical purposes, sits in a steel mill recovering an average 75 MMBtu/hr (120 million peak) and all it is doing is preheating boiler plant makeup water. Many a boiler plant can save a fair amount of energy in the winter if normal building exhaust can be trapped and used as combustion air. In those cases, the building and its occupants preheat the air.

BURNERS

Most boilers get their heat for the hot water or steam from the combustion of fuel, which requires a burner or firing system. There are some devices for combustion that are not called burners, including stokers and fluidized beds. All of them combine the fuel with air to form a combustible mixture in which the air and fuel react to produce combustion products and heat. The purpose of the burner, or firing system, is to control the mixing of fuel and air so that the combustion occurs smoothly and uniformly within the furnace of the boiler. There are several components of a burner and variations in construction that are designed to assist in this function. There are some of the important aspects of combustion that a burner design has to address.

The burner has to control the mixing of the fuel and air in a manner that ensures complete and stable combustion. Stability of combustion requires that the burner produce a fuel rich mixture right at the upper explosive limit, where the burning begins. That mixing point has to be stable, as described in the chapter on combustion. If the burner fails to produce a stable ignition point, the flame front will shift around in the furnace producing pulsations that disturb the process and make it worse.

The quality of the burner is indicated by noise. As the quality of mixing gets worse, the noise gets louder. Some burners are so bad that flame spurts out any open inspection port. Resolving that mixing problem is not a simple matter. It is a combination of engineering and art, with many solutions achieved solely by trial and error. It is not uncommon for a service technician to try several combinations of burner tips, diffusers, and burner adjustments to resolve an unstable ignition problem, sometimes taking days or weeks.

During startups, many owners and operators get frustrated with a new boiler because the startup takes so much time to resolve an unstable combustion problem. Despite the fact that the problem gets solved, they will never trust the boiler as much as they would if the problem never occurred. It is not uncommon and it is not something that is predictable. If it happens, don't blame the manufacturer and take a position that the boiler will always be a lemon. Unless the owner accepts something less than reliable operation out of a new boiler, it will always be more reliable than an older one.

It is the nature of burners that a deviation in any one part can produce several conditions inconsistent with good combustion, all of which can be due to several things. Many times, an operator unwittingly does something that alters a burner's performance, without being aware of it. The owner pays the price in higher fuel costs for long periods before the deviation is discovered and corrected. Understanding what the many adjustments on a burner do is one way of preventing such things from happening.

Air Supply and Distribution

The burner is normally fitted with some means of controlling the amount of air supplied to the fire. The means can vary from a simple single bladed damper to variable inlet vanes (VIVs) on the fan inlet and can include a variable speed drive (VSD) on the fan motor. To provide stable combustion, the dampers, or VSD, have to control the air flow without sticking or flopping around, which produces variations in air flow. The dampers also have to control the flow without producing distortions in the flow of air to the burner. If the dampers tend to shift air to one side of the burner inlet (or the fan inlet), it can shift the point of ignition to one side of the burner. That can produce instability. Sometimes, obstructions around fan inlets can produce unusual swirls that are carried through to the burner.

Installation of boilers that position building columns, pipes, racks of conduit, and similar obstructions within seven diameters of the fan inlet should be avoided. There are partial to total solutions to air distribution problems caused by such things, when it is impossible to move the boiler, or the obstruction. Of course, setting portable equipment, storage, and other things in front of a fan inlet can also cause problems with burner operation. Don't do it. The devices controlling the flow of air must present it at the burner throat in a manner that ensures mixing of the air and fuel to produce a mix in the flammable range, where the heat of the furnace will ignite it. To establish that ignition point where it is desired in the burner, there is always a primary air adjustment. It can be sectional dampers in a stoker, position of multiple burner registers, adjustment of cylindrical tubes in the burner that vary air flow, and, the most common, positioning of a diffuser.

A diffuser (Figure 10-43) contains slots or vanes that restrict air flow. Since the flow through the diffuser is restricted, the fuel–air mixture there will be richer in fuel than the mixture passing around the diffuser. The ignition point is usually aligned with the diffuser. It can be altered by changing the position of the diffuser. On a typical gas or oil fired burner, the diffuser normally has two positions: one for firing gas and one for firing oil. The reason for the different positions has nothing to do with the diffuser itself and everything to do with where the fuel enters the burner. In the typical burner, oil is admitted in the center from an oil nozzle. Gas is admitted through a gas ring, or spuds, on the outside of the burner. The diffuser positions must be switched to control the primary air ratio for each fuel. When that is the case, a semi-permanent marking should be applied

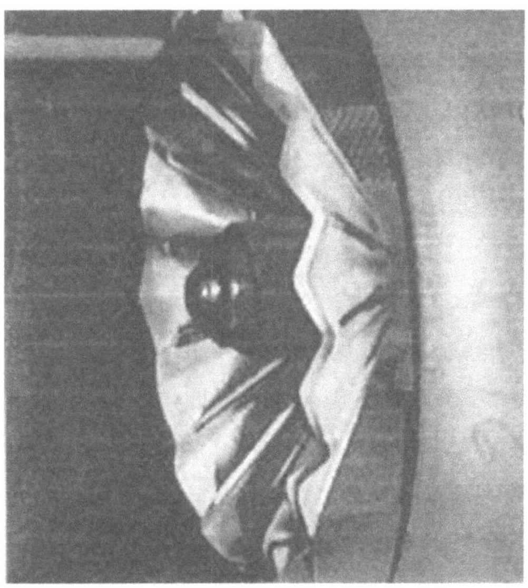

Figure 10-43. Burner diffuser.

to the adjustment for each fuel position so that an operator knows the diffuser is properly located. Paint a ring around the diffuser guide pipe with arrows pointing to the point where it enters the burner and label them for each fuel.

An inexperienced operator positioned a diffuser improperly on a boiler. The pipe was not marked. The operator knew that it was pushed in for firing oil, and pushed the diffuser guide pipe all the way in, as far as it would go. The burner failed to light several times, until enough oil had accumulated in the furnace to feed the explosion!

An ID oil or gas fired boiler does not have a FD fan and does not need any provisions to supply the air to the burner. It will still need means of controlling the distribution of air. Single burner boilers are typically fitted with a screen, or perforated plate, to provide uniform flow of air to the burner. Burners on multiple burner units are typically fitted with a register, a set of bent damper blades that form a circle around the burner inlet (Figure 10-44). Some are independently set with a locking bolt or screw on each blade, while others are fitted with linkage attached so that the blades turn uniformly and the flow of air to each burner can be adjusted while maintaining an even flow of air around the burner.

Burner registers will not only serve as a means to throttle air flow, they also deflect the air to create a swirl in the air. That produces additional turbulence for better mixing. Sometimes, two registers are employed: one to supply air around the outside of the fire and one for air supplied to the center, around the diffuser. When they are used, dual registers typically produce swirls in the opposite direction for better mixing. Another function of the burner registers and diffuser is flame shaping. Modern package boilers have very small furnaces and older sterling boilers have short furnaces. To prevent flame impingement on the furnace walls, the burner register and diffuser position combinations help shape the flame. On some boilers, the registers are modulated along with the air and fuel controls to alter the shape of the flame for different loads.

Figure 10-44. Burner register.

Rapid mixing makes for quick burning, hotter fires, and more NOx production. The register burner is being replaced by the axial flow burner, which is designed to minimize turbulence but ensures even distribution of air to the flame front. The original concept of the axial flow burner was developed in England, in concert with the Royal Air Force, to improve the performance of multiple burner boilers at the English Air Force Bases and included creation of a venturi throat for each burner that not only improved air flow distribution but also provided a means of measuring the air flow at each burner to allow final tuning of the air distribution through them (Figure 10-45(a)). One advantage of the venturi is that it creates a large static pressure to velocity pressure conversion at the burner inlet, most of which is recovered in the diverging section. The velocity conversion tends to balance the air flow through each burner to improve air distribution on multiple burner boilers. Control of air flow, including shutting off burners on axial flow units, is achieved by a damper that forms a sliding sleeve at the inlet of the venturi. Most low NOx burners applied to single burner boilers cannot benefit from the venturi design. Thus, other means are used to improve air distribution.

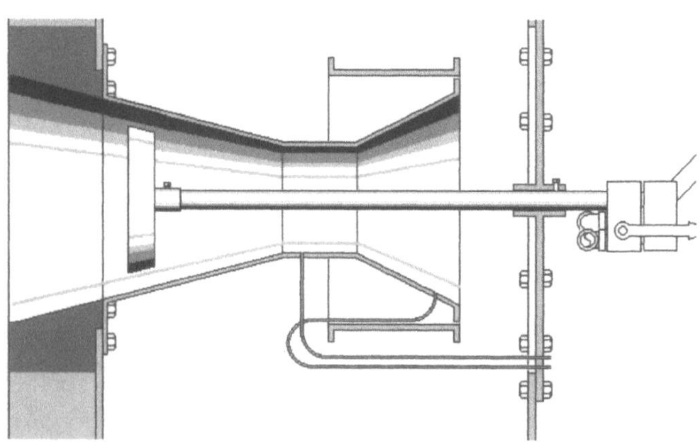

Figure 10-45(a). Venturi burner with flow sensing ports.

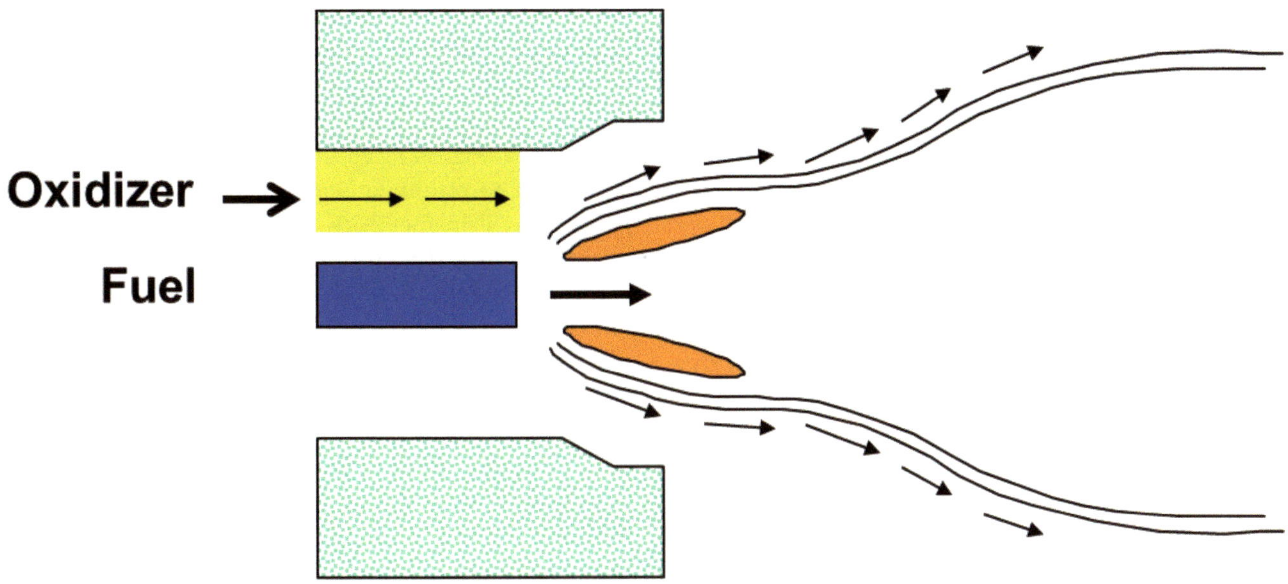

Figure 10-45(b). Radially stratified flame core burner.

One solution to this problem is the radially stratified flame core burner (RSFC) (Figure 10-45(b)). In this device, there are three air registers which can be adjusted to provide the proper amount of mixing to completely burn the fuel while reducing the NOx formation. The inner core is basically fuel rich. Under these conditions, the formation of NOx is minimized as the fuel competes with any nitrogen for oxidation. The second air register allows oxygen to diffuse into the inner core over a longer flame length, allowing the fuel to burn without raising the overall gas temperature to reduce the formation of NOx. The third register surrounds the flame with air and shapes it appropriately to provide the final burnout of the fuel without producing more NOx.

Large single burner and multiple burner boilers normally have one air supply, with the air flowing to the burner distributed within a windbox. The windbox receives the air from the FD fan and provides sufficient space around the burners for the air to be distributed evenly. At least that was supposed to be the idea. Several installations did not really provide adequate air distribution in the windbox, especially between burners. As a result, the fires were not truly uniform. A windbox has to be large enough to distribute the air and that is always larger than big enough to fit the burner. A burner manufacturer's dimensional tolerances for a burner are based solely on construction clearances. The minimum distance from the center of a burner to the inside of a windbox, as listed by the manufacturer, is just enough to prevent the register blades from hitting the windbox. If the blades are just clearing the inside of the windbox, there is no room for the air to get between them. With one of these windboxes, it will be difficult to get stack gas oxygen content down without generating a lot of carbon monoxide. Shrouds were developed to resolve the problem with limited room for registers within a burner windbox that would fit on the front of small package boilers. They consist of a cylinder of perforated plate, around 50% open area, larger than the open register (of course), but weighted for each application to achieve the most uniform distribution. The shroud is usually a couple of inches wider than the burner register.

In many single burner systems, proper placement of one or more 4 by 4 angles, mounted near the windbox air inlet, created sufficient turbulence and deflected the high velocity air from the fan which is enough to achieve good air distribution. Being cut long enough to be a press fit, they will hold position while testing for air velocity at the burner and can be moved to find the optimum position. One centered on the burner when the duct entering is centered, or at the point where the entering air velocity is highest, usually does most of the work. It prevents the direct blast of the air striking the shroud or register. It has to be far enough from the register so that some air can form an eddy behind it or a lot of air will immediately be lost behind it by deflection. Once the best distribution is achieved, be certain to weld them in place, as they will fall out when the windbox and boiler heat up in normal operation.

To check for uniform air distribution through a single burner (or each burner), measure it. First, do all the lock out and tag out protocols required for access into the boiler. Provide a means for operating the FD fan. Leave all the normal burner components in place, except for a center fired oil gun. Hang a manometer against the tubes in the furnace and connect flexible tubing at one end to a copper tube about 3 feet long that can be pointed at the burner. Take some paper on a clipboard and pen into the furnace to record the measurements. Be certain to wear good safety glasses. The breeze can do all sorts of strange things. With the fan in operation, point the tube directly at the burner while holding it so that the end is about flush with the face of the furnace wall and the tube is horizontal to get each reading. Begin with air flow consistent with low fire and record the total pressure read on the manometer at each point on the burner. Use clock positions (1 through 12) as a basis. Everyone will understand where the measurement was taken and the 12 readings provide reasonable resolution of the velocities around the burner. Since the point flush with the furnace wall and the open tube on the manometer are both exposed to the furnace, the static pressure is the same at both points. Velocity pressure is being read. Take readings at increasing air flow rates in steps of about 20% until reaching the top end, or the velocities get so high that standing up is difficult.

The actual velocity is reasonably estimated by multiplying the square root of the differential reading by 4005. That is done on any calculator by typing in the value of the differential (for example, 0.08 for that many inches of water column), pressing the square root key (√) to get the square root, then × (for multiply by) 4005, and the equals key to get the velocity in feet per minute (1132 in the example). There may be some argument about how much variation is permitted in the air flow around a burner. However, any deviation that exceeded 10% should get some attention. Take the sum of the readings (add them up) and then divide by 12 to get the average. Then multiply that result by 0.9 and 1.1 to get lower and upper limits, respectively. If any of the other readings are outside those limits, try ways to improve the air distribution in the burner including baffles, like the angles already mentioned. Then proceed to shrouds. Usually, corrections made at low fire do not alter air flow at higher firing rates. Correct the low fire variances first and repeat tests to determine their effect at higher flow rates.

That is a lot of work and is all after the fact. Yet, it does not cost as much as what is done for large utility boilers. Determining the best design of air distribution is such an art that utility boiler manufacturers will make models of the system and test them for proper air flow. That process is repeated to get it right before building the boiler. It is much easier to spend all the time on a model than to try to solve distribution problems on 24 or more burners in the field. Recent advances in computational modeling have shortened the time it takes to try various design options to arrive at a good design.

Speaking on large utility boilers, one firing system designed to minimize some of these issues was developed by CE in the 1920s. This system is called tangential firing. It is difficult to get uniform air flow from the ground level up to the windboxes on these large boilers. In tangential firing, the fuel and air are introduced from the corners of the furnace and are aimed tangentially to an imaginary circle in the center of the furnace. This arrangement causes the swirl and mixing to occur out in the furnace, instead of locally at the burner. In this way, the overall air and fuel mixture averages out any deviations in flow to each corner and each fuel introduction nozzle (Figure 10-45(c)).

A large number of burners were built for staged combustion in the last half of the 20th century. Some of those burners incorporated secondary air ports (openings in the refractory front wall around the circumference of the burner) with adjustment of the air flow to them, consisting of a piece of angle or other steel form surrounding the burner. Most of those provisions for

Figure 10-45(c). Tangential firing.

adjusting that air flow are so flexible that they do not provide uniform air flow around the burner. Some are so limber that they actually flop around in the air flow. To avoid that problem, the flows at the ports, adjust the flexible steel to equalize the flow through them and then tack weld the adjustment in position at each port. The goal of staged combustion is to reduce NOx formation. In the large utility units, the staging is accomplished by using overfire air. This term comes from stoker fired boilers, where air was introduced over the top of the fuel bed on the grate. For the larger boilers, separate ports at a higher elevation in the furnace are used to introduce the overfire air. With this approach, the lower part of the furnace can be slightly fuel rich, which reduces the formation on NOx. As the flue gas cools, the overfire air provides oxygen to complete the combustion without raising the gas temperature so high that additional NOx would be formed.

The mixture of fuel and air has to be heated to ignition temperature before it will start burning. For an individual burner, the burner has to provide means to heat the incoming air and fuel. The normal, and best means of heating the mix, is the application of a refractory throat. The radiant heat of the fire is reflected back into the entering fuel and air to heat them to the ignition temperature before they reach the proper mixture to provide stable combustion. The throat is also part of the insulating portion of the burner that protects the boiler front and burner housing from the heat of the furnace. There is a considerable variation in temperatures across that refractory during operation. Any large cracks, spalling, or shifting of pieces of throat tile can distort air flow at the burner to produce unstable combustion. For tangential firing, one flame impinges on the next one to provide flame stability. For startup, ignitors are used, firing oil or gas. These are essentially small burners that operate in the same manner as any other burner. Their job is to raise the overall temperature in the furnace zone to the point where ignition is guaranteed for the fuel and air entering through the fuel introduction nozzles.

Gas Burners

A gas burner can be premix or postmix. While most boiler burners are postmix, where the gas and oil mix after they enter the furnace, premix burners are available. Many operators think of a premix burner as hazardous. After all, a combustible mixture is made outside of the furnace. Many operators who moved from firing process equipment to the boiler plant are comfortable with premix burners, as they have a lot of experience with them. As long as the mixture is not heated above the ignition temperature, it cannot burn. Premixing permits a low cost arrangement of multiple burners which is frequently necessary for good heat distribution in processing equipment. There are not many boiler applications with premix burners. They are used in gas turbine combustors to provide low NOx emissions in that application. An understanding of combustion and a good reading of the instruction manual should be all that is needed to operate a premix burner.

Of the postmix gas burners, there are two choices which are normally identified as atmospheric burners or power burners. Atmospheric burners do not normally have fans or blowers to deliver the combustion air to the burner and seldom have ID fans. Lacking the power of the fan to introduce and help mix the fuel and air, atmospheric burners use some of the gas pressure for that process. The typical atmospheric burner has a "jet," which is a nozzle that the gas flows through on its way into the burner. That jet acts like an inducer to draw primary air in with it. The gas and primary air mixture is then distributed through the burner head (Figure 10-46) or flame runners (Figure 10-47) into the furnace. Secondary air is delivered by natural draft and mixes with the primary air and gas mixture as it burns. The several forms of flame runners shown in Figure 10-47 all seem to work well with no discernable difference in performance. Cracks between the holes and holes in the bottom, usually caused by rust, can produce distorted, inefficient, and dangerous fires.

Some gas furnaces can be subjected to very corrosive conditions between heating seasons. It is always a good idea to check an atmospheric burner right before the heating season and clean it if necessary. Large pieces of scale, from the heat exchanger lying on top of the

Figure 10-46. Gas burner head.

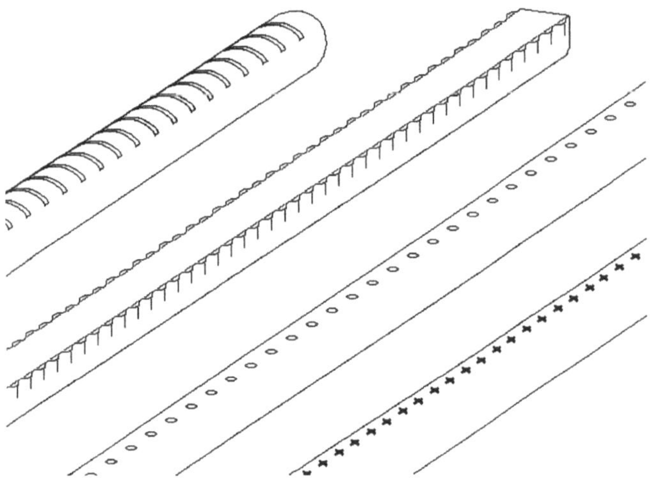

Figure 10-47. Flame runners.

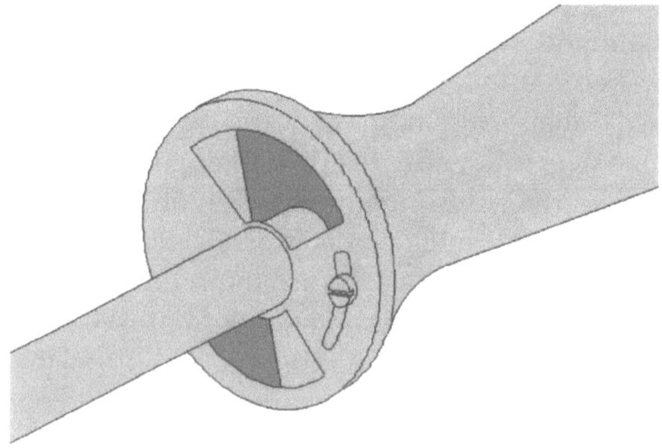

Figure 10-49. Primary air shutter.

runners, can form, as well as substantial quantities of rust. Replace any rusty, misshaped, or cracked furnace.

On atmospheric burners, primary air adjustment is accomplished by moving a sleeve (Figure 10-48) or rotating a shutter (Figure 10-49), thereby changing the opening for primary air. The gas nozzle (D in Figure 10-48) converts much of the static pressure in the gas to velocity pressure. The high velocity gas shoots into the distribution header (B), drawing primary air along with it through the opening that is adjusted by the sleeve (E). The mixture then flows into the flame runners (A) and out the ports, where heat from a spark or adjacent fire provides ignition energy to start the combustion. Always remember that additional air, secondary air, is required to complete the combustion and enters through openings like the one at (F).

The primary air and gas mixture is adjusted to produce a stable flame over the head, or flame runners, by adjusting the sleeve or shutter. Either one has a locking screw to ensure the piece stays where it was adjusted. The flame should burn clean and stable, just above the distribution ports. Lighting these burners can be interesting at times, especially during initial startup, because the pilot only lights one to four ports on the burner head or runner. The rest of the burner is ignited by flame at the adjacent port. Some atmospheric burners provide a degree of modulation and turndown by cutting out some of the jets, or controlling groups of jets, with individual shutoff valves and may be augmented by matching combustion air blowers. With this arrangement, it is very difficult to balance the fuel distribution to get them to burn cleanly and efficiently. They also tend to produce a considerable amount of CO.

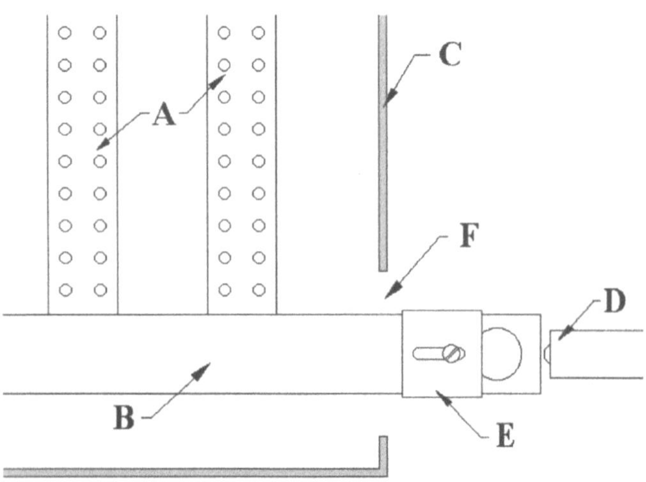

Figure 10-48. Primary air sleeve.

Atmospheric burners are only used on small boilers, hot air furnaces, and service water heaters for the most part. They are normally fixed fired and have very little control of overall excess air. If they serve a constant load for which they are well matched, there is some sense in their application. However, in any service with a varying load, the off cycle losses are very large. Controlling those losses with dampers that shut off air flow through the boiler when it is not firing can provide significant reductions in those losses. The dampers have to be confirmed to be open before the boiler fires. Modern high efficiency heating equipment, with pulse combustion or power burners, should be considered to replace most of that equipment. In some cases, a boiler with a power burner can pay for itself over an atmospheric fired unit in less than a year. Any medium to large boiler should have a power burner and, unless one is not available, it should be modulating.

Fuel gas can be introduced into a power burner via a gas ring, spuds, or a gas gun. A gas ring is a piece of pipe, fabricated steel, or a casting surrounding the burner right at the boiler front plate, with holes drilled in it to distribute the gas evenly around the outside diameter of the throat. Some gas rings are fitted with spuds, while other burners have spuds at the end of pipes, which deliver the gas from the front of the burner, or a gas ring located outside the front of the burner. Spuds are high temperature metal nozzles drilled with holes to admit the gas into the passing air stream. A gas gun consists of a pipe central to the burner, with a closed end drilled to admit the gas, very similar to an oil burner. Some gas guns consist of two concentric pipes that permit insertion of an oil gun down the center of them. The arrangement, distribution, and mix of holes drilled in gas rings, spuds, and guns vary with the manufacturer and, in many instances, are customized during startup to achieve smooth stable combustion.

Retaining data on the drilling of the gas burner is essential. That information may be the only accurate copy around. It is not unusual for a manufacturer to fail to update the records for changes made by the service technician. One element of the annual boiler inspection should be checking the diameter of the holes in the gas ring, nozzle, spuds, or combination thereof with matching drill bits. That is very important to do before closing a burner when refractory work is done. There is a tendency of masons to leave smears of refractory on and in the openings of gas rings.

The gas ring is usually bolted to the boiler front plate, that thick piece of steel that seals the front of the boiler, provides support, or is integrated with support of the refractory front wall. The front plate supports the burner throat to keep it centered and provides means of attaching the burner or windbox. If the gas ring or the boiler front plate is distorted, then air leakage around the gas ring at different points can produce very unstable firing conditions. The condition of the gas ring and clearances (if any) between the gas ring and boiler front plate should also be checked annually.

If the front plate is warped or there are other problems with irregular air spaces around a gas ring, they can be plugged with ceramic fiber rope. Always put the rope on the windbox side and be certain it is large enough that it will not blow through. It is very embarrassing to have someone ask about that thing fluttering in the fire. Gas rings can fail. Failure of adjustments of the firing rate, like linkage slipping, and other contributions, can produce situations where the heat of the fire is shifted into the throat, where it can overheat the gas ring to create cracks in it. Any cracked gas ring should be replaced before the boiler fires gas.

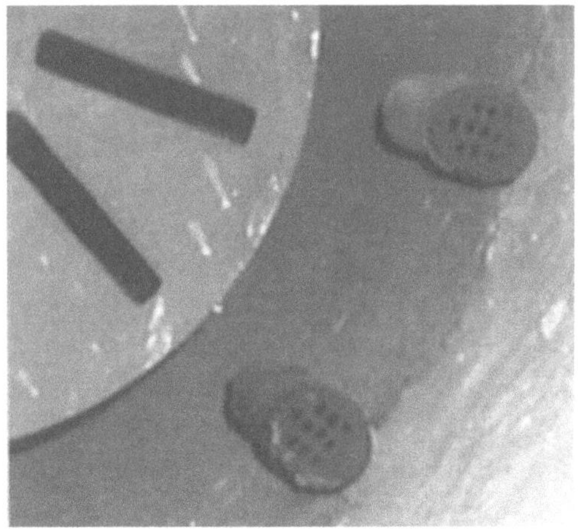

Figure 10-50. Gas spuds.

At one time, it was desirable to deliver the fuel gas into the flame in the furnace as uniformly as possible, to ensure complete mixing and permit low excess air firing. The discovery of NOx as a problem has resulted in irregular gas delivery schemes, usually using spuds (Figure 10-50) installed with pairs facing each other to produce alternating fuel rich and lean zones in the burner.

The fuel gas piping has to penetrate the windbox, or burner, to deliver the gas to the gas ring. A flanged connection was used on the gas ring to permit disassembly. However, that also placed a potential point for leakage of gas inside the burner windbox. There it could light off, producing heat in a windbox that was not designed to absorb that heat. There are also many variations in design of packing glands and other means of sealing the gas piping where it penetrates the windbox. If there are problems with air leakage at the gas line entrance, the best solution is welding it to the windbox. Normally, the windbox is flat and flexible enough at the gas line entrance that thermal expansion is accommodated by the windbox flexing. If there are problems with leaking piping joints inside the windbox, and the gas ring is not cast iron, cut the flanged joint out and weld the piping. Make sure it is free of gas before welding.

There are few options for the operator when it comes to gas rings. There is nothing to adjust. All the adjustments for fuel and air mixing have to be made by altering the combustion air flow. There are problems to watch out for. Gas rings can crack due to thermal shock, warping of the front plate, and improper repairs. The drilled openings for the gas can be blocked by dirt accumulation, careless application of refractory materials (a common one), and dirt when the burner port is used for furnace access. The ring can come loose from the boiler front plate when the mounting bolts vibrate loose. Any change in the appearance of a gas fire should be followed on the next shutdown by a careful examination of the gas ring.

There are a considerable number of gases that are fired in boilers in addition to natural gas (see the "FUEL GASES" section in Chapter 7). In many cases, they are burned only because the alternative is to waste them to atmosphere. Several are considered to be a source of pollution. The petroleum gases can, for the most part, be burned in equipment identical to natural gas burners, with adjustments in nozzle size or fuel supply pressure to compensate for the difference in the heating value and air to fuel ratio of the gases. Others, such as digester gas, can contain a large percentage of non-flammable gases and require special burners that can accommodate the larger volumes of fuel gas required to satisfy the heat input requirements. Digester gas (from sewage treatment plants) and landfill gas (tapped off a landfill) are noted for containing hydrochloric acid. The piping and burner has to be capable of handling the corrosive material. As always, take the time to seriously review the instruction manual to understand how those burners are to be operated.

Oil Burners

Fuel oil is introduced into a burner using a burner tip, which is normally mounted on the end of what is called an oil gun (Figure 10-51). The design and arrangement of the tip and gun is dependent on the type of atomizing system. Pressure atomizing burners have one or more tips on the end of a pipe positioned in the burner at the point where the oil has to be injected to develop the air/fuel mix. Pressure differential, air atomizing, and steam atomizing burners need two pipes: one to convey the oil to the tip and another to supply the air or steam or return the oil from the tip. Traditionally, the two pipes are concentric, with the oil supply down the center pipe and the annular space between the two providing the passage for air, steam, or return oil. Some manufacturers (like the one in the figure) provide two separate pipes running side by side. The tip introduces the oil into the furnace in a way that makes it possible for the oil and air to mix and burn. To ensure that the oil and air mix and burn completely, a fuel oil burner tip provides a means for "atomizing" the oil. Atomization is breaking the oil up into tiny droplets (not as small as an atom but small enough) so that the air can mix in between all the droplets for complete burning. If the oil is not atomized, it will not burn well. In some cases, it will not burn at all.

Don't accidentally leave the tip off an oil burner and try to start it that way. In one apartment house, the boiler operator did just that. The burner did not light the first few times. After several tries, enough oil was dumped in the furnace that the lighter portions, which evaporated, produced a flammable mixture that the ignitor managed to light. The resulting explosion did not destroy the apartment building, but it did manage to destroy the boiler.

Atomization is accomplished in different ways. All of them work. The principle difference between them is the degree of turndown they can accomplish. Pressure atomizing burners produce a fine spray pattern of oil. The quality of the atomization varies with the pressure drop across the burner tip. Many burner tips will have internal channels that divert the flow of oil (Figure 10-52). This causes the oil to accelerate as it approaches the central chamber and produces a whirling motion in the oil. As the oil flows out the tip, that spinning motion forces the oil to swirl out by centrifugal force and that causes the oil to tear apart into tiny droplets.

A similar principle was applied to a burner that is not legal to use anymore. Rotary cup burners used a brass cup mounted on the end of a pipe. The pipe and its cup were mounted on the shaft of the blower of the burner and centered in the burner throat. The oil entered the

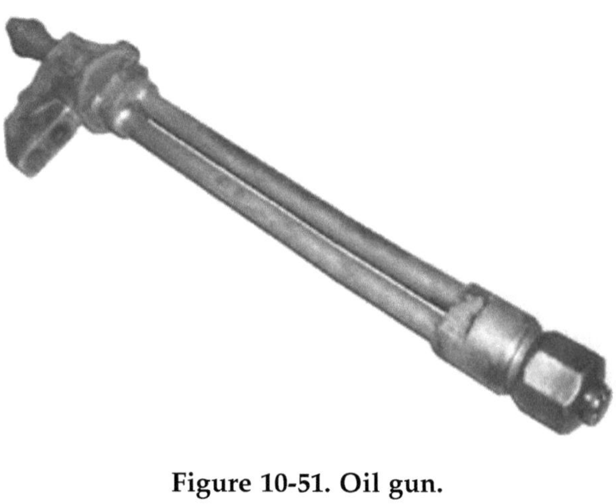

Figure 10-51. Oil gun.

Figure 10-52. Burner oil tip showing swirler pattern.

rotating pipe through a flexible connection and literally drizzled into the cup. The cup's rotary motion whirled the oil around the inside of the cup until it reached the top where it sheared off into the combustion air stream. This did not produce a very fine spray. The poor atomization was the reason the rotary cup burner is no longer legal.

Steam and air atomizing burners use one of two methods to atomize the fuel. Some of the burners introduce the oil into a jet of steam or air that cuts into and breaks the stream of oil up into tiny droplets and then transports them into the furnace. Most, however, simply mix the oil and steam or air in a chamber in the tip. When that mixture leaves the burner tip holes, the gas (steam or air) expands rapidly breaking the oil up into tiny droplets. Since the energy for atomization is provided by the air or steam, turndown of these burners is typically about 5 to 1, some as high as 10 to 1.

The typical pressure atomizing burner is limited in turndown capability to about 2 to 1. Once the flow is reduced to half (the pressure drop through the tip is one-fourth), the velocities are too low. The oil does not break up well. Oil return atomizers were produced as a solution to that problem. The full load flow of oil is delivered to the burner tip and flows through those slots to produce the spinning that breaks up the oil. To reduce the firing rate, some of the oil is returned from the tip to the suction of the fuel oil pumps. The turndown is generally a function of the differential pressure, where the turndown is equal to the delivered oil pressure divided by 100. A system firing oil supplied at 300 psig will have a 3 to 1 turndown and one with oil supplied at 800 psig will have an 8 to 1 turndown. The practical limit for those burners seems to be 1200 psig due to the price of the pumps, pumping, maintenance, and all the pressure containing components of the oil system getting very high.

With large boilers, additional turndown is accomplished by using multiple burners. Half the burners operating at maximum oil supply pressure will produce half the boiler load. One-fourth will produce one-fourth, and so on. If the boiler is limited to one to four burners, then other means of achieving additional turndown may be required. The typical solution is different sizes of burner tips. A smaller tip will pass a reduced amount of oil at the same pressures. The important thing to remember is that all the burners in a boiler should have the same size tips installed so that fuel delivery is uniform.

A boiler plant with steam atomizing burners needs a means for atomization on startup. There are two solutions. One is to use a small pressure atomizing tip to produce enough steam to get going. The other is to use compressed air temporarily. Temporary use means exactly that. A burner designed for steam atomizing will consume considerable amounts of compressed air at a high cost in electric power to generate it. A small control air compressor could be overloaded and damaged attempting to maintain fuel atomization.

Figure 10-53. Irregular drilling in oil tip.

As with the gas burners, oil guns have seen modifications in recent years to produce alternating zones of fuel rich areas in the flame for NOx reduction. Thus, irregular drilling of a burner tip (Figure 10-53) is now common.

The gun, in many cases, is nothing more than a piece of pipe connecting the fuel delivered to the burner on out to the tip. Many burner guns can be removed by breaking the connections outside the burner and pulling it out for cleaning and maintenance. There are guns which disconnect at a union (Figure 10-54) or simply break at tubing fittings as well as guns with elaborate quick connect capabilities, with many variations in between. Most arrangements are specific to a particular manufacturer, but the common arrangement is a yoke coupling (Figure 10-55), which is used by many manufacturers. A yoke with a set screw (Figure 10-56) clamps the two together.

With a yoke coupling, the gun has openings for oil and any atomizing medium that match with holes in the yoke coupling. To ensure alignment of the openings in the gun and yoke, there are usually ferrules (short smaller pieces of pipe or tubing) set in the yoke holes (Figure 10-56). The gun holes pass over them. The ferrules are removable because they can be damaged by pressing or throwing the gun against them. Then the holes in the

Figure 10-54. Oil gun quick assembly union connected.

Figure 10-55. Oil gun yoke coupling.

Figure 10-56. Yoke coupling clamp and set screw.

gun will not fit over them. The installation also includes some provision for sealing the joint of gun and yoke. Sometimes, it is a softer material, like brass, normally on the gun, that will deform under the pressure of the set screw to seal the joint. Frequently, it is a gasket. Most commonly, there is a thin layer of copper surrounding a fiber material that will conform to variations in the two surfaces to seal the joint. Being careful when inserting an oil gun prevents damage to the ferrules, which can prevent proper fit up of the yoke and gun.

Most manufacturers' instructions state that the gaskets should be replaced every time the burner is changed. Keep a set of gaskets handy to replace them when the existing ones fail or get damaged. There is no good reason to replace them every time an oil gun is changed. As long as there is no dirt or grit on them, they should still seal.

A skilled operator can remove one oil burner and install a fresh one in a matter of a few seconds. However, if the boiler has a single burner, the speed of the operator is not much of a consideration since the burner has to be shut down to remove the oil gun. To avoid shutting down the boiler, along with the processes of purging, low fire positioning, and trials for ignition, some burner manufacturers will provide single burners with the ability to accept two oil guns, while others provide as many as four. The typical two gun arrangement is designed to insert a temporary oil gun, transfer the fire to that gun, and then transfer back to the main oil gun. The fire may be lopsided or have voids when firing with the temporary gun. Other arrangements use guns with special tips that produce a uniform flame pattern when all the oil guns are in position and operating. Changing out the oil burner for cleaning is accomplished by switching guns one at a time. During the period that one of those oil guns is removed, there is a definite gap in the flame pattern. While changing guns, the operator should increase the air to fuel ratio so that the variations in fuel delivery do not produce fuel rich conditions in some portion of the furnace.

Of course, the reason for oil guns is that the tip gets dirty. They have to be removed for cleaning. Spare oil guns and tips are provided in order have a clean one ready to put back in the burner to permit continued firing. The need for cleaning is always "as soon as they get dirty." There are no hard and fast rules for cleaning burners. It depends on the oil, contaminants in the oil, the firing rates, and the condition of the burner itself. Change the guns and clean the tip of carbon before it builds up enough to start hampering atomization. Keep track of how long that is for the particular boiler, burner, and load combination.

The accumulation of unburned fuel that has been heated to drive off much of the lighter fractions and leave mostly carbon is called "carbon" by boiler operators. Carbon is a common problem when firing oil. It is less of a problem when firing light oils. There are many reasons for carbon buildup on burner tips, burner throats, and the floor and walls of a furnace when firing oil. The most common reason for carbon buildup is poor atomization. That can be produced by dirty oil that plugs burners or ties up the oil like glue so that it does not atomize properly. Other reasons are using tips too large for the load, worn tips, loose tips, and tips and other burner internals assembled improperly. One of the service engineers solved a poor atomization problem on a burner by assembling it improperly.

Steam and air atomizing burners can also suffer from condensate in the air or steam, the wrong pressure, and blockage of the atomizing medium piping. A problem encountered regularly with differential controls was a significant variation in the differential at the burner tip due to pressure drop in the oil or steam piping. Usually, the problem involved lighting the burner at low loads, where the differential was so high that the fuel–air mixture was always lean. The atomizing medium broke it up too much. The solution to that problem is adding an orifice nipple (steel bar simulating a piece of pipe with a hole drilled through it) which allows adjustment of the differential at low fire to get stable firing. As the load increases, the nipple introduces a pressure drop in the oil that increases the differential at the burner tip as load increases.

Every hole drilled in a burner tip comes from the factory fresh and sharp, with a pure 90 degree angle between the edges of each hole and the face of the tip. That is so that there is a sharp separation of the oil stream as it leaves the tip. Operators who get frustrated with the brass tools and wire brushes then resort to steel tools and wire brushes to round off those sharp edges. Now the oil stream does not make a sharp break with the tip. Some of the oil tends to follow the curved edges created by abrading the tip with steel tools. That oil forms carbon very quickly. Save a lot of trouble and stick with the brass tools.

Coal Stokers

There are many options for introducing coal into a furnace, and "coal burner" and "stoker" provide some differentiation. Stokers handle coal as a solid material. Coal firing can be as simple as a grate in the bottom of a furnace, with openings for the air and an individual opening a door in the side of the boiler to introduce the coal with a shovel. It can be as complex as a multi-tiered, tangentially fired furnace with overfire air ports and reburners.

Stokers come in a variety of forms and have basically been reduced to under feed, traveling grate, and over feed types. The difference in these is how the coal is introduced to the fire. An under feed stoker pushes the coal up into the furnace from below the grate. The coal is removed from storage or a hopper by a screw conveyor (Figure 10-57) or ram (Figure 10-58) which pushes the coal along through the "retort" and against the pile in the bottom of the furnace. As the coal is pushed up, it is mixed with air entering via the tuyeres (C in Figure 10-59), pipes, tubes, or slots in the grate that admit the air into the furnace. The mixture is ignited by coal already burning above the grate. The coal–air mixture partially burns on the grate and completes burning of hydrocarbons vaporized by the heat of the furnace in the space immediately above the grate. Air at the tuyeres and most active portions of the grate is considered primary air and is controlled by dampers supplying the air to the primary air zone (B).

As the hydrocarbons and sulfur in the coal are consumed, the remaining ash is pushed to the edge or sides of the grate, where it can be removed by hand or dumped (D) for removal by hand or screw conveyor. For final burnout and handling high loads temporarily, a controlled flow of air is supplied to the dump grate zone at (E), which must be reduced dramatically to permit removing ashes from that chamber manually.

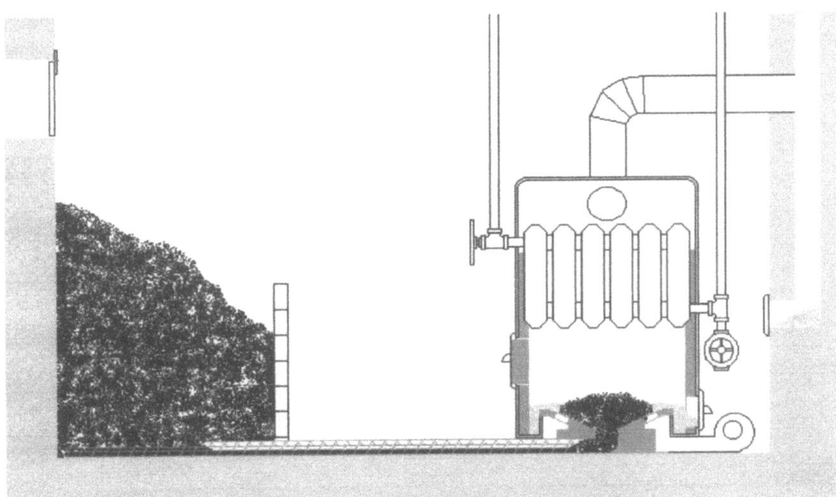

Figure 10-57. Coal screw conveyor.

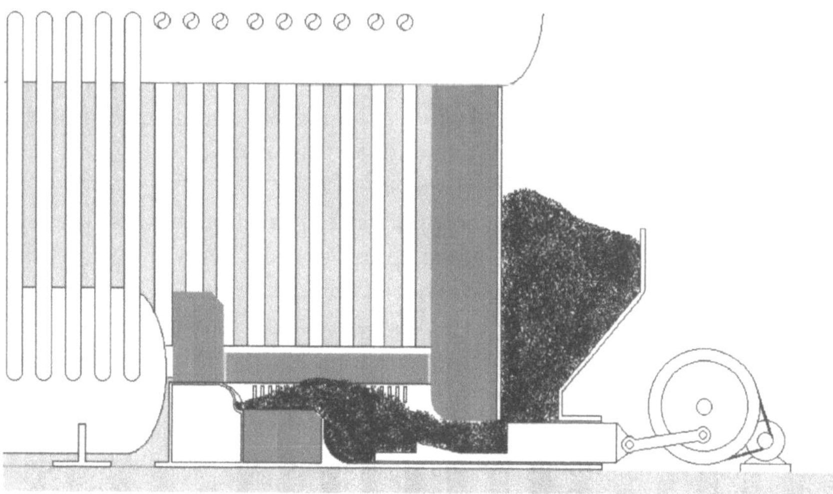

Figure 10-58. Underfeed stoker ram.

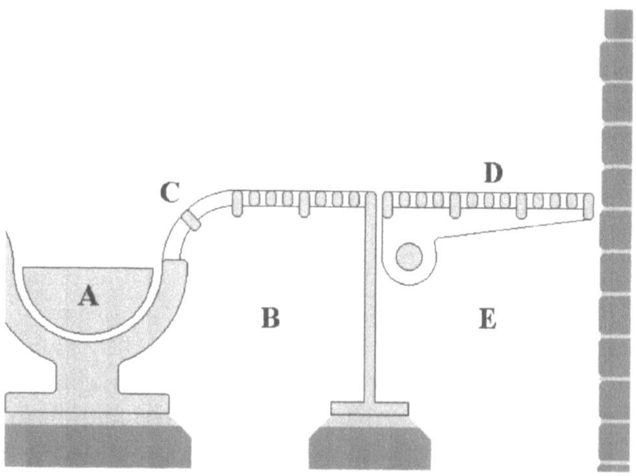

Figure 10-59. Dump grate.

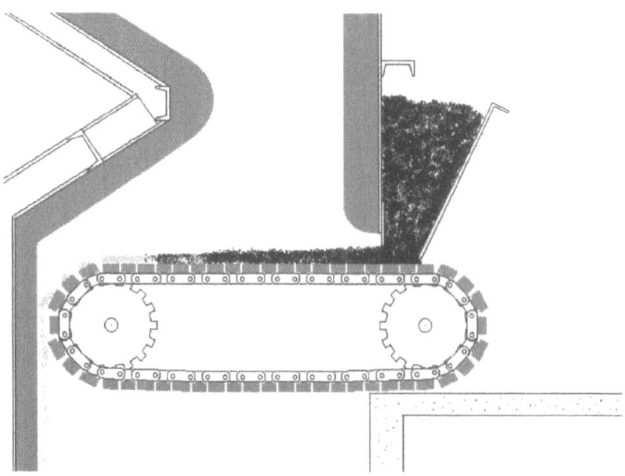

Figure 10-60(a). Traveling grate.

Under feed boilers with screw feeders like the one in Figure 10-57 are still found in homes in Pennsylvania, Ohio, and other coal states. Ram fed boilers can be powered by steam to eliminate the need for electricity. They are also available in sizes up to 100 MMBtu/hr by increasing the number of coal feed locations in a "multiple-retort" stoker. Some people might be surprised to learn that most of the nation's capitol was heated by those boilers up until the early 1990s. Under feed stokers are capable of burning a wide range of coals with different sizes. The common specification limits fines and particles smaller than one-half inch. The fines sift through the equipment and tend to compress and expand preventing proper operation of the feeder.

Traveling grate stokers burn coal particles in the range of one-eighth to three-quarters of an inch in size. The grate (Figure 10-60(a)) is a continuous belt of steel chain mounted between shafts spaced 10–16 feet apart, with lengths up to 20 feet. The steel is protected from the heat of the furnace by pieces of refractory which form an external layer on the grate with openings around each piece to admit the combustion air. Coal is stored in a hopper on the front of the boiler and is dragged into the furnace by the grate. The depth of coal over the bed is adjusted by a plate in the hopper at the front of the boiler. Proper control of air distribution in the zones below the grate and the ignition arch maintain combustion. As the coal burns down, the flaming particles under the ignition arch are blown up by the flow of combustion air and follow the flow of air and gas diverted by the arch so that they land on the entering coal to ignite it. That way, the coal burns from the top of the bed down to the bottom, eventually becoming ignition particles. Ash left over drops off the end of the grate as it makes the turn back toward the front of the boiler.

Over feed stokers have a grate just like the traveling grate stoker. The difference is in the way the fuel is introduced. Frequently, an over feed stoker is called a "spreader" stoker because the fuel is, to a degree, spread over the grate. Over feed stokers are further classified by the height of the feeders above the grate. "Low set" stokers will have feeders injecting the coal in the neighborhood of 3–5 feet above the grate. "High set" stokers can be as much as 80 feet above the grate. The grate on over feed stokers typically runs in the opposite direction of spreader stokers, delivering the ash to the front end of the boiler. The coal feeders come in a variety of forms, from plates connected to eccentrics on a shaft that toss the coal dropped on them into the furnace to rotating blades and rotary feeders with air blown into the feeder to transfer the fuel into the furnace. Over feed stokers are designed to fire a finer coal, from dust sized particles to pieces under one-quarter inch. The fines are burned in suspension over the grate and the heavy particles drop to the grate to complete burning.

Operation of stoker fired boilers normally requires more manpower than oil or gas fired boiler plants. The coal has to be received, moved to storage, and moved from storage to the "bunkers" that supply the coal to the stoker. The considerable amount of ash has to be removed from the boiler, moved to storage, and loaded into transports for final disposal. Occasional "dressing" of the fire is required to maintain uniform combustion over the bed of coal and to remove "clinkers," which are accumulations of carbon and ash that harden into solid deposits on the grate. Lighting a stoker fired boiler is accomplished by building a wood fire on the grate and then introducing coal to be ignited by the wood. Oil soaked rags are also used. Cleaning the plant of coal dust and equipment that accumulates the fines is an ongoing task. All those activities require more personnel. The lower cost of coal justified the added cost of personnel to handle it.

Coal can tend to "cake" before entering the stoker. The large pieces of compressed, usually wet, coal will not burn completely in the furnace unless it is broken up. Preventing caking is accomplished in the handling and preparation of the coal. Keeping the coal dry by unloading cars or trucks before it rains or snows and limiting exposure of the fuel to water will reduce caking. "Clinkers" is the name given to chunks of unburned coal and ash that form in the furnace. Those large particles can jam stokers and ash handling equipment. They are usually formed when low ash fusion temperature coal, or coal with a lot of dirt and other materials in it, is received. These materials melt at the normal furnace temperatures. They can also form when there is a hot spot in the furnace that is higher than the ash fusion temperature (See Fuels, Chapter 7). When they form, clinkers have to be broken up to prevent them from forming a blockage in the fire that reduces output and increases temperatures in other areas of the grate. The operator has to watch the coal bed and use special tools with one end inserted into the furnace to break up the clinkers.

Another operation that operators perform with coal stokers is "dressing" the fire. Despite all provisions, the coal never distributes perfectly evenly over the grate. Dressing the bed (the layer of coal on the grate) is accomplished with tools like those used for clinkers to move the coal around until the bed depth is uniform and burning evenly.

Breaking clinkers and dressing a coal fire are activities that require on-the-job training and experience to do it well. With the current concern for greenhouse gas emissions and the relatively lower prices for natural gas, the majority of stoker fired boilers burning coal have either converted to natural gas firing or shut down.

Coal Burners

Coal burners are principally designed to burn the fuel in suspension. It has to be pulverized before it is delivered to the burner. The bottom of a furnace, fitted with pulverized coal burners, will have means to remove the ash that drops out of the fire. This material is referred to as bottom ash. Much of the ash is transported through the boiler to be removed by dust collectors at the boiler outlet. This material is called fly ash. Pulverizers form an integral part of most coal burners. There are plants that can burn pulverized coal from storage (not many). Most plants have an integral pulverizer that grinds the coal to fine powder and mixes it with primary air to produce a fuel rich stream of air and coal, fed to the burners. The coal cannot be simply ground down. It has to be dried as well because it does contain water. The ground coal would become muddy without drying it. To dry the coal, the pulverizers are supplied with preheated combustion air from an air preheater or, in the case of some small plants, steam heated air.

One type of equipment that pulverizes the coal is a ball mill. It consists of a large drum mounted with its axis on the horizontal and is filled with cast iron balls. The trunions (extensions at the center of the heads of the drum which serve as a shaft) are hollow so that air and coal can be fed into one end and the pulverized mixture leaves the other. As the drum rotates, the balls are lifted and dropped on the coal to crush it. The finely ground particles are carried out with the heated air.

Bowl mills consist of a bowl spinning on a vertical shaft with rollers inside. The rollers are spring loaded and roll around on the inside of the bowl, crushing the coal that is dumped into the bowl. Most of the size reduction is achieved by coal particle to coal particle abrasion. The coal has to pass under the rollers to escape the bowl. Some use balls instead of rollers. The pulverized coal is carried away by heated air directed up around the bowl. The air flow proceeds to a classifier, which allows the very fine coal to pass through and rejects oversized coal back to the bowl. This primary air flow carries the pulverized coal up to the burners or fuel introduction nozzles. The standard particle size distribution is 75% through 200 mesh. The mesh size refers to screen with 200 holes to the inch. The average particle size is around 50 microns. For low NOx firing systems with staged fuel introduction, the particle size distribution is 90% through 200 mesh. These pulverizers can process up to 100 tons/hr of coal continuously.

Hammer mills use something comparable to several metal hammers that swing freely on a shaft connection. The metal hammers pound on an accumulation of coal to break it into fines that are carried away by the air.

Attrition mills are something like a combination of fan and grinder, with pins on the circumference of the fan wheel that strikes the coal particles to crush them. The attrition mills have stricter sizing requirements for feed than the others and mill capabilities vary with construction and manufacturer.

The fans or blowers that transport the coal and air mixture to the burners are called primary air fans, or exhausters, with the latter term reserved for those that move the coal laden air. Most installations use exhausters to limit potential leakage of powdered coal into the plant. The fuel and air mixture exits the mill into the exhauster inlet which discharges the mix under pressure to the burners. In smaller equipment, the pulverizer and exhauster are all in the same housing.

What is probably the most important part of a pulverizer and burner combination is the classifier. It is normally a static device (no moving parts) that separates large particles from the stream of coal dust and air heading to the burners and returns those particles to the mill for further grinding. The normal requirements for pulverized coal leaving a classifier are at 70%–75% of the coal through a 200 mesh sieve and no more than 2% over a 50 mesh sieve. Finally, the mixture of coal and primary air has to be fuel rich to provide a stable point for ignition of the fuel at the exit of the coal nozzle. The pulverized coal burner can be as simple as a pipe from the mill, or exhauster, pointed into the furnace, to a cast assembly with orifices, guide vanes, and other features that further mix the fuel and primary air and distribute it into the fire in the furnace. Over time, the coal flow can erode some of the more important parts of the burner to destroy baffles, etc., that produce the mix and, more importantly, provide that fuel rich concentration that is needed to get the fire started and stabilized.

Some utility boilers are equipped with cyclone furnaces, which use a pulverized coal with less size restriction than conventional pulverized fuel burners. The cyclone is a water cooled, refractory lined cylinder mounted horizontally at the side of the boiler. The coal and air is fired at very high heat release rates within the cyclone, with temperatures so high that all the ash is melted and removed as a liquid. The flue gases exit the cyclone furnace into the boiler furnace at temperatures around 3000 degrees. The primary purpose of the cyclone furnace is reduced size of the boiler. The initial problem with these units was the vaporization of alkaline materials in the ash (sodium and potassium). These vapors subsequently condensed on the heat transfer surfaces in the convection pass to form hard, glassy deposits that were extremely difficult to remove. With the advent of low NOx requirements, the high temperatures in the cyclone increased the NOx formation, which caused these units to be no longer offered.

Modern versions of coal burning boilers are fluidized bed boilers and circulating fluidized bed (CFB) boilers, where the entire furnace or the whole boiler is part of the burner. The coal is introduced as solid particles into a bed that is fluidized by the combustion air and flue gases passing up through it. Fluidizing is accomplished by distributing the air into the bed of coal over a broad area using special nozzles through a grid plate under the bed. The solid particles seem to boil just like water in a pot, as the air flows up between them. This type of fluid bed, with a defined surface, is called a bubbling bed, as the air moves through the bed in bubbles. In the case of a CFB, the smaller particles are carried out of the furnace with the flue gas to be captured in a cyclone and returned after they flow through the furnace section of the boiler (Figure 19-60(b)).

In addition to the coal, the fluid bed is fed finely ground limestone that reacts with and absorbs the sulfur dioxide. The reacted limestone and gas leave the boiler as part of the ash instead of emissions in the flue gas. CFB boilers actually allow considerable carryover of the bed into the initial passes of the boiler to prolong contact time of limestone and sulfur dioxide plus increased fuel and air mixing. Cyclone separators act like classifiers to remove the coal and limestone particles from the flue gases and return them to the furnace for additional reaction. A fluid bed heat exchanger can be utilized for a portion of the solids exiting the cyclone to independently control the bed temperature. The feed coal is typically crushed to 1/4 inch top size. The circulating mass is typically 200–800 microns. Larger particles are drained from the bed through a stand pipe. If the amount of large particles is rather high, these can be cooled in an ash cooler. Very fine particles (<60 microns) escape from the cyclone and

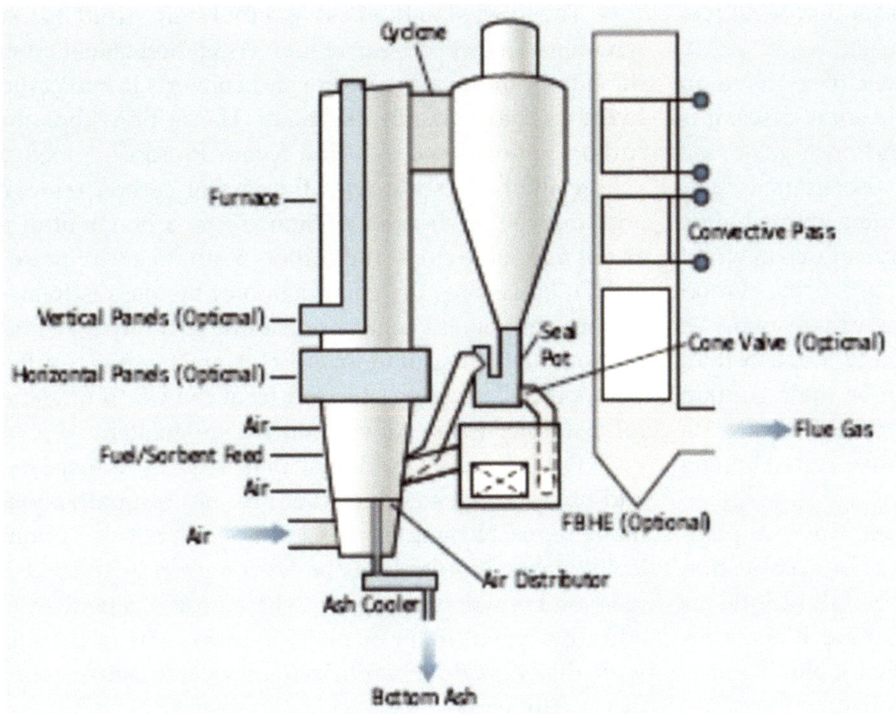

Figure 10-60(b). Circulating fluid bed boiler.

are carried along with the flue gas. These are removed in a fabric filter for disposal.

Coal firing requires consideration of the time it takes for fuel and air to mix and burn. A stoker fired boiler will hold the coal for several minutes, while the heat breaks each particle down, evaporating the lighter fractions of the fuel, and then converting the carbon. The furnace must be larger to hold the inventory of fuel. The fuel for a coal burner has to be pulverized because the particles have little time to burn in the furnace. In the CFB, the gas residence time is on the order of 5 seconds. Most of the solids stay much longer, as they are captured by the cyclone and recirculated back to the furnace for additional time to complete combustion.

One important element of coal firing is very low air flow purges. Operators, used to wide open damper full air flow purges for oil and gas, should be aware that it is possible to blow the boiler up if that is done with coal. There can be accumulations of coal or a mixture of coal and ash in the boiler which a full air flow purge would lift and stir to form a combustible (make that explosive) mixture. A purge should be conducted at low air flows to prevent that from happening. A high flow of products of combustion can stir that stuff up and move it without hazard because the flue gases are inert. They do not contain any air to mix with the fuel. Coal fired units take a lot more care and effort to operate and maintain. Be sure to read the instruction manuals provided by the manufacturer.

Wood Burners

Wood burners vary from a campfire to burners firing sander dust. On the one extreme, there are large pieces of wood which require long retention times in the furnace. On the other, there is wood so finely ground that it burns faster than fuel oil. There are a considerable number of different boiler, burner, and grate designs for burning wood, wood waste, and similar fuels.

Wood requires some special consideration if it is "green" or "wet." The moisture absorbs a considerable amount of heat and is capable of quenching the fire to the point it goes out. Dry wood from lumber operations (kiln dried), planing, sawing (of dried wood), trimming, and sanding burns readily and must be handled with care because it can easily produce an explosive atmosphere during air conveying or handling operations that mix the fuel with air. Most wood burning boilers serve industries that process that wood for such things as pine chemicals and furniture.

The fine materials, fine sawdust (some sawdust can be chips as large as one-half inch square), and sander dust are typically fed to a burner similar to a pulverized coal burner, where the material is burned in suspension like fuel oil. The furnace is usually also fitted with a grate, normally water cooled because there is no layer of fuel to protect the grate from the heat of the furnace. Larger materials are usually burned in a high set spreader stoker which allows for burning of the fine particles in suspension and the heavier pieces on the grate. A special consideration when firing wood is contamination with denser solids. Material cut specifically for firing can contain sand, rocks, and dirt. Sander dust can contain some of the abrasive material from the sanders. Those heavier and denser solids can seriously erode the fuel handling equipment, burners, grates, and the boiler tubes. Although some people do not consider it as wood fuel, paper plants burn large quantities of bark that is stripped from the wood used to make pulp for paper. Bark is usually burned over a high set, overfeed stoker, where the bark is introduced as much as 60 feet above the grate.

Wood and paper product manufacturers have an opportunity to convert waste to fuel. In some cases, it is less expensive to landfill the waste and burn gas or oil as a fuel. Environmental restrictions also limit its use. The increased cost of fuel and landfilling may change that in coming years. There are also innovations in wood burning systems, including fluidized bed firing. Wood firing problems and how an operator can respond to them vary with the fuel. There is usually less ash than with coal firing. However, the ash can be finer, contain more alkalis, and plug up equipment more. Keeping the systems clean and dressing the fire of stoker fired boilers are principal activities.

One thing that wood fired boilers have is plenty of air. Much of the air that is used for combustion comes with the wood. Wood tends to be full of little air spaces, especially if it is dry. There is so much air available that a pile of wood in a corner that looks like it is all burnt-out ash can have glowing embers underneath. The embers can also be there after several hours or even days. Always treat any accumulation of anything in a wood fired installation as a potential source of flame. Stir it up and mix it with a little air and an explosive mixture can result, same as with coal. Some processing of wood wastes have produced specialized wood fuels (pellets) and included material like leaves. Pellets are easier to ship and to handle, as long as they remain dry.

Biomass Burners

Wood is one of the many fuels identified as "biomass," a label applied to materials that are grown or produced from vegetation. Biomass is commonly restricted to mean a solid fuel. The burner may actually be a device that processes that fuel to convert it to a gas or liquid fuel that is then burned. Other biomass fuels are modified to produce a fuel in a different form. Charcoal is one. Also, the ethanol that is added to gasoline is made from corn. Unlike dried corn kernels, much of biomass has to be processed to prepare it for burning in conventional fuel burning equipment. A typical example is wood pellets. The equipment can also take larger particles and reduce them to dust (or nearly dust) such that the fuel can burn in suspension in a furnace like pulverized coal. A fairly common reduction creates a fuel labeled RDF (for refuse derived fuel). RDF can be a fairly fine material that is typically blended with pulverized coal for firing in combination with pulverized coal. Another form of RDF is a reduction to something less than 1/2 inch in dimension for firing in a spreader stoker.

The interest in biomass as a fuel stems from its designation as a carbon neutral fuel. Vegetation takes carbon dioxide from the atmosphere and converts it into carbohydrates that make up the plant. Harvesting stops this carbon removal process. The Kyoto Protocol agreed to charge the harvesting with the loss of carbon removal and treat the combustion of biomass as carbon neutral so as not to double count the carbon. With renewed interest in CO_2 emissions, the combustion of biomass is treated as zero emissions. Some countries have set up subsidies to promote the use of biomass. This practice has resulted in wood pellets being shipped from the US to England, for example, as the subsidies there are greater.

Producing a liquid fuel that is easily transported and burned, like standard fuel oils, has been attempted many times. No single method has stood out as economically viable. Normally, a conversion process performs a reaction known as pyrolysis, where heat is generated by adding oxygen (in air or pure) to burn some of the fuel, with the remainder vaporized into combustible gases, principally carbon monoxide. Another approach that is being considered is referred to as bio energy and carbon capture and storage (BECCS). In this process, the biomass is gasified to produce a synthesis gas (mainly CO and hydrogen). Any CO_2 that is produced is captured and stored in some way (most likely underground). The synthesis gas can be converted into liquid fuels by known processes. Again, any CO_2 that is produced in that processing can be captured and stored. Since the biomass is considered to be zero emissions and the CO_2 that is captured is stored (also called sequestration), the overall process can be considered to be a net reduction of CO_2 from the atmosphere (at least according to its proponents). In this way, the liquid fuels that are produced can be considered low carbon or even carbon neutral.

With biomass fuels, the instruction manual may not cover everything that would be good to know about burning that fuel. Advances in the technology are being made regularly. Looking through the equipment manufacturer's web site on a regular basis may provide awareness of changes and improvements in operation that have been developed. There are also trade organizations and user's groups, such as the Council of Industrial Boiler Owners (CIBO), that share problems, improvements, and solutions.

PUMPS

Pumps are used to move all of the liquids around a boiler plant. There is a diversity of designs and arrangements for pumping that provides many options. The word "application" means what the equipment is used for. Applications include feed water pumping, condensate pumping, fuel oil pumping, sewage pumping, etc. Over the years, the applications of pumps for boiler plants have singled out a particular pumping method and pump construction for each service. As a result, there will seldom be any deviations in the type of pump used for a particular fluid service.

High pressure feed water and condensate system pumps are usually centrifugal. Low pressure feed water and small volume condensate pumps are usually turbine type pumps. Fuel oil is moved with positive displacement progressive cavity pumps of the screw and gear types. There are other options, but their use is not very common. Technological advances could alter one or more of these general choices in the future. If only someone could come up with something better than a centrifugal pump, there could be dramatic reductions in electric power consumption, as many of the centrifugal pumps run at efficiencies less than 50%.

Pumps handle liquids, which are incompressible fluids. They are an essential part of the boiler plant. Modern pumps have become so reliable that operators tend to ignore them until something fails. In some plants, the pumps have been there operating for so long that the manufacturer's name that was formed in the casting of the pump had corroded until it could not be read. When asked, the operators could not produce an instruction manual or anything else that would identify the make and model of that pump. Pumps do not last forever. Their capacity and differential capability declines as they age. Their efficiency also declines with age. Pumps that are so old that the nameplate cannot be read may be using twice as much electricity as they did when they were new or, more likely, only pumping half of what they could originally. Monitoring the performance of the pumps is a wise thing to do.

Pumps are often oversized as well. Many times, boiler feed pumps were selected so that any one of them could run the plant at full capacity (all boilers on), which does not make sense when at least one boiler is usually a spare. Then, to compound the problem, multiple pumps were specified of the same size. It would be virtually impossible for an operator to select a pump that matches the load when they are all too big. In many instances, replacing a boiler feed pump with one that will just barely handle a spring or fall load will save enough electric power to pay for the pump in one summer. When there are more than two pumps, the capacity of each should be such that it takes all of them, less one, to carry the peak load. With three pumps, they should each handle half the peak load. With four pumps, they should each handle one-third of the peak load. Five pumps should be sized at one-quarter the peak load, etc. Since boiler feed pumps have to be capable of delivering water to the boiler when the safety valves are blowing (a code requirement), they are slightly oversized anyway because the capacity picks up as the differential is lowered to operating conditions.

In one plant with three boilers and four feed pumps, if two boilers were on line, the operator ran two pumps. During the winter, when there were three boilers on line, three pumps were running. It made no difference what the boiler load was. Run a boiler and run a pump. A quick look at the instruction manual revealed that any one pump could supply three boilers. Savings of electricity by only running one pump the year round was well over $50,000.00. Proper choices in the operation of pumps and monitoring their performance, as well as maintaining them, can make a significant difference in the cost of operating a plant.

With rare exceptions, pumps are powered by electric motors or steam turbines. The motor or turbine "drives" the pump and are called "drivers." They all serve to rotate or extend and contact the shaft of the pump. The energy is transmitted through a metal shaft that connects the driver to the pump. The rotating parts of a pump can be mounted directly on the driver's shaft or they can be mounted on their own shaft. When the pump has its own shaft, it is also fitted with bearings to maintain alignment of the shaft in the casing of the pump. Regardless of operation, rotating or extending, and retracting, the shaft moves and the design of the pump must allow it to move without allowing the liquid to leak out of the pump.

Packing seals the space along the shaft, where it penetrates the casing, to limit leakage. Some leakage through the packing is essential to lubricate the packing to shaft joint. If the packing is tightened enough to stop or reduce the leakage too much, then the packing and shaft rub with deterioration of each. Manufacturers started making rotating pumps with shaft sleeves to help with that problem. The sleeve was like a pipe, or tube, that slipped over the shaft. It was either clamped with other parts or threaded onto a matching thread on the shaft to make it removable. That way, if the pump was run with the packing dry, and the shaft sleeve was torn up, all that was needed was to replace the sleeve, not the entire shaft. If the annual replacement of shafts and shaft sleeves is common, it is because the operators consistently tightened the packing too much.

If the proper amount of leakage is unknown, try measuring the temperature of what leaks out and compare it with the temperature of the liquid inside the pump. It should not rise more than 5 degrees. That does not work for boiler feed pumps. The liquid in the boiler feed pump would flash. Usually, look for a tiny stream flowing out of the packing as a rule. A tiny stream means something no larger than a pencil lead. Over time, a relationship between the tightness of the packing and the right amount of leakage will be learned. The pump will usually indicate when it is too tight, as it will wear the sleeve or shaft until it gets enough flow. In that case, pay attention what the pump needs and allow that leakage.

A pump construction ensures that the packing, or seals described below, is not subjected to discharge pressures unnecessarily. By designing pumps with the packing, or shaft seal, at the lowest possible pressure point in the pump, wear and tear on them is reduced. Some pumps still have a seal, or packing, exposed to the highest pressure and construction is modified to reduce the effect of the high pressure. Packed pumps will have lantern rings (Figure 5-8), which permit bleeding off of the high pressure leakage to the suction of the pump, with the rest of the packing exposed only to the suction pressure. In the case of condensate pumps, and others that can operate with pressures below atmospheric, pressure on the suction side supplied by a connection from the discharge provides fluid to seal and cool the packing.

Modern rotating pumps are commonly supplied with a shaft "seal." It is a special construction, with very hard materials consisting of rotating and stationary parts that provide the liquid seal. Those materials are machined to very close tolerances. There are only a few hundred thousandths of an inch that separate them when operating. The two materials do not touch because a minute amount of liquid separates them. In many cases, the liquid forms a vapor between the two wearing surfaces. The vapor becomes the lubricant, with no leakage evident at the seal. Most of them require some flushing of the seal to keep it cool enough to operate properly, using a small line from the pump discharge to the seal to provide flushing liquid. When there is an opportunity for the liquid to contain small particles of rust, or other solids that could damage the seal, the flushing liquid is passed through a strainer to remove those solids, providing clean flushing liquid and extending the life of the shaft seal.

Sometimes the flushing liquid is improperly applied. If there is erosion around the area where the flushing liquid is admitted, check with the manufacturer. It is possible that the pump left the factory without a required orifice in the seal flushing piping. The seal materials have to be able to handle the temperatures under those vapor forming conditions. Some shaft seals require coolers to lower the temperature of the liquid. HTHW circulating pumps, for example, have strainers and coolers on the flushing liquid. Maintaining the coolers and strainers is an important factor in keeping those pumps in operation. Care is required to keep the temperature of the liquid within an acceptable range. It is possible to get too cool, producing thermal shock where it mixes with the fluid in the pump.

Alignment

When the pump and its driver are riding on separate bearings, the two shafts are connected with a coupling. Rotating shafts are equipped with flexible couplings which allow the two shafts to be centered in their own bearings. Shafts that extend and retract can be connected with rigid clamped couplings or a coupling containing a bearing that allows one shaft to swing like that for a chemical feed pump. Proper alignment of couplings is essential for long pump life. If the alignment is poor, the coupling will apply alternating forces to the shaft, constantly bending it back and forth until it finally breaks, assuming the bearings or packing do not fail first. The following discussion on aligning rotating pumps should provide all the clues needed to know about what has to be done for a reciprocating pump. The instruction manual should be read carefully to be sure to do it right.

The process of aligning a pump and driver begins with determining the differences between operating and cold conditions. A boiler feed pump, for example, will heat up when the pump is in operation. Its shaft can be higher when it is operating. A pump and turbine combination can have different changes in shaft position. It should not be necessary to correct for operating temperature on most pump and turbine combinations because both will be centerline supported. That means the pump and/or turbine is constructed with supporting feet that connect to the pump or driver near the centerline of the shaft. The temperature of the feet will not change much in operation. The shaft position will be the same regardless of the pump being hot or cold. When the pump or driver is not centerline supported, calculate the amount of growth or relative growth, given the operating temperatures and the material of the casing. Use that value in rough alignment. Then check the equipment when it is up to operating temperature.

Figure 10-61. Angular coupling alignment.

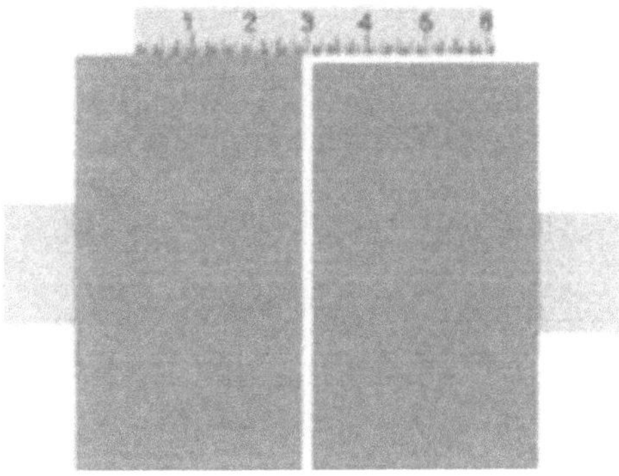

Figure 10-62. Coupling offset alignment.

Alignment should be performed in a particular order. Correct the vertical angular alignment (Figure 10-61) first and the vertical height (Figure 10-62) second. The horizontal angular alignment should be done next and horizontal alignment last. Those last two steps are done the same as the first two. They do not require shimming.

Shim stock of varying thicknesses will be needed. Commonly, shims are thin sheets of brass (preferably) or steel in varying thicknesses. Normally, some materials in 10, 5, 2, and 1 mil thicknesses will be needed (A mil being one thousandth of an inch). Occasionally, thicker pieces are required. Of course, this assumes that the pump was reasonably aligned in the factory. Sometimes, it takes some major pieces to rough in before dealing with the thinner pieces. Shims should be prepared as shown in Figure 10-63 so that they can be slipped under the supports of the driver (normally) and around its anchor bolts. It is important to make the slot at least a sixteenth larger than the anchor bolt. Be careful with their installation. Be sure that they do not interfere with bolting. When aligning a pump and turbine, it is sometimes easier to align the pump to the turbine. An electric motor does not have any connecting piping. That makes it easier to move the motor to achieve alignment.

Figure 10-63. Shims.

If aligning a pump to resolve some wear, or other problems that indicate misalignment, but do not find any problems with cold or hot alignment, be aware that a pump casing can be deformed by application of piping expansion stress at the pump nozzles. If that is the case, aligning the pump again is not going to solve the problem. The base that the pump and driver are mounted on also has to be firm. If the base can flex, it will allow vibrating misalignment, which usually results in coupling or bearing failure in a short period of time. In the case of a pair of condensate booster pumps, fairly large ones, their couplings were constantly failing. The base was suspended above the housekeeping pad by about an inch and held up only at the four anchor bolts in the corners of the base. A quick setup of a long ruler over a pivot next to the base showed how much it deflected when pressed by putting a foot on it. It was necessary to fill the base with grout, as specified by the manufacturer. The base has to be solid and not bent before worrying about alignment.

There are many different methods of pump alignment and selection is dependent on the speed, power requirements, and size of the pump. As speeds, power, and size increase, the precision of the alignment becomes more important. That does not mean that the smaller pumps should not be carefully aligned, only that the cost of the pump may be so low that the cost of precision alignment is higher. Replacing the pump more often may be less expensive than a precision alignment. The typical small pump is fitted with a coupling consisting of two metal halves with a rubber insert (Figure 10-64). The common method for aligning these pumps is to place a small metal ruler along the side of the coupling, as shown in the earlier figures, and adjusting until the rule shows the two coupling halves to be in line. Holding the rule as shown and holding a light behind it is the best way to see any gaps between the rule and the coupling halves. Turn the shafts 1/4 turn and repeat the reading three times when near the end. This does not correct for

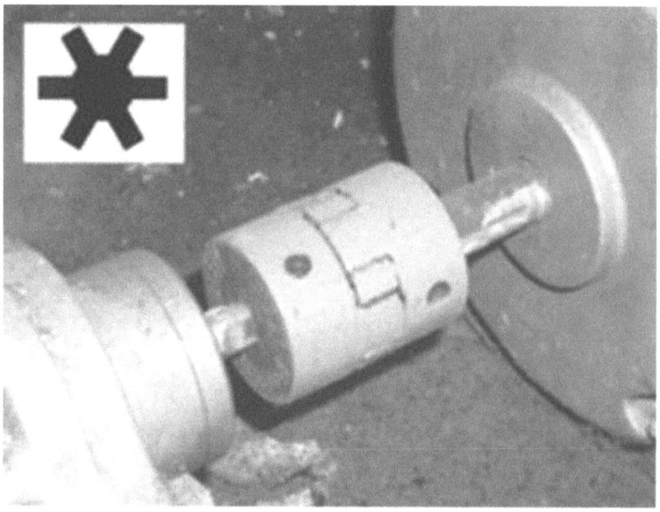

Figure 10-64. Small pump coupling.

couplings that are bored off center or where rough surfaces produce errors.

To determine how much angular adjustment is required, compare the length of the coupling half to the spacing between the motor mounts. Either eyeball the distance or slip varying thicknesses of shim stock in the gap between coupling half and ruler as shown in Figure 10-65. Then calculate the required adjustment by the ratio of the coupling half-length to the driver mount distance for vertical angular adjustments.

To correct the 2 mil difference over the coupling half, as shown in the figure, where the coupling is 1-1/2 inches long and the driver mounts are separated by 6 inches, adjust the shims at one end of the motor mount by 8 mils (2 × 6 ÷ 1.5). Be careful upon discovering a vertical angular misalignment. It can mean that some of the shims got knocked out from only one foot of the driver.

Sometimes, one mount is loosened and the shims are shaken out. When starting with a previously aligned pump, it is always a good idea to loosen all the anchor bolts of the pump and driver, and check to see if either rocks in any direction. Correct any rocking first. The pump could be distorted, or the driver frame, which is worse than misalignment for the bearings. It could even crack the motor housing. Once any vertical angular misalignment has been resolved, all the driver mounts should be level. Further adjustments involve adding or removing the same amount of shim stock under each of the feet. Be careful when performing the vertical center alignment. Different thicknesses of shims can be added or removed. The best thing to do is to use a micrometer (Figure 10-66) to measure the shims to be certain each foot is altered the same. If a micrometer is not available, use the ruler and light to compare the pieces of shim stock. Before working on the horizontal alignment, check the vertical with the driver bolted down on the shims. Sometimes, the shims can compress a little more or less to alter the alignment.

Once the vertical alignment is done, the job is simpler. The shim stock is no longer involved. However, it is hard to retain angular position horizontally while trying to correct centerline alignment. Once the pump is close to alignment, use a small hammer to tap the feet. With a little experience, the amount of tapping to get a movement of 1 mil will be discovered. Tapping both feet on one side consistently will shift the driver the same amount to retain angular displacement.

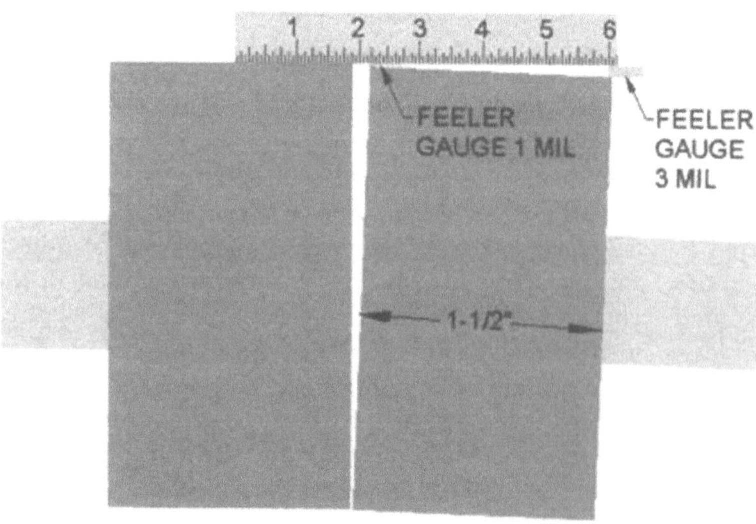

Figure 10-65. Aligning small coupling with ruler.

For better precision in aligning a coupling (pump and driver shafts), use a dial micrometer (Figure 10-67), which eliminates problems with poorly machined couplings and provides hard readings, instead of eyeballing it. Determine the error by clamping mounting bars furnished with the micrometer to hold it relative to one coupling half, while the micrometer stub (sticking out at the bottom left of the figure) rests against the half coupling attached to the other shaft, zeroing the micrometer, and then rotating the shafts to take a reading of 180 degrees from the original one. Zeroing the micrometer is accomplished by simply grabbing the dial and twisting it until the zero is centered under the needle. In this case, use twice the distance from the center of the shaft to the contact point of the

Figure 10-66. Measuring shim stock with micrometer.

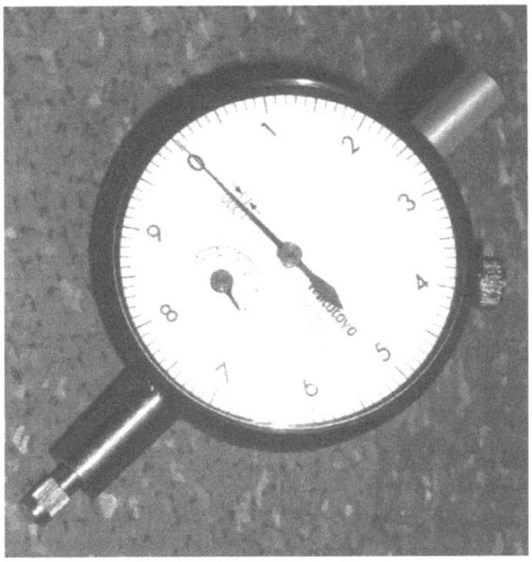

Figure 10-67. Dial micrometer.

micrometer instead of the length of the coupling to determine the ratio. Usually, the ratio is close to one, making life a little easier. Just use a shim matching the reading.

There are more precise methods using laser equipment and computers. That is best handled by a contractor who specializes in alignment. A lot of pumps would have to be aligned in order to justify the cost of a laser alignment system.

NPSH

NPSH stands for "net positive suction head." It is absolutely essential that an operator understand what it is and how it relates to the operation of pumps. In many a discussion, the term will mean one of two things: NPSHR is the "required" suction head and NPSHA is the "available" suction head. Suction head is the pressure at the inlet of the pump produced by two things: the height of the liquid above (below) the centerline of the pump and any pressure acting on the surface of that liquid. When the pump is running, the suction head has to account for the pressure drop in the suction piping. It will be a little lower when the pump is running. It will also decrease as the flow increases. When the available head is not adequate, the liquid in the pump will begin to boil. Small bubbles of gas will form in the suction. If enough of them form, the pump will become "vapor bound" and cannot pump any liquid. Once a pocket of vapor forms, the pump contains compressible gas, not incompressible liquid. The pump parts either spin in the vapor, producing no pressure, or the vapor will constantly compress and expand. The net result is the pump stops pumping. In some cases, this will cause a surge of discharged liquid back into the pump, which then gets pumped out again. That liquid surging back and forth damages the pump.

Sudden formation of vapor in a pump driven by a steam turbine will result in rapid overspeeding of the pump and turbine. Occasionally, that happens so fast that the turbine overspeed trip cannot respond before the turbine blades start flying out of the casing. When the bubbles start forming, they will collapse later when the pump increases the pressure in the liquid. In centrifugal and turbine pumps, the result is bubbles forming and then collapsing. The liquid rushing in, to fill the voids as bubbles collapse, hammer away on the parts of the pump. That is called "cavitation." It is evident by a small to fair amount of noise that can be heard. It is also evident when the pump is dismantled. There will be heavy wear, consisting of lots of tiny indentations where the bubbles collapsed. To prevent pump damage, be sure to have adequate NPSH.

The NPSHA is the difference between the suction head and the vapor pressure of the liquid. To get away from the math, assume a pump is submerged to its centerline in a tank of boiling water at sea level. Since the water level is right at the inlet, the suction head is zero gauge, 15 psia. Since the water is boiling, the vapor pressure is 15 psia and the NPSHA is zero. By submerging the pump in the tank so that its centerline is 4 feet below the surface and there is no suction piping to produce friction, the NPSHA is increased to 4 feet. If the water is colder, the suction head will increase. Assume the pump is in a tank of condensate at 162°F. The vapor pressure at that temperature (check the steam tables) is 5 psia. Subtract that from 15 psia to get an additional 10 psi of pressure that the suction can drop before the water boils. Checking the head tables, find that the 10 psi converts to about 23 feet. Now, that can be added to the 4 feet that the pump is submerged to get an NPSHA of 27 feet.

To help explain how a centrifugal pump can lift water out of a lake once it is flooded, the NPSH of water at 60°F is minus 14.5 psig, equal to 33.5 feet. A pump can lift 60° water that far before it will start boiling. Of course, the pump cannot pump the air out. A foot valve would have to be installed in the lake and fill the piping and pump casing with water to get it started. Once it is started, it will pump the water.

The NPSHR is specified by the pump manufacturer for the design operating condition and is usually shown on the pump curves. It is the required NPSH for that pump at the rate of flow. Some of the requirement is a function of how much the liquid has to accelerate at the inlet of the pump impeller. Some of the static pressure of the suction head has to be converted to velocity pressure to get the liquid into the impeller. The NPSHA is what is available, the actual NPSH at the inlet of the pump. That value always has to be higher than the NPSHR. Operating a pump when the level in a tank that it is taking suction on is too low can result in serious damage to the pump. Allowing a pump to continue operating when the suction head is inadequate does not make sense. If the tank is almost dry, there is nothing there for the pump to move anyway. Shut the pump down to prevent it from being damaged (priority number three). In some cases, a pump can be run to the point of losing liquid. Stripping a fuel oil tank before cleaning is one example. In that case, be prepared to stop the pump the instant it loses suction in order to limit the potential for damage.

Note that on the pump curves, the NPSHR increases as the flow through the pump increases. Throttling the discharge of a pump to reduce the flow will also reduce the NPSH required and can stop a pump from cavitating. Although this is occasionally required under unique operating conditions, it should not be the normal case. If the pump has to operate at the lower suction head, then it would be better to have the impeller turned down to reduce its capacity and horsepower requirement. Cutting down is a way of making a pump fit its application better. It cannot always be done. However, in many cases, it is something that should have been done and was not. If someone simply orders a new impeller, giving the manufacturer nothing but the pump model number, they could very easily get a full size impeller, not one that was trimmed for the application. Check the impeller diameter against the pump curve that came with the original instructions to determine if it is correct.

Pump Curves

Pump curves provide answers to a lot of questions about pumps. How much liquid can be pumped under varying differential pressures is the most important line on a pump curve. Any pump curve will normally have several of those depending on different construction and operating conditions. As stated above, the NPSH (NPSHR understood) will be shown when it is important. The pump will also have horsepower lines or efficiency lines or both. Either the horsepower or the efficiency will permit calculation of the other value. There is a standard formula for hydraulic horsepower.

The flow differential curve is the first one to look for. In many cases, they will be the darkest lines on the paper. The normal form of a curve lists the differential on the left side of the curve and the flow on the bottom. Differential is typically listed in feet, meaning head. That value has to be converted to psi to see how much pressure boost can be obtained from the pump. Some curves will show psi since the pump is not affected much by density. The rate of flow is normally listed in gallons per minute (gpm). Don't be surprised to see gallons per hour or hundreds of gpm. If there is no label, it is likely to be gpm. There are several lines on the curve. The pump can pump more than one type of liquid and some have variations in construction. The typical centrifugal pump, where the curves are almost always for cold water, will have different lines for the choices of impeller diameters. Normally, the curve is marked with the design point to show what diameter impeller was installed in the pump. Otherwise, look elsewhere in the manual to find out what size impeller was installed.

Once the proper line has been identified, the differential pressure for a given pumping volume can be determined. Sometimes, it is valuable for determining how much is being pumped based on the difference in pressure. Other curves will address characteristics of the liquid. Fuel oil pumps, for example, will have a number of lines on the curve for different viscosities of the oil. Unless specifically stated to the contrary, a pump curve is supplied to show the flow and differential characteristics for pumping cold water at 32°F and a density of 62.4 pounds per cubic foot. That provides a basis for determining the differential pressure at other fluid densities. Since ice water is seldom pumped, adjust the head characteristic of a pump curve to determine the actual differential pressure, which will always be lower than what the curve indicates. This gains some importance with water at high temperatures. It is important for things like boiler feed pumps.

Boiler feed water at 227°F (a common temperature) is not as dense as ice water. It only weighs about 59.4 pounds per cubic foot. While pumping that lighter feed water, the pump will only produce 95.3% of the discharge pressure that is produced when pumping ice water. That is enough to be significant when operating at high boiler pressures. It is also important to note that centrifugal pumps are volumetric machines. They pump so many gallons, not so many pounds. The 95.3% should also be applied to any calculation that converts the gpm to pph.

The horsepower, or efficiency, lines are primarily used by engineers in selecting pumps. They would like to buy the one with the lowest operating cost. The operator, on the other hand, can use those curves to get an idea of the best mix of pumps for a given operation or to provide answers to problems with the pump. There may be a choice of running one or two pumps. It could be that running one could be more efficient. While that may be a logical decision, it is not always the case. Running one large pump far out on its curve could be less efficient than running two smaller pumps because they are operating at a better efficiency.

At a major laundry, the owner complained that he was replacing the boiler feed pumps every six months. The pumps did not sound too bad, but it was obvious that they were cavitating during normal operation. A look at the pump curves and the installation revealed that inadequate suction head was the problem. New pumps were recommended for two reasons. One was that the NPSHR of the recommended pumps was less than what was available. The other reason was that the new pumps did the job at 3.5 horsepower and the existing pumps took 7.2 horsepower. There is that big a variation in pump efficiencies. The savings in motor horsepower was worth $1480.00 per year. If there is an opportunity to choose a pump, be conscious of power requirements in addition to NPSH. Another point to consider in using pump curves is the occasional use of a pump for a purpose other than originally intended. The curves can be used to see if the pump will work. Make certain to not overload the motor.

There is a standard formula for pump horsepower. It is called hydraulic horsepower or theoretical horsepower. It can be calculated by multiplying the flow, in gpm, by the head, in feet, and dividing by 3960. If the liquid is not water at 8.33 pounds per gallon, multiply by the specific gravity of the liquid. Note that it is theoretical horsepower. Divide by the pump efficiency to get brake horsepower, the amount the driver has to produce. If the efficiency is not known, use 33% (multiply the theoretical horsepower by 3) to be safe.

Reciprocating Pumps

Many boiler plant applications were predominantly served by reciprocating piston pumps until the middle of the 20th century when multi-stage centrifugal pumps displaced them. For that matter, most of the liquids in the plant were moved by the standard duplex reciprocating pump (Figure 10-68), which was the mainstay of the power plant at the beginning of that century. The pump, powered by steam from the boiler, was capable of producing very high pressures and, despite the reciprocating operation, produced a reasonably constant output.

The pressure differential of the pumped liquid is determined by the difference between the steam supply and exhaust pressures and the ratio of the cylinder areas. The maximum pressure that could be produced, an important consideration for selecting valves and piping materials, is the area of the face of the steam piston minus the area of the connecting rod times the difference in steam supply and exhaust pressures divided by the area of the fluid piston minus the area of the connecting rod (Figure 10-69).

There were, and still are, single piston pumps consisting of one steam cylinder and one fluid cylinder. It was difficult to adjust them, in which case they would operate continuously, occasionally hanging up at one end

Figure 10-68. Duplex reciprocating pump.

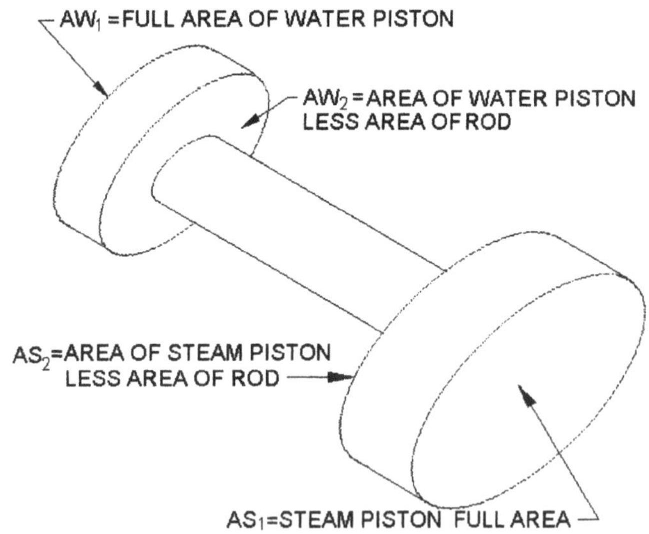

Figure 10-69. Areas of pistons for pump pressure.

Figure 10-70. Piston chemical feed pump.

of the stroke or another. Most of those were larger pumps used for fuel oil and ballast (water) transfer aboard the ships. The duplex pump practically eliminated problems with the pumps hanging up because the stroking of one piston tripped the valve to reverse the other. It is difficult to see that in the photograph. The linkage attached to one shaft operates the control valve for the other. A significant problem with these pumps was the lubrication, which tended to get into the condensate and then into the boiler. They also had a lot of sliding parts that would wear and required constant maintenance. Internal or external check valves also slammed open and shut with eventual wear and breakage.

Another form of reciprocating pump that can still be found, principally in boiler feed use, is a three piston, eccentric, cranked motor driven pump. The pistons are solid. They only pumped in one direction. Each of the three pistons operated off a different crank arm. Thus, the output was a little more uniform. The balance of pistons and a heavy counterweight on the shaft helped reduce the motor load from the imbalanced forces. A feature of the pump is control of the valves to vary capacity. The suction valve is held open on the discharge stroke (pushing the liquid back into the suction) for varying degrees of rotation to vary the amount of water pumped. The first nuclear merchant ship, the Savannah, had one of those pumps for boiler feed. The only reciprocating piston pump normally found in a modern boiler plant is a chemical feed pump. Usually, the piston is pumping a hydraulic fluid that transfers energy to the liquid being pumped using a diaphragm (Figure 10-70). The capacity of a reciprocating pump is easy to determine. It is equal to the area of the piston times the length of stroke times the revolutions per minute, if it is single acting. If it is double acting, where the liquid is admitted to and pushed out from both sides of the piston, it is twice that much less the cross-sectional area of the shaft times the length of stroke times the rpm (revolutions per minute).

Reciprocating pumps are positive displacement pumps. That is a term used to mean that plastic, wood, metal, or whatever the pump is made of displaces (moves into the space that was occupied by) the liquid to move it through the pump. The steam powered duplex pump had some balancing features, as the pressure on the liquid could not exceed the difference between the steam supply and exhaust pressures times the ratio of the areas of the pistons. That pump would simply stop if the pressure on the liquid got too high. Motor driven pumps seldom simply stop. They produce very high pressures because the motor's torque increases as it slows down. Usually, the motor starter will trip. Still, there are many reports where the pump or piping ruptured when someone accidentally started a pump without opening all the valves in the system.

To prevent damage of that nature and motor starters tripping, or motors burning up, a relief valve should always be installed at the discharge of a positive displacement pump. If it is reasonable to believe that the flow through the relief valve will always be of short duration, then the relief valve can dump the liquid back into the pump suction piping. It is always possible that the pump could be operated for some time pumping the same liquid and all the power will be diverted to heating up that liquid. It is better, whenever possible, to route that liquid back to a tank or sump where there is a larger mass of liquid to absorb the heat. A final note is appropriate before discussing specific types of pumps. Any of them can be run backwards. Some, like centrifugal pumps, can appear to operate, just not as well as with proper rotation. Gear and screw pumps will tend to pump the liquid in the opposite direction. If there is any doubt, check the instruction manual.

Centrifugal Pumps

The most common pump is the centrifugal pump. It achieved that position for reasons other than efficiency and energy costs. It is also misapplied more than any other pump. The range of efficiency of pumps in service runs from 30% to 70%. Now that has to be one serious variation. A pump at 30% efficiency will use 2.33 times as much energy as a pump operating at 70% efficiency. The main reason those significant differences are ignored is that the purchase decision is so intent on getting the lowest first cost. Then a pump is chosen that will chew up all the difference in first cost in comparison to an efficient pump in less than a year or two. There can be many reasons given for this choice (tight budget, someone else's problem, lack of analysis, inexperience, etc.). None are really compelling.

If a centrifugal pump was installed in connecting piping with no valves and stopped, the liquid would flow right back through the pump. There are no suction or discharge valves to block that flow. Some operators have a problem understanding how the pump even works. Centrifugal pumps simply grab the liquid and throw it. The impeller flings the liquid into the volute of the pump (Figure 10-71), where the velocity pressure is partially converted to static pressure and delivered to the discharge. Consider a pot or bowl half full of water. Start stirring it with a spoon. Stir the water in one direction (a pump only runs in one direction). Note that the level of the water in the bowl will vary from low in the middle to highest at the outside of the bowl. That difference in level is the head of that bowl pump at shutoff.

Figure 10-71. Centrifugal pump impeller and volute.

Stir faster and the head goes higher. Spin it fast enough and the water starts coming out of the bowl. Setting it under the spigot to add water and stirring it fast enough so that the water spills out at the same rate water is added and that is a simple version of a centrifugal pump. Do various things with the spoon and note what happens. That will provide a pretty good understanding of how a centrifugal pump operates.

A centrifugal pump does not move a fixed volume of liquid like a reciprocating pump. The amount of liquid moved varies with the differential. The flow of water pumped from a tank will vary with changes in the height of water in the tank or the discharge pressure at the outlet of the pump. If the spigot on the sink is opened up so that more water flows in and the rate of stirring is not changed, more water will flow, even though more work is not done. It might help to realize that a centrifugal pump simply boosts the pressure to a certain amount. That boost is related to the flow of water through the pump. Stop stirring the water in the bowl and it will still overflow once it has filled. The difference (head) that is created can be varied by changing the speed of stirring. There is no such thing as a limit on the flow through a centrifugal pump. The highest possible flow is much more than the design value and the minimum is zero. Without check valves in the discharge piping, a higher external differential pressure than the pump can handle will result in flow backwards through the pump. The actual flow rate is dependent on the performance of the pump itself and the difference in pressure between suction and discharge. There is a design point. It is a flow and differential that an engineer calculated for selecting the pump. It is usually indicated in the manual and on the pump curve. What an operator has to deal with is the actual flowing conditions. The odds that the actual conditions are precisely the same as the design conditions are between slim and none.

A feature of centrifugal pumps that is frequently forgotten is the use of wear rings (Figure 10-72). The space between the casing and the eye of the impeller is all that separates the suction and discharge pressure zones of the pump. Some water has to bleed back through that space because they cannot rub. As the pump is used, small particles in the liquid and the liquid itself can erode the material on either side of that gap. The provision of wear

Figure 10-72. Wear rings.

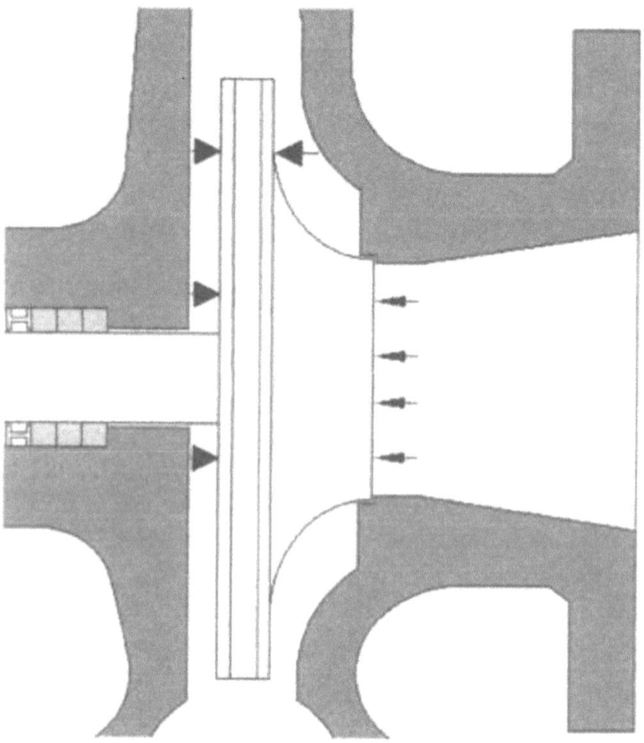

Figure 10-73. Axial forces on centrifugal pump.

rings makes it possible to restore a pump to a like-new condition by simply replacing the wear rings. The casing wear rings, the right one hanging loose in the photo, are keyed to set in the casing and not rotate. The impeller wear ring is heated and then inserted onto the end of the impeller, where it shrinks on for a tight fit. The strainer in the suction piping (a standard requirement for most pumps) does not remove the small particles that erode the wear rings. The strainer does remove pieces that would jam between them. Usually, a pump with wear rings will also have a shaft sleeve. Be cautious when replacing the wear rings. The outer wear ring can be distorted when the two halves of the pump casing bear down on it. Always make sure the pump rotates by hand as the bolts that hold the two casing halves together are drawn up.

Don't install a thicker gasket on a pump simply because the right thickness is not on hand. That will create gaps between the outer wear ring and casing, where erosion can cause problems. Too thin a gasket will normally bind the pump up.

There is a lot of variety in centrifugal pumps, depending on their application. The pressure differential they can produce depends on the density of the liquid being pumped and the speed of the tips of the vanes in the impeller. To make a pump operate at a higher differential pressure with the same liquid, the diameter of the impeller is increased. Once the impeller's maximum diameter is reached, a faster motor is used. As the impeller diameter and speed increases, the stress on the metal gets higher. There are practical limits on the pressure boost. If a larger differential pressure is required, the pump is supplied with additional impellers. These are called "multi-stage" pumps. The pressure is increased a little in each impeller which, along with its volute and share of the casing, constitutes a stage. That way, high pressures can be developed without making pumps of very large diameter.

Since the eye (inlet of an impeller) is exposed to suction pressure at that stage, and the rest is exposed to the discharge pressure of that stage, there is a difference in axial forces on the stage (Figure 10-73). In single-stage pumps, holes are drilled through the back plate of the impeller and a second set of wear rings added to balance the pressure (Figure 10-74). In multi-stage pumps, the stages are reversed on the shaft (Figure 10-75). The imbalance of one stage is opposed by the imbalance of another. Some pumps with vertical shafts are designed so that the axial thrust helps offset the weight of the shaft and impeller. Despite the best design, there is no guarantee that the pump will not see some axial forces. One end or the other is always fitted with a thrust bearing. If the pump is cantilevered off a single bearing, it is also the thrust bearing. As pumps wear, the direction of thrust can change. One excellent measure for pump condition is the axial position of the shaft, when accessible to measure it. Taking initial measurements of how much a shaft shifts along its axis (using a dial micrometer) before it is ever operated provides baseline measurements for bearing wear. Take them anyway if the pump is in good

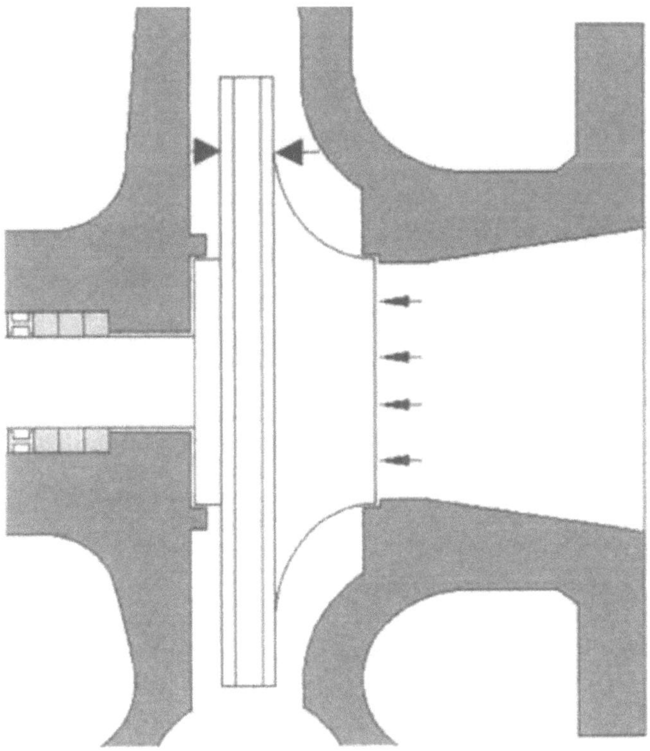

Figure 10-74. Back pressure with wear rings on a centrifugal pump.

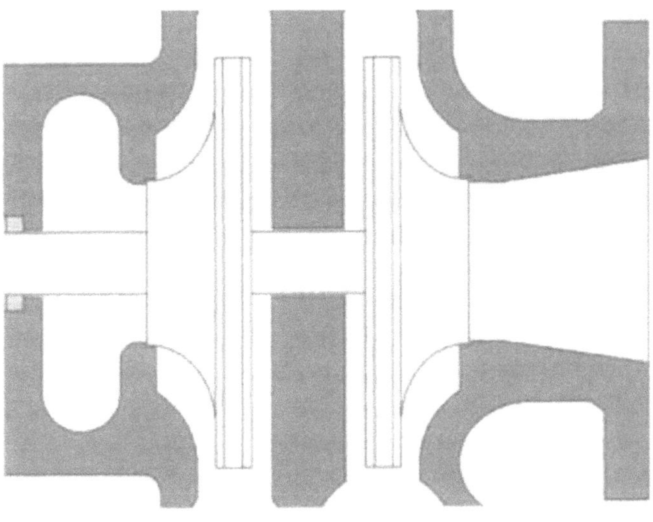

Figure 10-75. Opposing stages of centrifugal pump.

shape. Then compare them every year or two to check for wear. The first clue of potential operating problems with a pump is the shape of the curve. If the curve has a negative slope at all times, there should not be any operating problems with it under most circumstances. Slope is a value equal to the change in differential divided by the change in flow at any point on a curve, indicated by a line tangent to the curve at the point of observation. If the differential is always decreasing, the pump is easy to handle. A lot of pump curves have a positive slope as the flow approaches zero. The curve will have a hump in it where the slope is zero (differential does not change) at the top. The curve will have a positive slope (differential decreasing) to the left of the hump where flows are lower.

Anytime the pump is operating at a point close to or to the left of that hump, the pump's operation may be unstable. It may be unstable because, for one set differential across the pump, there are two possible flow rates. If the system somehow maintains a constant differential for those two flows, the pump will not align with one or the other, switching back and forth between the two points. When a pump does that, it is called "surging." It is usually accompanied by a lot of fluid noise in the pump and system. Multi-stage pumps can oscillate along the axis of the shaft when surging. That is another thing to look for when monitoring the operation of a centrifugal pump.

There are situations where the hump in the curve is not a problem. That is because the change in flow normally produces a change in pressure drop through the system. From the chapter on flow, the change in pressure drop is proportional to the square of the change in flow. With that knowledge and some actual operating conditions, the system flow curve can be spotted on a pump curve to see when the problem of surging will occur. Look at the difference in pressure when there is nothing flowing, a piece of data that is not always easy to measure. Then note differences in pressure in the system to find the loss due to flow at some point. Draw a system curve on the pump curve by starting with the difference in pressure when nothing is flowing and then add the pressure drop for the corresponding flows to continue it.

The curve in Figure 10-76 is a sample of a boiler feed pump curve with a couple of system curves plotted on it. The system curve "A" is for a normal plant. The system curve "B" is for a condition with very low system pressure drop between pump and boiler, one with a feed water control valve that is wide open for some reason. Note that there is no one flow rate where the slope of system curve "A" and the slope of the pump curve are close to each other. The slopes of the pump curve and system curve B are very similar. That is where things get unstable.

A change in flow increases the pressure drop in the system and also rides up the pump curve to increase the pump differential by the same amount. The rule of these curves is that the operating point is where the system curve and the pump curve intersect. It is the only point where both the pump and the system have the same characteristics. However, if one or the other did

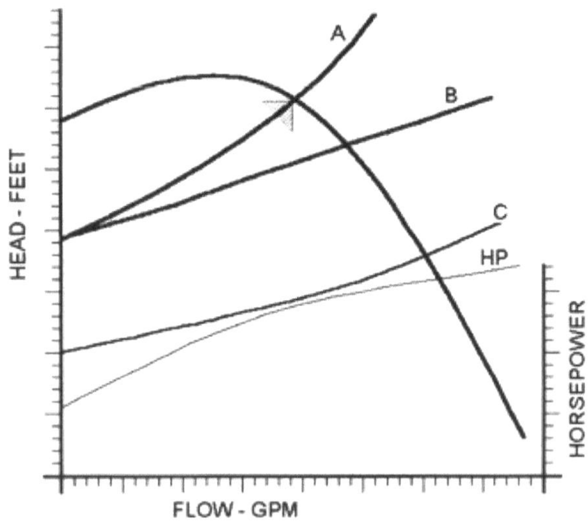

Figure 10-76. Boiler feed pump curve (A and B (no hump and hump) show horsepower).

not change, then the flow through the system would be constant and the water flow could not be controlled. A control valve somewhere in the system, or the differential at zero flow (the point where the system curves intersect the zero flow line), has to change to vary the flow. Picture the system curves being shifted up and down by the operation of the flow control valve. Then notice how a curve like the one labeled B can hit two points on the pump curve. If there is a problem with a surging pump, this should be a clue as to how to handle it. Simply increase system resistance when operating at the lower loads by throttling a valve someplace. Alternatively, open a bypass line to recirculate fluid so that the flow through the pump is beyond the hump of the curve where the slope is negative.

Recirculation of some fluid is typically recommended for centrifugal pumps that can be operated during periods of system flow stoppage to prevent overheating the pump or the fluid. If system flow is stopped, the water simply churns in the pump, soaking up all the motor horsepower that is used by the pump in that condition (all inefficiencies) to raise the temperature of the pump and fluid. If the fluid can take the high temperatures, it is possible that the heat will distort the pump or weaken the pump shaft until it springs off center or starts rubbing moving parts on stationary ones and fails dramatically. If the pump can take the heat, the next problem is the vapor pressure of the liquid in the pump. Once the temperature exceeds what matches the vapor pressure of the liquid, then the liquid will start vaporizing, creating cavitation first, then flooding the pump with vapor.

Operating under shutoff can happen regularly with boiler feed pumps. There is often a recirculating line on a centrifugal feed pump that returns some water to the deaerator or boiler feed tank. On most jobs, the line has an orifice between the connection at the pump discharge and an isolating valve on the recirculation line. The orifice is sized to bleed enough water off the pump to limit the temperature rise when the pump is operating in system shutoff conditions. If another orifice is installed in the piping before the deaerator or feed tank (included in the sizing to prevent pump and liquid overheating), there is an added advantage to these systems. The recirculating line of an idle pump can be used to bleed some liquid back through it and keep it hot. That way, it is ready to operate the moment it is started.

That represents a fair amount of water. If that flow was not included with the design capacity of the pump, there could be a shortage of pump capacity at high loads. However, that situation would be unusual, as the pumps are normally oversized. Engineers usually oversize pumps, including the recirculating flow, before applying a safety factor. What that flow represents is a lot of electrical energy to replace steam energy. The power used to pump that liquid heats it up. The electric power to do that costs a lot more than the fuel (generally 2–3 times as much). This is one place where an operator can reduce power costs. As long as the loads are such that the feed water valves should always be open, shut off the recirculating line. The pump will back up on the curve, producing a little higher feed water pressure, and the horsepower consumption will decrease. Open the valve when loads are low and periods of shutoff are possible. That will not save electricity at that time. There will be savings on demand. The FD fans and other equipment are at lower loads when the recirculating feed water pumping load is reinstated.

There are feed water systems that recirculate large quantities of water to maintain a constant feed water pressure or constant differential between feed water and steam pressure. Sometimes, it is nothing more than a concept of what should be done. The feed water pressure will get too high as the flow is reduced when the pumps have very steep curves. On the other hand, the pressure regulation is there to reduce pressure drop across the feed water control valves. They either cannot shut off at the higher differentials or they throttle so much that the valves wear dramatically. The feed valves can be changed to save electricity. Another solution is installing VSDs. The economics are not always there. The VSD will use less electricity as the load drops. That savings has to be compared with the added cost of the VSD.

If low load operation is infrequent, the payback period can be quite long.

When starting a centrifugal pump, it is common practice to open the suction valve, start the pump, and then open the discharge valve. The reason is that the pump cannot draw any more horsepower than what is used at shutoff during startup, reducing the load on the motor. In one instance, a discharge check valve had failed to close on a pump: the pump needed to be in operation. When the pump driver stopped, the fluid simply flowed backwards through the pump and tended to rotate it backwards. The additional motor load required to reverse the rotation before starting to pump resulted in a heavy starting current for too long. The starter tripped.

When the pump is operating under system startup conditions, the discharge valve may have to be throttled (they are normally gate valves and should not be throttled) until system pressure builds. Not all pumps are furnished with non-overloading motors. If a boiler feed pump is running when the boiler pressure is way below normal, and the feed water valve runs wide open, it is possible for the motor to overload. Look at that curve in Figure 10-76. Note that the horsepower at the design operating flow (indicated by the little triangle) is less than the maximum. Draw a vertical line at the design flow (point of the triangle) and a horizontal line from where it intersects the horsepower curve to the right to read the pump horsepower at that design condition. It is always possible to pick a motor smaller than the maximum horsepower of the pump. Even though the pump sees a higher pressure with curve A, it cannot pump as much volume as with curve B. Thus, the horsepower is less.

However, if the pressure in the boiler drops so that the system curve is C, then the flow can increase considerably. The motor horsepower requirement for the pump at that point is so much greater that it could overload the motor. If a pump has a limited horsepower motor, take action to prevent it from running out on the curve when the boiler pressure is low. Normal practice is throttling a valve down. Don't count on the throttling of the feed water valve. It could suddenly go wide open.

There are hundreds of variations in pump construction since there are many different applications. The shape of the vanes in the impellers can vary from highly efficient, backward curved (BC) to radial depending on the desired efficiency weighed against the solids in the liquid they pump. They can be rubber lined for purposes such as pumping a slurry of limestone or ash. They can be "canned" where the rotor of the motor is sealed in an enclosure with the pump to prevent the leakage of

Figure 10-77. Horizontal split case pump.

hazardous liquids. The most common arrangement is the horizontal split case pump (Figure 10-77). The American National Standards Institute (ANSI) pump (so called because the National Standard establishes fixed mounting dimensions so that all of manufacturer's pumps are interchangeable) is also gaining popularity. They are end suction pumps that require the piping be disconnected to get to the pump for maintenance.

Turbine Pumps

Mention turbine pumps and some people get the impression of a centrifugal pump powered by a steam turbine. That is not the case. A turbine pump is a type of pump. Although they exhibit some characteristics comparable to a centrifugal pump, they differ. The turbine pump grabs the liquid on the outer diameter of the impeller, spins it around inside the pump, and heaves it out the discharge. A turbine pump impeller looks like the one in Figure 10-78 with little slots all around the outside. The fins formed by those slots are what grabs

Figure 10-78. Turbine pump impeller.

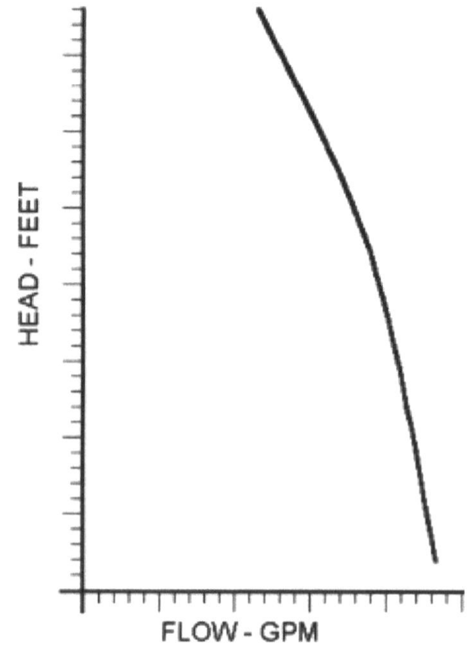

Figure 10-79. Turbine pump curve.

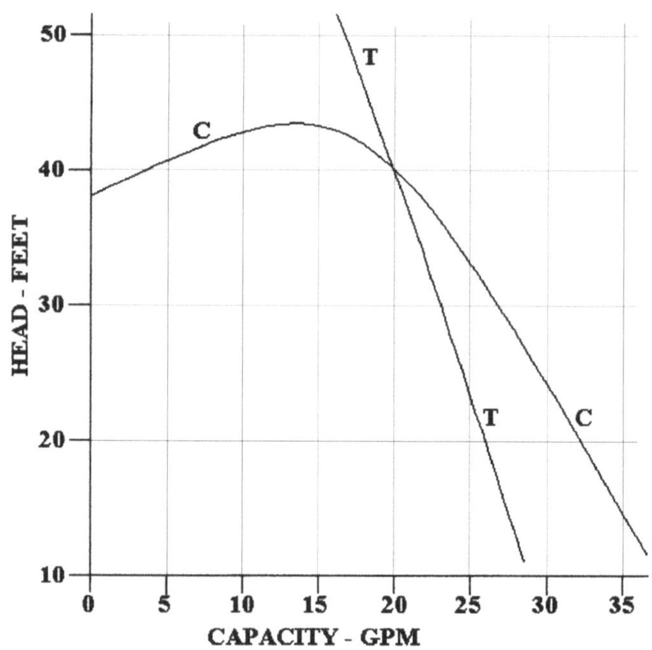

Figure 10-80. Centrifugal and turbine pumps on low pressure boilers.

the liquid and whirls it around inside the pump casing until it gets to the discharge.

Turbine pumps can produce very high differential pressures. They act more like a positive displacement pump than a centrifugal pump. The typical turbine pump curve (it is one that showed all conditions from zero flow) looks like a centrifugal pump curve. However, most of the turbine pump curves available look almost like a straight line, with a steep negative slope (Figure 10-79). Since they operate more like a positive displacement pump, treat them like one. Don't start a turbine pump with the discharge valve closed. Turbine pumps are commonly used as boiler feed pumps, especially on low pressure steam boilers. Their steep curves permit them to handle the significant variations in boiler pressure without any effect on pump capacity. Some plants have centrifugal pumps that also have curves so steep that their flow is not altered significantly by changes in boiler operating pressure. Centrifugal feed pumps in heating plants cannot handle the pressure variations.

Take the typical heating boiler plant. Both centrifugal and turbine pumps can be obtained to produce a design flow of about 31 gpm (15,500 pph) at the normal boiler operating pressure of 12 psig (31.7 feet). The density of water, for this example, is assumed to be 54.55 pounds per cubic feet, 175°F water, which means the head relationship is 2.64 feet per psi. There is a big difference in their operation as the pressure changes. They are selected for when the boiler runs up to the limit of the safety valves (15 psig or 39.5 feet). The concern with using centrifugal pumps is that any external pressure effects can result in total loss of water delivery (Figure 10-80). The curve as shown will deliver water to the boiler. However, changes in such things as level in the deaerator, or feed water tank, can prevent delivery. The values of head, used on this curve, assume that the pressure drop through the piping is negligible and that the level in the boiler is the same as the level in the feed water tank. From this curve, it is apparent that a drop in level at the feed water tank of a couple of feet will increase the head requirement for the pump to the point that the centrifugal pump cannot deliver any water, until the boiler water level or pressure drops enough to produce a differential that the pump can overcome.

On the other hand, any drop in boiler pressure will be accelerated by a centrifugal pump with a relatively flat curve. If the boiler pressure drops to 8 psi, a typical occurrence with a heavy load, the turbine pump output will only increase a little bit. The centrifugal pump will increase its delivery over twice as much. That additional water consumes more of the boiler's heat input, leaving less to make steam, meaning the pressure drops further. This shows that a poor choice in boiler feed pump selection on low pressure boilers can produce serious headaches for the boiler operator. Replacing those centrifugal pumps with turbine pumps can reduce the swinging pressure problems encountered in some plants and eliminate others because the pumps create the problem.

Screw and Gear Pumps

Screw and gear pumps are used principally for fuel and lubricating oils. They can be more efficient at moving liquids with viscosities higher than water than other types of pumps. They are capable of producing high differential pressures in a small package. Since they are positive displacement pumps, one running at 3500 rpm can be half the size of one running at 1750 rpm to pump the same amount of oil. Screw and gear pumps are positive displacement pumps and work pretty much alike. The pumps use two or more machined rotors that mesh closely together. The rotors produce a moving cavity as they rotate with each other. The cavity opens at the suction end and is sealed as the rotors turn. Then the cavity travels to the discharge end of the pump to deliver the liquid at the discharge pressure. The liquid serves to lubricate the rotors to prevent them from rubbing each other or the pump casing. The ends of the rotors are enlarged to increase bearing surfaces to balance the axial forces or shaft bearings take the thrust.

Some liquid is squeezed between rotors and casing in the opposite direction of the moving cavity, the amount depending on pump construction and wear. Smaller pumps usually have one rotor or gear that is driven and the rest of the rotating parts are driven by it in turn. Larger pumps, pumps that produce high differentials or pump very low viscosity liquids, can have external gearing so that each rotating element is driven. That reduces the amount of force that has to be transferred through the thin film of liquid between the rotating parts, replacing it with the lubrication of the external gears. The quality of internal lubrication is dependent on the differential pressure and pump speed. If the liquid is very viscous, it will maintain a stronger liquid film between metal parts to prevent them from rubbing. As the viscosity decreases, the film gets thinner and will break to allow the metal parts to touch. The fluid bleeding back through the spaces between the metal parts is what provides lubrication. The differential pressure between each adjoining cavity pushes the fluid through so that it wedges its way between each part. If pressure differentials are considerably lower than design, there may not be sufficient differential to force the lubrication of the pump. If the pump speed is too low, it will not generate that wedge effect as well. Other factors like the viscosity of the liquid have to aid in lubrication.

The typical pump used for pumping heavy fuel oil will not effectively pump light fuel oil. It may even fail if used to pump light fuel oil. Some people argue that a heavy oil pump is worn by the ash and sediment in the oil, thus increasing the gaps between rotors and casing. However, the truth of the matter is that the pump's design and speed were established for heavy oil and do not work well on light oil. The lower the viscosity, the faster the pump has to run.

Figures 10-81, 10-82, and 10-83 are the typical forms of screw and gear pumps used in boiler plants. A common gear pump consists of two gears in a casing. Usually, one is driven and the other is an idler. The term idler implies that it does not transmit power to anything else, not that it is lazy. The teeth of the driven gear engage in the teeth of the idler and they counter-rotate. Start with the gear pump in Figure 10-81. The liquid enters the pump where the gear teeth are disengaging, is trapped within the cavities formed between the teeth and casing,

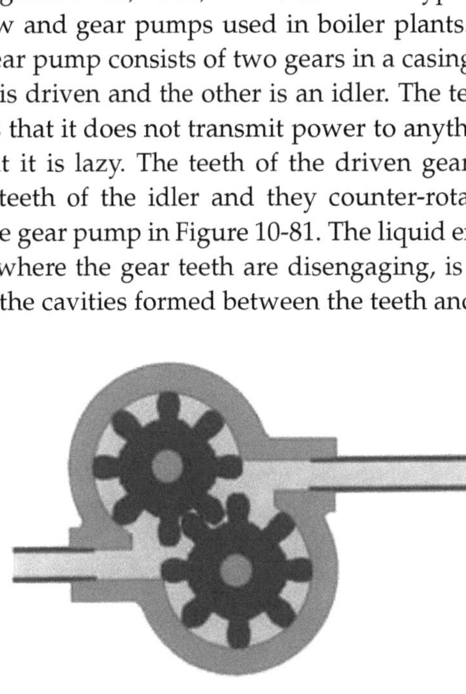

Figure 10-81. Gear pump.

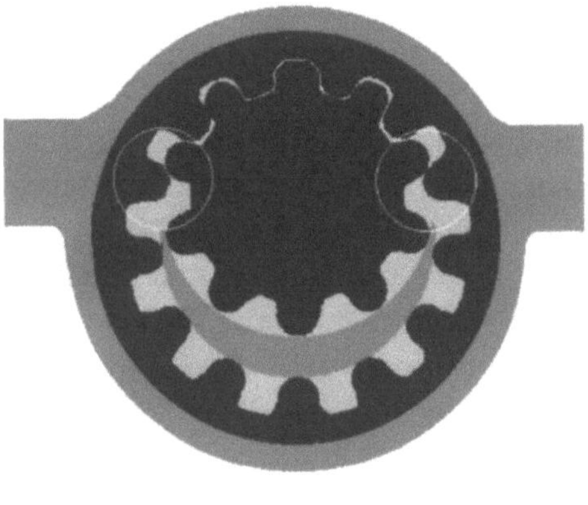

Figure 10-82. Crescent gear pump.

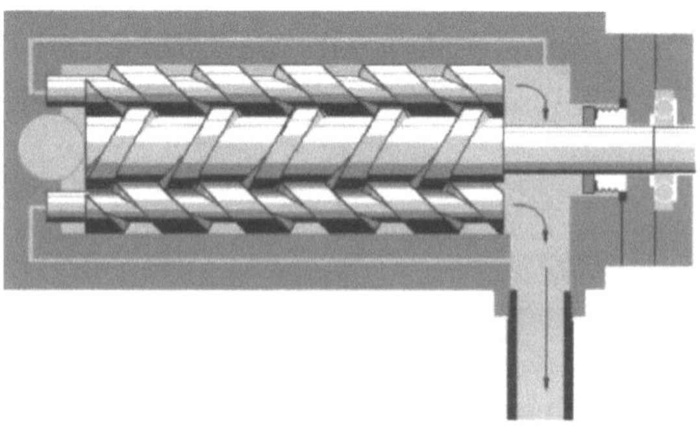

Figure 10-83. Screw pump.

and is carried to the discharge side of the pump, where it is forced out as the two gears engage, filling the cavity the liquid was in with a tooth of the other gear. A sectional view in the other direction would not reveal much. The sides of the gears are flat and just clear flat sides of the pump casing. The view in Figure 10-81 shows all that is relevant to the operation of the pump. Volumetric capacity of the pump is affected by the size, length of the gear teeth, and speed of rotation.

The crescent gear pump, in Figure 10-82, simply traps the liquid between the gears and the crescent shaped piece of the housing. The inlet and outlet ports are outlined. Either of these pumps will pump the fluid in either direction. The design capacity of a gear pump can be determined by calculating the area of the space between the casing and the root of the gear teeth and then multiplying that by the radius at the center of the teeth, the percent of the rotation where the liquid is trapped, and the rpm times two, to account for each side. The actual capacity will always be less because some of the liquid has to leak back past the teeth and the ends of the gears to lubricate the pump. In many of these pumps, the spacing between the casing and ends of the gears is adjustable, making them suitable for different viscosity fluids through adjustment.

The cavity in a screw pump (Figure 10-83) is formed by the intersection of the rotors and closed by the casing housing the rotors. The pump shown is supplied with two idler rotors that increase its capacity without an appreciable change in size. A smaller pump can be had with only two rotors. Liquid enters the pump at one end of the rotors, fills a cavity that opens as the grooves in the rotors separate, is trapped between the casing and rotors as the grooves engage, and then travels along the rotors to the discharge end of the pump. That movement, and the difference between suction and discharge pressures, produces an axial thrust on the pump that has to be opposed by the bearing of the driven rotor and the fluid film between the rotors, plus the end of the idler rotor bearing against the casing. Some manufacturers use an enlarged end on the rotors to increase the bearing surface. Other schemes include balancing lines between suction and discharge ends, applied to balance pressure forces. Another scheme is opposite hand ends of the rotors so that they draw liquid from both ends and discharge in the middle to balance the hydraulic pressures almost completely.

Screw and gear pumps do not do a very good job of pumping compressible fluids. An oil pump can easily get air bound when there is a sufficient volume of air, or vapors, at the discharge and inlet to expand and contract as each cavity between the rotors is opened and closed. This prevents any flow through the pump. The air also leaks back just like the oil. It is not uncommon for the pump to generate a loud audible roar when air or vapor is trapped in it. The air or vapor does not do a very good job of lubricating the pump and it forms bubbles in the oil as it leaks back. Operating a screw or gear pump with a vapor trap for any extended period of time will ensure complete breakdown of the film of lubricating oil on the rotating parts, with subsequent damage to the pump as the metal parts start rubbing. It is necessary to vent them to eliminate compressing air or vapors in the pump that will prevent liquid from entering. Properly vented, the pump will move air to eliminate it from the suction piping. When starting a dry pump (filled with air), it is important to ensure that lubricant film is maintained. Making certain that some of the piping is full of oil that will be drawn into the pump is important to limit wear. The oil is also a sealing film that helps the pump trap the air in its cavities and push it through. The best way to do it is to fill the suction strainer with oil, shutting down at regular intervals, and repeating the process, until all the air or vapor is pumped out. Once the suction line is full of liquid, the pump will work.

Pump Control

There was a time when the only control available over the operation of a pump was to turn it on or off. That is still a common means of controlling the pumping of liquids, used almost exclusively for feeding low pressure boilers and returning condensate. Modern technology has expanded the options. The wise operator should know how to utilize those methods. Note that this involves controlling the pump to control the flow. Before variable speed pump control, the fluid flow was controlled to maintain the operating parameters. Many times, that meant recirculating the liquid that was pumped, wasting the energy that was used to get the liquid up to pressure but a necessary means of controlling the flow. When dealing with on–off pump control, there are opportunities to improve that method of control to reduce energy costs and wear and tear on the pump. An attitude of limiting the number of starts by extending run time is one to adopt. That is not recirculating the liquid to keep the pump running. That saves starts but also wastes energy. It is an option to consider if the pump has extremely short off cycles, where it may run for 10 minutes and then shut down for 5–10 seconds. If that is the case, then recirculating some liquid to keep it running past those short off cycles will save on pump and motor wear and reduce wear and tear on the starter as well.

Every time the pump is started, the entire assembly is subjected to stresses above and beyond the normal operating conditions. Motor current is five to ten times the normal operating current when the pump is started. Those high currents produce rapid heating of the motor windings, with attendant thermal stresses, and also high magnetic forces that can dislodge the windings. Repeated startups and shut downs are also rough on the bearings, both in the motor and in the pump. The bearings will heat up and cool down, with some breathing that can increase the probability of air mixing into the grease or oil to corrode them (see Chapter 6, Lubrication). The pump always experiences pressure spikes when starting. The liquid in the connecting piping has to be accelerated from its stationary position and the check valve has to be lifted. By reducing starts, the strain on the equipment will be reduced to extend its life.

Reduce starts by stretching the pump's on–off settings as far as possible. Let the level get a little lower and run the pump until it is a little higher by adjusting the level controller. There are limits to this, including allowing the level in a boiler to get so low that a little upset results in operation of the low water cutoff. All that really does is provide the data on where the low limit is. As for controlling the flow through a pump automatically with modulating capability, it is not done consistently. The only pumps that can provide modulating control are chemical feed pumps, which use a reciprocating hydraulic pump acting on a diaphragm with an adjustment of the stroke of the reciprocating section. That is what the knob is for on the pump in Figure 10-70.

The centrifugal pump, which serves the majority of applications, is self-aligning. The flow through it is determined by the system. By throttling the flow in the system at some point, preferably after the pump discharge, the differential pressure required to force liquid through the system increases and the flow through the pump decreases, as it follows the differential up the pump curve. Applications with some centrifugal pumps and most screw and gear pumps normally incorporate recirculation control, where the flow through the pump does not change and a portion of that flow is diverted back to the pump suction to achieve a final control of delivery pressure.

Advances in motor speed control have made some pump control projects possible that were not possible before. They are limited. Varying the speed of a centrifugal pump with a relatively flat performance curve does not produce much of a savings in horsepower above what the pump automatically provides. A pump with a nearly flat curve will supply a reasonably constant differential pressure automatically. There would be no need for control. If the pump has a very steep curve, varying the speed will save power costs and allow differential pressure control. If the differential required for varying flow also varies with load, then some potential savings by controlling pump speed is possible, even with pumps with nearly flat curves.

Application of VSDs on pumps has resulted in significant operating cost savings. Typically, the pump is often oversized, like every other piece of equipment. Systems seldom operate at the design capacity, and the VSD permits tuning the pump to operate most efficiently under different load conditions. However, don't be surprised to see a pump with a VSD control that is simply operating at its highest speed. It is not unusual for the system to have problems that were not anticipated by the designer or there are other operators who simply do not understand the application and, therefore, solve problems by eliminating the automatic control and running the pump full out.

It is possible to control the drum level by controlling the speed of the boiler feed pump. Note that this requires an independent feed water pump and supply line for each boiler and will not work with a pump with a relatively flat curve. When the boiler is operating at low loads, recirculating some of the feed water back to the deaerator, especially on boilers with economizers, will eliminate problems with surging of the pump and flashing of the feed water in the economizer.

A more common means of boiler feed pump speed control consists of maintaining a constant header pressure. That can be enhanced, and should be, by actually maintaining a differential between the feed water pressure and the steam pressure. It is invaluable for good control during plant startup and in systems with varying steam pressure. The control system should also incorporate a set point adjustment relative to load to account for the pressure drop in the feed water piping and in the steam piping before the pressure sensing point.

In chilled water distribution systems, the major problem was piping friction losses. Applications to systems with three-way bypass valves on cooling equipment have to be changed to simple two-way valves to produce a change in demand that can be sensed. The three-way valves simply redirected the flow. There was no way to reduce system flow. Problems also appear when significant differences between actual operating conditions and design conditions are introduced by such things as buildings shutting down or varying significantly in load. A common significant shift in load is an auditorium or gym, where cooling is required for the high concentration of people and/or activity on a cool or cold night, when the other buildings do not require cooling. That is usually solved with application of a number of differential pressure transmitters in the system, artfully applied, so that the pump speed controls can operate to maintain a differential at the lowest measured differential in the system.

FANS AND BLOWERS

Fans and blowers are used to move gases (compressible fluids) around a boiler plant. In many cases, the terms "rotating equipment" or "fluid handling equipment" will be used to include pumps, fans, blowers, and compressors without regard to the fluid or the form of the equipment, as they all do basically the same thing. They move a certain volume of a fluid and add energy to it to permit it to flow through the rest of the system. For every design of pump, there is a comparable design of fan, blower, or compressor. Differences in the equipment are related primarily to the different densities, temperatures, and viscosities of the fluids the equipment handles and the effect the equipment has on the fluid. Fans and blowers are used to move compressible fluids, basically gases, not compress them. That is what makes fans and blowers differ from compressors.

Even though they are not designed to compress a gas, fans and blowers do manage to compress the fluid slightly. In most cases, the compressive effects are ignored. The density of the fluid does not change significantly. As the differential pressure of a fan or blower increases, compression becomes more significant. There is a very gray line between blowers and compressors, with no clear definition of when, specifically, one becomes the other. A fan, on the other hand, is almost never capable of compression. Still, gas handling equipment can be made slightly smaller when the fan is on the inlet side rather than the outlet, even though the pressure increase might be small.

The difference is principally intent. If it is intended to compress the gas, it is a compressor. If not, it is a fan or blower. As for whether a particular piece of centrifugal equipment is a fan or a blower, that is also a gray area. A centrifugal pump can pump a gas. It does not produce much differential, but it can do it. Look at any centrifugal pump, fan, or blower. Their construction is pretty much the same. The dynamics that allows them to move fluid is the same. These centrifugal devices will all perform according to their performance curve, regardless of the fluid that passes through them. The differential pressure they produce is directly related to the tip speed of the impeller and the density of the fluid. The impeller vanes throw the fluid and the pressure produced is related to the weight of the fluid flowing at a velocity related to the tip speed. Take a centrifugal pump curve and, realizing the differential head of the pump is feet of fluid, convert it to determine the inches of water differential pressure it would produce while pumping air. A fan curve could be used to calculate the differential it would produce if pumping water. The problem is that the denser water would produce so much load on the fan that it would break or the motor would overload before it actually pumped any water.

A lot of the rules for pumps apply just as well to fans and blowers. There are differences. Air leakage out of a fan is generally not a serious problem. There is seldom any kind of shaft seal. Since the density of the fluid is so low, fans and blowers can get a lot larger than pumps in order to handle enough volume to deliver the pounds of air, or other gases, that have to be moved. The typical application of a fan or blower also does not involve raising the pressure of the fluid to move it into a reservoir at a higher pressure. The differential pressure in a system at zero flow is typically zero for a fan or blower. The pressures at the far ends of the system are about the same. The system curve always starts at zero differential at zero flow. When it does not, the device is a compressor.

Propeller Fans

Propeller fans are not that efficient. Propeller fans have a niche in the world because a propeller can move air effectively as long as it does not have to produce any significant differential pressure. The blades of a propeller fan simply push the air along and add some spin to it (Figure 10-84). Housings around the propeller can redirect the flow to eliminate some spin and make them more efficient (Figure 10-85).

Propeller fans are primarily limited to ventilation services in a boiler plant, although they were used in the middle of the last century for FD and ID services when differentials were low. Some key things to know about propeller fans include the fact that they readily overload their motors if the system does not produce the design resistance. The differential pressure in the horsepower formula is total pressure, a combination of static and velocity pressure differences. A fan sitting on the floor and running has no static difference. The velocity pressure was there and a lot higher because, without the static resistance, the fan could force more air through to produce a higher velocity and, therefore, a higher velocity pressure. The increased flow and velocity pressure added up to produce a high horsepower. That will overload the motor. This is a lesson for testing any electrical device. If the fan was simply wired to a welding connection in the plant, with no starter and no overload device, the motor will be overloaded and burn up. Always check the instruction manual to determine the requirements for the device.

Fans, like pumps, have a theoretical horsepower. It is the total pressure across the fan that has to be used. The formula is cubic feet per minute (cfm) times the total pressure divided by 6356. If all that can be measured is the differential pressure, calculate the velocity pressure. Divide the cfm by the area of the fan discharge to get the velocity. Then look up the velocity pressure. Add the velocity and static pressure differentials to get the total pressure. The velocity is the capacity in cfm divided by the area of the outlet. Divide the velocity by 4005 and multiply the result by itself to get the velocity pressure. Add it to the static pressure to get total pressure.

Many fans and blowers are belt driven. The use of belts will allow an engineer to pick a fan for optimum speed for a given application because any speed can be established by the proper mix of motor speed and size of sheave (those pulleys the belts run on). In some cases,

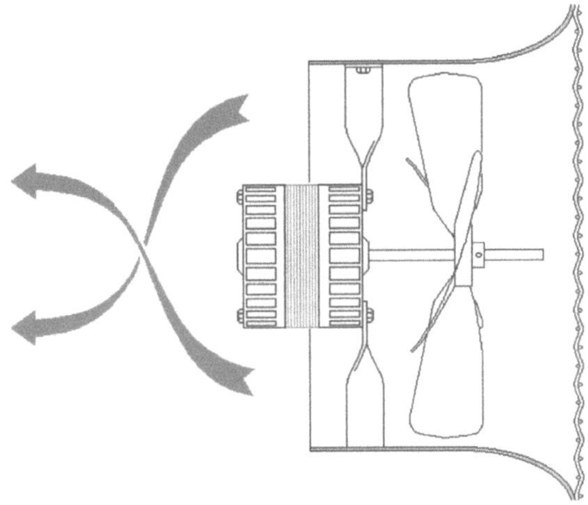

Figure 10-84. Propeller fan.

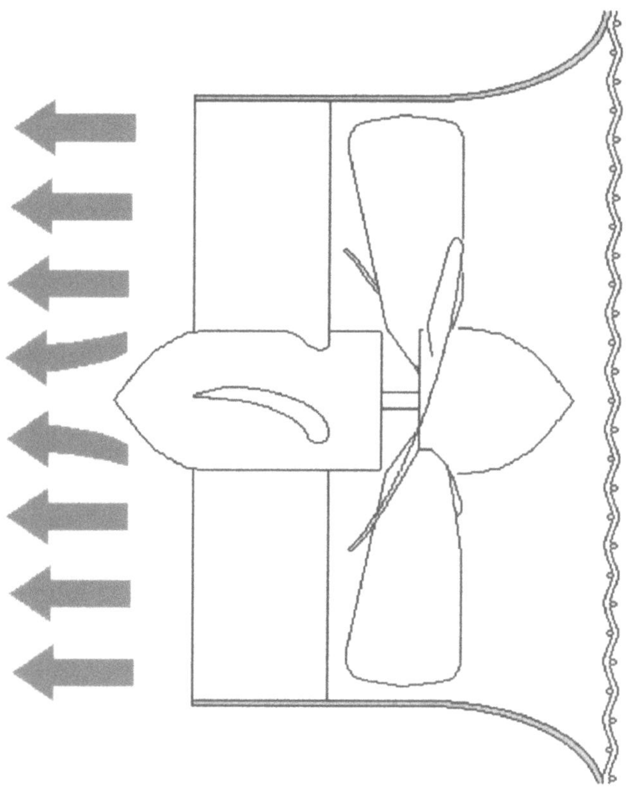

Figure 10-85. Propeller fan housing with flow redirected.

one of the sheaves is adjustable to permit field adjustment of the speed. All these features are very valuable for heating, ventilation, and air conditioning (HVAC) equipment, where the flow is constant and the fan can be tuned to achieve the precise required flow without chewing up added energy with dampers. They are not as valuable in a boiler application, where the air flow is varied. Another advantage of belts is that they can slip on startup to reduce the startup load on the motor, something to let go until the instruction manual is checked. Belts are typically provided to the degree that one belt can break and the rest can still carry normal loads. The problem is that the one belt that breaks usually gets tangled with the others with complete failure. With the growth of VSDs, where the fan can run at any speed, belts are not needed. Belts are a maintenance item and produce unnecessary radial loads on fan shafts and bearings.

Centrifugal Fans and Blowers

While there is not a hard and fast rule, when the width is as wide, or wider, than the center to scroll distance at the discharge, it is usually a fan. When the width is narrow, it is typically a blower. Some also use discharge pressure as a discriminator. When the discharge pressure is in modest inches of water, it is a fan. When it is in tens of inches of water, it is a blower. The two shapes in Figure 10-86 are fan on the left and blower on the right.

In more general terms, blowers produce significantly higher differential pressures than fans. Neither of those rules works every time. One other label that is common is the term "exhauster." When most of the pressure drop in the system is incurred before the fan inlet, they tend to be given that label. Primary air fans on pulverizers are commonly called exhausters.

Centrifugal fans are used in so many applications that standards have been developed to describe their construction. The different arrangements which relate to bearings and motor connections are defined in Figure 10-87. The motors for arrangement 1 and 3 fans are not left hanging in the air. The graphic only indicates that the fan manufacturer is not expected to provide anything to support the motor.

Discharge locations are shown in Figure 10-88. These are based on viewing the fan or blower as if one were sitting on the motor or the fan's sheave. Note that the rotation can be determined by simply looking at a fan's discharge position. Fans can operate with the wrong rotation. Centrifugal devices will work with either rotation. The only difference is that one way works better.

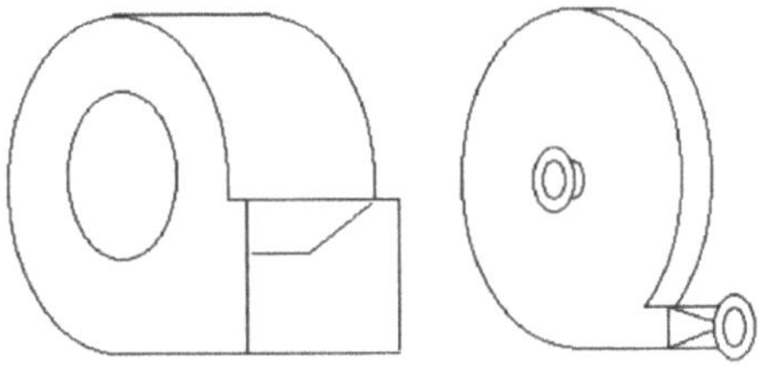

Figure 10-86. Centrifugal fan shape as opposed to blower.

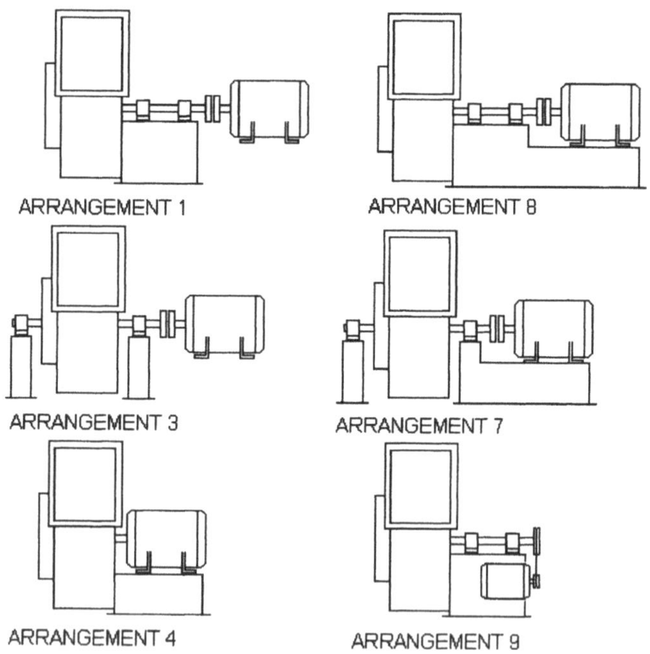

Figure 10-87. Fan arrangements.

There are single inlet fans, where air enters one side. There are also double inlet fans, where air enters both sides. They are defined by simple abbreviations, with SWSI (single width, single inlet) being the most common and DWDI (double width, double inlet), where the air can enter both sides, used in many applications from little convectors (those fan powered heating and cooling units mounted under windows in many buildings) to large FD fans. Instead of calling the primary rotating element an impeller, it is called a wheel. The term scroll is applied to the casing because the radius increases from the cutoff to the discharge. A casing is still a casing. Many other labels are consistent with what is used for pumps. The cutoff is the portion of the scroll that is closest to the outside diameter of the wheel. It is where the swirling fluid in the fan is cut off so that it heads out the discharge instead of riding around with the fan wheel. The inlet bell is that specially formed section that connects the fan inlet to the inside diameter of the wheel. Small fans will not have an inlet bell. They only have a hole in the casing that faces the wheel.

There are some additional gadgets that are not found on pumps. Fans usually do not have seals or packing glands, although they are used on occasion. There are "heat slingers" that are like little fan wheels located on the shaft outside the fan to draw cooling air over the bearings and protect them from hot gases and the heat that conducts along the fan shaft. Instead of strainers, a fan will be protected by inlet screens,

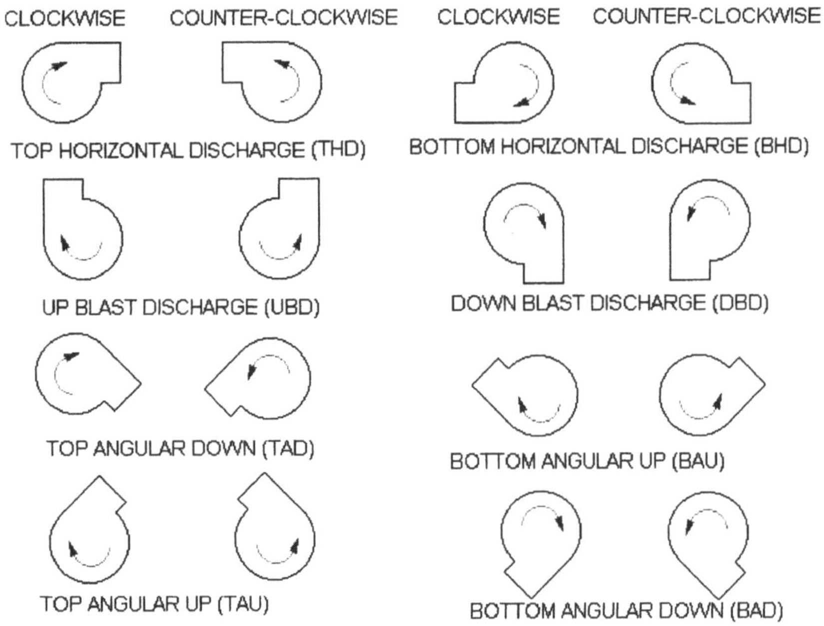

Figure 10-88. Fan discharge designations.

which keep sticks and stones out but not dust. Dust is, therefore, something an operator has to keep in mind. Keep it in mind for two reasons. First, it can damage the fan or hinder its performance. Dust can be converted from large harmless sizes into much finer particles that are injurious to human health after they pass through a boiler. A certain amount of dust will be struck by the blades on the fan wheel and trapped there, accumulating until they form a rather thick layer if they are not cleaned. The accumulation will tend to reduce the fan capacity. The bigger problem is that once it reaches a certain level, it will suddenly start breaking off. Losing a fair sized accumulation of dust on one blade will generate an imbalance in the fan wheel that adds load to the fan bearings, a variable shock load. If that is allowed to happen, there can be everything from shaft distortion to where the fan wheel hits the inlet bell, cutoff, or casing. Clean every fan, or have it cleaned, during the annual inspection. Some FD fans, or what is below them in the ductwork, cannot tolerate a water wash. Cleaning will be limited to brushing and vacuuming. Be sure to do the inside of the scroll too because the dust is thrown at it.

Centrifugal fans and blowers are used more than any other device for moving air. In order to accommodate a variety of applications, they are also supplied in a significant variety of configurations. Three principle variations involve the shape of the vanes, or blades, in the fan. A fan is called backward curved, forward curved (FC), or radial, depending on the shape of the blades as shown in Figure 10-89. These three shapes produce significantly different fan curves as shown in Figure 10-90.

Most applications in a boiler plant use BC or radial bladed fans. They are more efficient for the operating condition. BC and radial blades do not accumulate as much solids on the blades in operation. Radial bladed fans are used almost exclusively for application as ID fans and primary air fans for coal pulverizers. Most air conditioning and ventilation systems use FC fans. They are more efficient at delivering large volumes of air at low differential pressures. It is important to note that FC fans have a very stretched curve. It is not at all uncommon for the motors on those fans to be overloaded if nothing restricts the air flow. Blowers will have radial blades or BC blades, depending on the application, and can experience the same problems with surging that was discussed with centrifugal pumps. That surging will also occur for the same reasons. It is seldom encountered in

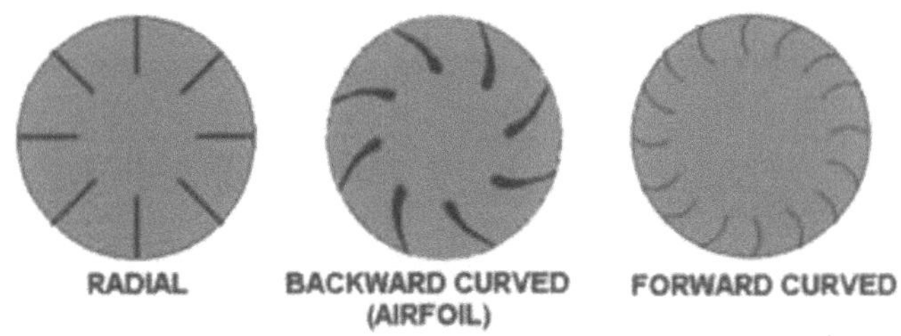

Figure 10-89. Different shapes of fan blades.

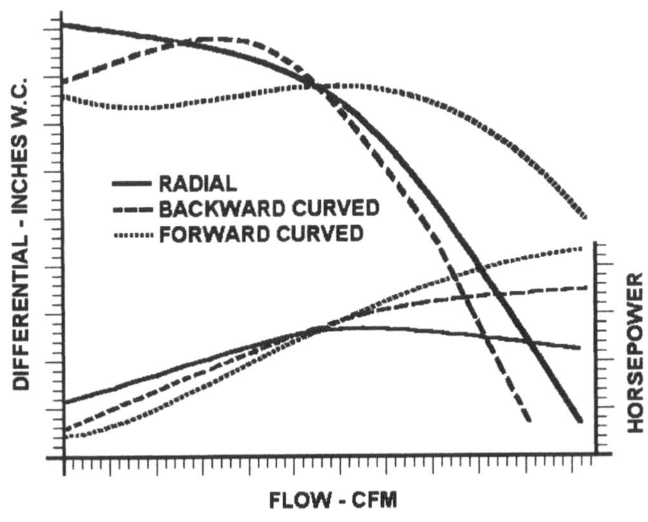

Figure 10-90. Fan curves: BC, radial, and FC.

fan and blower applications but is frequently encountered when compression is involved.

An important thing to remember about these fans is that they are centrifugal devices. The differential pressure that is produced by them is a function of the flow through the fan and the density of the gas flowing through. When the gas is colder, a fan will produce a higher differential pressure (in terms of inches of water) and, because it is moving denser air, more pounds of gas. When a fan is handling gases at higher temperatures, it will not produce as high a differential and move fewer pounds of gas. In many cases, the motor on an ID fan is not big enough to handle cold air. The power requirement is significantly higher when pumping cold air. That is why care is needed when starting a boiler with an ID fan to ensure that it is not overloaded. Once it is pumping hot flue gas, the load on it drops off. That is also justification for not getting excited when a boiler cannot produce full load in the summer time. If an FD fan is installed to collect the heated air in the top of the boiler plant, it will not pump as much air in the summer, when the temperatures are about 125–130°F. It will in the winter when those temperatures are 50–60 degrees lower. It is important to realize that the fan is moving less air (pounds of it) in the summer and excess air will be reduced. Take care not to be running fuel rich. If the boiler is summer tuned, then that excess air is higher in the winter because the air is denser than when the boiler was tuned.

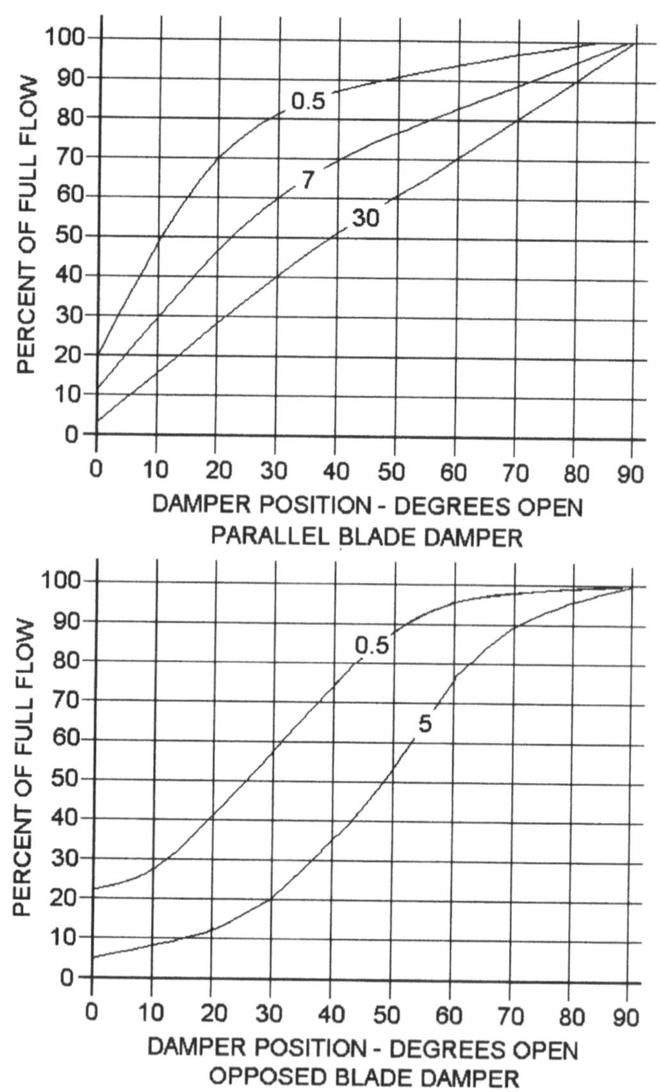

Figure 10-91. Resistance curves and diagrams of parallel and opposed blade dampers.

Rotary Blowers

Rotary blowers do not resemble fans. The same construction is used for compressors. The main reason to deploy rotary blowers is to produce high differentials that are necessary for material transport systems. Probably, the only time to use a rotary blower in a boiler plant is when it is used to provide air for ash or coal transport systems, which require some rather high differential pressures. The primary air for CFB boilers is one example, where there is a high pressure drop though the grid plate and the bed material itself. See the following discussion on rotary compressors for more information that would apply to blowers.

Fan and Blower Control

Control of the flow of gases in systems with fans and blowers is typically achieved using devices called dampers. These are a leaky version of valves. Sometimes, the system uses valves, or their equivalent, when leakage is not acceptable. Dampers are not the best method for controlling air flow. They are typically made to be inexpensive and there is not a linear relationship (See Controls, Chapter 11) between the damper position and the air flow. Opposed blade dampers (Figure 10-91) provide a better relationship than parallel bladed dampers. The different curves relate to the damper's wide open pressure drop divided by the maximum system differential pressure.

In the most common fan application that requires air flow control, FD fans, VIVs are typically used to reduce fan horsepower requirements. VIVs (Figure 10-92) on the inlet of an FD fan not only act as dampers but also put a swirl on the air as it enters the fan. By turning the vanes in a way that puts a twist on the air entering the fan, the air is rotated in the direction of fan wheel rotation. The inlet vanes reduce fan motor horsepower because they swirl the air so that the fan does not have to. The reduction of fan motor horsepower attributable to VIVs is indicated in the curve in Figure 10-93. Note that

Figure 10-92. Variable inlet vanes.

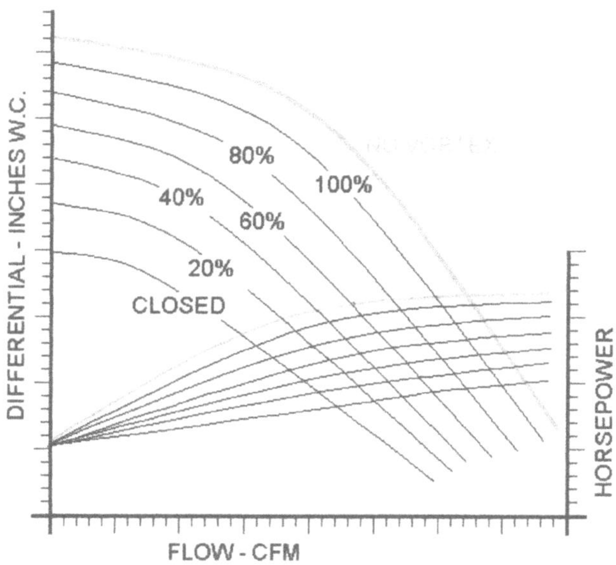

Figure 10-93. Fan curve, effect of variable inlet vanes.

the air has to be turned in the direction of fan rotation. If the vane positions are reversed when replacing that assembly, the horsepower could be much higher and the motor will overload. VIVs are fine for boilers operating with a maximum four to one turndown. They usually leak enough air when closed that they are not adequate for higher turndowns. Some applications use a discharge damper in addition to the VIVs to extend turndown.

Today, there are VSDs, sometimes called variable frequency drives (VFDs), that permit an almost infinite control of fan speed and, therefore, the air or gas flow. At an early installation, when the boiler was at low fire, the combination of 50 horsepower FD fan and 125 horsepower ID fan, along with all the controls and lights, pulled a total of 5 amps. That has to be compared to a full load motor rating of 218 amps. Any new installation should consider a VSD on the fan and a positive shutoff damper that is closed when the boiler is shut down to limit off cycle losses and rapid cooling of refractory by cold air.

Ejectors and Injectors

Using a water hose to sweep down a floor illustrates the principle of ejectors and injectors. The force of the fast moving water is capable of pushing a lot of additional water along. What happens is the high velocity is converted to pressure that pushes the rest of the water. When the motive fluid (the one going through at high velocity) is steam or air, it has less mass to contribute to the pressure. However, it is traveling at a much higher velocity. It can do almost as much work. These devices are also called jet pumps. Ejectors are used to produce lower pressures at their inlet (suction) by pushing a fluid along. The common use of an ejector is to produce a vacuum by pumping air, and sometimes water, out of a closed system. The typical condenser has a steam jet ejector to vent non-condensable gases from the system. Another common use is to remove condensate and rain water from underground vaults containing steam piping. An ejector with a float actuated steam shutoff valve is the least expensive means of automatically clearing water from underground piping vaults and they are quite reliable.

When ejectors are combined, or staged, as for a condenser ejector (Figure 10-94), they can produce an almost pure vacuum. The steam to the jets (C) entrains the air drawn from the condenser at (A), accelerating it through the venturi (B) to the first stage condenser (D), where the steam is condensed by the condensate pumped up from the condenser (J). Another jet draws the air from the first stage, accelerating it to a higher pressure through the venturi at (E), and then into the second stage condenser (F). After the steam from the second ejector is condensed, the air is vented into the boiler room at (G). The condensed steam drains from the second stage condenser through a liquid trap (H) into the first stage condenser. The liquid trap separates the two different pressures, the second stage being around atmospheric and the first stage being something in the range of 8–20 inches of mercury vacuum. The combined condensate in the first stage drains to the condenser through another liquid trap. A steam powered ejector can also lift water out of a vault,

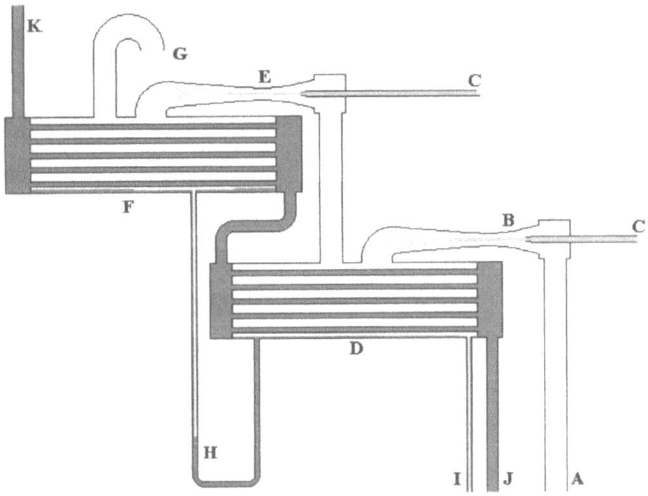

Figure 10-94. Dual jet ejectors for a condenser.

Figure 10-95. Feed water injector.

even when it is hot and flashing, because it will pump the flash steam.

Injectors are the same device but used to produce higher pressures at the discharge. They are found on a coal fired boiler (Figure 10-95) to provide an emergency means of feeding water to the boiler in the event that power is lost to the boiler feed pumps. The boiler's steam can be used to generate a higher feed water pressure to feed the same boiler. The heat energy of the steam is converted by the injector to mechanical energy to pump the water.

Ejectors and injectors have limited use because they use a considerable amount of energy compared to pumps, blowers, and compressors. They are only suitable for moving small volumes of fluid. Their use is limited to operations where there is little flow (condenser vacuum), or small flows, or no electricity available. An ejector or injector that does not boost the pressure or create a vacuum can be called an educator because it simply teaches the fluid where to go. They basically move water and are principally used to mix two fluids.

Compressors

Compressors are, of course, used to compress compressible fluids, mostly air and gases. It is possible to compress a liquid a little. However, most compressors will simply break if they try to compress a liquid. Compression is simply packing more pounds of a fluid into a certain volume. A simple example is pushing fluid into a container. Since none of the fluid leaves the container, and more is put in, each pound of fluid that is added has to share the space with what is already there. There are simply more pounds per cubic foot every second that the compressor is compressing the fluid into that space. A fluid can mean a liquid or a gas. Both can flow. The distinction for gases and liquids is that liquids are not compressible (or only very slightly compressible).

For most compressor operations, there is some fluid leaving the container as more is pressed in. The two flows do not have to match. A control air compressor may run for 5 minutes to fill a compressed air storage tank with enough air to supply the system after that tank for half an hour or longer. That should help explain why most compressor operations are on–off. The fluid stored under high pressure will expand to produce flow for the system. The fluid flows out of the container as it is used and very fewer pounds remain in the container. The container, or storage tank, serves as a reservoir for the fluid required by the system. The compressor refills the reservoir when the fluid level drops to a preset value.

Specific compressor operations require special consideration. The fluid being compressed may contain other fluids or contaminants that interfere with or require consideration in the process. When compressing air, there is also the moisture that is in the air, the humidity. By packing molecules of air into smaller and tighter spaces, the water vapor in that air is subjected to higher pressures. It will condense to form liquid water. Since compressors do not run well on liquids, that water has to be removed. Also, the combination of air and water is very corrosive. Water must be drained where it forms, and collected in the compressed air system, in between compressor stages, and in the storage tank. It also has to be drained at low points in the piping system, especially where the piping goes through a colder area (as in outdoors during the winter), where the water would be condensed by heat loss.

Coolers on compressors are not there to condense the water. As long as the water remains a vapor, it acts just like the air and does little harm to the compressed air system. The coolers are required because the compression is not efficient. Some of the energy that is used by the compressor does the work to compress the fluid. The inefficiency of the compressor is associated with simply heating the fluid. Since there is little mass in the fluid, the temperature of the fluid increases dramatically.

There is a relationship between the absolute temperature, the volume, and the absolute pressure which is called the Ideal Gas Law. It is called ideal due to the fact that there are additional forces that cause a fluid not to behave in an ideal manner. At relatively low pressures, most gases can be assumed to be ideal. The relationship is as follows:

$$P \times V = n \times R \times T$$

In this formula, n is the number of molecules and R is a constant (the ideal gas constant). When working on a fixed amount of gas, like compressing a gas from one pressure to a higher pressure, there is a simple formula for compression that says $P_1 \times V_1 \div T_1 = P_2 \times V_2 \div T_2$, which is derived from the Ideal Gas Law. It means that the pressure (P), volume (V), and temperature (T) are all related before and after compression. Pressure times volume divided by temperature, at one condition for a gas, will be equal to the pressure times temperature divided by the volume at another condition. If the pressure doubled, and the temperature remains the same, then the volume has to be half as much. It is important to note that the pressure and temperature have to be in absolute values. Add 15 to gauge pressure to get absolute pressure. Add 460 to the temperature reading to get absolute temperature. For volume, using cubic feet or cubic inches does not matter. Comparable metric units work just as well. It is the relationship, not the units that is determined. All that counts is using the same units on both sides of the equation. To eliminate any consideration of algebra, here are the solutions for each factor in the equation. To learn the second condition of any one of them, perform the math on the right of the equals sign:

$$P_2 = P_1 \times V_1 \times T_2 \div T_1 \div V_2$$
$$V_2 = P_1 \times V_1 \times T_2 \div T_1 \div P_2$$
$$T_2 = P_2 \times V_2 \times T_1 \div P_1 \div V_1$$

These expressions not only apply to compression but also to any change in the pressure, volume, or temperature of gases. It is most accurate with common diatomic gases, O_2, N_2, etc. It is the inefficiency of the compressor that produces the heat. The piping or the compressor head will be hot. It is a good thing to measure to monitor the health of the compressor system. The temperature will vary with load. Relate the temperature reading with one at a similar load at an earlier time to identify any pending problems.

Unless otherwise indicated, the capacity of a compressor is always described in standard cubic feet per minute (scfm). The standard is equal to air at 60°F, 0% relative humidity, and one atmospheric pressure. The actual cfm that may be flowing in a system is often referred to as acfm. This can cause some confusion. The scfm can also be called atmospheric cfm, also abbreviated acfm. Be careful as to what is meant for any particular application. Take a burner application using an air atomizer. The burner manufacturer's table indicates that 30 cfm of air at 80 psig is required. Call that 30 acfm. It is actually 190 scfm for a compressor ($30 \times 95 \div 15 = 190$), assuming that the air is at the same temperature. If the compressor manufacturer listed the capacity as atmospheric cfm, or acfm, the wrong sized compressor might be selected. Again, be sure to read and understand the manufacturer's information.

Normally, in a boiler plant, air is being compressed. It has its problems, but it is not as critical a process as compressing oxygen. In that application, the hydrocarbons from a fingerprint on one part can catch on fire in the compressor and do damage. Be aware of the hazards associated with any fluid that is being used. The best way, of course, is to read the instruction manual. Just because the fluid is flammable or hazardous does not mean that it cannot be handled with proper training and sensible operation. Gas compressors are used to boost the pressure of natural gas high enough to fire in boilers (or gas turbines). The key to their use is that the gas is all gas. It is so fuel rich that it cannot burn inside the unit. The concern will be with any leak that might form a flammable mixture and accumulate somewhere.

A unique feature of compressors, which is not associated with other fluid handling equipment, is the function of unloading. Unloading a compressor consists of bypasses, valves held open, or other methods built into the machinery, that prevents compression from occurring but does not require stopping the compressor. It is not efficient operation. The compressor is not doing anything but moving its parts around. The wear and tear of full blown starts and stops is eliminated, which makes life easier on the compressor and driver. Some equipment even has staged unloading, where part of the compressor is actually working, while the other part or parts are unloaded. The original purpose of unloading had nothing to do with continuing compressor operation. It still serves that purpose today. It allows the driver to bring the compressor up to speed before it starts compressing fluid. Even the smallest compressors have that feature.

Almost every boiler plant has a reciprocating compressor to produce compressed air for controls and actuators as well as pneumatic tools. That will probably be the case for a few more years, until microprocessor-based controls and electrically powered actuators are fully developed to eliminate both the compressor and all the compressed air distribution piping. Look at any other system in the plant. None will be more inefficient than the compressed air. Air is compressed to 80–120 psig and then used at 18–30 psig. Small compressors can be used to produce air at about 25 psig. That air can be distributed to all the controls. Then another one, with the higher pressure, can be used to serve actuators that need it as well as providing atomizing medium for emergencies. Replacing pneumatic controls with microprocessor-based controls in some plants has eliminated a lot of the waste. A wise operator can realize the opportunities for cost savings by locating and repairing leaks in air systems and eliminating wasteful use of compressed air. Waste can account for about 60%–80% of the consumption of compressed air.

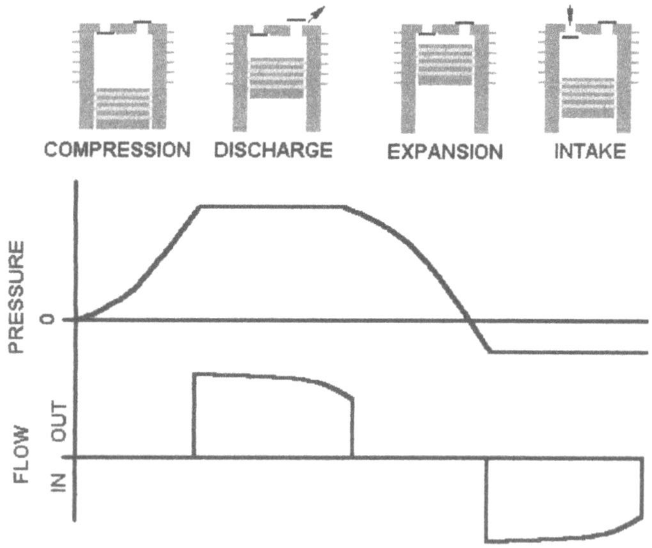

Figure 10-96. Reciprocating compressor operating stages.

Reciprocating Compressors

Just like reciprocating pumps, reciprocating compressors use a piston that changes the volume of a chamber to move the fluid. Intake valves are required to open as the piston moves down the chamber, increasing its volume, so that the air can enter the chamber. They close as soon as the flow stops. Unlike a reciprocating pump, the fluid does not start to leave the chamber as the piston moves up to reduce the volume. The fluid is compressed in the chamber instead. Not until the pressure is higher in the chamber than in the discharge piping connecting the compressor to its storage tank will the fluid begin to leave the chamber. When the piston reaches the end of its stroke, there is no difference in pressure. The discharge valves close. As the piston moves down the chamber to increase its volume, the fluid expands until the pressure in the chamber is lower than the pressure at the inlet. Then the fluid will flow into the chamber until the piston reaches the end of its stroke. The progression is depicted in Figure 10-96.

The typical air compressor valve looks something like a metal popsicle stick or tongue depressor. It is far more complicated than the typical liquid (incompressible fluid) pump, which fills and discharges. The capacity of the compressor cannot be calculated from the volume of the stroke because a good portion of the stroke is devoted to recompressing the fluid that expanded after the discharge valve closed. The less the fluid in the compressor at the end of its discharge stroke, the less that will be there to expand and get in the way of more fluid coming in. That is why compressors are built differently.

The piston and chamber are designed for minimum clearances at the end of the stroke. There is very little room devoted to passages between the chamber and the discharge valves. It is all those close clearances that create the problem when a little liquid gets into a compressor. It will pass out through the discharge valves. It will also make a lot of noise doing it. The hammering usually results in compressor damage. That is why it is so important to remove any liquid that forms between compressor stages. Staging in compressors is similar to staging in pumps. Let one part of the compressor do part of the job and another finish it (two stage), although more stages are common. Two-, three-, and four-stage compressors are all common, some with multiple intercoolers.

Compressors are fitted with intercoolers, which are heat exchangers used to cool the compressed air between the stages so that the next stage does not get too hot. An intercooler may be buried under the belt guard of the control air compressor. The sheave for the compressor has spokes, formed like fan blades, to force room air over the intercooler to remove the heat. If the screen and all the fins on that surface are kept clean, the compressor will run more efficiently. Usually, the compression is such that a control air compressor will not condense any water out of the air in the intercooler. There are no drains on it. Larger compressors will be cooled to the degree that water has to be separated, collected, and drained from the outlet of each intercooler. Most systems today are equipped with timed drain traps. These are solenoid valves connected to a timer that opens them at preset intervals to drain the liquid. Typically, they drain some liquid and a lot of air, another waste. The problem is that the demand on the compressor is not known. The timers are set for the worst (full load) condition. Occasionally, the compressor will be shut down or run unloaded between drain valve cycles. The only thing it is going to drain is air. They are reliable but wasteful. A wise operator would check the drains to see if they need to be opened.

Reciprocating compressors are designed to start unloaded. The typical scheme is use of lube oil pressure, where a small oil pump eventually builds up pressure as the compressor is started. That pressure is used to overcome the force of springs that hold the compressor's inlet valves open. During normal operation, that same oil pressure can be bled off to the crankcase to allow the springs to hold the inlet valves open for unloading. In compressors with multiple cylinders, it is possible to unload one set of valves while leaving others in operation to adjust the capacity of the compressor. That form of unloading is normally accomplished with a pressure switch that switches valves in the oil circuits. It can also be done with an electric switch and solenoid valves. Staged unloading is common in refrigeration compressors (see Chapter 5).

Be aware of that unloading scheme when adjusting anything in the system. One operator lowered the setting of a pressure control switch but was dumfounded to see that the compressor would run longer. He had simply reset the unloading setting. The compressor always ran with half the cylinders unloaded. Someone else lowered the setting of the on–off pressure control below the unloading value of a compressor with hydraulic unloading and could not understand why the motor burnt up. The compressor was constantly starting and stopping. The partial unloader, or unloaders, must operate within the span of the on–off control switch. If the unloading settings are not in the operating range of the compressor, they will not work.

Oil almost always requires attention in a reciprocating compressor. There are small compressors that use diaphragms, instead of pistons, to compress the air and others with synthetic rings that can operate without oil (oil-free compressors). Most of the ones in a boiler plant use oil. Always wait until the compressor has just shut down before checking the oil. Removing the cap on the oil reservoir while the machine is running will blow oil all over. Oil is required to lubricate the moving parts of a compressor and, except for oil-free units, serves to seal the space between piston and cylinder so that the air can be compressed. (By the way, some oil is needed in some oil-free compressors. It is only the air that has no oil in it.) Since the oil is coating the cylinder walls, is scraped by the piston rings, and is exposed to those parts heated by the inefficiency, some of it is vaporized. Some droplets form to leave the compressor with the air. As compressors age, they tend to load the air with oil more than when they were new. The system should have an oil separator to remove that oil so that it does not contaminate instruments, controls, and tools that use the air. Some systems are only used for tools and the oil helps lubricate them. In that case, the oil should be a non-hazardous type that does not form poisonous aerosols where it leaves the tool. In addition, there could be an oil coalescing filter that absorbs the oil. For the sake of the controls, please watch that coalescing filter. Change it when it is not quite saturated. Also make certain the separator is working to reduce the oil loading on the filter.

Other Types of Compressors

Centrifugal compressors were touted as the latest thing about 50 years ago. They quickly faded away because the tip speeds had to be so very high to develop the necessary pressure. The compressor required large speed, increasing gears to get that high tip speed. The stresses on the metals at those high speeds made them vulnerable to all sorts of problems. A reciprocating compressor, which runs at relatively low speeds, could take a small drop of water coming off the previous stage. A high speed, whirling impeller could not. One application that has seen more use is the compressor on a gas turbine.

Screw compressors function about the same as a screw pump. The important difference is that the screw is machined so that the cavity becomes smaller as it moves along the shaft. An added feature in the compressor world is a slide that bleeds air back to the suction to reduce capacity. Screw compressors are used extensively in the construction industry. They also need lubrication. The oil is what seals the cavities and keeps the metal parts from rubbing each other. Since most construction tools need lubrication, there is no problem with what is carried over with the air. A screw compressor in a plant is usually followed by an oil separator and a coalescing filter to provide the specified clean and dry air for boiler plant controls and actuators.

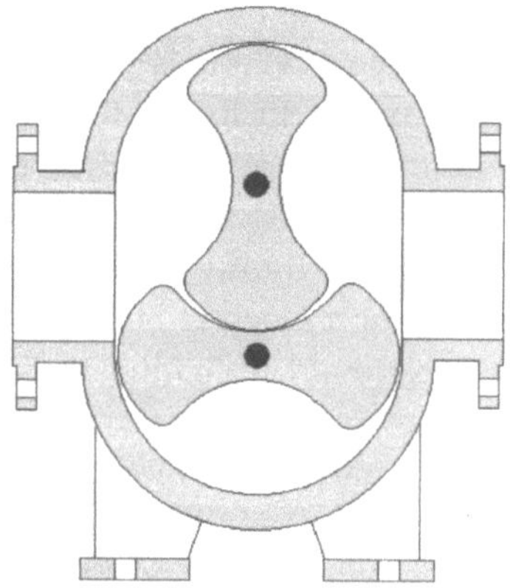

Figure 10-97. Lobe type rotary compressor.

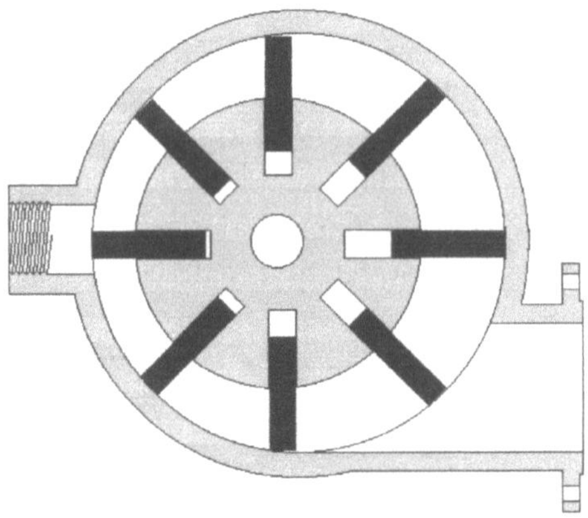

Figure 10-98. Vane type rotary compressor.

Some rotary compressors are very similar to gear pumps (Figure 10-97). They simply move air along with little concern for the fact that air rushes into the cavity as it opens to compress the air before it starts flowing out. Vane type rotary compressors (Figure 10-98) use the eccentrically positioned core to produce a cavity that changes volume to compress the air as the chamber rotates around the shaft.

These compressors must be lubricated and are typically used for low values of compression, producing air pressures in the range of 30–50 psig. A VSD for capacity control does not really work. The oil lubrication would be lost if the compressor was slowed down. Gas boosters are frequently found in a boiler plant. They are either the rotary or centrifugal type and cannot be turned down significantly. The gas has to be recirculated through the booster to reduce output to match the requirements of the boiler's burner when it is modulating. If the boiler shuts down, the booster must also shut down. During certain periods of boiler operation, the booster must run to produce pressure while the burner is not firing (to prove fuel pressure available). Full recirculating mode exists for a period of time. To prevent overheating the gas as it continues to recirculate in the booster, some means is required to cool it. An air cooled heat exchanger is recommended. Water cooled heat exchangers can waste a lot of good water and need so much that it all cannot be used for makeup. If a water cooled heat exchanger is used, and it is using city water, allow the water to get up to at least 140°F before discharging it. That way, as little as possible is wasted. Use whatever possible for makeup. Don't run a booster when it is not needed.

COGENERATION

The ability of the ordinary steam boiler plant to convert fossil fuel energy at an 80%–85% efficiency has allowed many facilities to incorporate power generation into their operations. This is called cogeneration (the generation of both heat and electric power). There is a lot of confusion surrounding cogeneration. Back at the start of the 20th century, cogeneration was very common. As improvements in central station power generation were made, the cost of purchased power declined. By the middle of the century, industrial facilities found it more convenient and more reliable to buy power than to generate it themselves. With the energy crisis of the 1970s, cogeneration was promoted as a means of improving overall efficiency. The Public Utilities Regulatory Policy Act of 1978 (PURPA) was enacted to promote cogeneration. Any plant that used more than 5% of its energy for heat and the rest for electric generation, or vice versa, was considered to be a qualified facility (QF). Note that these were guidelines. The Federal Energy Regulatory Commission (FERC) had jurisdiction over what was a QF and what was not. At that time, utility companies were regulated entities. Their rates were set by state agencies and they had an obligation to serve. If a facility was designated as a QF, a regulated facility was still obligated to provide backup power to the plant. Further, it could be forced to buy the power at its avoided cost (also determined by the state). In the 1990s, there was a move to deregulate the generation of electricity. Power lines were still regulated, but generation was opened up. About 75% of the country is deregulated. The goal was to provide competition to the generation of electricity. The US Department of Energy (DOE) and the US EPA still claim to promote cogeneration.

The economics of cogeneration are complex. The basic claim is that it is more efficient to generate power and thermal energy together than it is to generate each one individually. There is a lot of truth to that statement.

However, the claim that cogeneration is 85% efficient and the power grid is only 30% efficient is like comparing apples with oranges. The thermal energy portion of both plants is the same. It takes the same amount of energy to make steam for both plants. The difference comes on the power generation side. The traditional cogeneration plant used what is called a back pressure steam turbine for power generation. The back pressure steam turbine exhausted steam at the temperature and pressure needed by the industrial plant, rather than expanding the steam all the way down to condenser temperature. This practice provided a marginal heat rate for power generation that was better than a full blown power plant because the exhaust steam was used for thermal purposes, rather than being thrown away in the condenser. That is the source of the efficiency benefit. The marginal heat rate for this application is in the neighborhood of 5000 Btu/Kwhr (kilowatt-hour). This compares favorably to a conventional power plant at 9500 Btu/Kwhr. However, this benefit only applies to the power generation portion of the plant. To further complicate matters, a modern gas turbine combined cycle plant has a full load heat rate of around 6300 Btu/Kwhr. Thus, the improvement in efficiency is more modest than most proponents claim. Indeed, the major savings, from a cost point of view, come primarily from saving the transmission and distribution costs, which are now higher than the generation costs in many areas. There are other considerations such as equipment size, reliability, fuel cost, permitting requirements, the balance between electric demand and thermal demand, etc., that are not immediately obvious. Still, many industrial, commercial, and institutional plants are considering cogeneration. Many universities are using cogeneration because they have experimental work that needs continuous and reliable power supply. The extra considerations of power generation are outweighed by the need for self-generation. Understanding steam turbines and their operation is going to become more important for the wise boiler operator.

Operating a generator to minimize demand charges, when possible, can also be the responsibility of a boiler plant operator. With cogeneration, additional power can be produced, even if it is not particularly efficient to do so, to reduce a peak load and lower those demand charges. The degree to do so is dependent on the length of time a peak load is endured and the inefficiency associated with producing that extra power. If the peak is substantial and only occurs during a short period of time (like half an hour a week), it may pay to dump steam to atmosphere, as mentioned earlier, just to eliminate that peak. The cost to generate the power for that period of time and the overall savings on demand charges have to be evaluated. Thus, operation of an emergency generator can be a small cogeneration activity, as was explained in the section on reducing demand charges.

When capable of tri-generation (heating, cooling, and power generation), identify and develop standard operating procedures (SOPs) for spring and fall operation to balance wasted energy, by heating and cooling at the same time. The goal is to optimize the operating cost by generating more electricity and reducing thermal demand when there is normally insufficient thermal load at the generator exhaust. This requires knowledge of the electrical contract and how to manipulate it as well as good records on power generation and system loads. There are now consulting firms to help develop the program for such operations because it does get complicated. It will probably take a computer to guide the decision-making process because electricity costs will vary hourly. There are already situations where the cost of electricity varies with each hour of the day. Electricity costs as little as 2¢ per Kwhr at night and 26¢ per Kwhr in the early afternoon with hourly variations in between. The plant may be limited in controlling power usage to avoid the higher costs. Cogeneration provides the ability to potentially save some money on power.

There are several options for generating power with exhaust heat to be used for steam, hot water, service water, and absorption chillers. They include steam turbines and engines that have a long history in that service. Turbines require substantially less maintenance and operator attention than engines. Generating steam, or hot water, with exhaust from diesel generators, including those fired on natural gas, also have a long history. Like engines, these generators require a considerable amount of maintenance. Modern engines have improved on that maintenance requirement to the degree that they are being used. Modern devices include gas turbines and fuel cells. Once again, the instruction manual and other documentation is necessary to learn all that is required for operating them.

Steam Engines and Turbines

Take a good look at the photograph in Figure 10-99. It is a steam engine driven air compressor and it is probably one of the few that are still operating today.

Specific problems with steam engines have almost eliminated their use today. Lubrication oil getting into the boilers has just about been eliminated with provision of better materials that can seal the piston and shaft of a steam engine. The need for skilled workers to maintain and rebuild them and the high initial cost and cost of

Figure 10-99. Steam powered air compressor.

maintenance has pretty much priced them out of existence. It is not that they cost more than a motor over their lifetime. It is just that they cost more to begin with and their maintenance is not well understood. Direct conversion of steam to power should more than cover the maintenance costs. That investment cost may not pay back fast enough to justify these very efficient devices.

Steam turbines have also seen a decline in use mostly because electric power has become so reliable and there is such a low demand for steam turbines that they are only found in medium to large boiler plants which are willing to invest in them. There has been no argument for having turbines to ensure continued operation in the event of an electric power interruption because of power reliability and the ability to operate a generator to run motors. Auxiliary turbines (see earlier discussion) that exhaust steam to the deaerator are still a good application.

Steam turbines convert the heat energy in steam to mechanical energy. It is basically a matter of passing the steam through a nozzle from a higher pressure to a lower pressure in a manner that converts the static pressure in the steam to velocity pressure. Once the steam is moving at a high velocity, the mechanical energy is in the steam and the turbine has to transfer the energy in the steam to rotation of the shaft. Turbines use two methods to transfer the energy to the rotating shaft, either impulse or reaction. An impulse turbine works the same as a pinwheel. Blow on a pinwheel and the air strikes the surface of the pinwheel, giving up its velocity pressure to the blades of the pinwheel. A reaction turbine works more like a loose garden hose. The hose reacts to the water spraying out of the nozzle. Figure 10-100(a) shows the two types of turbine stages graphically and includes what is called a velocity compounded stage, where the back splash from an impulse stage is redirected to a second set of turbine blades to increase the performance. All steam turbines have a least one impulse stage because the steam is initially supplied through a set of stationary

Figure 10-100(a). Impulse and reaction stages of turbines.

nozzles in the turbine casing. Moving blocks are shown in black. Nozzles and stationary blades are in dark gray. Steam flow is in light gray.

By dropping the pressure in stages, better efficiency is achieved. Utility style turbines for power generation have 20 or more stages in the high pressure turbine and another 18 or more in the low pressure turbine(s). As the steam pressure drops from stage to stage, it expands. Check the volume of a pound of steam in the steam tables. Note that the volume of steam increases rather dramatically. The manufacturer of the turbine either has to make the latter stages of the turbine much larger or provide for bleeding off some of the steam. It is typical for a large power generating turbine to have at least two bleeds. High pressure bleed steam is regularly used for feed water heating between the deaerator and the economizer, or boiler. Some high pressure bleed steam is at a pressure high enough to be used as the steam supply for auxiliary turbines. Intermediate pressure steam can be used in the deaerator or for other purposes. Low pressure bleed steam can be used for plant services such as building heating and reheating condensate after it leaves the condensers. See the utility cycle (Figure 1-8).

Steam turbines, and engines, extract energy from the steam without condensing it. That is very important. The turbine would be severely damaged by droplets of condensate hammering the turbine blades. The energy that is removed from the steam to generate power is only enough to reduce the superheat. A turbine that runs on saturated steam is only extracting superheat. The steam contains the same amount of energy after it passes through the first nozzles of a turbine as it did at the inlet of the turbine. Since the pressure in the steam is lower, the steam has to be superheated. As long as the turbine does not draw too much energy from the steam, it will still be superheated (a little bit) at the outlet of the turbine.

Of course, to really generate power, the steam is superheated in the boiler. That allows for the use of a more efficient turbine that extracts more energy from the steam. In large power generation equipment, the steam is piped off the turbine and back to the boiler to be reheated before continuing its trip through the turbine. The reason to use reheat is that the superheat necessary to prevent condensation on a full path through the turbine would be so high that the superheater tubes and pipes would melt. There are no metals that could take those high temperatures. By reheating, the temperature can be increased at an intermediate stage in the turbine to about the same temperature as the steam at the turbine inlet, without requiring more exotic metals in the superheater and piping. The heat exchange surface in the boiler that does this is called the reheater and it requires special consideration in the startup and operation of a boiler that is equipped with one. Figure 10-100(b) shows

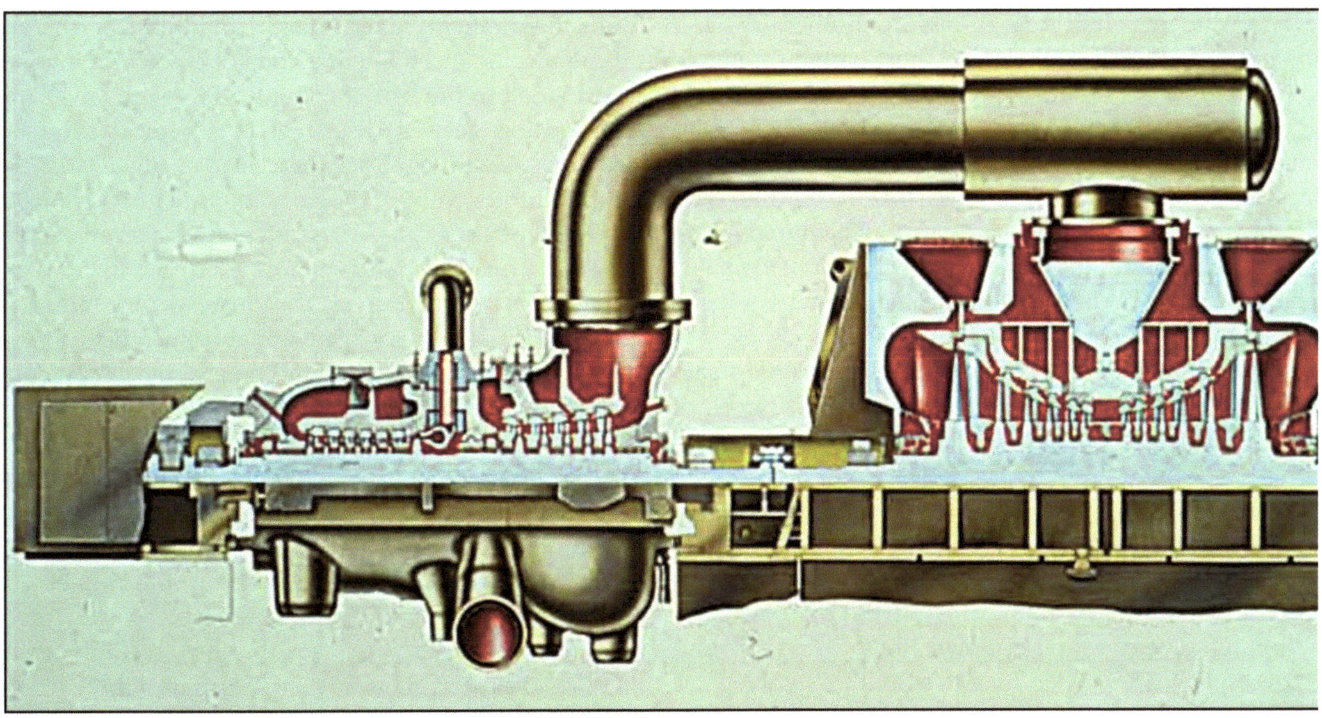

Figure 10-100(b). Cross section of a steam turbine showing high, intermediate, and low pressure sections.

a cross section of a steam turbine with a high pressure, intermediate pressure, and low pressure turbine sections.

Note that the blade length in the low temperature section (on the right) is much higher than in the blades in the high pressure and intermediate pressure sections (a consequence of the Ideal Gas Law). The turbine arrangement that will probably become more common with the development of distributed generation, or cogeneration, is the topping turbine, or high back pressure turbine, that will generate electric power. All the generated steam passes through the turbine for use in the facility. The steam will be produced at high pressure (600–900 psig being the most common) and superheated and then dropped through the power generation turbine to generate power. The pressure at the outlet of the turbine will match the pressures required in the facility served by the current boiler plant.

A couple of things that are important about turbines are that their lubrication is critical and the torque of a turbine is at maximum when it is not rotating and decreases as speed increases. Most power generation turbines have pressure lubrication. The oil is supplied to the bearings under pressure. The oil feed can be from a pump directly, in some cases, one attached to the turbine or from a head tank set well above the turbine, which is constantly refilled from the sump by pumps. In some power plants, there is a viewport in some piping where oil can be seen splashing through. That is the overflow from a head tank. As long as oil is spilling down that overflow, there is lubricating pressure for the turbine. If the oil cannot be seen, there is a short period of time to get that turbine shut down.

As for the torque, don't damage the turbine. Spinning open a steam valve on an idle turbine inlet can result in so much torque at the first stages that the plate holding the turbine blades, or the shaft, can be bent enough to cause the blades to hit the casing, with serious damage. The rapid acceleration of the turbine from zero speed can result in serious overspeed conditions. Just crack the steam valve to any turbine. It takes very little steam to get it moving. The marine turbines have a bunch of heavy gears, a 50 ton propeller, and long shaft holding it back. Gave it a bit of a blast to get it started, opening the valve a quarter turn or so and then quickly throttle back as it started moving. There is still considerable force on the turbine blades when a turbine is operating under load. If a power generator trips off the line, instantly stopping any generation of power, the turbine is bound to overspeed. There is nothing to stop it from taking off. That is why they all have overspeed trips. Some even have hydraulic brakes to limit the speed. Don't skip that very important function of testing the overspeed trip when starting up a turbine. If it does not work, it is possible to send turbine blades flying out of the casing and all over the place.

Power generating steam turbines are usually large pieces of equipment with very thick parts. The shaft of a turbine can be several inches thick. It is important to prevent quick temperature changes during the operation of those turbines. During startup, the turbine should be brought up to speed and loaded gradually. Always look for the instruction manual to see what the manufacturer recommends for startup. Then confirm that the facility's SOPs comply with those guidelines (they normally allow for slower heating). Stick with those rules.

The large shafts and heavy metal of steam turbine casings also retain heat. If the turbine is stopped, it should be given a bump to force a few rotations at regular intervals to prevent the shaft from buckling. The temperatures are higher in the top of the casing than in the bottom. On ships when docking, give the main shaft turbine a little steam to rotate it once or twice at regular intervals. It should be noted that the propeller only moves a couple of degrees during that operation.

When a power steam turbine (or a gas turbine) is shut down, it is normally connected to a "jacking gear" or "turning gear." A small electric motor, with (typically) a screw driver, is engaged to rotate the turbine at a low speed (normally less than 1 rpm) to provide for a balanced cool down. The turning gear is usually operated until the moment before steam is admitted to the inlet of the turbine throttle valve and some form of interlock is provided to prevent opening those steam valves until the turning gear is disengaged. A typical shipboard interlock was a sign "Jacking Gear engaged" hanging on the throttle valves. Modern equipment has more serious interlocks to prevent accidental admission of steam when the turbine shaft is connected to the turning gear (remember where torque is the greatest). On shut down, the turning gear is engaged and remains engaged until the temperature levels in the top and bottom are equal. Even then, for longer shutdown periods, the turning gear is used to rotate the shaft about 1/3 the way round at regular intervals to help prevent distortion of the shaft. In preparation for startup, the turbine is put back on the turning gear at least 4 hrs prior to startup. If there is any trip during the startup process, the turbine goes back on the turning gear for an additional 4 hrs.

All electric utility steam turbines, including those in nuclear plants, are condensing turbines. It means that at least some of the steam passes through the turbine to a condenser. The water from the condenser is then pumped back to the deaerator, usually through a number of heat exchangers. To condense the steam, all the heat of vaporization (the latent heat) has to be removed. That heat is transferred in the condenser to river water or cooling tower water. On rare occasions, it is dumped to air through air cooled condensers. Air cooled condensers are less efficient as they work at somewhat higher temperatures for condensation. This raises the vapor pressure of the condensing water, which, in turn, raises the back pressure on the steam turbine, making it less efficient.

For maximum power generation, the condenser must operate under a vacuum. Non-condensable gases, and any air that might leak in, must be removed from the condenser. That is typically done with a steam jet ejector. Motor driven vacuum pumps can also be used. The steam jet ejector (Figure 10-94) is usually two or more stages to pull as much vacuum as possible. The steam used to eject the air is then condensed in a heat exchanger using condensate. The actual vacuum achievable is dependent on the temperature of the cooling water, or air, which sets the condensation temperature. Around 27–28 inches of mercury vacuum is a typical value to shoot for. Steam will condense at 88–92°F to produce that vacuum. The cooling water will need to be around 80°F to achieve that. The ejector pulls the non-condensable gases out of the condenser in order to prevent them from raising the back pressure on the steam turbine.

Any condensing turbine requires special provisions to seal the shaft of the vacuum pressure stages to prevent drawing air into the turbine. That is usually accomplished with a special regulator that supplies steam from a high pressure bleed or a reducing station to keep pressure on the shaft seals. The regulator also dumps excess steam leaking from high pressure seals into the condenser during operation.

Maintaining a vacuum by providing adequate cooling water, or air, and keeping non-condensable gases and air out of the condenser is imperative for best power generation. A boiler plant that is converted to generate power in addition to heat, as opposed to one that generates power as well as heat, may have a condensing turbine, although most of the steam will be used in the plant. Various schemes, including bleeding steam and separating stages with piping and control valves, are used to maintain the pressure of the steam supply to the facility, while the steam to the turbine is controlled for power generation. One such scheme is a goggle plate inside the turbine, with slots that are opened and closed to control flow to the lower stages right after the facility steam is drawn off.

GAS TURBINES, ENGINES, AND HRSGs

Just because the fuel is not burned in a conventional boiler furnace does not mean a boiler operator cannot handle it. The combustion chemistry does not change. All the formulas stay the same. The new device still burns hydrocarbons to release heat. Gas turbines and gas engines burn the fuel to create a hot gas. Some of the thermal energy in that hot gas is extracted from it to generate power in a turbine, which operates on the same principles as a steam turbine. The gas turbine, by itself, is not a lot, if any, more efficient than a utility steam plant. That means that there is heat left over in the gas that can be used to make steam. There are many engine generator plants with waste heat boilers in this country that have been operating for more than 30 years. Gas turbines are not as new as some people think they are. In many cases, all that is really new is a way of putting equipment together. An HRSG is a prime example of that.

Most electric power generating engines work the same as an automobile engine, using either the Otto or Diesel cycle to convert energy in fuel to output power at the engine shaft, which drives the generator. Otto is the man who came up with the four-cycle engine, the scheme of intake, compression, ignition, and exhaust. A diesel engine can be two-cycle or four-cycle. Most are four-cycle with the only difference being that Otto used a spark to ignite the fuel. Fuel is injected into a diesel engine right before ignition and ignites spontaneously in the hot compressed air. The water jacket surrounding the engine's cylinders do not absorb much heat compared to a boiler. There is not much energy recovered by the water jacket. Some plants use the heat of the jacket for building heat and other purposes. Heat can also be collected from the lubricating oil and the turbocharger, if there is one. Most of the energy remains in the exhaust gases. Cogeneration plants using engines have a waste heat boiler that recovers the energy in the engine exhaust gases. The boiler or boilers are commonly manifolded to two or more engines so that steam generation can be maintained. Frequently, there is an auxiliary burner installed somewhere to provide additional heat or fire the boiler when the engines are shut down. The auxiliary burner in those applications should not be a conventional boiler burner. A proper burner for this application must match the actual conditions of temperature and pressure of the exhaust gas and have a higher turndown ratio.

The principle difference in engines and turbines, as far as combustion is concerned, is that engines are typically fired fuel rich to keep them cool and turbines are fired air rich. There is a catalytic converter on a car to remove or reduce the emissions created under fuel rich conditions (particularly when idle). By running the engine fuel rich, the combustion products are somewhat cooler. The combustion process is more stable. The catalytic converter passes the exhaust over two catalysts in the chamber. The first catalyst uses some of the CO created to react with oxides of nitrogen (NOx) to create nitrogen gas and carbon dioxide. The second catalyst then oxidizes any unburned hydrocarbons and any remaining CO with the air added by the air pump. The catalyst provides an active surface to allow these reactions to take place at the lower temperature of the exhaust gas and air mixture to ensure more complete combustion. It raises the gas temperature in the process to waste the heat out the tailpipe. It is not perfect or complete. A little CO still manages to sneak by a converter. The overall reduction is over 90% of a very small amount to start with. Some engine cogeneration plants are operating with catalytic converters, more for NOx reduction than CO reduction.

Gas turbines, on the other hand, use lots of excess air to absorb the heat of combustion and lower the operating temperatures so that the turbine blades do not melt. The gas turbine combustor has no boiler tubes to absorb heat. As a result, the gas temperatures would approach those seen in the burner throat. The second reason for temperature control in the combustor is to reduce the formation of NOx. At gas temperatures below 2800°F, the formation of NOx is greatly reduced. Since the gas temperature has to be reduced from traditional flame temperatures to acceptable temperatures for the turbine blades, most of that air is supplied to the combustor. Older gas turbines used to operate with as much as 400% excess air to produce gas temperatures of 1200°F. New techniques and construction are changing the form of gas turbines to allow lower excess air. Once scheme now used is to bleed air or steam through holes in the leading edge of the first row of turbine blades to create a film of cooling air or steam flowing back over the blades. Thermal barrier coating are applied to the blades to help protect them from the high gas temperatures. Current gas turbine temperatures are in the range of 2500°F, with excess air levels of 200%. Designs of power generating gas turbines are still evolving.

Gas turbines are not a new thing. The first gas turbine was demonstrated in 1898. It was a 2 Kw machine with 2% efficiency. Interest waned as steam turbines were more efficient. In the 1930s, interest was revived to create the jet engine for powering aircraft. A jet engine is basically a gas turbine mounted on an airplane. The first gas turbine powered ship, the Admiral Callahan, was powered by two airplane jet engines which exhausted to another turbine that drove the ship's propellers. Gas turbine plants that use that concept are now called "aero-derivative." The growing need for improved efficiency, fostered by the deregulation of electricity, has seen improvements in common shaft gas turbines (basically a jet engine with a shaft sticking out to drive the generator).

A gas turbine consists of three basic parts: a compressor, a combustor, and a turbine. The compressor draws in atmospheric air and compresses it before supplying it to the combustor. The combustor mixes the fuel with the compressed air and ignites it. There are usually multiple burners arranged around the shaft of the compressor to burn the fuel. The parts containing the burners are protected from the heat of the burning fuel by baffles cooled by some of the compressed air. The products of combustion and cooling air mix to provide a gas of the right temperature to enter the turbine. The turbine, a reaction type, converts the heat energy to shaft power to drive the compressor (a large portion of the turbine load) and a load, usually the generator. There are some gas turbine driven pieces of industrial equipment. Most of the time, they are used to power electricity generators.

It is the gas turbine, combined with an HRSG that feeds a steam turbine that makes the plants that are called combined cycle power plants. The HRSG could be modestly called a waste heat boiler but is much more than that. It consists of a combination of all elements of a modern boiler plant in a carefully matched and packaged combination designed for maximum efficiency. With combined cycles, utilities have been able to increase their efficiency to over 50% (higher heating value (HHV)). That compares favorably to a conventional steam-based power plant with 36%–40% efficiency (HHV). The basic arrangement of a combined cycle plant is a gas turbine followed by an HRSG which generates steam to power a steam turbine with both turbines generating electric power.

The typical HRSG is a combination of things, with the most common arrangement being a connecting duct for the turbine exhaust with a superheater, reheater, high pressure boiler, economizer, low pressure boiler for deaerator steam (which flows to the integral deaerator), and low pressure economizer. An inline duct burner can be used to generate additional steam, if required. These are primarily used for cogeneration plants to better match the steam demand with the electric demand. Typically, in those plants, the steam produced by the HRSG is used for thermal energy, as opposed to power generation. The burners would rarely be used in a power generating combined cycle plant since the heat generated by those duct burners does not go through the combined cycle. That reduces the overall efficiency of the system. The exception would be when deregulated power prices are very high. Then the extra revenue from the high priced electricity more than covers the extra fuel cost. The HRSG is designed to squeeze as much heat as possible at each section and then follow with a lower temperature boiler or heat exchanger that can absorb some of the heat that is left. As the flue gases cool in their path through the HRSG, they pass several "pinch points," where the flue gas temperature approaches the saturation temperature of the boiler or inlet temperature of the heat exchanger. The duct burners simply introduce fuel. The gas turbine exhaust already has plenty of air for the fuel. Some duct burners have air for the ignitors only and some have none.

Microturbines

Microturbines are very small gas turbine generators with some unique differences. Most generators are limited to a speed of 3600 rpm in order to generate 60 Hertz electricity. In Europe, the speed limit is 3000 rpm to generate 50 Hertz. A gas turbine is more efficient at higher speeds. Microturbines generate direct current and then invert the output with solid state electronics to produce alternating current. That way, there is no link between speed and frequency. The turbine can be operated at the most efficient speed for the power it is generating. Microturbines are an assembly line product, with common sizes being 30 and 60 Kw. The largest is around 500 Kw. They also produce hot exhaust which can be directed to a waste heat boiler. Many are used as emergency generators with no waste heat recovery. Microturbines still burn fuel. With the deregulation of the power industry, it was purported that distributed generation (as opposed to central station power plants) would take over the power industry. However, most entities found that being their own utility was not as easy as that.

Fuel Cells

Fuel cells do generate electricity without burning the fuel (i.e., combustion). That does not mean they run cool. Some of them operate at very high temperatures. The concept is one of hydrogen and oxygen combining to make water by a sort of reverse electrolysis. Electrolysis breaks down water by passing a current through two electrodes immersed in the water. Gas bubbles form at each electrode and rise up to be collected. One electrode produces oxygen. The other produces hydrogen. The process broke the water down into its two basic parts. A fuel cell does the opposite, using the reaction of hydrogen gas and oxygen to produce direct current electricity and water.

Fuel cells became the mainstay of electric power in the space program. They generated a lot of power with very little weight and produced water that could be consumed by the crew or jettisoned without degrading the environment. Their relatively low operating temperatures and lots of careful development produced a highly reliable electricity generator for space vehicles. When used in earthbound applications, the direct current produced has to be inverted to alternating current. They are used principally in plants where highly reliable backup electric power is required. The important thing to note is that they are designed for, and work well with, pure hydrogen. Since there are currently no hydrogen pipelines or storage tanks out there, a conventional hydrogen-based fuel cell is not the sort of thing that someone is ready to invest in. There was considerable hype around the development of fuel cells for automobiles as clean burning. The typical earth-based fuel cell installations currently burn a common hydrocarbon fuel such as natural gas with some modification. Natural gas can be reformed to generate hydrogen. The addition of an internal reformer reduces the overall efficiency. With the push for renewables for power generation, such as solar and wind, there is renewed interest in hydrogen as a storage medium. Excess power at, say, noon time for solar energy could be used to generate the hydrogen by electrolysis. The hydrogen can be stored for use at a time when the sun is not shining. This approach is still conceptual. However, if it does emerge, a fuel cell could be used to convert the hydrogen to electricity.

Currently, a fuel cell has to be dismantled and rebuilt on at least a five-year schedule. It is the sort of operation the manufacturer insists on doing, probably to retain secrecy regarding their methods of construction and other details. The schedule may be based on experience with the degradation of the electrolyte from contaminated fuel, particulates and stray gases in the air, etc. Programs of life extension based on chemical treatment or reconditioning of the electrolyte may be able to extend that operating period in the future. The actual operating temperature of a fuel cell depends on which electrolyte is used and can vary from very low (about 350°F) to high (about 1200°F). Thus, the temperature of the exhaust can vary considerably. The possible uses of the exhaust heat will vary as well. Currently, fuel cell applications are limited to sites where reliable electricity supply is all important. By installing several fuel cells, an owner can be reasonably certain that the power will never be interrupted. The boiler operator may be called on to monitor fuel cell operation. Once again, the important thing to do is read that instruction manual.

Chapter 11

Controls

There was a time when the operator was the only controlling influence on the operation of a boiler plant. Today, controls are an extension of the operator's eyes, ears, hands, and feet that help the operator keep the plant running in a safe and efficient manner. There is no question that less manpower is needed in a plant equipped with controls. However, unlike a human, they cannot always let someone know when they are not working correctly. The modern boiler plant operator uses the latest controls made available to manage the ever increasing cost of operating a boiler plant. With these tools, the operator can easily manage to reduce operating costs by as much as 15%. Controls are simply one of the things that help an operator to do a better job. The wise operator should know how to use them.

CONTROLS

There are two basic types of control: on/off and modulating. On/off control is not as simple as one might think. There are some unique conditions to be aware of. The following few paragraphs address the general elements of modulating controls which a boiler plant of any reasonable size will have. Recalling the section on flow, controls change the rate of flow in order to maintain a desirable condition such as pressure, level, or temperature. Despite the fact that these variables are not really controlled, a control loop can be identified using those parameters. Just keep in mind the fact that only the flow is actually being controlled.

Controls are not selective. The controller does not know if it is controlling to maintain a pressure or a level. That information is not even important to the controller. A parameter can be a level, a pressure, a temperature, a pH, or anything else that is chosen to be maintained with a controller. The controller does the same thing regardless. It adjusts something to maintain a target. The generic name for the target is a parameter.

There are a number of words used by the control designer and technician that should be understood. Controls personnel typically only know about controls and use words specific to their controls that do not differentiate among the hundreds or even thousands of different systems that can use those controls. A controller can be used for so many different applications that assigning names that are independent of the process being controlled was essential. Familiarity with these terms will provide a better understanding of controls.

The electronics of a controller really do not know what they are controlling or what parameters they are maintaining. They respond to control signals and produce control signals. The signals may represent air pressure, fluid pressure, electrical voltage, electrical current, or a bunch of electrical charges in a tiny microchip. Controls can be used in any application, not just boiler plants. The really wise boiler operator will be able to relate to how the controls work with the boiler and its auxiliaries.

Start with a parameter. It is a quantity, value, or constant whose value varies with the circumstances of the system. The controller does not know what the parameter is, and it does not care. It can be pressure, temperature, level, count, pH, oxygen content in percent, differential pressure, a flow of any fluid, a weight, etc. The controller basically deals with parameters that are called inputs. They are used to create an output or outputs. Inputs are assigned names that indicate what they are in relation to the controller. The two most important ones are process variable and set point. The process variable is a value representing the measurement of whatever it is that is trying to be maintained. If it is a pressure controller, it is the pressure. If it is a level controller, it is the level. It is the control system's representation of the actual value of the parameter that is trying to be controlled. The set point is a value representing what the process variable should be. If the boiler pressure should be 100 psig (pounds per square inch gauge), the set point must be adjusted until the parameter represents 100 psig. When properly applied, the controller will indicate that it is set at 100 psig. The actual value is an electronic signal. There is some math going on to convert that signal to a value that corresponds to 100 psig. The same is true for a digital readout on a sensor. The sensor actually measures milliamps. It has to convert that measurement into the required units to produce a readout that is meaningful. Set points are not always set by the operator. A set point can be the output of another controller.

A set point is termed "local" when it can be adjusted and "remote" when it is the output of another controller. Note that the terms do not relate to what is normally considered as local and remote. If one has to leave the boiler plant and go around the main building to the shed under the water tower to adjust the set point of the tower's water level controller, it is still a local set point, even though it may be far from the boiler plant. A remote set point can come from a controller right beside the next one in the control panel. Don't confuse local and remote with physical location.

The term loop is used to describe parts of a control system, as in a control loop. Each control loop is like a circle. There is no end to it. The parameter to be controlled (process variable) is sensed by the controller, which compares that value to the set point. It then adjusts its output accordingly. The change in output produces a change in the process variable. The controller compares that new value to the set point to change its output again. A control loop contains a controller, a device to measure the process variable, a source of set point, and an output device that controls the flow and anything else that changes the value of the process variable or the set point. A control loop can be as simple as a level controller, consisting of a controller with an internal set point adjustment, a level transmitter (LT), and a control valve. A complicated control loop could have similar devices in combination with a large number of computers located in different parts of the plant. The practical limit of a loop is at the devices that affect the process variable. Any one of those devices can be part of another control loop.

There is always a control range. The values the controllers use have an upper and lower limit. The range of transmitters has to be established to permit reasonable control and allow for the normal variations in the measured parameter. A range is selected by the applications engineer (the person who selects and specifies the controls to be used on a job) to ensure that the system will control properly. It is a question of accuracy and stability. When operating a low pressure steam plant, a transmitter will produce a signal in the range of 0–30 psig and can maintain a pressure of 10 psig ± minus 0.15 psi (pounds per square inch). The transmitter (which typically has an accuracy of ±1/2%) will produce a signal that accurate. On a plant operating at 3000 psig, a 0–4000 psi transmitter would be accurate to ±20 psi. That would not necessarily be considered accurate control. The engineer might select a transmitter that works in the range of 2500–3500 psig to get a transmitter accurate within 5 psi.

Control signals also have a range. Each system normally uses the same signal range for all the devices in the system. There are many standard ranges of control signals, with the most common ones being 3–15 psig (pneumatic), 0–5 volts (electric and electronic), and 4–20 milliamps (electronic). Several other signal ranges are also used. It is not uncommon to encounter a mix of these ranges within systems that are a mix of old and new instruments and controllers. Other signal ranges that may be encountered are 0–30 psig, 0–60 psig, and 3–27 psig pneumatic, 0–10 volt and –5 to +5 volt, 0–12 volt, and 0–24 volt values on electrical and electronic systems. There are others, but their use is industry specific and very limited.

The control signal range is representative of the value of the measured parameter, the process variable. Measure the control signal and, knowing the range of the transmitter, the actual value of the process variable can be determined. A simple example would be a loop to maintain 200 psig after a pressure control valve, where the transmitter range is 0–300 psig and the control signal is a 0–30 psig air pressure. The control signal value for the set point has to be 20 psig (or equal to it) and the actual value can be determined by multiplying the control signal by 10. If a remote indication of the pressure was needed, the transmitter output could be extended with 1/4 inch copper tubing to a 0–30 psig pressure gauge and adding a zero to each number on the gauge face. The tubing and lower pressure gauge would be considerably cheaper than running steel steam piping to the remote location with a high pressure gauge. This demonstrates the reason to use instrumentation. It saves money.

Instrumentation consists of devices that could be used in control loops but do not do any controlling. All they do is provide indications of the value of the process, including parameters such as pressure and temperature, and quantities like pounds, gallons, or cubic feet. The term controls and instrumentation are used to describe a complete system that not only maintains the desired parameters but also provides outputs that indicate how it is doing and what has been done. A common abbreviation for instrumentation and controls is "I&C."

There is another concept called "live zero." Live zero control signals are those for which the control signal value that represents zero is more than zero, such as in a 3–15 psig or 4–20 milliamp control range. The 3 psi or 4 milliamps represent zero. The main reason for a live zero is one of certainty. The pressure transmitter in the previous paragraph can be set at zero output with

Figure 11-1. Float control valve.

zero pressure applied to it. However, it is not certain that it will come off that zero properly. There may be slack in the linkage or stiffness in the bellows that has to be overcome. With a live zero, the signal value can be seen right where it is supposed to be with zero pressure at the process connection. The output can be adjusted while watching the signal approach the live zero from either direction. It is very hard to get a minus pressure, or minus electrical current, reading. The live zero solves that problem.

A common application is a simple level controller. Begin with a simple float control valve (Figure 11-1), which maintains the level (the process variable) by controlling the flow of water leaving the tank. With these float valves, the level has to vary. When there is no flow out of the tank, the valve has to shut off. Conversely, when water is drawn out of the tank at a high rate, the valve has to open fully. In order to change the position of the valve, the level in the tank has to change. When water use is low, the level is higher and the highest level is at shut off. The level has to drop for the valve to open fully. The level cannot be maintained at one precise point because the level has to change in order to control the flow.

The required change in the process variable to achieve control is called "droop." It is the difference between the value of the process variable at no flow and the process variable at maximum flow. The float controller is comparable to other "self-contained" devices that maintain desired pressures, temperatures, and other parameters. They work fine when the flows are low and the deviation in the process variable is acceptable. There are other factors that prevent all controls from being as simple as a float control valve. The pressure of the water supply can be so great, or the flow so great, that the float control valve simply will not work. If the pressure differential gets high enough, it will force the valve open regardless of the position of the float. The system in Figure 11-1 is obviously operating with very little pressure drop across the control valve. That is one of the few without a wire or cable led down to operating level to allow the operator to give it a yank to get it operating again.

It is possible to calculate the maximum supply pressure for a float valve controlling water supply to a tank. Calculate the volume of the float and multiply by the density of the water in the tank (62.4 pounds per cubic foot for cold water) and the equivalent length of the float arm from the pivot to the center of the float. That is the maximum torque that the float could impose on the valve since, at that point, the float is sinking. Divide the torque by the length of the pivot arm on the valve (from the pivot to the center of the valve disc) to determine the maximum force on the valve. Then divide that force by the area of the valve disc that is exposed to the difference between the supply pressure and the pressure in the tank. The result of this calculation is in psi. That is the maximum pressure difference for the float valve. If the drain leads to another tank at atmospheric pressure, the result is the maximum pressure (in psig) in the tank and the most that the valve can handle.

If the flow is high, the valve opening has to be large enough to handle the large flow and that requires the valve disc to be larger. Using the same procedure just described, it can be seen that, eventually, the disc will get so large that the water will force it open at very low pressures. A larger float could be used, but there are limits to float size imposed by the largest float chamber or, for floats in tanks, the tank opening. That is why some floats are cylinders, able to fit in the hole in the tank, but long enough to provide enough displacement volume to operate the control. Another problem with larger floats is that they will collapse when exposed to high pressures inside an enclosed tank, such as a boiler. The length of the float arm could be increased to increase the torque. However, there are limits to that imposed by the inside of the tank and the increased droop. Thus, simple float valves are seldom used to control water level in a boiler. Small residential boilers are frequently fitted with one, but it has a minimal water capacity and is limited to low pressure boilers.

A modulating controller that maintains a tank water level (on/off control is described later) can be compared to that simple float controller. A float operated valve can be used to produce the control. It can work just like the float valve, but control a much smaller volume of water with a very small valve. That way, it can handle the high

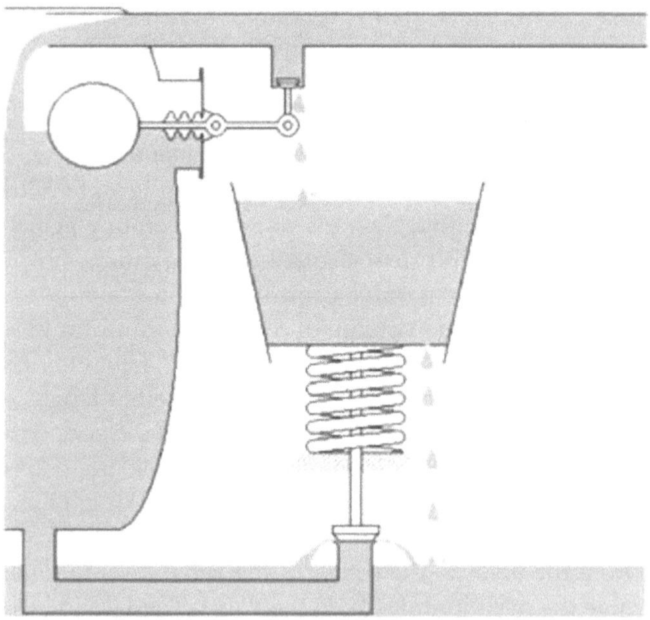

Figure 11-2. Bucket valve control.

differential. The controller in Figure 11-2 (which is a valve filling a bucket over the valve with an opposing spring) is uncommon. It does show some additional concepts of control. The valve controls flow to a bucket on top of the main control valve. When the water level drops, the float valve increases the water flow to the bucket to fill it. The heavier bucket overcomes the weight of the spring and closes the drain valve. The drain hole in the bucket lets the water out, somewhat essential since, without it, the main valve would close and never open until the water evaporated out of the bucket. Control is achieved by changing the level of the water in the bucket. It fills to close the valve and drains to open it. Note the differences between this system and the simple float control valve. An external source is used to power the system (weight of the water in this case). The transmitter and main control valve are separate, with no dramatic restrictions on the distance between them, which is another advantage of control systems. There is another notable difference in this control system. The float valve that is used as the controller is not the same as the typical float valve. It works backwards. Note that the flow of water through it decreases with level. This is just the opposite of the simple float valve. It happens because the pivot point is on the other side of the valve. It was necessary to make the control system work. It reveals one of the control concepts to get used to. There are direct acting controllers and reverse acting controllers.

A direct acting controller increases its output as the process variable increases. A reverse acting controller reduces its output as the process variable increases. Controllers like the one just described are seldom found today. There are a few problems with water. It is corrosive and contains solids that can eventually plug up the control orifices. In the prior example, dust from the atmosphere could get into the bucket and close the drain hole to prevent the valve from opening. There used to be hydraulic controls (which used oil instead of water in closed systems). However, their expense, and problems with corrosion and leaking, resulted in them having a short period of acceptability. They were replaced by pneumatic controls, which survived for several years before they were outstripped in price and function by microprocessor-based electronic controls, the current choice. Electrical and electronic controls saw some use and a share of the control market along with pneumatic controls as well.

The system just described consisted of a controller and a control valve. It is not consistent with modern control systems. The controller measured the process variable directly. A typical control system will have a transmitter, which produces a control signal proportional to the value of the measured variable, a separate controller, and a final element (control valve). The level of the water in the tank could be related to the level in the bucket. That will change as the drain hole plugs up or erodes and is also affected by the pressure drop through the valve and other factors. The float valve controller could be changed to a transmitter by drilling a hole in the outlet piping to let the water drain there and use the bucket as a reservoir. Installing a pressure gauge on the piping feeding the bucket provides an indication of the output of the transmitter. The problem is that the pressure transmitter cannot produce a control signal that is precisely proportional to the level in the tank. A variation in the water supply pressure, wear in the valve and drain orifice, and friction in the valve packing will all combine to generate changes in the signal that will produce errors.

A desire for accuracy and, more importantly, repeatability resulted in the development of precision transmitters by introducing feedback. Feedback comes from the output. It is used to test or correct the output. In the case of a transmitter, it is used to ensure that the output is really proportional to what is being measured, which is called the process value. Modify the float valve and use compressed air instead of water. There are two advantages to using air over water. One is that it has very little weight. The weight of the air does not alter the signal value when the signal is piped up or down two or three floors in the building. More importantly, nobody complains when it leaks out. People would complain about the water powered transmitter constantly leaking

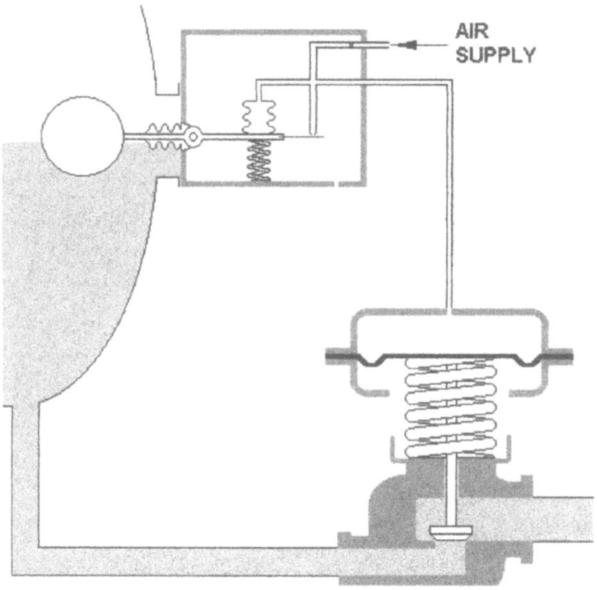

Figure 11-3. Pneumatic level transmitter and control valve.

water out. They will not even notice the air. The air leakage, normally undetected, proved to be a considerably costly part of control systems.

Change the valve and float arrangement so that the float arm compresses a spring. The spring force is opposed by a bellows that contains the output pressure. Now the transmitter looks like the one in Figure 11-3. The orifice has been moved from the bucket to the air supply and created another one consisting of a nozzle, the nozzle discharging against a baffle. That becomes the valve (less expensive than a valve). The valve moves with the float arm to make the output to accurately represent the level in the tank. Flow through the valve does not change based on the position of the float. It responds to differences between the position of the float and the balance of forces of the spring and the bellows.

This construction is typical of most pneumatic transmitters. As the level increases, the nozzle is moved away from the baffle. More air bleeds out at the nozzle. The pressure in the output bellows decreases. The spring pushes down on the float and up on the baffle, following the nozzle. When the level falls, the nozzle is pressed against the baffle. The pressure increases and the bellows compresses the spring to push the baffle down. The transmitter uses a pressure balance principle, where the output pressure of the transmitter is fed back (feedback) to restore the balance of the device, in this case the relative position of the nozzle and baffle.

This transmitter is reverse acting. The output increases as the level drops. In the figure, the valve is draining the tank and it drops the level. The same transmitter can be used for direct control of a makeup water valve supplying a boiler feed tank. The valve internals could be changed. The increasing air pressure would push the valve open.

The system shown is using the transmitter as a controller, and it would work. It is seldom done that way for several reasons, price and power predominating. By switching to compressed air, a much simpler valve could be made in the transmitter/controller. It could be made much smaller, lowering the cost of it dramatically. The reduction in size reduced air consumption as well. It costs less to operate. However, the small transmitter cannot move lots of air. It would take a very long time for it to pass enough compressed air to increase the pressure in the diaphragm casing of a large pneumatic control valve. If the transmitter was used as a controller, there would be a considerable time lag in the operation. It would have to pass all the air for the control valve, in addition to filling the feedback bellows and connecting tubing. The very limited output of transmitters prevents them being used as controllers for those reasons.

The simple transmitter would also have a droop, although not as noticeable as other methods. The distance between the nozzle and baffle would have to change to raise or lower the pressure in the output. That produces a difference between the control signal and the float position. Another important factor in the design of the transmitter also allows for increased droop. That is because the designer had to allow for something to go wrong (like loss of air pressure). The baffle is usually a flexible piece of spring steel that can bend without breaking when the level is low and there is no air pressure to compress the spring and keep the baffle to nozzle position. As the control signal increases, some of the pressure is used to bend the baffle slightly to introduce more droop. To reduce that effect on the transmitter, and save on even more air, the designers made the nozzle even smaller. The problem with that smaller nozzle was that it could handle even less air. Any leak in the tubing connecting the transmitter to other devices would introduce an error. The output would be lower than it should be.

To eliminate the problem of leakage loading down a transmitter, designers added boosters to their transmitters. The reduced size of the nozzle and baffle assembly and the savings in compressed air consumption allowed them to reduce the cost enough to justify adding the booster, which is a simple device. A booster installed in the transmitter eliminates any problems with tubing leakage loss, as the nozzle air only feeds the feedback bellows and the booster diaphragm (Figure 11-4). The large area

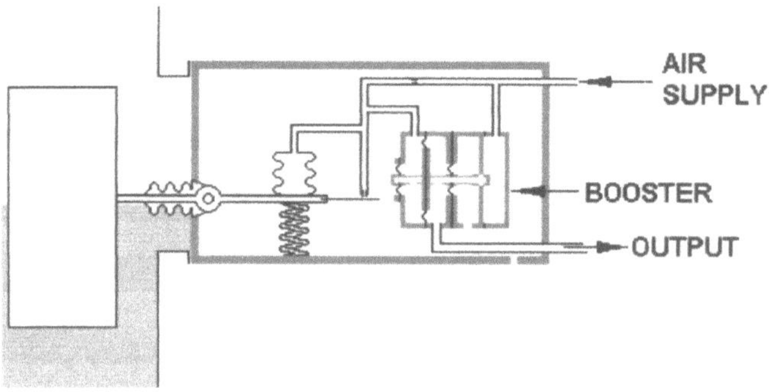

Figure 11-4. Booster for pneumatic transmitter.

of the diaphragm provides ample force to position the output valve so that the transmitter can pass enough air to compensate for small tubing leaks without a degradation in the value of the signal. It also allowed an operator to detect a leak by comparing a gauge connected to the output bellows and the tubing at another instrument. As designs of transmitters improved, the nozzles got even smaller, and, in some cases, a booster is used to feed the feedback bellows.

A displacement transmitter resolves some of the problems with floats. In a typical toilet, there will be a float valve to control the water filling the tank. Gently push down on the float and continue pushing it until it is completely under the water, noticing the force required. The force required to push the float down increased with depth. The additional force is equal to the difference between the weight of air in the float and the weight of water it displaces, the buoyancy principle. Displacement transmitters balance the force on the float with a force produced by a feedback bellows.

Pressure transmitters use the same principles of force balance to produce an output by using another bellows, or a diaphragm, sensing the pressure in the process and balancing that with an output feedback. Different pressures are accommodated by changing the size of the process bellows or diaphragm. Pressure transmitters would be very expensive if a special bellows had to be made for each pressure range. They are made adjustable within standard ranges by allowing adjustment of a pivot between two beams connected to the bellows and feedback.

Temperature transmitters work on the same principle. They just need a way to get motion or force proportional to the temperature and then convert it to a signal. Bimetallic sensors use the movement or force produced by the difference in thermal expansion of two metals. Fluid filled transmitters use the thermal expansion of the liquid to produce movement. Gas filled transmitters use the increase in pressure proportional to temperature.

Electronic pressure and differential transmitters sense process values using the same techniques as described for pneumatic transmitters, converting a force or movement to a voltage or current and generating a feedback force using an electromagnet. Temperature transmitters use a resistance to electric current, where the resistor's resistance varies with temperature. Another means of measuring temperature that has been around for years is a thermocouple. Two wires of different materials, connected at their ends, will produce an electric voltage when the two ends are subjected to different temperatures. Note that the reference temperature (one end of the two wires) has to be stable to get a reliable signal proportional to temperature at the other end. Digital transmitters use similar methods and then convert the analog signal to a digital one. For all practical purposes, all pneumatic, voltage, and current signals are analog signals. The signal represents (is analogous to) a process value. With a measurement of the signal, the process value can be determined from the value of the signal. That is all an analog signal is: a value that represents another one.

Digital signals are different. They change rapidly, commonly from a negative voltage to a positive voltage. There is no way to put a meter on the signal terminals and measure it. The value of a digital signal is a function of the number of changes in value and the time between each change, a complexity that requires a computer to read it. They are better because the actual value is not important. Any significant resistance in the signal wiring for a voltage signal, like a loose terminal, can alter the signal to produce an error. Digital signals represent zeros and ones, where a zero is considered anything between +5 and +15 volts and a one is considered anything between a −5 and −15 volts. That considerable range of voltage minimizes errors. The additional features of digital signal transmission provide more accuracy and reliability than analog signal transmission. All that and the lower cost, of both hardware and installation, of digital controllers have made them the controls of choice, replacing all other types of control.

Pneumatic control systems provide concepts that work with any type of controller and a pneumatic understanding will help with the comprehension of them. A controller that is no longer available (like most pneumatics) is a Hagan Ratio Totalizer as shown in Figure 11-5.

The totalizer has four diaphragm chambers. They could also be fitted with bellows. The totalizer was designed to provide universal use by adapting it. The output chamber and (A) input chambers are secured to the base of the transmitter. Sliding in the middle are clamps

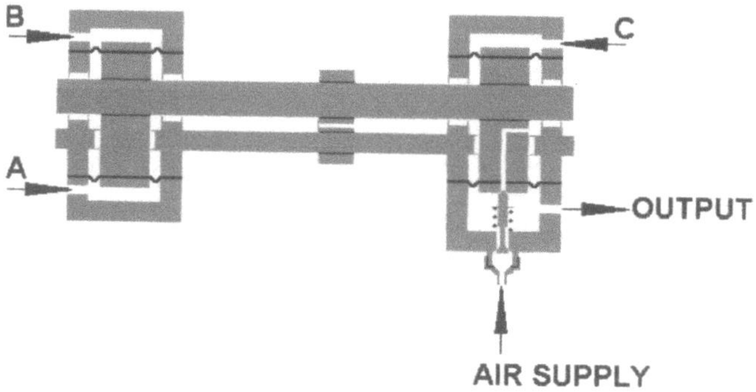

Figure 11-5. Hagan ratio totalizer.

that connect to the base and the beam. The beam floats in the middle of the assembly, attached to the diaphragms and the beam clamp. A very thin piece of spring steel connects the two clamps to form the pivot of the controller. The two clamps can be loosened and slide along the base and beam to positions right and left of center. The valve floats in the output chamber and will open to admit air if the beam is rotated clockwise or close off the air supply and stop while the beam continues to rotate counterclockwise. Further counterclockwise rotation will open the bleed end of the valve to dump air to atmosphere.

With proportional control, the output of a controller is proportional to the difference between the process value and the set point. Assume the LT covered earlier to produce the process value. In this case, the controller will be used for level control. Also assume that the level control valve is reverse acting. An increase in controller output will close the valve. When the water level in the tank increases, the control signal decreases. To make the system work, any increase in process value should result in an increase in output to close the valve. Now look at the ratio totalizer to see how to connect the process variable. The output bellows pushes up on the right side of the beam. Any increase in output will tend to rotate the beam around the pivot in a counterclockwise direction. It is a pressure balance system. The process variable has to create a tendency to rotate the beam in the opposite direction to balance the forces.

If the beam tends to rotate clockwise, more air is supplied to the output and output bellows to counter that rotation. If it tends to rotate counterclockwise, the vent valve opens to decrease the output. Connecting the signal from the transmitter to the bottom bellows (A) does the job. Now a change in the level will produce a change in the output of the controller to open or close the control valve. As shown, the controller acts pretty much like a signal booster because it produces a change in output that precisely matches a change in input. As the LT output changes from minimum to maximum, the controller output produces the same value because the bellows areas are identical. It works pretty much like the float controller, requiring the level change over the full range of the float to position the control valve between open and closed.

The whole reason for using a control system is to improve on the operation that is obtained with a float control valve. The change in level can be reduced by moving the pivot on the controller closer to the output end. It works just like a teeter totter. Adjust the controller so that the distance from the center of the process input to the pivot is twice the distance from the output bellows to the pivot (two-thirds of the beam length). Now, if the level varies to produce a 1 psi change in the transmitter output, the controller output has to change by 2 psi to maintain the force balance in the controller. There is a proportional difference in the change of the signals, where the output has to change twice as much as the input. That is the concept of proportional control. In this case, the controller has a gain of 2, which means the output has to change twice as much as the input. Now the controller will run the water valve from closed to open with half the change in the output of the LT, between 25% and 75% of the signal range.

The gain could be increased until there was very little change in the process value to produce a full stroke of the water control valve. The water level would not change much. That arrangement would not work too well. Any little ripple in the water level would produce a dramatic change in the valve position. There would be a lot of valve wear. There would also be controller "noise," where the output is jumping around with little relationship to the actual level in the tank. Conversely, the gain could be reduced to something less than 1, which would create another problem. The water valve would never close. It might work during normal plant operation, but when the plant is shut down, the controller output could not increase enough to close the valve. Too much gain produces a lot of noise and erratic operation, while too little gain can result in failure to operate at the extremes of load.

One problem with this controller arrangement is that there is no way to adjust the set point. For all practical purposes, the set point is the center of the transmitter's position. In order to have an adjustable set point, the B bellows of the controller is used. It is supplied with a control signal that is adjustable. The set point signal, in this case, is produced by a simple air pressure regulator. By connecting the regulator to the bellows opposite the one sensing the signal from the transmitter, a set point controller is created.

Now the output of the controller is proportional to the difference, which is called the error, between the set point and the transmitted level signal. Instead of acting only on the pressure from the float transmitter, the action is dependent on the difference between the set point and the process variable. The set point pressure acts on the diaphragm at B, pushing down on the right end of the beam opposite the process variable signal coming in at A. The force tending to rotate the beam is equal to the difference between the two pressures times the area of one diaphragm.

All modern controllers operate on the error, not the actual signal value. Now changes in output are proportional to changes in the error, not changes in the level. An important part of this to understand is that an error can be introduced by changing the set point. The gain will need to be set at much more than 2, in this case, or the output may not change enough to fully stroke the valve.

Creating a set point controller allows the use of something more than the level control range for the transmitter. The transmitter can be used for instrumentation as well as control. By putting a long arm on the float, output signal proportional to almost the full height of the tank can be produced. Then the level can be determined even when it is not in the control range. For example, the LT could be set to indicate levels from zero to 60 inches in the tank. Select the control range and adjust the controller gain accordingly. If the desired level to control is within 10 inches, set the gain of the controller to 6. If the set point is at 50 inches, the control valve will be fully closed when the level reaches 55 inches and open at 45 inches. The set point can be adjusted to anywhere from five inches to 55. Half the control range must be reserved to have control. That is why the set point cannot be anywhere within the range of the transmitter when using proportional control. If the set point was raised to, say, 58 inches, then the output valve will not be able to stroke completely.

The next question becomes how to use this controller in a boiler to keep the water level to within 1 inch of the set point. First, a maximum of 20 inches is used for the range of the boiler water LT (even if the boiler is over a 100 feet high). Then a reset is added to the controller. There are practical limits to an instrument's range when it is used for control. Reset control is a refinement that makes the set point realistic. To convert the controller to a reset controller, add some tubing, a needle valve, and a small volume chamber (Figure 11-6).

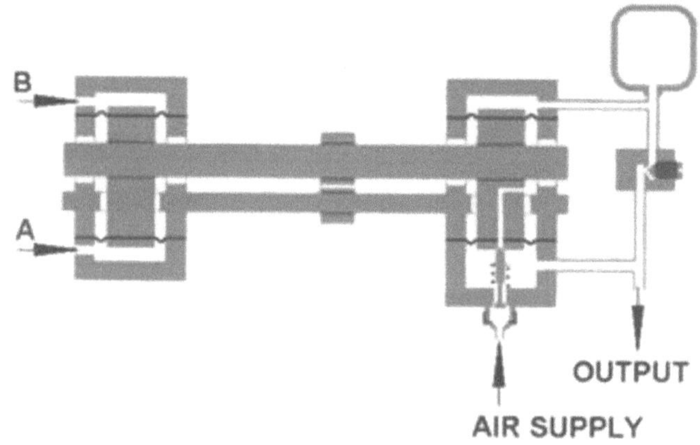

Figure 11-6. Totalizer with reset accessories.

It is these reset accessories that make the controller a reset controller. The controller has now acquired dynamic properties. The only time a reset controller will be in balance is when the set point and the process variable are precisely the same and the output has stopped changing. With the proportional controller, the system could be stable with the level holding at a value below or above the set point. Now, the left side of the controller is in balance only when the process variable and set point are precisely the same, when the error is zero. Even then, the controller output can be changing, when the pressures inside the output and reset bellows are different.

Operation of reset control is difficult to comprehend. Many technicians have an inappropriate perception of it because they think in terms of speed, not response to an error. The operation of the ratio totalizer provides a basis of understanding because the dynamic effects are apparent. Start with a steady-state condition, where the pressure in the output bellows matches the pressure in the reset bellows and the error is zero. Assume that the process variable drops a little. An error is generated. The proportional function of the controller responds immediately, changing the output an amount equal to the error times the gain. Also, assume that the error is not corrected immediately by the output change and holds. Since there is now a difference between the output and reset bellows, control air bleeds through the needle valve to (or from) the volume chamber and reset bellows. Since the error persists, the output will have to continue to change to balance the error. If the error continued to exist, the output would continue to change until it reached its practical limit (0 psig or supply pressure).

That does not happen often. Usually, the controller action results in the process variable returning to the set point. That is the beauty of reset control. It always works to return the process variable to the set point, not some value that is offset by the proportional value. It is real control. The only time the controller is not changing its output is when the process variable and set point are exactly the same and the pressures in the output and reset bellows equal each other. The controller can also be in balance with any pressure at the output. The output signal can be anything from zero to supply pressure balanced by the same pressure in the reset bellows. The controller will be satisfied as long as the set point and process variable pressures are the same. Unlike proportional control, a deviation in the process variable is not needed to get the required output.

It is reset control that keeps the boiler level right at the center of the glass, while changing the feed water control valve position from closed at low loads to almost wide open at high loads. It is also reset control that makes it possible to keep the steam header pressure at 120 psig, whether at low fire or high fire, and even when running five boilers instead of one. It is reset control that allows the burners to run air/fuel ratios so tight that oxygen in the flue gas can be held at one-half percent (for gas firing).

Tuning a reset controller is nowhere near as easy as tuning a proportional controller. The additional feature of the controller (there is P + I, proportional plus integral) allows more flexibility in matching the process. Integral is a mathematical term that means accumulating the average value. It is not important to understand mathematical integrals, only that it is another name for reset.

Tuning consists of changing the gain (proportional control) and reset (integral control) until the combination provides a response to an upset in process conditions, where the process variable returns to the set point within a short period of time and with only a little overshoot in response to the initial error. The text book curve shows that the error is plotted versus time. It starts as a big error with a rapid change in process variable quickly approaching the set point, overshooting it a bit, and then turning back toward the set point and falling in line with it. It is a pretty picture. Making it do that in the real world can be difficult at times.

Keep in mind that a simple proportional controller requires an error to do its job. Attempts to minimize that error can result in some pretty wild swings in the output of the controller. This is called instability or over control. Those swings are primarily associated with the fact that the process does not respond instantly to changes in the controller output. It can take a few hundredths of a second to several seconds before the full effect of a change in controller output is apparent by looking at the process variable. There is a time lag in the system. Consider a simple manual system. There is a flow control valve in the pipe feeding the tank. There is a drain valve that allows a constant flow leaving the tank. An operator adjusts the flow control valve in an attempt to maintain a constant level in the tank. The flow control valve is 20 feet away from the tank. The flow rate is 1 feet/second. It takes the water 20 seconds to get to the tank. The operator can see the level at the top of the tank. If the level is dropping, the operator will open the flow control valve to let more water into the tank. If the level is rising, the operator will close the valve to reduce the flow to the tank. If the operator sees that the level is going down, the valve can be opened. However, it will take 20 seconds for the increased flow to get to the tank. If the operator is anxious and sees that the level in the tank is still going down 5 seconds later and opens the valve some more, additional water will flow toward the tank. If the process is repeated every 5 seconds, there will have been five adjustments to the flow before the first adjustment is noticed at the tank. If the first adjustment was nearly sufficient to stop the level from declining, the other four will cause the level to start rising. The operator will start closing the flow valve. Again, it will take 20 seconds for that reduced flow to reach the tank. In the meantime, that additional adjustment to the flow to add more flow is arriving every 5 seconds. If the operator starts to make adjustments to close the valve at 5 second intervals, eventually too little flow will arrive at the tank. The system becomes unstable. This is a case of over control. The control system has to know not to make successive adjustments any more rapidly than the time lag in the system.

A reset controller is tuned to deal with those delays. The controller will have two adjustments, gain, and integral. Gain is the proportional part. The output is the error times the gain. The output changes when the error changes. Integral is the reset adjustment. It repeats the error multiplied by the value of the integral. Note that the reset effect is the error repeated. An integral adjustment is normally marked to indicate repeats per minute, meaning that is the number of times the error will be repeated in 1 minute. That does not mean the controller only repeats the error for a minute either. It continues repeating the error every minute. It also does not repeat it at the end of a minute. If the integral is set at 60 repeats per minute, it will increase or decrease the output by a value equal to the error every second. A proper combination of gain (proportional control) and integral (reset control) will make the process return to the set point quickly and smoothly.

Gain on modern controllers can be adjusted without affecting the output except for the difference in the gain (times the error). Increase the gain and the output changes more for a given error. The error is the difference between the set point and the process variable, what is wanted and what is there. Adjust the gain or reset to balance the system response. If a change in controller output produces an almost instantaneous change in process variable, then most of the control function can be left to proportional control. If, however, the process responds sluggishly to a change in controller output, then the integral adjustment is more critical. Watching what happens when an error is introduced will give a good idea of how to adjust the controller. A set point controller allows that to happen.

When starting up a new system, adjust the controller using some initial adjustments that are the average for comparable systems. Then switch it to automatic to see what happens. Quick swings in the process variable indicate instability. Reduce the gain immediately if they happen. Then slowly increase the gain until it starts getting a little unstable and back off some to eliminate the instability. If the process does not change due to external influences, introduce an error to see what happens. Actually, it is much easier to work with an error. It can intentionally be set to a value that can be meaningful, something simple like 1%, 5%, or 10%. Just change the set point, swinging it to the selected difference from the process variable. If the process overshoots the set point considerably, then reduce the gain. If it seems to take forever for the process to return to set point, then increase the integral. If the process returns to set point while swinging back and forth on either side of the set point, then reduce the integral. If the process slowly returns to set point, then increase the integral until the process overshoots the set point a little once.

Changes in one adjustment normally require an opposite change in the other when getting close to the desired characteristic of the controller (that curve where the process overshoots set point once then swings in to match it). An increase in gain will probably require a decrease in reset and vice versa. Figure 11-7 is a rendition of that popular graphic seen in all the instructions for tuning a controller. Hopefully, the previous discussion makes it meaningful now. It used to be rather difficult to get a graphic output on a recorder or other instrument that could be compared to something like Figure 11-7. Today, a recorder can be used. The chart speed can be adjusted or simply adjust the parameters for a trend screen. It is a graphic produced on a computer screen that draws a line between points of recorded values relative to time.

It is not always that simple. Some systems are set so that the data are only recorded every 5 seconds to every minute. In that case, any graphic can completely miss the swings that were generated by changing the set point. Be very aware of that potential limit on electronic data. When it comes to tuning controllers, there is no substitute for practice to gain experience. If practicing with a functioning controller, record the gain and integral adjustments before changing them so that they can be restored to the original settings. If it does not seem to work as well when the testing is done, keep in mind that hysteresis can have an effect. Restore the original settings by approaching them from the opposite direction.

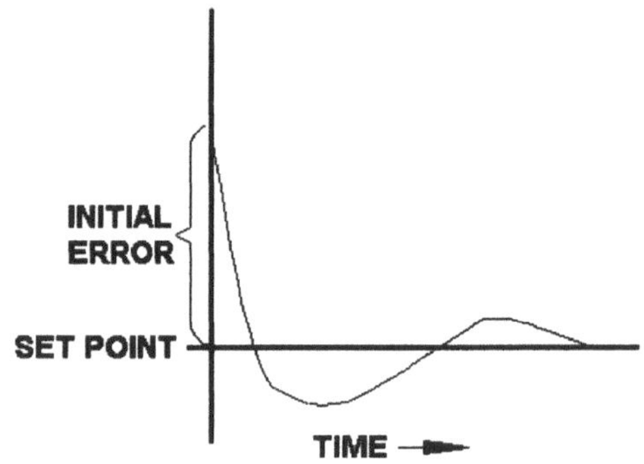

Figure 11-7. Controller error versus time, tuning guide.

Hysteresis has to do with friction in mechanical systems. It can occur in almost any situation. It often presents itself as not taking the same path going forward and coming back. The control valve in Figure 11-8 consists of a chamber over a rubber diaphragm, where the control pressure can push down on the valve stem and a spring that pushes up to resist the pressure forces. Without hysteresis, the position of the valve would be precisely proportional to the control pressure. The push down would be a force equal to the area of the diaphragm in square inches multiplied by the control pressure in psi. For a

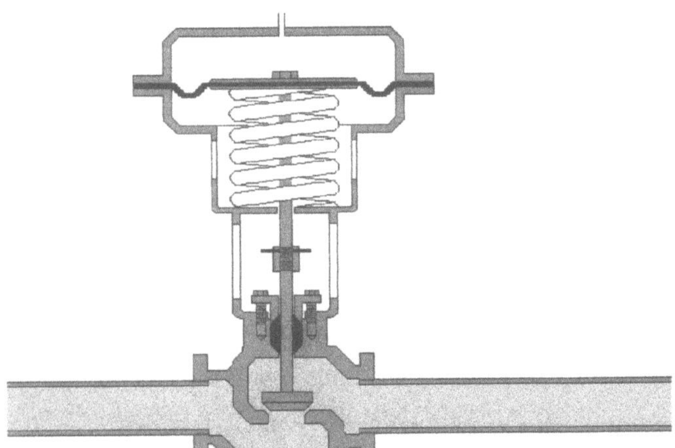

Figure 11-8. Simple pneumatic control valve diagram.

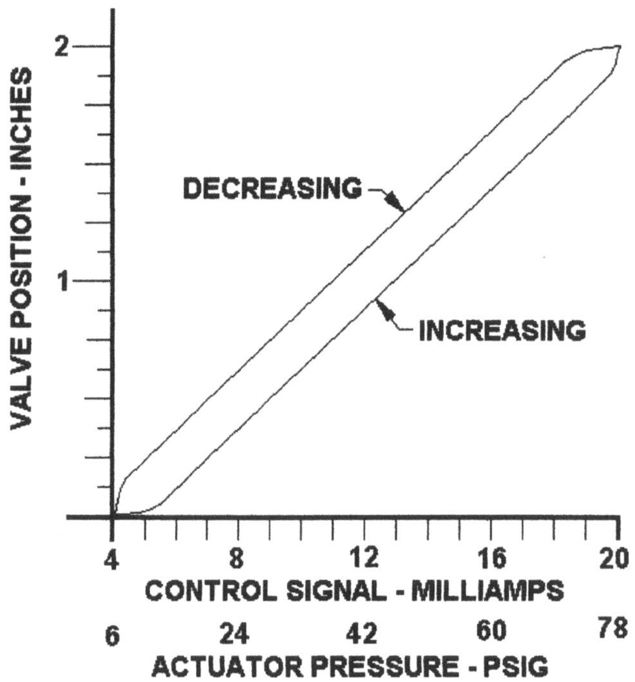

Figure 11-9. Hysteresis curve.

3–15 psig control signal and a 50 square inch diaphragm, the force would vary from 150 pounds at zero control signal (3 psi × 50 sq. in. = 150 pounds) to 750 pounds at 100% control signal. The spring would be compressed to balance the 150 pound force when the valve is closed and have a spring constant equal to 600 pounds divided by the stroke of the valve. If the valve stroked 1-1/2 inches, the spring constant would be 400 pounds per inch. The valve packing is tight on the valve stem to keep it from leaking. That tight packing tends to hold the valve where it is. The friction always acts in opposition to the travel of the stem. It will push against the diaphragm force when the valve is closing and oppose the spring when the valve is opening. It produces a difference in valve position for a given control signal, depending on whether the valve is opening or closing. The graph in Figure 11-9 is a typical hysteresis curve and it applies to the valve just described.

Mechanical hysteresis is not the only thing that creates a difference in position of a control valve operating on a control signal directly. There is a difference in the amount of air the controller must pass depending on the valve position. The volume of the diaphragm chamber increases and decreases with valve position to upset the performance of the controller. There are also the problems associated with the controlled fluid as well. When the valve is closed, the difference in valve inlet and outlet pressures act on the area of the valve opening, adding another force to the valve stem. If the valve is a boiler control valve, it can work perfectly fine when the boiler is operating but leak when the boiler is shut down because the pressure drop across the valve disc is so great that it overcomes the forces produced by control pressure. All these factors can be overcome by making sure the combination of diaphragm area and valve chamber pressure will keep the valve shut. Adding a positioner also helps because it can operate with higher actuator pressures using a separate air supply and match the valve position to the control signal.

A valve positioner is just another controller. It controls valve position by comparing the actual position (as a process variable) to the control signal (remote set point). The control signal becomes a remote set point because it is produced elsewhere. It is also a variable set point because it changes. A rather simple positioner is shown in Figure 11-10. The remote set point is the pneumatic signal coming to the positioner. The process variable is developed by the spring compressed by linkage attached to the valve stem. As the valve opens, it compresses the spring.

Changes in the control signal change the force on the diaphragm so that the spring is compressed or allowed to expand. That changes the position of the valve to divert air into or out of the diaphragm. The valve position is changed. The compression of the spring matches the control signal to return the valve to its center position. The pressure in the diaphragm is like the output of a reset controller. It is whatever it has to be to do the job. A positioner can also use a supply pressure higher than the control signal range to overcome high differential pressure on a valve and the friction of some packing that was tightened a little too much. Any control valve in a boiler plant should be equipped with a positioner. Today, with electronic control signals, the positioner has to adjust the air pressure to match an electronic signal. One simple positioner uses two solenoid valves: one to add air and one to bleed it off.

Valve positioners can experience windup. The feed water control valve mentioned earlier is a good example. A positioner was put on the valve and the pressure in the diaphragm of the valve actuator ran up, while the boiler cooled. It had to overcome the pressure of the feed water trying to open the valve. While the boiler came up to pressure, the actuator pressure did not change. When the boiler started making steam and the water level dropped, the level controller would have to raise the control signal a little, indicating that the valve should open a little. In normal operation, the valve would respond rather quickly. Yet this first time, after a shutdown, it will

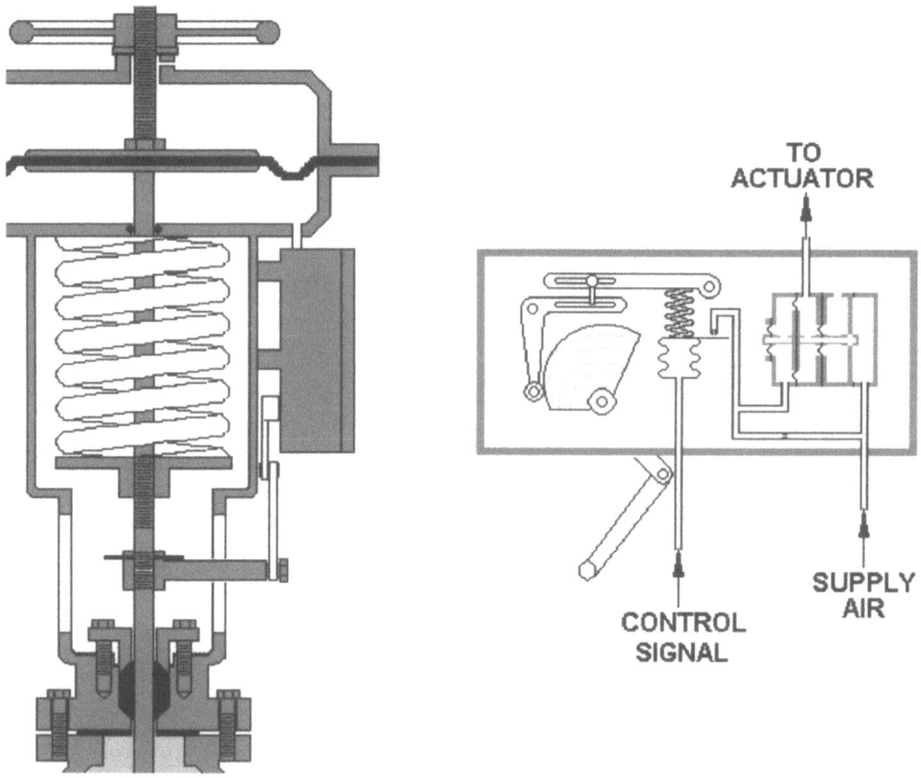

Figure 11-10. Simple valve positioner.

not. That is because the positioner has to bleed off all that air that was compressed into the diaphragm chamber to raise the pressure enough to keep the valve closed against the high feed water differential. That differential is now gone. That may explain why there is a lag in response when operating a valve manually.

The original pneumatic controllers did the same sort of thing. The introduction of live zero made it happen at both ends of the control signal range. A generic 3–15 psig controller could wind up to an output equal to the standard 18 psig supply pressure or wind down to zero output. When that happens, the system is out of control. The controller has done all it could to restore the process variable to set point. Unless that is an intentional condition (it could be), the output will eventually get the process variable going in the right direction. It will return to the set point. It will not stop there though. It will continue right on past because the output has not changed. With a reset controller, it cannot because the output is in windup or wind down.

Once the error is in the opposite direction, the output will start to change. During the period when the controller is building up from zero, or dropping down from supply pressure, the controlled device, a valve for example, does not respond. It only responds to signals within the control range. The result is always a long delay (seconds, not hours—although, sometimes, it seems like hours) before the output gets back into the normal control range. As a result, the process variable is swinging all over the place. That is the effect of reset windup. Most modern controllers avoid it. The control manufacturers have designed the controls to eliminate it. It may still be encountered with valve positioners and damper actuators.

The terms "procedureless" and "bumpless" have been applied to controllers. They are not common today but may be experienced on older units. Early pneumatic control systems that included hardware like the ratio totalizer had separate manual/auto stations, which were flush mounted on the panel and gave people the option of controlling the process by hand. When it was considered necessary to give people the option of changing the set point, the station also included that adjustment. Figure 11-11 shows what one of those stations looked like.

The set point adjustment was nothing more than a pressure regulator with the adjustment knob penetrating the front of the station. The set point was indicated on a pressure gauge mounted above the adjustment knob. The output of the controller was indicated on another pressure gauge and another pressure regulator produced the manual output signal. The valve handle in the middle of the station was used to switch from automatic to

Figure 11-11. Early pneumatic H/A station.

manual and back to automatic. However, it was not as simple as just turning that pointed knob. Simply twisting the valve knob from the automatic to manual position, and the automatic and manual pressures were not the same, produced a "bump." The output would jump from the output produced by the controller to the setting of the manual station. To transfer from auto to manual without a bump, the valve knob had intermediate positions at 1/4 turn for adjusting the outputs to match them up. When transferring from auto to manual, the gauge was switched to show the manual signal. That was adjusted until it matched the automatic output before turning the knob another 1/4 turn to manual. When transferring from manual to auto, it showed the automatic output. That was biased to match the manual output before switching to auto. It was necessary to perform a signal matching procedure during the transfer from hand to auto and vice versa to avoid a bump.

There are also controllers with a balance indicator. It consists of a clear plastic tube that is visible through a slot in the front plate of the controller. It contains a small ball that fits inside the tube with very little clearance. One side of the tube is connected to the manual output and the other to the automatic. When it is time to switch from one to the other, the manual output is adjusted until the ball floats to the middle. Then throw the auto/manual selector switch over. As pneumatic controls improved, the manufacturers included additional little controllers inside their devices so that the automatic signal automatically followed the manual output, and the manual output was automatically adjusted to match automatic to permit rapid and "procedureless" or "bumpless" transfer

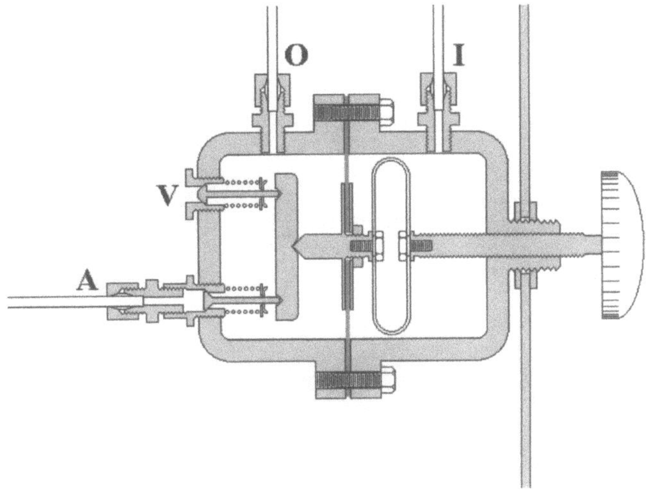

Figure 11-12. Bias regulator.

between manual and automatic operation. Electronic controls had similar procedures that were replaced by add-ins. Similar functions are understood to be included in modern controllers.

Bias is a control engineer's term for add or subtract. It is done a lot in controllers and, most of the time, is not seen. It became an integral part of the manual/auto stations to provide the ability to line up auto and manual signals. It was done in one control regulator, where the output of the regulator was a combination of the controller output and the pressure that opposed a spring. The manual adjustment loaded the spring. The assembly looked something like Figure 11-12. When the control designers noticed that the operators used that spring adjustment to produce a difference in the output of two manual/auto stations using the same control signal (like on coal pulverizers, where the primary air and coal feed could be biased), they simply manufactured another faceplate with that regulator on it and called it a bias station.

Derivative control is essentially "Rate" control. It is the "D" in a proportional, integral, and derivative (PID) controller. It is a helpful control feature in systems where the process is upset quickly and erratically by external influences all the time. When there is no relationship between what an output controls and something that upsets the process, a derivative control is almost a necessity. A ratio totalizer can be visualized as a rate controller. It will look like the diagram in Figure 11-13. With no change in the process variable, the output of the controller is equal to the output of the reset controller. The rate control occurs with changes in the process variable. If the process variable changes slowly, it will have very little effect on the output. The control air will bleed through

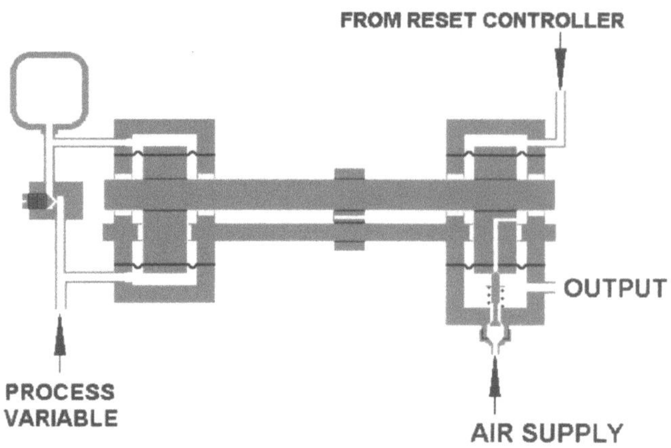

Figure 11-13. Ratio totalizer set up for rate control.

the needle valve fast enough that the pressure in the two bellows will remain about the same. If the process variable changes quickly, the air cannot bleed through fast enough. The difference between what it was and what it is produces an increase or decrease in the output on top of the reset controller signal.

Some manufacturers called the device a "pre-acting" controller because it changed the final output based on the action of the process variable. The output of this device changes according to the rate of change of the process variable. When the process variable stops changing, the output of the derivative element is always zero. It is called a derivative controller because the output is proportional to the rate of change of the input. Adjustments are in minutes per repeat.

To get an idea of where rate control can help, consider a system that maintains level in a small tank. This tank has an extra drain valve that is manually controlled. If the level is normal, and there is no flow out of the tank, the reset controller will wind down to shut off the water feed valve. Now someone opens the manual drain wide. With reset windup, or even a stable reset control situation, the level suddenly starts falling and the reset controller cannot respond fast enough to keep the level from dropping quickly. The rate control senses the rapid change and forces the output valve open quickly.

Another control term is "cascade." It is used to identify the use of the output of one controller as an input to another. Cascade controls are useful when the output of one process feeds into another. Changes in the first process, which are made by the output of one controller, create proportional changes in the second process. The impact on the control of the second process can be reduced by using the output of the controller for the first process as an input for the controller of the second process. Typically, drum level control and furnace pressure control on a boiler contain cascade control loops.

Control schematics are diagrams on paper that represent the elements of a control system for a process and how they are interconnected. Anything more elaborate than a simple proportional control system will normally have a diagram to show how the system is interconnected. That will help to figure out how it works. Control schematics or diagrams are not the same as process and instrumentation (P&ID) diagrams (See Documentation, Chapter 1). The P&ID shows the process itself and where the transmitters and control elements are in the process.

Control schematics show how the transmitters and control elements are linked in a control system to control the process. The better control schematics will show the transmitters across the top of the drawing and the controlled elements (like valves and dampers) at the bottom. In that way, there is a flow from inputs to outputs going down the drawing. Lines on the drawing show the flow of information and may or may not indicate how that information is transmitted. They may be process signals like 3–15 psig pneumatic or 4–20 milliamps of current. They can also be digital or something as unique as light (as used in fiber optics). It is not essential to know how the signal is transmitted to understand the system operation. That is needed only when it has to be fixed. Figure 11-14 is a simple, single loop, control schematic that can be used to explain some of the features of control diagrams. This control loop is the level control system, described earlier, presented in control schematic symbolism.

These diagrams use symbols comparable to those standardized by the Instrument Society of America (ISA) and the Scientific Apparatus Manufacturer's Association (SAMA). The LT produces the process variable signal that is fed to the controller. The line to the controller represents the level signal traveling from the transmitter to the controller by whatever means the control system employs. The set point of the controller (desired water level) is produced at the controller and is represented by the capital letter A in the diamond on the side. That symbol represents an analog output manually generated. The K in the box implies proportional control. (Engineers commonly use the letter K to represent a constant value and the gain of a proportional controller is a constant value that the error is multiplied by.) The funny looking symbol after the plus sign is the integral symbol. The integral symbol comes from mathematics as the symbol for the process of integration. On ISA drawings, the collection of symbols in the center is replaced by one circle with PID in it to represent the controller. This method

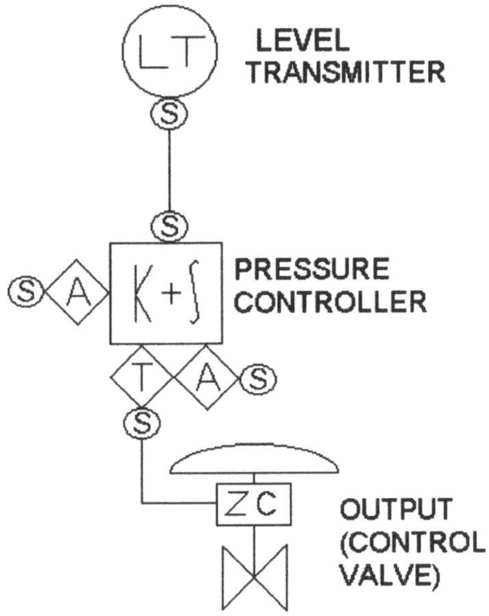

Figure 11-14. Single loop control diagram, SAMA symbols.

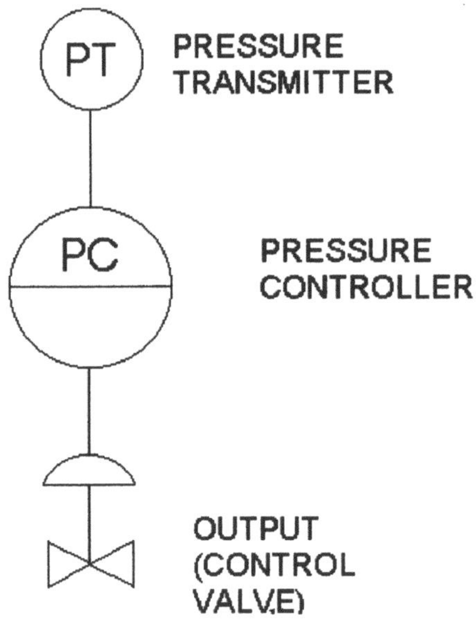

Figure 11-15. Single loop diagram, ISA symbols.

of diagramming shows more detail to get across the concept.

The diamond, with the T in it at the output of the controller, is the symbol for a transfer switch (like for hand to automatic). The analog output diamond next to it represents the manual output generator. The little circles with an S in them are used to indicate that the control signal value can be observed on a gauge, or other visible output, so that personnel can see its value. Note that it is shown at the LT. That is nice to have when the transmitter is at a long distance or several floors above or below the control panel. It depicts that the operator wants (or should want) to be able to see that signal value. This controller should let the operator see the process variable, set point, manual output setting, and controller output. It could use some switching device with only one display. It shows only one at a time. Note that the controller output is not shown at the valve positioner. It should be. Engineers use the letter Z to denote position. The ZC is defined here as a valve positioner, more clearly understood as a position controller.

Figure 11-15 shows the same loop in the simpler symbol method, the ISA methodology. It can be seen that there is a lot of detail missing. Other documents will be needed to clarify the diagram, but the control function is the same. One distinctive clarification is the line through the center of the symbol (or the lack of it).

The line through symbol indicates that it is panel mounted so that the controller is mounted in a control panel. Sometimes, double lines near the top and bottom of the drawing distinguish a separation between panel and field. The two figures could be modified to eliminate the lines representing logic flow by clustering symbols together. The controller in Figure 11-14 could be shown alongside the valve positioner, which would indicate that all those functions were in a controller and positioner mounted in one enclosure on the control valve. The PID controller and transmitter symbols could be put together to indicate that the controller and transmitter are in one package.

SELF-CONTAINED CONTROLS

Photographs of control rooms with long curved panels containing hundreds of knobs and buttons under a row of monitors that reveal graphic displays of the boiler systems can be impressive. Also impressive are some of the self-contained control devices, including some that have been around for years and proven themselves to be so reliable and economical that they may not be replaced by the end of the 21st century. A good example is the control valve on a little residential gas fired hot water heater. In one little box, it is a burner management system (BMS), pressure controller, and temperature controller. Simple versions use a thermostat to monitor the gas pilot for burner safety. A bulb, placed in the furnace over the pilot fire, contains a liquid which evaporates to generate an internal pressure conducted through a length of very small tubing to compress a spring opposite a bellows in the valve body. As long as the pilot burns, the bellows holds a latch, which holds up a disc in the valve body to admit gas to the pilot. To light one of these, hold in a button that opens the valve to admit gas to the pilot. Once the heat of the pilot generates enough pressure in the thermostat assembly, it holds the pilot valve open and the button can be released.

When the button is released, it allows a valve to open that admits pilot gas pressure to the main valve control. The temperature of the water in the heater is sensed by a bulb inserted into the side of the heater. That bulb can contain another liquid that expands to compress a spring, or linkage, that uses the difference in thermal expansion of metals for a mechanical movement relative to the temperature of the water in the tank. When the temperature knob on the side of the control valve is adjusted, it changes the relative position of the linkage, or spring, and another small valve to select the desired starting temperature. When the water cools, a switching valve opens to admit pilot gas pressure to a diaphragm, which opens the main valve. The main valve admits gas that is ignited by the pilot and heats the water. During that operation, the main valve also functions as a pressure regulator to maintain a constant gas pressure to the burner. When the temperature rises to a set amount above the starting temperature, the switching valve closes the pilot gas supply to the main valve diaphragm and drains the gas over the diaphragm to shut off the fire. The main valve closes until another operating cycle starts.

Modern self-contained hot water heater valves do not operate on a continuous pilot to save a little energy. They also eliminate the old problem of pilots blowing out. They include a piezoelectric starter that uses pilot gas flow to power a generator that creates a spark to light the pilot as the water temperature drops. The main valves have double seated valve discs to ensure safe operation. Look at the instruction manual for one of them to gain some appreciation of how something that appears to be so simple is rather sophisticated.

Some self-contained control valves are simple and effective. They do not have to be sophisticated. A gas pressure regulator, like the one in Figure 11-16, controls the flow of gas to maintain a constant outlet pressure. The position of the valve assembly is determined by the compression of the spring by the diaphragm. When the pressure at the outlet drops, the force on the diaphragm is less. The spring pushes the valve further open. When the outlet pressure goes up, the spring is compressed to close the valve.

The regulator contains an internal sensing tube that points downstream, allowing the velocity of the gas to produce a small venturi effect at the end of the tube, to effectively reduce the pressure in the diaphragm chamber as the flow increases. That helps open the valve at higher flows and reduce the droop. These valves have a limited operating range, as far as pressure drop is concerned. The spring has to have a low coefficient so that the valve can stroke completely. It does not have the strength to open the valve if, during shutdown when the valve is closed, a high differential pressure between inlet and outlet develops. If a regulator locks up after a no-flow situation, then the differential pressure across the valve is too high. Solve the problem, temporarily, by shutting off the supply to the inlet and bleeding off the pressure upstream of the regulator. Valves that lock up regularly need a larger diaphragm or should be replaced with an internal lever actuated, or pilot operated, valve. Internal lever actuated valves use a mechanical linkage to convert a longer motion at the spring and diaphragm to a shorter motion at the valve disc to allow higher pressure drops across the valve. The typical house regulator contains an internal lever. When self-contained, diaphragm actuated regulators are used for natural gas, the venting of the spring chamber requires special attention. If the diaphragm leaks, the vent of the spring chamber must bleed off the gas, or the spring will open the valve fully to raise the outlet pressure to unsafe levels. The gas bleeding out of that vent must be conveyed to a safe location (outside the building) to prevent a flammable

Figure 11-16. Gas pressure regulator.

mixture from forming near the valve or displacing air to asphyxiate someone in the room.

Occasionally, the valve is fitted with an internal pressure relief valve, which will drain gas to that vent in the event that the outlet pressure gets too high (from thermal expansion or main valve leaking). The vent piping location and size is very important. It is also important that vent lines for regulators on other boilers, and especially piping from the intermediate vent valves, are not connected to the regulator vent lines. Back in the 1960s, when many men had long hair and there were few restrictions on smoking, a steamfitter was replacing a diaphragm on a regulator, while the adjacent boiler was running. He was smoking and suddenly found his long hair on fire because his cigarette had ignited the gas leaking back down the vent line. Gas, fed from the regulator on the adjacent boiler, was also leaking.

Temperature control valves can use a probe mounted on the valve and penetrating the vessel or piping, where it can sense the temperature used to control flow through the valve (like the one on the hot water heater). That controls the location of the valve, which can require extra piping or create other problems with installation or maintenance. Using a probe connected to a bellows by a capillary allows the control valve and temperature sensor to be located separately. The capillary is a very small diameter tubing permanently connected to the bellows and probe assemblies. These consist of closed systems which are made up for a particular temperature range and valve actuating power. The contents of the system can be a liquid or a gas. Liquid systems are somewhat restrictive because the liquid expands and contracts with changes in temperature. They develop high pressures quickly if the expansion is restricted. Gas filled systems change pressure with variations in temperature and many of them contain mostly liquid that evaporates when heated to produce the pressure in the bellows.

Any of these systems rely on minimal changes in temperature at the capillary and bellows which interferes with control based on the temperature at the probe. The capillaries are also very narrow to minimize the amount of fluid they contain and the effect of heating or cooling them. Those small capillaries are easily pinched to block the transmission of pressure from the probe to the bellows or nicked, cracked, or cut to drain the fluid and eliminate control. Simple diaphragm operated valves and internal lever actuated valves have their limits when it comes to handling large pressure drops, large flow rates, or the need for low droop. Pilot operated self-contained control valves do a great job of handling those conditions. A pilot operated valve is basically a duplex valve, where the pilot controls the pressure by controlling the main valve.

Figure 11-17. Piloted gas pressure regulator.

The pilot valve is like a regular pressure regulator. However, its output is fed to the diaphragm chamber of the main valve (Figure 11-17). When the pressure at the outlet drops, the pilot feeds fluid into the main valve diaphragm chamber to compress the main valve spring and open the valve further to match the flow out of the system and restore the outlet pressure. The pilot cannot close the main valve. It can only close off its flow. In order to close the main valve, the diaphragm has a line connecting it downstream, with an orifice in it so that the fluid in the diaphragm chamber bleeds out to allow the valve to close. During normal operation, the balance between pilot fluid flow and the flow through the orifice holds the valve in position. These valves have a droop, but it is so small that it is not noticed. They require a minimum difference in inlet and outlet pressures. They actually work a little better as the pressure difference increases because the main valve operation is determined by the difference between inlet and outlet pressure.

A self-contained main flow control valve can be piloted by a small float valve, temperature element, or other devices to achieve control by using the difference between inlet and outlet pressure of the controlled fluid. Some important considerations for this control are filtering or installation of a strainer on the small stream of fluid used for control. Thus, it does not plug up the pilot valve or the orifice that bleeds the fluid downstream. The flow for the pilot is so low that many piloted gas pressure regulators do not have a vent line. There is a small orifice in the spring chamber that can bleed off enough gas to allow the valve to work when the diaphragm is leaking slightly but restricts the flow to limit gas entering the adjacent atmosphere. It is called a restrictor. It is important to be sure not to block the restrictor with paint.

CONTROL LINEARITY

A wise operator will understand what is meant by linearity and how important it is after reading this section. Regrettably, there are many control technicians who do not understand it and throw on more and more control features to correct the problems created by a non-linear output. A control loop is linear when any change in the controller output produces a proportional change in the process fluid flow. Remember that all that can be controlled is flow. Expect a 10% change in a controller output signal to produce a 10% change in flow in the controlled system. It should be consistent throughout the control range. If there is 20% flow with a zero output of the controller (typical for a boiler with 5 to 1 turndown), then the flow should change 0.8% for every 1% of control signal change. If a graph was plotted to compare control signal with flow, it should produce something close to a straight line.

The system's response to errors produces an output to correct that error. If the output produces a different change in flow at various loads, then the controller will overshoot at some loads and lag at others. Remember the joking comment "It always works fine when the serviceman is here?" The primary reason why that is true so often is that the serviceman is always there when the loads are the same as they were when he tuned the controller. Insist that the serviceman show up when the system is acting up. With good documentation, it should be possible to relate load and control problems. If, however, the technician tunes the system for those loads, it probably will not work well at the loads where it was originally tuned. If the system is linear, those problems will not occur.

To understand why linearity is difficult to achieve, consider a typical forced draft (FD) fan actuator. The fan can be equipped with a discharge damper or variable inlet vanes. Measure and plot the relationship of damper rotation and air flow. It will look something like the curve in Figure 11-18, hysteresis being ignored.

The flow at zero damper rotation is typical of leakage through a control damper. At high loads, the air flow does not change significantly. At low loads and in the middle, it does. There is a big difference between that curve and the straight line (which represents a linear flow

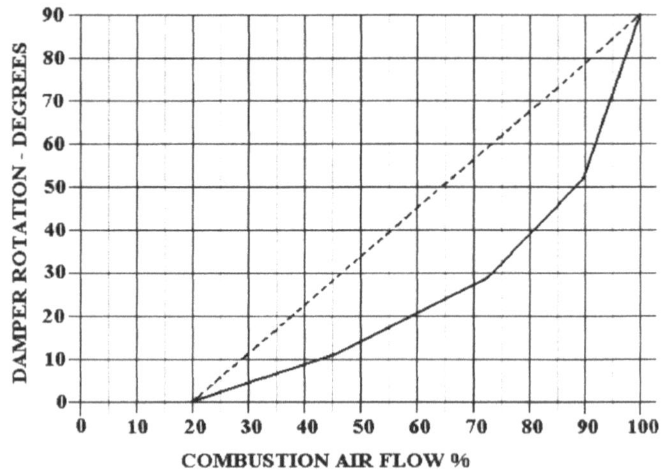

Figure 11-18. Non-linear air flow from damper.

characteristic). The modulating motor, or other actuator, that drives the damper cannot provide a linear response to controller output, unless something compensates for that variation in flow relative to damper position. Adjusting the mechanical linkage connecting the damper and its actuator can eliminate some of the non-linearity to produce a curve similar to the dotted line, which is the desired characteristic (linear). Pneumatic, hydraulic, and electric actuators with positioners can be fitted with cams to produce an excellent linear relationship between control signal (controller output) and flow.

One problem with microprocessor-based controllers is that technicians tend to avoid the rather laborious process of cutting a cam on a positioner. Instead, they simply program a function generator in the controller. That function generator produces an output that is a function of the controller output. The result is linear control. It works fine when the controls are in automatic. The function generator does not work when operating a boiler on manual. The linearity should be established at the final drive (damper actuator, fuel control valve, etc.) so that the response is consistent when operating on manual control. It is a lot nicer knowing that a 5% increase in firing rate will result when the fuel and air controller outputs are adjusted by 5% than tweaking each controller and watching the output changes to see what happens. Linearity adjustments are described in the chapter on maintenance.

STEAM PRESSURE MAINTENANCE

Steam pressure can be maintained by controlling the flow of steam from a higher pressure source into a system at a lower pressure or by the control of the operation of a boiler that generates steam. The steam pressure is used as the process variable to indicate how much steam is required and then control the pressure reducing valve or boilers accordingly. The control loop for a pressure reducing valve is identical to the control loop just described schematically for level maintenance. The difference is the use of pressure as the process variable instead of level. Controlling boilers to maintain steam pressure is accomplished in a variety of ways.

Regardless of the operating control method, all boilers have on/off controls. The boiler in a house and most hot water heaters use on/off as the only method of control. As sophistication and complexity of systems grow, on/off controlling seldom, if ever, happens but is always there. On/off control is normally achieved with one pressure sensing electrical switch that opens contacts to stop boiler operation and closes contacts to enable boiler operation, a pressure control switch. There is a lot more to that pressure switch than the light switch on the wall. This book is about operating wisely and the wise operator should know that the quality of the operation can be improved by adjusting that switch. It has two adjustments. One is the pressure at which it opens its contacts, as the pressure increases to stop operation. The other adjustment is the differential, which is the difference between the contact opening pressure and the pressure when the contacts will close again. The setting could also be the pressure it closes at and the differential determines when it opens. There are both types.

Set (stop) pressure minus differential equals start pressure. Many operators think they should set the differential as low as possible so that the pressure will not swing as much. The result is an increased cycling of the boiler and lower efficiency (See Cycling Efficiency, Chapter 3). To get the best performance out of the boiler, establish the widest possible operating range. For a simple on/off boiler operation, the operating range is the differential setting of that switch. The differential should be set as large as can be tolerated. It can be set larger in the summer than in the winter. The boiler will not start as often. It will run longer on each operation. That reduces the frequency of starts. Thus, there are less of them for higher overall operating efficiency and less wear and tear.

Enough pressure is needed so that all the heating equipment in the facility that the boiler is serving operates properly. Frequently, it is the pressure on the one that is the longest piping distance from the boiler. Sometimes, it is equipment at a shorter piping distance, but the pressure drop to that one is higher or it is not as oversized as everything else. The best way to determine it is to gradually drop the lowest pressure (increase switch differential) until someone complains. Then raise it a bit. If possible, wander around the facility to read pressure gauges and find the low one. Unless the equipment operates at full capacity summer and winter, and it has its own steam piping from the boiler plant, the same thing can be done in the summer. Summer loads are usually lower than winter loads. Piping pressure drops are less and steam demand on the equipment is less. The pressure can be reduced a little more at the boiler.

A typical heating plant, with a switch setting of 12 psig, can usually operate well in the summer with pressures lower than the maximum differential adjustment of the switch. Plants could operate as low as 2 psig. However, they had to install a special switch arrangement to get that spread. There is a caution here. Don't allow the starting pressure to go so low that the boiler will modulate above that setting. The main setting of the pressure control switch is the stop pressure. Many operators are instructed to set it as low as possible. That makes the steam and water temperature lower to cool flue gases more and to reduce stack losses. However, the small savings in lower stack temperature will be lost with more frequent cycling of the boiler. Set the switch as high as it can go and still prevent operation of the high steam pressure switch (See Burner Management, Chapter 11). There are two reasons. First, the larger the spread, the longer the run time for a boiler when it is cycling, and, second, the more room there is for continuous operation.

The most common modulating control is a simple electrical proportional control system. A Pressuretrol (trademarked name of Honeywell and still available) controller connects to the steam space in the boiler. It consists of a diaphragm or bellows connected to mechanical linkage that adjusts the position of a wiper on a coil of wire. The coil has a constant electrical voltage across it, supplied by a transformer. Voltage at any point on the coil is proportional to the position along the coil because the wire has a constant resistance. A matching coil is provided in the modulating motor that changes the firing rate of the boiler. The wipers are not exactly like an automobile windshield wiper but they operate similarly, swinging so that they touch the coil at any point from one end to the other. A schematic of the system is shown in Figure 11-19. When the steam pressure changes, it moves the wiper along the coil in the Pressuretrol. The voltage between ground and the wiper in the Pressuretrol will change, which produces an electrical current through the wiper to the balancing relay and the wiper on the coil in the modulating motor. The balancing relay is upset by the current when the two coils do not match so that it

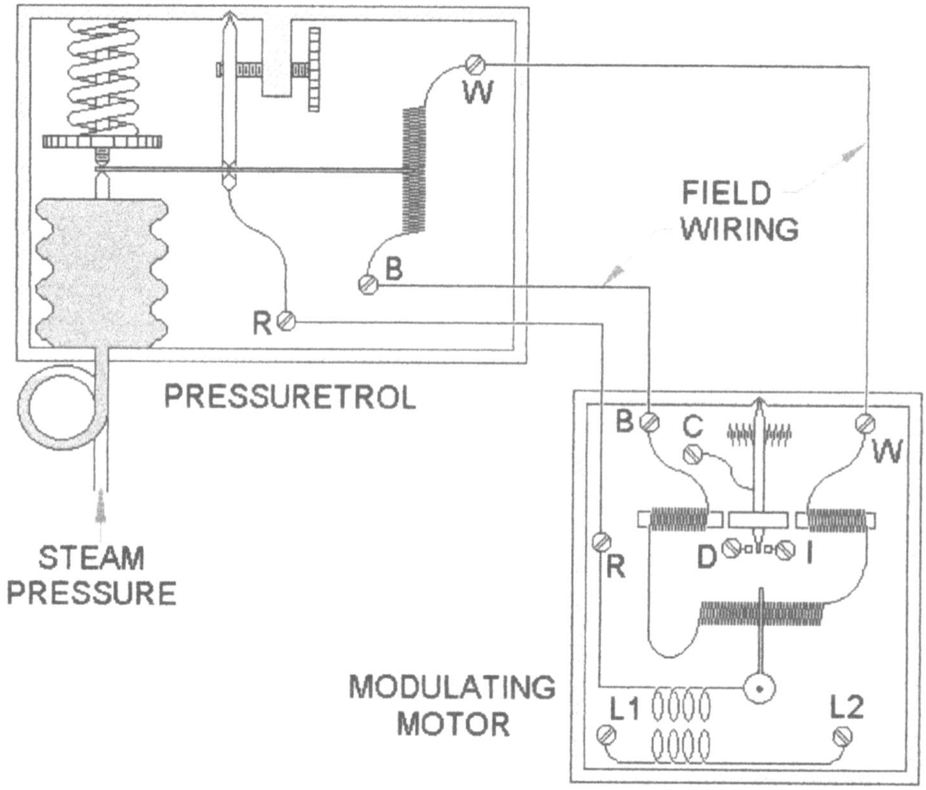

Figure 11-19. Pressuretrol—modulating motor schematic.

makes one of the electrical contacts that drive the modulating motor.

The direction of the motor is determined by the voltage imbalance. It runs in a direction that moves its wiper until it is at the same position as the wiper in the Pressuretrol. When the two wipers are in the same position, the voltage is the same and no current will flow through the balancing relay. It centers to stop the motor operation. The system rotates the modulating motor proportional to steam pressure. It is basically a proportional controller. The Pressuretrol has two settings, just like the pressure control switch. One establishes the center of the operating range (the steam pressure that will center the wiper in the middle of the coil). The other is the differential, which is the change in steam pressure necessary to drive the wiper from one end of the coil to the other. Tune it like any proportional controller, reducing the differential until the operation becomes erratic. Then increase it until it operates smoothly.

The setting of the center of the operating range of a Pressuretrol should always be such that the entire operating range is below the start pressure of the operating pressure switch. It should be below enough that the steam pressure after the boiler has started and purged at a load equal to low fire is slightly higher than the top of the operating range of the Pressuretrol. When a boiler is cycling on and off, the steam requirement is less than the boiler produces at low fire. At those loads, the boiler should not be modulating because that increases the input during the firing cycle to shorten it and increases the number of cycles (See Cycling Efficiency, Chapter 3). If there is only an operating pressure switch, its manufactured switch differential is the operating range. When there is a modulating control, the operating range is from the stop setting of the pressure control switch to the pressure that generates the maximum firing rate.

After tuning the modulating control to the minimum differential for smooth operation, adjust the differential of the pressure control switch and the setting of the Pressuretrol to establish an operating range as depicted in Figure 11-20. In many plants, it is possible to allow some of the differential of the Pressuretrol to fall below the minimum operating pressure. The boiler does not have to modulate to high fire to handle the maximum summer load.

Heating boiler plants may have more than one boiler and a need to control operations where two or more boilers are required to serve the needs of the facility. That requires a system that can stop and start each boiler, as needed, and may include modulating controls that

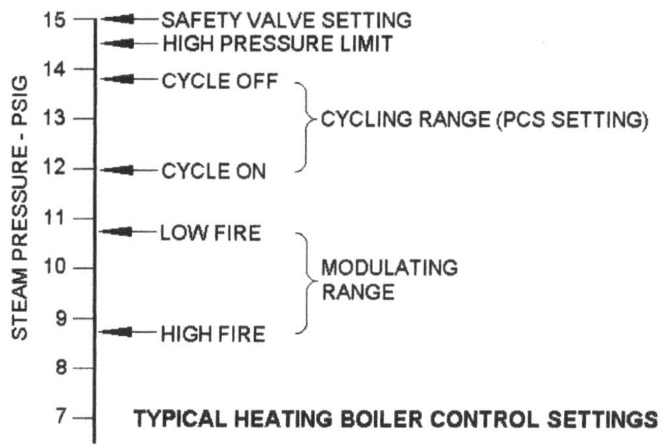

Figure 11-20. Range of control, modulating and on–off.

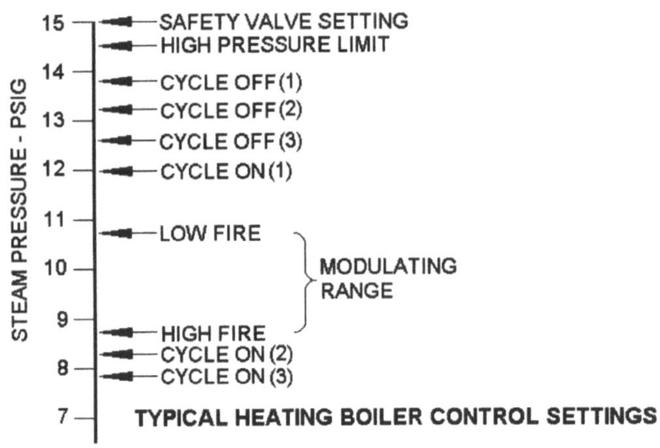

Figure 11-21. Settings for automatic three boiler control.

fire the boilers at different rates. Several methods, using complex arrangements of linkage, modulating motors that operate off another pressure control switch connected to the common steam header and powered by a shaft, which in turn controlled switches and multiple electric coils, like the one in a Pressuretrol, were provided. The principles described for a single burner control apply to them. They required a lot of maintenance and are, for the most part, replaced by modern digital controllers that simulate their functions. If one of those is encountered, read that instruction manual several times. Be sure to understand what it is supposed to do before making any adjustments. Then watch what happens when adjustments are made. They may not do what was intended. At the moment, there is no national standard applied to those devices. Their descriptions, labels, and settings may vary considerably.

With multiple boilers, there is no reason to change modulating control settings on the boilers, unless there is limited turndown (two to one or less) or the turndown was adjusted to the degree that it is very inefficient at low fire. That, by the way, is a normal thing to do. Boiler operators do not normally like to see a boiler shutting down regularly. Creating load and other unwise operations are not the proper way to deal with it though. Multiple boiler control is achieved by setting the operating pressure switches within the range that would be used for one boiler. It is easier to talk in terms of start and stop pressures, where the stop pressure is the setting of the boiler pressure control switch and the start pressure is the switch setting less the differential. Figure 11-21 shows the start and stop settings for three boilers to achieve automatic control. The difference between stop settings has to be enough that the residual energy in a boiler that was just shut down will not generate so much steam that the pressure rise associated with that steam generation trips another boiler.

The difference in start (cycle on) settings has to be sufficient to allow for the pressure drop that will occur while the boiler just started is purging and lighting off. Figure 11-21 also shows how to change the modulating range. Note that the setup requires a considerable swing in pressure to satisfy all the criteria. To change the order in which the boilers operate, it is necessary to change all the switch and Pressuretrol settings, which is a lot of work. It is much better to have one of those digital controllers to do it all.

Many plants have lead/lag controls as part of the package for controlling their boilers. The adjustment of settings in Figure 11-21 provides a form of lead/lag control since it varies the number of operating boilers, allowing Boiler 1 to carry the load until it cannot and then bringing on Boiler 2 and finally Boiler 3 to handle the maximum loads. All the boilers modulate together. Lead/lag controllers were designed to accomplish it in a slightly different manner. They would run the first boiler up to high fire and leave it there after starting the second boiler, which would modulate to carry the load until the load exceeds the capacity of two boilers. Then the third would start. Those controllers resolved a problem with the scheme of Figure 11-21, which provided different responses to load changes depending on how many boilers were on line. The lead/lag controller always had only one boiler responding to load changes. The others were either on at high fire or off. One controller was actually capable of controlling as many as 10 boilers.

High pressure boiler plants can operate with the same simple modulating control just described as long as there is no problem with the swinging pressure. Once a plant is large enough that someone installs a steam flow recorder, that will change. Normally, steam flow recorders require a constant pressure for accuracy (See Boiler Plant Instrumentation, Chapter 11).

When there is a need to keep producing steam at the same pressure, an integral controller is required. In the early days of controls, one of those controllers was an expensive item. One device was used to control all the boilers in a plant and called it the plant master pressure controller. It sensed the pressure in the common steam header so that it would not be affected by shutting a boiler down. It was close enough to the steam flow elements that it maintained a reasonably constant pressure at them for accuracy in recording. Those rules still apply today. The lower costs for instruments and controls have made it possible to have a controller at every boiler, if desired. A plant master pressure controller produces an output signal that is used by each set of boiler controls to adjust their firing rate so that they produce steam to satisfy the requirements of the facility, while maintaining the steam pressure in the header at the set point. There are several types of boiler firing rate control systems, but they all change the flow of steam out of their boiler proportional to the change in the plant master signal.

Always tune the plant master with the normal number of boilers on automatic. Most plants with multiple boilers in service and a plant master run one boiler on automatic and the rest on manual so that the plant master will operate properly regardless of the number of boilers in operation. Note that this works only when all the boilers are of the same size. If that is the case, and then two boilers are put on automatic, the pressure will jump around a bit when there is a change in the steam load. Under those conditions, the two boilers change their steam output. However, the master controller expected the change in output to alter the steam flow the same as the output of one boiler. With two boilers in automatic, the response to a controller action is doubled. If two boilers are run on automatic most of the time, and one occasionally, it is better to tune the master for two boiler operation and live with the slower response when one boiler is on. Plants with boilers of different sizes will also see a different response out of the plant master.

Prior to the days of digital controls, it was not practical to deal with that situation in the controls. The plant operator had to adjust the tuning of the master controller if the operation was erratic. Some plants added derivative control to help account for it. Complicated logic systems were created that adjusted the gain of the master controller according to the number of boilers on line in automatic. Modern digital controllers can use digital (on or off) inputs to determine which boilers are in automatic and calculate what the response will be to a controller action. A good digital control system should not be affected by the number of boilers in automatic or what size they are. In fact, most control companies, offering digital equipment, would prefer that everything be on automatic.

FLUID TEMPERATURE MAINTENANCE

Controls for heating fluids require special consideration that is not necessary for pressure controls. The largest single problem is making sure that the device that senses the temperature being used as a process variable is representative of the heat flow that is really being controlled. Always be aware that the sensor may be shielded by such things as air trapped above the fluid, or scale, or other material coating the sensor so that it cannot detect the temperature properly. It may be necessary to locate the sensor where it cannot detect changes in temperature when the flow is interrupted. Additional sensors and controls (like a flow switch) may be necessary to prevent hazardous operation under those circumstances. This chapter is focused on boiler plant controls, particularly hot water boilers for hydronic heating and similar applications. The control of boilers for service water heating (domestic hot water) is described in the chapter on water heating.

Most hot water boilers are supplied with a proportional control similar to that described for steam boilers. The only difference is that the temperature control switch and modulating controller sense boiler water temperature, not pressure. In many hydronic systems, the quantity of water in the boiler is large enough that it can operate much like a steam boiler, using temperature control instead of pressure. Simply convert the pressure values in the previous figures to the corresponding steam saturation temperature from the steam table in Appendix A and it is done.

The decisions for setting the start, stop, and modulating range for fluid temperature control are based on several considerations. The fluid has to be hot enough when it reaches the using equipment to transfer all the heat required. The fluid cannot be so cold that acids in the boiler flue gas condense on the boiler surfaces and corrode them. A normal low limit for natural gas is 170°F. Fuel oils used to cause corrosion at all temperatures below the maximum operating temperature for heating boilers (250°F). Thus, operation at 240°F was recommended. If oil is fired most of the time, ask the supplier for the normal acid dew point temperature of the oil and try to keep the water temperature above that. The lower the start temperature, the less the loss will be due to cycling. Review the section on steam pressure maintenance to get an understanding of how to set proportional fluid temperature controls. Also review the discussion on thermal shock.

For multiple boiler systems and large facilities, the setting of hot water controllers is a little different. The pressure maintained in a steam boiler pushes the heat out to the facility. In fluid systems, the heat is transferred by other means. There are basically two methods for transferring the heat and both rely on moving the heated fluid out of the boiler to the heat using equipment and returning the fluid, after it has given up some of that heat, back to the boiler to pick up more heat.

The simplest method is gravity. It relies on the difference in density of the fluid as it is heated. Most fluids expand when heated. They take up more space. The density of the fluid (number of pounds per cubic foot) decreases. The hotter fluid tends to float up in any pool of colder fluid, just like a block of wood floats on the top of water because it is lighter than the water. A boiler system with a proper piping arrangement can use this to force the heated fluid in a boiler to flow up through the pipes to radiators on the upper floors because the fluid cooled in the radiators fills the return lines to the bottom of the boiler. The colder water is heavier than the lighter, hotter water, producing a thermal siphon. That is a form of natural circulation.

Only simple small systems use natural circulation. Even most small residential systems use an electric pump to circulate the water. The pump can produce far more force to circulate the water than the thermal siphon effect. That allows pipes to be smaller and the system to cost less to install. Large hydronic heating systems for schools, office buildings, etc., simply cannot justify a system without pumps. They all include pumps to move the fluid around between boiler and heat user.

Unlike steam boilers, where load is balanced by the flow of steam from its source of generation to the load, hot water boilers cannot function with a plant master that controls the firing rate of all the boilers. Some systems have a master temperature controller. However, it does not control the firing rate of each boiler. There have been attempts to produce common control by operating the boilers in series (water flows through one boiler, then the next, and so on), although they do not always work well.

When fluid heating systems become so large that the volume of fluid in the boiler is a small part of the entire system, control of the water temperature becomes difficult. Another factor is the volume of water in the boiler. Fire tube boilers contain a large volume of water and can have long residence times (how long the water stays in the boiler). Water tube boilers can hold so little water that it is replaced every few seconds. Boilers like that (with short residence time) can have a problem because the temperature of the water at the sensor is not the same as the average temperature of the water in the boiler. Temperature maintenance of those units can get erratic. Another control method is required. The controls are very typical of high temperature hot water (HTHW) boilers, which operate at temperatures over 250°F and pressures over 160 psig.

For boilers heating water, the method is easy to understand. British thermal units (Btu's) are added to the water. The energy required is equal to the pounds of water going through the boiler and the temperature difference. The actual heating load is determined by multiplying the pounds of water flowing through the boiler by the difference in inlet and outlet temperatures. For other fluids, multiply by the average specific heat of the liquid. Control logic, which performs that calculation, provides a very responsive control. Any change of inlet temperature, or fluid flow rate, produces a change in the control signal, increasing or decreasing the firing rate of the boiler to compensate. Since multipliers were a problem in early controls, most plants relied on a constant fluid flow. Then only the temperature difference was needed to develop the control logic.

These systems cannot operate on that logic alone. There is no way to correct for changes in boiler efficiency or small errors in flow and temperature measurement that would produce an imbalance between the actual load and the firing rate. A temperature controller is used in these systems to provide a means of correcting for those differences. The typical HTHW boiler load control system is shown in the schematic in Figure 11-22. Refer to the following discussion on two-element boiler water level control, for an explanation of this particular type of control loop.

FLUID LEVEL MAINTENANCE

There are several locations where water level must be maintained in a boiler plant. The most important is the level in the boiler itself. The method of control varies significantly, depending on the size and complexity of

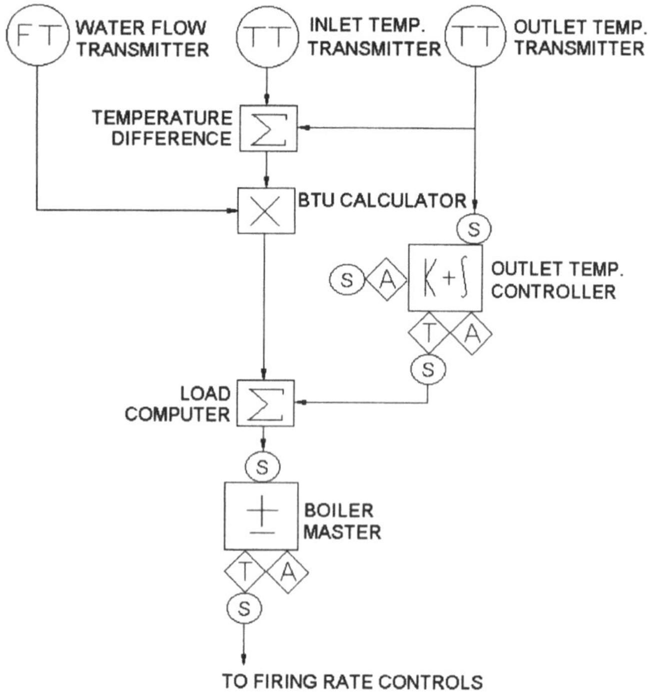

Figure 11-22. HTHW boiler control.

the plant. The simplest is a float controlled valve. The most complex (and expensive) is a three-element, boiler drum level control loop. Each has its place, advantages, and problems.

A float controller for level control is common in boiler water feed tanks, condensate tanks, makeup tanks, and other source tanks of water for the boiler plant. They are found in boiler feed service only on residential and small commercial boilers. They are not found on large, or high pressure, boilers. The float would have to be very large to produce the force needed to operate the control valve while operating on a very small change in level. As they get larger, they have to get stronger to prevent crushing them. They get heavier and the float chamber has to be thicker (See Strength of Materials, Chapter 9) and they become uneconomical. They do work fine for open tanks at small flow rates. One place where float controls have problems is in brine tanks used with water softeners. The salt tends to crystalize on the float and surrounding materials, usually a still pipe (a pipe placed around a float to prevent swinging operation due to wave action). It can be trapped in the brine crystals and fail to operate.

A float that only has to open and close an electrical contact can be quite small by comparison to something that has to open and close a valve. There are many systems controlled by float operated switches. The switch can energize a solenoid valve to open it and admit fluid to the tank or boiler. All the energy required to operate the valve is provided by the electricity (and, in many cases, the fluid itself; see pilot operated valves in the general discussion on controls). Thus, a small float can control any volume of fluid at any difference in pressure. The float still requires a change in level to function and only provides on/off control of the fluid flow. That is satisfactory in many situations. The switch can also be set to power a valve as the level rises to provide a system that allows fluid to flow on controller failure.

The typical heating steam boiler, and small commercial and industrial boilers, use float controls that start and stop the boiler feed pumps to control feed water flow for maintenance of the water level, instead of controlling a valve. These systems solve some of the problems with valve control by preventing operation of the feed pump when the control valve shuts off, a situation that would overheat the pump. It also eliminates feed water control valves as a maintenance item. Each boiler has to have its own pump for this control method to work. Operation of standby pumps is complicated because the electrical control has to be switched along with pump isolation valves. It is a simple and inexpensive method for level control and works well in many applications. However, it cannot be used with economizers. The higher electrical demand (pump and motor are normally sized at twice the boiler capacity) can create higher electrical power costs.

If a boiler is operated with very little reserve capacity, like most water tube boilers, and there is an economizer, or the swings in load associated with feed pump on/off control cannot be tolerated, a variable feed level control is required, one that modulates the feed water flow control valve to maintain the level. If a boiler has little reserve in it, the cold feed water rushing in at twice the boiler capacity can, for a short period of time, consume so much of the heat to simply heat up the feed water that some of the steam in the boiler is condensed. Then the water level drops suddenly every time the pump runs (see the following section on shrink and swell). Sometimes, it is enough to trip the low water cutoff. Considerable differences in boiler level are required for them to operate without false trips. Many of the new flexitube boilers are equipped with two level controls: one set for controlling level when the boiler is off and another for when the boiler is firing, which is set at a higher level. If the boiler has an economizer, the continuous flow of water is required to prevent generating steam in the economizer. The feed pump on/off operation produces a significant change in the output of a boiler, especially at low loads. That can cause bumps in the whole steam system. Anything larger than a small commercial boiler operation should have a better method of water level control.

Figure 11-23. Thermo-mechanical boiler level control.

There are two unique self-contained control systems to be aware of. They were used only on boilers, and can still be found in many locations. One is a thermo-mechanical system and the other is thermo-hydraulic. The key to these controls is that prefix, "thermal," which indicates the use of temperature to detect level and power the control valve. The thermo-mechanical systems (Figure 11-23) are manufactured by Copes-Vulcan. The thermo-hydraulic systems (Figure 11-24) are manufactured by Bailey (now ABB) and Swartout (now part of Ruskin) among others. Both systems use the difference in heat transfer rates between steam condensing and simple water heating.

They incorporate a tube connecting ends to the water space and steam space in the boiler. The water level in the boiler is repeated in the tubing. The tubing above the water level is exposed to steam and the tubing below the water level is exposed to boiler water (actually, it is mostly condensate from the steam condensing in the tube). Since steam condensing transfers heat much faster than hot water, the portion of the tube that is exposed to steam is hotter. Both systems arrange connecting piping so that the tube is at an angle. The slope of the thermo-mechanical tube is much shallower than the thermo-hydraulic to provide additional tube length and (as a result) heat exchange surface for better control. Since the heat transfer is much higher for steam condensing, the lower the level of the boiler water, the hotter the tube. The heat transfer from the finned water jacket of the thermo-hydraulic controller, or from the tube of the thermo-mechanical controller, to the surrounding air is increased slightly because of the hotter water jacket or tube. The expansion of the tube, or the water in the jacket, is converted to movement of the valve, opening it as the tube or jacket gets hotter.

The thermo-mechanical system uses a short pivot at the end of the tube, which consists of a lever point at the end of the tube and a pivot attached to the two steel channels on either side of the tube. The lever connected to the control valve moves as much as six inches from its end with a very small change in the length of the sensing tube. As the tube expands, the lever is pulled down by the weight to open the valve. The expanding water in the jacket of the hydraulic version acts on a diaphragm (Swartout) or bellows (Bailey) on the control valve, opening it. As the water rises in the tube, as a result of adding water, the tube or water jacket shrinks. The shrinking tube pulls the valve closed on the mechanical system. A spring pushing against the bellows, or diaphragm, of the hydraulic system closes the valve as the water in the jacket shrinks. Both systems will stabilize to maintain a constant water level. They do not respond rapidly to level changes and always open the valve fully as the boiler cools down. It is necessary to manually

Figure 11-24. Thermo-hydraulic boiler level control.

close off the water and manually control level on boilers equipped with these systems, until the boiler is at operating pressure.

The Copes-Vulcan system has another system with a feature to aid in response to changes in load. The control valve is fitted with a diaphragm connected to the feed water valve with sensing lines to the steam header at either side of an orifice. Increasing steam flow produces a higher pressure drop across the orifice, which produces a higher differential pressure on the valve diaphragm to force it further open. The lever of the thermo-mechanical tube is fitted with a chain extension that runs over a sprocket on the valve to the weight. The sprocket is connected to the valve stem, like a rack and pinion, to aid or restrict the diaphragm action for final water level control. This provides something comparable to two-element control. Modern controls and instruments have surpassed the use of those thermo-hydraulic or thermo-mechanical control valves. They are not the easiest things to work with. They do not control the level when the boiler is cold. They are relatively expensive. Now that LTs and controllers are so inexpensive that the cost of those older designs cannot justify their existence. They were fantastic controls years ago, but new controls can do so much more.

Shrink and Swell

A simple single loop control system, like the one covered in the beginning of this section, will satisfy the requirements of most heating boilers and commercial and industrial loads with fairly constant steam demands. If, however, the steam requirements change significantly, the control will actually operate in the wrong direction due to shrink and swell. "Shrink" and "swell" are terms used to describe what happens when the boiler load changes and feed water addition changes.

When the boiler is generating steam, some of the volume below the water surface has to consist of steam bubbles. The amount of bubbles depends on the load, the volume of the boiler below the water line in proportion to the capacity, the surface area of the water line, and the operating pressure. Many boilers, mostly fire tube boilers, contain so much water, in proportion to the steaming capacity, that the percentage of volume occupied by steam is small and the shrink and swell are not noticeable.

On the other hand, a low pressure, water tube boiler is most likely to show the most dramatic change because the steam density is low (volume of steam per pound is high). When a sudden increase in load occurs, the steam pressure in the boiler drops. The steam bubbles in the boiler water expand. Also, a small percentage of the water flashes to steam, adding to the number of bubbles. The result is an increase in the water level, which is called "swell" because the water level increases with no water being added to the boiler. A single element level control will react to the swell by closing down on the feed water valve, the opposite of what is needed. More water is required for the larger load. Closing of the feed water valve reduces the heat requirement for raising the temperature of the feed water. More heat is used to make steam (and more bubbles). That simply makes the water swell more.

When the opposite occurs, and the load decreases suddenly, pressure increases, and the bubbles are compressed. The water in the boiler is not up to the new saturated condition. Some of the steam condenses to heat it up. The water in the boiler shrinks and the level drops. A single element control senses the drop in water level and opens the control valve to increase the flow of feed water. The additional feed water requires heat to warm it to saturation condition. Some more of the steam is condensed to collapse more bubbles. Increasing the water flow is not required because the steam flow decreased. However, the shrinking water level makes the control open the feed water valve more.

Two-Element Control

To reduce the impact of shrink and swell, a water system that does not enhance the effect of it is required. Two- and three-element systems actually counter some of the effect by adding water when the level is swelling up to quench bubbles, which reduces the swell. Conversely, they reduce the addition of colder feed water when the level is shrinking.

Consider single element control operation. Single element feed water controls have a single process variable for control, i.e., water level. Two-element controls use another process variable (that is not maintained) and that is steam flow. Since the steam flow is not controlled as part of the feed water system, it is usually treated as a remote signal. The third variable for a three-element control is feed water flow. The two- and three-element systems act to maintain the balance of steam and feed water flow with adjustments for level. Both two- and three-element systems actually control the flow of water to match the flow of steam. It is given that every pound of steam that leaves a boiler must be replaced by a pound of feed water. That provides a logical way to do it.

These systems require a control element called a signal summer, which combines two or more control signals. The term "summer" is used instead of "adder" because a summer can subtract signals as well as add them. When mathematicians and control engineers use the word "sum," they mean to add up all the values and some of them can be negative. The ratio totalizer described earlier can be used as a signal summer. One input signal can be applied to the bellows opposite the output (port A in Figure 11-5). The output equals that signal, plus another signal be applied to port C of the totalizer for adding or port B for subtracting. A gain on the A and B values could be introduced by adjusting the pivot. A spring could also be added to the assembly. That would introduce a fixed bias (spring force) at either ends of the ratio totalizer. The mathematical equivalent of the summer output would be input C plus input A, minus input B, plus or minus a bias provided by a spring at their end, plus or minus the bias provided by a spring at the output end. The output equals (IA – IB ± KB) × G + IC ± KC), where the suffix identifies the port indicated in Figure 11-5, the letter "I" refers to input, "K" represents a spring attached to the pivot arm at that port, and "G" is the gain.

That is the basic concept of a summer. Most microprocessor-based controllers include the summer function inside the controller to eliminate the need for additional hardware. That way a two-element controller can be made out of a single element one by simply wiring the steam flow signal to the drum level controller. Actually, in many systems and any future system, it is simply a matter of telling the controller to get the steam flow signal. All the controllers have access to all the signals in a system. The two- and three-element systems control the feed water valve in proportion to steam flow, with an adjustment for drum level. A two-element feed water control system is shown in Figure 11-25. Two-element control is very common today because any boiler that needs the control is large enough to justify steam flow metering for monitoring the boiler demand and performance. Since the steam flow meter is there, it is simply a matter of adding, at most a little wiring, and, normally, just a few software instructions (for microprocessor-based controls), to make a two-element system out of a single element system. If the boiler has pneumatic controls, another device (summer) is required to create a two-element control and another hand automatic station may be necessary.

As steam flow increases, the output to the feed water valve increases. Provided the valve is selected, or its positioner is set to provide a linear output, the valve position for each value of steam flow will produce a feed water flow that matches the steam flow. To check if a two-element system is set up properly, note the output

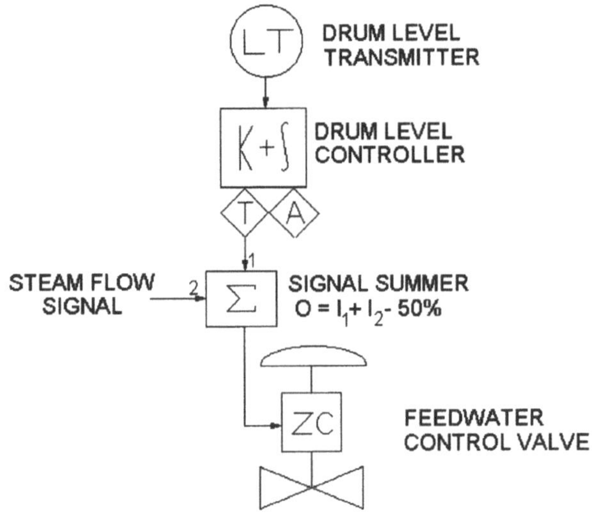

Figure 11-25. Two-element level control schematic.

of the level controller at different boiler loads when the level and steam flow are relatively stable. The output of the level controller should not change and should be about 50%. The setup of a boiler feed water control system usually ignores the blowdown (which requires a little more feed water). The valve position is set to handle the correct amount of feed water for the normal steam flow and blowdown. The technician setting up the system looks at the schematic and realizes that the level controller can add to the steam flow signal to increase flow and raise the water level. However, something has to be done to lower the level if it gets too high. A 50% bias opposed to the steam signal, and balanced by the 50% output of the level controller, has a net zero effect on the steam flow to valve position relationship but allows the level controller to modify the relationship up and down by 50%. Without that bias, the level controller output is around zero and it cannot do anything to lower the water level if there is a slight upset in operation (like lower steam pressure that will allow more water to flow into the boiler) that results in a high water level.

Three-element control accomplishes the same thing as the two-element control. It measures the feed water flow as a process variable. The feed water flow controller adjusts the feed water control valve until the feed water flow matches the steam flow, plus or minus any adjustment from the water level controller. Three-element systems are necessary whenever the feed water pressures are frequently changed or affected by heavy load swings such that the linearity of the control valve cannot be maintained. Of course, there is going to be a problem with a two-element system that is set up right if the steam flow signal is lost. The controller will not be able to open the valve more than 50%. Modern digital controls sort of solve that problem because their instructions include switching to single element control that can fully stroke the valve when the steam flow signal is lost or low. Failing that, and circumstances require the operation of a boiler without a steam flow signal, simply find the summer and adjust the bias from a –50% to zero and the valve will respond directly to the level controller signal.

BURNER MANAGEMENT

For years, the BMS was called the flame safeguard system, as the principal purpose of the system was to make sure that it was safe to light a fire and to continue the operation of the boiler. The BMS does two things: it "supervises" the operation to ensure that all operating parameters are within appropriate "limits" and it supervises, or performs and supervises, the procedures required to place a boiler and its burner (or burners) in operation. It can also provide a controlled shutdown of the boiler and its burner(s). There are several control systems manufacturers. Many are now part of larger companies. Fireye is now part of Carrier. Bailey is part of ABB. Their devices are competitively priced and do an excellent job of burner management for small and medium sized single burner boilers. The controller becomes a system when it is connected to a burner's controls, level, pressure, and temperature limit switches, and a flame scanner or flame rod.

Accurate detection of a fire is the most important function of the BMS. It has to ensure that there is no fire when it should not be there as well as ensure a fire is present when it is supposed to be, and respond accordingly. Detectors are flame rods, infrared sensing, ultraviolet sensing, or more modern multiple frequency sensing units. The detectors, with the exception of flame rods, are all called scanners.

The basis of flame rod operation is that a fuel and air mixture does not conduct electricity but a flame does. The rod has to be positioned where it is in the flame, a pilot flame on large burners. The flame also has to have a grounding electrode that touches the flame so that electricity can be conducted from one electrode to the other. The grounding electrode is normally connected electrically to the metal parts of the burner. It can also be another flame rod positioned at another point in the flame. The flame rod itself has to be constructed of a material that will not melt or oxidize in the fire. The insulation separating it from the metal parts of the burner has to be capable of withstanding the high temperatures. Normally, the rod is made of high chrome steel and the insulators are ceramic material. The portion of the BMS that identifies the presence of a fire has to produce a voltage adequate to push a detectable current through the flame and sense that current. It has to distinguish between no flame, false signals (like the rod shorting out on some metal in the burner), and a flame.

The flame scanner is a sensor that detects a fire in the burner by absorbing some of the light energy emitted by the fire. Scanners may detect infrared light or ultraviolet light, any frequency of light in between, or a combination. Some are called "self-checking." That label can be inappropriate. A scanner is self-checking when it contains a device that blocks the light from the sensor at intervals and the detector circuit has to sense a no-flame signal during that interval. Some scanners simply block the light at regular intervals so that the detector circuit can determine a flame is present because the signal from the scanner is constantly swinging to produce an alternating current. A constant signal from the scanner indicates no flame or scanner failure, including failure of the self-checker. Self-checking of scanners should not be confused with self-checking of the flame detector circuitry, which is different. If the scanner does not block the sensor's view of the fire, it is not a self-checking scanner. A self-checking circuit simply confirms a flame is not detected when it should not be. Of course, some systems were installed and connected in such a manner that the self-checking functions of the circuit were not allowed to work. If, on a burner that is turned off, the scanner does not detect a flame on a candle or cigarette lighter that is held in front of it (remove it from its mount), and it does not alarm and lock out as a result, that is one of those. Since many of them were only found after a boiler explosion, it would be wise to perform that test. If the system does not lock out, it is not safe.

The use of visible light frequencies coincided with the use of solid state electronics. Silicon carbide photodiodes convert light from the flame to an electrical signal that can be amplified, filtered, and analyzed to verify that flame is present. Visible light in the range of 400–750 nm is highly effective, as it is able to match the observations of trained operators. Improvements in sensitivity to coal flames extended the operating range of coal burners without the use of support fuel. Equipment reliability was also improved with solid state electronics. This provided the means to perform fault detection electronically rather than mechanically.

Regardless of the light spectrum being used, the speed and precision of digital processing brought improvements to plant safety and operation. Precise flame measurement in real time avoids delays in detecting unsafe firing conditions. Using microprocessors, more detailed information from the flame can be captured and analyzed. In addition to verifying the presence of a flame, analyzing narrow bands of light frequencies can allow users to determine the color of the flame and the shape of the flame curve. Algorithms for determining flame temperature resulting from localized fuel/air ratios have been developed.

The typical burner management system provides for automatic operation of the burner, performing all the steps described in the section on boiler startup. When a pressure control switch closes, the BMS should first determine that there is no flame in the burner. Provided operating limits like low water cutoff, high steam pressure, and low fuel pressure are all satisfied (contacts closed in a series circuit), it closes an output contact to start the burner fan or fans. When air flow is proven by closing contacts on an air flow switch, the firing rate control system is instructed to increase damper position to high fire. Some systems may include provisions to start an oil pump and prove it operating as well. The open damper, or a purge air flow switch, senses purge air flow to close a contact for another input to the BMS. The system then waits for the prescribed period of time for a purge.

Some are set with fixed timing. Modern units have provisions for setting the purge time to comply with the code requirements. The controller supervises the purge by requiring the damper open, or purge air flow switch, contacts to remain closed during the purge period. Some will simply restart the purge timing if the input is interrupted. Others will stop the startup. Once the purge timing is complete, the contacts for high fire are opened and another set close to instruct the firing rate controls to go to a low fire position for ignition. When the firing rate controls are at low fire, they close a low fire position (or ignition permissive where low fire is lower) switch contact to provide an input to the BMS.

During all of that portion of the startup sequence, the scanner should be looking for a flame. If it sees one, the system should lock out. The reasons can be anything from defective scanners to oil dripping out of a gun and lighting to glowing hot refractory from the previous firing. On more than one occasion, an operator figured out that the lockout could be prevented by pulling out the scanner and covering it with a glove during the purge and low fire positioning. Needless to say, there was not any accurate flame sensing and, eventually, an explosion occurred. If that scanner thinks it sees a flame where there is none, it is not safe to operate that boiler.

With low fire position proven, the controller closes a contact to energize an electric spark in the ignitor and another contact to energize the ignitor gas shut off valve(s) (if the burner is equipped with an ignitor). The controller then waits for 10 seconds to see if the valves open to admit gas that is lit by the spark to create an ignitor flame. If the flame is not detected in that time, it stops operation and energizes an alarm horn. If the flame is detected, it closes another contact to energize the main fuel valves. Some also de-energize the electric spark. At a prescribed time after the main fuel valves are open, it de-energizes the ignitor gas valves. If a flame remains detected, the controller opens the low fire contact and closes an automatic contact to permit automatic operation of the modulating controls to control the firing rate.

One of the most important elements of the burner management control is the checking provisions. The burner management controller had to include at least two relays: a power relay and a flame relay. The power relay could only be energized via a normally closed contact on the flame relay (proving the flame relay was de-energized). It closed a normally open contact to bypass the flame relay contact to continue operation. Once the power relay was up, the flame relay closed its contacts to power the main fuel valves and stop the timer motor. Any indication of a flame when there is not supposed to be one can mean the burner will continue to operate when there is no flame present, generating an explosive atmosphere in the boiler. Any component failure in the BMS should also act to safely shut down the boiler or, if its failure does not present an immediate danger, prevent a subsequent startup of the burner.

To meet international codes, the functions required to protect the boiler are provided in a standalone system that partitions protection functions from control functions. This protective portion of the overall boiler control system is called the boiler protection system (BPS). It is physically and functionally separated from the remainder of the boiler control systems. The system is based on usage of specific safety controllers holding the critical furnace functions, such as flue gas path protection, furnace purge, ignition time monitoring, ignition trials release, evaporator protection, oil burner trips, coal mill trips, and associated feeder trips. This separate control system is implemented in this manner due to the extensive control system redundancy requirements needed to meet international codes.

The key actions for a wise operator when it comes to BMSs are 1) to know what they are supposed to do, 2) to shut the boiler down when they do not do it, 3) to report inconsistencies in operation and regular interruptions in operation, and 4) to not change switch position settings without permission. That last one is a real key. Many states have adopted the ASME and National Fire Prevention Association (NFPA) standards that relate to burner management. Both standards are very exacting about the requirements for changes in BMSs.

No discussion of a BMS should be left without mentioning the important concept of fail-safe design. Every element of the system should be arranged such that its failure will not compromise the safety of the boiler operation. Each wire, relay, pressure switch, etc., should be evaluated for failure modes and analyzed for what will happen if the device fails. Only when every evaluation indicates the result will be safe should the system be considered fail-safe. Fail-safe concepts should be applied to all controls and applied in a sensible manner. Too many designers view fail-safe solutions as only resulting in a complete burner shutdown. That is not necessarily the safest thing to do. While that burner is operating, most of the furnace and boiler is full of inert gas (and less likely to explode). There are many other examples where a shutdown is not necessarily the safest solution to a failure.

There are always arguments as to what is safe as well. Is it better to have a feed water valve fail open so that the boiler will not run dry? Most of the time, the valve fails closed because there is no safety to prevent water flying down the steam lines and hammering them apart. However, it should be expected that the low water cutoff should safely shutdown the boiler. If a component of a control system is being replaced, and its operation is not exactly the same as the piece being replaced, consider what will happen if it fails. Much thought has gone into deciding if a particular component will fail in the safest manner. Replacing it with one of another action could reduce the safety and/or reliability of the plant.

FIRING RATE CONTROL—GENERAL

Firing rate controls regulate the flow of fuel and combustion air to the burner to produce a flame and heat input that satisfies the demand for heat at the boiler outlet. They are also called combustion controls. These are independent of the steam pressure controls on any system, except a simple jackshaft system. Typically, the terms "combustion controls" and "firing rate control" are not used with a jackshaft system. The heat input is primarily a function of the amount of fuel flowing to the fire. Control of air is also required to produce the heat input. In the chapter on fuels, the importance of maintaining an optimum air to fuel ratio was discussed. Part of the job of firing rate controls is to maintain an air to fuel ratio that is adequate for safe and efficient operation of the burner and the boiler. There are different control schemes for controlling the fuel and air in order to maintain the air to fuel ratio. Their ability to do the job varies with system cost and complexity.

The choice of control system for boilers will depend primarily on the size of the boilers. Size of the boilers implies a certain annual fuel consumption. The increasing cost of more refined controls has to be weighed against the savings that can be produced by improving the controls for better control of the air to fuel ratio. There is also the question of maintaining a certain steam or vapor pressure or a boiler outlet temperature that may, or may not, be critical to the facility served by the boiler plant. If the pressure or temperature is critical, the controls will be more refined. Finally, as the size increases, the plant will experience more and more emission limits that will need additional control features.

One plant had no pressure controls. They used the firing rate to swing the pressure around to deliver steam for heating. Swinging pressures will vary blowdown rates, increase the opportunities for carryover, and, if not caught at the right time, result in boiler shutdown or lifting of safety valves, which does not reflect well on the performance of the operators. The changes in temperature are adequate to define the operation as cycling. The standard boiler is constructed for a life of 7000 cycles. Swinging operation shortens boiler life. The boilers in question were equipped with firing rate controls. They were either inoperable due to no maintenance or not used. If the temperature swings, thermal stress problems can arise.

A low pressure steam plant can swing from a low of 8 psig to a high of 12 psig with a temperature swing of 9 degrees. Higher pressure plants have thicker boiler parts and swings of more than 4 or 5 degrees can cause problems with thermal stress in them. As a result, normal pressure swings should be held to less than 10 pounds. Another common trick, when maintenance is lacking, is to operate with the fan damper wide open. That way, there is always enough air to burn the fuel, right? Actually, that is wrong because at lower firing rates, the high excess air quenches the fire to produce combustibles, primarily carbon monoxide and, sometimes, unburned fuel products that are carcinogenic. Such careless operation is not only lacking concern for the cost of fuel but is potentially hazardous to the health of the operators as well as everyone within a one or two mile radius of the boiler. The combustion of CO becomes very slow at gas temperatures below 1800 °F. With relatively cold air flow, gas temperatures near the wall will certainly be below that temperature. Any CO that was formed will not burn, despite the high concentration of oxygen. As a wise operator, maintain adequate control of the air to fuel ratio.

Following are descriptions of the five of the more common methods of modulating a boiler's firing rate, followed by four possible enhancements to some systems. They run from the simplest to the most refined and complex. These systems are selected to provide optimum performance when they are working right. It is the job of the operator to ensure that they are working right, to keep them working right as much as possible, and to report it when they are not doing what they are supposed to so that they can be fixed by qualified technicians. A simple on/off boiler does not really have a firing rate control system. The first two simple systems are not a lot better. They do, however, change fuel and air flow rates and are, therefore, considered.

FIRING RATE CONTROL—LOW FIRE START

A low fire start control system only regulates the input of fuel and air to the furnace during the ignition period. The system limits fuel input for ignition. It then allows it to increase to the maximum firing rate which is maintained for the rest of the burner operating time. The controls for gas typically consist of a two-position fuel safety shut off valve, with a rack and pinion on its shaft connected to linkage that controls the position of the fan damper. The valve opens to a preset position during the main flame trial for ignition and the linkage limits opening of the fan damper to another preset position. Once a flame is established, and the ignitor is shut down, the valve opens the rest of the way and the fan damper opens with it.

For oil burners, the typical setup is a small hydraulic cylinder sensing the oil pressure at the burner. Two oil shut off solenoids are used to produce the two different oil flows or a solenoid is powered to bypass a manually set throttling valve for full fire. The cylinder contains a spring. It moves the damper according to the burner oil pressure, low then high. There is little advantage to a low fire start control system. Primarily, all it does is permit the use of a cheaper ignitor that would blow out if exposed to full load combustion air flow. A high/low firing rate control system is preferable if considering a low fire start. There is not enough difference in price that would prevent recovering the added cost of the modulating system in one or two heating seasons.

Adjusting low fire start controls is not easy. The manufacturer's instructions should be followed to the letter. Establish a suitable air to fuel ratio at the full load and ignition positions and ensure that the air to fuel ratio does not go too far astray as the controls swing from low fire to high fire. The process requires a thorough understanding of geometry to arrange the linkage so that the ratio is maintained.

FIRING RATE CONTROL—HIGH/LOW

High/low firing rate control is similar to the low fire start system (described above), except that the controls can switch between the low (ignition) position and the high firing position to vary the heat input to the boiler. Another pressure control switch is added to the boiler to control the positioning between high and low. Of course, if it is expected to work, it has to be set lower than the setting of the on/off pressure control switch to prevent pressure or temperature swings above the high/low switch settings shutting the boiler down. Setting of that pressure switch and the on/off pressure switch can be varied with the season as described for the on/off pressure switch and electric positioning control.

Maintaining a suitable air to fuel ratio during load swings is more important with the high/low system than the low fire start. The linkage has to maintain the ratio as the firing rate drops to low fire as well as when it increases to high fire. The burner may be frequently swinging from one to the other. The only reasonable way is to watch the fire as the control swings from high to low. It should not blow out at the high and it should not be smoking at the low. Preferably, it will be something close to a normal clean fire as it changes. Again, the process requires a thorough understanding of geometry to arrange the linkage so that a reasonable ratio is maintained.

FIRING RATE CONTROL—BURNER CUT OUT

Certain gas fired appliances incorporate this method of controlling heat input. It is not the same as having a multiple burner boiler. The application consists of installing multiple shut off valves (not safety shut offs necessarily) between the main safety shut off valves and parts of the burner. Oil burner cut out controls can shut down one or more burner nozzles, leaving the rest to continue supplying oil. Gas burner cut out controls typically shut down the gas to one or groups of flame runners. Sometimes, the combustion air is not changed (very inefficient operation). Several means of changing the air flow are available, including adjusting a damper, closing a valve in the air supply branch to the portion of the burner that is shut down, stopping a fan dedicated to that portion of the burner, or changing the fan speed. All of these systems are difficult, if not impossible, to adjust to achieve optimum combustion for each stage of operation.

FIRING RATE CONTROL—JACKSHAFT

This is a very common method for firing rate control for boilers equipped with modulating controls. The modulating motor described in the section on steam pressure control, or another form of actuator responding to a device that is attempting to maintain the pressure or temperature at the boiler outlet, is connected to a shaft (A in Figure 11-26) by mechanical linkage. The shaft is supported on the boiler by two or more bearings (B).

As the modulating motor shaft (C) rotates, or the actuator (not shown) changes position, the linkage (D) rotates the shaft. Some burners may not have a single central jackshaft, especially with small burners. The linkage may simply connect one device to the next. Most burners will have one. In Figure 11-26, the gas valve (not shown) is driven by a cam (E) which pushes on linkage (F) and the burner register is controlled by another link (G). Note that the linkage that controls the air, moving either a damper or register, is directly connected to the shaft without any adjustable cam. The jackshaft is connected by additional linkage to the fuel valves. Figure 11-27 shows the extension of the shaft (A), an end bearing (B), and the cam (H), which directly positions the fuel oil flow control valve. On this particular boiler, the cam for the gas valve is used to change the stroke of the linkage (Figure 11-28) for gas. Figure 11-29 shows another arrangement controlling a damper for air flow.

The controls are all linked to the one common shaft. Fuel and air flow controlling devices are all positioned together. To maintain a pressure, or fluid temperature, the modulating motor aligns its potentiometer with the Pressuretrol, or temperature controller, as described earlier. The movement of the motor changes the position of the fuel flow control valve to increase or decrease the quantity of fuel entering the burner and then the heat released in the furnace and transferred to the fluid and vapor inside the pressure vessel. This is commonly a proportional control. Occasionally, there will be a reset controller powering an actuator to position a jackshaft.

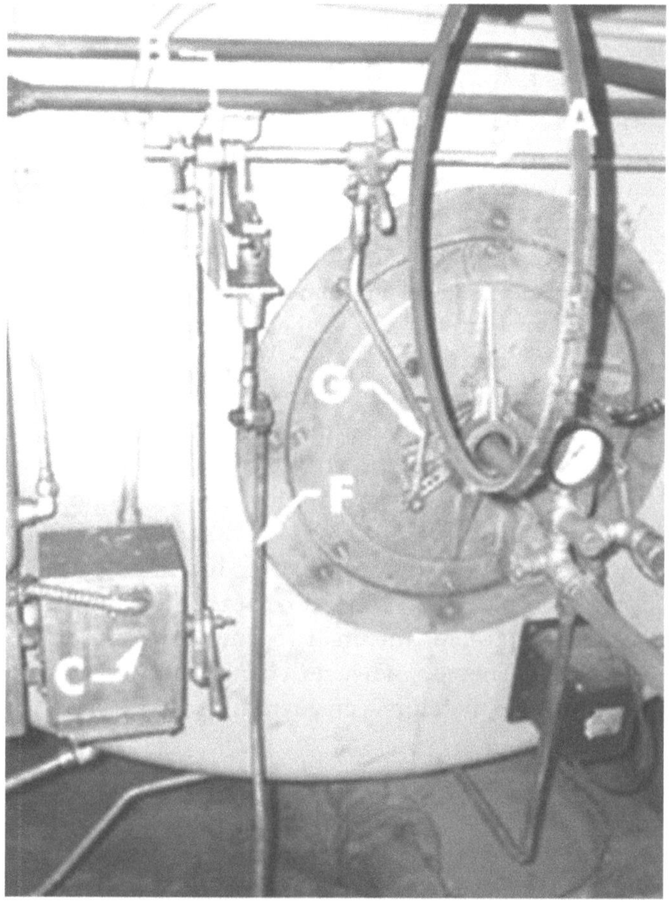

Figure 11-26. Jackshaft.

The first step in setting up controls with a jackshaft is to establish the linearity of air flow. That is all that is needed to get linear control. The fuel will be adjusted to match the air flow. With a simple linkage like that shown in Figures 11-26 and 11-29, establishing linearity can be very difficult, but it is an exercise that is essential to get consistent control. After establishing linearity, tuning consists of positioning the controls at each screw on the fuel valve (Figure 11-27 or 11-28) and then adjusting the screw to increase or decrease actual fuel flow at that position, until the desired air to fuel ratio is established.

That process should be repeated at each screw, although some technicians will do every other one or every third one. Then they adjust the ones in between to provide a smooth transition from screw to screw. Sometimes, the screws are not evident. They are concealed beneath a cover (Figure 11-30) to provide some tamper resistance. The series of screws form a cam that the roller on the fuel control valve shaft rides on as the jackshaft rotates. With some difficulty, it is possible to be positioned

Figure 11-27. Link to oil valve.

Figure 11-28. Link to gas valve.

Figure 11-29. Link to fan damper.

Figure 11-30. Linkage control valve with covered adjustments.

to see the shape of that cam. The transition should be smooth from low fire to high fire. If it is not, then the system is probably non-linear.

A jackshaft system provides simple, highly reliable control. Its performance is affected by external conditions and devices. Wise boiler operators need to be aware of how they can alter the air to fuel ratio independent of the jackshaft controls and maintain the plant accordingly. The flow that is the most susceptible to external influences is combustion air flow. Blocked or restricted air flow can cause the firing system to smoke due to insufficient air. Such conditions also aggravate the situation. The soot formed on the fire sides of the boiler from the smoke act to restrict the flow of flue gases through the boiler to block off the air flow even more. A wise operator knows the combustion air comes from the outdoors and makes sure the sources of that air flow are not blocked by leaves, snow, and other forms of debris. There are some offsetting conditions because a fan will deliver more pounds of cold air than hot air (see the section on centrifugal fans). The air to the burner actually increases as the boiler room gets colder. It tends to offset the additional friction as the operators start closing everything to keep warm. However, it cannot do it all.

Typically, it is better to tune the boiler in the winter, when the doors are shut and the air is cold. That is when the boiler is burning the most fuel and when the most efficient operation is needed. Any jackshaft controlled boiler should be tuned in cold weather with all doors, windows, etc., adjusted to winter positions. Some of that increased flow of colder air is required later in the winter, when the gas or oil gets colder. There is not a significant difference in the volume of oil as it cools, and the change in flow is not as measurable as it is with gas. Colder gas is denser. The boiler will burn more gas at each setting of the control valve. The colder air does not necessarily compensate for it. There are also variations in fuel and air flow associated with changes in atmospheric pressure. The pressure of fuel after a pressure regulator is equal to the sum of spring force and atmospheric pressure in the pressure reducing valve. The fuel gas pressure can vary a fair amount depending on where the regulator vent is. The pressure is higher when the vent is on the side the wind is hitting. A pressure below normal atmospheric is often produced on the downwind side of a building.

Wind forces can also affect the difference between the air inlets to the building and the stack to alter combustion air flow. Air density also varies slightly with atmospheric pressure. All these variations in temperature, wind, atmospheric pressure, and human generated interferences require that all burner adjustments have a cushion of excess air to absorb those variations. Accept a little loss in efficiency to ensure not to operate fuel rich, generating carbon monoxide and other hazardous and poisonous gases. A typical jackshaft system is adjusted for about 15% excess air at high loads, producing a flue gas with 3% oxygen remaining, to ensure that the boiler will always operate without going fuel rich. In testing, it can probably fire at 1/2% to 1% excess oxygen without combustibles. Almost any boiler will require some increase in excess air below 50% firing rate. The drop in velocity through the burner reduces the mixing of air and fuel. As the lower firing rates are approached, the excess air may go as high as 100% and, due to damper leakage, can go even higher.

The principal concern with the jackshaft control system itself is linkage slipping. It is not uncommon for one of the linkage connections to come loose. The best solution to loose connections on jackshaft linkage was provided by technicians at the Louisiana Army Ammunition Depot. They stopped at the auto supply store every fall to buy a different colored can of automotive spray paint. After making their adjustments, they sprayed all the connections with that paint. Any change in position was immediately apparent because the paint was cracked or a different color was showing. That does not mean that they will not slip–only that it is easy to notice if they do. Judicious use of lock tight or, preferably, star washers to prevent them from coming loose is also a wise thing to do.

A less common problem, but one to be aware of, is that the linkage rods can be bent to change their length and the relative position of the controls. That will not be evident with the paint trick described above. Some arrangements make this a difficult condition to spot because the rods are bent to begin with so that they can clear some obstruction on the burner. If there are such rods, the best thing to do is to mark their angle on a cardboard template and keep it for reference. Another problem, typical with fire tube boilers, is that the linkage gets disconnected when the boiler is opened for inspection or cleaning. The wise operator scratches match marks at all the connections before breaking them to open the boiler. That way, the linkage can be put back (almost) precisely where it was. Fresh paint after matching the scratches will restore confidence in the settings too.

ESTABLISHING LINEARITY

If systems are not linear, there will be problems with performance at different loads. Loss of linearity associated with repairs, rebuilding, and even other maintenance functions can suddenly create different operating characteristics for any system. To understand why, remember what has been discussed thus far on controls and the graph in Figure 11-31. Accept that the only thing that can be controlled in a boiler plant, refrigeration plant, and any process is flow. When a system load changes, the flows must change for it to be in balance. It is important that a change in flow is proportional to a change in other flows or controlling parameters. The most desirable relationship between flow and any control or instrument signal that is used to control that flow is described by a straight line drawn on a graph between zero and 100% of the flow and controlling parameter.

Two boilers were being converted from firing oil to gas/oil firing at a heating plant. One boiler had been finished. There was a two-week checkout period to complete on the first boiler before starting work on the second. Repeated calls from the plant regarding slow control response were resolved by a control adjustment that produced more calls regarding erratic and swinging operation. Adjusting the controls to respond properly at

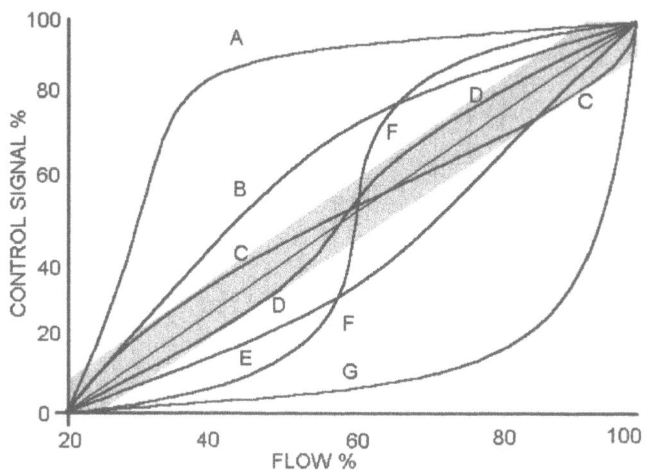

Figure 11-31. Flow characteristic curves.

Figure 11-32. Non-linear cam.

one load would produce slow or swinging operation at another load. Investigation revealed that the problem was due to the control of the boiler air flow, which had a characteristic similar to curve A in Figure 11-31. At low loads, a change in control signal (in response to a change in steam pressure) resulted in a significant change in air flow. At high loads, a large change in control signal was required to achieve a similar change in air flow. As analysis of the curve shows, there is inadequate change in air flow compared to lower loads. Steam pressure controls respond to a change in steam pressure by increasing or decreasing the control signal output in a consistent manner. The controls did not know that the controlled flow (air flow in this example) produces a varying change in flow. When the controlled flow changes vary considerably depending on load, the controls do not know it and cannot respond properly to it. The best performance will always be achieved when a change in controller output produces a consistently relative change in controlled flow. A 10% change in controller output should produce a 10% change in the controlled flow. Adjustment of the linkage achieved performance similar to curve C, which is close enough to eliminate most of the variation in performance. That resolved the problem for a while.

Some months later, the plant personnel pulled maintenance on the fan for that boiler and restored the original linkage positions which restored all the problems. Whenever working on a device that controls a flow, be certain to maintain the linearity of that device. There may be signs that one of the existing systems is not linear (note the curve of the cam in Figure 11-32) and resolves its non-linearity during a maintenance outing. Be prepared to have changes in related systems resolved after correcting linearity. If the air flow of the boiler was corrected to be linear, or close to it, then the fuel control cam, shown in Figure 11-32, has to be adjusted by a boiler technician (actually the technician should have established linearity of air flow first so that the cam would not look like it does).

Modern digital controllers are frequently adjusted to achieve linearity. The problem with them is that they correct it in automatic. There may be a system that responds to manual adjustments of the control output unsatisfactorily, but the system works well in automatic. Generally insist that every flow control device provide a linear response whether the control is in automatic or manual to ensure that when the signal is changed for a respective output to a flow control device, it does not matter whether it is in automatic or manual. Imagine trying to operate a feed water control valve in manual that has a flow characteristic similar to curve F in Figure 11-31. There is no reason for an operator to have to check a chart or curve to decide how much change in flow will occur with a given change in a controller output in manual. That requires the final element that actually changes the flow to produce the linear characteristic.

When working on anything that controls flow, the first thing to think about is retaining linearity. It is not at all uncommon for an operator, or maintenance man, to wonder why the linkage is arranged as it is and change it so that it "looks right." Adjusting the position of the linkage on a boiler can alter the fuel–air relationship. This has happened many times with different degrees of disastrous consequences. Before dismantling any linkage, including ones that are not on boilers, always mark it and restore its original position after the maintenance work is done. The correct position for linkage

Figure 11-33(a). Parallel linkage.

is seldom uniform, as shown in Figure11-33(a). Position on each shaft, length of shaft arms (from center of shaft to center of hole for connecting rod), and length of each connecting rod should be recorded on a sketch, along with labeling of the driving and driven shaft, that will be clear to another worker replacing the linkage, all before dismantling it.

On occasion, the linkage will have an unusual configuration. Figure 11-33(b) is one example. It is where the driving shaft on the left has no measurable effect on the driven shaft on the right for the last few degrees of rotation. Note that within the dashed lines, the rotation of the driven shaft will not change measurably, while the driving shaft rotates by almost 15 degrees. This would be consistent with a setup where the FD fan damper, attached to the driven shaft, is closed as much as possible and the linkage has to permit more rotation of the driven shaft to reduce the fuel flow rate controlled by another set of linkage connected to the same driven shaft. Don't expect this to be wrong if observed. If the link arm on the driven shaft is nearly aligned with the connecting rod, there is a danger of it flipping and trying to rotate opposite of the intended direction. When such an arrangement is used, it is advisable to have a stop mounted that will not permit the link arm and connecting rod to flip.

The rotation of the driven shaft will vary when the two link arms are not parallel when the link of the driving shaft is perpendicular to a line drawn through the center of the two shafts. This feature can result in nonlinearity or be used to produce it. Note that with the linkage shown in Figure 11-33(c), counterclockwise rotation of the driving shaft on the left will produce a varying rotation of the driven shaft on the right. As the driving shaft begins to rotate counterclockwise, the movement of its linkage connection is almost perpendicular to the connecting rod. There will be very little movement of the driven shaft. As the two shafts continue to rotate, the rotation of the driven shaft will accelerate until the driving link is perpendicular to a line between the two shafts and will then begin to decelerate. The graphic in the figure shows the relationship of the two links as they rotate. Note that the driving link could rotate clockwise, but the intent here is for it to rotate counterclockwise from the initial position shown. The driving shaft link cannot rotate more than 140 degrees in either direction because the driven shaft link and the connecting rod would be aligned. If the

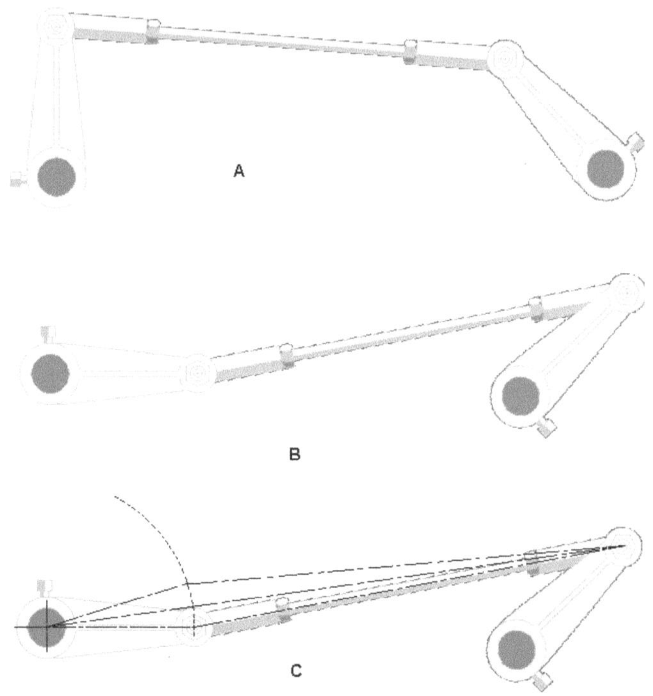

Figure 11-33(b). Linkage A.

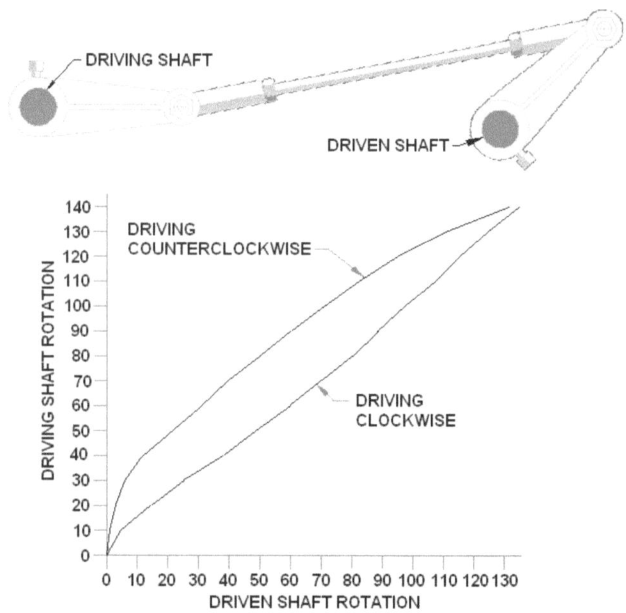

Figure 11-33(c). Offset linkage.

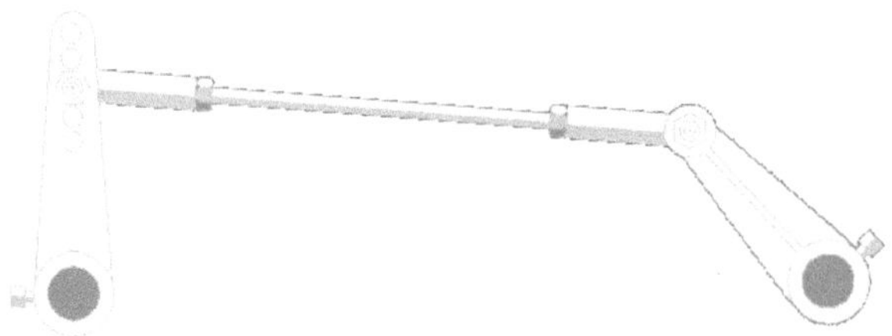

Figure 11-33(d). Typical linkage.

driving shaft rotates further, it will not be able to return to the original position.

With most linkage arrangements, there will be one link arm that has multiple holes in it for connecting the connecting rod at different distances from the attached shaft, as shown in Figure 11-33(d). Attaching the connecting rod at one of the holes that is further from the shaft than that of the link on the other shaft will result in more rotation of the other shaft than the one with the adjustable link. Conversely, attaching the rod at one of the holes closer to the shaft will produce less rotation in the other shaft. Combining that variation with those shown in Figure 11-33(c) should reveal why it is best left to the technicians to adjust linkage to achieve linearity.

Simply checking for linearity requires some care and understanding. Remember that flow is proportional to the square of pressure drop. There are two graphs in the Appendix that can be used to relate pressure drop and flow to get linear air flow characteristics. The easiest one to use is the square root graph paper in Appendix H. Measuring the pressure drop between furnace, or burner housing, and stack with a manometer, while the fan is running (no fire), will provide all the information necessary for establishing linear air flow. Setting the manometer on a slope (Figure 2-3) will allow the measurement of the pressure drop in hundredths of an inch. Extend tubing from the manometer, connecting the bottom end of it, into a hole in the stack. Be certain that the end of the tubing is not pointing toward or away from the direction of air flow to avoid getting any velocity pressure reading. Extend another piece of tubing through the observation port of the burner and connect it to the top end of the manometer. The end inside the burner has to be positioned to avoid velocity pressure as well. It is best to put a 90 degree bend in the end to make the end perpendicular to the flow. If there is an air flow measurement device, then plain graph paper can be used. Simply record the air flow. This exercise is useful, however, when there is any reason to question the air flow measurement. Compare the flow indicated at the differential, using the graph in Appendix G.

To ensure that the boiler will not fire while working on this, it is best to remove the burner management chassis. On small boilers, it may be necessary to jumper the fan starter to get it running independent of the BMS. Once the fan is running, locate the terminals that drive the modulating motor so that they can be jumped to control the position or, with other control systems, simply put it in manual and stroke the damper. Run the controls up to high fire to get the maximum air flow. Record the pressure differential on the manometer. On a jackshaft system, operate the modulating motor to lower air flow, stopping when even with each screw and recording the air differential. With more sophisticated controls, set the air flow controller output at maximum. Then decrease it and read the differential at 10% intervals (90%, 80%, 70%, etc.). With the differential pressure readings for all the flow values, the graph can be drawn.

Make a copy of the graph in Appendix H and sit down with it and a calculator. Write "air flow—%/100" on the bottom of the graph and "differential—%/100" on the left side. The chart values are 0–1. The "%/100" indicates that the range of the data is from 0 to 100%. If there were 10 cam positions, or the percentage scale of the air controllers output, then all that is needed is to use the scale on the bottom of the chart, remembering that each value indicated should have a zero after it and 1 is 100. On a typical cam with 12 positions, then 100% is 12 and 1 is zero adjustment with a span of 11. For each cam position (1 through 12), subtract 1 from it and then divide by 11. Note the result on the calculator, locate it on the bottom of the graph, draw a vertical line on the graph, and write the cam position under it. For each corresponding differential pressure reading, divide the reading by the maximum measured differential. Locate that value on the vertical scale of the graph and draw a light line horizontally until it intersects the corresponding cam position or controller output line. Mark a big dot there. Once the 10 or 12 dots are located, draw a line connecting them. The line should always extend to the upper right corner, where both values are 100%. Don't be surprised if a line from zero and zero is not appropriate. The lowest position, or controller output, is at low fire. The air flow at that point should be anywhere from 10% to 25% and the differential would be between 0.001 and 0.06. If the line is straight, or nearly so, the graph is done. If, however, the line is anything but straight (like the curves A, B, E, F, or G in Figure 11-31), adjust that linkage to get a more linear system. Something that falls in that narrow gray band on Figure 11-31 is needed for good control.

If dealing with a jackshaft, the position of the linkage will have to be changed. When possible, restore the original settings by the manufacturer. They should be linear. Otherwise, opt for changes that make sense. Then take some more readings to see how well they turned out, or did, repeating the process until something linear is obtained. For the best world, a damper actuator with a positioner, the data just collected will allow the production of a new positioner cam. Linear control should produce a straight line from low fire to 100%. Simply draw a straight line from the low fire point to the upper right corner. Draw horizontal lines through the data points until they intersect that line. The height of the existing cam at the data point is the height needed for the new cam at the position coinciding with the straight line.

Achieving linearity with a valve positioner, or an actuator, for dampers and other shaft operated devices, was simply a matter of plotting the control signal, process flow, and height of the positioner, or actuator, cam, along with its rotation. Then, take the blank cam and mark off the desired height at each angle of rotation, the desired height being determined by graphing the cam rotation, height, and flow. Then, use that data to note the required height. The data can be collected with whatever cam the manufacturer furnished.

Before collecting cam data for linearity, the stroke of the valve, or actuator, should be established to provide 120% of design flow. That extra flow provides an allowance for the system to catch up. Some valve manufacturers select their valves to produce 100% of the specified flow rate at 70% of the valve opening, which means that the valve could pass 143% of design flow when full open. That is likely to be too much. Check the selection of the trim of a control valve when ordering it. Once the stroke is determined, adjust the linkage of the positioner so that it rotates the cam by 100% of the design rotation at that stroke. Keep in mind that the valve stroke cannot be shortened excessively. It will have problems at the bottom end and may jump on and off the seat, making control sloppy. It is better to have the trim replaced to something smaller if the stroke is reduced by more than 30%.

The record of the valve characteristic, before and after linearizing, should look something like Figure 11-34(a), which is a sample alignment record sheet. The cam rotation relates to the valve stroke and degrees are simply values for scale.

The graph of actual measurements (normally on the back of the record sheet) and marking of the blank cam for a sample alignment are shown in Figures 11-34(b) and 11-34(c).

There are columns for zero position. For some applications, there has to be flow through the valve or damper, like minimum flow of fuel and air for a boiler. This positioner was set up for a feed water control valve and it has to shut for zero flow. The flow was read by the feed water flow meter and divided by the maximum flow to plot that curve in percent. The heights of the new cam positions were calculated using the desired flow

POSITIONER ALIGNMENT RECORD

VALVE / ACTUATOR SERVICE - BOILER 3 FEEDWATER CONTROL VALVE

VALVE / ACTUATOR AND POSITIONER TAG NO. - 1087

POSITIONER MFR. - FISHER

MODEL NO. - 667/503E SERIAL NO. - E503-347801

INSTALLED MEASUREMENTS: CAM NO. - 2

CONTROL SIGNAL	4MA										20MA
SIGNAL PERCENT	~~0~~	~~10%~~	~~20%~~	~~30%~~	~~40%~~	~~50%~~	~~60%~~	~~70%~~	~~80%~~	~~90%~~	~~100%~~
CAM ROTATION	0	30	60	90	120	150	170	190	210	240	270
CAM HEIGHT	0	10%	20%	30%	40%	50%	60%	70%	80%	90%	100%
PERCENT FLOW	0	26%	51.5%	68%	82%	94%	102%		108%	118%	120%
MEASURED FLOW	0	6,440	15,800	20,860	25,150	28,830	30,070		33,120	36,190	36,800

CAM CUT DATA AND FLOW VERIFICATION

CONTROL SIGNAL	4MA										20MA
SIGNAL PERCENT	0	10%	20%	30%	40%	50%	60%	70%	80%	90%	100%
CAM ROTATION	0	30	60	90	120	150	170	190	210	240	270
CAM HEIGHT	0	6.3%	10.3%	17.6%	26%	35.4%	49%		61.9%	88.5%	100%
PERCENT FLOW	0	11.1%	22.2%	33.3%	44.5%	55.6%	66.7%		77.4%	88.8%	100%
MEASURED FLOW	0	4,090	8,180	12,270	16,360	20,450	24,530		28,500	32,700	36,800

TECHNICIAN: KEN HESELTON BADGE NO. ______ DATE: 11 / 08 / 88

NOTES: NONE OF THE STANDARD CAMS PRODUCED LINEAR FLOW.

Figure 11-34(a). Record of linearizing a positioner.

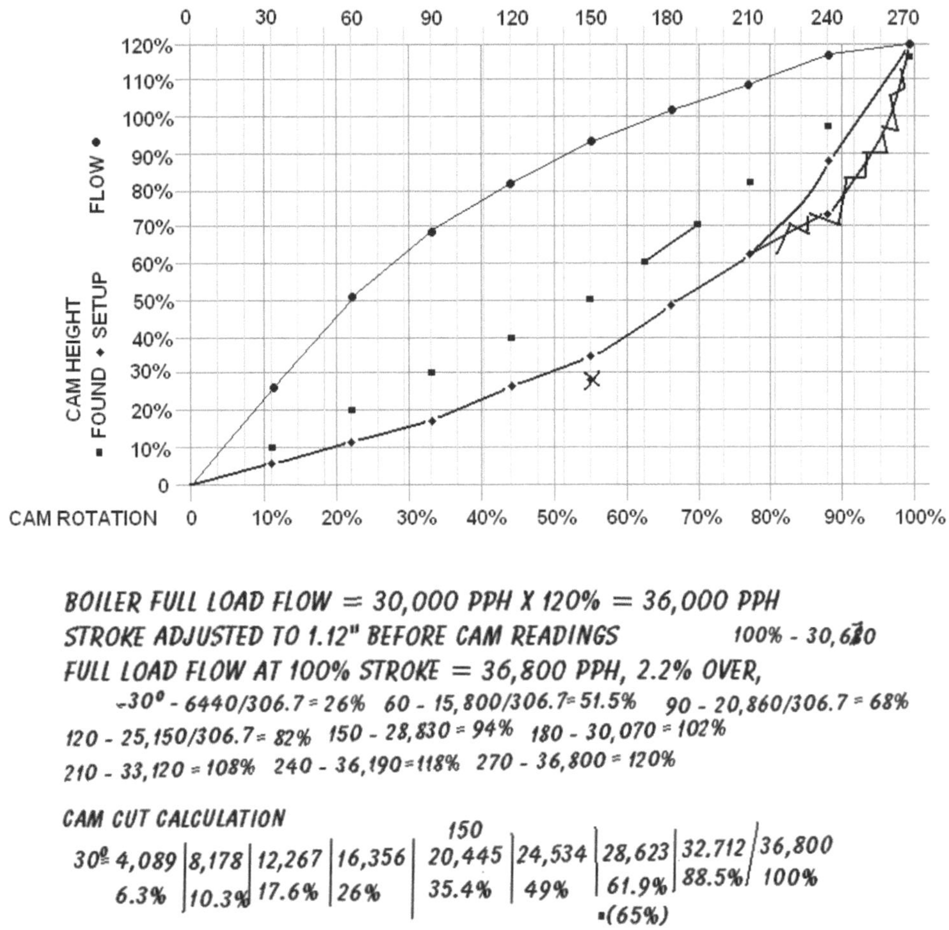

Figure 11-34(b). Graphs for positioner.

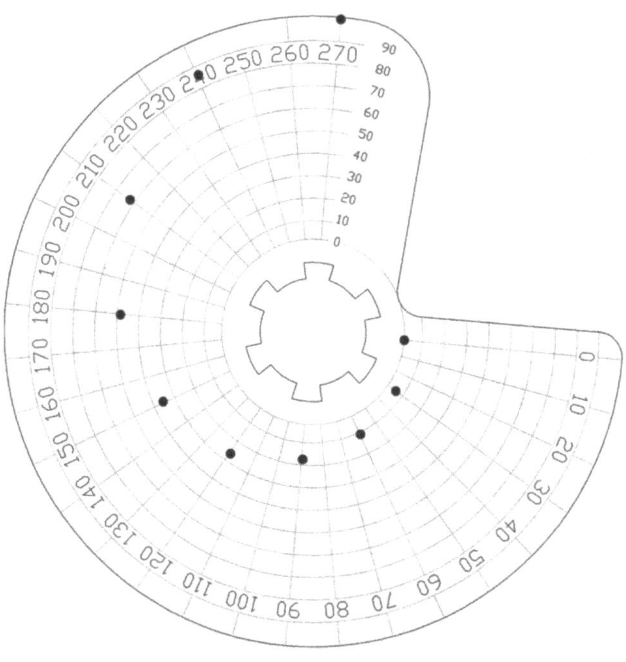

Figure 11-34(c). Blank positioner cam.

divided by the measured flow and multiplied by the original height (in percent divided by 100). The heights are shown plotted on the blank cam of Figure 11-34(c). A French curve (drafting instrument) is used to draw a smooth curve on the cam as a guide for cutting it.

This may seem like a lot of work, but it provides an accurate flow through the valve proportional to the signal. A change in the control signal always produces a desired flow through the valve (precise, provided the feed water header and steam pressures are near design). The controller will not see an unexpected response and control operation will be smooth. The added benefit is that a proportional change in flow can always be expected when operating in manual.

STARTUP CONTROL

The low fire hold control consists of an extra pressure or temperature switch that opens contacts to prevent automatic modulation of a burner when the pressure or temperature of the boiler is below the switch setting. Once the temperature, and temperature can be used even on steam boilers, or pressure rises to a level higher than the switch setting, the automatic controls can operate. What usually follows is a modulating control running on up to high fire. Now, supposedly, the boiler is warm enough that it will not experience any thermal shock, or excessive thermal stresses, in the process. That is not a certainty. On the larger boiler systems, which included an automatic startup provision, a ramping provision was used. Once the pressure exceeded the setting of the low fire hold switch, vent valves were automatically closed and the ramping system put in service. It allowed a very slow increase in the firing rate and prevented a more rapid increase, until it had completely ramped out.

The first ones were applied on pneumatic control systems and consisted of a low signal selector, three-way solenoid valve, and a volume chamber with a metering valve. The solenoid valve dumped the contents of the volume chamber to atmosphere, while the boiler was off, and applied supply air, usually at around 18 psig, to the chamber via the metering valve when the boiler had started and the low fire hold switch released. The pressure in the chamber was piped to the low signal selector along with the boiler master output or the plant master output. The low signal selector then fed either the fuel and air controls or the boiler master. Which one depended on the plant master operation. If all the boilers were operated off the plant master, then the output of the ramping control was fed to the boiler master. Otherwise, it went to the fuel and air controls so that the operators could set a firing rate manually at the boiler master. Once the low fire hold was released, the air bleeding into the volume chamber slowly increased the firing rate. Where possible, that ramp rate was set as slow as possible, sometimes to get it out in 2 hrs to reach high fire (and that is still faster than some boiler manufacturers specified). Once the firing rate exceeded the plant master, or boiler master, output, it was no longer the low signal and the boiler was operating on automatic.

The ramping control actually allowed a boiler to automatically come on line, generating steam and picking up load, without upsetting the operation of the other boilers. Steam traps to drain the boiler steam headers and many other features were required for fully automatic control. If a sudden increase in plant load required the boiler firing rate to increase once it was on automatic, the low signal selector would not allow the firing rate to increase any faster than the ramp rate. Once the pressure in the volume chamber bled up to supply pressure, the boiler operated automatically, as if it were not there. The automatic signals were always the low signals. Another nice feature of the system was that it drained off pressure from the volume chamber as slowly as it filled it up. If the boiler tripped for some reason and then started back up, the ramping simply swung back up but allowed the already hot boiler to fire at higher rates. With modern digital controls, the same feature can be added. It can be augmented to provide a ramp rate from a cold start and a different ramp rate from a hot or warm restart. All it takes is some additions to the software.

FIRING RATE CONTROL—PARALLEL POSITIONING

Parallel positioning controls perform about the same as a jackshaft control system because they simply establish the position of the fan damper and fuel valve. Large boilers and boilers with air heaters, or fans located away from the windbox, cannot easily use a jackshaft type of control. The weight of the linkage becomes a problem. A parallel positioning system allows the fan and fuel valve to be located conveniently for construction and for other reasons. The most commonly used parallel positioning system is an electric positioning system that uses potentiometers, like the modulating motor controls, to compare the position of the fan damper and fuel valve actuators and adjust them to match the position of a boiler master. Plants with parallel positioning control usually have a plant master for steam pressure control, which actuates a bunch of potentiometers that are matched in position by the boiler masters or fuel valve controllers on each boiler, plus or minus any bias introduced by adding resistance in the potentiometer loop.

Some advantages of parallel positioning controls include the ability to run on a plant master and bias boilers. It does not constrain the location of fuel valves and fan. It permits isolation of gas and oil control valves on two fuel boilers so that both are not operating to influence the operation of another boiler firing a different fuel. They permit maintenance on the valve and actuator for the alternate fuel. It also permits independent operation of the fuel and air controls so that a boiler operating on hand can be trimmed (adjusted by the operators) to reduce excess air.

A principle disadvantage of parallel positioning controls is variations in response of the actuators, especially for pneumatic actuators, which can produce temporary upsets in the air to fuel ratio during load changes. Unlike the jackshaft system, there is nothing to prevent the fuel valve actuator from moving faster than the fan damper actuator or vice versa. They also have all the disadvantages of the jackshaft system, with one provision to help offset the problems with maintaining air to fuel ratio, adjustment of the air to fuel ratio. By adjusting the resistance in the loop of a system where the air flow actuator follows the fuel flow actuator, the relative position of the two actuators can be varied to provide an excess air adjustment. The adjustment is a bias type adjustment in most systems. It does permit running the air to fuel ratio a little tighter, if the operating personnel choose to. It also allows the operators to compensate for soot accumulation in the boiler, something that is not easy to do with jackshaft controls. To help overcome the problems with the independent actuators, the controls can be enhanced to include a leading actuator provision so that the fan damper actuator follows the fuel on a decrease in load, and the fuel actuator follows the fan damper actuator on an increase in load.

In addition to problems with response, jackshaft and other modes of parallel control can suffer a degradation due to hysteresis associated with wear in the linkage connecting the devices used to position dampers and valves. To overcome the problems with response and hysteresis, manufacturers have developed systems utilizing synchronous actuators, which are, for the most part, mounted directly on the damper or valve shaft. The synchronous motors are basically electric actuators, with an integral positioner that responds to a current signal, or digital instructions, from a controller. These can work well but are sometimes erratic. A mechanical device that is constantly changing position back and forth is not exactly working right. Don't accept one that is doing it. For all practical purposes, a parallel positioning control system is set the same way as a jackshaft control system. There needs to be some way to set the fuel valve to match the air. Normally, a parallel positioning system will have cams just like a jackshaft system.

FIRING RATE CONTROL—ADD AIR METERING

The full title of this control logic is parallel positioning with air metering. The next evolution in control systems, after parallel positioning, was to add air flow metering. Since air flow is influenced by so many factors, it makes sense to measure the air flow and control it. The measured air flow provides feedback to the control system. Then the air flow controller can adjust the fan damper actuator to produce a repeatable air flow. Instead of simply positioning the fan damper, the control system adjusts the damper until the air flow signal matches the plant master position signal. The decision to measure air flow started the still-standing arguments about where it should be measured. The type of control system and boiler has some effect on the choice. Be aware of the variations. Air metering with measurement across the burner windbox to furnace, on multiple burner boilers, made it possible to compensate for the number of burners in operation. Because each burner throat is one orifice in the flow path, changing the number of burners does not change the differential measured at the air flow transmitter when the register on one closes, but it does change the air flow.

Measuring the air flow using a differential across the boiler itself is measuring the flue gas flow, not just the air flow. That is not a significant variation since the air is 93%–94% of the flue gas. The problem with using the boiler is that soot formation can change the differential relative to air flow. Other problems, like refractory seals breaking up, can also alter that differential without the operator being aware of it. The best measurement is in a suitable metering run, or venture, between the FD fan and burner windbox. Most boilers do not have enough room there for any kind of precision flow measurement. Using the inlet of the FD fan to measure air flow is always preferred, provided it is a single inlet fan. When a boiler is large enough to justify a double inlet fan, then a good metering element between the fan and the burner windbox is justified as well. Many operators do not understand fan inlet metering.

Within the boiler room, there is air movement. Typically, it is very slow except for right at the inlet of the FD fan or its silencer. In order for the air to accelerate from something very close to zero velocity in the boiler room to the speed it needs to get through the fan inlet, there has to be a difference in pressure between the two locations. The fan creates a lower pressure at the fan inlet by removing the air that enters it. It is that void, created by the fan removing that air, that the room air rushes into. The boiler room itself is nothing more than a big pipe that the combustion air flows through. The fan inlet is just like an orifice. By measuring the static pressure at the orifice and subtracting it from the pressure in the room, the velocity pressure is obtained, which indicates how fast the air is flowing into the inlet of the fan. There is one thing rather nice about this flow measurement. There is no orifice coefficient. There is no measurable friction applied to the airstream between the boiler room and the fan inlet. A good arrangement for this measurement of air flow is shown in Figure 11-35.

It requires a ring of half inch tubing forming a circle equal to two-thirds of the diameter of the inlet. The holes

Figure 11-35. Fan inlet flow measuring ring.

in the ring are drilled just a little past center to minimize plugging with dust from the air. The ring is mounted outside any screen, or other obstruction, in the fan inlet that could get dirty to vary the signal. The transmitter is mounted on at least five fan inlet diameters from the inlet of the fan and is independent of any obstructions that would produce air velocity near the high pressure sensing port of the transmitter. A drop leg is needed to prevent dirt from entering the high pressure connection of the transmitter. The transmitter should be mounted above the ring so that there is no way condensate can form and collect in the transmitter and sensing piping to block the signal. Any condensate that does form will run out of the holes in the sensing ring. There is only one caveat with this method of air flow measurement. Be certain that there is no way for the air being measured to go anywhere but to the burner. It will not work if there is air leakage, branch ducts, or the like between the fan and the burner.

The original systems were a little lax in producing a true flow signal. Recall that the pressure drop measured is proportional to the square of the flow. Some simply used the differential signal and counted on the screw cam type fuel flow control valve for setting the fuel air ratio. Others provided a flow signal somewhat related to actual air flow but still counted on the adjustment of a cam type fuel valve. The problem was developing an output proportional to flow from a differential pressure signal. Some of the original air flow transmitters used cams. Others used combinations of springs. Others used the stretching of the diaphragm used to sense the differential. They all gave way to differential pressure transmitters with panel mounted square root extractors, until microprocessor-based transmitters were developed. Modern microprocessor-based instruments either calculate the square root right in the transmitter so that the output is directly proportional to flow or the square root is calculated after the differential pressure signal is input to the controller.

An air metering addition to a parallel positioning controller allows tighter control of the air to fuel ratio and should permit operation at less than 15% excess air, in the range of 2-1/2%–3% oxygen in the flue gas and less on single burner boilers.

FIRING RATE CONTROL—INFERENTIAL METERING

Early inferential metering systems simply avoided the square root problem by comparing the differential signals. The reasoning was that if it is proportional to the square root for air flow, it must be for oil flow or gas flow. That is not quite right, but it can work up to a point. There are some differences between orifice coefficients and other factors that had to be taken into account and most control systems had provisions (adjustable cams) to compensate for it. Inferential control provided many of the features of metered control without the expense (and difficulty) of square root extracting. They also solved some problems that were principally associated with multiple burner boilers. There were a lot more multiple burner boilers in the middle of the 20th century. They were either converted from firing coal or designed to be convertible to coal. Coal fired designs use a reasonably square furnace, not the long skinny ones used on most package boilers today. The shorter furnace required the use of multiple burners.

Inferential metering is accomplished by considering the fuel delivery systems as an orifice, with a pressure drop that can be measured, and comparing that with the air side pressure drop. These systems were only applied to oil and gas fired boilers. They used the burner header pressure as a variable that equated to fuel flow. After all, the oil burner tip is an orifice, or group of them, and a gas ring or spud has orifices in it. The pressure in the furnace (downstream of the orifice) was relatively close to zero. Thus, it is reasonable to treat the burner header pressure as a value of differential. Some gas fired systems used gas at such low pressures that it was essential to include a furnace pressure input to the measuring device so that the changes in furnace pressure did not upset the flow signal although they did experience some difficulty with pressure fluctuations (See Draft Control, Chapter 11).

Modern instruments have erased the cost advantages of inferential metering systems. When inferential metering is used today, the differential is treated as a flow signal and the square root is extracted by the transmitter or controller to produce a linear flow signal. One of the more serious problems with inferential metering systems was their lack of linearity. The control response was normally tuned for the high end of the boiler operation and swings accepted at low loads. In dealing with those multiple burner boilers, they had a distinct advantage, even over today's full metering systems. The fuel flow based on the burner header did not account for the number of burners in service. The differential from the windbox to furnace did not account for the number of registers open. If a burner tripped, the control backed down to restore the header pressure, effectively decreasing the flow, and the air to fuel ratio at the other burners was restored. Later, the operator could close the register and the air control would restore the windbox to furnace differential to restore the air to fuel ratio again. The only problem came when someone put a burner in service and forgot to open the register.

FIRING RATE CONTROL—STEAM FLOW/AIR FLOW

Inferring fuel flow by pressure worked fairly well for oil and gas. It did not help with coal firing. Steam flow/air flow systems were developed for coal firing and are basically inferential metering systems because the steam flow could be equated to fuel flow. If the boiler efficiency and steaming conditions were constant, then a fixed relationship between steam flow and fuel flow would exist. The fuel would generate a proportional amount of steam. These systems eliminated the problems with, or impossibility of, measuring the coal flowing to the fire. Coal fired boilers larger than about 90,000 pph (pounds per hour) can justify the expense of metering the coal. Smaller units still use steam flow/air flow control. One problem with steam flow/air flow is the lag in response associated with load changes. If the plant master output increases, there is a delay associated with the inertia of the boiler. It takes a little time for the higher coal flow rate to heat up the boiler a little more and increase the steam flow rate. If the system was set up so that air flow followed fuel flow, the boiler would probably smoke on a load increase.

The systems normally use a parallel positioning control methodology, where the plant master changes produce a proportional change in fuel feed, primary air flow (on pulverized coal fired boilers), and combustion air flow and maintain the ratio of fuel and combustion air flow signals with the steam/air flow ratio on a slow reset. Some engineers refer to steam flow/air flow systems as parallel positioning with steam flow trim. The steam flow is used to trim the ratio between fuel and air.

It is the timing problem that dictates how tight the air to fuel ratio can be maintained with a steam flow/air flow system. Gas and oil fired systems could actually run a little tighter than a system with air flow metering added because changes in fuel input produced a rapid change in steam rate. Pulverized coal fired boilers have a delay in load changes associated with changes in coal inventory in the pulverizer and piping. They typically operate with excess air rates around 25%–30%. Stoker fired boilers have a larger inventory change effect and have to operate closer to 40%–50% excess air to eliminate fuel rich firing conditions (and smoking) during load changes. Of course, all of the boilers are not run at that rate. The wise operator will let one boiler take the load swing and set the others (if they are needed) to fire at a constant load and much tighter excess air rates. The steam flow/air flow controls can then respond to variations in fuel quality to maintain the appropriate air to fuel ratio.

FIRING RATE CONTROL—FULL METERING

Full metering control systems measure the flow of fuel and air. Similar to labeling the steam flow/air flow metering systems, some engineers will call the systems parallel positioning with flow tie-back. The advent of microprocessor-based controls (which have drastically reduced the cost of control systems) and continued reductions in device costs allow for smaller and smaller boiler control systems of the full metering type. Any oil or gas fired boiler that consistently operates at loads above 25,000 pounds of steam per hour (25 MMBtu/hr (million Btu's per hour)) should be equipped with full metering controls. They will return their cost in fuel savings in a matter of two or three years. Any step between a jackshaft system or parallel positioning and full metering (with the possible exception of adding oxygen trim which will be covered later) would be a waste of money.

The full metering system does use flow as feedback to the controls. The controllers control the flow of the fuel and the flow of the air to produce a heat flow into the boiler that matches the load. The plant master signal, which maintains a pressure at the common boiler header, is proportional to the heat load. The boiler masters in a steam system pass the plant master signal, plus or minus any bias at the boiler master, to the firing rate controls.

Hot water and fluid heating boilers each will have their own temperature control or, in large sizes, a load indication based on fluid flow and temperature differential to produce a boiler master signal for the firing rate controls.

The firing rate controls respond to the boiler master signal by changing their outputs until their respective fluid flow transmitters send back a signal that matches the boiler master. Modern full metering systems automatically include cross-limiting to prevent fuel rich firing conditions. There was a time when the term "cross-limiting" was added to the definition because it required additional control devices. Today, cross-limiting is simply a couple of extra instructions in the software.

The full metering system is shown in Figure 11-36 without the plant master controller. The lower of the master signal or air flow signal become the set point for the fuel flow controller. The symbol "<" in the diagram identifies a low signal selector. Its output is the lower of the two inputs. This is part of the cross-limiting. The fuel controller cannot see a demand for fuel flow greater than the air flow signal. The fuel flow controller adjusts its output, using PID algorithms, until the fuel flow signal matches the lower of the air flow or the master signal.

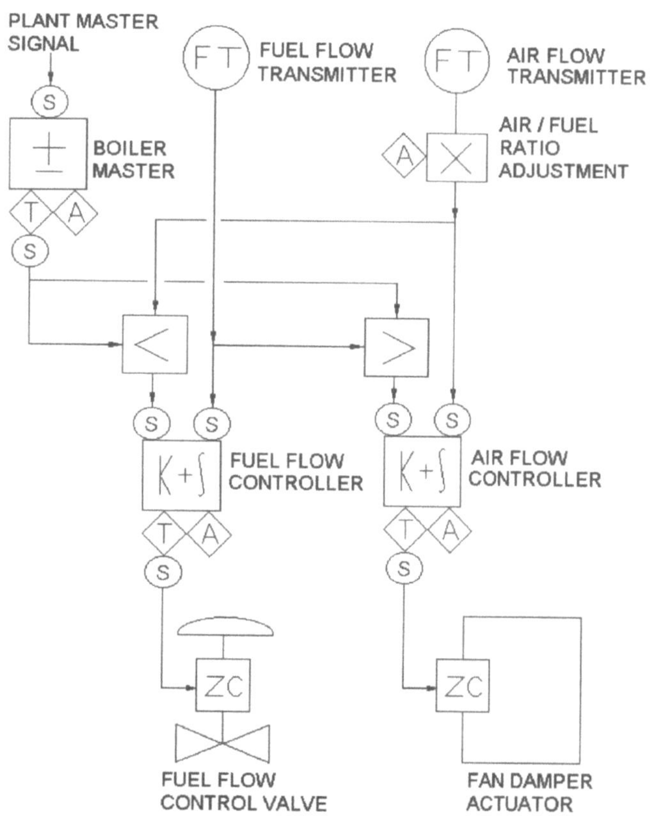

Figure 11-36. Full metering control schematic.

The air flow controller's set point is the higher of the master or fuel flow signal to provide the other part of cross-limiting. The symbol ">" in the diagram identifies selection of the higher signal. The air flow controller will adjust its output until the air flow signal coming back to it is equal to the higher of the fuel flow or the master.

Take a look at the system performance when a load change occurs. Say someone opened up a steam valve to a process in the facility so that there is an increase in load. The plant master will detect a drop in pressure and change to increase its output. The increase in plant master output is passed through the boiler master to the firing rate controls. Since the air flow signal matched the previous boiler master signal, it is lower than the master. The fuel flow controller does not see any change in its remote set point. The master signal is higher than the fuel flow signal. It passes through the high signal selector to become the remote set point of the air flow controller. The air flow controller then responds, changing its output to increase air flow. As the air flow increases, the transmitted flow signal increases to raise the set point of the fuel controller. If the master signal stops changing, the air flow signal will eventually come up to match the master and the fuel flow signal will follow it. Look at the diagram and see how the air will follow the fuel on a decrease in master signal. That is how the system with cross-limiting works.

On a decrease in load, the fuel flow goes down and the air flow controller follows the fuel flow signal (as its remote set point) down. Therefore, cross-limiting prevents a fuel rich condition. Some engineers try to think of these as lead/lag systems because the air leads the fuel going up and lags it going down. That is not correct. Lead/lag systems have been around for years and have nothing to do with fuel and air.

Since all the control signals have to match, there is a problem when the air to fuel ratio has to change. Any change in the master signal between air and fuel controllers will upset the cross-limiting. To resolve that problem, the air flow signal is modified to indicate an air flow that is less or more than what it actually is. A typical method is to insert a ratio control between the air flow transmitter and the fuel and air controls, with their signal selectors as shown in Figure 11-36.

Despite the fact that a full metering control eliminates many of the variables of pressure effects (people opening and closing windows and doors and other situations), there is one serious problem with full metering controls. If the fuel flow signal is lost, the controller will drive the fuel valve wide open almost instantly. If the air flow signal is lost, the air controller will drive the damper wide open and can blow the fire out or produce a lot of unburned fuel by quenching the fire. Either situation is hazardous. The loss of the fuel flow signal is the most dangerous. Many system designers incorporate differential sensing devices that will shut down the boiler if the fuel and air flow signals do not match within limits. Shutting the boiler down has other consequences. A better solution was to compare the fuel flow signal with a prescribed minimum and drive the boiler to low fire if the signal was less than that value. It does not result in a boiler shutdown. It does give the operator a chance to correct the situation, or fire the boiler in manual, rather than running around trying to get another boiler on line. The limit also prevents a shift above low fire in the event of the loss of the control signal after startup.

FIRING RATE CONTROL—DUAL FUEL FIRING

Dual fuel firing means firing two fuels at the same time and under control. Boilers that can fire gas or oil are two fuel boilers. They can fire gas or they can fire oil, but they cannot fire both at the same time. Low fire changeover systems are discussed in the section on operating wisely. They are not dual fuel firing either. In order to fire two fuels at once, a full metering system is required. In addition, a fuel flow summer is needed that combines the two fuel flow signals so that the total fuel flow is the feedback signal to the fuel controllers and to the high selector of the air flow controller. One of the two fuels has to be considered the primary fuel. The other fuel flow signal has to be adjusted with a gain such that it produces an output that equates to the air flow demand of the primary fuel. Some engineers call the summer a Btu summer because it takes about the same amount of air to release a Btu by combustion from firing oil or gas. The rest of the controls do not know that they are looking at two fuels. They operate normally.

The usual reason for dual fuel firing is switching fuels. There are other operating conditions that favor dual fuel firing, but the common one is switching fuels. Sometimes, emission limitations will drive the need for dual fuel firing. A dual fuel firing system is the ultimate in control for a boiler. It should be required unless only one fuel is fired or almost never switched. It is the best way to transfer fuels. The boiler is always operating with an inert furnace environment. It is safer than stopping and then restarting the boiler, and a lot safer than the low fire changeover systems.

The standby fuel is brought on the burners at low fire and then manually adjusted upward until the fuel flow controller output equals the manual output for the standby fuel. The controller will automatically reduce the firing rate of the leading fuel to compensate for the added standby fuel. When the two fuel flow signals are equal, switch the standby fuel controller to automatic and then switch the leading fuel to manual. It does not require an instant transfer since the controller will simply adjust the two fuels in parallel. Control action will not be smooth with both fuels in auto because every change in output produces twice the change in flow compared to firing one fuel. Don't leave both fuel controls in automatic unattended. It is possible to control two fuels in auto at once. Generally, it is not really necessary. If the unit is on dual fuel firing, there are other reasons, not one of which involves auto operation of both fuels to maintain steam pressure. To complete the transfer, reduce the firing rate of the lead fuel manually until it is at low fire. Then shut it down.

There is nothing preventing the firing of both fuels continuously as long as one is in manual control. It is convenient for burning down an oil tank, while still firing natural gas, or firing natural gas at the maximum rate allowed by the supplier. It is also possible to burn three fuels. For example, gas, oil, and a solid fuel can be burned. This might occur in a boiler system designed to fire municipal solid waste (MSW). Sometimes, the amount of MSW that is delivered or available is not enough to meet the steam demand. The supply of oil and gas may be temporarily limited. Typically, there are not a lot of opportunities to do this. If one does occur, just keep in mind that only one fuel can be operating on any one automatic control signal.

FIRING RATE CONTROL—CHOICE FUEL FIRING

It is possible to fire two fuels and have both on automatic control, just different automatic controls. Modern microprocessor-based controls allow dynamic changes in controller gain such that a fuel controller could fire oil and gas together. Choice fuel firing is the incorporation of additional controls to meet fuel supplier's criteria. There could be a system that calculates the rate that gas should be fired based on the amount of gas that the plant is allowed to burn (a monthly constraint) and the number of hours left in the month. The gas glut created by fracking (a process of fracturing gas laden rock in the earth that has produced the gas glut) will eventually diminish, and this type of firing may return. There may be permit limitations that constrain the use of one fuel or another. It is easy with computerized control. Input the amount of gas that is allowed each month and the controls do the rest. There are several parameters that the control system needs to make a decision about the gas firing rate at any time, including the need to burn a minimum amount of fuel oil. Some history of facility performance during that month can be used to predict situations when less gas could be burned and increase the current rate so that the gas is consumed by the end of the month.

Choice fuel firing can also be associated with burning other combinations of fuels, including a fuel produced by a process in the facility. Some of it may be sold to another facility. That case could involve burning the excess fuel the other facility does not use. When the other fuel is gas, it is handled by pressure control, sensing the pressure of the generated gas and varying the flow of gas to the fired equipment to maintain a constant pressure at the output of the process. It could also include a variable set point for the gas pressure developed by the process that generates the gas. When the other facility uses more of the generated gas, the pressure will attempt to drop. The controls reduce the amount fired in the fired equipment to maintain the pressure. When the other facility uses less generated gas, the pressure will attempt to rise with the result of burning more.

When neither facility needs more of the generated gas, the firing rate controls will first reduce the firing rate of the main fuel, which is the only one normally controlled to maintain boiler output. When the main fuel firing rate is at minimum, then the generated gas control is overridden to maintain the boiler output, reducing its use in the boiler. If there is still too much, the generated gas pressure will increase and a relief valve will send it to a flare to simply burn it and waste the energy. Note when this occurs (actually production should advise when it is going to happen). Look for ways to use the generated gas in other equipment to offset purchased fuel cost by utilizing all the generated gas that is possible to use. An adjacent landfill gas operation might provide an example of this situation. If the production generates a liquid fuel, the deviations in supply of the generated fuel can be absorbed in storage tanks. Automatic control for such applications is unusual. Production should be able to provide notification of any changes they make in the production of the fuel. Then determine an optimum firing rate for the generated fuel based on production data and the level in the storage tanks.

Burning of waste gases, oils, and solid fuels, generated by the plant process operations, can also require analysis of the fuel to determine its proper air to fuel ratio to ensure that it burns cleanly and efficiently. Liquid and solid fuels can be stored and tested with the ability to adjust the air to fuel ratio before firing them. Gas fuels, on the other hand, are not easily stored. At one refinery, they flared a waste gas rather than using it in the boilers. The waste gas varied considerably in its hydrogen to carbon ratio. An online analyzer could be used to automatically adjust the fuel/air ratio for the gas being burned and that could be incorporated into an equivalent fuel value and applied to the firing rate control. One tricky part of that application would be providing a means of accounting for the time delay between the analyzer's response to the waste gas values and the time it takes for the gas to get from the analyzer to the burner.

FIRING RATE CONTROL—OXYGEN TRIM

Oxygen trim controls actually measure and control excess air. The oxygen content of the flue gas is controlled. Basically, it is an indicator of excess air. An analyzer samples the flue gas in the furnace, or at the outlet of the boiler, to determine the amount of oxygen in the gas. The analyzer transmits a proportional signal to a controller, which then changes the firing rate controls to alter the fuel to air ratio to maintain the oxygen content at a set point.

In 2013, the US Environmental Protection Agency (EPA) finalized the Industrial Boiler maximum achievable control technology (MACT) rule, with a compliance date for existing units of 2016, with a limited extension to 2017 for some units. A facility is subject to the Boiler MACT rule if it owns or operates an industrial, commercial, or institutional boiler or process heater that is located at, or is part of, a facility that is classified as a major source of hazardous air pollutants (HAP). A "major source" HAP facility emits 10 or more tons per year of any single air toxic or 25 or more tons per year of any combination of air toxics. Note that it is the facility emissions as a whole that are counted, not just the boiler emissions. Any facilities that are not major sources of HAP are classified as area sources.

The following major source units are NOT subject to the Boiler MACT:

- An electric utility steam generating unit (EGU) covered by subpart UUUUU of part 63 (i.e., the mercury and air toxics (MATS) rule).
- Hot water heaters with a capacity of no more than 120 US gallons or a hot water boiler with a heat input capacity of 1.6 MMBtu/hr or less.

- Waste heat boilers, also known as heat recovery steam generators (these boilers recover traditionally unused energy and convert it to usable heat).
- Boilers that are used as control devices for other National Emission Standards for HAP (NESHAP), where at least 50% of the heat input to the boiler is provided by the NESHAP regulated gas stream.
- Research and development boilers.
- Boilers subject to other NESHAP standards, Section 129 standards, or hazardous waste boilers.
- A recovery boiler or furnace covered by subpart MM.
- Temporary boilers.
- Residential boilers.

Boilers that are subject to the Boiler MACT rule must apply oxygen trim for combustion controls. They must also have a tune-up, once per year in most cases.

Most oxygen trim systems will not have a simple oxygen set point. The amount of excess air required can vary with load. In most boilers, the excess air can be held constant at loads over 50% of maximum. Excess air has to increase almost exponentially as the load decreases (see the section on burners for reasons). The common approach is to generate an oxygen set point that, for all loads up to about 50%, is a function of the boiler master signal or the steam flow signal. Steam flow is generally preferred because it produces higher oxygen requirements when increasing the firing rate of a cold boiler. It is when more excess air is needed to complete combustion since the furnace and refractory are not as hot and the flame temperatures and gas temperatures are lower. The common approach is to use a function generator which allows the technician setting up the control to produce an output that bears no mathematical relationship to the input.

Data collected during firing tests on the boiler (to determine the necessary amount of excess air at each load) can be used to determine how to cut a cam in the function generator. Modern digital controllers have a similar application, except the numbers are entered, instead of measuring a plastic or aluminum plate and cutting it to get the desired shape to produce the output. It is another blessing of microprocessor-based controls. They can be easily changed. Cutting a new cam is not required if a mistake is made at one point.

On jackshaft controlled boilers, the trim is accomplished by adjusting the linkage connecting the fan damper to the shaft. Changes in the relative position of the damper and the jackshaft will alter the air to fuel ratio. The adjustment has to be made in a manner that maintains some relationship to firing rate because of a change in damper position near maximum fire that would be considered minimal can be a major change in air flow when the burner is at low fire. Once again, microprocessor-based controls serve to recognize those problems and correct for them. With an older system, be aware that the same correction at high fire has to have a much smaller effect on air flow at low fire. The same rules for linearity exist.

Oxygen trim control of parallel positioning systems (including steam flow/air flow, inferential and full metering) should use a multiplier to change the relationship of the fuel valve and fan damper position for oxygen trim control. That way, any change in the two signals is proportional to load. Multipliers are not an easy device to make for pneumatic systems. Many use a simple bias adjustment, adding to or subtracting from the signal to the damper positioner to trim the air to fuel ratio and maintain an oxygen set point. On inferential and full metering controls, the air flow signal is modified by the oxygen trim so that the output of the transmitter should be multiplied by the correcting output of the oxygen trim controller to change it proportionally over the load range.

Originally, only utility boilers would be equipped with oxygen trim control. The analyzers required almost constant maintenance and recalibration. Hot wire analyzers, which combined a flue gas sample with some hydrogen and heated it until the hydrogen burned, were the first analyzers to prove partially reliable and low enough in cost to be used in industrial plants. The paramagnetic analyzer, which used the difference in oxygen content of a gas to disturb a magnetic field, then followed. Both required drawing a sample of the flue gas from the boiler or stack and conditioning it before analysis. They used water systems to cool the gas, which always introduced a problem when there was any amount of oxygen in the water. The sampling systems had to operate at high velocity to reduce the time between analysis and a response to a change in burner operation. Leaks in the sample piping were always a concern.

The advent of the zirconium oxide analyzer made oxygen trim possible on even small commercial boilers. The analyzer can be mounted in the boiler or stack to achieve fast sampling and analysis. There were a few made with sampling systems, some integral to the analyzer. The *in-situ* zirconium oxide analyzer does not measure the oxygen content of the flue gas directly. The analyzer measures the difference between the oxygen in the flue gas and a reference gas. Often the reference gas is the air around the analyzer and, if the boiler casing, ductwork, or stack leaks, that reference can vary in its oxygen content. Many units still use a compressed air source as a reference gas. That can be complicated by particles or droplets of oil in the compressed air. To work the zirconium oxide cell (which is a ceramic substrate coated with the metal oxide), it must be heated to a temperature around 1500°F. At that temperature, any oil in the compressed air will burn and deplete some of the oxygen in the reference gas. If the analyzer(s) use compressed air, provide a separate compressor for them, one of the inexpensive, oil free compressors that only have to produce air at 10 psig or so. Besides, it is a real waste to dump air that was compressed all the way up to 100 or 150 psig for use as a reference gas. Size the little compressor to match the analyzer needs, plus a little for calibration. That provides far less expensive air, and it is oil free.

If oil is fired regularly, incorporate a procedure to prevent damage to the analyzer while blowing tubes. Steam soot blowers add a considerable amount of moisture to the flue gas when they are operating. The problem with that is that the steam has a much higher specific heat than air. The heater in a zirconium oxide analyzer has to really put out to push the gas temperature up to 1500°F. It is not the going up that is the problem. It is when the soot blower shuts off and all of a sudden that heat is no longer needed. The analyzer overheats and parts burn out. One way to avoid this problem is to insert a soot blower header pressure switch in the heater power circuit. The analyzer may not work very well while soot blowing the tubes and will indicate low oxygen, making the air flow run a bit high. However, the analyzers will quit failing every month.

Regular failures of the analyzers and drifting of the calibration led to the provision of an air fuel ratio adjustment independent of the oxygen trim control to really limit the trim control range. That way, an analyzer failure did not produce a hazardous situation or a lot of waste. Figure 11-37(a) shows a schematic of the air flow loop with this configuration. The summer is set to apply a gain of 0.1 to the input. The full range of output of the oxygen trim controller is reduced to a multiplier adjustment of ±5%. That not only limits the extent that the trim controller can adjust excess air, it also uses the

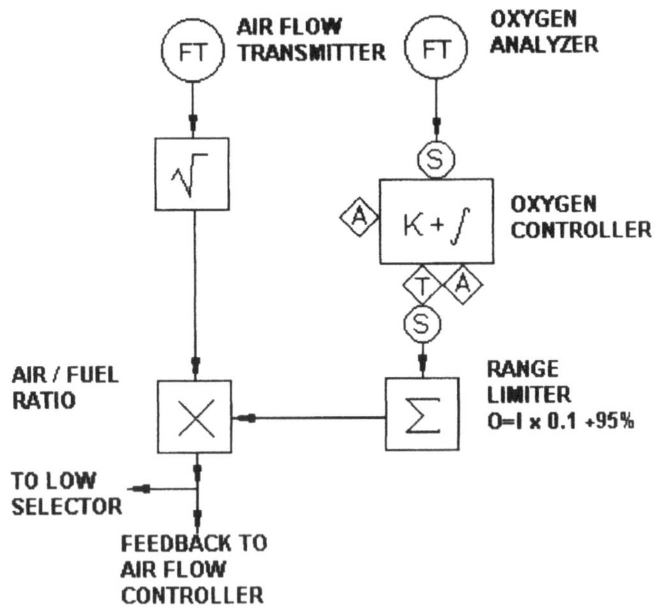

Figure 11-37(a). Air flow loop with limited oxygen trim.

full range of the trim controller output. When the oxygen controller output is at 50%, the multiplier for the fuel air ratio is 100%, basically one. The air flow signal flows directly to the controls without modification. That is where the output of the oxygen trim controller should be when everything is initially set up, right in the middle of the range. Then it can act to increase or decrease the excess air. It is something to look for after the controls are tuned. If the output is zero, the technician did not leave the control system any way to decrease the excess air. If it was set up in the summer, a reduction will probably be necessary in the winter, when the fan starts pumping colder air. On the other hand, if the output is well above 50%, it limits the amount the system can increase the excess air. Lacking any reasonable explanation from the technician, the output of the controller should be right at 50% at any firing rate immediately after it is set up.

Recall that the oxygen content has to increase as firing rate decreased because of the loss of turbulence. An oxygen trim control has to provide for that. It does so by introducing a variable set point that is fed to the oxygen controller. Two methods are used. One adjusts the oxygen set point as a function of firing rate (basically, the boiler master signal). The other adjusts the set point as a function of load using steam flow. Using the steam flow is generally preferred. A boiler needs more excess air during cold starts because the fire does not get the benefit of the heat reflected back from a hot burner throat and any furnace refractory. Using steam flow to establish a variable oxygen set point produces a higher set point for the trim controls while the boiler is warming up. Once steam flow is established, tighter control can be accommodated. Then the result is the same as using the master.

The variable set point is generated by a control device called a function generator. A function generator permits a control technician, or a savvy boiler operator, to produce preset outputs at multiple points of input. Some function generators provide for four points, while others allow eleven. Its setup is simple with new microprocessor controls. It is possible to set an output for each input by simply entering the input and output values into the controller. In the past, this was done with a cam. The results were a little better because the cam was smooth. A function generator that is set with specific points has sharp changes at the points that can introduce instability in the control. In addition to the function generator, the oxygen set point generator should include a bias adjustment, or independent summer, that allows the operator to increase the oxygen set point to accommodate upsets, or unusual fuel conditions, and decrease it when the boiler is firing at a fixed rate (manually set) to optimize operation under that condition. The result is a set point that follows the same slope for excess oxygen that is produced by the offset of the air flow zero.

With this method, the oxygen trim controller is limited. It could easily wind up or down to the end of its output range if there was a considerable change in the fuel or some other factor. If the operator notes that condition, the first action should be to check the fires to see if their appearance indicates a condition indicated by the analyzer output. If the fires appear normal, then the analyzer should be checked for calibration. It is uncommon to need any more adjustment than that ±5%. With the addition of liquefied natural gas (LNG), there was no guarantee that 5% would always be enough. Rather than allowing the trim controls more latitude, add a manual station so that a boiler operator could put another signal on the summer. Giving that input a gain of 0.1, and changing the summer's bias from 95% to 90%, allows the trim control a ±5% and the boiler operator ±5%. If the fires indicate that the analyzer is right, the operator can adjust the manual air to fuel ratio adjustment (slowly) to restore the trim controller's output to 50%.

A gain was used on those inputs. Then the controller, or operator, could adjust their respective outputs over the full control range, from 0 to 100%. A large number of trim controls, and similar control loops, use limits that will not allow that. The output may be adjusted from 0 to 100%, but the limits only allow it to work from 40% to 60%. The rest of the time, nothing happens. Operators complained when they were exposed to limits on the output. The control parameters were modified to get the full range. A control technician should not set up a control system to impose limits that do not make sense. A 2-1/2% swing in air to fuel ratio is not something that is expected in the short term. Regular checks and adjustment of the manual ratio adjustment (to restore the trim output to 50%) are not likely to happen. Any significant changes should be discussed with the control technician. It may indicate problems with the controls or (more likely) one of the flow transmitters.

A boiler with a high stack temperature (over 500°F) will benefit from oxygen trim control. The ability to operate with somewhat lower excess air will reduce the total amount of flue gas. That, in turn, will reduce the stack loss and improve efficiency. Low pressure heating boilers, and boilers with economizers or air heaters, already have a lower stack temperature and will see less fuel savings. Single burner boilers can be fired with stack oxygen content of 1/2%–1% with a combination of full metering and oxygen trim. Also, don't expect oxygen trim to be a cure all. There is a definite lag in time between the change of an air to fuel ratio on a burner and the appearance at the analyzer cell of a gas sample that is the result of that change. If the controls are not aligned properly to maintain an air to fuel ratio with load changes, don't expect the oxygen trim controls to correct that. If the trim controller output changes with load, that is what it is trying to do. Be aware that the oxygen measured at the analyzers did not necessarily come through the burners. This is particularly true on induced draft (ID) and balanced draft boilers, where the furnace and boiler passes are at a lower pressure than atmospheric, and air can leak in at several points after the burners.

One solution to problems with oxygen analyzers sensing oxygen that did not come through the burners is to control based on another parameter. Carbon monoxide, the result of incomplete combustion and a gas that is always present in some minute quantities in flue gas, can be sensed and controlled. The original oxygen analyzer problems still hold for analyzing CO. There is also the problem of its minute quantity. Typical requirements are control at about 50 ppm (parts per million). That is 0.005%, a very little amount of gas and hard to measure accurately. Some states are requiring CO limits down to 10 ppm. Large utility boilers frequently use CO measurement to resolve the problems with casing and ductwork leakage. The operating modes are the same as for oxygen trim. Utility units have to install continuous emissions monitoring systems (CEMs) and report their emissions to the EPA on a regular basis.

Utility boilers, and other very large boilers, have some special considerations. In particular, they are connected to steam turbines with electric generators. They are also at very high temperatures and pressures. The size contributes to longer time lags. The high temperatures and pressures put more stresses on the thick walled pressure parts, requiring slower response times. The steam turbine and generator put additional requirements on the boiler. The steam turbine and generator set have rotating parts. In North America, the rotational speed is 3600 rpm (revolutions per minute), which provides 60 cycle (60 Hertz) alternating current. In Europe, and much of the rest of the world, the rotational speed is 3000 rpm or 50 cycle power. The 50 cycle speed was chosen as that is the speed below which the human eye can detect the fluctuations in an incandescent light bulb. These turbine and generator sets have to be synchronized with the power grid. Those fluctuations, of 60 cycles per second, must be exactly the same for all the light bulbs and electrical machinery in North America. This is called automatic frequency control (AFC). In North America, deviations of more than 0.1 cycles are not tolerated. Problems with frequency control on the generator can lead to plant shut downs.

The generator works by moving a conducting wire through a magnetic field. The magnetic field is created by an electromagnet. In commercial generators, the electromagnet is rotated, while three sets of windings around the rotating magnet interact with the magnetic field to induce a current. From an electrical point of view, this produces three-phase current flow. Think of it as positive, negative, and ground, although that is not precisely correct. That current flow goes to a transformer that increases the voltage for transmission to the end users. At the user end, there are transformers that step down the voltage to the required levels for industrial plants, commercial buildings, and home use. The voltage is maintained constant at the generator, while the current increases or decreases to meet the demand. There is a governor on the generator that detects the demand level and increases or decreases the strength of the magnetic field through the electromagnet.

The governor provides a signal to the steam turbine to increase or decrease steam flow accordingly. The rotation of the steam turbine shaft is what drives the generator. An increase in steam flow will try to increase the speed of rotation of the turbine. The governor on the generator increases the strength of the magnetic field, which provides a resistance to the steam turbine rotation, keeping the speed constant. When the demand calls for more power, the signal from the governor goes to the steam turbine to open the valves that allow steam flow into the turbine. Opening the valves calls for more steam flow. When that happens, the pressure in the steam drum starts to drop. A pressure transmitter in the steam drum senses the drop in pressure and translates that into a request for more fuel. As more fuel is added to the boiler, the oxygen content of the flue gas will begin to decrease. The oxygen sensor will pick up the decrease and send a signal to the windbox dampers to open up to allow more air to flow into the boiler to burn the additional fuel. The rate of change, or ramp rate, is generally limited to 2% per minute to avoid undue stress on the thick wall pressure parts. Note that this system does not necessarily need to know the precise values of the various flow rates. The system only needs to know that there is enough to satisfy the appropriate set points for the steam drum pressure and the oxygen concentration. The control is entirely a feedback control system. For example, if some fuel leaked out before getting to the firing system, the controller would request more fuel flow until the drum pressure was restored or the feed system was maxed out. Likewise, if some air leaked out of the ductwork prior to getting to the boiler, the system would attempt to open the dampers further until the dampers were full open and the fan was at highest setting. Air that leaks into the boiler due to locations of negative pressure have to be accounted for by increasing the oxygen set point.

Currently, the predominant technique for implementing the boiler protection system (BPS), the BMS, and the burner control system (BCS) is with a microprocessor-based distributed control system (DCS) or programmable logic control (PLC) system. These devices make it possible to program the control system directly from the logic diagrams. They are configured to be strong in logic capability. They can be packaged ruggedly enough to be used in a power plant environment. This last feature is especially important in a coal fired power plant. The logic for a large steam system is commonly distributed to multiple controllers that are arranged in a fault tolerant configuration. With the reliability and capability of modern microprocessors, logic controllers are typically arranged in a "hot standby" configuration. This configuration is arranged where two controllers simultaneously execute logic instructions. In the event that the primary controller fails, control to the input/output (I/O) and communication is automatically transferred to the redundant, or back up, controller. To improve overall safety, additional fault tolerant configurations are in use, which are arranged in a "two out of two" or "two out of three" redundant architecture. The basic design criterion is that the failure of any single controller, with all of its logic capability, will not jeopardize the boiler and turbine safety or availability.

One of the reasons for this emphasis on reliability is that it takes time to cool down these large systems and then takes more time to start them back up again. For a large steam turbine, the temperature increase limits are typically 200°F per hour. With steam temperatures in the neighborhood of 1000–1100°F, it takes 5–6 hrs to bring the unit up to full temperature after the steam turbine has been "rolled" (i.e., steam has been admitted to turn the turbine) and then synchronized. Synchronization is carried out at as low a load as possible in order to minimize the forces put on the turbine and generator shaft when the final circuit breaker is closed and the generator is connected to the grid. The current grid (December 2020) has over 1.2 million Mw of installed capacity. Even a large 500 Mw unit is small when compared to the grid. Any error in timing or frequency during the synchronization process produces a force on the turbine and the generator shaft that attempts to push the shaft back away from its bearings. Since the grid is larger, the grid always wins. The grid always imposes its current frequency onto the equipment that is starting up. When this process was carried out manually, it was possible to actually break the turbine shaft. Now there is instrumentation that adjusts the turbine speed to match up with the grid and make the final connection electronically. This minimizes the forces exerted on the shaft. There is always some small difference between the turbine being started and the grid system.

With all of these considerations, it is clearly preferable to keep a plant on line once it has been started up. The operator and the control system cooperate with each other to try to avoid any unnecessary shutdowns. Plant operators still retain the ultimate responsibility for the operation and protection of both utility and industrial boilers. There are a number of conditions that would cause an emergency shutdown, or trip, of a large boiler. They include the following:

Low airflow

- Loss of all FD fans
- Loss of all ID fans
- Loss of all primary air fans (pulverized coal units)
- Turbine trip
- Inadequate water wall circulation
- High furnace gas side pressure
- Low furnace gas side pressure
- Low drum level (drum units)
- High water wall outlet temperature (once-through units)
- High separator level (once-through units)
- High superheater outlet pressure (once-through units)
- Flame failure
- Loss of logic power to the BPS/BMS
- Loss of primary and redundant BPS/BMS
- Operator's emergency trip push button is depressed

Awareness of all of these potential issues combined with the knowledge of how the system and controls actually work provides the wise operator with the tools to manage these larges systems safely and economically.

DRAFT CONTROL

Many small boilers use natural draft and a natural means of draft control. The gas fired hot water heater in a house is one. Most use a draft hood, nothing more than an open box over the outlet of the boiler. Natural draft up the stack produces a difference in pressure between the bottom of the hood and the rest of the room. The space under the hood is negative with respect to the room. Check to ensure that there is a negative pressure there by holding a match or lighter near the bottom edge of the hood. If there is a draft, air flowing from the room into the hood will pull the flame into the hood. The air that is pulled into the draft hood from the room goes up the stack. It cools the stack gases, which lowers the natural draft until there is not a significant difference between the pressure in the room and the pressure under the hood. Since the boiler outlet is under the hood, the pressure at the outlet and the boiler inlet differs by the natural draft through the boiler. If the hood was not there, the pressure at the boiler outlet could vary so much that it could blow out the fire, as on cold days, or be so high that there would not be enough air for clean combustion. That draft hood stabilizes a fixed fire operation to ensure maintenance of the air to fuel ratio.

Instead of a draft hood, a barometric damper can be used. That is a single bladed, usually round and pivoted just above its center, damper that separates the boiler room and the stack. These usually have stops on them such that the damper will not swing out at the bottom into the boiler room. As the stack draft increases, the difference in boiler room pressure and the stack base forces the damper open and cold air from the boiler room slips into the stack to cool the stack gases and reduce the draft. If the draft gets too low, the damper closes down to restrict the flow of cold boiler room air into the stack. The stack temperature then increases to raise the draft. Those dampers usually have a weight mounted on a stud to adjust them. By screwing the weight in and out, the pressure required to open the damper can be adjusted.

Barometric dampers, or some other means of controlling draft, are essential on systems with two or more boilers attached to the same stack. Draft is always a balance between the differential pressure produced by natural draft and the resistance to gas flowing up the stack. Double the flow of gas, by firing two boilers instead of one, and the resistance to flow up the stack will increase by a factor of 4. It is obvious that the pressure at the base of the stack will differ considerably. That means that the flow of combustion air and flue gases through the boilers will too. It is impossible to maintain air to fuel ratios in boilers with a common stack unless there are draft or metering controls. Barometric dampers do a fair job. They also require a lot of additional air supplied to the boiler room. In the winter, that air has to be heated or worry about freezing some pipes. If exhaust air from other sources is not available and there is a lot needed to control that draft, other means of draft control, more expensive means, will lower the operating cost and pay for that expensive control.

In addition to accounting for a variation in flue gas flow, draft controls can maintain a parameter in the boiler, such as furnace pressure, a requirement for balanced draft boilers. Many operators believe that is the only place to control the pressure with draft controls. Nothing could be further from the truth. If there are two or more boilers capable of pressurized firing, the draft can be controlled anywhere between the furnace and the outlet for individual boiler control. If controlling the common draft (at all boiler outlets), it has to be controlled there, or at a central point in the breaching, where there is little difference in pressure as flows change. A common control is not recommended because it can fail to prevent operation of all the boilers, and it is very difficult to get the large damper in a common stack to handle all the turndown that is required of it.

Balanced draft boilers require a means of controlling the ID fan to keep a constant pressure in the furnace, something slightly less than atmospheric pressure. A typical control loop looks no different than any other control loop. A transmitter senses the furnace pressure and sends a signal to a controller, which alters its output to an actuator for a boiler outlet damper or a damper at the inlet or outlet of the ID fan. It can also vary the speed of the ID fan. The control is not that simple. There are a number of factors that influence it. First, the pressure in the furnace of the boiler should only be slightly less than the pressure in the boiler room outside the furnace. That way, any air that leaks into the furnace and boiler passes is kept at a minimum. That air is heated to stack temperature and thrown away, just like excess air. It is a loss that should be minimized. The furnace pressure transmitter is really a differential pressure transmitter, comparing furnace and boiler room pressure. It should have a maximum range of 6 inches water column and, preferably, have a range of 1 inch (for smaller boilers). Don't mount the transmitter on the control panel in another building and sense the pressure using draft gauge piping. The boiler will pressurize regularly, blowing smoke and soot out into the operating area.

The differential that the transmitter measures is so low that it needs a large diaphragm to accurately measure it in the required range (less than 2 inches of water column). The larger diaphragm transmitter costs a lot more than the standard differential pressure transmitter (like about 3–4 times as much). Many plants are fitted with one that saved the contractor a lot of money but really does not work. The desired operating point for a furnace pressure is 0.05–0.2 inches of water column below the boiler room pressure. Transmitters with a wide range, like 50 inches or so, become too unstable for good control. Also, at the low pressures encountered for draft control, any pressure fluctuations due to a noisy fire create a very noisy pressure signal. Considerable filtering is necessary to get a steady output. Frequently, the location of a furnace pressure sensing connection has to be moved because the selected spot just happens to be where heavy pressure waves from combustion noise strike it. Of course, there is also the problem of incomplete combustion that can create a noisy signal.

Once any problems with the furnace pressure signal are resolved, there is the problem of load changes. In the old days, when controls were expensive, that was accepted unless the boiler loads were constantly changing more than 10% or so. When necessary, cascade control was added, where the output to the boiler outlet damper became the output of the air flow controller plus or minus the output of the furnace pressure controller. The summer, which combined the air flow controller output and the furnace pressure controller output, also needed a bias spring to subtract 50% so that the furnace pressure controller output would end up at midrange, just like two-element and three-element feed water controls. Those draft control systems got tuned with changes in gain applied to the air flow controller input, along with the bias to satisfy control requirements, which varied between boiler startup and operating conditions. Modern microprocessor controls can use the stack temperature as an input to help compensate for the variation in conditions.

The use of balanced draft allows the boiler manufacturer to use open inspection doors and joints that are not exactly gas tight in furnace construction. The result is that there are plenty of places for atmospheric air to enter the furnace. The leakage at two tenths of an inch is three times as much as the leakage at five hundredths, where the set point should be. Operating at five hundredths will allow an occasional puff of furnace gases into the boiler room, especially during startup, but will provide far more efficient operation. With modern microprocessor-based controls, using the stack temperature could permit varying the set point to minus two tenths for startup and increasing to minus five hundredths for normal operation.

FEED WATER PRESSURE CONTROLS

Feed water pressure control is unique to boiler plants. For pumps, the differential follows the pump curve. As long as the discharge pressure is less than the maximum pressure rating of the pump and piping, there is no way the pressure can get too high. Some pumps do have a rather steep curve. That may cause a concern about the pressure getting too high. Most of the time, the problem is with the feed water control valves. A pneumatic, or electro-hydraulic, actuated feed water control valve can be selected with an adequate diaphragm, or enough hydraulic pressure, to keep the valve closed under conditions of the maximum feed pump discharge pressure and no pressure in the boiler. The thermo-hydraulic and thermo-mechanical valves, described earlier, had limited power and, in most cases, could not operate with a pressure differential greater than 30–50 psi.

Another reason for pressure control was to improve operating efficiency. Turbine driven boiler feed pumps could be controlled using feed water header pressure, or the difference between feed water and steam header pressures, to throttle the steam to the turbine. It reduced steam flow through the turbine to save energy. Actually, it saved by allowing operation of more auxiliary turbines to eliminate motor operating costs. The normal practice for maintaining a constant feed water header pressure, or a differential between feed water and steam headers, consisted of the installation of a control valve that dumped feed water back into the deaerator or boiler feed water tank. That method maintains a header pressure or differential, but it also wastes a lot of energy. In one industrial plant, the operators had two feed pumps running (in case one failed) in the summertime, when one was four times larger than the actual load. Maintaining the header pressure by recirculating the water ensures that the pump runs at full capacity (maximum horsepower) all the time. That industrial plant was running one pump at 30 horsepower for nothing but the mental comfort of the operators, in case the other one failed. Total cost for that pump operation was 52.5 horsepower more than necessary, equal in 1997 energy costs to about $5000 per year. That is money that never became a bonus for the operators.

If it is necessary to control feed water header pressure with electric motor driven feed pumps, try to get an evaluation of the application of one or more variable speed drives (one for each pump preferably). They can be used to maintain pressure by slowing the pump down and saving on the horsepower. There is a practical limit to how slow they can go, but, most of the time, they will provide all the pressure control that is necessary. As with other things, technology improvements and manufacturing cost reductions has made such controls a wise investment.

All constantly operating boiler feed pumps have a potential problem with overheating, cavitation, and pump damage that can occur if all the feed water control valves shut off. Temporary upsets in plant operations can result in high water levels in all the boilers, meaning that can happen. If the water does not flow through the pump, then it just sits there and churns. All of the energy put in by the motor is converted to heat that raises the temperature of the water. The water that left the deaerator was nearly at saturation conditions. Thus, the additional heat will most certainly result in steam generation, cavitation, and pump damage (see the discussion on pumps). To prevent the damage from such an incident, some feed water circulation is provided. The recirculation provided by pump discharge pressure control solves this problem but at a very significant operating expense. The standard practice was to install a small recirculating line on each pump that returned enough water to the deaerator to prevent overheating of the water.

An orifice nipple is made for those recirculating lines. The recirculating lines were usually 3/4 inch pipe size. The orifice could be made out of a 1 inch diameter steel rod. One was installed on each pump to provide protection in the event that a pump was started with the discharge valve shut and then forgotten to open it. Of course, there were times when someone forgot to open an isolating valve on that recirculating line as well. They should have been left open. Another feature of those recirculating lines that was used in later designs was combining all the recirculating lines into one line returning to the deaerator with another orifice (sized for all pumps). That way, some of the recirculating water would flow backwards through the other, idle, pumps to keep them warm (the recirculating line was connected before the discharge check valve). That may also have prevented damage to the pumps due to lower stress from the piping.

Higher pressure boilers require substantial feed pump energy. The recirculating flow represents a significant amount of extra horsepower. It can also represent a significant reduction in the amount of water that could be supplied to the boiler. Plants operating at pressures of 250 psig and higher have had a solution for this problem. It is a self-contained check and recirculation valve, which consists of a spring loaded check valve that checks the main fluid flow and an integral recirculating valve that opens as the main flow decreases. The problem with those valves is that they are very expensive. Regardless, they work and they pay for themselves in power savings. The advent of microprocessor controls allows the addition of more and more features to a control system. These provide ways to help the boiler operator. Limit circuits prevent the system from trying to operate beyond the current capability of the plant, thus maintaining safe operation. With all of the information that is now available to the operator, there is a danger of "information overload." I&C permit the operation of plant systems from central location. Control room screens, meters, gauges, and lights display equipment status. Distributed controls systems permanently chronicle plant performance. Diagnostic systems monitor and evaluate plant status and performance.

With these more complex systems, an alarm is typically sounded to warn the operator of an impending problem. In view of all of the information that is being taken and analyzed, there is the risk of perpetual alarms going off. An events and alarms viewer allows the operator to get information about operational problems immediately and monitor whether corrective actions are being taken automatically by the system. Events and alarms can be sorted and filtered. A hierarchical system provides a priority assignment to the alarm event and guides the operator in monitoring the situation while analyzing the problem. A well-designed operator interface (Figure 11-37(b))

Figure 11-37(b). Modern operator interface.

provides the operator with all of the information needed for safe operation of the boiler.

INSTRUMENTATION

Instruments are there to provide the operator with information on the status of the plant and provide a history of the plant's performance. The wise operator knows how to use those instruments. Instrumentation varies in sophistication and precision from an indicating light to a fully compensated fuel gas flow recorder. Some, like the indicating light, give an immediate perception of the status of the plant. Others, like a flow totalizer reading, have to be subjected to study before the status is determined. One key to the use of instrumentation is that it is not worth anything if it is not recorded. Many of the reasons for recording data are explained in the section on boiler logs. The purpose of this section on instrumentation is to convey some points on interpreting readings and understanding the effect of other conditions on the instruments. For those plants that must report their emissions to the EPA and the state permitting agency, the plant control system must have a data historian that collects and saves all of the data from the plant instrumentation.

An indicating light provides information on two states or conditions: on and off. That presumes that it is working properly. If the light is off, it could be because the bulb is blown or the power is shut down to that piece of equipment. If the light is on, there is voltage and current at the indicating light. That does not necessarily mean that the status it is indicating really exists. That is very true for pumps, fans, etc., that are powered out of a motor control center with a common control power transformer or where there is a motor area disconnect. It is possible for the motor starter to pull in, making a contact and energizing a motor running indicating light, while the motor is sitting there powerless, because the power circuit breaker, or the disconnect, at the motor is open.

Meters and other electrical devices are directional. Recently, some operators blew up a rather expensive set of electrical switchgear because the phases were not realigned after some maintenance. They thought "alternating current goes both ways, meaning that there is not any direction." It is important to remember that it is different for each phase. Single phase power can come from a transformer on one of the other two phases and they are not parallel. When power lines to the plant are downed during a storm, it is very important to be sure that when the lines are reconnected, all three phases are connected up. With only two phases connected, the voltage will double, causing serious damage to electrical equipment.

Pressure gauges show pressure. Yet, a steam pressure gauge that is mounted at the operating level and has connecting piping to a steam drum 20 feet or more above the gauge also has a standing leg of water on it. To properly indicate the pressure in the boiler, the gauge has to be calibrated to read zero when it has that standing head of water. Thermometers read the temperature at their bulb. That does not mean that the fluid is at the same temperature just a few inches away from the bulb. Use the steam tables in the Appendix to find the temperature of the steam in the boiler. Thermometers in the top of a pipeline can fail to indicate the temperature of the liquid flowing underneath the bulb. Similarly, air in the top of piping or a vessel can insulate the thermometer from the heat of the liquid. Part of using instrumentation properly is realizing when a reading has to be wrong.

Steam flow recorders, unless compensated, are calibrated for a certain operating pressure. If the header pressure is higher or lower than the recorder, then the readings are wrong. If the steam pressure varies at the recorder (more than plus or minus 2 or 3 psi), and it should be accurate, it needs to be compensated. Compensated recorders for steam use a steam pressure and/or temperature input that allows calculation of the density of the steam at the orifice for accurate measurement. Superheated steam flow recorders need both pressure and temperature inputs to determine the density of the steam. Saturated steam only needs one of them.

Fuel gas flow recorders are subject to the same errors from pressure and temperature fluctuations as steam flow recorders. By maintaining the pressure constant, there is usually little variation between actual and recorded flow, in which case it is suitable to use a simple recorder. Normally, fuel gas flow is recorded at each boiler since there are flow instruments to provide a control signal for the firing rate controls and it does not cost much more to add the recorder. For purposes of control, little errors in the gas flow recordings are not significant. Compare the readings of the fuel gas recorders with the gas company's meter to be sure that the fuel billing is correct.

Track inventory and manage it. When burning heavy fuel oil, it has to be heated. It will look like the unit burned more oil than what was available. The oil in the tanks can be checked by sounding the fuel oil tanks. The oil in the tanks is maintained at a temperature much lower than the temperature at which the oil is burned (to save energy). The oil that is burned is measured by a fluid meter, after the oil is heated for firing. The oil expanded as it was heated. A gallon of fuel burned will be less mass than the gallon in the tank and less mass than the gallon of oil that was delivered. In order to get the proper heat input, more gallons of oil will flow through the burner. Correct for temperature to keep a good accounting of the oil inventory. The oil in the tanks should match a calculation of what was there, plus what was delivered, minus what was burned. If it is a little less or more, simply show an inventory adjustment. If the calculation shows that there should be a lot more than what is in the tanks, there is a leak or an oil thief. If it is a leak, the local Coast Guard office must be called to inform them. That is federal law. The state environmental agency will also have to be called to make sure any leak is properly contained, collected, and disposed of.

Fuel oil or gas is measured by the supplier. The user has to pay for what they measured. The plant meter readings, values from the instrumentation, should be corrected to match the supplier's numbers so that the data are considered accurate. Divide the combined fuel meter readings for all boilers by the fuel supplier's number to produce a correction factor. Then multiply that result by the meter readings at each boiler to get the actual fuel consumed in the each boiler. Keep track of the correction factor and ask "why?" if it changes significantly. When firing oil, verify each delivery with a sounding.

A sounding is a measurement of the depth of liquid in a tank. The term comes from taking an ullage reading. To measure the depth of heavy fuel oil in a tank, a probe is used at the end of a tape, which looks like a brass rod with an upside down cup on the end. When that is lowered into the tank, it makes a plop sound when it hits the surface of the oil. Using the tape measurement from the top of the pipe and subtracting the depth of the tank from the reading gives the depth of the oil. Since the process involves making a sound (the probe going "plop" when it hits the oil), it is called sounding the tank. The actual measurement is called an "ullage," when it is the distance down to the top of the oil.

The sounding of light oil storage tanks does not require wiping off a lot of black sticky oil. Then the soundings are taken where the probe is simply a pointed brass rod or wood stick that drops to the bottom of the tank. Read the level where the liquid coats the rod or stick and wipe the thin coat of oil off the rest of it. That stick that is dropped into the oil tank is an instrument too. The tip can be torn off (there is usually a brass button on the bottom) or someone can need a piece of wood about that size and cut a few inches off. Also, just like the meter readings, strange results can occur when the temperature of the oil in the tank and the oil delivered differ considerably. Sometimes, it pays to take another reading on a tank a day later to ensure the change in volume is accounted for.

One of the most valuable and important instruments in any steam plant is the drum level gauge on the boiler. It is also one that can go wrong, with disaster close on its heels. Either leg can plug and present a false water level indication. Keep in mind that the only force that produces the level indication in that gauge glass is the level in the boiler. The measurement is in inches of water column. The steam side can be plugged to the point that only a small opening remains and the steam condenses in the glass faster than it can get through that small opening. The result is that the level rises, compared to the level in the boiler, until the condensing matches the amount that can get through the opening. If there is nothing but a small opening in the water leg, the level in the glass may rise to produce the additional pressure needed to force the condensate through the small opening. Any leak on the steam side of the glass has to be fed by steam flowing from the boiler. There is a pressure drop in the connection and piping associated with the friction of that steam flowing, making the pressure in the glass lower than it is in the drum. The result is a false high level indication. Note that all those potential problems produce a false high level. It can look pretty normal but can be wrong. Only a liquid side leak in a gauge glass assembly will produce a false low level indication. For large utility boilers, there is often a three-element system. Whichever measuring device is selected for use, there are three of them. The control system takes all three readings and selects the best 2 out of 3. That way, if one is inoperative, or otherwise inaccurate, the other two are relied upon to provide the control signal. That system minimizes shutdowns from faulty instrumentation.

A common instrument that does not get the attention it deserves is the draft gauge. Many plants do not even have them. Typical vertical draft gauges provide an indication of the pressures in the air and flue gas flow streams of a boiler and are valuable for indicating soot formation and damage to baffles, seals, and dampers. If installed properly draft lines will not plug. The best connection for sensing draft with a draft gauge is shown in Figure 11-38. The large pipe is sloped where it penetrates the boiler wall so that soot and dust that tries to accumulate in it can roll out. The cap at the end allows easy access to clean the boiler penetration when necessary.

Every change in direction of the sensing piping is made with a cross, closed with nipples and cap. Plugs

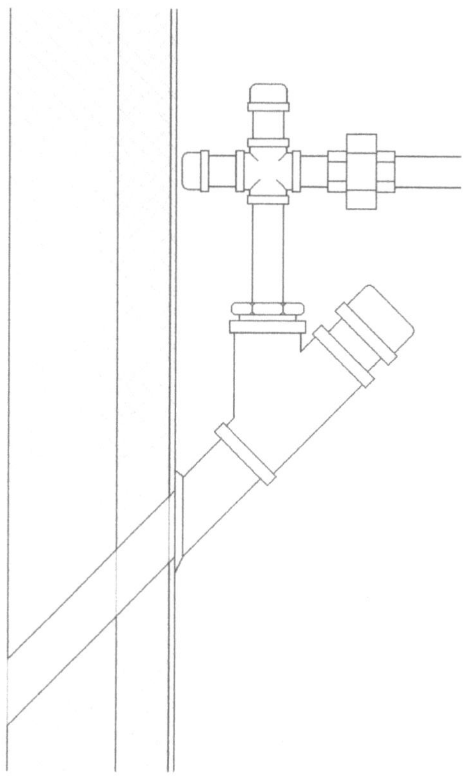

Figure 11-38. Draft sensing connection.

in those crosses will be next to impossible to remove after a year or two. Note that pipe is shown, usually no smaller than 3/4 inch. That is to allow a lot of room for dust to pile up before it fouls up the indication. In addition to allowing for removal of the piping, a union close to the sensing point is a great place to insert an orifice. There are always problems with draft gauges because they are measuring such low pressures and the flame can make a lot of noise. In some cases, a connection will need to be relocated because it is looking right at the fire, which can produce a very noisy pressure signal. Use a thin (1/16 inch) piece of copper with a small (1/8 inch) hole drilled off center in it and insert it in the union closest to the boiler. If the signal at the draft gauge is still noisy, take the piece to a vise and hammer around the hole to close it down some and then try it again. If, on the other hand, the indication seems sluggish, ream the hole out some. It is always a good idea to hang a tag on the union with this orifice to make it the first thing somebody checks if the gauge line acts like it is plugged.

Sensing lines for pressure gauges can affect the quality of their reading and, in some cases, can produce some operating problems if not installed and maintained properly. First, there is the matter of the size of the sensing connection. None should be smaller than 1/2 inch nominal pipe size (NPS). A 1/2 inch schedule 80 pipe nipple and valve is strong enough for most people to stand on without damage. Anything smaller is simply looking for trouble. Sensing connections should be made at the side or top of process lines to limit any debris settling into the smaller line and blocking it. The connection should be isolated with a valve as close as reasonable. Only provide enough room for a hand to get at the valve handle and make allowances for insulation. After the isolating valve, smaller piping or tubing can be installed from the connection to the gauge. If it gets broken, quickly shut the valve.

If it is not heated, heavy fuel oil does not flow well. Below a certain temperature, it becomes quite solid. To prevent blockages in sensing lines for heavy fuel oil, don't put heavy fuel oil in them. There are two approaches to the problem of sensing pressure of heavy fuel oil. They are dependent on the fill liquid. A light fuel oil, like Number 2, can be used or a heavy mineral oil such as Nujol. One is lighter than (floats on) the heavy fuel oil and one is heavier. When using light oil, the process connection, and all pipe and tubing connected to the process line, has to be flooded in such a manner that the light oil is trapped above the heavy oil. When using a heavy mineral oil, the process connection should be on the side of the piping and turn down immediately into a separating chamber. Thereafter, the sensing piping can be routed however it is needed.

With both systems, the separating oil must be injected into the sensing lines at regular intervals to refresh it because it will gradually mix with the heavy fuel oil. Since both burn, it is best to inject the separating oil while the burner is in operation. Most heavy fuels are fired with steam atomizing. The atomizing steam differential control valves have a chamber filled with oil to sense the burner oil pressure. It is best to inject the separating oil at the valve chamber to flush the piping and tubing all the way to the process line. A valve for that purpose should be provided at the chamber or at the sensing line connection to the chamber. Pump it slowly to avoid blowing the fire out.

A fuel oil sensing line can produce a hazardous condition. In a recent situation, the piping from the burner manifold to a pressure gauge in the control room was not properly vented. Since the line was full of air, it compressed every time the burner operated, allowing more than half the line to fill with fuel oil. When the burner shut down, the air expanded, forcing the contents of the sensing line into the furnace through the burner tip. In most instances, the oil simply burns off. However, keep in mind that a tablespoon of fuel oil, properly atomized and mixed with air to form an explosive mixture, can blow a boiler casing off.

Always bleed the air out of the piping when the accumulating effect of air is not desirable. Provide vent valves at the high points of the piping. Keep a piece of the appropriate sized pipe, bent with a 180-degree turn, to insert in the outlet of those vent valves to allow for cleanly and safely bleeding the air and catching any liquid spill in a bucket.

On the other hand, some sensing lines and gauges are protected by air trapped in the sensing lines. The air can serve as a cushion to limit the impact of noise on the gauge. A gauge line, for a heavy fuel gear pump, can use the air to quiet the effect of the bump each time a gear squeezes out its oil. Centrifugal pumps can produce fluctuations in the line that are associated with the vanes passing the cutoff. Some acid and caustic processes provide for the air to separate a process fluid and a pressure gauge that would be destroyed by that fluid. When there are situations where it is desirable to have the gauge sensing piping full of air, the sensing lines should be fitted with vent and drain valves to allow removal of any liquid that may absorb the air.

Pressure and flow transmitters, or any transmitter, should be installed where it is convenient to get at them for checking and calibration. With today's computer tools, it is possible to model the physical arrangement of the plant and have the program check for interferences and convenient access. Back in the day, for large systems, plastic models were made to check for interferences, platform and ladder locations, access doors, and instrumentation locations. These scale models then became a showcase for visitors to get an appreciation of the plant. A good location for transmitters, or other process instrumentation, is an elevation 4 feet above a floor or platform and readily accessible to a person standing on that floor or platform. Sometimes, it requires extra piping and installations where the operator may have to blow down the sensing lines a little more frequently.

Pressure and differential flow transmitters require piping connecting them to the process line. Some of those lines require long runs of sensing lines. They should be installed in a manner that limits problems with the instruments. The most common problem (Figure 11-39) is a transmitter installed at the bottom of a sensing line. Any scale, rust, or sediment that comes drifting down the line ends up inside the transmitter.

Liquid pressure and differential pressure transmitters should be installed as shown in Figure 11-40 so that the only thing that flows to the transmitter is liquid. The

Figure 11-39. Improper transmitter installation.

rust and sediment ends up in the drop leg, where it can be removed by blowing down the line through the drain valve. It is almost impossible for the dirt to get up into the transmitter (it will if the transmitter is vented too fast). Despite some arguments to the contrary, steam will not get into the transmitter when a steam pressure sensing unit is blown down.

The location of the transmitter and the type of fluid to be sensed have a lot to do with how a transmitter is piped. The diagram in Figure 11-41 is recommended for dirty liquid systems, making it more difficult for solids and debris in the system getting into the transmitter.

The piping routed to the process sensing connection should always run vertically or at least slope up to the connection so that any gas that may form in the sensing piping will naturally rise to the process connection and be replaced by liquid. A little air in a liquid sensing line for flow measurement will introduce a considerable error. If the transmitter is sensing a non-condensing gas

Figure 11-40. Proper transmitter installation.

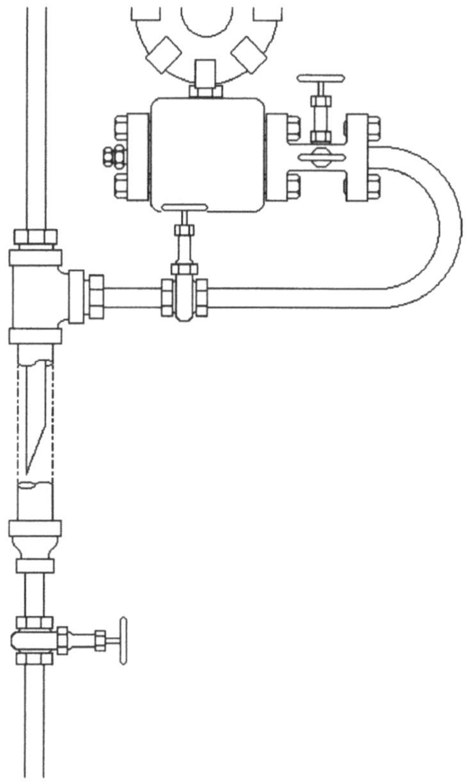

Figure 11-41. Steam and liquid transmitter piping.

(just about anything but steam), the transmitter should be mounted above the process sensing connection and run in such a manner that anything condensing out of the gas will run back out of the sensing lines into the process line. When it is absolutely necessary to install the transmitter below gas piping (especially for compressed air and, in some parts of the country, fuel gas), the arrangement shown for liquids should be used, and a schedule prepared for regular draining of the dirt legs. Otherwise, install it above the line to let everything drain away.

Installation of oxygen analyzers, and their sampling locations, has varied with the type of instrument over the years. The *in-situ* analyzers eliminate problems with sampling lines but introduce other problems. The analyzer has to be installed where it senses a representative sample of the flue gas. It also has to be where the wiring will not be overheated and in a manner that ensures the reference gas is not contaminated. See the discussion under the section on oxygen trim. Some *in-situ* analyzers have been installed at the furnace outlet, which will work well on boilers with low heat release rates. If the temperature of the flue gas at the sampling point is above 1500°F, then the gas will be too hot to control its temperature and the analyzer will produce erroneous signals. A good location for the installation of the analyzer is such that the probe is centered in the upper third of the smallest gas passage (in cross section) at the boiler outlet. If the boiler is equipped with an economizer, or air heater, it should be installed before that equipment. On very large boilers, the economizer outlet is a common location.

Thermometers and temperature transmitters are occasionally installed in such a manner that they are nearly useless. The temperature sensing portion of the instrument must be in the process fluid where it is flowing. One measurement that is always a problem is boiler stack temperature. A short stem will not penetrate the stack. On others, the thermometer bulb was located in a zone where the flue gas was idle, a stagnant zone where the gas was much cooler than the flowing flue gas. Here is a case where modern technology has created some problems. The two common instruments used for temperature detection, resistance temperature detectors (RTDs) and thermocouples, are point instruments. They only sense temperature at one point. A probe can be traversed back and forth across the stack, or boiler outlet, several times in a pattern to ensure that a good average reading of the gas temperature is obtained. The problem is that they were filled with mercury. An RTD has temperature limitations and should not be used for the higher temperature locations.

A multipoint thermocouple, with an element that spans the stack and has several terminations in it, along with several reference junctions outside the stack, will provide an average reading. A single point bimetal thermometer can be used for the local instrument. The large dial makes it easy to read from the floor or access platform level. Make certain the stem is long enough to always be in the center of the gas flow. A boiler stack should have 3″–4″ of insulation and a nipple and coupling extending through to mount the transmitter. A standard 2-1/2″ bulb is too short. With the possible exception of stack and air duct temperature measurements, all thermometers and temperature transmitter elements should be installed in thermowells. That way, if there is a question of an instrument's accuracy, it can be removed and have its calibration checked.

Stacks and air ducts may simply contain air at ambient temperatures or be under negative pressure. There is no hazard associated with removal of the thermal element and a thermowell is not necessary. Sometimes, however, the well is essential to support the thermal element. Thermowells tend to slow the response of the instrument to changes in temperature. They have to heat up before the thermal element. Thus, there is no reason to install them where they are not necessary. Some process applications do not use thermowells to achieve faster response time. Many thermowells are filled with grease or other compounds to improve heat transfer between the well and the element. Both RTDs and thermocouples require more expensive wiring than the typical twisted shielded pair required for a transmitter. Exposing that wiring to electromagnetic fields in the plant can also produce erroneous outputs.

By installing local transmitters, the inventory of special wire can be eliminated as well as a lot of running back and forth when trying to check the calibration of the instrument. A local reading of what the transmitter is sensing can be provided by adding a relatively inexpensive meter on the transmitter. The only caveat with local transmitters is that they are not designed to be mounted on hot ductwork and piping. Mount the temperature transmitter away from the probe on another support attached to the building structure. That requires the temperature element to be fitted with extension leads long enough to reach the transmitter. Three feet should be a requirement for extension leads (except stack temperature elements where double that is needed). There should be enough lead to conveniently locate the transmitter at a platform or grade, where it is readily accessible, 4 feet above just like for pressure and flow transmitters.

Data Analytics

With the cost of sensors and microprocessors declining, and the amount of data increasing, there is the question of what to do with all of that data. Data is of little use if it is not analyzed to find out what the data really means. There are now software programs that can take that data and analyze it to determine its significance. "Data analytics" is the term used to describe this multivariate type of analysis. Multivariate software analysis looks at all of the plant variables, typically over a two-year period. It then produces an output that shows which variables impact the operation the most. If reduced emissions are the goal, the software will examine all of the measured data to determine which ones impact the emission the most, either positively or negatively. The data are typically represented in a single bar graph, arranged from the most positive to the most negative. The most significant five or six variables at the ends of the graph are highlighted as the ones to concentrate on for improvement. The same thing can be done for efficiency or other outcomes. This information now provides a guide as to which items in the plant should be looked at further to see how plant operations can be improved. Some may be obvious, such as the impact of excess air on efficiency. Others may be less obvious such as a minor burner instability. With this initial data, it is possible to construct a real-time model of the plant. The model can then be operated in real time to optimize the operation. Historical data is used to create the model. At one plant, data from 18,000 sensors over a four-year period were used to set up the model. The software focuses on a limited number of outputs that are plotted in real time. If something is amiss, the operator is alerted so that the source and the cause can be identified and fixed. The model can also be run in predictive mode to look ahead at potential improvements. The system is designed to reduce the number of screens down to one or two. The problem points can be isolated and amplified to identify the contributors to that problem. The software provides a reasonably fast return on investment as there is no permit required. Of course, if there are changes to the plant configuration or the fuel being fired, the process has to be started over again to allow the software to "learn" from the new data.

The ultimate goal of these types of software programs is to create the "digital twin." The digital twin is supposed to be a model of the plant that is so representative and so accurate that one cannot tell the difference between the actual plant and the digital plant model. The software should be capable of running fast enough that it is essentially running in parallel to the actual plant itself. In this manner, the software can not only observe the actual plant for trends and procedures that are moving away from the best results but also suggest to the operator the proper course of action to achieve a desired result. This type of software, if successful, would lead to a truly automatic control of the plant. Even so, there are equipment failures and potential for software corruption that still require the presence and judgment of the wise operator.

Chapter 12

Why They Fail

When a boiler or related equipment fails, it is usually due to a lack of attention. While modern control systems normally manage to ensure that a failure is handled in a safe manner, i.e., a shutdown, the news media often have headlines involving catastrophic failures. Some of those catastrophes involve human suffering and death. Although not at the frequency and numbers of a century ago, it still generates grave concerns when an incident does occur.

WHY THEY FAIL

A Little Bit of History

A boiler explosion holds the record for the most deaths from a single accident. It happened in 1865, at the end of the Civil War, shortly after Lee surrendered to Grant. Over 1900 union soldiers clambered aboard the riverboat Sultana heading north to Cincinnati. Shortly after leaving the dock, the boilers exploded. Some died immediately. Others suffered from burns and shrapnel wounds until succumbing weeks later. About 1800 people, including women and children, died in that accident. With little left of the ship, the actual cause remains undetermined.

In the early 1900s, thousands died each year from boiler accidents. That is why the ASME proceeded to produce the boiler construction codes at the beginning of the 20th century. The dramatic improvements that produced continued up until the end of that century. The situation seemed to reverse in 1999. Data from 1996 to 2002 is depicted by the charts in Figures 12-1 and 12-2, which show the swing in primary cause from low water to operator error and poor maintenance plus an increase in incidents. Note that the data are very old. That is because the lawyers for National Board of Boiler Inspectors advised the board in 2003 to not publish that information. Fortunately, incidents reported to Occupational Safety and Health Agency (OSHA) appear to indicate a much lower trend in the last decade with one explosion reported in 2011, one in 2012, two in 2013, and two in 2017.

Boilers seldom wear out. The effects of wear that are associated with machinery and automobiles are nowhere near as significant with boilers. Most of the time, the boiler just sits there. There is rubbing associated with movement as it heats up and cools down. In a normal plant, it does not happen often enough to be important. Don't confuse the boiler with the burner. They really are separate items. Burners wear because there are moving parts associated with most of them. Pulverizers wear due to the grinding of the coal. Pumps and fans can wear, as they have rotating machinery. The boiler itself basically has no moving parts.

There are many boilers that are over 50 years old. Many have no evidence that they are nearing the end of their life. Boilers usually fail by incident. The most common incidents have to do with lack of, or improper, water treatment.

Water Treatment

Improper water treatment, or the lack of it, contributes to most of the failures. The boiler fails because scale builds up until some metal overheats. The metal fails to allow the steam and water to escape, where the water then flashes into steam so quickly that it violently blows the boiler apart. There is a whole chapter in this book on water treatment. There are opportunities to learn more at the treatment supplier's school or other sources. If a boiler operator is comfortable with the water treatment, the likelihood that the boiler will fail is very low.

LOW WATER

For years, the reports of boiler failures would show low water as the primary reason that the boiler failed. Even today, with special systems and all current knowledge, low water always stands out as a significant cause for boiler failures. Taking all the precautions, and conducting the regular testing, should prevent them, but they continue to occur. Occasionally, someone even tries to fire the boiler without putting any water in it.

RECENT HISTORY OF BOILER ACCIDENTS ACCORDING TO NATIONAL BOARD DATA					
POWER BOILERS	1996	1999	2000	2001	2002
SAFETY VALVE	1	1	1	4	8
LOW WATER	356	67	183	161	137
LIMIT CONTROLS	16	27	22	8	4
IMPROPER INSTALLATION	5	14	15	2	5
IMPROPER REPAIR	6	24	16	1	14
DESIGN OR FABRICATION	8	22	8	2	6
OPERATOR ERROR OR POOR MAINT.	125	140	193	82	90
BURNER FAILURE	40	27	10	29	16
HEATING BOILERS - STEAM					
SAFETY VALVE	5	2	14	2	2
LOW WATER	490	397	437	519	359
LIMIT CONTROLS	27	33	66	17	17
IMPROPER INSTALLATION	14	10	22	10	5
IMPROPER REPAIR	7	36	23	11	2
DESIGN OR FABRICATION	14	33	34	31	54
OPERATOR ERROR OR POOR MAINT.	125	258	412	406	262
BURNER FAILURE	59	20	19	29	16
HOT WATER HEATING BOILERS					
SAFETY VALVE	5	5	7	6	7
LOW WATER	112	221	258	195	96
LIMIT CONTROLS	24	68	69	19	23
IMPROPER INSTALLATION	15	31	68	13	11
IMPROPER REPAIR	3	87	28	10	2
DESIGN OR FABRICATION	20	67	40	30	60
OPERATOR ERROR OR POOR MAINT.	221	314	406	260	215
BURNER FAILURE	70	31	30	26	28

Figure 12-1. Chart of reasons for boiler failures in prior years.

BOILER ACCIDENTS, INJURIES AND DEATHS ACCORDING TO NATIONAL BOARD					
	1996	1999	2000	2001	2002
SAFETY VALVE	11	8	22	12	17
LOW WATER	958	685	878	875	592
LIMIT CONTROLS	67	128	157	44	44
IMPROPER INSTALLATION	34	55	105	25	21
IMPROPER REPAIR	16	147	67	22	18
DESIGN OR FABRICATION	42	122	82	63	120
OPERATOR ERROR OR POOR MAINT.	471	712	1011	748	567
BURNER FAILURE	169	78	59	84	60
TOTAL INCIDENTS	1768	1935	2381	1873	1439
INJURIES	56	63	24	66	16
DEATHS	4	15	8	8	3

Figure 12-2. Chart of accidents, injuries, and deaths, from late 1990s to 2002.

It does not matter if it is a hot water boiler or a steam boiler. It should have a low water cut off. Steam boilers should have two. In the last century, the most consistent reason for a boiler failure, accounting for about one-third of the incidents, was loss of water. Check the cut offs as often as possible and under different situations to be certain they are reliable. Low water cut offs come in two basic forms: float and conductance. Float operated cut offs, as their name implies, use a float to detect the water level and a lever connected to the float keeps the float in position and actuates the electrical contacts that open to stop burner operation. Conductance cut offs use probes, looking something like a spark plug, to detect water level by the difference in conductivity of water and steam or air. Low water cut offs should be installed to prevent burner operation in the event that the boiler water drops below a safe level, where the heating surfaces are exposed to steam. Normally, the lowest safe operating level in a boiler is the bottom of the gauge glass. The cut off should prevent burner operation near it. Cut offs are installed in two forms: external and internal. There are arguments for each installation and some boilers have both.

The failures of boilers, due to low water, continue despite the provisions of extra low water cut offs and regular testing of them. Perhaps, one principle reason is the failure to test them regularly so that a problem is detected before a failure occurs. Never fail to test the low water cut offs immediately after arriving on the job. They can fail because mud builds up in the piping connecting the cut off to the boiler or an accumulation of mud in the cut off housing. The mud is dirt that enters with the makeup and accumulates in the boiler water. It is usually suspended in the boiler water by the rapid circulation but will settle out in the water column and cut off piping and chambers because the water moves slowly in that equipment.

Float operated low water cut off failures include the normal problems of mud collecting in the piping between boiler and the float housing. The float chamber cannot drain. Then the level is higher than that in the boiler. (This happens if either the water leg or the steam leg is plugged. The chamber fills with condensate and cannot drain.) Mud filling the bellows and hardening to resist transmission of the float position, friction preventing operation of magnet actuated switches, stiffening with age of wiring connected to magnet actuated switches, fusing of contacts due to excessive electrical current, freezing of the switch actuating mechanism due to corrosion from boiler water leaks, or leakage into the switch housing, all can cause failures.

Probe types, using conductance, can fail because deposits coat the probe to simulate the presence of water. The opportunities for a low water cut off to fail are so many that regular testing (to detect problems) is the most important thing that can be done.

Remember that, despite the many schemes for testing the low water cut off, the only sure proof that the low water cut off works is gradually dropping the water level with the burner operating until the cut off shuts the burner down. Do it as often as possible, while keeping a close eye on the water level. Other tests to check it, explained in the normal operation description, should be performed with the recommended frequency. Always watch the level until cut off occurs since the odds are rather high that it will not work.

Since incorporating the timing of low water cut off testing into the burner management systems, there have been almost no failures of the boilers with those systems. There were a few incidents of the testing, revealing a problem with a low water cut off.

THERMAL SHOCK

Of all the modes of boiler failure, thermal shock seems to be the one that can happen at any time. In some cases, boilers did not make it past their initial week's operation without failing as a result of thermal shock. Boilers also failed after years of operation due to an incident of thermal shock. It is important to understand exactly how thermal shock destroys a boiler. There are several situations that are called thermal shock that are not consistent with the normal perception. Thermal shock can destroy a boiler in a single incident or it can take several shocks to produce evident damage. There is a specific combination that must exist for thermal shock damage. First, the metal of the boiler (or refractory) must be exposed to a change in temperature that is enough to produce a range of stress in the material.

The best example of thermal shock is pouring water over ice cubes fresh out of the freezer. The ice cubes crack. Even if cold water stored in the refrigerator is used, they crack. Steel is only about 7% stronger than ice. Similarly, thermal shock can destroy a boiler. The reason for the ice cracking can be explained by noting how the cracks form. When the water hits the ice, there is a rapid transfer of heat from the water to the surface of the ice. Keep in mind that ice contracts as it is heated. The operation is just the opposite for steel. The inside of the cube remains cold because the heat does not transfer through the ice as fast as the outside is warmed by the water flowing over it. Because it is warmed and tends to shrink, the outer layer of the ice cube is placed in tension, as if something was trying to pull it apart. The result is that it is pulled apart. Cracks form and, as the rest of the cube shrinks, the crack continues.

The second important element of thermal shock is thickness of the material. Shaved ice does not crack when cold water is poured over it. When the metal is thin enough, the difference in temperature across it is not adequate to produce enough stress to produce cracking. The thicker parts of a boiler, tube sheets, shells, headers, and drums, are more susceptible to thermal shock than the tubes. The third element is frequency. One violent shock may not be good for a boiler. Hundreds of little ones, repeatedly occurring, will eventually result in failure because of tiny micro fissures (very little cracks) that form in thinner metals or, where the temperature differences are not dramatic, will, if constantly bombarded with thermal shock conditions, eventually grow into large cracks that finally result in boiler failure.

Many people do not realize that thermal shock does not have to happen on the water side of a boiler. Any boiler that trips, while running at high fire, and immediately goes into a purge is subjected to thermal shock. The metal of the boiler's heating surface is immediately subjected to contact with cold purge air right after it was exposed to the hottest flue gas of normal operation. Add to that, the trip occurring near the maximum operating temperature (and related pressure) and there is the potential for failure.

The most common failure in this mode occurs with the ends of the fire tubes at the inlet of the second pass of a fire tube boiler. The reason they fail is because they are sticking out into the hot flue gas, where their temperature is elevated by the high fire flue gases. Then they suddenly encounter the cold purge air. That failure is usually one that results in gradual growth of micro fissures in the ends of the tubes and will even happen in tubes that are welded to the tube sheet. The primary reason for this type of failure is improper adjustment of the firing rate controls such that the boiler cycles off while the modulating controls are still at high fire or just left high fire.

Hydronic heating systems can operate to produce significant thermal shock by returning water from idle sections of the system (where the water got very cold) to the hot boiler. A slug of cold water is directed against the boiler heating surfaces. In some cases, this can be caused by automatic controls operation, especially day/night controls. Sources of the problem are usually close to the boiler because any slug of cold water in a remote system will be heated by the metal in the piping as it returns to the boiler. Service water heating with a hydronic boiler has a high potential for thermal shock if the heating water to the service water heater is cycled on and off. It is better to use a constant flow to the service water heat exchanger, with other provisions to prevent overheating the service water. See the section on service water heating for more on thermal shock in that application.

CORROSION AND WEAR

Nothing lasts forever, and that is very true for boilers. There are not too many boilers in operation that are more than 70 years old. There are a few, but they are few and far between. Nominal design life for boilers designed to the ASME codes is 30 years. Boilers that are well maintained and have good water chemistry can last longer. The normal end of a well-maintained boiler's life is almost always due to a decision to replace them, not wear. There are areas in a boiler that cannot be reached to monitor and prevent corrosion. Sometimes, they are due to installation and sometimes to manufacturing, but they are there. In many cases, the only way to address those spots is a major rebuild of the boiler to reach them and clean, protect, and recover them to extend the boiler's life. That is a sound decision in many cases. Most of the time, the problems with wear are all at the burner.

When the control valve on a boiler has run from high fire to low fire six or seven times a day, 365 days a year, for 10 years, it has run over 21,000 cycles. The ASME code has factors for operating cycles, with no additional allowance for boilers that are expected to cycle less than 7000 times in their lifetime. Except when the specification calls for it, most design engineers do not allow for more than that many cycles. Exactly how long can that system expect to run without failing? A major revamping of a boiler's burner and controls on a five-year cycle should prevent failures due to wear.

OPERATOR ERROR AND POOR MAINTENANCE

Regrettably, the National Board statistics, which are quoted here, do not provide enough breakdown to clearly indicate why trends exist or to detect reasons for trends. In many locations, central boiler plants have been replaced with multiple low pressure heating plants that are maintained by individuals without a license. This can increase the likelihood of operator error. Typically, problems that are attributable to operator error or poor maintenance are usually accompanied by an attitude, on the part of the plant management, that promotes or enforces the improper action or lack of action. Training can be useful for plant management as well as plant operators. Additional requirements under OSHA regulations include safety training and risk management for all personnel.

Frequently, it is not the operator that contributes to poor maintenance. The operator manages to keep the plant running by a growing mountain of temporary fixes that accumulate until nothing can keep the boiler running. The reason is management's attitude about maintenance. In some cases, operators simply have to allow the boiler to fail or shut down due to unsafe operating conditions. One of the advantages of a license is that a license provides the authority to do just that. Shut it down and refuse to operate it.

Ideally, the plant has a culture of safety. Safety must be a core value. The objective is to ensure that all employees and workers arrive home safely each and every day. Work conditions must be kept safe. All personnel are expected to behave safely. Safety is everyone's responsibility. It is not just about regulations and paperwork. Complacency is the enemy of a good culture of safety. Just because someone did not get hurt in a situation, it does not mean that the situation was safe. Under the OSH Act of 1970, employers are responsible for providing a safe and healthful workplace. OSHA's mission is to assure safe and healthful workplaces by setting and enforcing standards and by providing training, outreach, education, and assistance. Employers must comply with all applicable OSHA standards. Employers must also comply with the General Duty Clause of the OSH Act, which requires employers to keep their workplace free of serious recognized hazards.

It is true that a licensed boiler operator could make a mistake with disastrous consequences. A license is no guarantee and neither is training. However, the lack of the discipline involved in training and preparing for the exam leaves lots of room for error. Part of the business of acquiring a license includes the development of respect for the profession and a greater understanding of the responsibility. Attempt to get a license even if one is not needed. It is more a matter of attitude than the actual license. When a state licensing program exists, the wise operator seeks to obtain the license to support a professional perception of the role. Attitude and perception seem to be the key to operator error. When a boiler is damaged, any failure in operation is usually attributable to an attitude. It is the ones who seem to believe that they can get away with doing the minimum and that the company should be happy that they even show up, who eventually make the mistakes that result in damage. It is important to respect the potential for a boiler or furnace explosion and act accordingly. It is the people without fear, with an attitude that they are infallible, that take unnecessary risks with everything from shortening purge periods to skipping boiler water analysis, which eventually result in a failure. Recall from Chapter 1 that the wise operator's top priority should be personal safety. A word to the wise should be sufficient.

Appendix A

Properties of Water and Steam

Vacuum in. Hg	TEMP. °F	CU. FT. PER LB.		HEAT IN BTU PER POUND		
		LIQUID	STEAM	LIQUID	LATENT	STEAM
29.75	40	.01602	2423.7	8	1071	1079
29	79	.01608	652.3	47	1049	1096
25	134	.01626	143.2	102	1018	1119
20	161	.01640	74.8	129	1001	1131
15	179	.01650	51.1	147	991	1138
10	192	.01658	39.4	160	983	1143
5	203	.01666	31.8	171	976	1147
PRESS. PSIG	TEMP. °F	CU. FT. PER LB.		HEAT IN BTU PER POUND		
		LIQUID	STEAM	LIQUID	LATENT	STEAM
0	212	.01672	26.8	180	970	1150
2	218	.01675	24.1	186	966	1152
5	227	.01682	20.1	196	961	1156
8	234	.01688	17.9	202	955	1158
9	237	.01690	17.2	205	954	1159
10	240	.01692	16.6	207	953	1160
11	241	.01693	16.0	209	951	1161
12	243	.01695	15.4	212	950	1162
15	250	.01700	13.9	218	945	1163
15	250	.01700	13.9	218	945	1163
20	258	.01707	12.1	227	940	1167
25	266	.01715	10.6	235	934	1169
30	274	.01721	9.5	243	929	1172
45	281	.01727	8.59	250	924	1174
50	298	.01743	6.68	257	911	1179
60	307	.01753	5.89	277	904	1182
70	316	.01761	5.18	286	898	1184

PROPERTIES OF WATER AND STEAM (*Continued*)

PRESS. PSIG	TEMP. °F	CU. FT. PER LB.		HEAT IN BTU PER POUND		
		LIQUID	STEAM	LIQUID	LATENT	STEAM
80	324	.01770	4.65	294	892	1186
90	331	.01778	4.25	302	886	1188
100	338	.01785	3.90	309	881	1189
110	344	.01792	3.59	315	875	1191
120	350	.01800	3.34	322	871	1192
125	353	.01802	3.22	325	868	1193
150	366	.01819	2.74	339	857	1195
175	377	.01833	2.40	350	847	1197
200	388	.01847	2.13	362	837	1199
225	397	.01860	1.92	372	828	1200
250	406	.01873	1.74	382	820	1201
275	414	.01885	1.59	390	812	1202
300	422	.01897	1.47	399	804	1203
400	448	.01940	1.11	428	756	1204
600	489	.02020	0.73	474	728	1203
750	513	.02070	0.61	500	700	1200
900	534	.02130	0.49	529	665	1195
1200	574	.02233	0.39	587	624	1183

Notes:

- Data is for gauge pressure at one standard atmosphere (14.696 psia)
- Entropy and internal energy shown on standard steam tables is not included for clarity.
- Only pressures commonly used are shown, use of tables from another source is recommended if precision is desired.

PROPERTIES OF SUPERHEATED STEAM

28" Hg	Temperature	**150**	**200**	**250**	**350**	**600**
(101)	Volume	**2.01**	**2.05**	**2.08**	**2.10**	**2.27**
	Heat	**1128**	**1150**	**1173**	**1219**	**1336**
26" Hg	Temperature	**150**	**200**	**250**	**350**	**600**
(126)	Volume	**1.94**	**1.97**	**2.01**	**2.06**	**2.19**
	Heat	**1125**	**1150**	**1172**	**1227**	**1336**
11" Hg	Temperature	**250**	**300**	**350**	**450**	**700**
(200)	Volume	**1.80**	**1.84**	**1.87**	**1.92**	**1.99**
	Heat	**1169**	**1193**	**1216**	**1263**	**1335**
0 psig	Temperature	**250**	**300**	**350**	**450**	**700**
(212)	Volume	**1.78**	**1.81**	**1.84**	**1.90**	**2.02**
	Heat	**1169**	**1193**	**1216**	**1263**	**1383**
5 psig	Temperature	**250**	**300**	**350**	**450**	**700**
(227)	Volume	**1.75**	**1.78**	**1.82**	**1.87**	**1.98**
	Heat	**1167**	**1192**	**1134**	**1263**	**1383**
10 psig	Temperature	**300**	**350**	**400**	**500**	**750**
(240)	Volume	**1.75**	**1.78**	**1.81**	**1.87**	**1.98**
	Heat	**1186**	**1214**	**1238**	**1286**	**1407**
15 psig	Temperature	**300**	**350**	**400**	**500**	**750**
(250)	Volume	**1.73**	**1.76**	**1.79**	**1.85**	**1.96**
	Heat	**1209**	**1213**	**1238**	**1286**	**1400**
60 psig	Temperature	**350**	**400**	**450**	**550**	**800**
(307)	Volume	**1.67**	**1.70**	**1.73**	**1.78**	**1.89**
	Heat	**1207**	**1233**	**1268**	**1307**	**1430**
120 psig	Temperature	**400**	**450**	**500**	**600**	**850**
(350)	Volume	**1.61**	**1.65**	**1.67**	**1.72**	**1.83**
	Heat	**1222**	**1259**	**1276**	**1327**	**1453**

PROPERTIES OF SUPERHEATED STEAM (*Continued*)

150 psig	Temperature	400	450	550	600	850
(366)	Volume	1.58	1.62	1.67	1.70	1.81
	Heat	1217	1245	1299	1324	1452
200 psig	Temperature	450	500	550	650	900
(388)	Volume	1.58	1.61	1.64	1.69	1.80
	Heat	1241	1267	1304	1346	1476
250 psig	Temperature	450	500	550	650	900
(406)	Volume	1.55	1.59	1.62	1.66	1.77
	Heat	1231	1262	1301	1344	1474
300 psig	Temperature	450	500	550	650	900
(422)	Volume	1.53	1.56	1.59	1.64	1.75
	Heat	1236	1256	1286	1340	1472
400 psig	Temperature	500	550	600	700	950
(448)	Volume	1.52	1.56	1.58	1.64	1.74
	Heat	1243	1271	1306	1362	1495
600 psig	Temperature	550	600	650	750	1000
(489)	Volume	1.49	1.53	1.56	1.61	1.71
	Heat	1254	1289	1320	1380	1516
750 psig	Temperature	550	600	650	750	1000
(513)	Volume	1.45	1.49	1.52	1.57	1.68
	Heat	1238	1274	1312	1370	1512
900 psig	Temperature	600	650	700	800	1050
(534)	Volume	1.47	1.49	1.53	1.58	1.68
	Heat	1258	1297	1331	1393	1536
1200 psig	Temperature	600	650	700	800	1050
(574)	Volume	1.40	1.44	1.48	1.54	1.64
	Heat	1220	1268	1310	1378	1537

Note: Value in parenthesis is temperature of steam at saturation for that pressure.

Appendix B

Water Pressure per Foot Head

HEAD	Psi produced by water at				HEAD	Psi produced by water at			
(Feet)	60°F	140°F	212°F	240°F	(Feet)	60°F	140°F	212°F	240°F
1	0.433	0.456	0.415	0.410	36	15.59	15.35	14.95	14.78
2	0.866	0.853	0.831	0.821	38	16.45	16.20	15.78	15.60
3	1.299	1.279	1.246	1.231	40	17.32	17.05	16.61	16.42
4	1.732	1.705	1.661	1.642	45	19.48	19.18	18.69	18.47
5	2.165	2.132	2.077	2.052	50	21.65	21.32	20.77	20.52
6	2.598	2.558	2.492	2.463	55	23.81	23.45	22.84	22.57
7	3.031	2.984	2.907	2.873	60	25.98	25.58	24.92	24.63
8	3.464	3.410	3.323	3.283	65	28.14	27.71	27.00	26.68
9	3.987	3.837	3.738	3.694	70	30.31	29.84	29.07	28.73
10	4.329	4.263	4.153	4.104	75	32.47	31.97	31.15	30.78
11	4.762	4.689	4.569	4.515	80	34.64	34.10	33.23	32.83
12	5.195	5.116	4.984	4.925	85	36.80	36.24	35.30	34.89
13	5.628	5.542	5.399	5.336	90	38.97	38.37	37.38	36.94
14	6.061	5.968	5.815	5.746	95	41.13	40.50	39.46	38.99
15	6.494	6.395	6.230	6.156	100	43.29	42.63	41.53	41.04
16	6.927	6.821	6.645	6.567	110	47.62	46.89	45.69	45.15
17	7.360	7.247	7.061	6.977	120	51.95	51.16	49.84	49.25
18	7.793	7.673	7.476	7.388	130	56.28	55.42	53.99	53.36
19	8.226	8.100	7.891	7.798	140	60.61	59.68	58.15	57.46
20	8.659	8.526	8.307	8.209	150	64.94	63.95	62.30	61.56
22	9.525	9.379	9.137	9.029	200	86.59	85.26	83.07	82.09
24	10.391	10.231	9.968	9.850	250	108.24	106.58	103.83	102.61
26	11.257	11.084	10.799	10.671	300	129.88	127.89	124.60	123.13
28	12.122	11.936	11.629	11.492	350	151.13	149.21	145.37	143.65
30	12.988	12.789	12.460	12.313	400	173.18	170.52	166.14	164.17
32	13.854	13.642	13.291	13.134	450	194.83	191.84	186.90	184.69
34	14.720	14.494	14.121	13.955	500	216.47	213.15	207.67	205.21

Appendix C

Nominal Capacities of Pipe

Size	Sch.	Water gpm	Air - scfm @ 30 psig	Air - scfm @ 100 psig	Steam pph @ 12 psig	Steam pph @ 30 psig	Steam pph @150 psig	Steam pph @ 250 psig
3/4	S	6.70	0.15	0.38	35	45	116	198
	XS	5.43	0.09	0.23	28	36	94	112
1	S	12.70	0.50	1.28	66	89	233	730
	XS	10.57	0.32	0.81	55	74	194	445
1-1/4	S	26.00	1.98	5.06	138	199	523	3,188
	XS	22.30	1.35	3.44	118	171	449	2,111
1-1/2	S	39.50	4.28	10.9	210	309	813	2,921
	XS	34.28	3.00	7.67	182	268	706	2,536
2	S	75.00	14.9	38.1	410	627	1,650	4,815
	XS	65.99	10.8	27.7	361	552	1,452	4,237
2-1/2	S	120.0	36.3	92.7	660	1,033	2,430	6,870
	XS	106.2	26.7	68.3	584	914	2,151	6,082
3	S	210.0	69.3	177.1	1,160	1,880	4,210	10,608
	XS	187.6	61.9	158.2	1,036	1,679	3,760	9,478
	LW	416.3	125.2	320.0	2,518	4,197	8,814	19,168
4	S	396.8	119.3	305.0	2,400	4,000	8,400	18,268
	XS	358.3	107.8	275.4	2,168	3,614	7,588	16,498
5	S	623.6	187.6	479.3	4,250	7,390	15,000	28,708
	XS	567.1	170.6	435.9	3,863	6,718	13,636	26,107
	LW	937.1	281.9	720.3	7,284	12,633	26,224	43,141
6	S	900.5	270.8	692.2	7,000	12,140	25,200	41,457
	XS	812.5	244.4	624.5	6,322	10,964	22,758	37,405
	LW	1,641	493.5	1,261	15,048	26,570	52,614	75,541
8	S	1,559	469.0	1,199	14,300	25,250	50,000	71,787
	XS	1,423	428.1	1,094	13,070	23,079	45,700	65,526
	LW	2,602	782.7	2,000	27,527	49,654	95,285	119,798
10	S	2,458	739.3	1,889	26,000	46,900	90,000	113,153
	XS	2,327	700.0	1,789	23,660	42,680	81,901	107,137
	LW	3,674	1,105	2,824	41,684	77,115	161,526	169,124
12	S	3,525	1,060	2,710	40,000	74,000	155,000	162,291
	XS	3,380	1,017	2,598	38,338	70,925	148,559	155,599
	LW	4,461	1,342	3,429	50,608	93,625	196,106	205,400
14	S	4,298	1,293	3,304	48,751	90,189	188,910	197,863
	XS	4,137	1,244	3,180	46,929	86,818	181,848	190,466
	LW	5,881	1,769	4,521	66,714	123,420	258,515	270,767
16	S	5,693	1,712	4,376	64,579	119,471	250,243	262,103
	XS	5,508	1,657	4,234	62,479	115,586	242,106	253,580
	LW	7,497	2,255	5,763	85,041	157,325	329,533	345,150
18	S	7,284	2,191	5,599	82,628	152,862	320,185	335,359
	XS	7,075	2,128	5,438	80,251	148,464	310,971	325,709
	LW	9,308	2,800	7,155	105,589	195,340	409,159	
20	S	9,071	2,728	6,973	102,899	190,364	398,735	417,632

Size	Sch.	Water gpm	Air - scfm @ 30 psig	@ 100 psig	Steam pph @ 12 psig	@ 30 psig	@150 psig	@ 250 psig
	XS	8,837	2,658	6,793	100,244	185,451	388,445	406,854
	LW	11,316	3,404	8,698	128,359	237,465	497,393	
22	S	11,054	3,325	8,497	125,392	231,975	485,893	508,920
	XS	10,796	3,247	8,298	122,459	226,549	474,527	497,016
	LW	13,519	4,066	10,392	153,351	283,699	594,235	
24	S	13,233	3,980	10,172	150,106	277,695	581,659	609,225
26	S	15,607	4,694	11,997	177,041	327,526	686,034	718,546
28	S	18,178	5,468	13,973	206,198	381,466	799,016	836,883
30	S	20,944	6,300	16,099	237,576	439,516	920,607	964,237
32	S	23,906	7,191	18,376	271,176	501,675	1,050,806	
34	S	27,064	8,140	20,803	306,997	567,945	1,189,614	
36	S	30,418	9,149	23,381	345,040	638,324	1,337,029	
42	S	41,654	12,529	32,018	472,497	874,119	1,830,925	
48	S	54,653	16,439	42,010	619,947	1,146,902	2,402,295	

As stated in the title, these are nominal capacities. The pipe can always handle less than the indicated flow and will handle much more than the indicated flow with increasing pressure drop. These capacities are approximately what a piping designer would allow through the pipe.

Appendix D

Properties of Pipe

Schedule / Weight	Thickness Inches	I.D. Inches	Surface Inside sq. ft.	Cross-sectional Metal Area	Cross-sectional Flow Area	Weight of Pipe pounds	Weight of Water pounds	Weight of Pipe & Water
0.405" O.D.				**1/8" NPS**		Outside surface area 0.106 sq. ft.		
40 / S	0.068	0.269	0.070	0.072	0.057	0.245	0.025	0.270
80 / XS	0.095	0.215	0.056	0.092	0.036	0.314	0.016	0.330
0.540" O.D.				**1/4" NPS**		Outside surface area 0.141 sq. ft.		
40 / S	0.088	0.364	0.095	0.125	0.104	0.425	0.045	0.470
80 / XS	0.126	0.302	0.079	0.157	0.072	0.535	0.031	0.566
0.675" O.D.				**3/8" NPS**		Outside surface area 0.177 sq. ft.		
40 / S	0.091	0.493	0.129	0.167	0.191	0.568	0.083	0.651
80 / XS	0.126	0.423	0.111	0.217	0.140	0.739	0.061	0.800
0.840" O.D.				**½" NPS**		Outside surface area 0.220 sq. ft.		
40 / S	0.109	0.622	0.163	0.250	0.304	0.851	0.132	0.983
80 / XS	0.147	0.546	0.143	0.320	0.234	1.088	0.101	1.189
160	0.187	0.466	0.122	0.384	0.171	1.304	0.074	1.378
XX	0.294	0.252	0.066	0.504	0.050	1.715	0.022	1.737
1.050" O.D.				**3/4" NPS**		Outside surface area 0.275 sq. ft.		
40 / S	0.113	0.824	0.216	0.333	0.533	1.131	0.231	1.362
80 / XS	0.154	0.742	0.194	0.434	0.432	1.474	0.187	1.661
160	0.218	0.614	0.161	0.570	0.296	1.937	0.128	2.065
1.315" O.D.				**1" NPS**		Outside surface area 0.344 sq. ft.		
40 / S	0.133	1.049	0.275	0.494	0.864	1.679	0.374	2.053
80 / XS	0.179	0.957	0.250	0.639	0.719	2.172	0.311	2.483
160	0.250	0.815	0.213	0.836	0.522	2.844	0.226	3.070
XX	0.358	0.599	0.157	1.076	0.282	3.659	0.122	3.781

PROPERTIES OF PIPE (*Continued*)

Schedule / Weight	Thickness Inches	I.D. Inches	Surface Inside sq. ft.	Cross-sectional Metal Area	 Flow Area	Weight of Pipe pounds	 Water pounds	 Pipe & Water
1.660" O.D.				**1 1/4" NPS**		Outside surface area 0.434 sq. ft.		
40 / S	0.140	1.380	0.361	0.668	1.496	2.273	0.648	2.921
80 / XS	0.191	1.278	0.334	0.881	1.283	2.997	0.555	3.552
160	0.250	1.160	0.304	1.107	1.057	3.765	0.458	4.223
XX	0.382	0.896	0.234	1.534	0.630	5.215	0.273	5.488
1.900" O.D.				**1 ½" NPS**		Outside surface area 0.497 sq. ft.		
40 / S	0.145	1.610	0.421	0.799	2.036	2.718	0.882	3.600
80 / XS	0.200	1.500	0.939	1.068	1.767	3.632	0.765	4.397
160	0.281	1.337	0.350	1.431	1.404	4.866	0.608	5.474
XX	0.400	1.100	0.288	1.885	0.950	6.409	0.411	6.820
2.375" O.D.				**2" NPS**		Outside surface area 0.622 sq. ft.		
40 / S	0.154	2.067	0.541	1.074	3.356	3.653	1.453	5.106
80 / XS	0.218	1.939	0.508	1.477	2.953	5.022	1.278	6.300
160	0.343	1.689	0.442	2.190	2.240	7.445	0.970	8.425
2.875" O.D.				**2 ½" NPS**		Outside surface area 0.753 sq. ft.		
10S	0.120	2.635	0.690	1.039	0.545	3.531	2.361	5.892
40 / S	0.203	2.469	0.646	1.704	4.790	5.794	2.073	7.867
80 / XS	0.276	2.323	0.608	2.254	4.240	7.662	1.835	9.497
160	0.375	2.125	0.556	2.945	3.550	10.01	1.540	11.550
XX	0.552	1.771	0.464	4.028	2.460	13.70	1.070	14.770
XX	0.436	1.503	0.393	2.656	1.774	9.030	0.768	9.798

PROPERTIES OF PIPE (*Continued*)

Schedule / Weight	Thickness Inches	I.D. Inches	Surface	Cross-sectional		Weight of		
			Inside sq. ft.	Metal Area	Flow Area	Pipe pounds	Water pounds	Pipe & Water
3.500" O.D.				**3" NPS**		Outside surface area 0.916 sq. ft.		
10S	0.120	3.260	0.853	1.274	8.35	4.33	3.61	7.94
40 / S	0.216	3.068	0.803	2.228	7.39	7.58	3.20	10.78
80 / XS	0.300	2.900	0.759	3.016	6.60	10.25	2.86	13.11
160	0.437	2.626	0.687	4.205	5.42	14.30	2.35	16.65
XX	0.600	2.300	0.602	5.466	4.15	18.58	1.80	20.38
4.500" O.D.				**4" NPS**		Outside surface area 1.178 sq. ft.		
10S	0.120	4.260	1.115	1.651	14.25	5.61	6.17	11.78
LW	0.188	4.124	1.080	2.550	13.36	8.66	5.78	14.44
40 / S	0.237	4.026	1.054	3.170	12.73	10.79	5.51	16.30
80 / XS	0.337	3.826	1.002	4.410	11.50	14.99	4.98	19.97
120	0.437	3.626	0.949	5.580	10.33	18.96	4.47	23.43
160	0.531	3.438	0.900	6.620	9.28	22.51	4.02	26.53
XX	0.674	3.152	0.825	8.100	7.80	27.54	3.38	30.92
5.563" O.D.				**5" NPS**		Outside surface area 1.456 sq. ft.		
40 / S	0.258	5.047	1.321	4.30	20.01	14.62	8.66	23.28
6.625" O.D.				**6" NPS**		Outside surface area 1.734 sq. ft.		
5S	0.109	6.407	1.667	2.23	32.2	7.58	13.96	21.54
10S	0.134	6.357	1.664	2.73	31.7	9.29	13.74	23.03
LW	0.219	6.187	1.620	4.41	30.1	14.99	13.02	28.01
40 / S	0.280	6.065	1.588	5.58	28.9	18.98	12.51	31.49
80 / XS	0.432	5.761	1.508	8.40	26.1	28.58	11.29	39.87
120	0.562	5.501	1.440	10.7	23.8	36.40	10.29	46.69
160	0.718	5.189	1.358	13.32	21.1	45.30	9.16	54.46
XX	0.864	4.897	1.282	15.64	18.8	53.17	8.16	61.33
80 / XS	0.375	4.813	1.260	6.11	18.19	20.78	7.88	28.66

PROPERTIES OF PIPE (*Continued*)

Schedule / Weight	Thickness Inches	I.D. Inches	Surface Inside sq. ft.	Cross-sectional Metal Area	Cross-sectional Flow Area	Weight of Pipe pounds	Weight of Water pounds	Weight of Pipe & Water
8.625" O.D.				**8" NPS**		Outside surface area 2.258 sq. ft.		
5S	0.109	8.407	2.201	2.92	55.5	9.91	24.04	33.95
10S	0.148	8.329	2.180	3.94	54.5	13.4	23.59	36.99
LW	0.219	8.187	2.143	5.78	52.6	19.66	22.94	42.60
20	0.250	8.125	2.127	6.58	51.8	22.37	22.45	44.82
30	0.277	8.071	2.113	7.26	51.2	24.7	22.15	46.32
40 / S	0.322	7.981	2.089	8.40	50.0	28.56	21.68	50.24
60	0.406	7.813	2.045	10.48	47.9	35.6	20.8	56.4
80 / XS	0.500	7.625	1.996	12.76	45.7	43.4	19.8	63.2
120	0.718	7.189	1.882	17.84	40.6	60.6	17.6	78.2
XX	0.875	6.875	1.800	21.30	37.1	72.4	16.1	88.5
10.750" O.D.				**10" NPS**		Outside surface area 2.81 sq. ft.		
5S	0.134	10.482	2.74	4.47	86.3	15.2	37.4	52.6
10S	0.165	10.420	2.73	5.49	85.3	18.7	36.9	55.6
20	0.250	10.250	2.68	8.25	82.5	28.0	35.7	63.7
30	0.307	10.136	2.65	10.07	80.7	34.2	34.9	69.1
40 / S	0.365	10.020	2.62	11.91	78.9	40.5	34.1	74.6
60 / XS	0.500	9.750	2.55	16.10	74.7	54.7	32.3	87.0
80	0.593	9.564	2.50	18.92	71.8	64.3	31.1	95.4
160	1.125	8.500	2.23	34.02	56.7	115.7	24.6	140.3
12.750" O.D.				**12" NPS**		Outside surface area 3.34 sq. ft.		
5S	0.156	12.438	3.26	6.17	121.5	21.0	52.6	73.6
10S	0.180	12.390	3.24	7.11	120.6	24.2	52.2	76.4
20 / LW	0.250	12.250	3.21	9.82	117.9	33.4	51.0	84.4
40S / S	0.375	12.000	3.14	15.58	113.1	49.6	49.0	98.6
40	0.406	11.938	3.13	15.74	111.9	53.5	48.5	102.0
30	0.330	12.090	3.17	12.88	114.8	43.8	49.7	93.5

PROPERTIES OF PIPE (*Continued*)

Schedule / Weight	Thickness Inches	I.D. Inches	Surface Inside sq. ft.	Cross-sectional Metal Area	Cross-sectional Flow Area	Weight of Pipe pounds	Weight of Water pounds	Weight of Pipe & Water
12" NPS (continued)								
80S/XS	0.500	11.750	3.08	19.24	108.4	65.4	47.0	112.4
80	0.687	11.376	2.98	26.04	101.6	88.5	44.0	132.5
160	1.312	10.126	2.65	47.14	80.5	160.3	34.9	195.2
14" O.D.				**14" NPS**		Outside surface area 3.67 sq. ft.		
5S	0.156	13.688	3.58	6.78	147.2	23.1	63.7	86.8
10S	0.188	13.624	3.57	8.16	145.8	27.7	63.1	90.8
10 / LW	0.250	13.500	3.53	10.80	143.1	36.7	62.0	98.7
20	0.312	13.375	3.50	13.44	140.5	45.7	60.8	106.5
30 / S	0.375	13.250	3.47	16.05	137.9	54.6	59.7	114.3
40	0.438	13.125	3.44	18.66	135.3	63.4	58.6	122.0
XS	0.500	13.000	3.40	21.21	132.7	72.1	57.5	129.6
80	0.750	12.500	3.27	31.22	122.7	106.1	53.1	159.2
120	1.093	11.814	3.09	44.32	109.6	150.7	47.5	198.2
160	1.406	11.188	2.93	55.63	98.3	189.1	42.6	231.7
16" O.D.				**16" NPS**		Outside surface area 4.19 sq. ft.		
5S	0.169	15.670	4.1	8.21	192.9	27.9	83.5	111.4
10S	0.188	15.624	4.09	9.34	191.7	31.8	83.0	114.8
10 / LW	0.250	15.500	4.06	12.37	188.7	42.1	81.7	123.8
20	0.312	15.375	4.02	15.4	185.7	52.4	80.4	132.8
30 / S	0.375	15.250	3.99	18.41	182.7	62.6	79.1	141.7
40 / XS	0.500	15.000	3.93	24.35	176.7	82.8	76.5	159.3
60	0.656	14.688	3.85	31.62	169.4	107.5	73.4	180.9
80	0.843	14.314	3.75	40.14	160.9	136.5	69.7	206.2
120	1.218	13.564	3.55	56.56	144.5	192.3	62.6	254.9

PROPERTIES OF PIPE (*Continued*)

Schedule / Weight	Thickness Inches	I.D. Inches	Surface	Cross-sectional		Weight of		
			Inside sq. ft.	Metal Area	Flow Area	Pipe pounds	Water pounds	Pipe & Water
16" NPS (continued)								
160	1.593	12.814	3.35	72.1	129.0	245.1	55.8	300.9
18" O.D.				**18" NPS**		Outside surface area 4.71 sq. ft.		
5S	0.165	17.670	4.63	9.24	245.2	31.4	106.2	137.6
10S	0.188	17.624	4.61	10.52	243.9	35.8	105.6	141.4
20	0.312	17.375	4.55	17.36	237.1	59.0	102.7	161.7
S	0.250	17.250	4.52	20.76	233.7	70.6	101.2	171.8
30	0.438	17.124	4.48	24.17	230.3	82.2	99.7	181.9
XS	0.500	17.000	4.45	27.49	227.0	93.5	98.3	191.8
40	0.562	16.876	4.42	30.79	223.7	104.7	96.9	201.6
60	0.750	16.500	4.32	40.64	213.8	138.2	92.6	230.8
80	0.937	16.126	4.22	50.23	204.2	170.8	88.4	259.2
120	1.375	15.250	3.99	71.81	182.7	244.2	79.1	232.3
160	1.781	14.438	3.78	90.75	163.7	308.5	70.9	379.4
20" O.D.				**20" NPS**		Outside surface area 5.24 sq. ft.		
5S	0.188	19.624	5.14	11.7	302.5	39.8	131.0	170.8
10S	0.218	19.564	5.12	13.55	300.6	46.1	130.2	176.3
10 / LW	0.250	19.500	5.11	15.51	298.6	52.7	129.3	182.0
20 / S	0.375	19.250	5.04	23.12	291.0	78.6	126.0	204.6
30 / XS	0.500	19.000	4.97	30.6	283.5	104.1	122.8	226.9
40	0.593	18.814	4.93	36.2	278.0	122.9	120.4	243.3
60	0.812	18.376	4.81	48.9	265.2	166.4	114.8	281.2
80	1.031	17.938	4.70	61.4	252.7	208.9	109.4	318.3
120	1.500	17.00	4.45	87.2	227.0	296.4	98.3	394.7
160	1.968	16.064	4.21	111.5	202.7	379.1	87.8	466.9

PROPERTIES OF PIPE (*Continued*)

22" O.D.				**22" NPS**		Outside surface area 5.76 sq. ft.		
10 / LW	0.250	21.500	5.63	17.1	363	58.1	157.2	215.3
S	0.375	21.250	5.56	25.5	355	86.6	153.6	240.2
XS	0.500	21.000	5.50	33.8	346	114.8	150.0	264.8
24" O.D.				**24" NPS**		Outside surface area 6.28 sq. ft.		
5S	0.218	23.564	6.17	16.3	436	55.4	188.8	244.2
10 / LW	0.250	23.500	6.15	18.7	434	63.4	187.8	251.2
20 / S	0.375	23.250	6.09	27.8	425	94.6	183.8	278.4
XS	0.500	23.000	6.02	36.9	415	125.5	179.9	305.4
40	0.687	22.626	5.92	50.3	402	171.1	174.1	345.2
60	0.986	22.064	5.78	70.0	382	238.1	165.6	403.7
80	1.218	21.564	5.65	87.2	365	296.4	158.1	454.5
120	1.812	20.376	5.33	126.3	326	429.4	141.2	570.6
160	2.343	19.314	5.06	159.4	293	542	126.9	668.9
26" O.D.				**26" NPS**		Outside surface area 6.81 sq. ft.		
S	0.375	25.250	6.61	30.2	501	102.6	216.8	319.4
XS	0.500	25.000	6.54	40.1	491	136.2	212.5	348.7
28" O.D.				**28" NPS**		Outside surface area 7.33 sq. ft.		
S	0.375	27.250	7.13	32.5	583	110.7	252.5	363.2
XS	0.500	27.000	7.07	43.2	573	146.9	247.9	394.8
30" O.D.				**30" NPS**		Outside surface area 7.85 sq. ft.		
5S	0.250	29.500	7.72	23.4	683	79.4	296.0	375.4
10 / LW	0.312	29.376	7.69	29.1	678	98.9	293.5	392.4
S	0.375	29.250	7.66	34.9	672	118.7	291.0	409.7
20 / XS	0.500	29.000	7.59	46.3	661	157.6	286.0	443.6
30	0.625	28.750	7.53	57.7	649	196.1	281.1	477.2
32" O.D.				**32" NPS**		Outside surface area 8.38 sq. ft.		
S	0.375	31.250	8.18	37.3	767	126.7	332.1	458.8
XS	0.500	31.000	8.12	49.5	755	168.2	326.8	495.0

PROPERTIES OF PIPE (*Continued*)

Schedule / Weight	Thickness Inches	I.D. Inches	Surface Inside sq. ft.	Cross-sectional Metal Area	Cross-sectional Flow Area	Weight of Pipe pounds	Weight of Water pounds	Weight of Pipe & Water
34" O.D.				**34" NPS**		Outside surface area 8.9 sq. ft.		
S	0.375	33.250	8.70	39.6	868	134.7	376.0	510.7
XS	0.500	33.000	8.64	52.6	855	178.9	370.3	549.2
36" O.D.				**36" NPS**		Outside surface area 9.42 sq. ft.		
S	0.375	35.250	9.23	42.0	976	142.7	422.6	565.3
XS	0.500	35.000	9.16	55.8	962	189.6	416.6	606.2
42" O.D.				**42" NPS**		Outside surface area 11.0 sq. ft.		
S	0.375	41.250	10.80	49.0	1336	166.7	578.7	745.4
XS	0.500	41.000	10.73	65.2	1320	221.6	571.7	793.3
48" O.D.				**48" NPS**		Outside surface area 12.56 sq. ft.		
S	0.375	47.250	12.37	56.1	1753	190.8	759.2	950.0
XS	0.500	47.000	12.30	74.6	1735	253.7	751.2	1004.9
LW	0.219	10.310	2.7	7.28	83.5	24.7	36.1	60.8

Notes applicable to properties of pipe:

- Metal area and flow area are in square inches.
- Surface area and weights of pipe and water are per foot of length
- There are other standard sizes of pipe based on schedule numbers that are not shown and other wall thicknesses. Check with your local pipe supplier for more information if it is needed.
- LW, S, XS, XX refer to Light Wall, Standard, Extra Strong, and Double Extra Strong commercial sizes.
- 5S and 10S sizes apply specifically to corrosion resistant materials.

Appendix E

Secondary Ratings

SECONDARY RATINGS OF JOINTS, FLANGES, VALVES, AND FITTINGS

SOLDER JOINTS - MAXIMUM WORKING PRESSURE				
Solder used in joints	Maximum working temperature	1/4" thru 1"	1-1/4" thru 2"	2-1/2" thru 4"
50-50 Tin - lead	100	200	175	150
	150	150	125	100
	200	100	90	75
	250	85	75	50
95 - 5 Tin - Antimony	100	500	400	300
	150	400	350	275
	200	300	250	200
	250	200	175	150
These ratings apply to any valve or fitting joined by soldering. They are also limited to water and other non-flammable liquids and gases. For steam, the pressure cannot exceed 15 psig saturated (maximum temperature of 250°F). These values limit all other ratings.				

BRONZE VALVES AND FITTINGS						
Pressure Class & Conn	125	150		200	300	
	Screwed	Screwed	Flanged	Screwed	Screwed	Flanged
-20 to 150°F	200	300	225	400	1000	500
200°F	185	270	210	375	920	475
250°F	170	240	195	350	830	450
300°F	155	210	180	325	740	425
350°F	140	180	165	300	650	400
400°F				275	560	375
406°F	125	150	150			
450°F				250	480	350
500°F				225	390	325
550°F				200	300	300

Note: The 406°F value, saturation temperature for 250 psig steam, is normally the limit for all bronze valves according to the boiler construction code.

SECONDARY RATINGS OF JOINTS, FLANGES, VALVES, AND FITTINGS (*Continued*)

IRON VALVES AND FITTINGS							
Temp.	125 LB CLASS			150 LB	250 LB CLASS		300 LB
	2" thru 12"	14" to 24"	30" to 36"	DUCTILE	2" thru 12"	14" thru 24"	DUCTILE
-20 to 100	200	150	150	250	500	300	500
150	200	150	150	250	500	300	500
200	190	135	115	235	460	280	480
225	180	130	85	230	440	270	470
250	175	125	65	225	415	260	460
275	170	120	50	220	395	250	450
300	165	110		215	375	240	440
325	155	105		212	355	230	430
350	150	100		208	335	220	420
375	145			204	315	210	410
400	140			200	290	200	400
425	130			190	270		390
450	125			180	250		380
500				170			360
600				140			320
650				125			300

Note: The values in the table are restricted to valves that are rated for steam at the class pressure. There are certain constructions that cannot use these secondary ratings.

SECONDARY RATINGS OF JOINTS, FLANGES, VALVES, AND FITTINGS (*Continued*)

CARBON STEEL VALVES AND FLANGES						
Temperature °F	PRESSURE CLASS					
	150	300	400	600	900	1500
-20 TO 100	275	720	960	1440	2160	3600
200	240	675	930	1400	2100	3500
300	210	655	910	1365	2050	3415
400	180	635	890	1330	2000	3330
500	150	600	835	1250	1875	3125
600	130	550	740	1110	1660	2770
650	120	535	690	1030	1550	2580
700	110	535	640	960	1440	2400
750	95	505	590	890	1330	2220
800	80	410	545	815	1225	2040
850	65	270	495	745	1115	1860
900	50	170	450	670	1010	1680
950	35	105	400	600	900	1500
1000	20	50	285	425	635	1065

Notes:

- The secondary ratings can vary by manufacturer, valve material, and valve construction. These are the lowest of secondary ratings listed by two manufacturers and can be considered reasonable for most valves. When operating a system with pressures close to these ratings the manufacturer's ratings for the valves and fittings being used should be consulted.
- When piping is designed for operation at temperatures less than saturated steam temperature these tables should be consulted to ensure the hydrostatic test pressure does not exceed the rating of the valves and flanges.

Appendix F

Pressure Ratings for Various Pipe Materials

PIPE PRESSURE RATINGS							
SIZE NPS	PVC 75°F	CPVC 180°F	COPPER	BRASS	STEEL		
					STD	XS	XX
1/8	100	100		1,989	330	2,058	
1/4	100	100	1,530	1,984	1,178	3,115	
3/8	100	100	1,212	1,712	1,049	2,400	
1/2	100	100	954	1,625	1,390	2,626	8,485
3/4	100	100	900	1,355	1,200	2,248	6,903
1	100	100	691	1,188	1,344	2,288	6,575
1-1/4	70	100	560	1,079	1,164	1,980	5,460
1-1/2	60	100	524	961	1,080	1,844	4,969
2	50	100	460	793	954	1,658	4,318
2-1/2	50		425	784	1,221	1,895	4,731
3	50		409	751	1,091	1,722	4,202
4	40		380	663	960	1,539	3,659
5	35		365	531	868	1,412	3,291
6	35		367	444	809	1,401	3,215
8			391	425	740	1,269	2,445
10			391	409	691	1,010	
12			391	343	600	847	
14					545	769	
16					476	671	
18					423	595	
20					380	534	
24					316	444	
36					210	295	
48					157	220	

Notes:

- These are nominal pressure ratings for the common materials and standard wall thicknesses
- Steel pipe data includes a 1/16 inch corrosion allowance.

Appendix G
Square Root Flow Curve

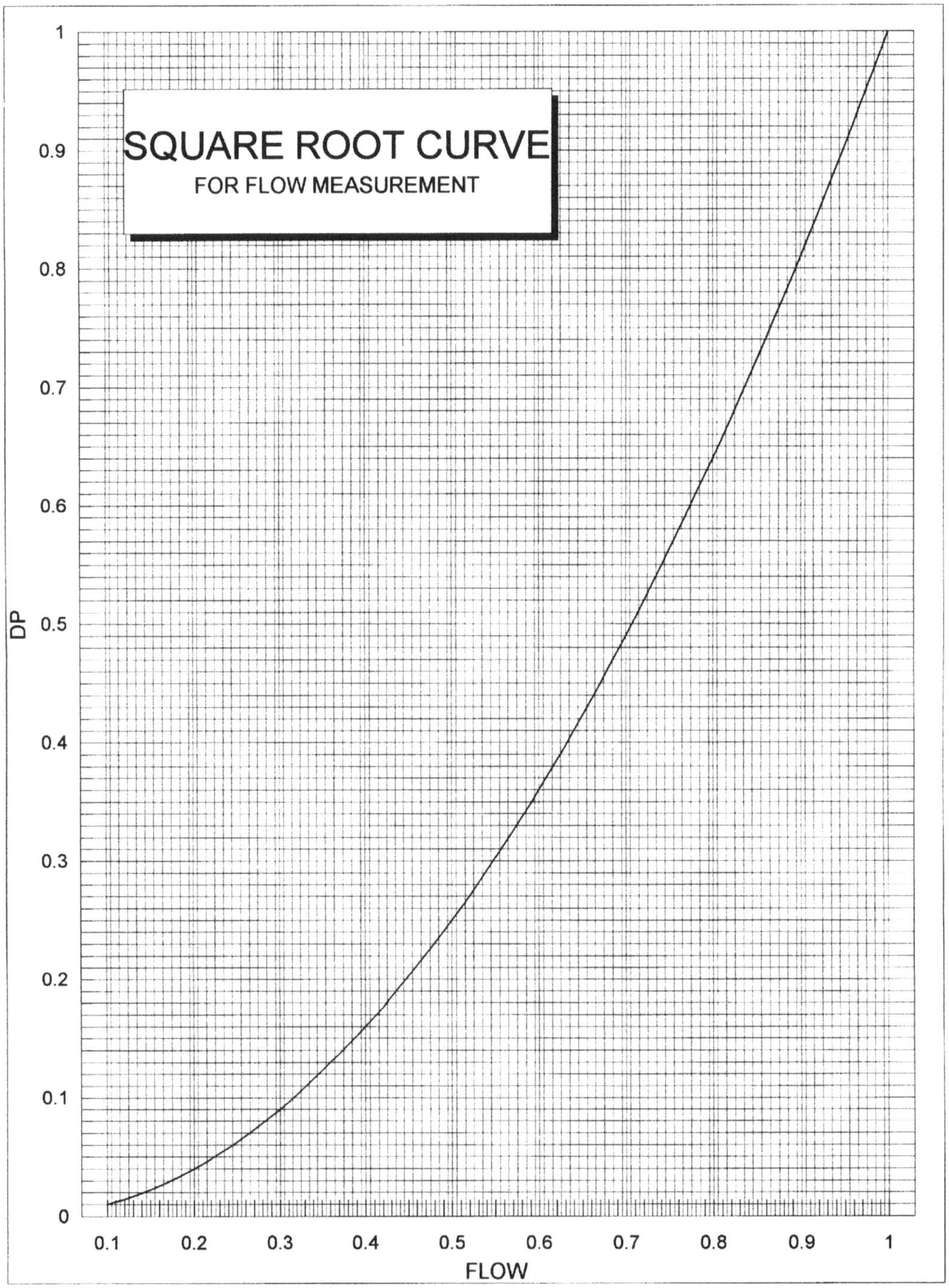

Appendix H

Square Root Graph Paper

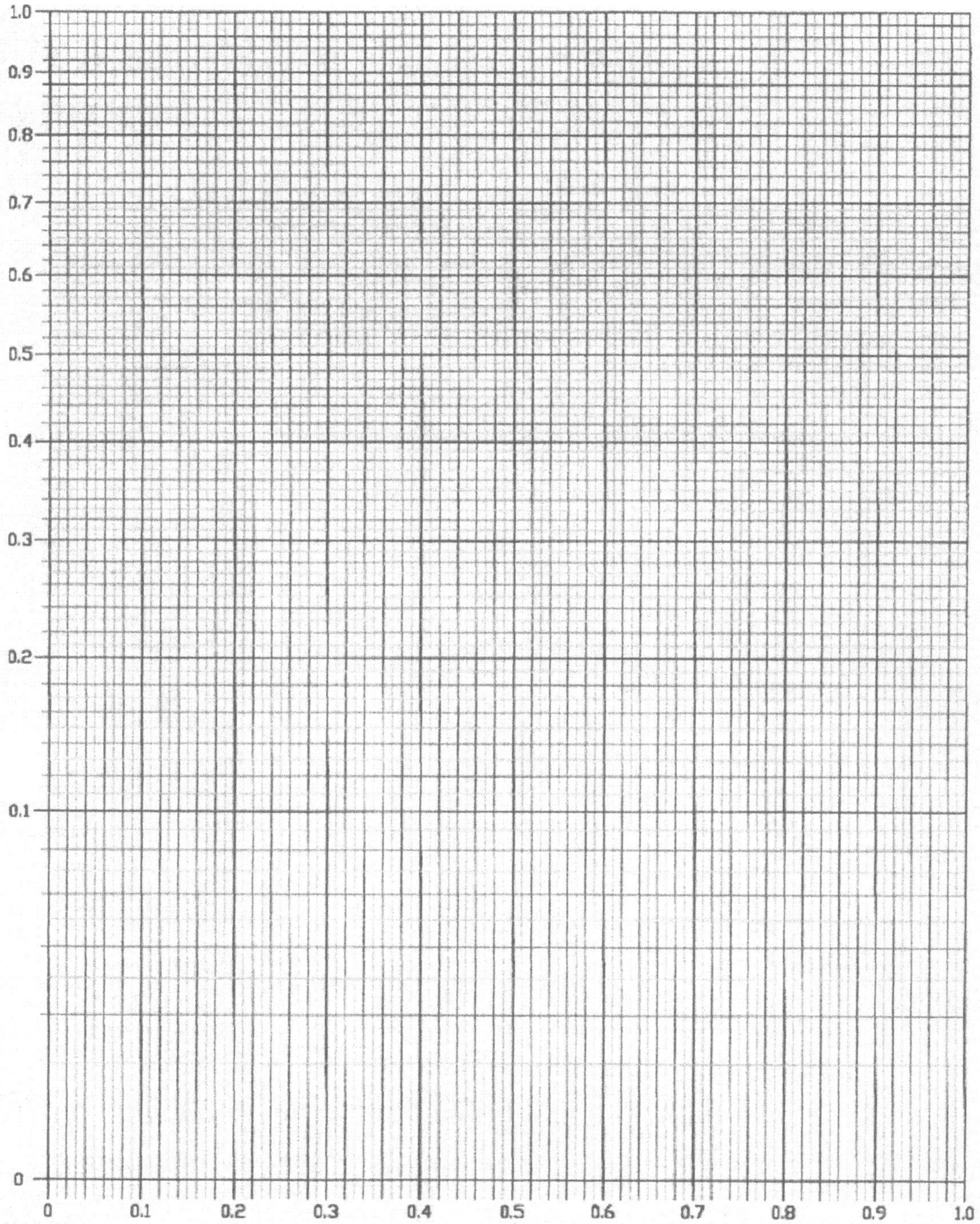

Appendix I

Viscosity Conversions

VISCOSITY CONVERSIONS					
SSU	SSF	Kinematic	SSU	SSF	Kinematic
40	6.4	0.042	1100	112	2.419
50	7.2	0.074	1110	113	2.441
60	8.2	0.103	1120	114	2.463
70	9.1	0.130	1130	115	2.485
80	10	0.156	1140	116	2.507
90	10.9	0.182	1150	117	2.529
100	11.9	0.207	1160	118	2.551
110	12.8	0.230	1170	119	2.573
120	13.7	0.253	1180	120	2.595
130	14.6	0.276	1190	121	2.617
140	15.6	0.298	1200	122	2.639
150	16.5	0.321	1300	132	2.859
160	17.5	0.344	1400	142	3.079
170	18.4	0.366	1500	152	3.299
180	19.4	0.389	1600	162	3.519
190	20.3	0.411	1700	172	3.739
200	21.3	0.433	1800	182	3.959
250	26.2	0.545	1900	192	4.179
300	31.2	0.656	2000	202	4.399
400	41.2	0.877	2100	212	4.619
500	51.2	1.097	2200	222	4.839
600	61.4	1.318	2300	232	5.059
700	71.5	1.538	2400	242	5.279
800	81.7	1.758	2500	252	5.499
900	91.8	1.979	2600	262	5.719
1000	102	2.199	2750	277	6.050

Notes: Conversion is for viscosity in Saybolt Universal Seconds at 100°F, Saybolt Seconds Furol at 122°F, and Kinematic viscosity in Centistokes.

FUEL FIRING TEMPERATURE CALCULATOR

To determine correct burning temperature draw a diagonal line parallel to the one on the chart through the viscosity at temperature reported for the oil. Note the temperature for where that line intersects the line for the correct viscosity for firing

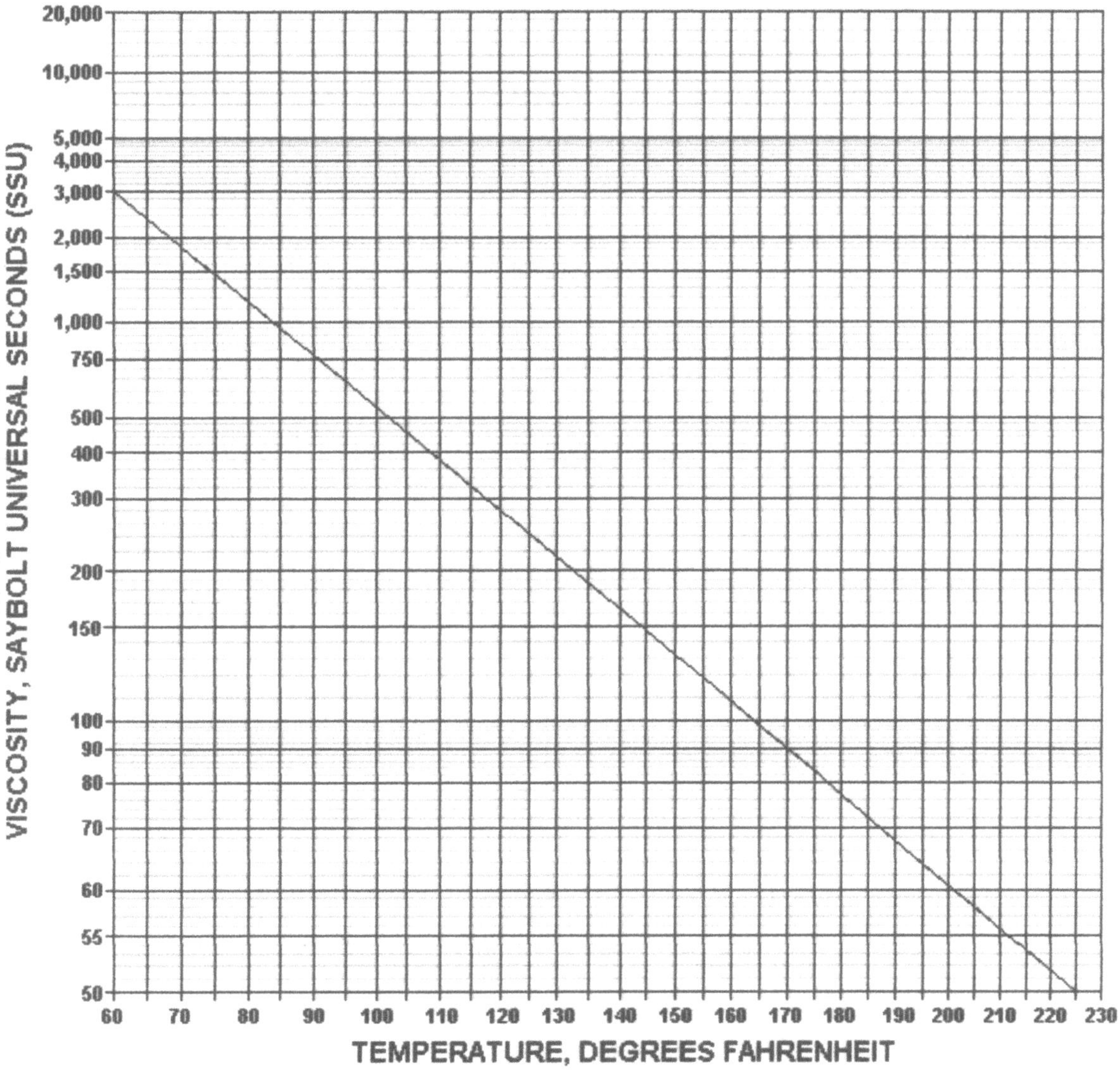

Appendix J

Thermal Expansion of Materials

EXPANSION OF MATERIALS, INCHES PER 100 FEET									
	Temperature range, from 70°F to:								
Material	200	300	400	500	600	700	800	900	1000
Air	22.08	39.06	56.04	73.02	90.00	107.0	124.0	140.9	157.9
Aluminum	2.00	3.66	5.39	7.17	9.03				
Austenitic stainless	1.46	2.61	3.80	5.01	6.24	7.50	8.80	10.12	11.48
Brass	1.52	2.76	4.05	5.40	6.80	8.26	9.78	11.35	12.98
Brick	0.78	1.44	2.14	2.87					
Bronze	1.56	2.79	4.05	5.33	6.64	7.95	9.30	10.68	12.05
Carbon steels	0.99	1.82	2.70	3.62	4.60	5.63	6.70	7.81	8.89
Concrete	1.26	2.31	3.43						
Chrome steels	0.94	1.71	2.50	3.35	4.24	5.14	6.10	7.07	8.06
Cast iron	0.90	1.64	2.42	3.24	4.11	5.03	5.98	6.97	8.02
Copper-nickel	1.33	2.40	3.52						
Glass	0.78	1.44	2.14	2.87					
Glass, pyrex	0.28	0.52	0.77	1.03	1.31	1.61	1.91		
High Chrome stainless	0.86	1.56	2.30	3.08	3.90	4.73	5.60	6.49	7.40
Titanium	0.52	0.96	1.42	1.90	2.42	2.95			
Water (liquid)	42.6	104	193	324	563	1557			
Wood	0.47	0.31	0.46	1.72					
Wrought iron	1.14	2.06	3.01	3.99	5.01	6.06	7.12	8.26	9.36

Values are approximate, for information only, and do not indicate the material is suitable for use at the temperatures indicated.

Appendix K
Value Conversions

To obtain	multiply	by	To obtain	multiply	by
atmospheres	ft. of water	0.0295	knots	miles per hour	0.8684
atmospheres	in. mercury	0.0334	liters	cubic feet	28.316
atmospheres	psi	0.0680	liters	gallons (US)	3.7853
barrels	gallons (US)	0.0238	horsepower	Btu/hr	0.00039
Btu	calories	252	horsepower	kW	1.341
Btu	hp-hr	2545	meters	feet	0.3048
Btu	kW-hr	3413	meters	inches	0.0254
Btu	watt-hr	3.413	meters	nautical miles	1852
Btu/hr	horsepower	2545	meters	miles	1609.34
Btu/hr	kW	3413	microns	inches	25.4
Btu/hr	refrigeration ton	12,000	miles	feet	5280
centimeter	inches	2.54	miles	meters	0.00062
cubic feet	gallons (US)	0.1337	miles	nautical miles	1.151
cubic meters	cubic feet	0.0283	miles, nautical	kilometers	0.54
feet of H_2O	atmospheres	33.899	miles, nautical	miles	0.8690
feet/minute	miles per hour	88	milliliters	microns	0.001
feet/second	gravity	32.174	mils	centimeters	393.7
foot-pounds	Btu	778	mils	inches	1000
foot-lbs/min.	horsepower	33,000	mils	microns	0.03937
gallons (US)	barrels	42	ounces	grains	0.00228
gallons (US)	cubic feet	7.4805	ounces	grams	0.03527
gallons (US)	Imperial gallons	1.201	ounces, liquid	gallons (US)	128
gallons (US)	Liters	0.2642	parts/million	grains/gallon	17.118
grains	grams	15.432	percent grade	ft. per 100 ft.	1.0
grains	ounces	437.5	pounds	grains	0.00014
grains	pounds	7000	pounds	grams	0.00220
grains/gallon	parts per million	0.0584	pounds	kilograms	2.2046
grams	grains	0.0648	pounds	long tons	2240
grams	ounces	28.35	pounds	metric tons	2204.6
Grams	pounds	453.59	pounds	short tons	2000
Inches	centimeters	0.3937	lbs.ice melt/hr.	refrigeration ton	83.711
inches	microns	0.00004	pounds/cu.ft.	grams/cu.cm.	62.428
in. mercury	feet of water	0.88265	pounds/cu.ft.	pounds/gallon	7.48
inches water	psi	27.673	psi	atmospheres	14.696
kilograms	pounds	0.45359	psi	feet of water	0.43352
kilometer	mile (US)	1.6093	psi	inches water	0.0361
km/hr	mph	1.6093	quarts	cubic feet	29.922
kW	Btu/minute	0.01758	quarts	liters	1.057
kW	horsepower	0.7457	Tons, metric	Tons, short	0.9072
kW-hour	Btu	0.00029			

Appendix L

Excess Air/O_2 Curve

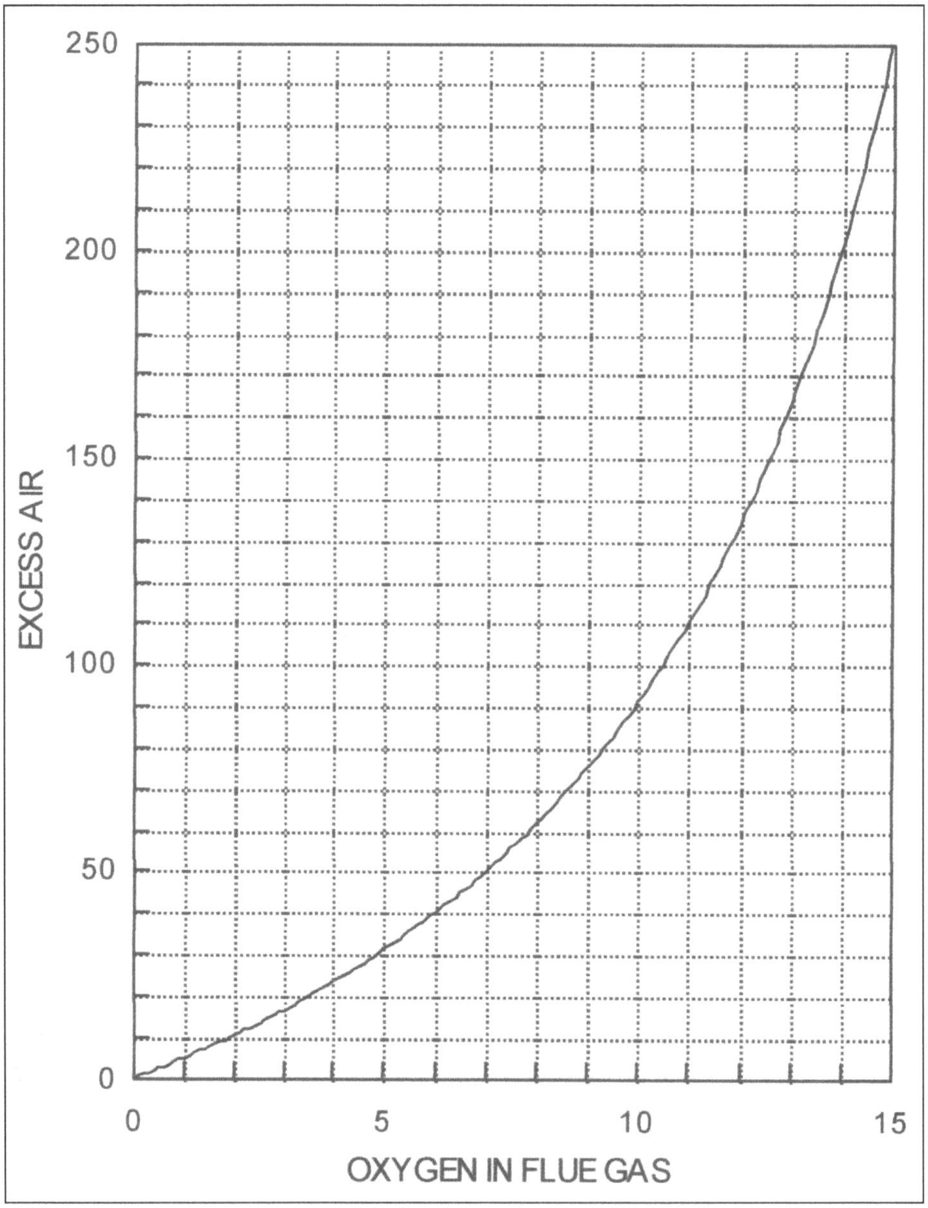

Appendix M

Properties of Dowtherm A

Vacuum Pressure	TEMP. °F	LB. PER CU. FT.		HEAT IN BTU PER POUND		
		LIQUID	VAPOR	LIQUID	LATENT	VAPOR
29.92 in. Hg	60	66.54	0.0000	2.4	174.4	176.8
29.92 in. Hg	80	65.82	0.0000	9.9	172.0	181.9
29.92 in. Hg	100	65.27	0.0000	17.6	169.6	187.2
29.92 in. Hg	120	64.72	0.0001	25.5	167.2	192.7
29.91 in. Hg	140	64.16	0.0002	33.5	164.9	198.4
29.90 in. Hg	160	63.60	0.0004	41.6	162.7	204.3
29.87 in. Hg	180	63.03	0.0007	49.9	160.4	210.3
29.83 in. Hg	200	62.46	0.0012	58.3	158.3	216.6
29.74 in. Hg	220	61.88	0.0021	66.9	156.2	223.1
29.60 in. Hg	240	61.30	0.0034	75.7	154.0	229.7
29.40 in. Hg	260	60.71	0.0055	84.5	152.0	236.5
29.09 in. Hg	280	60.11	0.0086	93.6	149.9	243.5
28.65 in. Hg	300	59.50	0.0129	102.7	147.9	250.6
27.97 in. Hg	320	58.89	0.0191	112.1	145.8	257.9
27.06 in. Hg	340	58.28	0.0274	121.5	143.8	265.3
25.80 in. Hg	360	57.65	0.0385	131.2	141.7	272.9
24.11 in. Hg	380	57.02	0.0532	140.9	139.8	280.7
21.87 in. Hg	400	56.37	0.0720	150.9	137.6	288.5
19.00 in. Hg	420	55.72	0.0959	160.9	135.6	296.5
15.31 in. Hg	440	55.06	0.1258	171.1	133.5	304.6
10.71 in. Hg	460	54.38	0.1626	181.5	131.3	312.8
5.01 in. Hg	480	53.70	0.2076	192.0	129.1	321.1
0 psig	494.8	53.18	0.2470	199.9	127.4	327.3
0.00 psig	494.8	53.18	0.2470	199.9	127.4	327.3
0.95 psig	500	53.00	0.2618	202.7	126.9	329.5
5.08 psig	520	52.29	0.3267	213.5	124.5	338.0
10.01 psig	540	51.56	0.4037	224.5	122.1	346.6
15.85 psig	560	50.82	0.4943	235.7	119.5	355.3
22.68 psig	580	50.06	0.6003	247.1	116.9	364.0
30.64 psig	600	49.29	0.7237	258.6	114.1	372.7
39.81 psig	620	48.49	0.8667	270.2	111.3	381.5
50.33 psig	640	47.67	1.032	282.0	108.3	390.4
62.30 psig	660	46.82	1.223	294.0	105.2	399.2
75.86 psig	680	45.94	1.442	306.1	102.0	408.1
91.10 psig	700	45.03	1.695	318.3	98.6	416.9
108.3 psig	720	44.08	1.988	330.7	95.0	425.8
127.4 psig	740	43.09	2.327	343.4	91.2	434.6
152.5 psig	760	42.04	2.723	356.2	87.1	443.3
198.6 psig	800	39.74	3.749	382.7	77.6	460.2

Reference state for heat of the fluid is zero at the freezing temperature of 53.6°F

Appendix N

Properties of Dowtherm J

Vacuum / Pressure	TEMP. °F	LB. PER CU. FT.		HEAT IN BTU PER POUND		
		LIQUID	VAPOR	LIQUID	LATENT	VAPOR
29.92" Hg	0	55.64	0.0000	-32.8	175.1	142.3
29.92" Hg	20	55.14	0.0000	-24.5	172.3	147.8
29.92" Hg	40	54.64	0.0001	-16.0	169.6	153.6
29.90" Hg	60	54.13	0.0003	-7.4	167.0	159.6
29.88" Hg	80	53.61	0.0006	1.3	164.5	165.8
29.81" Hg	100	53.09	0.0011	10.2	162.1	172.3
29.72" Hg	120	52.56	0.0021	19.3	159.7	179.0
29.55" Hg	140	52.02	0.0037	28.6	157.3	185.9
29.29" Hg	160	51.47	0.0063	38.0	155.0	193.0
28.86" Hg	180	50.92	0.0103	47.6	152.7	200.3
28.19" Hg	200	50.36	0.0161	57.4	150.4	207.8
27.21" Hg	220	49.79	0.0246	67.4	148.1	215.5
25.81" Hg	240	49.21	0.0364	77.6	145.7	223.3
23.83" Hg	260	48.62	0.0525	87.9	143.5	231.4
21.16" Hg	280	48.01	0.0739	98.5	141.1	239.6
16.48" Hg	300	47.40	0.1018	109.3	138.6	247.9
12.90" Hg	320	46.77	0.1375	120.3	136.2	256.4
6.91" Hg	340	46.13	0.1826	131.4	133.6	265.0
0 psig	358.4	45.53	0.2340	141.9	131.2	273.1
0.32 psig	360	45.48	0.2385	142.8	131.0	273.8
4.93 psig	380	44.80	0.3072	154.4	128.3	282.7
10.57 psig	400	44.11	0.3907	166.3	125.4	291.7
17.39 psig	420	43.40	0.4911	178.4	122.4	300.7
25.55 psig	440	42.67	0.6110	190.7	119.2	309.9
35.19 psig	460	41.91	0.7533	203.2	115.9	319.1
46.50 psig	480	41.12	0.9215	215.9	112.5	328.4
59.63 psig	500	40.30	1.1198	228.9	108.9	337.7
74.79 psig	520	39.44	1.3533	242.0	105.0	347.1
92.16 psig	540	38.53	1.6285	255.4	101.0	356.4
111.96 psig	560	37.58	1.9542	269.1	96.6	365.7
134.42 psig	580	36.56	2.3425	283.1	91.8	374.9
159.80 psig	600	35.46	2.8110	297.4	86.6	383.9

Appendix O

Chemical Tank Mixing Table

CHEMICAL TANK MIXING TABLE					
Inches tank Diameter	Gallons per inch of depth	Chemicals to be added for % solutions in pounds per inch			
		1%	2%	5%	10%
12	0.49	0.04	0.08	0.20	0.41
18	1.10	0.09	0.18	0.46	0.92
24	1.96	0.16	0.33	0.82	1.63
30	3.06	0.25	0.51	1.27	2.55
36	4.41	0.37	0.73	1.84	3.67
42	6.00	0.50	1.00	2.50	5.00
48	7.83	0.65	1.31	3.26	6.53
54	9.91	0.83	1.65	4.13	8.26
60	12.24	1.02	2.04	5.10	10.20
66	14.81	1.23	2.47	6.17	12.34
72	17.63	1.47	2.94	7.34	14.68
78	20.69	1.72	3.45	8.62	17.23
84	23.99	2.00	4.00	9.99	19.98
90	27.54	2.29	4.59	11.47	22.94
96	31.33	2.61	5.22	13.05	26.10
102	35.37	2.95	5.89	14.73	29.47
108	39.66	3.30	6.61	16.52	33.03
114	44.19	3.68	7.36	18.40	36.81
120	48.96	4.08	8.16	20.39	40.78
132	59.24	4.93	9.87	24.67	49.35
144	70.50	5.87	11.75	29.36	58.73
192	125.34	10.44	20.88	52.20	104.41
216	158.63	13.21	26.43	66.07	132.14
240	195.84	16.31	32.63	81.57	163.13

Use the data on this table to create one's own table for a specific size of chemical tank and solution strength to be maintained. On a fresh piece of paper, write or print the range of levels of the tank from bottom to top and then multiply the values of those levels by the pounds of chemical per inch numbers from the table above. The levels and chemical requirements in multiple columns can be listed to produce a table like the one on the following page.

For this example, the tank is 48 inches deep. Thus, 48 different level readings and the matching quantity of chemical to produce a 5% solution are needed. After laying out the table so that all 48 inches are accounted for, each level is multiplied by the 1.27 pounds per inch to determine the number of pounds that must be added to produce a 5% solution at each level.

Here is the table that was made. Once the table is made, it would pay to find some laminating plastic and cover it and then mount it next to the tank. With this table, the level in the tank is subtracted (before it is filled with water) from the level after it is filled and then the pounds of chemical to add from the table is found.

Level	Chemical	Level	Chemical
1	1.27	25	31.75
2	2.54	26	33.02
3	3.81	27	34.29
4	5.08	28	35.56
5	6.35	29	36.83
6	7.62	30	38.1
7	8.89	31	39.37
8	10.16	32	40.64
9	11.43	33	41.91
10	12.7	34	43.18
11	13.97	35	44.45
12	15.24	36	45.72
13	16.51	37	46.99
14	17.78	38	48.26
15	19.05	39	49.53
16	20.32	40	50.8
17	21.59	41	52.07
18	22.86	42	53.34
19	24.13	43	54.61
20	25.4	44	55.88
21	26.67	45	57.15
22	27.94	46	58.42
23	29.21	47	59.69
24	30.48	48	60.96

A chart could also be created based on the level before it is filled when the tank is filled to a consistent level. If the level is always raised to the 48 inches, subtract the level values in the table from 48 and replace them with the result.

Appendix P

Suggested Mnemonic Abbreviations

The use of mnemonic abbreviations to simplify communications and labeling of devices in a boiler plant is a common practice. This is a recommended list for identifying plant devices on logs, equipment lists, maintenance records, reports, etc. A plant can have many identical devices that are numbered sequentially (although earlier numbers may no longer exist) as indicated by the pound sign (#). Also, some devices are redundant (such as two safety shutoff valves in series). Thus, the number can be followed by a letter, indicated by the asterisk (*). The two symbols (# and *) are shown only where the inclusion of a number/letter is common. This list does contain duplicate abbreviations where it is necessary to determine which one is correct by how and where it is used.

3VBP............ Three valve bypass
AACV.......... Atomizing air control valve
AAPS........... Atomizing air pressure switch
ABC .... Automatic blowdown control
ABCV .... Automatic blowdown control valve
AFT.............. Air flow transmitter
AIS............... Automatic Interruptible System
ALWCO ...... Auxiliary low water cutoff
ASBOV......... Atomizing steam blow-out valve
ASCV........... Atomizing steam control valve
ASPS............ Atomizing steam pressure switch
ASSV .... Atomizing steam shutoff valve
ASV .... Anti-siphon valve
AT .... Analysis transmitter
AW............... Acid waste
BD................ Blowdown (piping)
BDFT .... Blowdown flash tank
BDHX.......... Blowdown heat exchanger
BDQT .... Blowdown quench tank
BF................. Boiler feed water (piping)
BFP# .... Boiler feed pump
BFV .... Butterfly valve
BGV#* .... Burner gas safety shutoff valve
BLR#............ Boiler
BO................ Blow off (piping)
BOQT .... Blow off quench tank
BOS.............. Blow off separator
BV .... Ball valve
BVV# .... Burner gas vent valve
CAFS .... Combustion air flow switch
CF .... Chemical feed
CHX............. Condensing heat exchanger
CO................ Carbon monoxide
CO_2.............. Carbon dioxide
COND .... Condensate
CP .... Circulating pump
CP# .... Condensate polisher
CPMP .... Condensate pump
CPMS .... Circulating/condensate pump motor starter
CR................ Control relay
CW............... City water (piping)
DA. .... Deaerator
DEGAS......... Degassifier
DBB.............. Double block and bleed (valve arrangement)
DI .... Draft indicator
DI ....Demineralized (water)
DLT.............. Drum level transmitter
DT .... Draft transmitter
FC .... Flow controller
FD .... Forced draft
FDF.............. Forced draft fan
FFT............... Feed water flow transmitter
FIC .... Flow indicating controller
FMS .... Fan motor starter
FOR.............. Fuel oil return
FOP#............ Fuel oil pump
FOS .... Fuel oil supply/suction
FOT#............ Fuel oil tank
FPSC .... Frost proof sill cock
FR................. Flame relay
FR................. Flow recorder
FW .... Boiler feed water (piping)
FWCV.......... Feed water control valve
FWHTR........ Feed water heater
FY................. Flow totalizer
GCV............. Gas flow control valve
GFT.............. Gas flow transmitter
GOS .... Gas - off - oil selector (switch)
GPR .... Gas pressure regulator
GT#.............. Gas turbine
GV................ Gate valve
H_2................. Hydrogen
HFPS .... High furnace pressure switch
HGP............. High gas pressure (limit switch)
HIGP .... High ignitor gas pressure (limit switch)
HOT............. High oil temperature (limit switch)

HPC .... High pressure condensate
HPS .... High pressure steam
HPS .... High pressure switch
HTS .... High temperature switch
ID .... Induced draft
IDF .... Induced draft fan
IGV#* .... Ignitor gas safety shutoff valve
IPS .... Intermediate pressure steam
IT .... Ignition timer
IT .... Ignition transformer (see IX)
IVV# ....Ignitor gas vent valve
IX .... Ignition transformer
LAAD .... Low atomizing air differential pressure (limit switch)
LAAP .... Low atomizing air pressure switch
LAF .... Low air flow (limit switch)
LASD .... Low atomizing steam differential pressure (limit switch)
LASP .... Low atomizing steam pressure (limit switch)
LC .... Level controller
LDS .... Low draft switch
LG .... Level glass
LGP .... Low gas pressure (limit switch)
LI .... Level indicator
LIC .... Level indicating controller
LIGP .... Low ignitor gas pressure (limit switch)
LOP .... Low oil pressure (limit switch)
LOT .... Low oil temperature (limit switch)
LPC .... Low pressure condensate
LPHTR .... Low pressure heater
LPS .... Low pressure switch
LPS .... Low pressure steam
LR .... Level recorder
LS .... Level switch
LSH .... Level switch, high level
LSL .... Level switch, low level
LT .... Level transmitter
LTS .... Low temperature switch
LWCO .... Low water cutoff
LWFS .... Low water flow switch
LWL .... Low water level
MAFS .... Minimum air flow switch (limit switch)
MBDI .... Mixed bed demineralizer
MGPR .... Minimum gas pressure regulator
MGV#* .... Main gas safety shutoff valve
MOPR .... Minimum oil pressure regulator
MS .... Motor starter
MU .... Makeup water (piping)
MVV# .... Main gas vent valve
N_2 .... Nitrogen
NG .... Natural gas
NRV .... Non-return valve
O_2 .... Oxygen
O_2T .... Oxygen transmitter
OCV .... Oil flow control valve
OF .... Overflow
OFT .... Oil flow transmitter
OPMS .... Oil pump motor starter
OPR .... Oil pressure regulator
OV#* .... Oil safety shutoff valve
PC .... Pressure controller
PC .... Pumped condensate
PI .... Pressure indicator
PIC .... Pressure indicating controller
PPT .... Post purge timer
PR .... Pressure recorder
PR .... Pressure regulator
PRV .... Pressure reducing valve (station)
PT .... Purge timer
PT .... Pressure transmitter
PV .... Plug valve
RO .... Reverse osmosis
ROW .... Reverse osmosis water (permeate)
ROV .... Recirculating oil valve
RV .... Recirculating valve
RV .... Relief valve
SAN .... Sanitary sewer
SOFT# .... Softener
SPT .... Steam pressure transmitter
STM .... Steam
STRNR .... Strainer
SV .... Safety valve
SW .... Softened water
TC .... Temperature controller
TE .... Temperature element
TI .... Temperature indicator
TIC .... Temperature indicating controller
TR .... Temperature recorder
TSTAT .... Thermostat
TT .... Temperature transmitter
TV .... Globe valve (throttling valve)
VC .... Vent condenser
VTR .... Vent through roof
ZC .... Position controller (valve positioner)

NOTE: A mnemonic is a device to help someone remember. The letters used in an alphabetic abbreviation help one remember the device that is referred to.

Appendix Q

Specific Heats of Some Common Materials

It takes less heat to raise the temperature of most substances than it does to raise the temperature of water. To determine how much steam or hot water is needed to heat another substance, multiply the temperature rise of the substance by its specific heat and the quantity in pounds. The result is the number of Btus needed. For heating products continuously use pounds per hour of the substance to get the result in Btu/hr.

SPECIFIC HEATS OF COMMON MATERIALS[14]

Acetic Acid	0.51	Hydrochloric acid	0.60
Acetone	0.54	Ice	0.47
Air	0.24	Iron oxide	0.17
Alcohol	0.58	Kerosene	0.50
Alumina (aluminum oxide)	0.18	Lead	0.03
Aluminum	0.23	Lead oxide	0.06
Asbestos	0.20	Limestone	0.22
Ashes	0.20	Magnesia	0.22
Bakelite	0.35	Marble	0.21
Basalt (Lava)	0.20	Nickel	0.07
Benzol	0.40	Nickel steel	0.10
Borax	0.23	Oil, fuel	0.50
Brass, Bronze	0.09	Oil, machine	0.40
Brick	0.22	Oil, olive	0.40
Carbon	0.20	Petroleum	0.50
Chalk	0.22	Rubber	0.37
Charcoal	0.20	Salt, rock	0.21
Chloroform	0.23	Sand	0.19
Cinders	0.18	Sandstone	0.22
Coal	0.30	Silica	0.19
Concrete	0.16	Silver	0.10
Copper	0.09	Soil	0.44
Copper oxide	0.11	Solders	0.04
Cork	0.48	Steel	0.11
Dolomite	0.22	Sulfur	0.18
Ether	0.54	Sulphuric acid	0.34
Ethylene glycol	0.60	Talc	0.21
Flint glass	0.12	Titanium	0.13
Gasoline	0.50	Toluene	0.40
Glass	0.20	Turpentine	0.42
Glycerin	0.58	Water, sea	0.94
Gold	0.03	Wax	0.69
Granite	0.20	Wood, oak	0.57
Graphite	0.20	Wood, pine	0.67
Gypsum	0.26	Zinc oxide	0.12

Appendix R

Design Temperatures and Degree Days

Design outdoor winter temperature and the number of degree days are provided below for a number of North American cities.[15] More precise values should be available for your plant from the local weather service.

Alabama		
Anniston	5	2806
Birmingham	10	2611
Mobile	15	1566
Montgomery	10	2071
Alberta		
Calgary	-29	9520
Edmonton	-33	10320
Lethbridge	-32	8650
Medicine Hat	-35	8650
Arizona		
Flagstaff	-10	7242
Phoenix	25	1441
Yuma	30	1036
Arkansas		
Fort Smith	10	3226
Little Rock	5	3009
British Columbia		
Prince George	-32	9500
Prince Rupert	8	6910
Vancouver	11	5230
Victoria	15	5410
California		
Eureka	30	4758
Fresno	25	2403
Los Angeles	35	1391
Sacramento	30	2680
San Diego	35	1596
San Francisco	35	3137
San Jose	25	2823
Colorado		
Denver	-10	5839
Grand Junction	-15	5613
Pueblo	-20	5558
Connecticut		
Hartford	0	6113
New Haven	0	5880
Delaware		
Wilmington	0	
District of Columbia		
Washington	0	4561
Florida		
Apalachicola	25	1252
Jacksonville	25	1185
Key West	49	59
Miami	35	185
Pensacola	20	1281
Tampa	30	571
Tallahassee	25	1463
Georgia		
Atlanta	10	2985
Augusta	10	2306
Macon	15	2338
Savannah	20	1635
Idaho		
Boise	-10	5678
Lewiston	5	5109
Pocatello	-5	6741
Illinois		
Cairo	0	3957
Chicago	-10	6282
Peoria	-10	6004
Springfield	-10	5446
Indiana		
Evansville	0	4410
Fort Wayne	-10	6232
Indianapolis	-10	5458
Iowa		
Davenport	-15	6252
Des Moines	-15	6375
Dubuque	-20	6820
Keokuk	-10	5663
Sioux City	-20	6905
Kansas		
Concordia	-10	5425
Dodge City	-10	5069
Topeka	-10	5075
Wichita	-10	4664
Kentucky		
Lexington	0	4792
Louisville	0	4417
Louisiana		
New Orleans	20	1203
Shreveport	20	2132

Maine

Eastport	-10	8445
Presque Isle		9644
Portland	-5	7377

Manitoba

Brandon	-32	10930
Churchill	-42	16810
Winnipeg	-29	10630

Maryland

Baltimore	0	4487

Massachusetts

Boston	0	5936
Fitchburg	0	6743

Michigan

Alpena	-10	8278
Detroit	-10	6560
Escanoba	-15	8777
Grand Rapids	-10	6702
Lansing	-10	7149
Marquette	-10	8745
Sault St. Marie	-20	9307

Minnesota

Duluth	-25	9723
Minneapolis	-20	7966
Saint Paul	-20	7985

Mississippi

Meridian	10	2330
Vicksburg	10	2069

Missouri

Columbia	-10	5070
Kansas City	-10	4692
Saint Louis	0	4596
Saint Joseph	-10	5596
Springfield	-10	4569

Montana

Billings	-25	7213
Havre	-30	8416
Helena	-20	7930
Kalispell	-20	8032
Miles City	-35	7981
Missoula	-20	7604

Nebraska

Lincoln	-10	5980
North Platte	-20	6384
Omaha	-10	6095
Valetine	-25	7197

Nevada

Reno	-5	5621
Tonopah	5	5812
Winnemucca	-15	6357

New Brunswick

Fredericton	-6	8830
Moncton	-8	8700
Saint John	-3	8380

Newfoundland

Corner Brook	-1	9210
Gander	-3	9440
Goose Bay	-26	12140
Saint Johns	1	8780

New Hampshire

Concord	-15	7400

New Jersey

Atlantic City	5	5015
Newark	0	5500
Sandy Hook	0	5369
Trenton	0	5256

New Mexico

Albuquerque	0	4517
Roswell	-10	3578
Santa Fe	0	6123

New York

Albany	-10	6648
Binghamton	-10	6818
Buffalo	-5	6925
Canton	-25	8305
Ithaca	-15	6914
New York City	0	5280
Oswego	-10	7186
Rochester	-5	6772
Syracuse	-10	6899

North Carolina

Asheville	0	4236
Charlotte	10	3224
Greensboro	10	3849
Raleigh	10	3275
Wilmington	15	2420

North Dakota

Bismark	-30	8937
Devils Lake	-30	10104
Grand Forks	-25	9871
Williston	-35	9301

Northwest Territories

Aklavik	-46	17870
Fort Norman	-42	16020

Nova Scotia

Halifax	4	7570
Sydney	1	8220
Yarmouth	7	7520

Ohio

Cincinnati	0	4990
Cleveland	0	6144

Columbus	-10	5506
Dayton	0	5412
Sandusky	0	6095
Toledo	-10	6269
Oklahoma		
Oklahoma City	0	3670
Ontario		
Fort William	-24	10350
Hamilton	0	6890
Kapuskasing	-30	11790
Kingston	-11	7810
Kitchener	-3	7380
Ottawa	-15	8830
Toronto	0	7020
Oregon		
Baker	-5	7197
Portland	10	4353
Pennsylvania		
Erie	-5	6363
Harrisburg	0	5412
Philadelphia	0	4739
Pittsburgh	0	5430
Reading	0	5232
Scranton	-5	6218
Prince Edward Island		
Charlottetown	-3	8380
Quebec		
Arvida	-10	10440
Montreal	-9	8130
Quebec City	-12	9070
Sherbrooke	-12	8610
Rhode Island		
Providence	0	5984
Saskatchewan		
Prince Albert	-41	11430
Regina	-34	10770
Saskatoon	-37	10960
Swift Current	-33	9660
South Carolina		
Charleston	15	1866
Columbia	10	2488
Greenville	10	3059
South Dakota		
Huron	-20	7940
Rapid City	-20	7197
Tennessee		
Chattanooga	10	3238
Knoxville	0	3658
Memphis	0	3090
Nashville	0	3613
Texas		
Abilene	15	2573
Amarillo	-10	4196
Austin	20	1679
Brownsville	30	628
Corpus Christi	20	965
Dallas	0	2367
El Paso	10	2532
Fort Worth	10	2355
Galveston	20	1174
Houston	20	1315
Palestine	15	2068
Port Arthur	20	1352
San Antonio	20	1435
Utah		
Modena	-15	6598
Salt Lake City	-10	5650
Vermont		
Burlington	-10	8051
Virginia		
Cape Henry	10	3538
Lynchburg	5	4068
Norfolk	15	3364
Richmond	15	3922
Roanoke	0	4075
Washington		
North Head	20	5367
Seattle	15	4815
Spokane	-15	6138
Tacoma	15	5039
Tatoosh Island	15	5857
Walla Walla	-10	4910
Yakima	5	5585
West Virginia		
Elkins	-10	5800
Parkersburg	-10	4928
Wisconsin		
Green Bay	-20	7931
La Crosse	-25	7421
Madison	-15	7405
Milwaukee	-15	7079
Wyoming		
Cheyenne	-15	7536
Lander	-18	8243
Sheridan	-30	7239
Yukon Territory		
Dawson	-56	15040

Appendix S

Code Symbol Stamps

Your boiler or boilers will have one or more of these ASME Code Symbol stamps applied to the construction.

The letter within the symbol identifies the product and quality of construction. These stamps can only be applied by manufacturers authorized by ASME to use their respective stamp. Under no circumstances should the symbol stamp and the lettering next to it (which is also required by the Code) be removed, altered, or obliterated. The definition of the stamps and the general scope of the authorization, including those that will be found on pressure vessels (not shown above) are as follows:

- A - Assembly, to assemble boilers.
- E - Electric boiler, to manufacture electric boilers.
- H - Heating boiler, to manufacture heating boilers.
- M - Miniature boiler, to manufacture miniature boilers.
- PP - Power piping, to manufacture boiler external piping.
- S - Steam boiler, to manufacture power boilers, high temperature hot water and organic fluid heating boilers.
- U - Unfired pressure vessel, to manufacture pressure vessels.
- UM - Miniature unfired pressure vessel, to manufacture small pressure vessels.
- UV - Safety valves, manufacture of safety valves for unfired pressure vessels.
- V - Safety valves, manufacture of safety valves for high pressure boilers.

Note that the manufacturer's certificate will also define the locations where the manufacturing can be done, either in the shop named on the Certificate of Authorization, or (also) in the field.

BIBLIOGRAPHY

1. ASME Boiler & Pressure Vessel Code Section IV, *Rules for Construction of Heating Boilers.*
2. ASME Boiler & Pressure Vessel Code Section VIII, *Rules for Construction of Pressure Vessels.*
3. ASME Codes & Standards, B31.1, *Power Piping.*
4. Alstom Power, *Clean Combustion Technologies*, 5th ed., edited by Carl Bozzuto, 2009.
5. C. B. Oland, *Guide to Combined Heat and Power Systems for Boiler Owners and Operators.* Oak Ridge, TN, USA: Oak Ridge National Laboratory, 2004.
6. C. B. Oland, *Guide to Low Emission Boiler and Industrial Equipment Selection.* Oak Ridge, TN, USA: Oak Ridge National Laboratory, 2002.
7. Council of Industrial Boiler Owners (CIBO), *Steam System Efficiency Handbook*, 1997.

Abbreviations

CHAPTER 1

AC alternating current 30
BHP boiler horsepower 16
BPVC Boiler and Pressure Vessel Code 6
Btu British thermal unit 6
CDR compact disk read only memory 40
CEMS continuous emissions monitoring systems 43
CSD controls and safety device 6
DC direct current 30
E.D.R. equivalent direct radiation 16
EPA Environmental Protection Agency 27
FGR flue gas recirculation 43
FMS fan motor starter 31
gpm gallons per minute 6
H.P. high pressure 20
HTHW high temperature hot water 42
I.P. intermediate pressure 20
kpph thousand pounds of steam per hour 9
Kwhr kilowatt-hour
LEL lower explosive limit 26
L.P. low pressure 20
MACT maximum achievable control technology 27
MBtu/hr thousand Btu's per hour 9
MCC motor control center 32
MMBtu/hr million Btu's per hour 9
MSL mean sea level 7
NFPA National Fire Protection Association 6
O&M operating and maintenance 35
OSHA Occupational Safety and Health Agency 34
P&IDs process and instrumentation diagrams 36
PPE personal protective equipment 5
pph pounds per hour 9
PRV pressure reducing valve 19
psi pounds per square inch 9
psia pounds per square inch absolute 10
psig pounds per square inch gauge 10
SOP standard operating procedure 4
SRV safety relief valve 19
TDS total dissolved solids 45
UEL upper explosive limit 26
UST underground storage tank 44
VSD variable speed drive 38
W.C. water column 10

CHAPTER 2

AIS automatic interruptible system 76
BMS burner management system 56
Btu British thermal unit 50
CEMS continuous emissions monitoring systems 88
CFB circulating fluid bed 56
DCS distributed control system 68
EPA Environmental Protection Agency 88
gpm gallons per minute 53
HAP hazardous air pollutant 88
HRSG heat recovery steam generator 47
HTHW high temperature hot water 60
MACT maximum achievable control technology 88
MFTI main flame trial for ignition 56
MIC microbe induced corrosion 73
MMBtu/hr million Btu's per hour 76
pph pounds per hour 77
ppm parts per million 86
psi pounds per square inch 60
PTFI pilot trial for ignition 56
PTV pressure and temperature relief valve 71
rpm revolutions per minute 93
SASB Sustainability Accounting Standards Board 81
SOP standard operating procedure 51
TDS total dissolved solids 69

CHAPTER 3

ABMA	American Boiler Manufacturers Association 104
AQCS	Air Quality Control System 105
Btu	British thermal unit 98
cfm	cubic feet per minute 99
EIA	Energy Information Agency 97
EPA	Environmental Protection Agency 106
ESCO	Energy Service Company 109
gpm	gallons per minute 103
HHV	higher heating value 107
HRT	horizontal return tubular 107
kW	kilowatts 109
kWhrs	kilowatt-hours 102
LHV	lower heating value 107
MMBtu/hr	million Btu's per hour 106
NAPE	National Association of Power Engineers 111
pph	pounds per hour 100
psig	pounds per square inch gauge 101
SOP	standard operating procedure 104

CHAPTER 4

ASHRAE	American Society of Heating, Refrigerating, and Air Conditioning Engineers 122
Btu	British thermal unit 115
CDPF	catalyzed diesel particulate filter 128
DEF	diesel exhaust fluid 128
EGR	exhaust gas recirculation 128
EPA	Environmental Protection Agency 128
gpm	gallons per minute 116
GT	gas turbine 127
HHV	higher heating value 134
HL	high limit 124
HRSG	heat recovery steam generator 127
HTHW	high temperature hot water 119
IC	internal combustion 130
ISO	International Standards Organization 134
LHV	lower heating value 134
MACT	maximum achievable control technology 128
OL	operating limit 124
PCV	positive crankcase ventilation 128
pph	pounds per hour 116
ppm	parts per million 116
psi	pounds per square inch 114
psig	pounds per square inch gauge 119
PTV	pressure and temperature relief valve 124
RICE	reciprocating internal combustion engine 128
SOP	standard operating procedure 113
ST	steam turbine 128
TDS	total dissolved solids 116
TV	throttling valve 116

CHAPTER 5

ADP	apparatus dew point 174
AHU	air handling unit 162
ASHRAE	American Society of Heating, Refrigeration, and Air Conditioning Engineers 170
Btu	British thermal unit 135
CFC	chlorofluorocarbon 135
cfm	cubic feet per minute 171
CT	(queried for expansion)
DB	dry bulb 170
DX	direct expansion 181
EPA	Environmental Protection Agency 136
GHG	greenhouse gas 136
gpm	gallons per minute 166
HCFC	hydrochlorofluorocarbon 135
HEPA	high efficiency particulate air 177
HVAC	heating, ventilation, and air conditioning 168
Kwhrs	kilowatt-hours 135
psig	pounds per square inch gauge 137
RH	relative humidity 171
RTU	roof top unit 177
RV	recreational vehicle 164
SCADA	Supervisory Control and Data Acquisition 135
SCBA	self-contained breathing apparatus 136
SNAP	Significant New Alternatives Policy 136
SOP	standard operating procedure 160
TAB	testing, adjusting, and balancing 181
TDS	total dissolved solids 151
TXV	thermostatic expansion valve 154
UV	ultraviolet 136
UPS	uninterruptible power supply 163
VAV	variable air volume 177
W.C.	water column 178

CHAPTER 6

ANSI	American National Standards Institute 208
ASJ	All Service Jacket 192

BPVC	Boiler and Pressure Vessel Code 203
EPA	Environmental Protection Agency 207
GFCI	ground fault circuit interrupter 199
GTAW	Gas Tungsten Arc Weld 205
LED	light emitting diode 198
MAWP	maximum allowable working pressure 206
NFPA	National Fire Prevention Association 202
OSHA	Occupational Safety and Health Agency 188
PCB	polychlorinated biphenyl 201
PE	protective equipment 201
PM	preventive maintenance 185
psig	pounds per square inch gauge 204
RT	radiation testing 208
SMAW	shield metal arc weld 207
SOP	standard operating procedure 207
TIG	tungsten inert gas 205
UPS	uninterruptible power supply 198
WPS	welding procedure specification 208

CHAPTER 7

API	American Petroleum Institute 217
ASTM	American Society for Testing and Materials 217
BS&W	bottom sediment and water 213
Btu	British thermal unit 215
CIBO	Council of Industrial Boiler Owners 225
EPA	Environmental Protection Agency 214
gpm	gallons per minute 225
HHV	higher heating value 213
HMIWI	hospital, medical, and infectious waste incinerator 224
IMO	International Maritime Organization 217
LNG	liquefied natural gas 214
LOI	loss on ignition 223
LPG	liquefied petroleum gas 214
MACT	maximum achievable control technology 224
NPDES	National Pollutant Discharge Elimination System 226
PID	proportional–integral–derivative 220
ppm	parts per million 217
PRB	Powder River Basin 222
PSEG	Public Service Electric & Gas 225
psig	pounds per square inch gauge 215
SOP	standard operating procedure 221
SSU	seconds Saybolt universal 217
TDS	total dissolved solids 225
UST	underground storage tank 214
WOTUS	waters of the US 206

CHAPTER 8

ABMA	American Boiler Manufacturers Association 238
AVT	all volatile treatment 242
DC	direct contact (queried whether to include) 240
gpm	gallons per minute 238
HRSG	heat recovery steam generator 242
ppm	parts per million 232
ppb	parts per billion 238
psi	pounds per square inch 233
psia	pounds per square inch absolute 247
psig	pounds per square inch gauge 241
RO	reverse osmosis 236
TDS	total dissolved solids 234

CHAPTER 9

ASTM	American Society of Testing and Materials 253
BPVC	Boiler and Pressure Vessel Code 253
I.D.	inside diameter 254
MAWP	maximum allowable working pressure 256
psi	pounds per square inch 251
psig	pounds per square inch gauge 255
WOG	water, oil, or gas 255

CHAPTER 10

acfm	actual cfm/atmospheric cfm 339
ANSI	American National Standards Institute 327
BC	backward curved 327
BECCS	bio energy and carbon capture and storage 314
BPVC	Boiler and Pressure Vessel Code 257
Btu	British thermal unit 261
B&W	Babcock and Wilcox 276
CE	Combustion Engineering 274
CFB	circulating fluidized bed 312
cfm	cubic feet per minute 333
CHP	combined heat and power 279
CHX	condensing heat exchanger 266
CIBO	Council of Industrial Boiler Owners 314
DOE	Department of Energy 342
DWDI	double width, double inlet 334

E.D.R.	equivalent direct radiation 261
EGU	electric generating unit 257
EOR	enhanced oil recovery 270
EPA	Environmental Protection Agency 257
FC	forward curved 335
FD	forced draft 273
FERC	Federal Energy Regulatory Commission 342
FRP	fiberglass reinforced plastic 298
gpm	gallons per minute 320
HHV	higher heating value 348
HRSG	heat recovery steam generator 296
HRT	horizontal return tubular
HTHW	high temperature hot water 258
HVAC	heating, ventilation, and air conditioning 333
ID	induced draft 273
kpph	thousand pounds of steam per hour 279
Kwhr	kilowatt-hour 343
LWCO	Low Water Cut Off 288
MMBtu/hr	million Btu's per hour 257
NPSHA	NPSH available 319
NPSHR	NPSH required 319
NPSH	net positive suction head 319
pph	pounds per hour 258
psi	pounds per square inch 279
psia	pounds per square inch absolute 258
psig	pounds per square inch gauge 257
PTV	pressure and temperature relief valve 284
PURPA	Public Utilities Regulatory Policy Act 342
QF	qualified facility 342
RDF	refuse derived fuel 279
rpm	revolutions per minute 322
RSFC	radially stratified flame core burner 301
scfm	standard cubic feet per minute 339
SOP	standard operating procedure 343
SWSI	single width, single inlet 334
VFD	variable frequency drive 337
VIV	variable inlet vane 299
VSD	variable speed drive 299

CHAPTER 11

AFC	automatic frequency control 400
Btu	British thermal unit 373
BCS	burner control system 400
BMS	burner management system 365
BPS	boiler protection system 399
	boiler protection system 400
CEM	continuous emissions monitoring system 399
DCS	distributed control system 400
EGU	electric utility steam generating unit 396
EPA	Environmental Protection Agency 396
FD	forced draft 368
HAP	hazardous air pollutants 396
HTHW	high temperature hot water 373
I&C	instrumentation and controls 352
ID	induced draft 399
I/O	input/output 400
ISA	Instrument Society of America 364
LNG	liquefied natural gas 399
LT	level transmitter 352
MACT	maximum achievable control technology 396
MATS	mercury and air toxics 396
MMBtu/hr	million Btu's per hour 393
MSW	municipal solid waste 395
NESHAP	National Emission Standards for HAP 397
NFPA	National Fire Prevention Association 380
NPS	nominal pipe size 407
P&ID	process and instrumentation 364
PID	proportional, integral, and derivative 363
PLC	programmable logic control 400
pph	pounds per hour 393
ppm	parts per million 399
psi	pounds per square inch 352
psig	pounds per square inch gauge 351
rpm	revolutions per minute 400
RTD	resistance temperature detector 409
SAMA	Scientific Apparatus Manufacturer's Association 364

CHAPTER 12

OSHA	Occupational Safety and Health Agency 413

Index

A
absolute pressure 10
absorption chillers 164
acceptance testing 62
accounting of your oil inventory 406
accumulators 120
acfm 339
acid dewpoints 295
acid washing 204
actual cubic feet per minute 339
actual flowing conditions 323
adjustable orifice 153
ADP (apparatus dew point) 174
aero-derivative 348
air atomizing 306
air bound 330
air can serve as a cushion 408
air changes 54
air conditioning 168
  control 181
air-conditioning system, functions of 169
air cushion 48
air drying 195
air ejectors 94
air filter sensors 129
air-fuel ratio 23
air handling units (AHU) 178
air in a sensing line 198
air preheaters 296
air ventilation 171
aligning a coupling 318
aligning a pump and driver 316
alignment 316
alkalinity 235
allowable stress 253
all valves do leak 49
alternating current 30
American Boiler Manufacturer's Association 104
ammonia 135
analog 356
anion exchange resin 237
annual inspection 82
annual tests 69
ANSI (American National Standards Institute) 208
Anthracite 221
API gravity 217
apparatus inlet 172
arc chutes 33
arc flash protection 33
area 7
arrangements 334
arrangements of hydronic boilers 117
asbestos 192
  bearing insulation 192
  insulation 191
ash fusion point 214
ASHRAE 122
ASJ 192
ASME B31.1 Power Piping Code 208
ASME B31.9 Building Services Piping Code 208
ASME Boiler and Pressure Vessel Codes (BPVC) 6
ASME CSD-1 202
ASME P-4 294
ASME PTC-4.1 62
asphyxiate 367
atmosphere 10
atmospheric burners 303
atmospheric cubic feet per minute 339
atomic weight 25
atomization 306
attrition mills 312
automatic blowdown control 292
automatic expansion valve 153
automatic interruptible gas service 76
auxiliary burner 347
auxiliary turbine operation 89, 92
axial flow burner 300
axial measurement 8

B
Babcock and Wilcox 276
back pressure control valves 159
back pressure regulator 220
back pressure turbines 20, 93
backward curved 335
backwash 236
bagasse 224
Bailey Standard Line controls 351
balanced draft boilers 273
ball mill 311
bank 260
barometric damper 401
battery 30
bellows joint 256
belts 333
bending stress 253
Bernoulli principle 14
bias 363
bicarbonate 237
bill of material number 35
binary refrigerants 136
biomass 213, 223
  burners 314
Bituminous 221
black sky effect 260
bleeds 345
bleed steam 21
blister 260
blow by 128
blowdown heat recovery 226
blowdown safety valve 283
blowdown transfer 78
blowing down 116
blowing sediment out 116
blowoff 244
  schedule 226
blow-thru apparatus 177
BMS 379
Boiler and Pressure Vessel Code (BPVC) 257
boiler efficiency 104
boiler external piping 293
boiler feed pumps 22, 315
boiler feed tanks 239
boiler horsepower (BHP) 16
boiler on-line 66
boiler operating efficiency 109
boiler trim 283
boiler tube cleaning 203
boiler vent valve 63
boiler warm-up 60
boiler water circulation 263
  pumps 128

boil-out 58
bonding and grounding 32
bonding jumpers 199
bottom blowoff 79
valves 292
bowl mills 311
box header boiler 272
break 26
breakdown maintenance 185
break in period 61
brick 264
brick or tile laid up dry 195
British thermal unit (Btu) 15
Btu/hr 16
brushing 261
bubble point 136
bulges 260
bull ring 194
bumpless 362
bunker 222
buoyancy 15
principle 356
burner cutout control 381
burner ignition cycle 56
burner management 378
burner register 300
burners 298
wood 313
burner throat 193
bus bars 200
butane 215
bypass factor 174

C
calcium ions 236
California Energy Commission 28
cam contacts 379
cam positioner 388
capillary 153, 291
capillary type temperature transmitter elements 409
carbon 121, 309
carbon dioxide 23, 136
carbonic acid 246
carbon monoxide 24
detectors 177
cardboard 224
Carrier 135
carryover 243
cascade 364
casing 265
castable 195, 265
cast iron boilers 265
catalysts 25
catalytic converter 347
caustic embrittlement 246
cavitation 319
CDPFs 128
central boiler plant 28
central plant 28
centrifugal compressors 341
refrigerant 147
centrifugal devices 332
centrifugal feed pumps 328
centrifugal pumps 323
ceramic fibers 193
CFCs 135
change the light bulbs 198
channeling 103, 238
charting data 12
checking the oil 341
checklist 187
chelate 248
chemical treatment 244
chilled water design conditions 162
chillers 162
chloride 235
choice fuel firing 395
CHX 298
circuit breaker 32
circuit test 31
circulating fluidized bed boilers 312
circulators 115
classifier 311
Clean Air Act 137
clean dry air 197
cleaning 186
cleaning waterside scale 204
clinkers 311
closed circuit 30
CO_2 24
coal 24, 221
coal and oil slurry 221
coal burners 311
Coast Guard 406
cogeneration 342
collecting performance data 61
combined cycle power plants 348
combustion 22
chemistry 23
controls 380
efficiency 106
engineering 274
optimization 27, 88
comfort for occupants 182
comfort zones 170
common units of measure 9
compact fluorescent 198
compressing air 338
compressing oxygen 339
compressive stress 252
compressor refrigerant 142
compressors 338
hermetic 143
open 143
other types of 341
semi-hermetic 143
condensate 226
polisher 237
pumps 22
condenser 21
condenser pressure control valves 151
condenser, refrigerant 148
condensing heat exchangers 298
condensing units 149
conduction 259
conductive heat transfer 259
conduit covers 197
constant friction 178
contamination of the oil 190
continuous blowdown 226, 243
continuous blowdown piping 243
continuous blowdown valve 292
continuous duty 34
contractors 3
contractor's log 41
control air compressor 340
control linearity 368
controlling flow 13
control range 352
control schematics 364
control signals 352, 361
controls 352
boiler 352
maintenance 197
refrigerant 159
convection section 260
convection superheater 279
convective heat transfer 259
convectors 182
conversion of velocity pressure 15
conveyors 222
coolers on compressors 339

cooling coil 173
cooling towers 166
Copes valves 376
corn 224
corn cobs 224
corrosion 232
    and wear 416
corrugated cardboard 224
cost differential 29
cost of electricity varies 343
cost of failure 203
CO trim 396
coupling 316
    guard 202
crack 25
crack a valve 48
crescent gear pump 330
cross drum sectional header boiler 271
cross-limiting 394
crude oil 25
CSD-1 6
culm 221
current 29
    imbalance 201
custom log book 41
cycling efficiency 109
cyclone furnaces 312

D
damper wide open 380
data to record 42
day tank 219
dead plant start-up 62
dealkalizers 237, 238
decarbonators 239
defrosting 139
defrost valves 158
degassifiers 239
degree day ratio 110
degree days 97
delivery rate 110
demand charges 102
demineralizers 237
density 8
desuperheater 21, 75
diaphragm actuated regulators 366
diatomic gases 339
differential setting 369
diffuser 180, 299
    guide pipe 300
digester gas 215, 306
digital signals 356
diked areas 74
dimensional analysis 10
direct acting controllers 354
direct current 30
disaster plans 34
discharge 34
disconnects 37
discrimination 56
displacement transmitters 356
distance 7
distributed generation 346
documentation 34
doubler 255
downcomers 263
draft control 401
draft gauges 198, 406
draft hood 401
drainable superheaters 279
drain pan 174
drain traps 341
draw-thru apparatus 177
dressing of the fire 311
drift 166
drilling of your gas burner 305
drip pan ell 285
drivers 315
droop 353
drum level gauge 406
drums 263
dry back design 268
dry bulb (DB) temperature 170
dry lay-up 87
dry-out 195
D type boilers 275
dual fuel firing 77, 395
duct burners 134, 348
ductwork 178
duplex oil strainer 187
DX 181

E
economics 342
economizers 294
economy cooling 171
eddy current testing 202
eductor 338
efficiency 104
ejectors and injectors 337
electricity 29
elevation 7
emergency boiler start-up 67
enthalpy 16
environmental testing 189
EPA 106
equipment number 35
error 358
ESCO (energy service company) 109
establishing linearity 384
establish proper firing conditions 58
ethylene or propylene glycol 114
evaporation rate 110
    curves 103
evaporative condensers 150
evaporators 137, 139
evase 130
excess internal reinforcement 209
exhausters 312
expanding liquid 73
expansion cracks 193
expansion joint 256
expansion tanks 17, 114, 120
    on the boiler 86
explosion of steam and boiling hot water 85
explosive range 26
extraction steam 21
eyeballing 11

F
face and bypass dampers 173
fail-safe concepts 380
false high level 406
fan and blower control 336
fan horsepower 336
fan housing 177
fan inlet metering 391
fan powered VAV boxes 179
fans and blowers 332
federal law 406
feedback 354
feed water circulation 403
feed water piping 292
feed water pressure controls 403
feet MSL 7
fiberglass tanks 218
field tanks 219
fill 167
    liquid 408
    systems 51
film 261

filter-dryer 156
filters 172
firebox boiler 268
fire side cleaning 203
fire triangle 22
fire tube boiler 266
Fireye 378
firing aisle 222
firing rate control 380, 393
fit up 208
flame impingement 261
flame rod 378
flame runners 303
flame scanner 378
flame sensors can deteriorate 64
flame shaping 300
flammability limits 26
flammable range 26
flare 216
flash point 218
flash steam 227
flash tank 19, 226
flash type deaerators 240
flexible couplings 316
flexitube boiler 278
flick of the switch 76
float valve 154
floor drains 218
flow 13, 384
flow hood 180
fluid handling 332
fluid heater 121
fluid heating systems 373
fluidized bed boilers 312
fluid level maintenance 373
fluid temperature maintenance 372
FM Cock 49
foaming 70, 142
force 9
forced draft boiler 273
forward curved fans 335
fossil fuels 22, 213
four-pass fire tube boilers 269
free blow drain 291
free cooling 148
freezing 141
Freon™ 135
from and at 212°F 16
FRP 298
fuel analysis 106
fuel cells 215, 343, 349
fuel oil 217
fuel oil pumps 219
fuel oil sensing line can produce a hazardous condition 407
full load or 100% heating load 97
full metering control systems 393
function generator 368, 399
furnace 188
    controller setpoint 403
    pressure 403
fusible plug 269
future service connections 116

G
gain 357
gallon 8
garbage 224
gas boosters 342
gas compressors 339
gas, digester 306
gas engines 347
gas gun 304
gas holders 216
gas, landfill 306
gas pressure regulator 366
gas ring 305
gas turbine 129, 347
gauge faces 202
gauge glass 406
gauge set 138
gear pump 329
GFCI 199
goggle plate 21
grain 170
grapes 209
graphite tape 288
grate 309
grease 189
grease fitting 190
grease gun 189
grease lubricated bearings 189
grounded conductor 31
ground fault interrupter 199
ground grid 32
grounding 199
ground source heat pumps 161
ground wire 31

H
Hagan Ratio Totalizer 356
hammer mills 312
hand holes 272
hardness 235
harmonics 197
hay 224
head 9
headers 271
header temperature control 118
head tank 346
heat and energy wheels 177
heat balance 89
heat drying 195
heat exchangers 177
heating boiler control settings 371
heating coil 175
heating season 97
heat loss efficiency 105
heat pipes 148, 297
heat pump 161
heat slingers 334
heat transmittance 261
heat traps 294
heavy oils 217
HEPA filter (high efficiency particulate arresting) 177
Hg 240
high and low oil temperature switches 160
higher heating value (HHV) 107
high-low firing rate control 381
high pressure boilers 257
high pressure switch 159, 290
high pressure washers 204
high pressure water wash 204
high set firebox boiler 268
high temperature hot water 119, 258
high temperature switch 290
Honeywell 369
horizontal split case pump 327
hospital waste 224
hot gas bypass 146
hot gas line 156, 159
hot pass 210
hot water heating load 100
hot water heating systems 114
hot wire analyzers 397
HRSGs 131, 348
HRT boiler 267
HTHW 258
    boiler control 374
    design conditions 119
    generator 119

humidifiers 176
HVAC 168
Hydrazine 247
hydrocarbons 22
hydrogen 349
  as a fuel 215
  -carbon 221
  heating 114
hydrostatic testing 84
hyperbolic cooling towers 167
hysteresis 360, 391

I
ice expands 141
ice storage 141
ideal gas law 346
idle systems 71
ignition arch 310
ignition cycle 56
ignition permissive 379
imbalance 335
impeller turned down 320
impending emergencies 66
implied measures 10
implosion 273
impulse turbine 344
inches of water 10
incomplete combustion 23
incomplete penetration 209
individuals without a license 416
induced draft boiler 273
inert gas 189
inferential metering 392
infrared thermometer 200
inlet bell 334
inlet conditions 171
inlet screens 334
inlet vanes 147
input-output efficiency 105
in-situ analyzers 409
inspection and access doors 174
inspector's gauge connection 82
instability 128
installing packing 196
instantaneous hot water heaters 123
instructions 187
instrumentation 352, 405
instrument maintenance 197
Instrument Society of American 364
insulation inventory 192
insulation studs 192
insurance companies 82
integral 359
intercoolers 340
intermediate supported units 265
intermittent duty motors 34
interruptible gas 28, 75
interstitial space 219
in. W.C. 10
ion 231

J
jacking gear 93, 346
jackshaft control 382
jet pumps 337

K
key caps 272
keyed in 195
kiln dried wood 223
King Valve 157
know your plant 101
kpph 9

L
lack of a ground 32
landfill gas 216, 306
lantern ring 196
large hydronic heating systems 373
lawn sprinkler example 14
law of conservation of mass 24
lay-up 86
lead-lag controls 371
leakage in a long duct run 182
leakage is necessary 316
leak testing of fuel oil safety shut-off valves 69
LEDs 198
Legionella 125
LEL 26
levels 7
licensed individuals 416
life cycle cost 264
life of electrical equipment 200
lift test 60
light-off conditions 51
light-off position 54
limit and operating controls 159
limit switches 159
  boiler 290
linear air flow 387
linearity 387
linkage adjustment 385
liquid line 156
liquid observation port 156
list of disasters 39
lithium bromide 164
little bits 192
live zero 352
LNG 214
load 10
load control valves 158
local boilers 28
local setpoint 352
local transmitters 410
lock-out, tag-out 188
locomotive boiler 267
log book 39
log calculations 46
logs 40
longitudinal welds 275
loop 352
loop trap 94
losing calibration 198
lower explosive limit 26
lower heating value (LHV) 107
low fire changeover 76
low fire hold 59, 389
low fire position 77
  switches 55
low fire start 381
low hydrogen electrodes 210
low load 97
low oil pressure switches 159
low pressure boilers 257
low pressure drop check valve 292
low pressure heat exchangers 21
low pressure switch 159
low set firebox boiler 268
low temperature switch 160
low water cutoff 288
  fail 82
low water flow switches 119
LPG 214
lubricated plug valve 49
lubricating system 190
lubrication 189
Lungstrom air preheater 297

M
magnesium 236
main flame trial for ignition (MFTI) 56

maintaining a vacuum 347
maintaining pneumatic controls 197
maintenance and repair
    history 35
    log 41
    of the oil systems 190
    operating during 83
makeup water pumps 120
management's attitude 416
manholes 292
manometer 52
matching equipment to the load 101
MAWP 266
maximum allowable pressure 253
MBtu/hr 9
measurements 6
measures of a rate 6
measures of quantity 6
meniscus 234
mercaptans 25
mercury switches 291
meter on the makeup 116
methane 25
methyl orange 235
methyl purple 235
MIC (microbe induced corrosion) 73
microturbines 349
Mill Test Certificates 294
minimum fire pressure regulator 55
minimum stop 55
misting 131
mixing box 171
moderator 23
modernizing and upgrading 111
modulating 13
    controls 369
    motor 369
Morrison tubes 270
motor control centers (MCCs) 33
motor speed control 331
motor starter 31
multiple boilers in service 372
multiple-retort 310
multiplier 397
multi-stage pumps 324

N

nameplates 202
NAPE (National Association of Power Engineers) 111
national board data 82
National Board R-1 forms 294
National Board "R" Symbol Stamp 84
National Board statistics 416
National Fire Protection Association (NFPA) Codes 6
natural circulation 373
natural convection 259
natural draft 15
net positive suction head 319
new startup 50
New York Telephone Company 200
NFPA 85 202, 380
nitrogen 17, 136
non-overloading motors 327
non-return valve 292
NPSHA 319
NPSHR 319
number of pumps in operation 103

O

Ohm's law 30
oil 190
oil burner tip 306
oil field boilers 270
oil filled transformers 201
oil for refrigeration 137
oil gun 306
oil maintenance service 190
oil pressure for light-off 52
oil transfer pumps 219
once through boilers 128, 277
on-off boiler 381
on-off control 369
open 30
operating modes 47
operating unit 202
operator's log 40
operator's narrative 41
operators 111
opposed blade dampers 173
order of operations 37
organic fluid 121
orifice in the seal flushing piping 316
orifice nipple 309
Otto 347
O type boiler 275
outdoor conditions 170
output 104
outsized 109
over feed stokers 310
overload a motor 34
over-lubrication 189
over speed trip 95, 319
oxygen pitting 246
oxygen trim 215
ozone depletion 136

P

package boiler 275
packing 195
painters 200
parallel positioning controls 390
parallel positioning with air metering 391
parallel positioning with flow tieback 393
parallel positioning with steam flow trim 393
paramagnetic analyzer 397
parameter 351
partial alkalinity 235
partial combustion 23
parts per million 232
pass 269
PCBs 201
peaking generators 129
peak load 97
pendant type superheaters 279
percent humidity 170
percent makeup 235
perception 358
permeate 238
perpendicular 8
personal protective equipment (PPE) 5
person in charge of lock-out, tag-out 188
petcock 286
petro-tite leak test 219
pH 231
phenopthalein 235
phosphate 248
PIDs 36
pigtail 290
pilot operated valve 367
pilot trial for ignition (PTFI) 56
pilot turndown test 57
pinch points 132, 349
piping flexibility 255
plan for the failure of every utility 203
plans for fire 38
plant efficiency 107

plant master 372
plant rate 111
plastic 193, 264
plugged economizer 295
plugging tubes 204
Plumbing Code 81
pneumatic testing 85
pneumatic transmitters 355
poisonous carbon monoxide
pop testing 71
pop valves 284
positioner 388
postmix 303
potato peels 247
pour point 217
Powder River Basin 222
Power Magazine 134
power test code 62
power turbines 93
pph 9
ppm 232
predicted performance 105
predictive maintenance 185
premix 303
preparing for operation 50
preprinted log 41
preserving historical data 3
pressure 8
pressure atomizing burners 306
pressure balance principles 355
pressure differential atomizing 306
pressure drop and flow 14
pressure gauges 290
pressure swings 380
pressure, temperature relief valves 284
pressure testing 84
pressuretrol 369
pretreatment 236
prevent failures due to wear 416
preventing scale formation 247
preventive maintenance 185
primary air adjustment 299
primary air fans 312
primary air shutter 304
priming 70
priorities 1
procedureless 362
process of combustion 23
process variable 351
production loads 100
propeller fans 333
proper grease 190
proper rotation 323
proportional control 364
prove combustion air flow 53
provisions for thermal expansion 114
psia 10
psig 10
psychrometrics 169
puff 27
pulse combustion or power burners 304
pulverized coal burner 312
pulverizers 311
pump and heater set 219
pump control 331
pumps 314
pump set 219
purge the boiler 58
purge timing 54, 379
purging 189

Q

qualified, experienced boiler operators 111
questions an operator should have answers to 101
quill 248

R

R-22 136
R-134a 136
radial 8
    bladed fans 335
radiant heat transfer 259
radiant superheater 278
radiation loss 106
rain load 100
ramping controls 59, 77
rates 9
reacting to changing loads 104
reactions 23
receiver 157
reciprocating compressors 340
reciprocating pumps 321
recirculate oil 73
recirculating control valves 220
recirculating line 326
recommended rules 6
recorder charts 41
recover condensate 227
recycling the water 225
Redler conveyor 222
reformer 349
refractory 193, 264
    anchor 195
    dry-out 56
    maintenance coating 194
    repair 194
    throat 303
refrigerant compressor heater 142
refrigerant oils 137
refrigerants 135
refrigeration compressors 341
refrigeration cycle 137
regeneration cycle 237
registers 179
regulations for lock-out, tag-out 188
reheat coils 179
reheaters 21, 279
remote setpoint 352
removing a tube 205
repeats per minute 359
replacement of filters 173
replacements 202
representative sample 232
requirements for combustion air 50
reset 358
reset accessories 358
reset push-button 27
reset windup 364
residual 245
resin bed 236
resistance 30
restoring insulation 192
retort 309
reverse acting controllers 354
reverse osmosis (RO) 236
right-sizing 109
risers 263
riveted boiler 270
Roman baths 264
room conditions 171
root pass 208
rotary blowers 336
rotary compressors 342
rotary cup burners 306
rotate a pump 72
rotating boilers 78
rotating equipment 332
roughness on light off 27
RTUs (roof top units) 181
rule of thirds 78

rupture 85
rupture disc 161
RV refrigerator 164

S
safety 5
factor 253
relief valves 283
salt 227
salt elutrition test 237
sample cooler 233
sander dust 223
sand filters 236
saturation card 140
saturation condition 16
saturation point 16
saturation temperature 18
sawdust 223
SCADA 170
scale formation 232
SCBA 136
scfm (standard cubic feet per minute) 339
schools 42
Scientific Apparatus Manufacturer's Association 364
screen tubes 272
screw and gear pumps 329
screw compressors 341
refrigerant 146
screw pump 330
scroll 334
scroll compressors 145
scrubber type of deaerator 241
seatless blowoff valves 80, 293
secondary air ports 302
secondary ratings 255
self-contained controls 365
sensing connections 407
sensing lines 407
sentinel valve 91
separating fluid 198
separating oil 407
sequester 247
service stub 169
service water 122
Servidyne Systems 28
setpoint 352
setting 264
for automatic three boiler control 371
severe duty motors 34
shaft seal 316
on a power turbine 94
sheave 333
shim stock 317
shortening purge 417
short off cycles 331
shot feeders 228
shrink 376
shrouds 301
single element control 376
single loop control 365
single phasing 33
siphon 290
slag 210
sling psychrometer 182
slope 325
slow startup 64
sludge 247
sludge conditioners 248
small tools 228
soda-phosphate 248
sodium 236
sodium hydroxide 246
sodium sulfite 228, 246
soot blowers 204
operation 104
sounding 405
spalling 193
sparge line 239
specific gravity 8
specific volume 9
split case pump 327
spray type deaerators 240
spreader stoker 313
spuds 304
Sq. Ft. E.D.R. 16
square 11
stable combustion 303
stack thermometers 409
staged combustion 302
staged unloading 341
stage of a turbine 20
standard operating procedures 36
standard ranges of control signals 352
standards 255
standby operation 78
starting a boiler with an induced draft fan 336
starting a dry pump 330
startup control 389
startup sheet 124
static electricity 30
static pressure 14
static regain 178
staybolts 270
steam 291
steam air heaters 296
steam and water cycle 19
steam atomizing burners 306
steam drum 271
internals 281
steam explosions 18
steam flow/air flow 393
steam flow recorders 405
steam generating units 62
steam humidifiers 176
steam pressure maintenance 368
steam quench 121
steam tracing 121
steel tanks 219
Sterling boilers 273
still pipe 374
stoichiometric 23
storage water heaters 123
stored fuel oil 202
stratosphere 136
stress 251
stress-strain diagram 251
strike three 27
stringers 210
sub-cooling 138
suction accumulator 143
suction line 156
sulfite corrosion 242
sulfur dioxide 24
summer 376
load 97
supercritical boiler plants 20
superheat 75, 138
superheated gas 138
superheated steam 18
boiler 257
superheater 20, 278
vent first 75
surface blowdown 243
surface tension 243
surging 325
sweep 194
swell 376
switching fuels 75

synthetic oils 189
synthetic replacements 191
system curves 325

T
TAB report 182
tack welds 208
tall stack 273
tangent 8
TDS 234
teapot 264
temperature control 372
    switch 372
    valves 366
temperature glide 136
temperature limit switches 290
temperature piloted pressure control valve 220
temperatures of casing surfaces 62
tensile stress 251
tensile test specimen 251
test safeties 60
test stand 234
test the low water cutoff properly 53
test water temperature 84
thermal bulb mounting 155
thermal shock 126, 415
thermocouple 356
thermo-hydraulic 375
thermo-mechanical 375
thermometers read 405
thermostat 365
thermostatic expansion valves 154
thermosyphoning 115
thermowells 410
third party inspection 83
throttling device 152
throttling the vent valve 240
Tier 4 regulations 128
TIG (GTAW) 205
tiles 184, 193
timer motor 379
ton of refrigeration 135
topping turbine 346
top supported boilers 265
total alkalinity 235
total dissolved solids 225
tramp air 273
transformers 201
transmitter installation 408
transmitter, mounted 408
trash burners 224
traveling grate stokers 310
tray type deaerator 241
trends 10, 360
tribology 190
tri-generation 343
trim controller 397
troubleshooting burner management 31
tubeless boiler 265
tube roller 206
tune ups 88
tuning firing rate controls 198
turbine pumps 327
turbining 204
turndown 10, 306
tuyeres 309
two element control 376
TXV 154

U
U bend 256
UEL 26
ullage 406
ultimate analysis 22, 26, 213
ultraviolet lights 175
under-feed stoker 310
uniform air distribution 301
units 6
universal solvent 231
unloading 339
    pumps 219
    systems 144
untested pipe 188
upper explosive limit 26
UPS (uninterruptible power supply) 198
USTs 218

V
vacuum 16, 21
vacuum breakers 17
vacuum deaerator 239
vacuum pumps 94
vacuum systems 113
value of documentation 4
valve manipulation 47
valve packing 197
valve positioner 361, 388
valves 291
    blowdown and blowoff 292
    feedwater 292
valve wrench 49
vanadium 193
vapor bound 319
vaporizer 121, 258
variable frequency drives, fans 337
variable inlet vanes 337
variable speed drives 331, 337
    pumps 331
VAV boxes 179
VAV systems 179
velocity pressure 14
velocity regain 179
vent condenser 240
ventilation air 171
ventilation loads 99
venting of refrigerants 136
vents and drains 48
venturi throat 300
vertical fire tube boiler 267
viscosity 10, 217
    control 221
visitor's log 41
voltage 30
    imbalance 201
volume 7
VR (valve repair) symbol stamp 82
VSD (Variable Speed Drive) 299, 337

W
WADITW 11
warm up bypass valves 48
warped front plate 305
waste heat service 127
waste paper 224
water 225
water column 285
water flow switches 160
water sample cooler 225
water softener 236
water, steam and energy 15
water treatment chemicals 227
water treatment consultant
water treatment log 41
water tube boilers 270
water walls 271
wear rings 323
weave pass 210
weekend load 97
weigh feeders 222
weigh lorries 222

weld efficiency 254
welding machine 197
Welding Procedure Specification 208
wet back arrangement 269
wet bulb temperature 170
wet layup 86
wheel 335
why they fail 413
Willians line 90
windage 90
windbox 301
window air conditioner condenser 149
window unit 168
window weld 205
winter load 97
wire drawing 66
wood burners 313
wood chips 223
wrong bolts or nuts 187

Z

zirconium oxide analyzer 397
zone dampers 177

Biography

Carl Bozzuto received the bachelor's and master's degrees in chemical engineering from Massachusetts Institute of Technology (MIT), Cambridge, MA, USA, and the master's degree in management science from the Hartford Graduate Center, Hartford, CT, USA.

He is a retired vice president of Technology from Alstom Power's Boiler and Environmental Sector. He is currently an independent consultant (currently with CIBO and EUCI). He is also a judge for the Carbon XPRIZE. Mr. Bozzuto's background and experience have been in the areas of new technology development and commercialization. At ALSTOM Power, he has been associated with a wide variety of new process technologies including CCS, pressurized fluidized bed combustion systems, atmospheric fluidized bed combustion systems, flue gas desulfurization systems, NOx control and removal technologies, firing systems, coal gasification combined cycle systems, coal water mixtures, coal conversion technologies, solar thermal power systems, iron production systems, and paper mill recovery systems. He has also been responsible for marketing, market planning, strategic planning, and business development for utility steam generation systems and related products and the profit and loss of the Fluidized Bed System Business Unit.

Mr. Bozzuto was the Editor-in-Chief of the award winning textbook entitled "Clean Combustion Technologies" published by Alstom in 2009. He was the technical leader for the National Coal Council Report on DOE's Technical Progress in CCS Technology in 2015. He received the Lifetime Achievement Award from Marquis Who's Who in 2020.